Food Bioactive Compounds: Chemical Challenges and Opportunities

Food Bioactive Compounds: Chemical Challenges and Opportunities

Editor

Smaoui Slim

MDPI • Basel • Beijing • Wuhan • Barcelona • Belgrade • Manchester • Tokyo • Cluj • Tianjin

Editor
Smaoui Slim
Laboratory of Microbial and
Enzymatic Biotechnologies
and Biomolecules
Center of Biotechnology of
Sfax
Sfax
Tunisia

Editorial Office
MDPI
St. Alban-Anlage 66
4052 Basel, Switzerland

This is a reprint of articles from the Special Issue published online in the open access journal *Molecules* (ISSN 1420-3049) (available at: www.mdpi.com/journal/molecules/special_issues/BioFood).

For citation purposes, cite each article independently as indicated on the article page online and as indicated below:

LastName, A.A.; LastName, B.B.; LastName, C.C. Article Title. *Journal Name* **Year**, *Volume Number*, Page Range.

ISBN 978-3-0365-6793-8 (Hbk)
ISBN 978-3-0365-6792-1 (PDF)

Contents

About the Editor

Smaoui Slim

Slim SMAOUI is currently working as an Associate Professor in the Centre of Biotechnology of Sfax (CBS-Tunisia). In 2010, He received Ph.D.s in "Process Genius and Environment"from National Polytechnic Institute of Toulouse (France) and in "Biological Sciences"from Faculty of Sciences of Sfax (Tunisia). He received the accreditation to supervise research in 2020 from National School of Engineers of Sfax (Tunisia).

In 2013, he integrated the "Laboratory of Microorganisms and Biomolecules"of CBS as a Senior Research Scientist. He conducted his researches by studying the biomolecule production processes and their applications in meat products preservation. Slim's research interests are centered on the Development of new healthier meat products; use of agro-food industries by-products as potential sources to obtain valuable and bioactive compounds and incorporation of bioactive compounds and development of functional foods to protect foods against oxidative and microbial degradation. He deals with meat quality and safety, analytical chemistry applied to food analysis, research and development and analytical methods validation by the application of chemometric techniques.

He has published more than 115 articles in peer-reviewed and Academic Journals, and books. The H index of Slim SMAOUI has reached 22 and citations exceed 1600 (source: Scopus).

He is also an Associate Editor in Quality Assurance and Safety of Crops & Foods, an Academic Editor in: *Journal of Food Quality, Journal of Food Processing and Preservation, Biomed Research International* and *Evidence-Based Complementary and Alternative Medicine*. He was a Guest Editor in *Molecules, Frontiers in Microbiology* and *Frontiers in Nutrition* (section: Food Microbiology). In addition, Slim is a reviewer in many international journals. He has participated in many research programs as a coordinator or scientific member.

molecules

MDPI

Article

Cupressus sempervirens Essential Oil: Exploring the Antibacterial Multitarget Mechanisms, Chemcomputational Toxicity Prediction, and Safety Assessment in Zebrafish Embryos

Sarra Akermi [1], Slim Smaoui [1,*], Khaoula Elhadef [1], Mariam Fourati [1], Nacim Louhichi [2], Moufida Chaari [1], Ahlem Chakchouk Mtibaa [1], Aissette Baanannou [2], Saber Masmoudi [2] and Lotfi Mellouli [1]

1 Laboratory of Microbial Biotechnology and Engineering Enzymes (LMBEE), Center of Biotechnology of Sfax (CBS), University of Sfax, Road of Sidi Mansour Km 6, P.O. Box 1177, Sfax 3018, Tunisia; sarahakermi221@gmail.com (S.A.); elhadefkhawla@gmail.com (K.E.); mariamfourati@ymail.com (M.F.); moufida.chaari97@gmail.com (M.C.); ahlemchakchouk@yahoo.fr (A.C.M.); lotfi.mallouli@cbs.rnrt.tn (L.M.)

2 Laboratory of Molecular and Cellular Screening Processes, Center of Biotechnology of Sfax, University of Sfax, Road of Sidi Mansour Km 6, P.O. Box 1177, Sfax 3018, Tunisia; nacim.louhichi@gmail.com (N.L.); aissette.baanannou@yahoo.fr (A.B.); saber.masmoudi@cbs.rnrt.tn (S.M.)

* Correspondence: slim.smaoui@yahoo.fr; Tel.: +216-53759478

Citation: Akermi, S.; Smaoui, S.; Elhadef, K.; Fourati, M.; Louhichi, N.; Chaari, M.; Chakchouk Mtibaa, A.; Baanannou, A.; Masmoudi, S.; Mellouli, L. *Cupressus sempervirens* Essential Oil: Exploring the Antibacterial Multitarget Mechanisms, Chemcomputational Toxicity Prediction, and Safety Assessment in Zebrafish Embryos. *Molecules* **2022**, *27*, 2630. https://doi.org/10.3390/molecules27092630

Academic Editor: Junhu Cheng

Received: 11 March 2022
Accepted: 18 April 2022
Published: 19 April 2022

Publisher's Note: MDPI stays neutral with regard to jurisdictional claims in published maps and institutional affiliations.

Abstract: Nowadays, increasing interest has recently been given to the exploration of new food preservatives to avoid foodborne outbreaks or food spoilage. Likewise, new compounds that substitute the commonly used synthetic food preservatives are required to restrain the rising problem of microbial resistance. Accordingly, the present study was conducted to examine the chemical composition and the mechanism(s) of action of the *Cupressus sempervirens* essential oil (CSEO) against *Salmonella enterica* Typhimuriumand *Staphyloccocus aureus*. The gas chromatography analysis revealed α-pinene (38.47%) and δ-3-carene (25.14%) are the major components of the CSEO. By using computational methods, such as quantitative structure–activity relationship (QSAR), we revealed that many CSEO components had no toxic effects. Moreover, findings indicated that α-pinene, δ-3-carene and borneol, a minor compound of CSEO, could inhibit the AcrB-TolC and MepR efflux pump activity of *S. enterica* Typhimurium and *S. aureus*, respectively. In addition, our molecular docking predictions indicated the high affinity of these three compounds with active sites of bacterial DNA and RNA polymerases, pointing to plausible impairments of the pathogenic bacteria cell replication processes. As well, the safety profile was developed through the zebrafish model. The in vivo toxicological evaluation of (CSEO) exhibited a concentration-dependent manner, with a lethal concentration (LC_{50}) equal to 6.6 μg/mL.

Keywords: *Cupressus sempervirens* essential oil; antibacterial activity; membrane permeability; replication and transcription inhibition; computational toxicology; molecular docking; zebrafish

1. Introduction

It is clear that the long-term use of antibiotics has provoked the mass production of genetically resistant bacteria [1]. As a result, some pathogenic bacteria have become resistant to entire antibiotics classes [1–3]. For instance, some bacterial species usually susceptible to carbapenems and colistin, such as *Enterobacteriaceae*, *P. aeruginosa* and *K. pneumoniae*, have gained the ability to hydrolyze β-lactams and make them highly resistant to most β-lactam antibiotics [1]. In addition, methicillin-resistant *Staphylococcus aureus* (MRSA) and extended-spectrum β-lactamase producers are resistant not only to methicillin and cephalosporin, but to tetracycline, aminoglycosides, macrolides, and chloramphenicol [4].

By virtue of the extensive bacterial resistance to numerous drugs and antibiotics, bacterial infections have become great health challenges, creating expanded concern in the search and the development of new antimicrobial agents. In this regard, during the past 20 years, the number of new drugs that have reached the marketplace has significantly decreased and the number of new antibiotics approved for marketing is in continuous decline. Thus, there is an urgent need for new discovery strategies to control antibiotic-resistant bacteria. One of the main approaches is the identification and exploitation of new targets in pathogens. Recently, with the dramatic reduction in the cost of bacterial genome sequencing and the development and evolution of bioinformatic tools, it has become possible to compare several bacterial genome sequences including those of pathogenic bacteria, to understand the interactions between targets and active compounds and consequently, to predict novel therapeutic targets against pathogenic microorganisms.

One of the easily manageable origins of such compounds are medicinal plants that offer a wide kind of phytochemicals with antimicrobial activity [5,6]. Plants and their derivatives, including essential oils (EO), have presented a broad range of secondary metabolites that are commonly reported to prevent or delay the growth of bacteria, yeasts, and molds [7–9]. In this respect, the antibacterial mechanism of EOs could be related to the phenolic constituents and their interaction with minor constituents [10–12]. However, the compounds' hydrophobicity, presented in EOs, permitted them to transfer throughout the cell wall and cytoplasmic membrane, perturb the structure of their different layers of polysaccharides, fatty acids, and phospholipids, and, eventually, permeabilize them [10,11]. Additionally, EOs can inhibit diverse enzyme systems covering the enzymes responsible for managing energy and synthesis of structural components [12]).

Containing twelve plant species, the genus *Cupressus* is distributed in different parts of the world, such as the Mediterranean regions [13]. *Cupressus sempervirens* is the sole species of this genus indigenous from Tunisia [14] and has been customarily utilized for influenza and rheumatism treatments, as an antiseptic, and against the inflammation, colds, curing diabetes [15]. From a chemical point of view, previous studies conducted on *C. sempervirens* EO (CSEO) have revealed that it contains various bioactive substances such as α-pinene [14,15], β-caryophyllene, and germacrene D [15], which are described to have considerable antimicrobial potentials [15–17].

On the other hand, the toxicological profiles of the majority of medicinal plant EOs have not been greatly elucidated. In this way, toxicity challenging in a varied range of in vitro studies using animal models is crucial and comprises experimental screening methods for determining the safety profile of EOs. By way of illustration, zebrafish embryos are helpful for assessing vertebrate development of endpoint morphological changes in toxicity studies [18,19]. Indeed, each growth stage (from fertilization, embryogenesis, and organogenesis to larva hatching) matches other higher vertebrates' embryogenesis, including humans [19–21].

The objective of our paper was to highlight the updated focus on the (CSEO) application in food preservation. By skillfully using in silico and software tools, here, molecular docking interactions of all (CSEO) compounds with bacterial DNA and RNA polymerases and DNA topoisomerase II (DNA gyrase) of the two pathogenic bacteria *Staphylococcus aureus* and *Salmonella enterica* Typhimurium, as well as molecular docking interactions of the major (CSEO) constituents with the cell membrane of these two pathogenic bacteria, were investigated. In addition, the in vivo innocuity of acute exposure of the efficient concentration of *C. sempervirens* EO was explored.

2. Results and Discussion

2.1. Chemical Composition Analysis of CSEO

(CSEO) GC-MS analysis exhibited 27 different components (Table 1). The main components were α-pinene, the most abundant compound (38.47%), δ-3-carene (25.14%), D-limonene (5.84%), and citronellal (5.33%). Additionally, four components were present in more than 2% of the (CSEO), which are α-terpinyl acetate (2.82%), β-myrcene (2.78%), cedrol (2.24%), and β-pinene (2.04%).

Table 1. Chemical composition of (CSEO).

Compound	Molar Mass (g/mol)	Molecular Formula	Retention Time (min)	EO (%)
β-terpinene	136.23	$C_{10}H_{16}$	4.80	0.11
Tricyclene	136.23	$C_{10}H_{16}$	5.25	0.22
α-pinene	136.23	$C_{10}H_{16}$	5.76	38.47
α-fenchene	136.23	$C_{10}H_{16}$	5.95	1.36
Sabinene	136.23	$C_{10}H_{16}$	6.67	1.18
β-pinene	136.23	$C_{10}H_{16}$	6.75	2.04
β-myrcene	136.23	$C_{10}H_{16}$	7.22	2.78
δ-3-carene	136.23	$C_{10}H_{16}$	7.91	25.14
D-limonene	136.23	$C_{10}H_{16}$	7.99	5.84
P-cymene	134.22	$C_{10}H_{14}$	8.23	0.86
Linalool	154.25	$C_{10}H_{18}O$	10.62	0.38
Isopulegol	154.25	$C_{10}H_{18}O$	11.96	0.88
Citronellal	154.25	$C_{10}H_{18}O$	12.23	5.33
Borneol	154.25	$C_{10}H_{18}O$	12.58	1.37
Terpinen-4-ol	154.25	$C_{10}H_{18}O$	12.92	1.55
α-terpineol	154.25	$C_{10}H_{18}O$	13.39	0.54
β-citronellol	156.26	$C_{10}H_{20}O$	14.54	0.21
α-fenchyl acetate	196.29	$C_{12}H_{20}O_2$	15.95	1.28
Camphene	136.23	$C_{10}H_{16}$	16.29	0.29
α-terpinyl acetate	196.29	$C_{12}H_{20}O_2$	17.73	2.82
α-zingibirene	204.35	$C_{15}H_{24}$	19.31	0.52
α-carophyllene	204.35	$C_{15}H_{24}$	19.49	0.83
α-humulene	204.35	$C_{15}H_{24}$	20.36	0.18
Germacrene D	204.35	$C_{15}H_{24}$	20.60	0.83
α-amorphene	204.35	$C_{15}H_{24}$	20.94	0.18
δ-cadinene	204.35	$C_{15}H_{24}$	20.09	0.30
Cedrol	222.37	$C_{15}H_{26}O$	24.02	2.24
Monoterpenes hydrocarbons				**78.29 (%)**
Oxygenated monoterpenes				**17.28 (%)**
Sesquiterpens				**2.84 (%)**
Total				**98.41 (%)**

α-Pinene [2,6,6,-trimethylbicyclo(3.1.1)-2-hept-2-ene] is a natural and active monoterpene that is used in flavorings, fragrances, insecticides, fine chemicals, and pharmaceuticals [22]. Several studies have attributed interesting biological activities to α-pinene, including antimicrobial [23], hypertensive [24], antinociceptive [25], and anti-inflammatory [26]. In addition, the US Food and Drug Administration [27] approved this compound as a food additive generally recognized as safe. δ-3-carene, a bicyclic monoterpene, is widely known for its antimicrobial activity, notably against *Aspergillus* and *Candida* species [28], antitussive and expectorant properties [29], and its activity against acute inflammation [30]. D-limonene, a simple monocyclic monoterpene, has been used as a flavor and fragrance additive in cleaning and cosmetic products, food, beverages, and pharmaceuticals [31] and has demonstrated potential chemo-preventive and anticancer activity in preclinical and clinical studies [32]. The monoterpenoid citronellal presents many interesting clinical activities like its central nervous system depressant and anticonvulsant [33] and its potential benefit in managing inflammatory disorders and correlated damage caused by oxidant agents [34].

In our case, the major constituents of (CSEO) were α-pinene, δ-3-carene, D-limonene, and citronellal. Nevertheless, it should be noted that the essential oil composition of plants is closely related to the location of the plant and the method used to extract and isolate essential oils. In fact [35], have studied the aerial parts of the same plant *Cupressus sempervirens* collected from Makther in Tunisia, using the same hydrodistillation technique obtained a significant variation in (CSEO) composition and the major compounds were α-pinene (31.61%), α-cedrol (13.50%), δ-3-carene (9.50%) and germacrene D (8%). This difference in both composition and compounds percentage of (CSEO) can be explained by the fact that the region of Sfax, Tunisia, is characterized by semi-arid climatic conditions with an annual average precipitation of 230 mm, while Makther, Tunisia, is known for its Mediterranean climate and about 450 mm of precipitation falls annually in this region World widely, the literature revealed wide variations in the (CSEO) collected from different locations [14,17].

2.2. Antibacterial Activity

(CSEO) has been screened for its antibacterial activity against four bacterial strains (two Gram-positive and two Gram-negative bacteria). The antibacterial activity was assessed by evaluation of the growth inhibition zones and the determination of MIC values. As compiled in Table 2, (CSEO) showed antibacterial activity against the four tested strains with inhibition zones of 21 and 15 mm against Gram-positive and Gram-negative bacteria, respectively. However, it should be noted that the monoterpene family is known by its antibacterial activity [36], and in our case, this family represented a high level (74.78%) of the (CSEO) composition with α-pinene (38.47%), δ-3-carene (25.14%), D-limonene (5.84%), and citronellal (5.33%). The obtained MIC values (Table 2), indicated that (CSEO) was most effective against Gram-positive bacteria (MICs = 6.25 μg/mL) than Gram-negative bacteria (MICs = 12.5 μg/mL).

Table 2. Antibacterial activity of (CSEO); zones of growth inhibition expressed in (mm) and minimum inhibitory concentrations (MICs) expressed in (μg/mL) of the (CSEO) and the standard antibiotic gentamicin. All tests were performed in triplicate; values with a different letter (a,b) within a row for each antibacterial test are significantly different ($p < 0.05$).

Bacterial Strains	Inhibition Zones Diameters (mm)		MIC (μg/mL)	
	(CSEO)	Gentamicin	(CSEO)	Gentamicin
S. aureus ATCC 6538	21 ± 1.00 [a]	20 ± 0.83 [a]	6.25 ± 0.00 [a]	12.5 ± 0.00 [b]
L. monocytogenes ATCCC 19117	21 ± 0.83 [a]	20 ± 0.66 [a]	6.25 ± 0.00 [a]	12.5 ± 0.00 [b]
S. Typhimurium ATCC 14028	15 ± 0.5 [a]	25 ± 1.00 [b]	12.5 ± 0.00 [b]	2.5 ± 0.00 [a]
E. coli ATCC 8739	15 ± 0.5 [a]	25 ± 1.25 [b]	12.5 ± 0.00 [b]	2.5 ± 0.00 [a]

These data were in concordance with previous findings and it has been reported that Gram-positive bacteria were more sensitive to plant essential oils than Gram-negative bacteria. The resistance of these latter bacteria to plant essential oils was attributed to the presence of external lipopolysaccharide surrounding the peptidoglycan cell wall, which acts as a hydrophobic barrier to essential oils [37]. Concerning the antibiotic gentamicin used as standard, the antibacterial activity was higher against Gram-negative bacteria than Gram-positive bacteria with an inhibition zone of 25 mm and an MIC value of 2.5 μg/mL against Gram-negative bacteria, and an inhibition zone of 20 mm and an MIC value of 12.5 μg/mL against Gram-positive bacteria (Table 2). It should be mentioned that gentamicin is an aminoglycoside bactericidal and is a broad-spectrum antibiotic active against a wide range of bacterial infections; mostly Gram-negative bacteria [38].

2.3. Compounds Toxicity Evaluation by In Silico Tools

2.3.1. Computational COMPOUND Toxicity Prediction by VEGA HUB Software

By using VEGA HUB software: the QSAR (quantitative structure–activity relationship) approach, the toxicity of some selected food preservatives recommended by FDA (Food and Drug Administration), approved antibiotics: rifamycin SV and ciprofloxacin, efflux pump inhibitors (EPIs): cathinone and thioridazine, and (CSEO) compounds were elucidated (Table 3).

Table 3. Toxicity predictions of (CSEO) compounds and selected antibiotics using Vega QSAR model.

Toxicity Measurements / Compound	Mutagenicity (Ames Test) Model (CAESAR) 2.1.13	Carcinogenicity Model (CAESAR) 2.1.9	Developmental Toxicity Model (CAESAR) 2.1.7	Developmental/Reproductive Toxicity Library (PG) 1.1.0	Estrogen Receptor Relative Binding Affinity Model (IRFMN)	Androgen Receptor-Mediated Effect (IRFMN/COMPARA) 1.0.0	Thyroid Receptor Alpha Effect (NRMEA) 1.0.0	Thyroid Receptor Beta Effect (NRMEA) 1.0.0	In Vitro Micronucleus Activity (IRFMN/VERMEER) 1.0.0
β-terpinene	-	-	-	-	-	-	-	-	-
Tricyclene	-	-	-	-	+	-	-	-	-
α-pinene	-	-	+	-	-	-	-	-	-
α-fenchene	-	-	-	-	-	-	-	-	-
Sabinene	-	-	-	-	-	-	-	-	-
β-pinene	-	-	+	-	-	-	-	-	-
β-myrcene	-	+	-	-	-	-	-	-	+
δ-3-carene	-	-	-	-	-	-	-	-	-
D-limonene	-	+	-	-	-	-	-	-	-
P-cymene	-	-	-	+	-	-	-	-	-
Linalool	-	-	-	-	-	-	-	-	+
Isopulegol	-	+	+	+	-	-	-	-	-
Citronellal	-	-	-	-	-	-	-	-	-
Borneol	-	-	-	-	-	-	-	-	-
Terpinen-4-ol	-	-	+	-	-	-	-	-	-
α-terpineol	-	-	+	-	-	-	-	-	-
β-citronellol	-	-	-	-	-	-	-	-	-
α-fenchyl acetate	-	-	-	-	-	-	-	-	-
Camphene	-	-	-	-	-	-	-	-	-
α-terpinyl acetate	-	-	-	-	-	-	-	-	-
α-zingibirene	-	-	-	-	-	-	-	-	+
α-carophyllene	-	-	-	-	-	-	-	-	+
α-humulene	-	-	-	-	-	-	-	-	+
Germacrene D	-	-	-	-	-	-	-	-	+
α-amorphene	-	+	+	-	-	-	-	-	+
δ-cadinene	-	+	+	-	-	-	-	-	-
Cedrol	-	-	+	-	-	-	-	-	-
Ciprofloxacin	+	-	+	+	-	-	-	-	+
Rifamycin SV	+	-	+	-	-	-	-	-	+

(-): Non toxicant/inactive; (+): Toxicant/active.

The selected antibiotics are predicted to be toxic in different assays. In this regard rifamycin SV and ciprofloxacin are found to be mutagenic in the mutagenicity test/model (Ames test) and predicted to be toxic in the developmental toxicity model. These two antibiotics also produce active genotoxicity signals in the in vitro micronucleus activity model Similar trends have been observed in recommended food preservatives. The latter showed some toxicity measurements and none of the selected preservatives were detected to be nontoxic (Table 4). A previous study conducted by Damayanti et al. (2015) evaluated the toxicity of some food preservatives in silico using Toxtree and OECD QSAR Toolbox software [39]. These authors reported that ascorbic acid is slightly toxic; butylhydroxyanisole (BHA) was predicted to be moderately toxic, carcinogen, and could engender reproduction toxicity; and citric acid was demonstrated to be slightly toxic and carcinogenic.

Table 4. Toxicity assessment of food preservatives and efflux pumps inhibitors (controls) using the Vega QSAR model.

Toxicity Measurements	Citric Acid	Butylated Hydroxyanisol (BHA)	Ascorbic Acid	Propionic Acid	Benzoic Acid	Cathinone	Thioridazine
Mutagenicity (Ames test) model (CAESAR) 2.1.13	-	-	-	-	-	-	-
Carcinogenicity model (CAESAR) 2.1.9	-	+	-	-	-	-	-
Developmental Toxicity model (CAESAR) 2.1.7	-	-	-	+	+	+	+
Developmental/Reproductive Toxicity library (PG) 1.1.0	-	-	-	+	-	+	+
Estrogen Receptor Relative Binding Affinity model (IRFMN)	-	-	-	-	-	-	+
Androgen Receptor-mediated effect (IRFMN/COMPARA) 1.0.0	-	-	-	-	-	-	-
Thyroid Receptor Alpha effect (NRMEA) 1.0.0	-	-	-	-	-	-	-
Thyroid Receptor Beta effect (NRMEA) 1.0.0	-	-	-	-	-	-	-
In vitro Micronucleus activity (IRFMN/VERMEER) 1.0.0	+	-	+	Not predicted	-	+	+

(-): Non toxicant/inactive; (+): Toxicant/active.

Additionally, EPIs were predicted to cause reproductive/developmental toxicity and produce genotoxic signals. For instance, cathinone's toxicity was previously evaluated in vitro and experiments confirmed that this EPI is hepatotoxic and causes huge damage in the liver [40]. Similarly, thioridazine has been associated with liver toxicity [41]. Therefore, these molecules are quite unsafe and can be harmful to human health.

Interestingly, many (CSEO) components such as β-terpinene, α-fenchene, sabinene, δ-3-carene, citronellal, borneol, β-citronellol, α-fenchyl acetate, camphene and α-terpinyl acetate showed no toxicity effect. This predicts that these molecules can be safe antimicrobial agents, economically low-cost choice as compared to synthetic antibiotics and could also be used as bio-preservative agents in a view to extend the shelf life of stored products, enhance the nutritional quality and guarantee the safety of food for future consumption.

2.3.2. Rodent Oral Toxicity and Cytotoxicity of (CSEO) Compounds Predicted by PROTOX II Tool

The oral toxicity of (CSEO) compounds in rodents was predicted using the webserver PROTOX II. This tool divided the compounds into different classes based on their lethal dose upon swallowing [42,43]. None of the (CSEO) compounds were found to cause cytotoxicity. In addition, except for P-cymene, LD_{50} values of the 27 (CSEO) compounds and the two controls did not reveal any fatal or toxic molecule which belongs to Class 1, 2, and 3. Remarkably, the two (CSEO) major compounds, α-pinene and δ-3-carene, were predicted to belong to Class 5, which means that they can be harmful if swallowed ($2000 < LD50 \leq 5000$). Further, tested food preservatives were predicted to belong to class 3 and 4, meaning that these substances could be toxic if swallowed ($50 < LD50 \leq 300$) and harmful if swallowed ($300 < LD50 \leq 2000$), respectively, except L-ascorbic acid, which was categorized into class 5 (Table 5). Finally, cathinone and thioridazine were predicted to belong to class 4 (harmful if swallowed ($300 < LD50 \leq 2000$). The same model was used to evaluate the toxicity of chlorogenic acid in order to use it as a promising efflux pump inhibitor against AcrB protein of *E. coli* TG1. It was demonstrated to have a high LD50 value indicating that it is nontoxic [44]. Results of toxicity evaluation by the use of in silico tools confirmed that (CSEO) compounds could be safely used as antibacterial agents and biopreservatives as compared to synthetic food preservatives and FDA-approved antibiotics.

Table 5. Rodent oral toxicity and cytotoxicity of (CSEO) compounds, food preservatives, and efflux pumps inhibitors (EPI) predicted by the PROTOX II tool.

Compound	Cytotoxicity	Probability	LD 50 (mg/kg)	Toxicity Class
β-terpinene	Inactive	0.80	4400	5
Tricyclene	Inactive	0.77	15,380	6
α-pinene	Inactive	0.75	3700	5
α-fenchene	Inactive	0.74	5000	5
Sabinene	Inactive	0.71	5000	5
β-pinene	Inactive	0.71	4700	5
β-myrcene	Inactive	0.75	5000	5
δ-3-carene	Inactive	0.71	4800	5
D-limonene	Inactive	0.82	4400	5
P-cymene	Inactive	0.89	3	1
Linalool	Inactive	0.82	2200	5
Isopulegol	Inactive	0.93	5000	5
Citronellal	Inactive	0.82	2420	5
Borneol	Inactive	0.88	500	4
Terpinen-4-ol	Inactive	0.88	1016	4
α-terpineol	Inactive	0.64	2830	5
β-citronellol	Inactive	0.86	3450	5
α-fenchyl acetate	Inactive	0.73	3100	5
Camphene	Inactive	0.76	5000	5
α-terpinyl acetate	Inactive	0.80	4800	5
α-zingibirene	Inactive	0.82	1680	4
α-caryophyllene	Inactive	0.79	3650	5
α-humulene	Inactive	0.79	3650	5
Germacrene D	Inactive	0.83	5300	5
α-amorphene	Inactive	0.76	4400	5
δ-cadinene	Inactive	0.69	4390	5
Cedrol	Inactive	0.87	2000	4

Table 5. *Cont.*

Compound	Cytotoxicity	Probability	LD 50 (mg/kg)	Toxicity Class
Rifamycin SV	Inactive	0.60	2120	5
Ciprofloxacin	Inactive	0.92	2000	4
Citric acid	Inactive	0.73	80	3
BHA	Inactive	0.83	700	4
L-ascorbic acid	Inactive	0.65	3367	5
Propionic acid	Inactive	0.75	300	3
Benzoic acid	Inactive	0.86	235	3
Cathinone	Inactive	0.82	400	4
Thioridazine	Inactive	0.68	360	4

Class 1: fatal if swallowed (LD50 ≤ 5); Class 2: fatal if swallowed (5 < LD50 ≤ 50); Class 3: toxic if swallowed (50 < LD50 ≤ 300); Class 4: harmful if swallowed (300 < LD50 ≤ 2000); Class 5: may be harmful if swallowed (2000 < LD50 ≤ 5000); Class 6: nontoxic (LD50 > 5000).

2.4. In Vivo Toxicity Assessment Using Zebrafish Model

By using the zebrafish model in toxicity screening, findings showed that the use of DMSO at 0.1%, used as a positive control, did not display any toxicological effect on zebrafish embryonic development (Figure 1). This result came to approve previous studies achieved by Hoyberghs et al. (2020) [45] and Thitinarongwate et al. (2021) [21]. For concentrations superior to 6 µg/mL, a significant ($p < 0.05$) increase in mortality rate in zebrafish embryos was detected. In addition, no viable zebrafish embryos were observed for the groups treated with 8 µg/mL of (CSEO); therefore, (CSEO) toxicity is a concentration-dependent effect. On the other hand, the LC_{50} obtained in the present study was 6.6 µg/mL. It should be noted that we studied the toxicity of all 27 components of (CSEO) separately by using the Vega QSAR model and, according to Table 3, we have demonstrated that some components are totally safe and others possess a limit of toxicity. Among these later, α-pinene developed toxicity only in the developmental toxicity model (CAESAR) 2.1.7 out of nine toxicity measurements. Concerning the in vivo study (LC50), we used the whole (CSEO), and consequently, we must find a certain limit of toxicity.

Similar assays were conducted using the zebrafish model to assess different EOs' toxicities. Thitinarongwate et al. (2021) evaluated the toxic effects of *Zingiber ottensii* Valeton (ZO) EO [21]. These authors concluded that the LC_{50} value was equal to 1.003 µg/mL, meaning that ZO EO showed more toxic effects on zebrafish embryos as compared to the query EO.

2.5. Antibacterial Mechanisms of (CSEO)

2.5.1. Alteration of Bacterial Cell Permeability: Inhibition of Efflux Pumps by (CSEO)

To better understand the mechanism of membrane permeability alteration by inhbiting efflux pumps against *S. enterica* Typhimurium and *S. aureus*, two major foodborne pathogen bacteria, docking studies of all (CSEO) components were carried out. In this approach, the AcrB efflux pump protein model of *S. enterica* Typhimurium and the cristallographic structure of MepR of *S. aureus* (PDB ID: 3ECO) were established. Homology modeling results displayed that AcrB model identity was 94.47% with a QMEAN value equal to −3.04 and Ramachandran Plot values of favored regions and allowed regions were >90%.

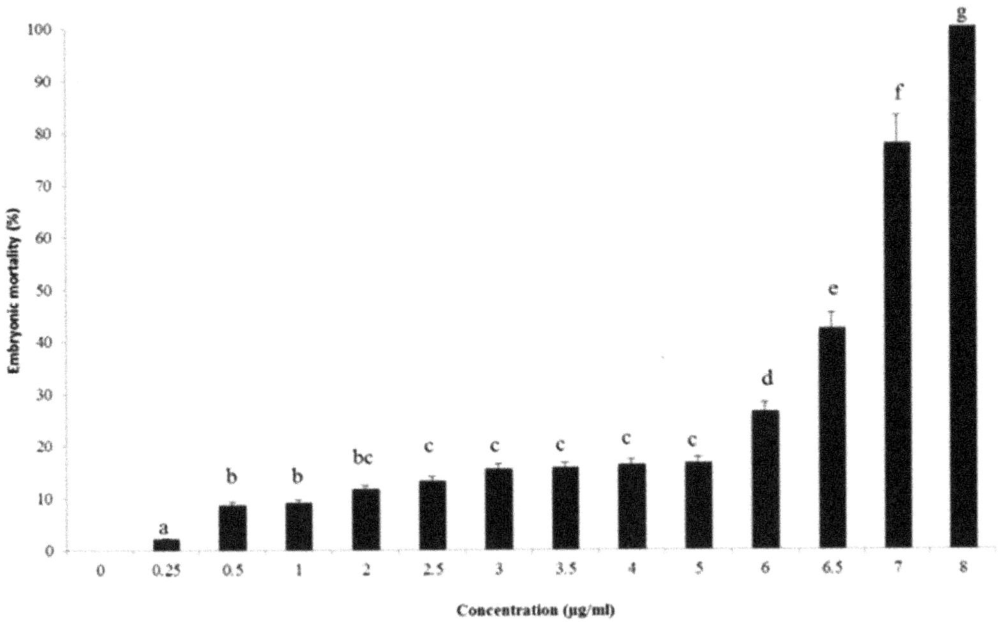

Figure 1. The embryotoxicity of different concentrations of (CSEO) in embryonic zebrafish mortality. All tests were performed in triplicate; Values with a different letter (a–g) are significantly different ($p < 0.05$).

It should be noted that the AcrAB-TolC is the major RND (resistance–nodulation–division) efflux system providing the *S. enterica* Typhimurium resistance to many antibiotics [46]. The AcrAB-TolC system is specifically formed by the AcrB efflux pump associated with an outer membrane protein (TolC). This complex pump out therapeutic molecules accumulated in bacterial periplasmic space after binding to them, resulting to a broad substrate specificity against several classes of antibiotics (phenicols, cyclins, β-lactams, fluoroquinolones) [47]. On the other hand, MepR is a multidrug binding transcription regulator that represses the expression and the activity of the multidrug efflux pump MepA of *S. aureus* [48]. MepA is a transporter that belongs to the MATE family (multidrug and toxic compound extrusion). This efflux pump is able to engender antibiotic resistance in *S. aureus* by extruding hydrophilic antibiotics such as fluoroquinolones and aminoglycosides [49]. MepR was chosen for molecular docking to understand the mechanism of MepA efflux pump inhibition.

Docking results displayed that the two major compounds of (CSEO), α-pinene and δ-3-carene showed free energy of binding at −6.7 Kcal/mol and −6.4 Kcal/mol, respectively (Table 6). These later free energies of binding are better than cathinone. This alkaloid, basically used as a dopamine stimulator in the central nervous system (amphetamine), showed a free energy of binding equal to −4.8 Kcal/mol and employed against *S. enterica* Typhimurium by inhibiting the function of AcrABTolC efflux pump [50,51]. Moreover, it was reported that the repression of AcrB efflux function can induce loss of virulence in *S. enterica* Typhimurium by the reduction of bacterial factors involved during infection, leading to an alteration of noxious molecules retention inside the bacterium [46].

Table 6. Molecular docking results for complexes between compounds of (CSEO) and efflux pump targets of *S. aureus* and *S. enterica* Typhimurium by using Autodock Vina (kcal/mol).

Compound	AcrsB Efflux Pump	MepR
β-terpinene	−5.1	−5.5
Tricyclene	−5.8	−4.7
α-pinene	**−6.7**	**−6.5**
α-fenchene	−5.3	−4.8
Sabinene	−5.0	−5.2
β-pinene	−5.0	−4.9
β-myrcene	−4.8	−4.5
δ-3-carene	**−6.4**	**−6.2**
D-limonene	−5.3	−5.3
P-cymene	−6.2	−5.6
Linalool	−4.8	−4.4
Isopulegol	−5.2	−5.1
Citronellal	−4.2	−3.9
Borneol	**−7.9**	**−7.7**
Terpinen-4-ol	−5.3	−5.0
α-terpineol	−5.3	−5.3
β-citronellol	−4.4	−4.3
α-fenchyl acetate	−6.4	−4.3
Camphene	−5.4	−4.8
α-terpinyl acetate	−6.2	−5.2
α-zingibirene	−5.4	−5.6
α-carophyllene	−6.0	−6.3
α-humulene	−6.5	−6.3
Germacrene D	−7.0	−6.4
α-amorphene	−6.5	−6.6
δ-cadinene	−6.6	−6.7
Cedrol	−6.3	−6.5
Cathinone	−4.8	-
Thioridazine	-	−6.7

Previous studies reported that the α-pinene detected in *Alpinia Katsumadai* seeds EO can inhibit the activity of Gram-negative efflux pumps [52]. It also showed an inhibitory effect against *Campylobater jejuni* (Gram negative bacterium) by reducing CmeABC and Cj1687 efflux pumps activities in order to increase bacterial susceptibility to ciprofloxacin, erythromycin, and triclosan [52]. Another study confirmed the existence of a synergitic effect between different monoterpenes hydrocarbons compounds present in *Citrus aurantium* L EO such as pinene, δ-3-carene and D-limonene in a view to act as bio-enhancer of antibiotics and to limit the emergence of drug-resistant infections [53].

However, it's important to mention also that the borneol, an oxygenated monoterpene, showed an important inhibitory effect against AcrB efflux pump of *S.enterica* Typhimurium with the best free energy of binding value (−7.9 Kcal/mol) (Table 6). A previous study indicated that *Thymus* species such as *(T. broussonetii, T. marocanus, T. riatarum)* were proven to inhibit the AcrAB-TolC efflux system of some *Enterobacteriaceae* strains due to their high content in monoterpenes (carvacrol and borneol) which could act as efflux pump substrates and to disrupt bacterial membranes [54].

On the other hand, regarding anti-*S. aureus*, α-pinene and δ-3-carene revealed a great inhibitory effect on MepR, the transcription regulator of the MepA efflux pump, with free energies of binding equal to −6.5 Kcal/mol and −6.2 Kcal/mol, respectively (Table 6). Similar free energy of binding was detected for thioridazine (−6.7 Kcal/mol). This molecule had an inhibitory effect on efflux pumps of methicillin-resistant *Staphylococcus aureus* (MRSA) [55–58]. The previous study conducted by De Medeiros et al. (2017) showed that the presence of α-pinene, as a main component of *Croton growioides* EO, may modulate and reduce the activity of another efflux pump (NorA) of *S. aureus* in order to overcome bacterial resistance to antibiotics [59].

Remarkably, borneol showed an interesting affinity towards MepR (−7.7 Kcal/mol), as indicated in Table 6. To the best of our knowledge, none of the previous studies has reported the inhibitory effect of borneol on the MepR or MepA efflux pump. However, it has been generally described that some oxygenated monoterpenes such as nerol and 3,7-dimethyl-1-octanol were able to potentiate the antibiotic activity of norfloxacin and act as efflux pump inhibitors of NorA of *S. aureus* [60]. These in silico results confirmed the fact that (CSEO) compounds have a potential inhibitory effect which consists of the disruption of bacterial membrane permeability by inhibiting the function of efflux pumps.

The results of the interaction profiles between (CSEO) major compounds and the selected efflux pumps proteins of the two pathogens bacteria are represented in Figures 2 and 3. Results indicated that AcrB efflux pump receptor of *S. enterica* Typhimurium complexed with α-pinene showed Pi-alkyl and alkyl-type interactions with LEU972, LEU976, PHE1020, VAL1016, ILE 1019, and Van der Waals interaction with THR1015 (Figure 2A). The complex of δ-3-carene with AcrB efflux pump receptor presented alkyl and Pi-alkyl interactions with MET1008, ALA915, LEU914, and Van Der Waals interactions with GLY1009, GLY911, and THR1013 (Figure 2B). Additionally, borneol complexed with the AcrB efflux pump of *S. enterica* Typhimurium and displayed the existence of 5 types of interactions with nonpolar amino acids which have hydrophobic character. It showed Pi-alkyl and alkyl interactions with ALA553, LEU88, VAL905; also Van der Waals interactions with ALA873, PRO874, ILE382, Val909, LEU931, PRO906, ILE935, MET552, VAL557; one conventional hydrogen bond with ALA878, and Pi-stacked interactions with a polar amino acid TYR 877 and nonpolar aromatic amino acid PHE 556 (Figure 2C).

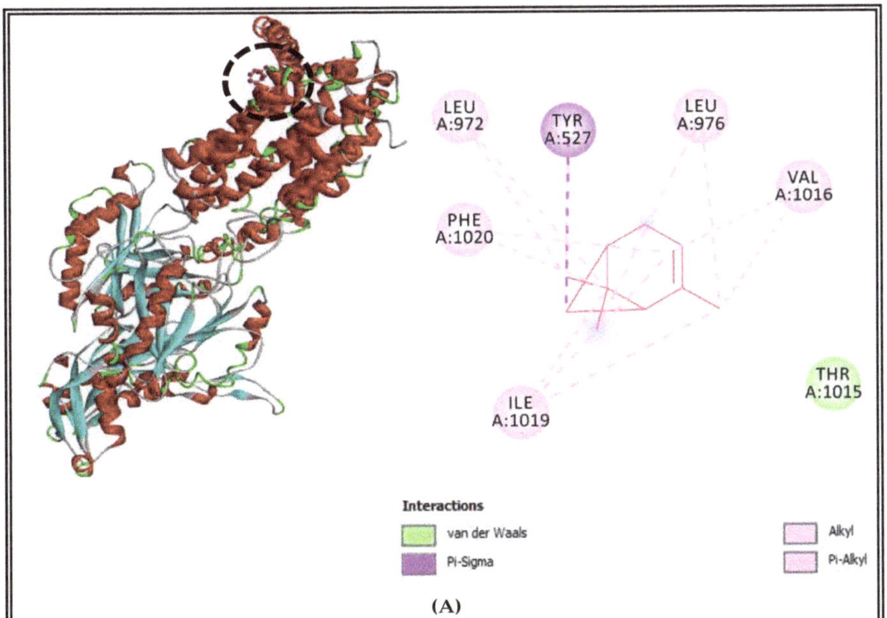

(A)

Figure 2. *Cont.*

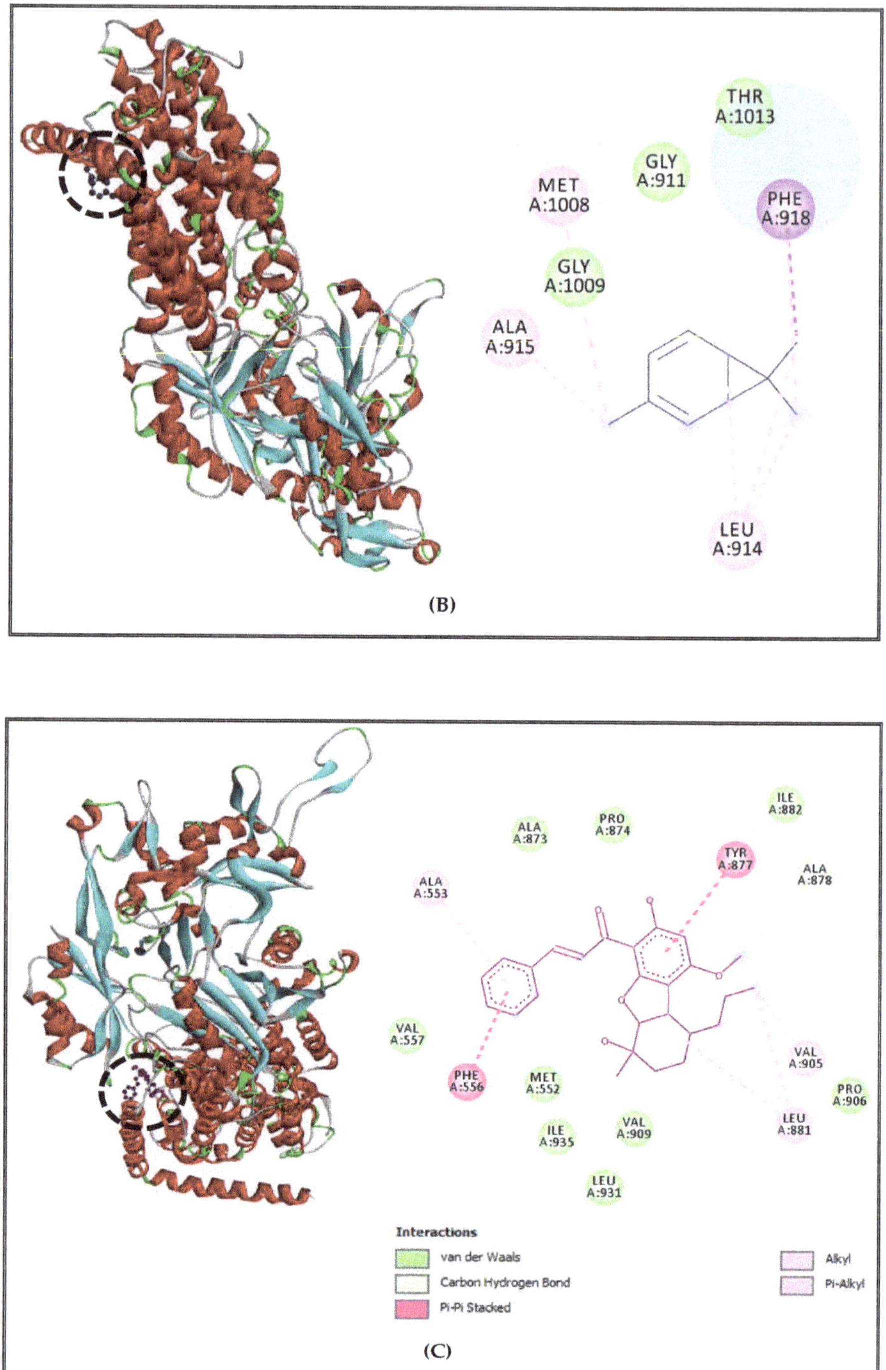

Figure 2. AcrB efflux pump receptor of *S. enterica* Typhimurium complexed with α-pinene (**A**), δ-3-carene (**B**), and borneol (**C**).

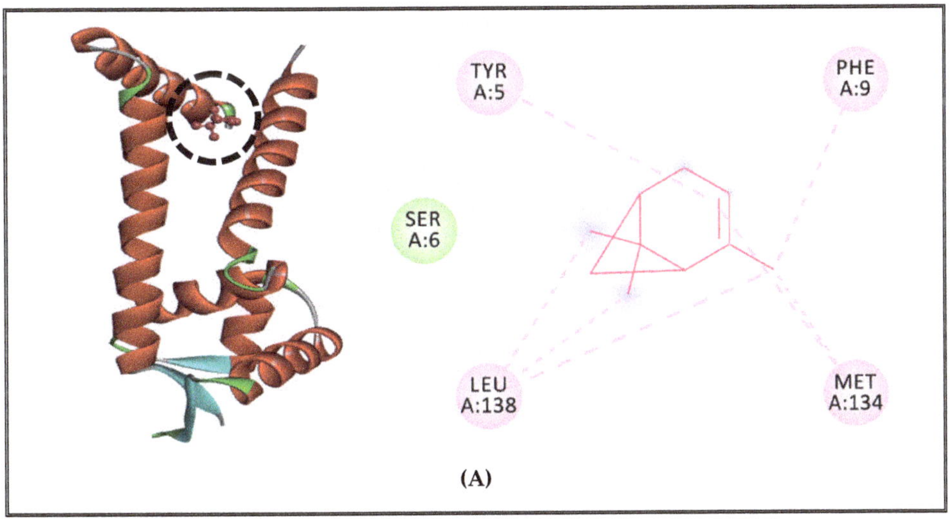

(A)

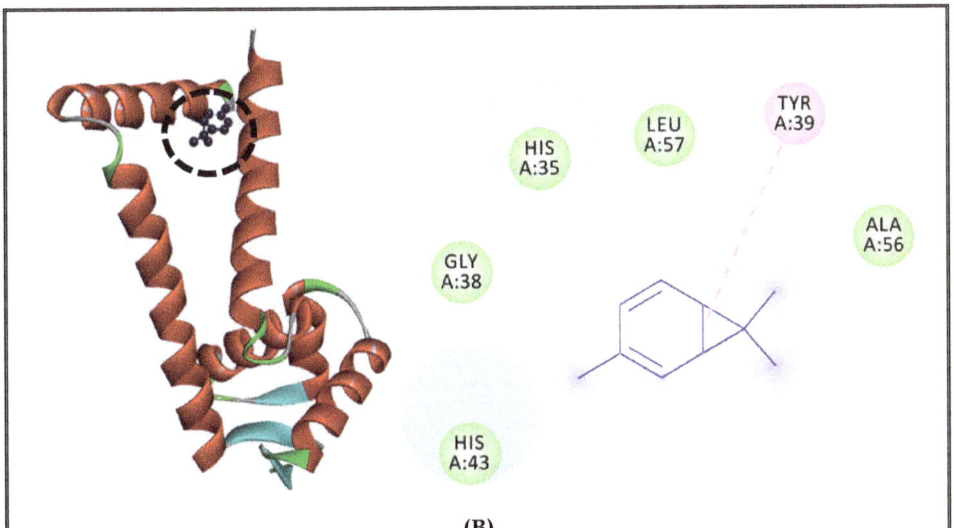

(B)

Figure 3. *Cont.*

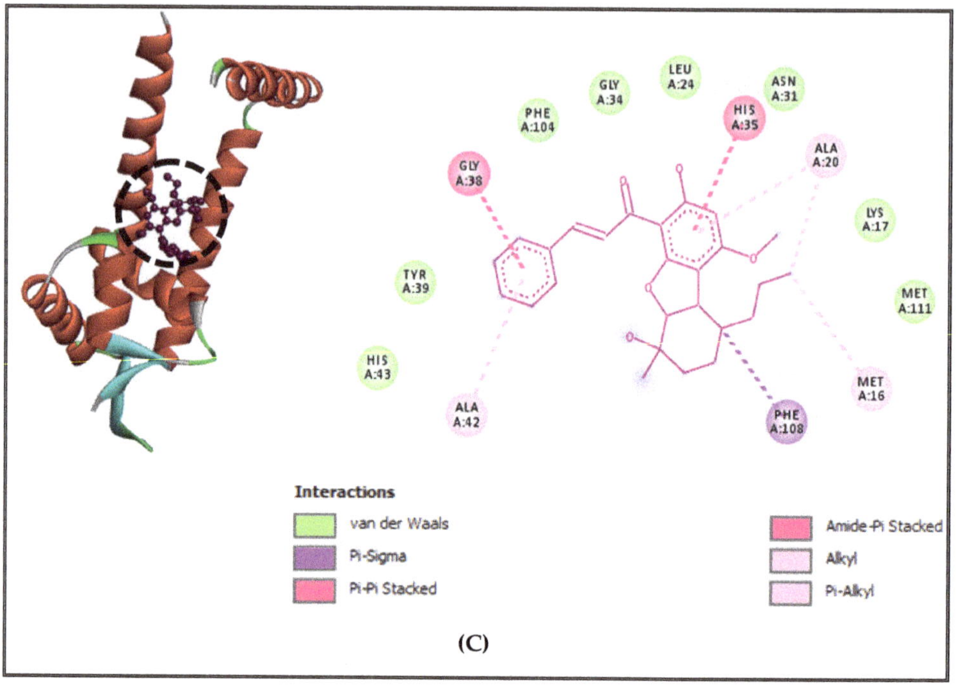

Figure 3. MepR receptor of *S. aureus* complexed with α-pinene (**A**), δ-3-carene (**B**), and borneol (**C**).

MepR transcription regulator of the MepA efflux pump complexed with α-pinene showed Pi-alkyl and alkyl interactions with TYR5, PHE9, LEU138, MET134 and a Van der Waals interaction with SER6 (Figure 3A). The complex of δ-3-carene with the MepR receptor showed Pi-alkyl interaction with TYR39 and Van der Waals interactions with HIS35, LEU57, GLY38,ALA56, and HIS43 (Figure 3B). Finally, borneol complexed with MepR, indicating the existence of 6 types of interactions as compared to both major compounds, which showed just 2 types of interactions. It showed the existence of alkyl and Pi-alkyl interactions with the nonpolar amino acids, which have hydrophobic characteristics such as MET16, ALA42, ALA20; Van der Waals interactions with polar amino acids such as HIS43, TYR39,ASN31, LYS17 and nonpolar amino acids such as GLY34, PHE104, LEU24, and MET111; Pi-sigma with nonpolar aromatic amino acid PHE108, Pi–Pi stacked and amide stacked with nonpolar amino acid GLY38 and a polar amino acid HIS35 (Figure 3C).

The docking scores and (CSEO) compounds' interaction profiles with target proteins confirmed that α-pinene, δ-3-carene, and borneol can bind to multiple targets involved in the efflux pump inhibition process and have a potential to block the AcrAB-TolC pump of *S. enterica* Typhimurium and the MepR transcription regulator of MepA of *S. aureus*. It might also be employed in other efflux pumps mechanisms.

These findings support the fact that monoterpenes present naturally in EO have the ability to inhibit efflux pumps mechanisms due to their hydrophobic character [61]. Monoterpenes could alter the membrane's permeability by moving to the bacterial membrane and interacting with polysaccharides, phospholipids, and fatty acids [62].

It is very interesting to pass from in silico methods to in vitro analysis, and it is already among our perspectives. Therefore, we project in subsequent work to perform further in vitro assays either by analyzing efflux of k+ and extracellular nucleotide leakage or by observing microscopic changes in cell structure and simulation experiments of artificial cell membranes.

2.5.2. Interactions between CSEO Molecules and Bacterial Topoisomerase II, DNA and RNA Polymerases

In order to understand the mechanism of the interactions between the (CSEO) compounds and pathogenic bacteria, some bacterial receptors were selected as possible targets. In this part, we investigated the inhibitory effect of (CSEO) on Topoisomerase II (DNA gyrase) and DNA and RNA polymerases of two foodborne bacteria *Staphylococcus aureus* and *Salmonella enterica* Typhimurium. Molecular homology results of the target- templates, their identity percentages, and their corresponding Ramachandran plot and QMEAN values are presented in Table A1. Templates can be used for homology modeling when their identities (%) are higher than 30% [63]. Hence, the identities of the selected templates were over 30% and Ramachandran Plot values of favored regions and allowed regions were, together, over 90%, therefore, predicted models can be used for molecular docking simulation [64].

Nevertheless, it is important to mention that DNA and RNA polymerases are two crucial enzymes which are playing an imperative role in DNA replication, transcription and translation as well as influencing the nucleic acid formation in bacterial cells [65]. Without forgetting that topoisomerase II (DNA gyrase) is also another pivotal enzyme in DNA transcription and translation. This enzyme can catalyze the unwinding of supercoiled DNA strands and is implicated in DNA replication and transcription processes [66]. For better understanding, we choose to perform molecular docking to simulate the binding mode and to estimate the inhibitory potential of (CSEO) on pathogenic bacteria by predicting the interaction energies between each compound and Topoisomerase II, DNA and RNA polymerases.

The main components of the tested (CSEO) were complexed with different bacterial receptors, and the results of binding affinity were elucidated in Table 7. Docking results showed that the two major compounds of (CSEO), α-pinene and δ-3-carene, showed a good inhibitory effect on Topoisomerase II, RNA polymerase, and DNA polymerase of both analyzed bacteria.

Previous studies reported that α-pinene has an important antibacterial activity against several Gram-negative and Gram-positive bacterial strains and activity against methicillin-resistant *staphylococcus aureus* (MRSA) [67,68]. It was also disclosed that it can cause damage to DNA and to bacterial membranes by increasing its permeability [69]. Moreover, α-pinene present in rockrose essential oil (39.25%) was employed to develop films used in food packaging in order to extend product shelf-life [70]. On the other hand, δ-3-carene could interrupt biofilm formation and cause damage to bacterial biosynthetic pathways [71]. Generally, EOs rich in α-pinene and δ-3-carene exhibit a stronger antimicrobial activity and could be employed as a good source of natural food preservatives [72].

At a percentage equal to 1.37, borneol showed the lowest free energy of binding (Kcal/mol) and the best inhibitory potential with topoisomerase II, DNA and RNA polymerase in both analyzed bacteria (Table 7). Previous studies have revealed the important antibacterial activity of borneol [73,74]. It was used as an antibacterial agent, showing excellent bactericidal activity via membrane disruption mechanism, especially against MRSA [75]. In addition, in other studies, borneol was reported to have anti-adhesion effects by minimizing bacterial attachment and biofilm formation [76,77]. This alcohol monoterpene exhibited great antibacterial activity by causing damage and impairment to the bacterial cell membrane. It presented broad-spectrum activity against both Gram-positive and Gram-negative bacteria and it was encapsulated and used as general surface disinfectants and as antiseptics for food preservation due to its effectiveness and safety [78].

The results of interaction profiles between α-pinene and topoisomerase II, DNA and RNA polymerases of *S. aureus* and S. Typhimurium are presented in Figures 4 and 5. α-pinene made a complex with DNA polymerase receptor via alkyl and Pi-alkyl interactions with ARG270, TYR273, LEU333, TYR634 and Van der Waals interactions with GLU335 and PHE334 (Figure 4A). In addition, RNA polymerase complexed with α-pinene showed Alkyl interactions with PRO142, VAL139, ILE163 and Van der Waals interactions with ARG140, ARG407, GLY492, ARG409, ASN165 and PRO164 (Figure 4B). Likewise, it interacted with

Topoisomerase II via Alkyl and Pi-alkyl interactions with ALA640, TYR192 and it made Van derWaals interactions with ASN636, ASP635,ASP218, ARG223, ASP215, VAL638, ARG217, TYR 190 and ASN191 (Figure 4C).

Table 7. Molecular docking results for complexes between compounds of *C. sempervirens* EO and protein targets of *S. aureus* and S. Typhimurium using Autodock Vina (kcal/mol).

	S. aureus (Strain Mu50/ATCC 700699)			S. Typhimurium (strain LT2/SGSC1412/ATCC 700720)		
Compound	**DNA Polymerase**	**RNA Polymerase**	**Topoisomerase II**	**DNA Polymerase**	**RNA Polymerase**	**Topoisomerase II**
β-terpinene	−5.2	−5.8	−5.5	−5.3	−6	−5.4
Tricyclene	−5.2	−5	−5.4	−5.2	−5	−5.2
α-pinene	−5.2	−5.1	−5.3	−5.4	−5.4	−5.2
α-fenchene	−5.4	−5.6	−5.2	−5.2	−5.1	−5.5
Sabinene	−5.2	−5.4	−5.4	−5.2	−5.2	−5.1
β-pinene	−5.5	−5.1	−5.3	−5.4	−5.4	−5.2
β-myrcene	−4.6	−4.3	−5.3	−4.7	−4.3	−4.3
δ-3-carene	−5.1	−6.1	−5.2	−5.6	−5.4	−5.6
D-limonene	−4.9	−4.6	−5.4	−4.9	−5.2	−5.1
P-cymene	−5.4	−5.2	−5.6	−5.4	−5.5	−5.2
Linalool	−4.8	−4.8	−5.2	−5.1	−4.5	−4.2
Isopulegol	−5.3	−5.8	−5.2	−5.4	−5.4	−5
Citronellal	−4.5	−4.4	−4.8	−4.6	−4.8	−4.2
Borneol	−7.7	−8.2	−8.8	−7.7	−7.3	−7.4
Terpinen-4-ol	−5.2	−5.3	−5.7	−5.7	−5.7	−5.3
α-terpineol	−5.3	−5.9	−5.6	−5.7	−5.7	−5.2
β-citronellol	−4.8	−4.3	−5	−5.1	−4.6	−4.7
α-fenchyl acetate	−5.7	−5.4	−5.6	−5.2	−5.3	−5.1
Camphene	−5.2	−5.4	−5.1	−5.5	−5.5	−5.2
α-terpinyl acetate	−5.4	−5.1	−6.4	−6	−6	−5.3
α-zingibirene	−5.8	−5.1	−6.1	−6.1	−5.2	−5.8
α-carophyllene	−6.2	−5.9	−6.8	−6.1	−6.1	−6.3
α-humulene	−6.1	−6.3	−6.8	−6.1	−6.1	−6.3
Germacrene D	−6.5	−7.1	−6.7	−6.2	−6.2	−6.5
α-amorphene	−6.8	−6.2	−6.9	−6.8	−6.2	−6
δ-cadinene	−6.3	−6.7	−6.7	−6.5	−6.7	−6.1
Cedrol	−6.6	−6.7	−6.8	−6.9	−6.2	−6.7
Rifamycin SV	−9.2	−9.8	-	−8.4	−8.6	-
Ciprofloxacin	-	-	−6.7	-	-	−6.3

The inhibition of bacterial DNA replication mechanism by the use of EOs was previously confirmed by studies reported by Dai and al. (2020) [79]. These authors analyzed the inhibitory effect of *Litsea cubeba* EO on topoisomerase, DNA and RNA polymerases of *E.coli*. De Souza-Moura et al. (2020) studied the antibacterial activity of *Siparuna guianensis* EO and revealed the inhibitory effect of germacrene B against bacterial DNA and RNA polymerases of *E.coli, P. aeruginosa, S. aureus,* and *S. pyogenes* [80]. Therefore, these findings revealed that (CSEO) has a potential inhibitory effect on pathogenic bacteria based on the inhibition of genetic material synthesis.

Interactions details of α-pinene, δ-3-carene, and borneol with the active sites of selected bacterial targets are summarized in (Table 8).

Table 8. Interaction details of (CSEO) compounds and the active sites of selected bacterial targets.

Bacteria	Compound	Targets	Number of Residues Interacting	Residues with H-Bond
S. aureus (strain Mu50/ATCC 700699)	α-pinene	DNA polymerase	6	-
		RNA polymerase	6	-
		Topoisomerase II	6	-
	δ-3-carene	DNA polymerase	8	-
		RNA polymerase	4	-
		Topoisomerase II	4	-
	Borneol	DNA polymerase	4	Ser 901, Lys 901
		RNA polymerase	11	Glu 79, Ala 672, Gly 670, Gln 725
		Topoisomerase II	5	Thr 194
S. Typhimurium (strain LT2/SGSC1412/ATCC 700720)	α-pinene	DNA polymerase	3	-
		RNA polymerase	9	-
		Topoisomerase II	8	-
	δ-3-carene	DNA polymerase	7	-
		RNA polymerase	6	-
		Topoisomerase II	2	-
	Borneol	DNA polymerase	8	Gly 640
		RNA polymerase	8	Asp81, Glu 963
		Topoisomerase II	8	Leu 509

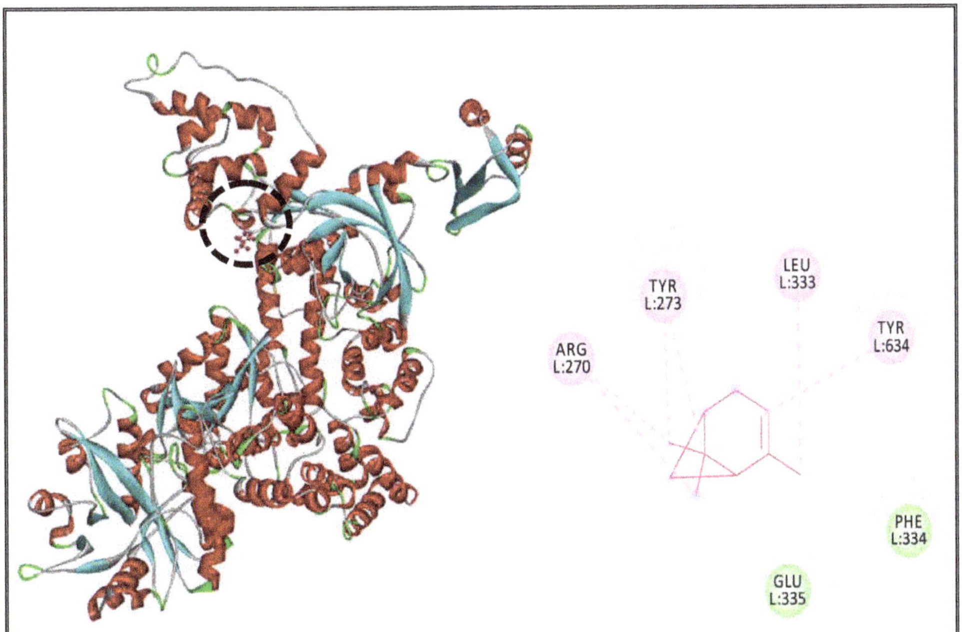

(A)

Figure 4. *Cont.*

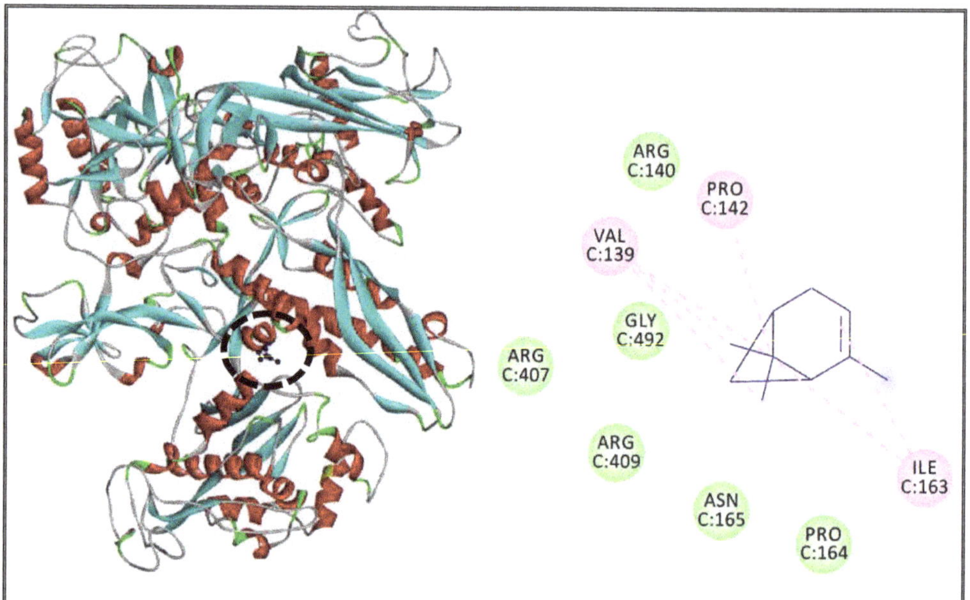

(B)

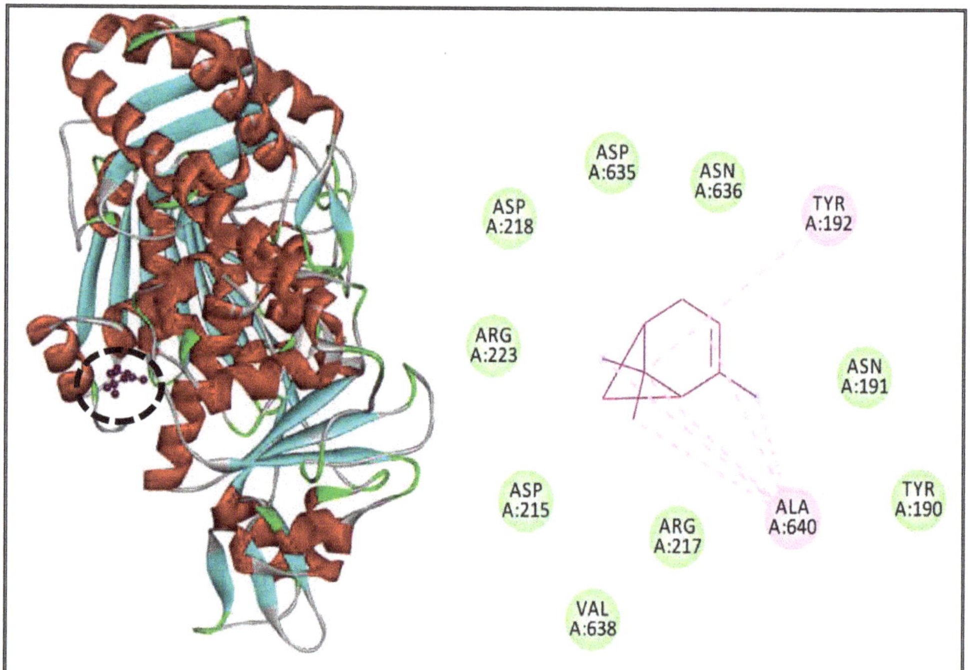

(C)

Figure 4. α-Pinene complexed with DNA polymerase enzyme (**A**), RNA polymerase (**B**), topoisomerase II (**C**) of *S. aureus*.

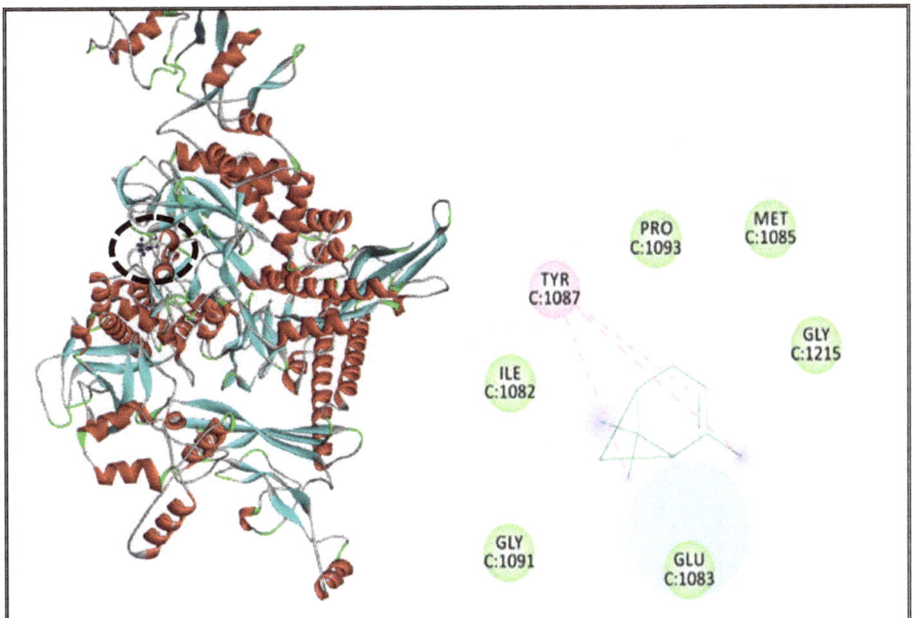

(A)

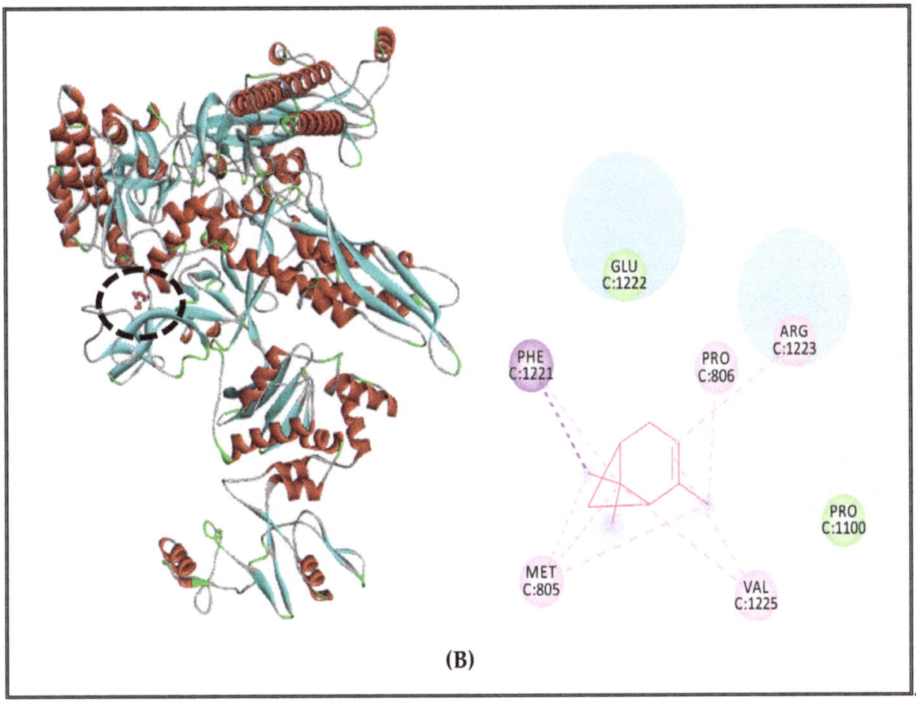

(B)

Figure 5. *Cont.*

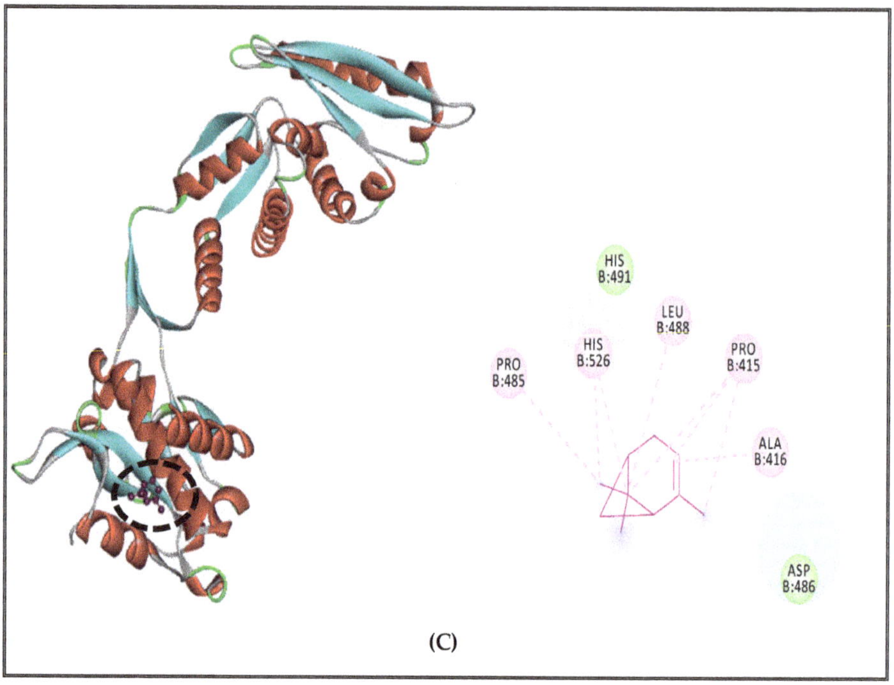

Figure 5. α-Pinene complexed with DNA polymerase enzyme (**A**), RNA polymerase (**B**), topoisomerase II (**C**) of *S. enterica* Typhimurium.

α-Pinene complex with *S. Typhimurium* DNA polymerase showed interactions with active site amino acids and the ligand: Pi-alkyl interaction with TYR1087 and Van der Waals interactions with GLY1091, GLU1083, ILE1082, PRO1093, MET1085, and GLY1215 (Figure 5A). For the RNA polymerase, we found alkyl and Pi-alkyl interactions with PRO806, ARG1223, VAL1225, MET805, and only two Van der Waals interactions with PRO1100 and GLU1222 (Figure 5B). Concerning topoisomerase II, α-pinene presented alkyl and Pi-alkyl interactions with PRO485, HIS526, LEU488, PRO415, ALA416, and Van der Waals with HIS491 and ASP486 (Figure 5C).

On the other hand, Interaction profiles between δ-3-carene and Topoisomerase II, DNA and RNA polymerases of *S. aureus* and S. Typhimurium are outlined in Figures 6 and 7). The complex between δ-3-carene and DNA polymerase of *S. aureus* displayed the existence of Alkyl interactions with ILE507, MET732, VAL722, ILE718, and Van der Waals interactions with LYS728, THR545, GLY546, ASN721, and ARG503 (Figure 6A). Moreover, it showed interactions with RNA polymerase via alkyl and Pi-alkyl interactions with VAL536, TRP39, and Ver der Waals with GLU413, GLU538, GLY540, ASN537, SER410, and SER36 (Figure 6B). Concerning topoisomerase II of *S. aureus*, δ-3-carene interacted with the receptor via alkyl interactions with VAL189, ALA640, and one Van der Waals interactions with ASN636, GLU193, ARG217, TYR192, TYR190, ASN191, TYR639, and VAL638 (Figure 6C).

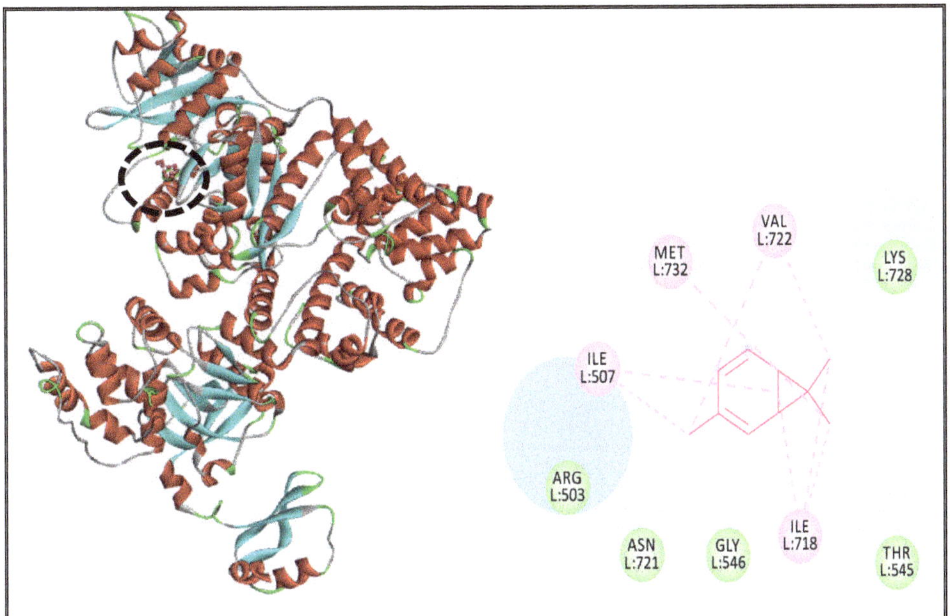

(A)

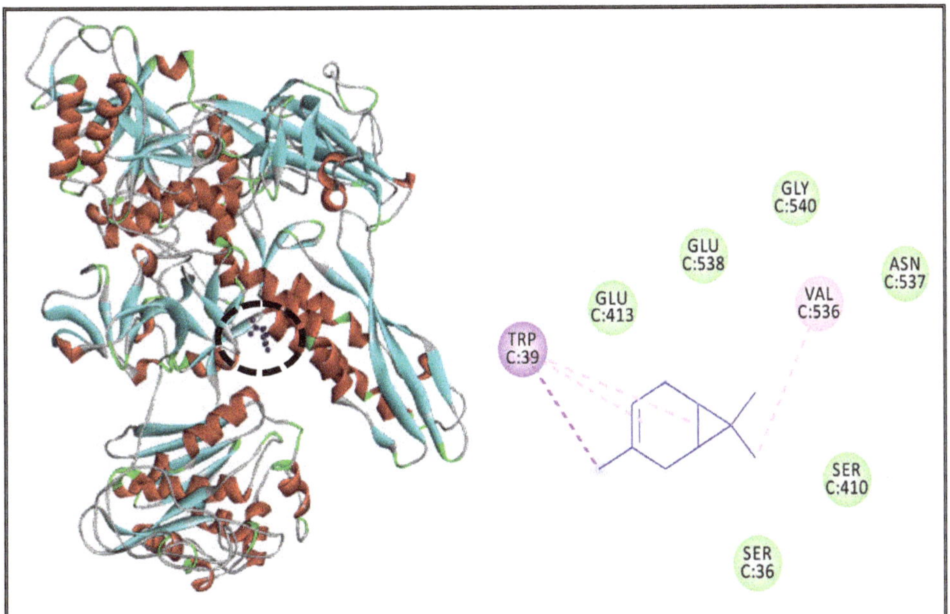

(B)

Figure 6. *Cont.*

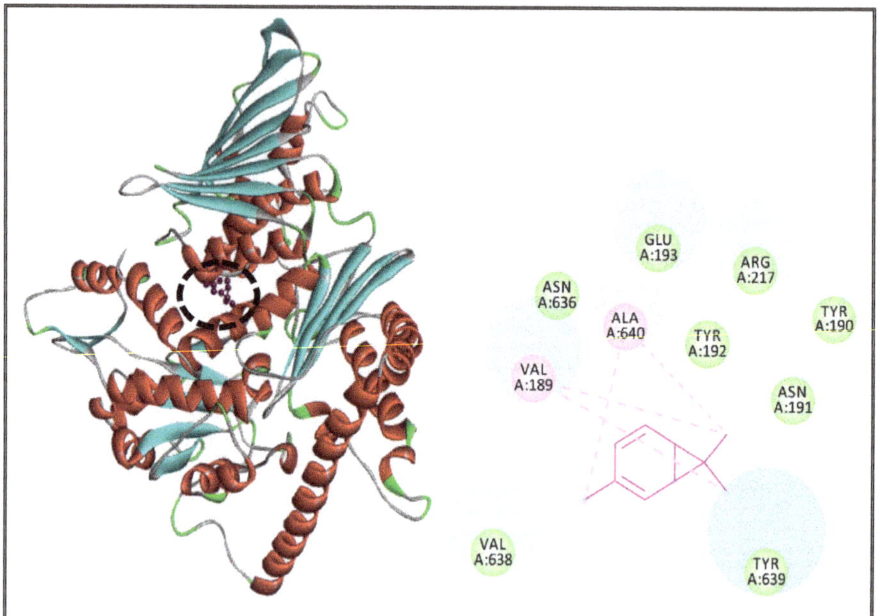

(C)

Figure 6. δ-3-Carene complexed with DNA polymerase enzyme (**A**), RNA polymerase (**B**), topoisomerase II (**C**) of *S. aureus*.

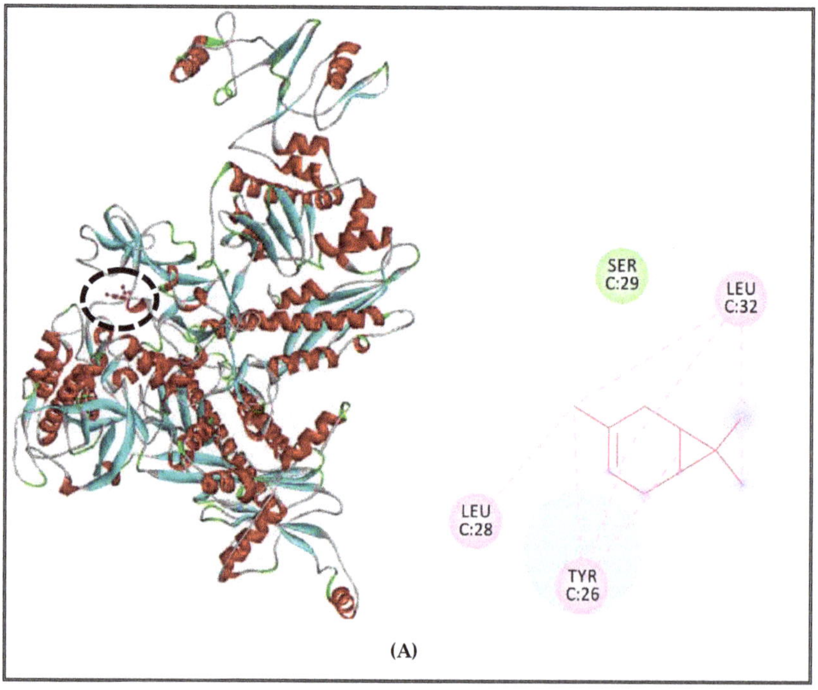

(A)

Figure 7. *Cont.*

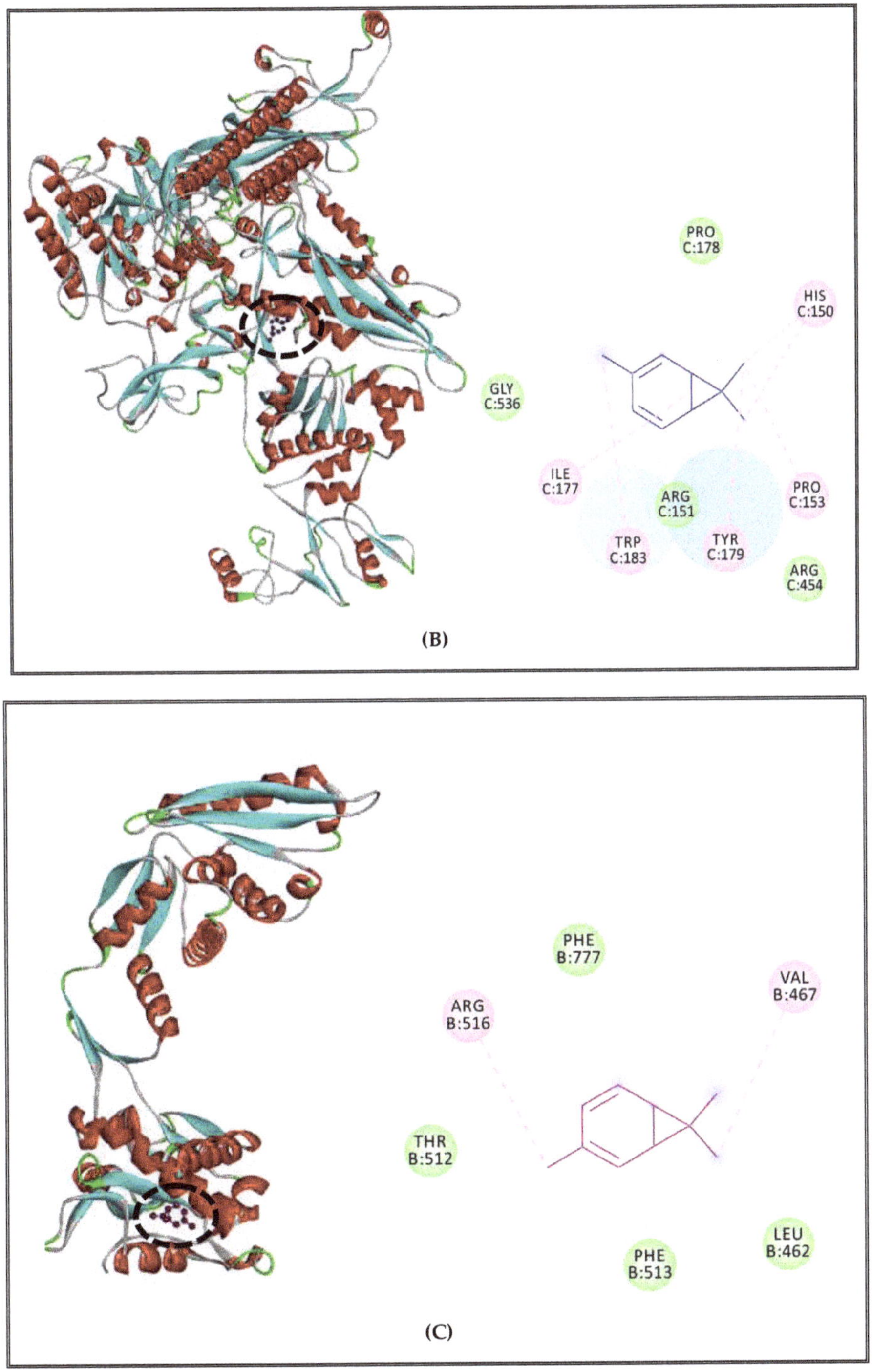

Figure 7. δ-3-carene complexed with DNA polymerase enzyme (**A**), RNA polymerase (**B**), topoisomerase II (**C**) of *S. enterica* Typhimurium.

DNA polymerase receptor of *S. enterica* Typhimurium complexed with δ-3-carene showed alkyl and Pi-alkyl interactions with LEU28, TYR26, LEU32 and only one Van der Waals interaction with SER29 (Figure 7A). Moreover, when complexed with RNA polymerase of S. Typhimurium, it revealed the existence of alkyl and Pi-alkyl interactions with ILE177, TRP183, TYR179, PRO153, HIS150 and Van der Waals with GLY536, ARG151, ARG454 and PRO178 (Figure 7B). Finally, the complex between δ-3-carene and the topoisomerase II of *Salmonella* exhibited the presence of alkyl interactions with VAL467, ARG516 and Van der Waals interactions with PHE777, LEU462, PHE513 and THR512 (Figure 7C).

Results also indicated that DNA polymerase of *S. aureus* receptors complexed with borneol, showing 3 types of interactions including alkyl-type interaction with nonpolar amino acid LEU941, Van der Walls interactions with polar amino acids GLN731, GLN975, GLN974, THR940, GLU939, ASN904, GLU735 and nonpolar amino acids LEU902, ILE899,PHE900, ILE938, and one conventional hydrogen bond with polar amino acid SER903 (Figure 8A). On the other hand, the complex with RNA polymerase presented 4 types of interactions including Alkyl and Pi-alkyl interactions with a polar amino acid TYR709 and nonpolar amino acids PRO710, ILE673, ALA712, Van der Walls interactions with nonpolar amino acid LEU78, ALA671 and polar amino acids GLN131, GLN472, LYS82, LYS715, THR122, LYS676, ASP121 and a conventional hydrogen bonds with two polar amino acids GLU79, GLN725 and two nonpolar amino acids GLY670 and ALA672 (Figure 8B). Additionally, the complex between borneol and topoisomerase II of *S. aureus* indicated the presence of 4 different interactions, including Pi-alkyl interactions with nonpolar amino acids ILE532, Leu521, ALA614; Van der Walls interactions with polar amino acids GLU613, ASP610, THR617, ASN171, GLU41, GLU609, ARG42, HIS46, HIS45, ARG198, TYR525 and nonpolar amino acids TRP49, LEU608, VAL606; one conventional hydrogen bond with a polar amino acid THR194; and Pi–Pi interaction with nonpolar aromatic amino acid PHE618 (Figure 8C).

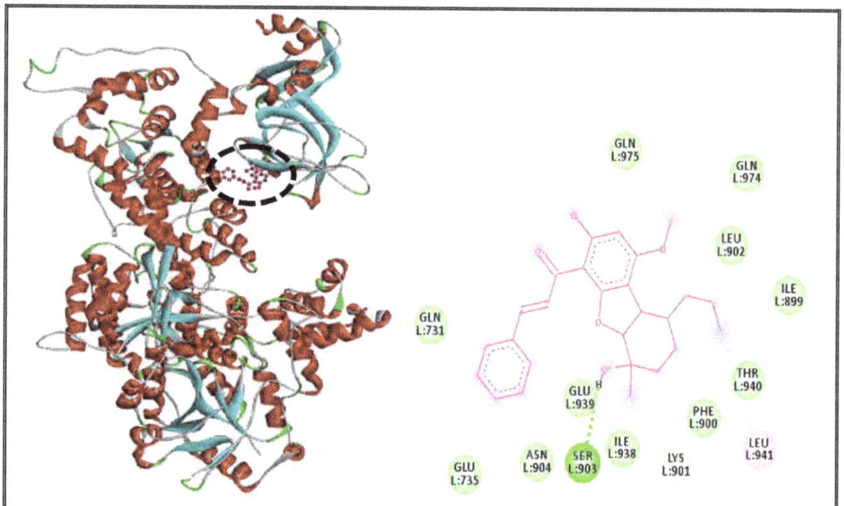

(A)

Figure 8. *Cont.*

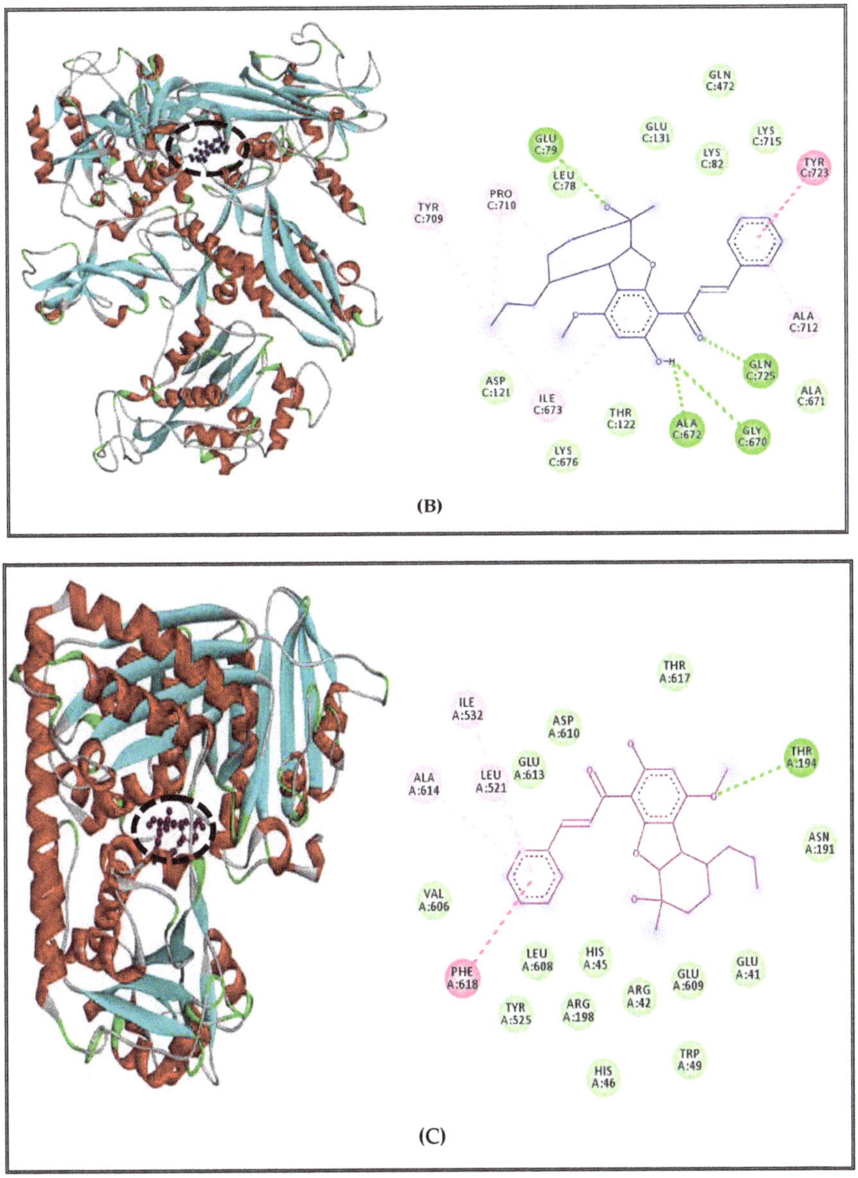

Figure 8. Borneol complexed with DNA polymerase enzyme (**A**), RNA polymerase (**B**), topoisomerase II (**C**) of *S.aureus*.

Borneol complex with *S. enterica* Typhimurium DNA polymerase showed different interactions with active site amino acids and the ligand: alkyl and Pi-alkyl interactions with nonpolar amino acids PRO552, PRO560, VAL550, VAL660, ALA619; Van der Walls interactions with polar amino acids such as TYR555, HIS554, GLN618, GLU641, ARG637, THR657; and only one conventional hydrogen bond with nonpolar amino acid GLY640 (Figure 9A). For the RNA polymerase, we found alkyl interactions with nonpolar amino acids ALA956, ALA1031 and a polar amino acid LYS1028; Van der Walls interactions with nonpolar amino acid LEU960 and polar amino acid ASN752, Pi-cation with polar amino acid LYS1032 and two conventional hydrogen bond with polar amino acid ASP81 and GLU963 (Figure 9B). Finally, the complex between borneol and topoisomerase II showed 5 types of interactions, including alkyl and Pi-alkyl with nonpolar amino acids such as LEU780, MET461, VAL467, PHE513, and a polar amino acid LYS460; Van der Walls interactions with nonpolar amino acids MET781, LEU462, LEU510 and polar amino acids such as SER464, THR512; one conventional hydrogen bond with nonpolar amino acid LEU509 and Pi-sigma interaction with nonpolar aromatic amino acid PHE777 (Figure 9C).

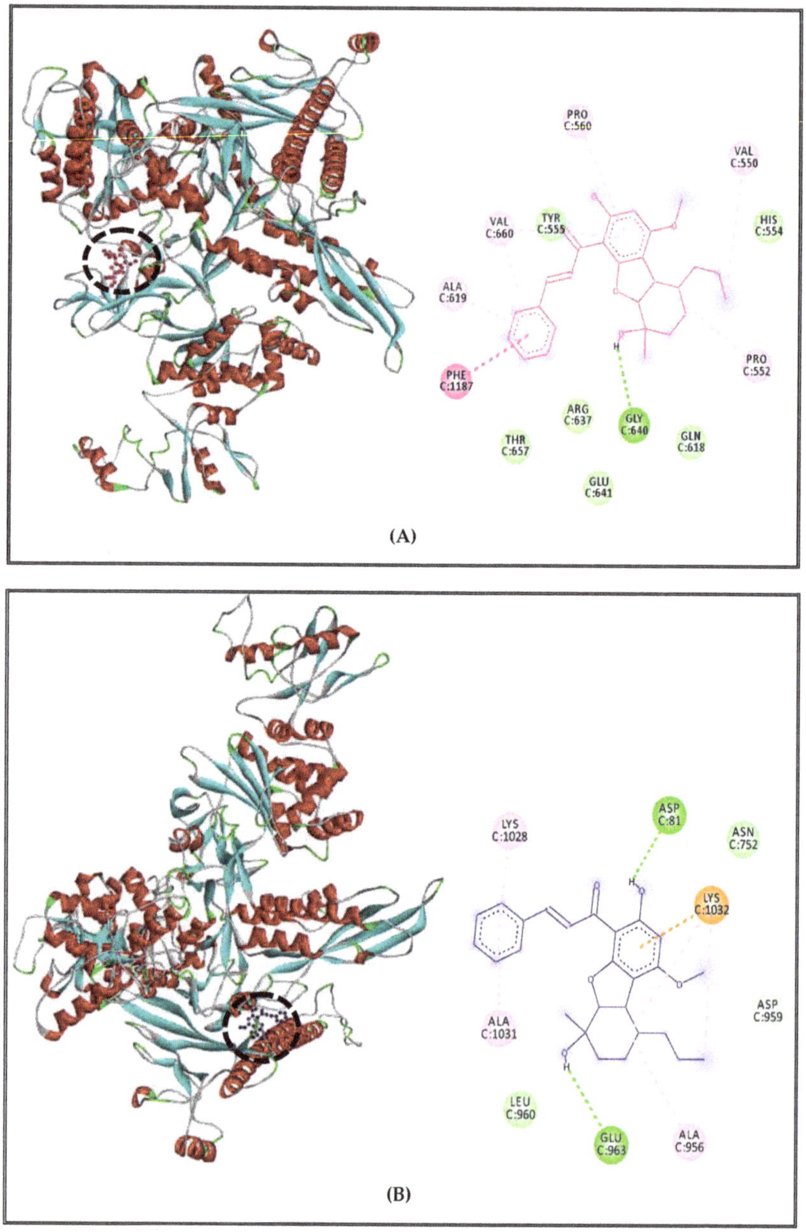

Figure 9. *Cont.*

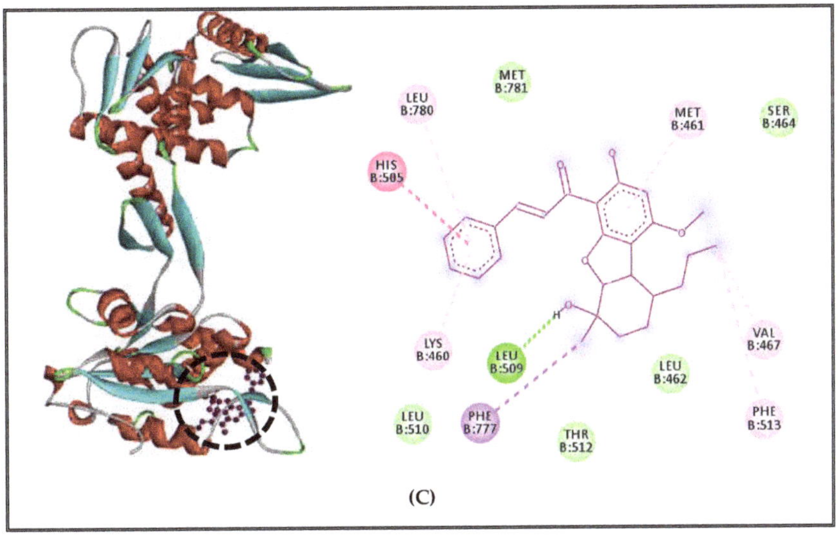

Figure 9. Borneol complexed with DNA polymerase enzyme (**A**), RNA polymerase (**B**), topoisomerase II (**C**) of *S. enterica* Typhimurium.

3. Materials and Methods

3.1. Plant Material, Essential Oil Extraction, and Gas Chromatography–Mass Spectrometry (GC–MS) Analysis

3.1.1. Plant Material

Aerial parts of *C. sempervirens* were collected from Sfax, Tunisia (N: 34.4426°, E: 10.4537°) which is characterized by semi-arid climatic conditions. Aerial parts were harvested at the vegetative stage and were air-dried in obscurity at room temperature.

3.1.2. Extraction and Analysis of (CSEO)

The EO of dried samples of *C. sempervirens* aerial parts was hydrodistilled for 3 h by using a Clevenger apparatus. The obtained (CSEO) was collected and dried over anhydrous sodium sulfate and maintained at 4 °C until analysis.

The analysis of the (CSEO) was carried out on a GC/MS HP model 6980 inert MSD, equipped with an Agilent Technologies capillary HP-5MS column (60 m × 0.25 mm, 0.25 mm film thickness) and coupled to a mass selective detector (MSD5973, ionization voltage 70 eV, all Agilent, Santa Clara, CA, USA). The carrier gas was helium and was used at 1.2 mL/min flow rate. The oven temperature program was as follows: 1 min at 100 °C ramped from 100 to 280 °C at 5 °C/min and 25 min at 280 °C. The chromatograph was equipped with a split/split less injector used in the split less mode. Identification of components was appointed by matching their mass spectra with Wiley Registry of Mass Spectral Data 7th edition (Agilent Technologies) and National Institute of Standards and Technology 05 MS (NIST) library data.

3.2. Antibacterial Activity

3.2.1. Microorganisms and Growth Conditions

For antibacterial activity determination, bacteria used as indicator cells were obtained from international culture collections (ATCC); two Gram-positive bacteria: *Staphylococcus aureus* (*S. aureus*) ATCC 6538 and *Listeria monocytogenes* (*L. monocytogenes*) ATCC 19117, and two Gram-negative bacteria: *Salmonella enterica* Typhimurium ATCC 14028 and *Escherichia coli* (*E. coli*) ATCC 8739. According to the previous work of [81], the bacterial cultures were performed in Luria-Bertani (LB) agar medium composed of (g/L): peptone, 10; yeast extract, 5; NaCl, 5; and agar, 20 at pH 7.2, then the bacterial strains were incubated at 37 °C. Bacterial cultures were prepared by inoculating a loopful of each test bacteria in 3 mL of LB broth.

3.2.2. Agar Diffusion Method

Antimicrobial activity of the essential oil of *Cupressus sempervirens* plant (CSEO) was evaluated by agar-well diffusion assay according to [82]. Fifteen milliliters of the molten agar (45 °C) were poured into sterile Petri dishes (Ø 90 mm). Working cell suspensions were prepared and 100 µL were evenly deposited onto the surface of plates containing LB agar medium. Plates were aseptically dried and then 5 mm wells were punched into the agar with a sterile Pasteur pipette. The (CSEO) was dissolved in dimethylsulfoxide (DMSO)/water (1/9; *v*/*v*) to a final concentration of 1 mg/mL and then filtered through 0.22 µm pore-size black polycarbonate filters. 100 µL of this filtered solution were placed into the wells and the plates were incubated at 37 °C.

3.2.3. Minimal Inhibitory Concentration (MIC)

MIC of the (CSEO) against the four tested bacteria was determined using the microdilution method with serial dilution described by Chandrasekaran and Venkatesalu (2004) [83]. The final volume in each tube was 100 µL. The cell suspension was added to each test, to the final inoculum concentration of 10^6 CF/mL of the corresponding indicator bacterium. The contents of the tubes were mixed by pipetting and were incubated for 24 h at 37 °C. The MIC was defined as the lowest concentration that inhibits the visible growth of the used indicator microorganism.

For the antibacterial activity determination (inhibition zones and CMIs), each experiment was carried out simultaneously three times under same conditions. The obtained diameters of inhibition zones reported in mm and the MIC values reported in mg/mL were quite similar and the reported results are the average of the two experiments.

3.3. *Chemo-Computational Toxicity Evaluation Using In Silico Tools*

3.3.1. Toxicity Prediction of Compounds by VEGA HUB Software Using QSAR Method

The compounds from (CSEO): five recommended food preservatives (citric acid, BHA, L-ascorbic acid, propionic acid, and benzoic acid [84], rifamycin, and ciprofloxacin) were selected previously for molecular docking, as positive controls to inhibit DNA polymerase, RNA polymerase [85], and topoisomerase II [86] and efflux pumps inhibitors (cathinone and thioridazine) which are usually employed as controls to tackle multidrug-resistance to several antibiotics, were selected respectively. All compounds were subjected to 9 toxicity tests/measurements, including the genotoxicity/mutagenicity Caesar hybrid model for bacterial reverse mutation (Ames test), which consists of the detection of substances that could cause genetic mutations [87], the carcinogenicity CAESAR model for carcinogenicity which is based on spotting the ability of a molecule to induce tumors depends on its molecular structure [88]. The developmental toxicity model (CAESAR) and developmental/reproductive toxicity library (PG) help us to specify if the query compounds could be developmental toxicants and cause reproductive problems or not [89]. Besides, we selected another toxicity endpoint that facilitated the detection of endocrine-disrupting chemicals that interfere with the biosynthesis, metabolism, or action of endogenous hormones by the activation of their receptors. In this case, 3 models were employed: the estrogen receptor relative binding affinity model using 17-estradiol as the androgen receptor-mediated effect and thyroid receptor alpha and beta effects [90,91]. Finally, we used in vitro micronucleus activity model to evaluate the ability of an agent to cause DNA damage as an alteration in the structure or information content of genetic material in cells [92]. All toxicity endpoints measurement were performed by VEGA software version 1.1.5 using the QSAR (quantitative structure–activity relationship) approach [93].

3.3.2. Rodent Oral Toxicity and Cytotoxicity of Selected Compounds Predicted by PROTOX II

PROTOX II is an in silico tool (Charite University of Medicine, Institute for Physiology, Structural Bioinformatics Group, Berlin, Germany) which is generally employed to evaluate multiple types of toxicity, such as acute toxicity, hepatotoxicity, cytotoxicity, carcinogenicity, mutagenicity, immunotoxicity, different toxicological pathways and targets, according to preliminary saved data obtained from both in vitro and in vivo assays [94].

In this research paper, PROTOX II was employed to predict rodent oral toxicity and cytotoxicity of *C. sempervirens* EO compounds, FDA approved drugs, food preservatives and EPIs in order to classify those compounds into several classes of toxicity using a globally harmonized system (GHS) of chemical labeling classification [95]. The SMILES (simplified molecular input line entry systems) of these compounds were introduced into the software for more chemo-computational toxicology evaluations.

3.4. *In Vivo Toxicity Assessment Using Zebrafish Model*

3.4.1. Zebrafish Maintenance and Embryos' Collection

Danio rerio, a tropical freshwater fish, was used as a test species at the Laboratory of Molecular and Cellular Screening Processes (LPCMC), Center of Biotechnology of Sfax (CBS, Tunisia). Adults were maintained in culture under controlled conditions in a custom-made flow-through system. Each breeding group consisted of 7 females and 7 males, which were kept in 13 L circulation tanks at 26 ± 1 °C under continuous aeration. 14:10 h photoperiod cycle (light:dark) was maintained. Adult fish were fed twice a day with a commercially available dry food and fresh Artemia larvae.

The eggs were collected after a period of one hour of natural mating of 6 adult fish in a female/male ratio of 2:1. Then the eggs were washed thoroughly and rinsed several times with water. Healthy and developing embryos were selected within 1 hpf for exposure testing using a Stemi 2000-C stereomicroscope (Zeiss, Göttingen, Germany), transferred to crystallizing dishes and briefly stored in an incubator (26 ± 1 °C) until exposure. Screening the eggs before assaying ensured the exclusion of unfertilized eggs, injured, or deformed embryos. All experiments were approved by the Animal Ethics Committee of National School of Veterinary Medicine, IACUC, ENMV- Sidi Thabet, Tunisia (Permit No. CEEA-ENMV 44/22, 1 February 2022).

3.4.2. Zebrafish Embryonic Toxicity Test and Determination of LC_{50}

Embryotoxicity measurement in the zebrafish was elaborated by evaluating the mortality rate of the zebrafish embryos. Exposure of the embryos to the EO was performed based on the OECD Guideline for Testing of Chemicals 236—Acute Fish Embryo Toxicity Test (FET) [96]. Generally, for each concentration treatment, 20 fertilized eggs (at 1hpf) were used and placed in individual wells of 12-well plates.

The embryos were exposed to multiple concentrations of (CSEO) containing 0.1% dimethyl sulfoxide (DMSO) diluted in 2 mL of embryo water (embryonic medium). EO was serially diluted to produce 12 increasing concentrations (0.25; 0.5; 1; 2; 2.5; 3; 3.5; 4; 5; 6; 7; 8 μg/mL). The control (untreated group) was exposed only to 2 mL of embryo water. After treatment, the embryos were placed into at 27 °C and they were continuously examined every 24 h, using a stereomicroscope (Zeiss, Göttingen, Germany). Coagulation, absence of hatching and /or heartbeats were used as criteria to differentiate viable embryos from dead ones [97]. Finally, in order to LC_{50} values for (CSEO), the number of dead embryos was calculated in each concentration. The experiment was triplicated.

3.5. *Interaction Study between the (CSEO) Molecules and Bacterial Protein Targets by Molecular Docking*

3.5.1. Homology Modeling of the Proteins

We performed homology modeling of proteins from two different microorganisms (1): *Staphylococcus aureus* (strain Mu50/ATCC 700699) and (2): *Salmonella* Typhimurium (strain LT2/SGSC1412/ATCC 700720) using the Swiss model server [98,99]. We selected DNA polymerase (Uniprot ID: P63979), RNA polymerase (Uniprot ID: Q932F8), and topoisomerase II (DNA gyrase) (Uniprot ID: P66936) from *Staphylococcus aureus* (strain Mu50/ATCC 700699). Similarly, DNA polymerase (Uniprot ID: P14567), RNA polymerase (Uniprot ID: P06173) and topoisomerase II (DNA gyrase) (Uniprot ID: P0A213) from *Salmonella* Typhimurium (strain LT2/SGSC1412/ATCC 700720). We also selected the AcrB efflux pump protein of *S.* Typhimurium (PDB ID: 5FFZ) from the NCBI database. The FASTA sequences of proteins were obtained from the UNIPORT server [100,101] and submitted to the Swiss Model server for Automatic Homology Modeling using the default parameters. The finding of the best template proteins was performed by the BLASTp program [102]. The predicted homology models are ranked based on the target-template protein sequence identities, QMEAN Z score, and GMQE score.

3.5.2. Validation of Protein Models

The obtained protein models from the Swiss model server were subjected to model validation using the Profunc Server [103]. The Profunc server has an integrated protein analysis tool known as Procheck, which analyzes the overall quality of the models using the Ramachanderan plot based on the distribution of dihedral angles of the amino acids backbone Phi (Φ) and Psi (Ψ) angles [104]. We also considered the QMEAN Z score and GMQE score from the Swiss model server for the selection of the best homology models [105].

3.5.3. Binding Site Prediction

All selected protein models were subjected to binding site prediction using the Profunc server [103]. Our query protein structures are searched against the protein database with known binding site residues using "reverse template comparison vs. structure in PDB approach" integrated in the SiteSeer program. The binding site residues are extracted from best hits with E-value 0.

3.5.4. Selection of the Compounds

Based on the available literature, two antibiotics, Rifamycin SV and Ciprofloxacin, were used as controls and were downloaded from the Drug Bank database [106]. Twenty-seven compounds of *C. sempervirens* and two efflux pump inhibitors (cathinone and thioridazine) were downloaded from the Pubchem database [107]. Their Smiles strings were obtained from the Pubchem database and converted into a 3D structure via the Corina server [108]. All files were saved in the pdb file format.

3.5.5. Molecular Docking by Autodock Vina

The virtual screening of the compounds was performed against proteins using the Autodock Vina [109]. First, the compound pdbqt files were prepared by Autodock.4.2 software. All hydrogen atoms were added to the compound's structures, followed by the merging of the nonpolar hydrogen atoms. Subsequently, Gasteiger charges were added to each atom. The number of rotatable bonds is set to be maximum according to the torsional bonds in the compound. All proteins files were also prepared by Autodock.4.2 software. Like compounds, all hydrogen atoms were added to the protein's structures, followed by merging the nonpolar hydrogen atoms. Subsequently, Gasteiger charges were added to each atom and Kollman united atom charges were assigned to the receptor atom. The grid box was built around the binding sites of proteins by making a grid box size of $126 \times 126 \times 126$ with a grid spacing of 0.375 Å. This grid box dimension covered the whole binding site and provided enough space for translation and rotation of ligands. The corresponding grid center coordinates were set according to the respective binding site residues of the proteins. The conFigure file (*conf*) of the Autodock vina was set with name of protein.pdbqt, information about *center_x, center_y, center_z* and box size of *xyz*. The exhaustivness value was set to 10 with number of modes (*num_modes*) of 200 and *energy_range* to 4. All autodock vina executable files were put in same folder, where the compounds and proteins files were present. We used open babel tool for converting all compounds pdb files in to pdbqt format using the command "*obabel *.pdb -opdbqt -m*". We have developed our own docking script for automation of docking simulation by Autodock vina. All compound pdbqt files were added to the ligand.txt file for our docking perl script. Our perl script was set to take each ligand and screen against the binding site of the protein and all outputs were stored in the docking folder. The single docked conformation was selected from each docking round based on the clustering RMSD and lowest binding energy.

The most stable conformations of the ligand molecule were selected based on the lowest binding energy and their binding mode at the active site of proteins and analyzed by discovery studio software for h-bond analysis and non-bonded interactions between compounds and proteins [110].

3.6. Statistical Analysis

All tests were assayed in triplicate and expressed as the mean ± standard deviation of the measurements. The statistical program SPSS version 21.00 for Windows (SPSS Inc., Chicago, IL, USA) was used to analyze data. Variance was analyzed by one-way ANOVA and Tukey's multiple range tests were calculated for the significant data at $p < 0.05$.

4. Conclusions

The tendency towards the application of EOs as safer antimicrobial agents has increased. Results of this study evidenced the anti-food-borne bacterial activities of (CSEO) against *S. enterica* Typhimurium and *S. aureus*. Interestingly, α-pinene, δ-3-carene, and borneol, belonging to monoterpenes, can increase the cell wall permeability and inhibit DNA and RNA polymerases and topoisomerase II of bacteria. Predicted molecular docking analysis showed that these three compounds were highly reactive molecules when compared to the reference antibiotics. These findings indicate that (CSEO) can be used in targeted drug development to combat antibiotic resistance associated with efflux pump expression, modulation of DNA topology, and DNA and RNA synthesis. In addition, mutagenic, toxicological, and carcinogenic properties of all (CSEO) compounds vs. some recommended food preservatives commonly used in the food industry were investigated by applying in silico tools and software. Notably, we revealed an absence of toxic effects of many (CSEO) components. Throughout the zebrafish model, we confirmed the safety of (CSEO). Overall, experimental validation by studying in vitro the process of impairment of membrane permeability and replication of pathogenic bacteria insured by (CSEO) would be required for conclusive confirmation in order to be applied to reduce the impact of the diseases caused by such pathogenic microorganisms.

Author Contributions: Conceptualization, S.S. and L.M.; methodology, S.A., S.S.; software, S.A.; validation, S.A., S.S. and L.M.; formal analysis, K.E.; investigation, M.F., N.L., M.C., A.C.M.; resources, S.S., N.L., A.B., S.M.; data curation, S.A., S.S.; writing—original draft preparation, S.A., S.S.; writing—review and editing, S.S., L.M.; visualization, S.A., S.S.; supervision, L.M., S.S.; project administration, S.S.; funding acquisition, S.S. and L.M. All authors have read and agreed to the published version of the manuscript.

Funding: This research was funded by the Tunisian Ministry of Higher Education and Scientific Research (LR15CBS06).

Institutional Review Board Statement: Not applicable.

Informed Consent Statement: Not applicable.

Data Availability Statement: All data generated or analyzed during this study are included in this published article.

Conflicts of Interest: The authors declare that they have no known competing financial interests or personal relationships that could have appeared to influence the work reported in this paper.

Sample Availability: Samples of the compounds are not available from the authors.

Appendix A

Table A1. Receptors models of the two pathogenic bacteria used to analyze the molecular docking with the major constituents of the (CSEO).

Bacterial Strain	Bacterial Target	Receptor (UniprotKB)	Template	Identity (%)	Ramachandran Plot		QMEAN
					Favoured Regions (%)	Additional Allowed Regions (%)	
***Staphylococcus aureus* (strain Mu50/ATCC 700699)**	DNA polymerase	P63979	4IQJ.1.L	34.77	87.9	10.1	−3.26
	RNA polymerase	Q932F8	6WVK.1.C	81.09	85.5	12.7	−2.36
	Topoisomerase II	P66936	6GAV.1.A	54.42	88.4	10.7	−1.69
***Salmonella* Typhimurium (strain LT2/SGSC1412/ATCC 700720)**	DNA polymerase	P14567	5FKU.1.A	96.72	88.0	10.0	−2.35
	RNA polymerase	P06173	4LLG.2.C	98.66	88.0	11.0	−1.11
	Topoisomerase II	P0A213	4TMA.2.B	95.41	90.2	9.2	−1.82

References

1. Nicoloff, H.; Hjort, K.; Levin, B.R.; Andersson, D.I. The high prevalence of antibiotic heteroresistance in pathogenic bacteria is mainly caused by gene amplification. *Nat. Microbiol.* **2019**, *4*, 504–514. [CrossRef] [PubMed]
2. Ogawara, H. Comparison of antibiotic resistance mechanisms in antibiotic-producing and pathogenic bacteria. *Molecules* **2019**, *24*, 3430. [CrossRef] [PubMed]
3. Breijyeh, Z.; Jubeh, B.; Karaman, R. Resistance of gram-negative bacteria to current antibacterial agents and approaches to resolve it. *Molecules* **2020**, *25*, 1340. [CrossRef] [PubMed]
4. Ngaruka, G.B.; Neema, B.B.; Mitima, T.K.; Kishabongo, A.S.; Kashongwe, O.B. Animal source food eating habits of outpatients with antimicrobial resistance in Bukavu, DR Congo. *Antimicrob. Resist. Infect. Control* **2021**, *10*, 124. [CrossRef]
5. Kebede, T.; Gadisa, E.; Tufa, A. Antimicrobial activities evaluation and phytochemical screening of some selected medicinal plants: A possible alternative in the treatment of multidrug-resistant microbes. *PLoS ONE* **2021**, *16*, e0249253. [CrossRef]
6. Vaou, N.; Stavropoulou, E.; Voidarou, C.; Tsigalou, C.; Bezirtzoglou, E. Towards advances in medicinal plant antimicrobial activity: A review study on challenges and future perspectives. *Microorganisms* **2021**, *9*, 2041. [CrossRef]
7. Kosakowska, O.; Węglarz, Z.; Pióro-Jabrucka, E.; Przybył, J.L.; Kraśniewska, K.; Gniewosz, M.; Bączek, K. Antioxidant and antibacterial activity of essential oils and hydroethanolic extracts of Greek oregano (*O. vulgare* L. subsp. hirtum (Link) Ietswaart) and common oregano (*O. vulgare* L. subsp. vulgare). *Molecules* **2021**, *26*, 988. [CrossRef]
8. Zaharieva, M.M.; Zheleva-Dimitrova, D.; Rusinova-Videva, S.; Ilieva, Y.; Brachkova, A.; Balabanova, V.; Gevrenova, R.; Kim, T.C.; Kaleva, M.; Georgieva, A.; et al. Antimicrobial and Antioxidant Potential of Scenedesmus obliquus Microalgae in the Context of Integral Biorefinery Concept. *Molecules* **2022**, *27*, 519. [CrossRef]
9. Capatina, L.; Napoli, E.M.; Ruberto, G.; Hritcu, L. *Origanum vulgare* ssp. hirtum (Lamiaceae) Essential Oil Prevents Behavioral and Oxidative Stress Changes in the Scopolamine Zebrafish Model. *Molecules* **2021**, *26*, 7085. [CrossRef]
10. Smaoui, S.; Hlima, H.B.; Tavares, L.; Ennouri, K.; Braiek, O.B.; Mellouli, L.; Abdelkafi, S.; Khaneghah, A.M. Application of essential oils in meat packaging: A systemic review of recent literature. *Food Control* **2022**, *132*, 108566. [CrossRef]
11. Ni, Z.J.; Wang, X.; Shen, Y.; Thakur, K.; Han, J.; Zhang, J.G.; Hu, F.; Wei, Z.J. Recent updates on the chemistry, bioactivities, mode of action, and industrial applications of plant essential oils. *Trends Food Sci. Technol.* **2021**, *110*, 78–89. [CrossRef]
12. Jugreet, B.S.; Suroowan, S.; Rengasamy, R.K.; Mahomoodally, M.F. Chemistry, bioactivities, mode of action and industrial applications of essential oils. *Trends Food Sci. Technol.* **2020**, *101*, 89–105. [CrossRef]
13. Arsalani, M.; Griessinger, J.; Pourtahmasi, K.; Bräuning, A. Multi-centennial reconstruction of drought events in South-Western Iran using tree rings of Mediterranean cypress (*Cupressus sempervirens* L.). *Palaeogeogr. Palaeoclimatol. Palaeoecol.* **2021**, *567*, 110296. [CrossRef]
14. Fadel, H.; Benayache, F.; Chalchat, J.C.; Figueredo, G.; Chalard, P.; Hazmoune, H.; Benayache, S. Essential oil constituents of *Juniperus oxycedrus* L. and *Cupressus sempervirens* L. (Cupressaceae) growing in Aures region of Algeria. *Nat. Prod. Res.* **2021**, *35*, 2616–2620. [CrossRef] [PubMed]
15. Argui, H.; Youchret-Zalleza, O.B.; Suner, S.C.; Periz, Ç.D.; Türker, G.; Ulusoy, S.; Ben-Attia, M.; Büyükkaya, F.; Oral, A.; Coskun, Y.; et al. Isolation, Chemical Composition, Physicochemical Properties, and Antibacterial Activity of *Cupressus sempervirens* L. Essential Oil. *J. Essent. Oil-Bear. Plants* **2021**, *24*, 439–452. [CrossRef]

16. Rguez, S.; Djébali, N.; Slimene, I.B.; Abid, G.; Hammemi, M.; Chenenaoui, S.; Bachkouel, S.; Daami-Remadi, M.; Ksouri, R.; Hamrouni-Sellami, I. *Cupressus sempervirens* essential oils and their major compounds successfully control postharvest grey mould disease of tomato. *Ind. Crops Prod.* **2018**, *123*, 135–141. [CrossRef]
17. Selim, S.A.; Adam, M.E.; Hassan, S.M.; Albalawi, A.R. Chemical composition, antimicrobial and antibiofilm activity of the essential oil and methanol extract of the Mediterranean cypress (*Cupressus sempervirens* L.). *BMC Complement. Altern. Med.* **2014**, *14*, 179. [CrossRef]
18. Alimi, D.; Hajri, A.; Jallouli, S.; Sebai, H. Phytochemistry, anti-tick, repellency and anti-cholinesterase activities of *Cupressus sempervirens* L. and *Mentha pulegium* L. combinations against *Hyalomma scupense* (Acari: Ixodidae). *Vet. Parasitol.* **2022**, *303*, 109665. [CrossRef]
19. He, Y.L.; Shi, J.Y.; Peng, C.; Hu, L.J.; Liu, J.; Zhou, Q.M.; Guo, L.; Xiong, L. Angiogenic effect of motherwort (*Leonurus japonicus*) alkaloids and toxicity of motherwort essential oil on zebrafish embryos. *Fitoterapia* **2018**, *128*, 36–42. [CrossRef]
20. Haddad, J.G.; Picard, M.; Bénard, S.; Desvignes, C.; Desprès, P.; Diotel, N.; El Kalamouni, C. Ayapana triplinervis essential oil and its main component thymohydroquinone dimethyl ether inhibit Zika virus at doses devoid of toxicity in zebrafish. *Molecules* **2019**, *24*, 3447. [CrossRef]
21. Thitinarongwate, W.; Mektrirat, R.; Nimlamool, W.; Khonsung, P.; Pikulkaew, S.; Okonogi, S.; Kunanusorn, P. Phytochemical and Safety Evaluations of Zingiber ottensii Valeton Essential Oil in Zebrafish Embryos and Rats. *Toxics* **2021**, *9*, 102. [CrossRef] [PubMed]
22. Niu, F.X.; He, X.; Wu, Y.Q.; Liu, J.Z. Enhancing Production of Pinene in *Escherichia coli* by Using a Combination of Tolerance, Evolution, and Modular Co-culture Engineering. *Front. Microbiol.* **2018**, *9*, 1623. [CrossRef] [PubMed]
23. Gomes-Carneiro, M.R.; Viana, M.E.; Felzenszwalb, I.; Paumgartten, F.J. Evaluation of beta-myrcene, alpha-terpinene and (+)- and (−)-alpha-pinene in the Salmonella/microsome assay. *Food Chem. Toxicol.* **2005**, *43*, 247–252. [CrossRef] [PubMed]
24. Kamal, E.H.; Al-Ajmi, M.F.; Abdullah, M.B. Some cardiovascular effects of the dethymoquinonated *Nigella sativa* volatile oil and its major components α-pinene and p-cymene in rats. *Saudi Pharm. J.* **2003**, *11*, 104–110.
25. Him, A.; Ozbek, H.; Turel, I.; Oner, A.C. Antinociceptive activity of alphapinene and fenchone. *Pharmacologyonline* **2008**, *3*, 363–369.
26. Orhan, I.; Küpeli, E.; Aslan, M.; Kartal, M.; Yesilada, E. Bioassay-guided evaluation of anti-inflammatory and antinociceptive activities of pistachio, *Pistacia vera* L. *J. Ethnopharmacol.* **2006**, *105*, 235–240. [CrossRef]
27. FDA. *Code of Federal Regulations Title 21*; Food and Drug Administration: Washington, DC, USA, 2015.
28. Cavaleiro, C.; Pinto, E.; Gonçalves, M.J.; Salgueiro, L. Antifungal activity of Juniperus essential oils against dermatophyte, *Aspergillus* and *Candida* strains. *J. Appl. Microbiol.* **2006**, *100*, 1333–1338.
29. Niccolini, P.; Pasotti, V.; Caliari, W. New Pharmacological Properties of Delta-3-Carene. Antibacterial and Expectorant Effects. *Boll. Chim. Farm.* **1964**, *103*, 598–608.
30. Gil, M.L.; Jimenez, J.; Ocete, M.A.; Zarzuelo, A.; Cabo, M.M. Comparative study of different essential oils of *Bupleurum gibraltaricum* Lamarck. *Pharmazie* **1989**, *44*, 284–287.
31. Ravichandran, C.; Badgujar, P.; Gundev, P.; Ashutosh Upadhyay, A. Review of toxicological assessment of d -limonene, a food and cosmetics additive. *Food Chem. Toxicol.* **2018**, *120*, 668–680. [CrossRef]
32. Chebet, J.J.; Ehiri, J.E.; McClelland, D.J.; Taren, D.; Hakim, I.A. Effect of d-limonene and its derivatives on breast cancer in human trials: A scoping review and narrative synthesis. *BMC Cancer* **2021**, *21*, 902. [CrossRef] [PubMed]
33. Quintans-Junior, L.J.; Souza, T.T.; Leite, B.S.; Lessa, N.M.N.; Bonjardim, L.R.; Santos, M.R.V.; Alves, P.B.; Blank, A.F.; Antoniolli, A.R. Phythochemical screening and anticonvulsant activity of *Cymbopogon winterianus* Jowitt (Poaceae) leaf essential oil in rodents. *Phytomed* **2008**, *15*, 619–624. [CrossRef] [PubMed]
34. Melo, M.S.; Guimarães, A.G.; Santana, M.F.; Siqueira, R.S.; De Lima, A.C.; Dias, A.S.; Santos, M.R.V.; Onofre, A.S.C.; Quintans, J.S.S.; De Sousa Almeida, D.P.J.; et al. Anti-inflammatory and redox-protective activities of citronellal. *Biol. Res.* **2011**, *44*, 363–368. [CrossRef] [PubMed]
35. Amri, I.; Hamrouni, L.; Hanana, M.; Gargouri, S.; Jamoussi, B. Chemical composition, bio-herbicidal and antifungal activities of essential oils isolated from Tunisian common cypress (*Cupressus sempervirens* L.). *J. Med. Plant Res.* **2013**, *7*, 1070–1080.
36. Moo, C.L.; Osman, M.A.; Yang, S.K.; Yap, W.S.; Ismail, S.; Lim, S.H.E.; Chong, C.M.; Lai, K.S. Antimicrobial activity and mode of action of 1,8-cineol against carbapenemase-producing Klebsiella pneumoniae. *Sci. Rep.* **2021**, *11*, 20824. [CrossRef]
37. Alamoti, M.P.; Gilani, B.B.; Mahmoudi, R.; Reale, A.; Pakbin, B.; Di Renzo, T.; Ata, A. Essential Oils from Indigenous Iranian Plants: A Natural Weapon vs. Multidrug-Resistant *Escherichia coli*. *Microorganisms* **2022**, *10*, 109. [CrossRef]
38. O'Sullivana, M.E.; Songa, Y.; Greenhousea, R.; Lina, R.; Pereza, A.; Atkinsona, P.J.; MacDonaldb, J.P.; Siddiquib, Z.; Lagascab, D.; Comstockc, K.; et al. Dissociating antibacterial from ototoxic effects ofgentamicin C-subtypes. *Proc. Natl. Acad. Sci. USA* **2020**, *117*, 32423–32432. [CrossRef]
39. Damayanti, S.; Permana, J.; Tjahjono, D.H. The use of computational chemistry to predict toxicity of antioxidants food additives and its metabolites as a reference for food safety regulation. *Pharma Chem J.* **2015**, *7*, 174–181.
40. Valente, M.J.; Araújo, A.M.; Bastos, M.D.L.; Fernandes, E.; Carvalho, F.; Guedes de Pinho, P.; Carvalho, M. Editor's highlight: Characterization of hepatotoxicity mechanisms triggered by designer cathinone drugs (β-Keto amphetamines). *Toxicol* **2016**, *153*, 89–102. [CrossRef]

41. Eftekhari, A.; Ahmadian, E.; Azarmi, Y.; Parvizpur, A.; Fard, J.K.; Eghbal, M.A. The effects of cimetidine, N-acetylcysteine, and taurine on thioridazine metabolic activation and induction of oxidative stress in isolated rat hepatocytes. *Pharm. Chem. J.* **2018**, *51*, 965–969. [CrossRef]
42. Drwal, M.N.; Banerjee, P.; Dunkel, M.; Wettig, M.R.; Preissner, R. ProTox: A web server for the in silico prediction of rodent oral toxicity. *Nucleic Acids Res.* **2014**, *42*, 53–58. [CrossRef] [PubMed]
43. Zhang, Q.L.; Fu, B.M.; Zhang, Z.J. Borneol, a novel agent that improves central nervous system drug delivery by enhancing blood–brain barrier permeability. *Drug Deliv.* **2017**, *24*, 1037–1044. [CrossRef] [PubMed]
44. Samreen Qais, F.A.; Ahmad, I. In silico screening and in vitro validation of phytocompounds as multidrug efflux pump inhibitor against E. coli. *J. Biomol. Struct.* **2022**, 1–13. [CrossRef] [PubMed]
45. Hoyberghs, J.; Bars, C.; Pype, C.; Foubert, K.; Hernando, M.A.; Van Ginneken, C.; Van Cruchten, S. Refinement of the zebrafish embryo developmental toxicity assay. *MethodsX* **2020**, *7*, 101087. [CrossRef] [PubMed]
46. Wang-Kan, X.; Blair, J.M.; Chirullo, B.; Betts, J.; La Ragione, R.M.; Ivens, A.; Ricci, V.; Opperman, T.J.; Piddock, L.J.V. Lack of AcrB efflux function confers loss of virulence on *Salmonella enterica* serovar Typhimurium. *MBio* **2017**, *8*, e00968-17. [CrossRef] [PubMed]
47. Takatsuka, Y.; Chen, C.; Nikaido, H. Mechanism of recognition of compounds of diverse structures by the multidrug efflux pump AcrB of *Escherichia coli*. *Proc. Natl. Acad. Sci. USA* **2010**, *107*, 6559–6565. [CrossRef] [PubMed]
48. Kumaraswami, M.; Schuman, J.T.; Seo, S.M.; Kaatz, G.W.; Brennan, R.G. Structural and biochemical characterization of MepR, a multidrug binding transcription regulator of the *Staphylococcus aureus* multidrug efflux pump MepA. *Nucleic Acids Res.* **2009**, *37*, 1211–1224. [CrossRef] [PubMed]
49. Seukep, A.J.; Kuete, V.; Nahar, L.; Sarker, S.D.; Guo, M. Plant-derived secondary metabolites as the main source of efflux pump inhibitors and methods for identification. *J. Pharm. Anal.* **2020**, *10*, 277–290. [CrossRef] [PubMed]
50. Kaizaki, A.; Tanaka, S.; Numazawa, S. New recreational drug 1-phenyl-2-(1-pyrrolidinyl)-1-pentanone (alpha-PVP) activates central nervous system via dopaminergic neuron. *J. Toxicol. Sci.* **2014**, *39*, 1–6. [CrossRef]
51. Ghosh, A.; Roymahapatra, G.; Paul, D.; Mandal, S.M. Theoretical analysis of bacterial efflux pumps inhibitors: Strategies in-search of competent molecules and develop next. *Compul. Biol. Chem.* **2020**, *87*, 107275. [CrossRef]
52. Kovač, J.; Šimunović, K.; Wu, Z.; Klančnik, A.; Bucar, F.; Zhang, Q.; Možina, S.S. Antibiotic resistance modulation and modes of action of (-)-α-pinene in *Campylobacter jejuni*. *PLoS ONE* **2015**, *10*, e0122871. [CrossRef] [PubMed]
53. Nidhi, P.; Rolta, R.; Kumar, V.; Dev, K.; Sourirajan, A. Synergistic potential of Citrus aurantium L. essential oil with antibiotics against Candida albicans. *J. Ethnopharmacol.* **2020**, *262*, 113135. [CrossRef] [PubMed]
54. Fadli, M.; Chevalier, J.; Hassani, L.; Mezrioui, N.E.; Pagès, J.M. Natural extracts stimulate membrane-associated mechanisms of resistance in Gram-negative bacteria. *Lett. Appl. Microbiol.* **2014**, *58*, 472–477. [CrossRef] [PubMed]
55. Thorsing, M.; Klitgaard, J.K.; Atilano, M.L.; Skov, M.N.; Kolmos, H.J.; Filipe, S.R.; Kallipolitis, B.H. Thioridazine induces major changes in global gene expression and cell wall composition in methicillin-resistant *Staphylococcus aureus* USA300. *PLoS ONE* **2013**, *8*, e64518. [CrossRef]
56. Rana, T.; Singh, S.; Kaur, N.; Pathania, K.; Farooq, U. A review on efflux pump inhibitors of medically important bacteria from plant sources. *Int. J. Pharm. Sci. Rev. Res.* **2014**, *26*, 101–111.
57. Ruth, M.M.; Pennings, L.J.; Koeken, V.A.; Schildkraut, J.A.; Hashemi, A.; Wertheim, H.F.L.; Hoefsloot, W.; van Ingen, J. Thioridazine is an efflux pump inhibitor in Mycobacterium avium complex but of limited clinical relevance. *Antimicrob. Agents Chemother.* **2020**, *64*, e00181-20. [CrossRef]
58. Dabul, A.N.G.; Avaca-Crusca, J.S.; Van Tyne, D.; Gilmore, M.S.; Camargo, I.L.B.C. Resistance in in vitro selected Tigecycline-resistant methicillin-resistant *Staphylococcus aureus* sequence type 5 is driven by mutations in mepR and mepA genes. *Microb. Drug Resist.* **2018**, *24*, 519–526. [CrossRef]
59. de Medeiros, V.M.; do Nascimento, Y.M.; Souto, A.L.; Madeiro, S.A.L.; de Oliveira Costa, V.C.; Silva, S.M.P.; dos SantosFalcão, V.; de Fátima Agra, M.; de Siqueira-Júnio, J.P.; Tavares, J.F. Chemical composition and modulation of bacterial drug resistance of the essential oil from leaves of Croton grewioides. *Microb. Pathog.* **2017**, *111*, 468–471. [CrossRef]
60. Coêlho, M.L.; Ferreira, J.H.L.; de Siqueira Júnior, J.P.; Kaatz, G.W.; Barreto, H.M.; Cavalcante, A.A.D.C.M. Inhibition of the NorA multi-drug transporter by oxygenated monoterpenees. *Microb. Pathog.* **2016**, *99*, 173–177. [CrossRef]
61. Agreles, M.A.A.; Cavalcanti, I.D.L.; Cavalcanti, I.M.F. The Role of Essential Oils in the Inhibition of Efflux Pumps and Reversion of Bacterial Resistance to Antimicrobials. *Curr. Microbiol.* **2021**, *78*, 3609–3619. [CrossRef]
62. Limaverde, P.W.; Campina, F.F.; da Cunha, F.A.; Crispim, F.D.; Figueredo, F.G.; Lima, L.F.; de, M. Oliveira-Tintino, C.D.; de Matos, Y.M.L.S.; Siqueira-Júnior, J.P. et al. Inhibition of the TetK efflux-pump by the essential oil of Chenopodium ambrosioides L. and α-terpinene against *Staphylococcus aureus* IS-58. *Food Chem. Toxicol.* **2017**, *109*, 957–961. [CrossRef] [PubMed]
63. Xiang, Z. Advances in homology protein structure modeling. *Curr. Protein Pept. Sci.* **2006**, *7*, 217–227. [CrossRef] [PubMed]
64. Giacoppo, J.D.O.; Carregal, J.B.; Junior, M.C.; Cunha, E.F.D.; Ramalho, T.C. Towards the understanding of tetrahydroquinolines action in Aedes aegypti: Larvicide or adulticide? *Mol. Simul.* **2017**, *43*, 121–133. [CrossRef]
65. Robinson, A.; Van Oijen, A.M. Bacterial replication, transcription and translation: Mechanistic insights from single-molecule biochemical studies. *Nat. Rev. Microbiol.* **2013**, *11*, 303–315. [CrossRef] [PubMed]
66. Cui, H.; Bai, M.; Sun, Y.; Abdel-Samie, M.A.S.; Lin, L. Antibacterial activity and mechanism of Chuzhou chrysanthemum essential oil. *J. Funct. Foods* **2018**, *48*, 159–166. [CrossRef]

67. Utegenova, G.A.; Pallister, K.B.; Kushnarenko, S.V.; Özek, G.; Özek, T.; Abidkulova, K.T.; Voyich, J.M. Chemical composition and antibacterial activity of essential oils from Ferula L. species against methicillin-resistant Staphylococcus aureus. *Molecules* **2018**, *23*, 1679. [CrossRef]
68. Yang, C.; Hu, D.H.; Feng, Y. Antibacterial activity and mode of action of the Artemisia capillaris essential oil and its constituents against respiratory tract infection-causing pathogens. *Mol. Med. Rep.* **2015**, *11*, 2852–2860. [CrossRef]
69. Melkina, O.E.; Plyuta, V.A.; Khmel, I.A.; Zavilgelsky, G.B. The Mode of Action of Cyclic Monoterpenees (−)-Limoneneand (+)-α-Pinene on Bacterial Cells. *BioMolecules* **2021**, *11*, 806. [CrossRef]
70. Luís, Â.; Ramos, A.; Domingues, F. Pullulan films containing rockrose essential oil for potential food packaging applications. *Antibiotics* **2020**, *9*, 681. [CrossRef]
71. Aghoutane, Y.; Moufid, M.; Motia, S.; Padzys, G.S.; Omouendze, L.P.; Llobet, E.; El Bari, N. Characterization and analysis of Okoume and aiele essential oils from Gabon by gc-ms, electronic nose, and their antibacterial activity assessment. *Sensors* **2020**, *20*, 6750. [CrossRef]
72. Anka, L.; Rammal, H.; Kobeissi, A.; Saab, H.B. Chemical composition and biological potentials of Lebanese *Cupressus sempervirens*L. leaves extracts. *J. Med. Plant Res.* **2020**, *14*, 292–299.
73. Lee, S.Y.; Kim, S.H.; Hong, C.Y.; Park, M.J.; Choi, I.G. Effects of (−)-borneol on the growth and morphology of Aspergillus fumigatus and Epidermophyton floccosom. *Flavour Fragr. J.* **2013**, *28*, 129–134. [CrossRef]
74. Xiong, Z.Y.; Xiao, F.M.; Xu, X.; Wu, Y.F.; Jiang, X.M. Studies on pharmacological activity of borneol. *China J. Chin. Mater. Med.* **2013**, *38*, 786–790.
75. Yang, L.; Zhan, C.; Huang, X.; Hong, L.; Fang, L.; Wang, W.; Su, J. Durable Antibacterial Cotton Fabrics Based on Natural Borneol-Derived Anti-MRSA Agents. *Adv. Healthc. Mater.* **2020**, *9*, 2000186. [CrossRef] [PubMed]
76. Luo, L.; Li, G.; Luan, D.; Yuan, Q.; Wei, Y.; Wang, X. Antibacterial adhesion of borneol-based polymer via surface chiral stereochemistry. *ACS Appl. Mater. Interfaces* **2014**, *6*, 19371–19377. [CrossRef] [PubMed]
77. Sun, X.; Qian, Z.; Luo, L.; Yuan, Q.; Guo, X.; Tao, L.; Wei, Y.; Wang, X. Antibacterial adhesion of poly (methyl methacrylate) modified by borneol acrylate. *ACS Appl. Mater. Interfaces* **2016**, *8*, 28522–28528. [CrossRef] [PubMed]
78. Wang, W.; Ren, Z.; Wang, L.; Cai, Y.; Ma, H.; Fang, L.; Su, J. Nanoparticle-stabilized encapsulation of borneol and citral: Physicochemical characteristics, storage stability, and enhanced antibacterial activities. *J. Food Sci.* **2021**, *86*, 4554–4565. [CrossRef]
79. Dai, J.; Li, C.; Cui, H.; Lin, L. Unraveling the anti-bacterial mechanism of Litsea cubeba essential oil against E. coli O157: H7 and its application in vegetable juices. *Int. J. Food Microbiol.* **2021**, *338*, 108989. [CrossRef]
80. De Souza Moura, W.; de Souza, S.R.; Campos, F.S.; Cangussu, A.S.R.; Santos, E.M.S.; Andrade, B.S.; Gomes, C.H.B.; Viana, K.F.; Haddi, K.; Oliveira, E.E.; et al. Antibacterial activity of Siparuna guianensis essential oil mediated by impairment of membrane permeability and replication of pathogenic bacteria. *Ind Crops Prod.* **2020**, *146*, 112142. [CrossRef]
81. Sellem, I.; Chakchouk-Mtibaa, A.; Zaghden, H.; Smaoui, S.; Ennouri, K.; Mellouli, L. Harvesting season dependent variation in chemical composition and biological activities of the essential oil obtained from *Inula graveolens* (L.) grown in Chebba (Tunisia) salt marsh. *Arab. J. Chem.* **2020**, *13*, 4835–4845. [CrossRef]
82. Güven, K.; Yücel, E.; Cetintaş, F. Antimicrobial activities of fruits of Crataegus and Pyrus species. *Pharm. Biol.* **2006**, *44*, 79–83. [CrossRef]
83. Chandrasekaran, M.; Venkatesalu, V. Antibacterial and antifungal activity of Syzygium jambolanum seeds. *J. Ethnopharmacol.* **2004**, *91*, 105–108. [CrossRef] [PubMed]
84. Lennerz, B.S.; Vafai, S.B.; Delaney, N.F.; Clish, C.B.; Deik, A.A.; Pierce, K.A.; Ludwig, D.S.; Mootha, V.K. Effects of sodium benzoate, a widely used food preservative, on glucose homeostasis and metabolic profiles in humans. *Mol. Genet. Metab.* **2015**, *114*, 73–79. [CrossRef] [PubMed]
85. Mosaei, H.; Zenkin, N. Inhibition of RNA polymerase by Rifampicin and Rifamycin-like Molecules. *EcoSal Plus* **2020**, *9*, 1–16. [CrossRef]
86. Al-Wahaibi, L.H.; Amer, A.A.; Marzouk, A.A.; Gomaa, H.A.; Youssif, B.G.; Abdelhamid, A.A. Design, synthesis, and antibacterial screening of some novel heteroaryl-based ciprofloxacin derivatives as DNA gyrase and topoisomerase IV inhibitors. *Pharmaceuticals* **2021**, *14*, 399. [CrossRef]
87. Ferrari, T.; Gini, G. An open source multistep model to predict mutagenicity from statistical analysis and relevant structural alerts. *Chem. Cent. J.* **2010**, *4*, S2. [CrossRef]
88. Fjodorova, N.; Vračko, M.; Tušar, M.; Jezierska, A.; Novič, M.; Kühne, R.; Schüürmann, G. Quantitative and qualitative models for carcinogenicity prediction for non-congeneric chemicals using CP ANN method for regulatory uses. *Mol. Divers.* **2010**, *14*, 581–594. [CrossRef]
89. Benfenati, E.; Chaudhry, Q.; Gini, G.; Dorne, J.L. Integrating in silico models and read-across methods for predicting toxicity of chemicals: A step-wise strategy. *Environ. Int.* **2019**, *131*, 105060. [CrossRef]
90. Roncaglioni, A.; Piclin, N.; Pintore, M.; Benfenati, E. Binary classification models for endocrine disrupter effects mediated through the estrogen receptor. *SAR QSAR Environ. Res.* **2008**, *19*, 697–733. [CrossRef]
91. Bruning, J.B.; Parent, A.A.; Gil, G.; Zhao, M.; Nowak, J.; Pace, M.C.; Smith, C.L.; Afonine, P.V.; Adams, P.D.; Katzenellenbogen, J.A.; et al. Coupling of receptor conformation and ligand orientation determine graded activity. *Nat. Chem. Biol.* **2010**, *6*, 837–843. [CrossRef]

92. Baderna, D.; Gadaleta, D.; Lostaglio, E.; Selvestrel, G.; Raitano, G.; Golbamaki, A.; Lombardo, A.; Benfenati, E. New in silico models to predict in vitro micronucleus induction as marker of genotoxicity. *J. Hazard. Mater.* **2020**, *385*, 121638. [CrossRef] [PubMed]
93. Available online: https://www.vegahub.eu/ (accessed on 27 January 2022).
94. Banerjee, P.; Eckert, A.O.; Schrey, A.K.; Preissner, R. ProTox-II: A webserver for the prediction of toxicity of chemicals. *Nucleic Acids Res.* **2018**, *46*, 257–263. [CrossRef] [PubMed]
95. Available online: https://tox-new.charite.de/protox_II/ (accessed on 27 January 2022).
96. No, O.T. 236: Fish embryo acute toxicity (FET) test. *OECD Guidel. Test. Chem.* **2013**, *2*, 1–22.
97. De Guzman, M.C.; Chua, P.A.P.; Sedano, F.S. Embryotoxic and teratogenic effects of polyethylene microbeads found in facial wash products in Zebrafish (Danio rerio) using the Fish Embryo Acute Toxicity Test. *bioRxiv* **2020**. [CrossRef]
98. Available online: https://swissmodel.expasy.org (accessed on 28 January 2022).
99. Schwede, T.; Kopp, J.; Guex, N.; Peitsch, M.C. SWISS-MODEL: An automated protein homology-modeling server. *Nucleic Acids Res.* **2003**, *31*, 3381–3385. [CrossRef] [PubMed]
100. Available online: https://www.uniprot.org (accessed on 28 January 2022).
101. UniProt Consortium. The universal protein resource (UniProt) in 2010. *Nucleic Acids Res.* **2010**, *38*, 142–148. [CrossRef]
102. Martí-Renom, M.A.; Stuart, A.C.; Fiser, A.; Sánchez, R.; Melo, F.; Šali, A. Comparative protein structure modeling of genes and genomes. *Annu. Rev. Biophys. Biomol. Struct.* **2000**, *29*, 291–325. [CrossRef]
103. Available online: http://www.ebi.ac.uk/thornton-srv/databases/ProFunc (accessed on 29 January 2022).
104. Sharma, A.D.; Kaur, G.R.I.; Kocher, G.S. Molecular modeling and in-silico characterization of alkaline protease from bacillus circulans MTCC 7906. *Online J. Bioinform.* **2015**, *16*, 61–87.
105. Waterhouse, A.; Bertoni, M.; Bienert, S.; Studer, G.; Tauriello, G.; Gumienny, R.; Heer, F.T.; de Beer, T.A.P.; Rempfer, C.; Bordoli, L.; et al. SWISS-MODEL: Homology modelling of protein structures and complexes. *Nucleic Acids Res.* **2018**, *46*, 296–303. [CrossRef]
106. Available online: https://www.drugbank.ca/ (accessed on 26 January 2022).
107. Available online: https://pubchem.ncbi.nlm.nih.gov/ (accessed on 26 January 2022).
108. Available online: http://www.molecular-networks.com/online_demos/corina_demo (accessed on 26 January 2022).
109. Trott, O.; Olson, A.J. AutoDock Vina: Improving the speed and accuracy of docking with a new scoring function, efficient optimization, and multithreading. *J. Comput. Chem.* **2010**, *31*, 455–461. [CrossRef]
110. Pawar, S.S.; Rohane, S.H. Review on discovery studio: An important tool for molecular docking. *Asian J. Res. Chem.* **2021**, *14*, 86–88. [CrossRef]

molecules

MDPI

Review

Soluble Dietary Fiber, One of the Most Important Nutrients for the Gut Microbiota

Zhi-Wei Guan [1,2,†], En-Ze Yu [1,†] and Qiang Feng [1,3,*]

1 Shandong Provincial Key Laboratory of Oral Tissue Regeneration, Shandong Engineering Laboratory for Dental Materials and Oral Tissue Regeneration, Department of Human Microbiome, School of Stomatology, Shandong University, Jinan 250012, China; guanzhiwei1109@163.com (Z.-W.G.); yuenze31415@163.com (E.-Z.Y.)
2 School of Life Science, Qi Lu Normal University, Jinan 250200, China
3 State Key Laboratory of Microbial Technology, Shandong University, Qingdao 266237, China
* Correspondence: fengqiang@sdu.edu.cn
† These authors contributed equally to this work.

Abstract: Dietary fiber is a widely recognized nutrient for human health. Previous studies proved that dietary fiber has significant implications for gastrointestinal health by regulating the gut microbiota. Moreover, mechanistic research showed that the physiological functions of different dietary fibers depend to a great extent on their physicochemical characteristics, one of which is solubility. Compared with insoluble dietary fiber, soluble dietary fiber can be easily accessed and metabolized by fiber-degrading microorganisms in the intestine and produce a series of beneficial and functional metabolites. In this review, we outlined the structures, characteristics, and physiological functions of soluble dietary fibers as important nutrients. We particularly focused on the effects of soluble dietary fiber on human health via regulating the gut microbiota and reviewed their effects on dietary and clinical interventions.

Keywords: soluble dietary fiber; resistant oligosaccharide; viscous fiber; gut microbiota; human health

Citation: Guan, Z.-W.; Yu, E.-Z.; Feng, Q. Soluble Dietary Fiber, One of the Most Important Nutrients for the Gut Microbiota. *Molecules* **2021**, *26*, 6802. https://doi.org/10.3390/molecules26226802

Academic Editor: Smaoui Slim

Received: 17 September 2021
Accepted: 2 November 2021
Published: 11 November 2021

1. Introduction

The dietary pattern is closely related to human health. Hu et al. identified two major dietary patterns, which are the Prudent and Western diets [1]. The Prudent diet is considered healthy and is characterized by higher intake of vegetables, fruit, legumes, whole grains, fish, and poultry. Compared to the Prudent diet, the Western diet is characterized by abundant red meat, fat, and refined carbohydrates [2,3]. The Western diet contributes to increased risk of non-communicable diseases, especially gastrointestinal diseases and metabolic diseases which are partially attributed to the deficiency of dietary fiber in the Western diet [4–10]. According to the widely accepted definition derived from Codex Alimentarius Alinorm in 2009, dietary fiber is considered as edible carbohydrate polymers with three or more monomeric units that are resistant to endogenous digestive enzymes and thus are neither hydrolyzed nor absorbed in the small intestine [11]. Based on their structures, dietary fibers can be classified into non-starch polysaccharides, resistant starches (RSs), and resistant oligosaccharides. Moreover, the non-carbohydrate polymer, lignin, which coexists with cellulose in plant cell walls, is also considered in this definition as dietary fiber. Based on the source, dietary fibers are mainly comprised three subgroups: (i) Carbohydrate polymers naturally existing in edible plants and consumed as vegetables, fruits, seeds, cereals, and tubers; (ii) edible carbohydrate polymers obtained from raw foods by physical, enzymatic, and chemical means that have a proven physiological benefit (e.g., resistant oligosaccharides, inulin, and psyllium); and (iii) synthetic carbohydrate polymers with a proven physiological benefit (e.g., methylcellulose). [11,12].

Dietary fibers from food pass through the small intestine into the large intestine, where they play physiological roles. Dietary fibers contain a variety of organic polymers, with different monomers linked by different glycosidic bonds, showing complex and heterogeneous structure [13]. To help correlate physicochemical characteristics of dietary fiber with their physiological functions, many ways in classifying dietary fiber were established, which include solubility, viscosity, and fermentability [14]. Depending on solubility, dietary fiber can be categorized as insoluble or soluble (SDFs) [15]. The sugar chains in insoluble dietary fiber associate with each other by dense hydrogen bonds, forming a hydrophobic and crystalline structure, which can resist the hydrolysis of exogenous glucosidases. As the most widely distributed and abundant insoluble fiber in nature, cellulose is a polysaccharide with high molecular weight, composed of β-glucose. It is the main structural component of plant cell walls, which usually combines with hemicellulose, pectin, and lignin [16]. A schematic diagram of molecular structure of cellulose is shown in Figure 1A. Most insoluble dietary fibers, such as cellulose, hemi-cellulose, and lignin, have an effect on bulking fecal material, but are not or just slowly utilized by gut bacteria. On the contrary, SDFs can be readily and quickly metabolized by gut bacteria, in the process of which SDFs significantly influence the abundance and diversity of the human gut microbiota [17]. Studies confirmed that dietary fibers, especially SDFs, can positively regulate the gut microbiota and be metabolized to beneficial products, mainly short-chain fatty acids (SCFAs), thus providing many advantages to human health, such as reducing the risk of gastrointestinal diseases including irritable bowel syndrome (IBS), inflammatory bowel disease (IBD), diverticular disease, functional constipation, fecal incontinence, and colorectal cancer (CRC) [18–20].

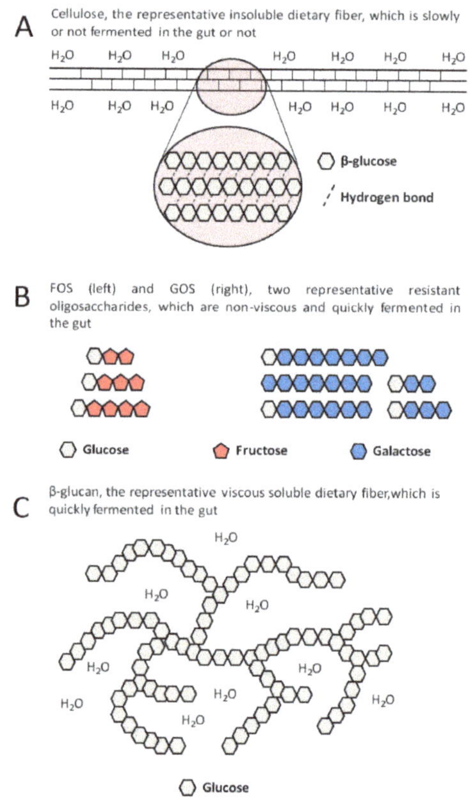

Figure 1. Schematic diagram of the structures of representative insoluble dietary fiber and SDF. (**A**). The representative insoluble dietary fiber, cellulose. (**B**). The representative resistant oligosaccharides, FOS and GOS. (**C**). The representative viscous SDF, β-glucan.

Here, we summarized the structures and characteristics of SDFs as well as their effects on the gut microbiota, by which they help to improve human health. We hope that this review can help the readers learn about SDFs and their influence on the gut microbiota, and provide some advice for future research in related fields.

2. SDFs Include Many Different Substances

Solubility means the ability of dietary fibers to dissolve in water. Compared to insoluble dietary fibers, SDFs have a high affinity for water. SDFs includes various active substances with different structures, which are mainly composed of resistant oligosaccharides and viscous dietary fibers with a high molecular weight [21]. It is worth noting that solubility can be quite variable depending on not only the structures of fibers but also the external factors such as temperature and pH value [13]. For example, the solubility of pectin improves with the increase of the abundance of its side chains [22]. In addition, some viscous fibers are obtained via chemically modifying insoluble fibers, such as methylcellulose and type IV RS (RS-4), which were produced from cellulose and type II RS, respectively [23,24]. Significantly, SDFs include a large amount of active substances showing the structural complexity, which can be administered in the form of natural foods or dietary supplements [25]. In consideration of the fact that certain dietary fibers, including most SDFs, have the effect of promoting the proliferation of specific probiotics, they are also called microbiota-accessible carbohydrates (MACs) or prebiotics [26]. The definition of prebiotic published in 2017 is a substrate that is selectively utilized by host microorganisms conferring a health benefit [27]. The structures, sources, and physicochemical characteristics of common SDFs are shown in Table 1.

Resistant oligosaccharides, which are also called functional oligosaccharides, refer to oligosaccharides with a degree of polymerization (DP) from 3 to 9 that have prebiotic effects. Fructo-oligosaccharides (FOS) and galacto-oligosaccharides (GOS) are the most studied and typical resistant oligosaccharides. They are easily fermented in the gut because of their low molecular weight and high solubility [28]. A schematic diagram of molecular structures of FOS and GOS is shown in Figure 1B. The preparation of resistant oligosaccharides can be achieved by enzymatic methods, such as hydrolysis and isomerization, and chemical methods [29,30].

Inulin is a chain polysaccharide composed of D-fructose linked by a β-glycosidic bond with a glucose at the end, and with a DP from 2 to 60 [31]. Inulins with a DP less than 10 belong to resistant oligosaccharides, while the others belong to viscous dietary fibers. Due to their large molecular weight and strong hydrophilicity, viscous fibers, including β-glucan, pectins, and gums, can dissolve in water as well as forming a gelatinous structure at the required critical concentration with a water-holding capacity, which can inhibit the absorption of glucose and lipids in the gut [17]. β-glucan is composed of D-glucose linked by a β-glycosidic bond, which is the main structural component of plant cell walls and can also be synthesized by enzyme technology [32]. A schematic diagram of the structure of viscous β-glucan is shown in Figure 1C. Pectin, a hetero-polysaccharide widely existing in the middle lamella and primary cell walls of the plants, is mainly composed of D-galacturonic acids linked by an α-1,4-glycosidic bond and contains a small amount of rhamnose, arabinose, and galactose [33]. Gums are high-molecular-weight carbohydrates that at low concentrations can combine with water to form gels. There are many types of gums, such as exudate gums (e.g., acacia gum), mucilage gums (e.g., psyllium), and microbial gums (e.g., xanthan gum). Exudate gums are the exudates of certain trees and shrubs, which are wildly applied in the food and other industries because they are easy to produce [34].

Most viscous dietary fibers are rapidly and completely fermented in water with the gelatinous structure disappearing, but there are exceptions, such as psyllium. Psyllium, a widely concerned and studied dietary gum, is a powder ground from the seeds of Plantago ovate. Psyllium is rich in mucilage, which is a mixture of polysaccharides consisting of pentoses, hexoses, and uronic acids [35]. The results of in vitro experiments showed that psyllium is a kind of fermentable fiber, but in vivo experiments confirmed that psyllium can hardly be fermented [36]. Therefore, compared with other fermentable viscous fibers, psyllium can always retain its water-holding capacity in the gut, which is the reason why psyllium is used as a laxative in clinics [35].

Methylcellulose is the simplest cellulose derivative. It is made by alkalizing cellulose and then methylating alkaline cellulose by chloromethane, in the process of which methyl groups substitute the hydrogens of the hydroxyls at C-2, C-3, and/or C-6 positions of β-D-glucose units in cellulose. Finally, the product is purified and ground to a powder. Methylcellulose is approved as a thickener and gelling additive in food processing in many countries [37]. As a viscous dietary fiber, methylcellulose is completely non-fermentable in the gut, but has the function of relaxing the bowel [38].

RS is a kind of starch, which can escape digestion in the small intestine and then arrives in the colon where it can be utilized by certain specialized bacteria [39]. RS is divided into five subtypes. Although all types of RSs exhibit prebiotic activity, only RS-4 is soluble. RS-4 is a group of starches that are chemically modified by conversion, substitution, or cross-linking in order to lower their digestibility in the small intestine [40].

Table 1. The structures, sources, and physicochemical characteristics of common SDFs and their effects on the gut microbiota.

Type	Structure	Source	Viscosity	Fermentability	Changes Related to the Gut Microbiota
FOS	Sucrose combines with 1 to 3 fructoses linked by a β-glycosidic bond	Vegetables, fruits, produced by enzyme-catalyzed synthesis	No	Yes	Increased α-diversity, Increased *Bifidobacteria* and *Lactobacilli* [41]
GOS	Galactose or glucose combines with 1 to 7 galactoses linked by a β-glycosidic bond	Milk, produced by enzyme-catalyzed synthesis	No	Yes	Increased β-diversity, Increased Lactobacillaceae and Lachnospiraceae, Decreased Ruminococcaceae [42]
Inulin	D-fructose linked by a β-glycosidic bond with glucose at the end	Vegetables, fruits, grains	DP < 9: No DP ≥ 9: Yes	Yes	Increased β-diversity, Increased Prevotellaceae [43] Increased α-diversity, Increased *Bifidobacteria*, Decreased *Desulfovibrio* [44]
β-glucan	High polymer composed of D-glucose linked by a β-glycosidic bond	Grains	Yes	Yes	Increased *Bifidobacteria* and *Lactobacilli*, Decreased *Enterobacteriaceae* [45]
Pectins	Polysaccharides with complex structures containing D-galacturonic acid, rhamnose, arabinose, and galactose	Vegetables, fruits, beans	Yes	Yes	Increased *Bifidobacteria*, *Lactobacilli*, and *Faecalibaculum* spp. [46] Increased β-diversity, Inhibited *Citrobacter rodentium* [47]
Gums	Polysaccharides with complex structures containing mannose, galactose, glucose, and D-galacturonic acid	Leguminous plants, nuts, seaweeds	Yes	Yes	Increased *Bifidobacteria* and *Lactobacilli* [48] Inhibited *Clostridium histolyticum* [49]
Psyllium	Mixture of polysaccharides consisting of arabinose, xylose, galactose, rhamnose, and D-galacturonic acid	*Plantago ovate*	Yes	No	No
Methylcellulose	Long-chain substituted cellulose, in which about 30% of the hydroxyl groups exist in the form of methoxyl	Synthesized	Yes	No	No
RS-4	Chemically modified starch, such as acetyl starch, hydroxypropyl starch, heat-modified starch, and phosphorylated starch	Synthesized	Yes	Yes	Increased *Bacteroides*, *Bifidobacteria*, *Lactobacilli*, *Coprococcus*, and *Allobaculum* [50–52]

FOS: fructo-oligosaccharides, GOS: galacto-oligosaccharides, DP: degree of polymerization, RS-4: type IV resistant starch.

3. Mechanism in the Utilization of SDFs by the Gut Microbiota

The human body is occupied by trillions of microorganisms, most of which are bacteria [53]. They are closely related to human health and the gut is their most densely populated habitat [54]. In 2020, Almeida et al. established the Unified Human Gastrointestinal Genome (UHGG) collection by editing and analyzing previous human gut microbiome datasets. UHGG consists of 204,938 non-redundant genomes from 4644 gut prokaryotes and is the most comprehensive sequence database of the human gut microbiome so far [55]. Firmicutes and Bacteroidetes are the dominant phyla in the gut microbiota [56]. The digestive enzymes of the human body cannot degrade SDFs. When SDFs in foods enter the colon though the small intestine, the gut bacteria can degrade SDFs into oligosaccharides or monosaccharides through different degradation systems, and then absorb them by specific transport systems for energy source [57]. Although each organism contains relatively few cellulolytic enzymes, the intestinal microbes in total contain about 130 glycoside hydrolase (GH) families, 22 polysaccharides lyase (PL) families, and 16 carbohydrate esterase families, which provide the microbiota with the flexibility to switch between different fiber energy sources [58].

In terms of the degradation of long-chain dietary fibers, in vitro studies by co-culturing with human fecal samples indicated that the species from the phylum Firmicutes (e.g., *Ruminococcus bromii, Eubacterium rectale, Clostridium* spp., and *Roseburia* spp.), *Bifidobacterium* spp., and *Bacteroides* spp., are the major degraders of RSs [59,60]. However, bioinformatic analysis of a model human microbiome constructed from 177 human microbial genomes revealed that the bacteria from the phylum Bacteroidetes are the likely primary degraders of the various complex polysaccharides in the plant cell walls [61]. Many gut microbes evolved diverse strategies to utilize SDFs. For example, *Bacteroides*, the typical Gram-negative bacteria, is the most widely studied microbe in the field of polysaccharide transport and utilization of the gut microbiota due to its efficient polysaccharide degradation system, which may be the reason that *Bacteroides* is the dominant bacteria in the gut microbiota. About 20% of the genes in the genome of *Bacteroides* are involved in the utilization of polysaccharides [57]. Its transportation and decomposition of polysaccharides require the participation of multiple functional proteins and the system containing these proteins is called the starch utilization system (Sus). Sus consists of eight components, SusA-SusG and SusR, which are responsible for the detection, binding, hydrolysis, and transport of exogenous polysaccharides, respectively [62]. Notably, all *Bacteroides* possess the orthologous components of the Sus system, many of which have been demonstrated to be responsible for the absorption and utilization of specific SDFs such as inulin and pectin [63]. Moreover, certain Gram-positive bacteria in the gut microbiota are also prominent in the process of glycan utilization. The most representative Gram-positive intestinal saccharolytic microbe is the *Bifidobacterium* genus from the Actinobacteria phylum [64]. Bioinformatics analyses showed that the genes encoding modules involved in carbohydrate utilization account for nearly 13% of the identified genes in the bifidobacterial genome [65]. Unlike Gram-negative bacteria, Gram-positive *Bifidobacteria* have no periplasmic space. Therefore, complex polysaccharides first need to be digested into oligosaccharides extracellularly by the GHs anchored on the cell surface. Then, the produced oligosaccharides are transported into the cytoplasm for further degradation or shared by other members in the gut microbiota as nutritional sources [66]. The bacterial ability to degrade fibers is relative to the chain length of fibers. Resistant oligosaccharides are usually easier to degrade than polysaccharides. Many bacteria in the gut microbiota can utilize short-chain fibers. Besides the degraders of long-chain fibers, *Faecalibacterium prausnitzii* and *Lactobacillus* spp. were found to be able to utilize FOS in vitro [67]. Except for the decomposition of SDFs, the subsequent production of SCFAs is also necessary for human health. SCFAs are the fatty acids with 1 to 6 carbon atoms, mainly including acetate, propionate, and butyrate. Succinate and lactate are also produced by intestinal microbes, but they are usually regarded as intermediates due to their small amounts and the conversion of them to SCFAs by other microorganisms. The current findings have demonstrated that the health benefits of SDFs depend more on SCFAs [15,20]. Many intestinal bacteria have been proved to be specific SCFA producers, although some of them have a poor ability to degrade SDFs (e.g., *Anaerostipes* spp., *Coprococcus* spp., *Dialister* spp., *Veillonella* spp., and *Salmonella* spp.) [20,68]. The process of degrading and metabolizing SDFs by the gut microbiota is illustrated as Figure 2.

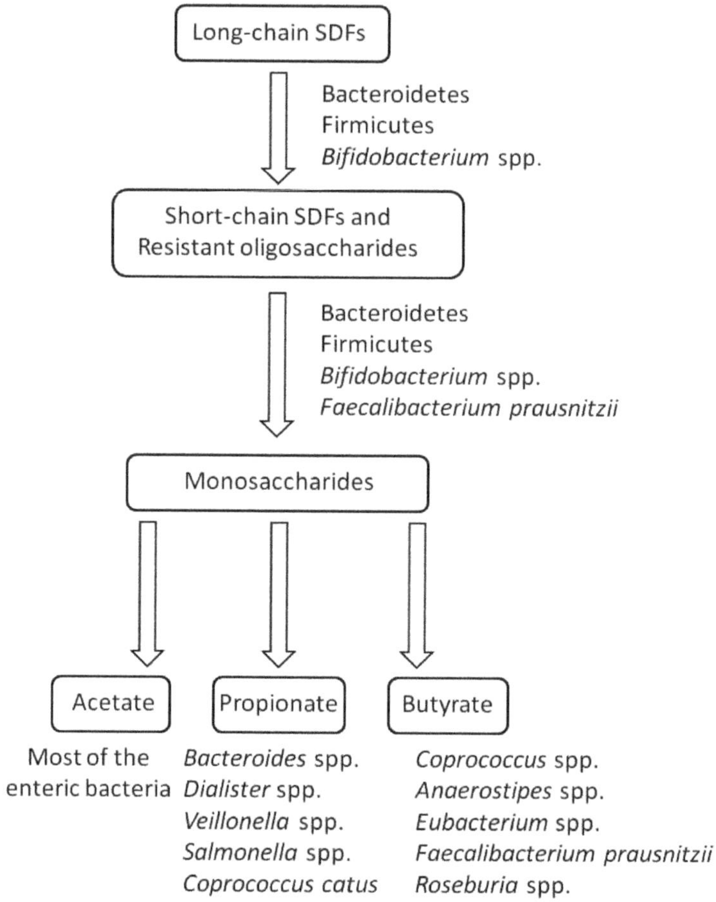

Figure 2. Schematic diagram of the process of degrading and metabolizing SDFs by the gut microbiota.

The above data are indicative of the link between the ability of the intestinal bacteria in glycan utilization and its corresponding ecological niche in the human gut. It also makes SDF metabolism closely related to human health.

4. Effects of SDFs on the Gut Microbiota

Dietary fibers, especially SDFs, provide the main carbon and energy source for the gut microbiota. SDFs have prebiotic effects by increasing the beneficial bacteria and improving the intestinal environment [69].

For example, FOS can effectively increase the bacterial diversity of the human gut microbiota and improve the abundance of *Bifidobacteria* and *Lactobacilli* [41]. The results of the study by Yang et al. showed that the administration of GOS is conductive to improving neuro-inflammatory and cognitive dysfunction in a rat model of abdominal surgery. Furthermore, in-depth fecal microbiota analysis by 16S rRNA sequencing revealed that the administration of GOS can induce a significant increase in β-diversity of the gut microbiota and the proliferation of *Bifidobacterium* and other potentially anti-inflammatory bacteria [42]. Inulin was confirmed to be beneficial to health by regulating the gut microbiota. In *ob/ob* mice, it was found that an inulin-containing diet can improve fat accumulation and glucose intolerance. Further fecal microbiota analysis displayed that the β-diversity of the *ob/ob* mice tended to the level of the wild type mice and the Prevotellaceae UCG 001 family was significantly enriched, which positively affected the leptin-related pathways in the *ob/ob* mice [43]. In a randomized controlled trial on healthy adults, the addition of agave inulin bettered the diversity and activity of the gut microbiota including increasing *Bifidobacteria* and depleting *Desulfovibrio* [44]. Cereal is the main dietary source of β-glucan. A study in rats showed that the administration of cereal β-glucan can promote the growth of *Bifidobacteria* and *Lactobacilli*, whereas it can reduce the abundance of Enterobacteriaceae in a dose-dependent manner [45]. Pectin rich in rhamnogalacturonan-I which is refined from citrus segment membranes was demonstrated to significantly promote the production *Bifidobacteria*, *Lactobacilli*, and *Faecalibaculum* spp. in C57BL/6J male mice, and to be metabolized to SCFAs, which reduces the pH value in the intestine [46]. Another experiment with C57BL/6J female mice showed that administration of pectin extracted from orange peel relieved the development of *Citrobacter rodentium*-induced colitis, which may be because of the increase in the diversity of the gut microbiota [47]. Acacia gum, also known as gum Arabic, a branched-chain polysaccharide exudate gum mainly produced from *Acacia senegal*, is composed of arabinose and galactose [34]. Calame et al. found that the intake of acacia gum at 10 g/d was effective in increasing the abundance of *Bifidobacteria* and *Lactobacilli* in healthy human volunteers [48]. The results of an in vitro study by Rawi et al. showed that acacia gum significantly increased the abundance of *Bifidobacteria* while it inhibited that of *Clostridium histolyticum*, the bacterium which gives rise to gut dysbiosis. Moreover, acacia gum promoted the production of butyrate, which may also contribute to ameliorating the gut microbiota [49]. RSs, both insoluble and soluble, can be fermented and utilized by colonic microorganisms. They can increase the abundance of *Bacteroides*, *Bifidobacteria*, *Lactobacilli*, *Coprococcus*, and *Allobaculum* [50–52] and cooperatively perform their prebiotic effects with other active constituents such as chitosan oligosaccharide [70]. The effects of SDFs on the gut microbiota are summarized in Table 1.

Except for the different types and amounts of SDFs and other polysaccharides from food, dietary and endogenous proteins and mucins that come from intestinal epithelial cells are also nutrients for the gut microbiota [66,71]. The intestinal epithelium is covered and protected by a mucous layer to keep bacteria isolated from the mucosa [72]. SDFs and SCFAs stimulate mucus production and secretion [73]. Inadequate intake of SDFs will reduce the number of probiotics and transfer the metabolism of the gut microbiota to utilize amino acids in the other substrates, which can give rise to injury of the intestinal mucosa by accumulation of harmful metabolites, such as branched-chain fatty acids, ammonia, amines, N-nitroso compounds, and phenolic compounds [74]. Therefore, a dietary pattern with high sugar, fat, and protein but low fiber may result in the development of chronic inflammatory diseases such as IBD, CRC, allergies, cardiovascular disease, and obesity [75–77]. Sufficient intake of SDFs can protect the intestinal mucosa from degradation by intestinal bacteria and to a certain extent prevent these diseases.

Studying the impact of SDFs on health via the gut microbiota is sophisticated due to certain interferences. First, animal experiments often administer refined or modified SDFs in the form of feed additives. However, the physicochemical properties of SDFs such as viscosity or fermentability may be different when SDFs are eaten in the form of natural foods, thus altering their physiological functions. Second, natural foods rich in SDFs also contain other beneficial nutrients, such as minerals and phytochemicals, which make it difficult to determine the precise effects of SDFs alone on human health. Therefore, whether the results from animal experiments can be applied to humans requires further study.

5. SDFs and Their Metabolites Display Important Physiological Effects on Human Health

Although SDFs hardly directly provide energy for humans, SDFs per se exhibits specific physiological functions as recognized nutrients. In addition to stimulating the production and secretion of mucus, SDFs and SCFAs have other important physiological functions. In this section, we summarized the physiological functions of SDFs, including those that are attributed to SCFAs.

5.1. Increase Satiety and Reduce Energy Intake

Viscous SDFs retard the hydrolysis and absorption of energetic nutrients from food in the small intestine, such as starch and triglyceride. So, SDFs can significantly reduce the total intake of energy as well as glucose and cholesterol, so they contribute to slowing down the process of obesity, type 2 diabetes mellitus (T2DM), hyperlipidemia, and related metabolic diseases [78,79]. Then, the chyme gets to and stimulates the terminal ileum, where the mucosa responses and releases glucagon-like peptide-1 (GLP-1). The results from the experiments in human and pigs indicated that GLP-1 can inhibit gastric emptying and reduce intestinal peristalsis [80,81]. Therefore, viscous SDFs are helpful to control appetite, improve insulin sensitivity, and reduce weight.

5.2. Promote the Metabolism and Absorption of Active Substances

Many viscous SDFs can provide a platform for the metabolism of active substances. Take the bile acids as an example. Bile acids are produced in the liver and metabolized by enzymes from intestinal bacteria, which not only promote the absorption of dietary fat but also play indispensable roles in maintaining the healthy gut microbiota, balanced lipid and carbohydrate metabolism, insulin sensitivity, and innate immunity [82]. In the upper segment of the ileum and colon, conjugated primary bile acids combine with SDFs, where they are hydrolyzed to free primary bile acids by the bile salt hydrolase (BSH) from the intestinal bacteria, mainly *Bacteroides* and *Lactobacilli* [83]. Then, 7α-dehydroxylase also from the intestinal bacteria, such as *Clostridium* spp. and *Eubacterium* spp., catalyzes the free primary bile acids to secondary bile acids [84,85].

In addition to organic substrates, viscous SDFs can bind with inorganic nutrients such as metal ions. It was reported that after being bound with SDFs, calcium, magnesium, iron, copper, and zinc are transported to the distal colon. With the degradation of SDFs by the local bacteria, the ions are released and exhibit specific effects including resisting pathogens, increasing the diversity of the intestinal bacterial community, and protecting the gut from infection [76]. Furthermore, SCFAs produced via the gut microbiota fermenting SDFs combine with ions to form soluble salts, which are more prone to absorption by the colon [86,87].

5.3. SCFAs Act as Histone Deacetylase (HDAC) Inhibitors

Intracellular butyrate and propionate inhibit the activity of HDACs in colon cells and immune cells, leading to histone hyperacetylation, which in turn affects gene expression and cell differentiation, proliferation, and apoptosis [88]. Many studies have shown that SCFAs have important anti-inflammatory effects due to HDAC inhibition. For example, SCFAs can down-regulate proinflammatory cytokines such as interleukin-6 (IL-6) and IL-12 in colonic macrophages and differentiate dendritic cells from bone marrow stem cells [89,90]. Moreover, SCFAs can induce colonic regulatory T cells (Tregs) in mice [91]. There is also evidence that butyrate and propionate can induce the differentiation of Tregs, which can express the transcription factor Foxp3 via increasing the acetylation at the gene locus of foxP3. Foxp3 is found to play a crucial role in controlling intestinal inflammation in mice [92]. In addition, butyrate and propionate activate the AP-1 signaling pathway in human epithelial cells, which plays an important role in controlling proliferation and inducing apoptosis of colon cancer cells [93].

In brief, SCFAs, especially butyrate, not only provide the most energy for colon cells, but also aid to a large extent in the prevention of inflammation and CRC due to HDAC inhibition.

5.4. SCFAs Are Important Ligands for Specific G-Protein Coupled Receptors (GPCRs)

In addition to acting as HDAC inhibitors, SCFAs also exert important physiological functions as ligands for GPCRs. Three GPCRs (GPR41, GPR43, and GPR109A) involved in immune regulation were proven to specifically respond to free fatty acids. Therefore, GPR43 and GPR41 were also named FFAR (free fatty acid receptor) 2 and FFAR3, respectively [94].

In mice, butyrate can increase the secretion of Tregs, IL-18, and T cells producing IL-10 in intestinal epithelial cells via stimulating GPR109A [95]. Additionally, the study by Macia et al. in mice indicated that SCFAs derived from a high-fiber diet stimulated GPR43 and GPR109A to activate the NLRP3 inflammasome, which produces IL-18. This effect maintains intestinal homeostasis by decreasing the inflammatory response of the gut and maintaining the integrity of the mucosal barrier, which prevents bacterial invasion and infection [96]. In the intestine and white adipose tissue (WAT) of mice, SCFA-dependent GPR43 stimulation (especially acetate and propionate) displays beneficial effects in ameliorating the metabolism of glucose and lipids by GLP-1 secretion and anti-lipolytic activity, respectively [97]. GPR41 also plays a role in the regulating appetite. Samuel et al. reported that by binding with GPR41, SCFAs induce the production of peptide YY, which inhibits gastrointestinal motility and gastric acid secretion in mice [98].

The above studies indicated that SCFAs have significant immunologic and metabolic functions as ligands for GPCRs, affecting the incidence of IBD, CRC, and other cancers as well as chronic metabolic diseases.

The identified physiological effects of SDFs and SCFAs on human health are summarized in Figure 3.

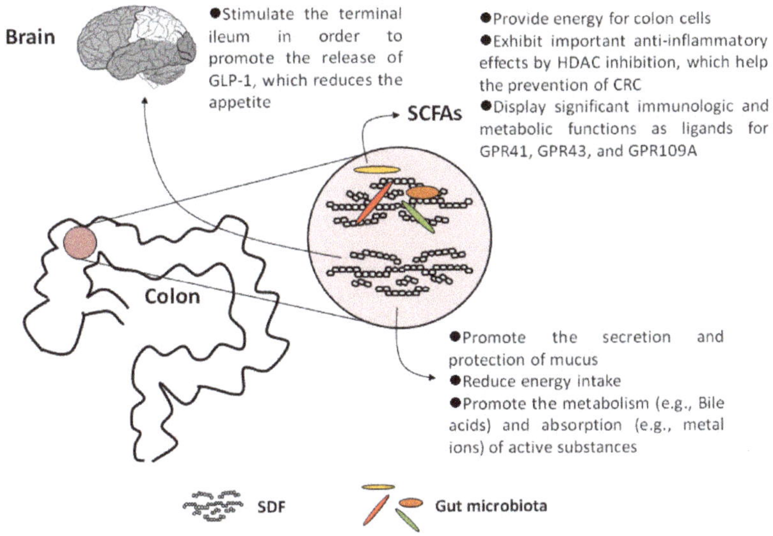

Figure 3. Summary of identified physiological effects of SDF and SCFAs on human health.

6. Safety of SDFs

Although SDFs show excellent health effects, it cannot be ignored that inappropriate SDF intake may lead to certain health hazards, which depend on the type and quantity of SDFs as well as the physiological background of the host.

First, a sudden increase of SDF intake, even when consumed judiciously, may lead to abdominal distension, flatulence, constipation, diarrhea, and other syndromes of IBS [99,100].

Second, as mentioned above, the binding by SDFs can promote to a certain extent the absorption of certain micronutrients such as some metal ions in the colon. However, the study on six healthy young women by Riedl et al. indicated that the bioavailability of β-carotene, lycopene, and lutein, was markedly reduced by three different kinds of SDFs, pectin, guar gum, and alginate [101]. This suggests that excessive SDF intake may be disadvantageous to certain people with micronutrient deficiencies. In addition, Bruggencate et al. reported that the rapid fermentation of FOS by endogenous microbiota damaged the intestinal mucosal barrier and increased intestinal permeability, which caused pathogen infection in rats [102]. Moreover, the study by Singh et al. found that formulating a diet rich in refined inulin or other SDFs to feed TLR5-deficient (*T5KO*) mice with obesity caused by dysregulated gut microbiota, about 40% of mice had a lower weight than before. However, many mice suffered from icteric hepatocellular carcinoma (HCC). Further research revealed that SDF-induced HCC in mice developed via elevation of secondary bile acids in the systemic circulation, cholestasis, and hepatocyte death, followed by neutrophilic inflammation of the liver. Furthermore, fecal microbiota analysis by 16S rRNA sequencing showed that the HCC-prone mice exhibited gut dysbiosis characterized by a loss in species richness and diversity and an increase in the Proteobacteria phylum and the fiber-utilizing microbes including *Clostridium* spp. However, such HCC in T5KO mice cannot be induced by cellulose, the insoluble and non-fermentable fiber, and it was not observed in germ-free nor antibiotics-treated mice [103]. Although acting as a kind of metabolite with important physiological functions, certain bile acids, especially secondary bile acids, were proven to be cytotoxic and cancer-promoting, which have adverse effects on the structure and function of the colonic epithelium by many mechanisms including DNA oxidative damage, inflammation, NF-κB activation, reducing apoptosis, and differentiation, as well as enhancing cell proliferation [104]. Apart from HCC, the changed bile acid profile derived from dysregulated gut microbiota was demonstrated to be associated with a variety of digestive diseases. For example, many patients suffering from CRC exhibited abnormal bile acid metabolism characteristic of redundant secondary bile acids. These redundant secondary bile acids, including deoxycholic acid, lithocholic acid, and taurochenodeoxycholic acid, originate in the aberrant elevation of certain gut bacteria expressing BSH and 7α-dehydroxylase, whose proportion is significantly higher than that in the gut microbiota of healthy people [82]. These facts hinted that, despite notable physiological benefits, fortification of diets with SDFs should be done with great caution as it may cause severe digestive disorders, especially under the background of dysregulated gut microbiota.

7. Conclusions and Prospects

Studies have confirmed that the gut microbiota is closely related to human health. The results from experiments with mice showed that the genetic background contributed to less than 12% of the difference of the gut microbiota, but dietary structure and habits contributed to 57% [105]. SDFs have a significant advantage in improving the gut microbiota due to their high fermentation efficiency. Furthermore, many prospective cohort studies showed that the diversity of the gut microbiota is negatively correlated with the incidence rate of many chronic diseases, including IBD, CRC, obesity, and T2DM [106–110]. In view of the fact that specific SDFs can promote the proliferation of specific intestinal bacteria, a diet rich in various SDFs is beneficial to health. A reasonable combination of various SDFs by raw food is undoubtedly helpful to improve the gut microbiota. In addition, studies also showed that the host response to SDF intervention is personalized, and the results to a significant extent depend on the individual's intestinal ecology before treatment [15]. Therefore, based on the precise analysis of host gut ecology, the personalized treatment of dietary intervention with SDFs combined with antibiotic therapy and/or fecal microbial transplantation (FMT), may be effective in improving health, especially in the prevention and treatment of intestinal diseases. SDFs have a promising future by increasing the community of beneficial microorganisms and their products of growth and metabolism in the host. Notably, certain prospective cohort studies showed that insoluble fibers in diet is more protective against some metabolic diseases (e.g., T2DM) than SDFs [111]. At the same time, considering the side effects of SDFs via dysregulated microbial fermentation, precise SDF intake as well as a reasonable association with insoluble dietary fiber on the basis of the background of the gut microbiota in hosts is of great importance in achieving the goal of disease prevention and cure. With the development of microbiome analysis, the functions of intestine ecology will be clearer. The precise understanding of metabolic pathways and end products involved with the utilization of SDFs and other nutrients by the gut microbiota is the hotspot of the present research. There is an urgent need to elucidate the interactions between bacterial strain levels and specific types of SDFs, which may undoubtedly help improve human health by accurately determining the diversity and functions of the gut microbiota.

Author Contributions: Conceptualization, Q.F.; writing, Z.-W.G. and E.-Z.Y.; Figures and table made by Z.-W.G.; supervision, Q.F. All authors have read and agreed to the published version of the manuscript.

Funding: We gratefully acknowledge the financial support from The National Natural Science Foundation of China (Grant No. 81630072, 82071122), Program of Taishan Young from Shandong Province, Oral Microbiome Innovation Team of Young Scientist Project of Shandong Province (Grant No. 2020KJK001), The National High-level Young Scientist Project Foundation (2019), Excellent Young Scientist Foundation of Shandong Province (Grant No. ZR202102230369), and Higher Educational Science and Technology Program of Shandong Province (Grant No. J1LLE53). The funders had no role in the study design, data collection and analysis, decision to publish, or preparation of the manuscript.

Institutional Review Board Statement: Not applicable.

Informed Consent Statement: Not applicable.

Data Availability Statement: Not applicable.

Acknowledgments: Thanks to Edward C. Mignot, Shandong University, for linguistic advice.

Conflicts of Interest: The authors declare no conflict of interest.

References

1. Hu, F.B.; Rimm, E.B.; Stampfer, M.J.; Ascherio, A.; Spiegelman, D.; Willett, W.C. Prospective study of major dietary patterns and risk of coronary heart disease in men. *Am. J. Clin. Nutr.* **2000**, *72*, 912–921. [CrossRef] [PubMed]
2. van Dam, R.M.; Rimm, E.B.; Willett, W.C.; Stampfer, M.J.; Hu, F.B. Dietary patterns and risk for type 2 diabetes mellitus in U.S. men. *Ann. Intern. Med.* **2002**, *136*, 201–209. [CrossRef] [PubMed]
3. Strate, L.L.; Keeley, B.R.; Cao, Y.; Wu, K.; Giovannucci, E.L.; Chan, A.T. Western Dietary Pattern Increases, and Prudent Dietary Pattern Decreases, Risk of Incident Diverticulitis in a Prospective Cohort Study. *Gastroenterology* **2017**, *152*, 1023–1030.e2. [CrossRef] [PubMed]
4. Christ, A.; Lauterbach, M.; Latz, E. Western Diet and the Immune System: An Inflammatory Connection. *Immunity* **2019**, *51*, 794–811. [CrossRef]
5. Sdona, E.; Georgakou, A.V.; Ekström, S.; Bergström, A. Dietary Fibre Intake in Relation to Asthma, Rhinitis and Lung Function Impairment—A Systematic Review of Observational Studies. *Nutrients* **2021**, *13*, 3594. [CrossRef]
6. Larrosa, S.; Luque, V.; Grote, V.; Closa-Monasterolo, R.; Ferré, N.; Koletzko, B.; Verduci, E.; Gruszfeld, D.; Xhonneux, A.; Escribano, J. Fibre Intake Is Associated with Cardiovascular Health in European Children. *Nutrients* **2020**, *13*, 12. [CrossRef]
7. O'Keefe, S.J. Diet, microorganisms and their metabolites, and colon cancer. *Nat. Rev. Gastroenterol. Hepatol.* **2016**, *13*, 691–706. [CrossRef]
8. Zhao, L.; Zhang, F.; Ding, X.; Wu, G.; Lam, Y.Y.; Wang, X.; Fu, H.; Xue, X.; Lu, C.; Ma, J.; et al. Gut bacteria selectively promoted by dietary fibers alleviate type 2 diabetes. *Science* **2018**, *359*, 1151–1156. [CrossRef]
9. Canfora, E.E.; Meex, R.C.R.; Venema, K.; Blaak, E.E. Gut microbial metabolites in obesity, NAFLD and T2DM. *Nat. Rev. Endocrinol.* **2019**, *15*, 261–273. [CrossRef] [PubMed]
10. Dinan, T.G.; Cryan, J.F. The Microbiome-Gut-Brain Axis in Health and Disease. *Gastroenterol. Clin. N. Am.* **2017**, *46*, 77–89. [CrossRef]
11. Stephen, A.M.; Champ, M.M.; Cloran, S.J.; Fleith, M.; van Lieshout, L.; Mejborn, H.; Burley, V.J. Dietary fibre in Europe: Current state of knowledge on definitions, sources, recommendations, intakes and relationships to health. *Nutr. Res. Rev.* **2017**, *30*, 149–190. [CrossRef]

12. Davani-Davari, D.; Negahdaripour, M.; Karimzadeh, I.; Seifan, M.; Mohkam, M.; Masoumi, S.J.; Berenjian, A.; Ghasemi, Y. Prebiotics: Definition, Types, Sources, Mechanisms, and Clinical Applications. *Foods* **2019**, *8*, 92. [CrossRef] [PubMed]
13. Lovegrove, A.; Edwards, C.H.; De Noni, I.; Patel, H.; El, S.N.; Grassby, T.; Zielke, C.; Ulmius, M.; Nilsson, L.; Butterworth, P.J.; et al. Role of polysaccharides in food, digestion, and health. *Crit. Rev. Food Sci. Nutr.* **2017**, *57*, 237–253. [CrossRef] [PubMed]
14. Williams, B.A.; Mikkelsen, D.; Flanagan, B.M.; Gidley, M.J. "Dietary fibre": Moving beyond the "soluble/insoluble" classification for monogastric nutrition, with an emphasis on humans and pigs. *J. Anim. Sci. Biotechnol.* **2019**, *10*, 45. [CrossRef]
15. O'Grady, J.; O'Connor, E.M.; Shanahan, F. Review article: Dietary fibre in the era of microbiome science. *Aliment. Pharmacol. Ther.* **2019**, *49*, 506–515. [CrossRef] [PubMed]
16. Takahashi, T.; Karita, S.; Ogawa, N.; Goto, M. Crystalline cellulose reduces plasma glucose concentrations and stimulates water absorption by increasing the digesta viscosity in rats. *J. Nutr.* **2005**, *135*, 2405–2410. [CrossRef]
17. McRorie, J.W., Jr.; McKeown, N.M. Understanding the Physics of Functional Fibers in the Gastrointestinal Tract: An Evidence-Based Approach to Resolving Enduring Misconceptions about Insoluble and Soluble Fiber. *J. Acad. Nutr. Diet.* **2017**, *117*, 251–264. [CrossRef]
18. Gill, S.K.; Rossi, M.; Bajka, B.; Whelan, K. Dietary fibre in gastrointestinal health and disease. *Nat. Rev. Gastroenterol. Hepatol.* **2021**, *18*, 101–116. [CrossRef]
19. Staller, K.; Song, M.; Grodstein, F.; Whitehead, W.E.; Matthews, C.A.; Kuo, B.; Chan, A.T. Increased Long-Term Dietary Fiber Intake Is Associated with a Decreased Risk of Fecal Incontinence in Older Women. *Gastroenterology* **2018**, *155*, 661–667.e1. [CrossRef] [PubMed]
20. Koh, A.; De Vadder, F.; Kovatcheva-Datchary, P.; Bäckhed, F. From Dietary Fiber to Host Physiology: Short-Chain Fatty Acids as Key Bacterial Metabolites. *Cell* **2016**, *165*, 1332–1345. [CrossRef]
21. Chawla, R.; Patil, G. Soluble dietary fiber. *Compr. Rev. Food Sci. Food Saf.* **2010**, *9*, 178–196. [CrossRef]
22. Ralet, M.C.; Dronnet, V.; Buchholt, H.C.; Thibault, J.F. Enzymatically and chemically de-esterified lime pectins: Characterisation, polyelectrolyte behaviour and calcium binding properties. *Carbohydr. Res.* **2001**, *336*, 117–125. [CrossRef]
23. Nasatto, P.L.; Pignon, F.; Silveira, J.L.M.; Duarte, M.E.R.; Noseda, M.D.; Rinaudo, M. Methylcellulose, a Cellulose Derivative with Original Physical Properties and Extended Applications. *Polymers* **2015**, *7*, 777–803. [CrossRef]
24. Birt, D.F.; Boylston, T.; Hendrich, S.; Jane, J.L.; Hollis, J.; Li, L.; McClelland, J.; Moore, S.; Phillips, G.J.; Rowling, M.; et al. Resistant starch: Promise for improving human health. *Adv. Nutr.* **2013**, *4*, 587–601. [CrossRef]
25. Hamaker, B.R.; Tuncil, Y.E. A perspective on the complexity of dietary fiber structures and their potential effect on the gut microbiota. *J. Mol. Biol.* **2014**, *426*, 3838–3850. [CrossRef] [PubMed]
26. Payling, L.; Fraser, K.; Loveday, S.M.; Sims, I.; Roy, N.; McNabb, W. The effects of carbohydrate structure on the composition and functionality of the human gut microbiota. *Trends Food Sci. Technol.* **2020**, *97*, 233–248. [CrossRef]
27. Gibson, G.R.; Hutkins, R.; Sanders, M.E.; Prescott, S.L.; Reimer, R.A.; Salminen, S.J.; Scott, K.; Stanton, C.; Swanson, K.S.; Cani, P.D.; et al. Expert consensus document: The International Scientific Association for Probiotics and Prebiotics (ISAPP) consensus statement on the definition and scope of prebiotics. *Nat. Rev. Gastroenterol. Hepatol.* **2017**, *14*, 491–502. [CrossRef]
28. Sanders, M.E.; Merenstein, D.J.; Reid, G.; Gibson, G.R.; Rastall, R.A. Probiotics and prebiotics in intestinal health and disease: From biology to the clinic. *Nat. Rev. Gastroenterol. Hepatol.* **2019**, *16*, 605–616. [CrossRef] [PubMed]
29. Wilson, B.; Whelan, K. Prebiotic inulin-type fructans and galacto-oligosaccharides: Definition, specificity, function, and application in gastrointestinal disorders. *J. Gastroenterol. Hepatol.* **2017**, *32* (Suppl. 1), 64–68. [CrossRef]
30. Wu, J.; Yang, R.; Gao, M.; Zhang, H.; Zhan, X. Synthesis of functional oligosaccharides and their derivatives through cocultivation and cellular NTP regeneration. *Adv. Appl. Microbiol.* **2021**, *115*, 35–63. [CrossRef]
31. Man, S.; Liu, T.; Yao, Y.; Lu, Y.; Ma, L.; Lu, F. Friend or foe? The roles of inulin-type fructans. *Carbohydr. Polym.* **2021**, *252*, 117155. [CrossRef]
32. Bulmer, G.S.; de Andrade, P.; Field, R.A.; van Munster, J.M. Recent advances in enzymatic synthesis of β-glucan and cellulose. *Carbohydr. Res.* **2021**, *508*, 108411. [CrossRef] [PubMed]
33. Schmitz, K.; Protzko, R.; Zhang, L.; Benz, J.P. Spotlight on fungal pectin utilization—From phytopathogenicity to molecular recognition and industrial applications. *Appl. Microbiol. Biotechnol.* **2019**, *103*, 2507–2524. [CrossRef]
34. Barak, S.; Mudgil, D.; Taneja, S. Exudate gums: Chemistry, properties and food applications—A review. *J. Sci. Food Agric.* **2020**, *100*, 2828–2835. [CrossRef] [PubMed]
35. McRorie, J.W.; Daggy, B.P.; Morel, J.G.; Diersing, P.S.; Miner, P.B.; Robinson, M. Psyllium is superior to docusate sodium for treatment of chronic constipation. *Aliment. Pharmacol. Ther.* **1998**, *12*, 491–497. [CrossRef]
36. McRorie, J. Clinical data support that psyllium is not fermented in the gut. *Am. J. Gastroenterol.* **2013**, *108*, 1541. [CrossRef]
37. Souza, P.R.; de Oliveira, A.C.; Vilsinski, B.H.; Kipper, M.J.; Martins, A.F. Polysaccharide-Based Materials Created by Physical Processes: From Preparation to Biomedical Applications. *Pharmaceutics* **2021**, *13*, 621. [CrossRef]
38. Topping, D.L.; Oakenfull, D.; Trimble, R.P.; Illman, R.J. A viscous fibre (methylcellulose) lowers blood glucose and plasma triacylglycerols and increases liver glycogen independently of volatile fatty acid production in the rat. *Br. J. Nutr.* **1988**, *59*, 21–30. [CrossRef] [PubMed]
39. Bello-Perez, L.A.; Flores-Silva, P.C.; Agama-Acevedo, E.; Tovar, J. Starch digestibility: Past, present, and future. *J. Sci. Food Agric.* **2020**, *100*, 5009–5016. [CrossRef]

40. DeMartino, P.; Cockburn, D.W. Resistant starch: Impact on the gut microbiome and health. *Curr. Opin. Biotechnol.* **2020**, *61*, 66–71. [CrossRef]
41. Tandon, D.; Haque, M.M.; Gote, M.; Jain, M.; Bhaduri, A.; Dubey, A.K.; Mande, S.S. A prospective randomized, double-blind, placebo-controlled, dose-response relationship study to investigate efficacy of fructo-oligosaccharides (FOS) on human gut microflora. *Sci. Rep.* **2019**, *9*, 5473. [CrossRef] [PubMed]
42. Yang, X.D.; Wang, L.K.; Wu, H.Y.; Jiao, L. Effects of prebiotic galacto-oligosaccharide on postoperative cognitive dysfunction and neuroinflammation through targeting of the gut-brain axis. *BMC Anesthesiol.* **2018**, *18*, 177. [CrossRef]
43. Song, X.; Zhong, L.; Lyu, N.; Liu, F.; Li, B.; Hao, Y.; Xue, Y.; Li, J.; Feng, Y.; Ma, Y.; et al. Inulin Can Alleviate Metabolism Disorders in ob/ob Mice by Partially Restoring Leptin-related Pathways Mediated by Gut Microbiota. *Genom. Proteom. Bioinform.* **2019**, *17*, 64–75. [CrossRef]
44. Holscher, H.D.; Bauer, L.L.; Gourineni, V.; Pelkman, C.L.; Fahey, G.C., Jr.; Swanson, K.S. Agave Inulin Supplementation Affects the Fecal Microbiota of Healthy Adults Participating in a Randomized, Double-Blind, Placebo-Controlled, Crossover Trial. *J. Nutr.* **2015**, *145*, 2025–2032. [CrossRef] [PubMed]
45. Shen, R.L.; Dang, X.Y.; Dong, J.L.; Hu, X.Z. Effects of oat β-glucan and barley β-glucan on fecal characteristics, intestinal microflora, and intestinal bacterial metabolites in rats. *J. Agric. Food Chem.* **2012**, *60*, 11301–11308. [CrossRef]
46. Mao, G.; Li, S.; Orfila, C.; Shen, X.; Zhou, S.; Linhardt, R.J.; Ye, X.; Chen, S. Depolymerized RG-I-enriched pectin from citrus segment membranes modulates gut microbiota, increases SCFA production, and promotes the growth of Bifidobacterium spp., Lactobacillus spp. and Faecalibaculum spp. *Food Funct.* **2019**, *10*, 7828–7843. [CrossRef] [PubMed]
47. Beukema, M.; Akkerman, R.; Jermendi, E.; Koster, T.; Laskewitz, A.; Kong, C.; Schols, H.A.; Faas, M.M.; de Vos, P. Pectins that Structurally Differ in the Distribution of Methyl-Esters Attenuate *Citrobacter rodentium*-Induced Colitis. *Mol. Nutr. Food Res.* **2021**, *65*, 2100346. [CrossRef]
48. Calame, W.; Weseler, A.R.; Viebke, C.; Flynn, C.; Siemensma, A.D. Gum arabic establishes prebiotic functionality in healthy human volunteers in a dose-dependent manner. *Br. J. Nutr.* **2008**, *100*, 1269–1275. [CrossRef]
49. Rawi, M.H.; Abdullah, A.; Ismail, A.; Sarbini, S.R. Manipulation of Gut Microbiota Using Acacia Gum Polysaccharide. *ACS Omega* **2021**, *6*, 17782–17797. [CrossRef]
50. Kawakami, S.; Han, K.H.; Araki, T.; Ohba, K.; Wakabayashi, T.; Shimada, K.; Fukushima, M. Potato powders prepared by successive cooking-process depending on resistant starch content affect the intestinal fermentation in rats. *Biosci. Biotechnol. Biochem.* **2017**, *81*, 359–364. [CrossRef]
51. Li, T.; Teng, H.; An, F.; Huang, Q.; Chen, L.; Song, H. The beneficial effects of purple yam (*Dioscorea alata* L.) resistant starch on hyperlipidemia in high-fat-fed hamsters. *Food Funct.* **2019**, *10*, 2642–2650. [CrossRef] [PubMed]
52. Zeng, H.; Zheng, Y.; Lin, Y.; Huang, C.; Lin, S.; Zheng, B.; Zhang, Y. Effect of fractionated lotus seed resistant starch on proliferation of Bifidobacterium longum and Lactobacillus delbrueckii subsp. bulgaricus and its structural changes following fermentation. *Food Chem.* **2018**, *268*, 134–142. [CrossRef] [PubMed]
53. The Human Microbiome Project Consortium. Structure, function and diversity of the healthy human microbiome. *Nature* **2012**, *486*, 207–214. [CrossRef] [PubMed]
54. Sender, R.; Fuchs, S.; Milo, R. Revised Estimates for the Number of Human and Bacteria Cells in the Body. *PLoS Biol.* **2016**, *14*, e1002533. [CrossRef]
55. Almeida, A.; Nayfach, S.; Boland, M.; Strozzi, F.; Beracochea, M.; Shi, Z.J.; Pollard, K.S.; Sakharova, E.; Parks, D.H.; Hugenholtz, P.; et al. A unified catalog of 204,938 reference genomes from the human gut microbiome. *Nat. Biotechnol.* **2021**, *39*, 105–114. [CrossRef]
56. Costea, P.I.; Hildebrand, F.; Arumugam, M.; Backhed, F.; Blaser, M.J.; Bushman, F.D.; de Vos, W.M.; Ehrlich, S.D.; Fraser, C.M.; Hattori, M.; et al. Enterotypes in the landscape of gut microbial community composition. *Nat. Microbiol.* **2018**, *3*, 8–16. [CrossRef]
57. Singh, R.P. Glycan utilisation system in Bacteroides and Bifidobacteria and their roles in gut stability and health. *Appl. Microbiol. Biotechnol.* **2019**, *103*, 7287–7315. [CrossRef]
58. Flint, H.J.; Scott, K.P.; Duncan, S.H.; Louis, P.; Forano, E. Microbial degradation of complex carbohydrates in the gut. *Gut Microbes* **2012**, *3*, 289–306. [CrossRef]
59. Leitch, E.C.; Walker, A.W.; Duncan, S.H.; Holtrop, G.; Flint, H.J. Selective colonization of insoluble substrates by human faecal bacteria. *Environ. Microbiol.* **2007**, *9*, 667–679. [CrossRef]
60. Ze, X.; Duncan, S.H.; Louis, P.; Flint, H.J. Ruminococcus bromii is a keystone species for the degradation of resistant starch in the human colon. *ISME J.* **2012**, *6*, 1535–1543. [CrossRef]
61. El Kaoutari, A.; Armougom, F.; Gordon, J.I.; Raoult, D.; Henrissat, B. The abundance and variety of carbohydrate-active enzymes in the human gut microbiota. *Nat. Rev. Microbiol.* **2013**, *11*, 497–504. [CrossRef]
62. Foley, M.H.; Cockburn, D.W.; Koropatkin, N.M. The Sus operon: A model system for starch uptake by the human gut Bacteroidetes. *Cell. Mol. Life Sci.* **2016**, *73*, 2603–2617. [CrossRef]
63. Martens, E.C.; Lowe, E.C.; Chiang, H.; Pudlo, N.A.; Wu, M.; McNulty, N.P.; Abbott, D.W.; Henrissat, B.; Gilbert, H.J.; Bolam, D.N.; et al. Recognition and degradation of plant cell wall polysaccharides by two human gut symbionts. *PLoS Biol.* **2011**, *9*, e1001221. [CrossRef]
64. Shortt, C.; Hasselwander, O.; Meynier, A.; Nauta, A.; Fernandez, E.N.; Putz, P.; Rowland, I.; Swann, J.; Turk, J.; Vermeiren, J.; et al. Systematic review of the effects of the intestinal microbiota on selected nutrients and non-nutrients. *Eur. J. Nutr.* **2018**, *57*, 25–49. [CrossRef]

65. Milani, C.; Lugli, G.A.; Duranti, S.; Turroni, F.; Bottacini, F.; Mangifesta, M.; Sanchez, B.; Viappiani, A.; Mancabelli, L.; Taminiau, B.; et al. Genomic encyclopedia of type strains of the genus Bifidobacterium. *Appl. Environ. Microbiol.* **2014**, *80*, 6290–6302. [CrossRef] [PubMed]
66. Koropatkin, N.M.; Cameron, E.A.; Martens, E.C. How glycan metabolism shapes the human gut microbiota. *Nat. Rev. Microbiol.* **2012**, *10*, 323–335. [CrossRef] [PubMed]
67. Holscher, H.D. Dietary fiber and prebiotics and the gastrointestinal microbiota. *Gut Microbes* **2017**, *8*, 172–184. [CrossRef]
68. Baxter, N.T.; Schmidt, A.W.; Venkataraman, A.; Kim, K.S.; Waldron, C.; Schmidt, T.M. Dynamics of Human Gut Microbiota and Short-Chain Fatty Acids in Response to Dietary Interventions with Three Fermentable Fibers. *mBio* **2019**, *10*, e02566-18. [CrossRef]
69. Tap, J.; Furet, J.P.; Bensaada, M.; Philippe, C.; Roth, H.; Rabot, S.; Lakhdari, O.; Lombard, V.; Henrissat, B.; Corthier, G.; et al. Gut microbiota richness promotes its stability upon increased dietary fibre intake in healthy adults. *Environ. Microbiol.* **2015**, *17*, 4954–4964. [CrossRef] [PubMed]
70. Shang, W.; Si, X.; Zhou, Z.; Li, Y.; Strappe, P.; Blanchard, C. Characterization of fecal fat composition and gut derived fecal microbiota in high-fat diet fed rats following intervention with chito-oligosaccharide and resistant starch complexes. *Food Funct.* **2017**, *8*, 4374–4383. [CrossRef]
71. Larsson, J.M.; Karlsson, H.; Sjovall, H.; Hansson, G.C. A complex, but uniform O-glycosylation of the human MUC2 mucin from colonic biopsies analyzed by nanoLC/MSn. *Glycobiology* **2009**, *19*, 756–766. [CrossRef]
72. Johansson, M.E.; Phillipson, M.; Petersson, J.; Velcich, A.; Holm, L.; Hansson, G.C. The inner of the two Muc2 mucin-dependent mucus layers in colon is devoid of bacteria. *Proc. Natl. Acad. Sci. USA* **2008**, *105*, 15064–15069. [CrossRef]
73. Wrzosek, L.; Miquel, S.; Noordine, M.L.; Bouet, S.; Chevalier-Curt, M.J.; Robert, V.; Philippe, C.; Bridonneau, C.; Cherbuy, C.; Robbe-Masselot, C.; et al. Bacteroides thetaiotaomicron and Faecalibacterium prausnitzii influence the production of mucus glycans and the development of goblet cells in the colonic epithelium of a gnotobiotic model rodent. *BMC Biol.* **2013**, *11*, 61. [CrossRef]
74. Windey, K.; De Preter, V.; Verbeke, K. Relevance of protein fermentation to gut health. *Mol. Nutr. Food Res.* **2012**, *56*, 184–196. [CrossRef] [PubMed]
75. Schroeder, B.O.; Birchenough, G.M.H.; Stahlman, M.; Arike, L.; Johansson, M.E.V.; Hansson, G.C.; Backhed, F. Bifidobacteria or Fiber Protects against Diet-Induced Microbiota-Mediated Colonic Mucus Deterioration. *Cell Host Microbe* **2018**, *23*, 27–40.e7. [CrossRef]
76. Makki, K.; Deehan, E.C.; Walter, J.; Backhed, F. The Impact of Dietary Fiber on Gut Microbiota in Host Health and Disease. *Cell Host Microbe* **2018**, *23*, 705–715. [CrossRef] [PubMed]
77. Zou, J.; Chassaing, B.; Singh, V.; Pellizzon, M.; Ricci, M.; Fythe, M.D.; Kumar, M.V.; Gewirtz, A.T. Fiber-Mediated Nourishment of Gut Microbiota Protects against Diet-Induced Obesity by Restoring IL-22-Mediated Colonic Health. *Cell Host Microbe* **2018**, *23*, 41–53.e4. [CrossRef]
78. Burton-Freeman, B. Dietary fiber and energy regulation. *J. Nutr.* **2000**, *130*, 272S–275S. [CrossRef]
79. Howarth, N.C.; Saltzman, E.; Roberts, S.B. Dietary fiber and weight regulation. *Nutr. Rev.* **2001**, *59*, 129–139. [CrossRef]
80. Ratanpaul, V.; Williams, B.A.; Black, J.L.; Gidley, M.J. Review: Effects of fibre, grain starch digestion rate and the ileal brake on voluntary feed intake in pigs. *Animal* **2019**, *13*, 2745–2754. [CrossRef]
81. Benton, D.; Young, H.A. Reducing Calorie Intake May Not Help You Lose Body Weight. *Perspect. Psychol. Sci.* **2017**, *12*, 703–714. [CrossRef]
82. Jia, W.; Xie, G.; Jia, W. Bile acid-microbiota crosstalk in gastrointestinal inflammation and carcinogenesis. *Nat. Rev. Gastroenterol. Hepatol.* **2018**, *15*, 111–128. [CrossRef] [PubMed]
83. Ridlon, J.M.; Kang, D.J.; Hylemon, P.B. Bile salt biotransformations by human intestinal bacteria. *J. Lipid Res.* **2006**, *47*, 241–259. [CrossRef]
84. Chiang, J.Y. Bile acids: Regulation of synthesis. *J. Lipid Res.* **2009**, *50*, 1955–1966. [CrossRef] [PubMed]
85. Chiang, J.Y.L.; Ferrell, J.M. Bile Acids as Metabolic Regulators and Nutrient Sensors. *Annu. Rev. Nutr.* **2019**, *39*, 175–200. [CrossRef]
86. Baye, K.; Guyot, J.P.; Mouquet-Rivier, C. The unresolved role of dietary fibers on mineral absorption. *Crit. Rev. Food Sci. Nutr.* **2017**, *57*, 949–957. [CrossRef]
87. Bosscher, D.; Van Caillie-Bertrand, M.; Deelstra, H. Effect of thickening agents, based on soluble dietary fiber, on the availability of calcium, iron, and zinc from infant formulas. *Nutrition* **2001**, *17*, 614–618. [CrossRef]
88. Fung, K.Y.; Cosgrove, L.; Lockett, T.; Head, R.; Topping, D.L. A review of the potential mechanisms for the lowering of colorectal oncogenesis by butyrate. *Br. J. Nutr.* **2012**, *108*, 820–831. [CrossRef]
89. Chang, P.V.; Hao, L.; Offermanns, S.; Medzhitov, R. The microbial metabolite butyrate regulates intestinal macrophage function via histone deacetylase inhibition. *Proc. Natl. Acad. Sci. USA* **2014**, *111*, 2247–2252. [CrossRef]
90. Singh, N.; Thangaraju, M.; Prasad, P.D.; Martin, P.M.; Lambert, N.A.; Boettger, T.; Offermanns, S.; Ganapathy, V. Blockade of dendritic cell development by bacterial fermentation products butyrate and propionate through a transporter (Slc5a8)-dependent inhibition of histone deacetylases. *J. Biol. Chem.* **2010**, *285*, 27601–27608. [CrossRef]
91. Furusawa, Y.; Obata, Y.; Fukuda, S.; Endo, T.A.; Nakato, G.; Takahashi, D.; Nakanishi, Y.; Uetake, C.; Kato, K.; Kato, T.; et al. Commensal microbe-derived butyrate induces the differentiation of colonic regulatory T cells. *Nature* **2013**, *504*, 446–450. [CrossRef]

92. Arpaia, N.; Campbell, C.; Fan, X.; Dikiy, S.; van der Veeken, J.; deRoos, P.; Liu, H.; Cross, J.R.; Pfeffer, K.; Coffer, P.J.; et al. Metabolites produced by commensal bacteria promote peripheral regulatory T-cell generation. *Nature* **2013**, *504*, 451–455. [CrossRef]
93. Nepelska, M.; Cultrone, A.; Beguet-Crespel, F.; Le Roux, K.; Dore, J.; Arulampalam, V.; Blottiere, H.M. Butyrate produced by commensal bacteria potentiates phorbol esters induced AP-1 response in human intestinal epithelial cells. *PLoS ONE* **2012**, *7*, e52869. [CrossRef] [PubMed]
94. Brown, A.J.; Goldsworthy, S.M.; Barnes, A.A.; Eilert, M.M.; Tcheang, L.; Daniels, D.; Muir, A.I.; Wigglesworth, M.J.; Kinghorn, I.; Fraser, N.J.; et al. The Orphan G protein-coupled receptors GPR41 and GPR43 are activated by propionate and other short chain carboxylic acids. *J. Biol. Chem.* **2003**, *278*, 11312–11319. [CrossRef] [PubMed]
95. Singh, N.; Gurav, A.; Sivaprakasam, S.; Brady, E.; Padia, R.; Shi, H.; Thangaraju, M.; Prasad, P.D.; Manicassamy, S.; Munn, D.H.; et al. Activation of Gpr109a, receptor for niacin and the commensal metabolite butyrate, suppresses colonic inflammation and carcinogenesis. *Immunity* **2014**, *40*, 128–139. [CrossRef]
96. Macia, L.; Tan, J.; Vieira, A.T.; Leach, K.; Stanley, D.; Luong, S.; Maruya, M.; McKenzie, C.I.; Hijikata, A.; Wong, C.; et al. Metabolite-sensing receptors GPR43 and GPR109A facilitate dietary fibre-induced gut homeostasis through regulation of the inflammasome. *Nat. Commun.* **2015**, *6*, 6734. [CrossRef]
97. Kimura, I.; Ozawa, K.; Inoue, D.; Imamura, T.; Kimura, K.; Maeda, T.; Terasawa, K.; Kashihara, D.; Hirano, K.; Tani, T.; et al. The gut microbiota suppresses insulin-mediated fat accumulation via the short-chain fatty acid receptor GPR43. *Nat. Commun.* **2013**, *4*, 1829. [CrossRef] [PubMed]
98. Samuel, B.S.; Shaito, A.; Motoike, T.; Rey, F.E.; Backhed, F.; Manchester, J.K.; Hammer, R.E.; Williams, S.C.; Crowley, J.; Yanagisawa, M.; et al. Effects of the gut microbiota on host adiposity are modulated by the short-chain fatty-acid binding G protein-coupled receptor, Gpr41. *Proc. Natl. Acad. Sci. USA* **2008**, *105*, 16767–16772. [CrossRef]
99. Eswaran, S.; Muir, J.; Chey, W.D. Fiber and functional gastrointestinal disorders. *Am. J. Gastroenterol.* **2013**, *108*, 718–727. [CrossRef]
100. El-Salhy, M.; Gundersen, D. Diet in irritable bowel syndrome. *Nutr. J.* **2015**, *14*, 36. [CrossRef]
101. Riedl, J.; Linseisen, J.; Hoffmann, J.; Wolfram, G. Some dietary fibers reduce the absorption of carotenoids in women. *J. Nutr.* **1999**, *129*, 2170–2176. [CrossRef] [PubMed]
102. Ten Bruggencate, S.J.; Bovee-Oudenhoven, I.M.; Lettink-Wissink, M.L.; Van der Meer, R. Dietary fructooligosaccharides increase intestinal permeability in rats. *J. Nutr.* **2005**, *135*, 837–842. [CrossRef]
103. Singh, V.; Yeoh, B.S.; Chassaing, B.; Xiao, X.; Saha, P.; Olvera, R.A.; Lapek, J.D., Jr.; Zhang, L.; Wang, W.B.; Hao, S.; et al. Dysregulated Microbial Fermentation of Soluble Fiber Induces Cholestatic Liver Cancer. *Cell* **2018**, *175*, 679–694.e22. [CrossRef] [PubMed]
104. Bernstein, H.; Bernstein, C.; Payne, C.M.; Dvorak, K. Bile acids as endogenous etiologic agents in gastrointestinal cancer. *World J. Gastroenterol.* **2009**, *15*, 3329–3340. [CrossRef]
105. Zhang, C.; Zhang, M.; Wang, S.; Han, R.; Cao, Y.; Hua, W.; Mao, Y.; Zhang, X.; Pang, X.; Wei, C.; et al. Interactions between gut microbiota, host genetics and diet relevant to development of metabolic syndromes in mice. *ISME J.* **2010**, *4*, 232–241. [CrossRef]
106. Mozaffarian, D.; Hao, T.; Rimm, E.B.; Willett, W.C.; Hu, F.B. Changes in diet and lifestyle and long-term weight gain in women and men. *N. Engl. J. Med.* **2011**, *364*, 2392–2404. [CrossRef] [PubMed]
107. Wang, T.; Cai, G.; Qiu, Y.; Fei, N.; Zhang, M.; Pang, X.; Jia, W.; Cai, S.; Zhao, L. Structural segregation of gut microbiota between colorectal cancer patients and healthy volunteers. *ISME J.* **2012**, *6*, 320–329. [CrossRef]
108. O'Keefe, S.J.; Li, J.V.; Lahti, L.; Ou, J.; Carbonero, F.; Mohammed, K.; Posma, J.M.; Kinross, J.; Wahl, E.; Ruder, E.; et al. Fat, fibre and cancer risk in African Americans and rural Africans. *Nat. Commun.* **2015**, *6*, 6342. [CrossRef]
109. Nilsson, A.C.; Johansson-Boll, E.V.; Bjorck, I.M. Increased gut hormones and insulin sensitivity index following a 3-d intervention with a barley kernel-based product: A randomised cross-over study in healthy middle-aged subjects. *Br. J. Nutr.* **2015**, *114*, 899–907. [CrossRef]
110. Nicolucci, A.C.; Hume, M.P.; Martinez, I.; Mayengbam, S.; Walter, J.; Reimer, R.A. Prebiotics Reduce Body Fat and Alter Intestinal Microbiota in Children Who Are Overweight or With Obesity. *Gastroenterology* **2017**, *153*, 711–722. [CrossRef]
111. Davison, K.M.; Temple, N.J. Cereal fiber, fruit fiber, and type 2 diabetes: Explaining the paradox. *J. Diabetes Its Complicat.* **2018**, *32*, 240–245. [CrossRef] [PubMed]

molecules

MDPI

Article

Olive Mill Wastewater Fermented with Microbial Pools as a New Potential Functional Beverage

Paola Foti [1], Paride S. Occhipinti [1], Nunziatina Russo [1,2], Antonio Scilimati [3], Morena Miciaccia [3], Cinzia Caggia [1,2,4,*], Maria Grazia Perrone [3], Cinzia L. Randazzo [1,2,4] and Flora V. Romeo [5]

1 Department of Agriculture, Food and Environment (Di3 A), University of Catania, Via Santa Sofia 100, 95123 Catania, Italy
2 ProBioEtna srl, Spin-Off of University of Catania, Via Santa Sofia 100, 95123 Catania, Italy
3 Department of Pharmacy-Pharmaceutical Sciences, University of Bari "Aldo Moro", Via E. Orabona 4, 70125 Bari, Italy
4 CERNUT (Interdepartmental Research Centre in Nutraceuticals and Health Products), University of Catania, Via le A. Doria 6, 95125 Catania, Italy
5 Consiglio per la Ricerca in Agricoltura e l'Analisi dell'Economia Agraria (CREA), Centro di Ricerca Olivico-Tura, Frutticoltura e Agrumicoltura, Corso Savoia 190, 95024 Acireale, Italy
* Correspondence: cinzia.caggia@unict.it

Abstract: Olive mill wastewater (OMWW) represents a by–product but also a source of biologically active compounds, and their recycling is a relevant strategy to recover income and to reduce environmental impact. The objective of the present study was to obtain a new functional beverage with a health–promoting effect starting from OMWW. Fresh OMWW were pre–treated through filtration and/or microfiltration and subjected to fermentation using strains belonging to *Lactiplantibacillus plantarum*, *Candida boidinii* and *Wickerhamomyces anomalus*. During fermentation, phenolic content and hydroxytyrosol were monitored. Moreover, the biological assay of microfiltered fermented OMWW was detected versus tumor cell lines and as anti-inflammatory activity. The results showed that in microfiltered OMWW, fermentation was successfully conducted, with the lowest pH values reached after 21 days. In addition, in all fermented samples, an increase in phenol and organic acid contents was detected. Particularly, in samples fermented with *L. plantarum* and *C. boidinii* in single and combined cultures, the concentration of hydroxytyrosol reached values of 925.6, 902.5 and 903.5 mg/L, respectively. Moreover, biological assays highlighted that fermentation determines an increase in the antioxidant and anti–inflammatory activity of OMWW. Lastly, an increment in the active permeability on Caco-2 cell line was also revealed. In conclusion, results of the present study confirmed that the process applied here represents an effective strategy to achieve a new functional beverage.

Keywords: olive mill wastewater; microfiltration; fermentation; microbial pool; hydroxytyrosol; functional beverage

Citation: Foti, P.; Occhipinti, P.S.; Russo, N.; Scilimati, A.; Miciaccia, M.; Caggia, C.; Perrone, M.G.; Randazzo, C.L.; Romeo, F.V. Olive Mill Wastewater Fermented with Microbial Pools as a New Potential Functional Beverage. *Molecules* **2023**, *28*, 646. https://doi.org/10.3390/molecules28020646

Academic Editor: Smaoui Slim

Received: 2 December 2022
Revised: 1 January 2023
Accepted: 6 January 2023
Published: 8 January 2023

1. Introduction

Olive oil by-products, while representing a management problem for olive oil companies, may actually represent a source of high value–added compounds that can be used in pharmaceuticals, food, feed and cosmetics for their health properties [1–3]. In fact, olive mill wastewater (OMWW) is a resource rich in phenols including hydroxytyrosol (HT) and tyrosol (TYR), characterized by high antioxidant, anti–inflammatory, antimicrobial and anticarcinogenic activities [4]. The scientific community has proposed several strategies for the valorization of this by–product including solvent extraction techniques, selective resins, membrane filtration or enzymatic applications [5]. These techniques allow the extraction and/or concentration of bioactive compounds in order to increase the nutraceutical component and produce new products or functional ingredients, thus, responding to the demand of consumers who are now aware of the beneficial role that these natural products play in human and animal diets. In the food industry, OMWW have been proposed as an added functional ingredient in meat, dairy, fish, bakery products and juices [6–8]. As a matter of fact, the addition of such phenolic components in food matrices has been shown not only to fulfil a technological function (i.e., to extend the shelf life) but also to improve the health and safety properties of the food. Although the interest of the scientific community in the use of microorganisms in the bioprocessing of agro-industrial waste has grown in recent years [9], only a few microbial applications have been proposed for the valorization of this matrix. Authors have reported that the use of live microorganisms increases the content and bioavailability of the phenolic compounds, and especially of HT and TYR [10,11]. In addition, the driven microbial fermentation provides several advantages by preserving and improving food safety and shelf life due to the formation of organic acids, such as lactic, acetic, formic, propionic acids, etc. [12]. The diversities of acids are dynamic among different alcoholic beverages and fermented food, as are the synergistic effects of abiotic and biotic factors [13]. Functional microorganisms, such as lactic acid bacteria (LAB) and yeasts, are responsible for the metabolism of organic acids. Therefore, the use of selected microorganisms, especially yeasts and LAB isolated from spontaneous similar fermented matrices such as table olives, could represent a low–cost strategy to stabilize and improve the nutraceutical and sensory traits of OMWW. In detail, *Lactiplantibacillus plantarum* strains from fermented olives have been largely associated with the metabolism of phenolic compounds as they can produce degradation enzymes, such as β-glucosidase, esterase, tannase, decarboxylase [14]. Moreover, some of them have been proposed due to their potential probiotic activity [15]. With regard to yeasts, several species show β-glucosidasic, lipasic and esterasic activity and have been used for their ability to improve sensorial profile through production of esters from fatty acids and free fatty acids. Among yeasts, *Candida boidinii* and *Wickerhamomyces anomalus* are the most commonly used as starters [16]. In addition, yeasts isolated from oil matrices, especially strains of *W. anomalus*, have demonstrated several probiotic characteristics, among which the most known is the in vitro cholesterol removal capacity [17].

Today, the functional beverage sector is steadily increasing worldwide thanks to its high nutritional value and the possibility to add flavors. Furthermore, nutraceutical beverages with added probiotics and prebiotics are of considerable interest to the consumer [18], as this matrix was shown to inhibit proliferation and induce apoptosis in several tumor cells, prevent DNA damage and exert anti–inflammatory activity [19].

The aim of this study was to set up a process to obtain a new functional beverage with a health–promoting effect starting from OMWW. For this purpose, OMWW were pre–treated through filtration and microfiltration and then subjected to fermentation with selected microbial pools, isolated from spontaneously fermented table olives. During fermentation, the biotechnological aptitude of the different strain combinations, their effect on the fermentation parameters, the increase of the phenolic content, especially as HT increase, were evaluated. Furthermore, a biological characterization to evaluate the safety profile and the antioxidant activity was performed on treated OMWW samples. Finally, the ability to cross Caco-2 cell monolayers, as a model of gastrointestinal tract absorption, was performed.

2. Results

2.1. Chemico-Physical Characterization of Sample of Different Trials

The OMWW belonging to Trial I were monitored at different times (0, 8, and 30 days), through the detection of pH, total soluble solids (TSS), total phenol content and single phenols by HPLC (Table S1). Regarding pH, any significant difference was observed at the beginning of fermentation, and the lowest pH value (4.45) was reached at T8 in sample inoculated with *C. boidinii* in single culture. The TTS at the beginning of fermentation showed values between 7.08 and 8.32, reaching values between 5.60 and 6.34 at T30. During the fermentation, the total phenol content showed, to some extent, a constant trend, reaching the highest concentration at T30 in samples fermented with *W. anomalus* in single culture, with a value of 3241.9 mg/L. The results obtained by HPLC confirmed this increase, as samples treated with *W. anomalus* showed the highest concentration of HT, equal to 2630.4 mg/L. Regarding TYR, an increase during fermentation was observed, reaching, after 30 days, values between 508.6 and 679.4 mg/L in all treated samples. The chemical analyses performed on Trial I were repeated on Trial II (Table S1). The pH decreased during fermentation, showed the lowest values at T8. In detail, all inoculated samples showed a lower pH than the control sample. In particular, the lowest value was found in the samples with *L. plantarum* and *W. anomalus* in single culture, but also in the combination *L. plantarum* and *C. boidinii* and, finally, with the mix of the three strains with values ranging from 3.97 to 3.99. With regard to TSS, the greatest decrease occurred with the combination of *L. plantarum* in association with *W. anomalus,* going from a value of 7.84 at T0 down to 5.50 at T8. At T30, almost all samples maintained the value showed at T8 of fermentation. In addition, total phenols at T8 and T30, in all inoculated samples, showed a higher content over time compared with the control sample. In detail, the samples with significantly higher phenolic content were the three–strain association (3379.5 mg/L) and *W. anomalus* in single combination (3261.9 mg/L) at T8, while at T30 was the sample inoculated with *L. plantarum* with a value of 3577.6 mg/L. The results obtained by HPLC showed a decrease at T8 of HT in all samples except in the samples with *L. plantarum* where there was an increase of 115 mg/L of HT, and the sample inoculated with *L. plantarum* with *W. anomalus* which showed an increase equal to 262 mg/L of HT. All samples inoculated up to T30 had higher HT content. Opposingly, the TYR decreased during fermentation from an average range of values from 319.7 mg/L to 136 mg/L.

In Trial III, microfiltration resulted in a clear and sterile matrix. Before starting the final fermentation, a preliminary test was carried out in a reduced volume (100 mL) to ascertain if any difference could be revealed between trials with the addition of glucose, peptone and yeast extract (added at the same concentrations) and the trials without any additions. The results showed the same pH values and cell density during fermentation. Moreover, the addition of these compounds made the OMWW turbid (data not shown). For these reasons, to improve the acceptability of the product to consumers, the thesis without additions was chosen for the final test. During fermentation, pH, TSS and total phenol content were monitored (Table 1). Regarding pH, no significant difference was found at the beginning of fermentation. The pH at T0 was in a range of 5.12 and 5.19. Fermentation stopped at T21 for all samples examined. The end of fermentation was revealed by the stabilization of the pH value that was evaluated every three days of fermentation (data not shown). In particular, the samples inoculated with *W. anomalus* in single culture and in association with *L. plantarum* reached a pH value of 4.54 and 4.49 at T21, respectively. In addition, although slower than the previous theses, the theses containing *L. plantarum* and *C. boidinii* in single culture also reached at T21 a pH of 4.65 and 4.60, respectively. Total soluble solids showed no significant difference at any of the fermentation times. Initial values ranged from 8.30 to 10.85 °Brix, while values between 5.32 and 8.17 °Brix were reached at the end of fermentation. The sample used as a control during fermentation maintained its pH and TSS values. Regarding the content of total phenols, the highest values at the beginning of fermentation were found in the sample containing the *L. plantarum* and *W. anomalus* combination, a value that decreased during the fermentation process. In contrast, the sample with the three–strains combination showed an increase in total phenol content up to T14 with a value of 4015 mg/L, and then decreased at T21 reaching a value of 1543 mg/L.

Table 1. Chemical parameters detected in samples of Trial III.

Sample	Time	pH	TSS (°Brix)	Total Phenol (mg/L)
Control	0	5.18 ± 0.01	8.30 ± 0.77	3627.4 ± 0.54 [c]
L. plantarum	0	5.16 ± 0.01	10.60 ± 0.78	3711.2 ± 4.89 [b]
C. boidinii	0	5.18 ± 0.01	10.28 ± 0.70	3539.8 ± 0.54 [d]
W. anomalus	0	5.12 ± 0.08	9.76 ± 1.53	3172.9 ± 1.63 [f]
L.p + W.a	0	5.13 ± 0.06	8.56 ± 0.80	4135.0 ± 4.89 [a]
L.p + C.b	0	5.19 ± 0.02	10.04 ± 1.13	3474.7 ± 1.09 [e]
L.p + W.a + C.b	0	5.18 ± 0.01	8.88 ± 1.39	2967.1 ± 2.18 [g]
		n.s.	n.s.	**
Control	8	5.17 ± 0.01 [a]	8.30 ± 0.78	1985.8 ± 3.26 [g]
L. plantarum	8	5.04 ± 0.03 [b]	10.30 ± 0.98	3032.5 ± 2.18 [b]
C. boidinii	8	4.97 ± 0. 01 [bcd]	8.88 ± 2.18	3020.6 ± 0.54 [c]
W. anomalus	8	4.87 ± 0.02 [e]	9.14 ± 1.77	2395.3 ± 1.63 [f]
L.p + W.a	8	4.88 ± 0.04 [de]	7.52 ± 1.17	2897.2 ± 0.01 [e]
L.p + C.b	8	5.00 ± 0.01 [bc]	9.50 ± 0.32	2991.0 ± 5.44 [d]
L.p + W.a + C.b	8	4.94 ± 0.01 [cde]	8.24 ± 1.29	3268.2 ± 1.63 [a]
		**	n.s.	**
Control	14	5.18 ± 0.02 [a]	8.30 ± 0.78	1809.5 ± 0.54 [f]
L. plantarum	14	4.68 ± 0.01 [cd]	9.90 ± 0.99	3282.9 ± 0.54 [c]
C. boidinii	14	4.77 ± 0.01 [b]	7.99 ± 2.82	2443.8 ± 1.63 [d]
W. anomalus	14	4.67 ± 0.04 [cd]	8.28 ± 1.44	3539.7 ± 0.54 [b]
L.p + W.a	14	4.62 ± 0.09 [d]	6.64 ± 1.52	2199.2 ± 3.81 [e]
L.p + C.b	14	4.89 ± 0.01 [b]	9.03 ± 1.12	3545.1 ± 53.84 [b]
L.p + W.a + C.b	14	4.82 ± 0.02 [bc]	7.62 ± 1.36	4015.0 ± 2.72 [a]
		**	n.s.	**
Control	21	5.19 ± 0.01 [a]	8.20 ± 0.61	1009.4 ± 0.54 [f]
L. plantarum	21	4.65 ± 0.03 [c]	8.15 ± 1.20	3392.1 ± 0.54 [a]
C. boidinii	21	4.60 ± 0.01 [d]	5.60 ± 0.57	3005.2 ± 0.54 [b]
W. anomalus	21	4.54 ± 0.04 [e]	6.36 ± 0.37	2394.2 ± 1.09 [c]
L.p + W.a	21	4.49 ± 0.01 [de]	5.32 ± 0.04	1914.6 ± 1.63 [d]
L.p + C.b	21	4.84 ±0.02 [b]	8.00 ± 1.41	3403.2 ± 10.88 [a]
L.p + W.a + C.b	21	4.82 ± 0.01 [b]	5.99 ± 0.01	1543.6 ± 1.09 [e]
		**	n.s.	**

Data are expressed as mean ± standard deviations. Mean values with different letters within the same column at the same time interval are statistically different. n.s. not significant; ** Significance at $p \leq 0.01$.

2.2. Microbiological Analyses

Results on microbiological analyses (Table S2) are referred at the same sampling times reported for chemical analyses. Overall, for samples of Trial I, high microbial densities were detected for aerobic mesophilic bacteria, enterobacteria, and yeast in all sampling times. Regarding LAB, an increase of 1 Log unit at T8 was detected, and the values were quite constant until T30, with some exceptions. In detail, at the beginning of fermentation, the sample treated with *L. plantarum* showed a significantly higher cell density, with a value of 5.85 log CFU/mL, whilst at T30 the highest LAB densities were detected in samples treated with the combination of the three strains, namely of *L. plantarum* and *C. boidinii*, *W. anomalus* in samples inoculated with *W. anomalus* in single culture, and in samples treated with *L. plantarum* and *C. boidinii* in mixed cultures. Yeasts and molds also showed a similar trend in all samples. In fact, cell density increased at T8 of fermentation and then decreased at T30, when an average value of 6.45 Log CFU/mL were detected. Aerobic mesophilic bacteria counts showed only a slight variation during fermentation, reaching a final mean value of 6.25 Log CFU/mL, whereas Enterobacteriaceae and staphylococci showed a significant decrease during fermentation. At the beginning of fermentation, the latest microbial groups showed an initial average value of 6.63 and 3.05 Log CFU/mL, respectively. These values decreased significantly during fermentation in the inoculated samples, reaching values under the detection limit. In Trial II, the LAB and yeast counts increased during fermentation (Table S2). In detail, the LAB mean value starting from 4.27 Log CFU/mL reached, after 30 days, a mean value of 7.34 Log CFU/mL, whilst in samples inoculated with *W. anomalus* it reached the lowest cell density. A similar trend was observed for yeasts that at the 30th day exhibited a mean cell density of 9 Log CFU/mL in the sample inoculated with *L. plantarum* and *C. boidinii* in mixed culture. Aerobic mesophilic bacteria were found at high density, until the end of fermentation when a final average value of 7.52 Log CFU/mL was counted. Different trends were observed for Enterobacteriaceae and staphylococci, for which after a slight increase a significant decrease was detected after 30 days in all samples.

Regarding Trial III, before starting fermentation, the microfiltered OMWWs were subjected to microbiological analyses to confirm the achieved sterility. The following microbial groups were searched: LAB, yeasts, staphylococci, total mesophilic aerobic bacteria, Enterobacteriaceae and *Clostridium perfringens*. All used media and conditions are reported in Section 4.

Once the OMWWs were analyzed, the selected strains were inoculated at a cell density of 10^8 and 10^7 CFU/mL for *L. plantarum* and yeasts, respectively. As shown in Figure 1, a different growth pattern between the two yeasts and the LAB strains was observed during fermentation. In fact, while in all inoculated samples LAB showed an initial decrease, during the first 14 days they increased until the 21st day; the yeasts increased their cell density during the first 18 days, when they reached values between 7.03 and 7.78 Log CFU/mL.

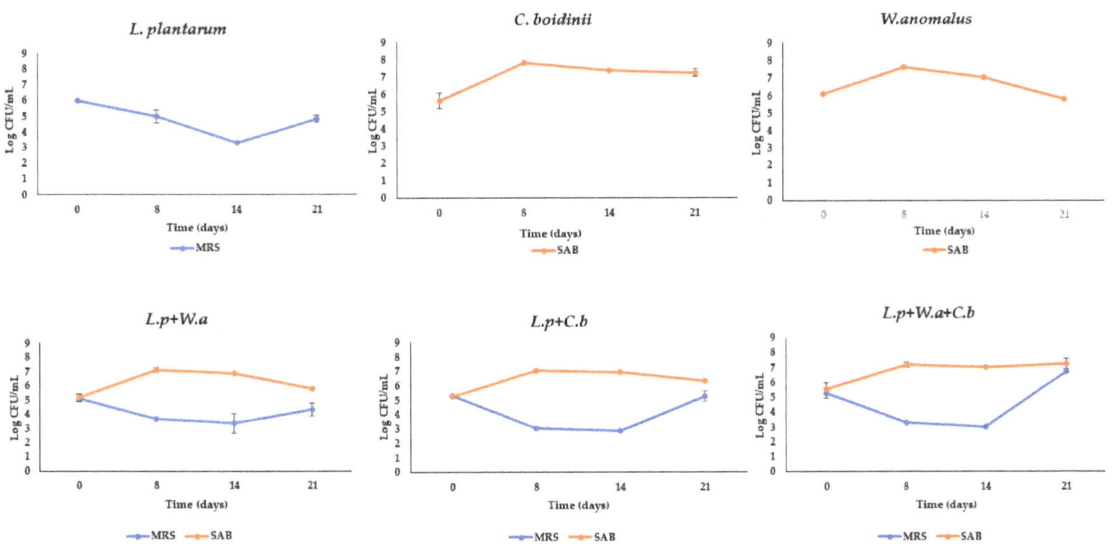

Figure 1. Microbial counts detected in MRS and SAB during fermentation in microfiltered OMWW differently inoculated. Data are expressed as means of Log CFU/mL ± standard deviations.

2.3. Phenol and Organic Acid Detection

Regarding phenolic content, HT and TYR were the main detected compounds, found at high concentration by HPLC during fermentation (Figure 2). As for HT, at the beginning of fermentation a concentration between 341.7 and 469.1 mg/L was found. At the end of fermentation, an exponential increase of HT in all inoculated samples was observed. Particularly in the samples treated with *L. plantarum* and *C. boidinii* in single and in combined cultures, the HT concentration was found as 925.6, 902.5 and 903.5 mg/L, respectively. A slowly increase in concentration of TYR was observed during fermentation, reaching values between 315.6 and 544.7 mg/L in all inoculated samples. In contrast, the control samples showed a significant decrease in HT along fermentation, reaching values of 170.6 mg/L and a slight increase in TYR, reaching final value of 303.7 mg/L.

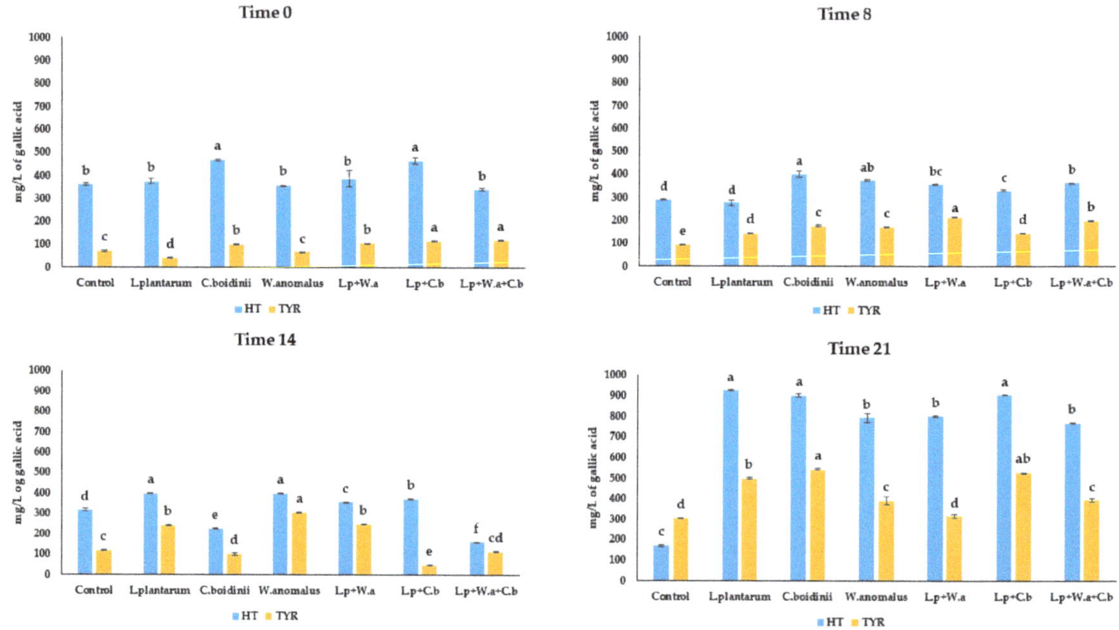

Figure 2. Concentration of HT and TYR during fermentation in microfiltered OMWW differently inoculated. Different letters indicate statistical differences within the columns for the same compound (significance at $p \leq 0.01$).

In addition, organic acids were evaluated at the end of fermentation in all samples. The control sample, at the beginning of fermentation, was used as an initial control (Table 2). The control, analyzed at both the beginning and at the end of fermentation, showed a constant value of acids except for butyric, for which a concentration of 566.4 mg/L was detected only at T21. For all samples inoculated with the different microbial combinations, on the other hand, an acid increase during fermentation was observed, except for isobutyric acid that decreased in sample inoculated with *W. anomalus* and in all the inoculated combinations. In detail, the sample inoculated with *L. plantarum* showed the highest increase for all the detected acids.

Table 2. Organic acids (mg/L) detected by HPLC.

Sample	Time (Days)	Citric Acid	Lactic Acid	Acetic Acid	Propionic Acid	Isobutyric Acid	Butyric Acid
Control	0	4172.9 ± 96.54	1606.6 ± 99.00	416.8 ± 97.31	3865.9 ± 268.47	3136.9 ± 188.31	0.00 ± 0.00
Control	21	4529.3 ± 100.00 ^{de}	1219.2 ± 18.03 ^{f}	326.3 ± 78.89 ^{e}	3743.1 ± 34.21 ^{g}	1654.9 ± 15.21 ^{d}	566.4 ±48.79 ^{d}
L. plantarum	21	7033.4 ± 15.76 ^{a}	4512.6 ± 18.07 ^{a}	7212.8 ± 82.59 ^{a}	9802.4 ± 12.82 ^{a}	3235.3 ± 5.51 ^{a}	4666.4 ± 103.03 ^{a}
C. boidinii	21	6624.4 ± 87.69 ^{b}	4123.3 ± 20.03 ^{b}	4568.4 ± 58.78 ^{c}	9153.8 ± 19.41 ^{b}	3202.7 ± 27.72 ^{a}	4393.3 ± 44.23 ^{a}
W. anomalus	21	5214.4 ± 121.00 ^{c}	3774.5 ± 99.00 ^{c}	6214.7 ± 168.83 ^{b}	8219.2 ± 41.95 ^{c}	2072.0 ± 77.52 ^{b}	4239.4 ± 176.96 ^{a}
L.p + W.a	21	4126.9 ± 106.79 ^{f}	2846.0 ± 35.53 ^{e}	6188.4 ± 85.52 ^{b}	6831.2 ± 10.08 ^{f}	1626.7 ± 1.41 ^{e}	3682.8 ± 26.00 ^{b}
L.p + C.b	21	4744.1 ± 16.31 ^{d}	3167.5 ± 33.49 ^{d}	4366.7 ± 132.82 ^{c}	7913.0 ± 24.25 ^{d}	2096.8 ± 16.35 ^{b}	2995.9 ± 54.51 ^{c}
L.p + W.a + C.b	21	4381.7 ± 20.88 ^{ef}	3075.3 ± 31.95 ^{d}	3334.0 ± 8.10 ^{d}	7342.0 ± 116.15 ^{e}	1810.2 ± 5.28 ^{c}	3550.1 ± 25.58 ^{b}

Data are expressed as mg/L of means ± standard deviations. Different letters indicate statistical differences within the same column (Significance at $p \leq 0.01$).

2.4. Biological Assay

2.4.1. Transepithelial Transport through Caco-2 Cell Monolayers

The intestinal permeability values, estimated with the Caco-2 cell experimental model, correlate well with human in vivo absorption data for many drugs and chemicals. Caco-2 cells are a human colon epithelial cancer cell line that, when cultured as a monolayer, differentiate to form tight junctions between cells to serve as a model of paracellular movement of compounds across the monolayer. The monolayer represents the human intestinal epithelial cell barrier and by this assay, the measured endpoint is intestinal permeability (expressed as apparent permeability—P_{app} value) (Table 3).

Table 3. Apparent permeability of different samples of OMWW.

Samples	Concentration of HT (mg/L)	P_{app} BA (nm/s) Passive Transport	P_{app} AB (nm/s) Active Transport	BA/AB	λ (nm)	ε
Control	1.70	2581	457	4.22	275	0.80
L. planturum	9.50	4015	1014	3.95	285	0.20
C. boidinii	9.00	2540	575	4.41	275	0.09
W. anomalus	7.90	1958	367	5.34	285	0.22
L.p + W.a	8.00	1905	335	5.67	284	0.19
L.p + C.b	9.00	2912	1125	2.58	275	0.19
L.p + W.a + C.b	7.70	2587	522	4.95	283	0.20

All samples were tested at a dilution of 1:100. BA indicates basolateral to apical transport; AB indicates apical to basolateral transport; BA/AB values are from P_{app} AP–BL/P_{app} BL–AP.

The flux from the apical part of the monolayer to the basolateral side (BA) is referred to the passive transport, while the measurement of active transport is obtained by measuring the reversed flow (AB), since Caco-2 cells express efflux pumps in the apical side. The smaller the BA/AB ratio value, the greater the contribution of the active transport to the membrane crossing. In all tested samples, the contribution of active transport to the membrane crossing was always lower than that due to passive diffusion, as demonstrated by high values of P_{app} AB. This occurs mainly for samples inoculated with *L. plantarum* and *C. boidinii*, which showed the highest value (as 1125 nm/s). This value could be related to a synergic effect between the LAB and the *C. boidinii* strains, that also in single cultures showed P_{app} values of 1014 and 575 nm/s, respectively. This result is confirmed by the lowest BA/AB value (2.58), detected in samples fermented with *L. plantarum* and *C. boidinii*.

OMWW samples and HT pure (used as a control) at the opportune dilution (1:25) have been evaluated on different cell lines, normal (HepG2) and tumoral (Caco-2), in order to evaluate their toxicological profile [20]. Since no cytotoxic effect was detected (data not shown), they resulted to be safe at a dilution of 1:25, while with higher concentrations (as such and 1:10) a cytotoxic effect was registered. These results are in agreement with data reported by Di Mauro et al. [21], confirming that the use of higher concentrations (as such and 1:10) induced a reduction in cell viability in a dose–dependent manner, while lower concentrations did not affect cell viability.

2.4.2. Evaluation Activity on COX-1 and COX-2 Isoenzymes

COX, also called Prostaglandin H synthase (PGHS), is a key enzyme in the inflammatory cascade. It catalyzes the conversion of arachidonic acid (AA) in prostanoids, bioactive lipids mediating numerous physiological and pathological processes in the body. Prostanoids include thromboxane A2 (TXA2), prostaglandins (PGD2, PGE2, PGF2α) and prostacyclin (PGI2). Two COX isoforms are known, COX-1 and COX-2, encoded by different genes. The two isoforms show 60% homology in their amino acid sequence. COX-1 is the isoform constitutively expressed in most tissues and responsible for maintaining normal physiological functions such as gastric protection, modulation of platelet function, and renal homeostasis. COX-2, differently from COX-1, is the inducible isoform upon pro–inflammatory stimuli. The possibility of finding anti–inflammatory properties in nutraceutical compounds would make the products under study extremely interesting, thus, the OMWW fermented sample inhibition of *ovine*COX-1 (*o*COX-1) and *human*COX-2 (*h*COX-2) enzyme activity was investigated and HT was used as positive control. Pure HT showed, at a concentration of 40 mg/L, inhibition activity on *o*COX-1 and *h*COX- 2 with a percentage of 6.41 and 26.11, respectively (Figure S1). The control OMWW sample did not show any anti–inflammatory activity, while low anti–inflammatory activity was found for the different OMWW samples. In detail, samples fermented with *L. plantarum*, *C. boidinii* and *W. anomalus* in single culture showed a moderate inhibitory activity towards both isoforms (Table S3). In particular, the sample inoculated with *L. plantarum* in single culture showed a selective inhibition of *o*COX-1, whereas samples treated with *C. boidinii* showed an inhibition towards both *o*COX-1 and *h*COX-2 with the percentage of inhibition reaching 15.96% and 12.95%, respectively. In addition, the sample inoculated with a combo of *L. plantarum* and *C. boidinii* preserves a selective inhibition towards *o*COX-1, with an inhibition of 8.20%. It could be hypothesized that *C. boidinii* produces some metabolites with a greater affinity and selectivity towards *o*COX-1 isoform.

2.4.3. Antioxidant Activity

Diabetes, cardiovascular diseases, arthritis and joint diseases, allergies and chronic obstructive pulmonary diseases are classified, according to the World Health Organization, as specific inflammation–mediated chronic diseases. The processes underlying these diseases are many, but oxidative stress is undoubtedly involved in their pathogenesis and in the development and establishment of a sustained inflammatory state. All selected samples were evaluated for their antioxidant activity by measuring their reactivity with 1,1-diphenyl-2-picrylidrazyl (DPPH), a purple–colored stable radical that strongly absorbs at λ = 517 nm, in order to determine their efficacy as scavengers of stable free radicals. Testing was carried out to compare the effect of fermented OMWW samples with the known antioxidant activity of HT (Figure 3). The data showed that the OMWW control exhibited lower antioxidant capacity, at all dilutions tested, compared with both that exerted by HT and fermented samples. In particular, at the lower tested volume (12.5 µL), the best antioxidant activity was obtained in the sample inoculated with *C. boidinii*, reaching a % RSA value higher than pure HT. The same behavior was observed for samples inoculated with *L. plantarum* and *W. anomalus*, in single culture.

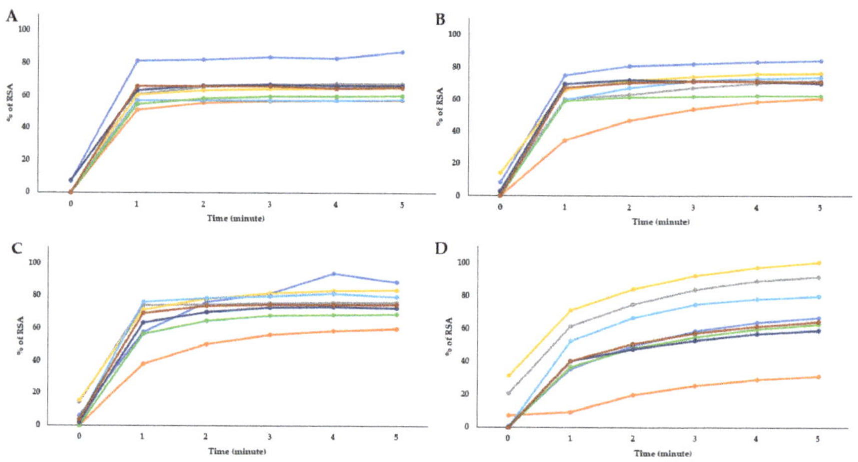

Figure 3. Evaluation of antioxidant activity expressed as % RSA. Each graph corresponds to a volume (µL) used for each sample: (**A**) 50 µL of samples; (**B**) 37.50 µL of samples; (**C**) 25 µL of samples; (**D**) 12.5 µL of samples.

3. Discussion

Fermentation is widely considered a low–cost strategy to recovery and valorize agro-industrial by–products [22]. In this study, in order to obtain a suitable matrix to be fermented with selected microbial pools, different Trials were set up. For this purpose, fresh OMWWs were collected at two successive seasonal years. Samples obtained from Trial I, untreated fresh OMWW, appeared very turbid and rich in unwanted solids at both the beginning and end of fermentation. Therefore, in Trial II, the OMWWs were subjected to on farm filtration using carton filters with different porosity. To date, such a technique is used to remove unwanted solid components from the matrix, maintaining the nutritional compounds as phenolic fraction (Figure S2) [23].

Results of Trial II showed that although the OMWWs (filtered through cardboard filters) visually appeared as clear from a physical point of view, at both the beginning and end of fermentation, they were not microbiologically suitable, in relation to the high total aerobic mesophilic bacteria densities. According to the European Regulation (EC) No. 1441/2007, the absence of pathogens, such as *Salmonella* spp. and *L. monocytogenes*, is considered an essential criterion for the microbiological safety of vegetable products, while no mandatory microbiological criterion is fixed for total aerobic mesophilic bacterial count. However, some guidelines include *Escherichia coli* and total aerobic mesophilic count as quality parameters, fixing the following thresholds (as CFU/g): *E. coli* < 10 for satisfactory; between 10 and $\leq 10^2$ for acceptable; and $>10^2$ as not acceptable [24]. The same authors, for total aerobic mesophilic count, proposed the following thresholds: $\leq 10^4$ for satisfactory, between 10^4 and lower or the same of 10^6 for acceptable, and $>10^6$ not acceptable, respectively [24]. Therefore, OMWW obtained through the last cardboard filter, with a porosity between 0.20 and 0.40 μm, were afterwards subjected to microfiltration (0.22 μm) in the laboratory. This procedure resulted in a microbiologically sterile, clear matrix mainly composed of phenols (Figure S3).

To date, the microfiltration technique is successfully applied in food industries, such as the dairy industry, as it induces an improvement in the microbial quality of the final product [25]. In the present study, the application of such a strategy allowed the evaluation of the biotechnological aptitude of the strains, used as single or mixed cultures, and enabled an understanding of how they interact with the matrix. The results showed that the use of microbial starters drove fermentation by lowering the pH to values as low as 4.49 and inducing an increase in the phenolic compounds. In detail, the combinations of *L. plantarum* and *C. boidinii*, both in single and in mixed cultures, resulted, at the end of fermentation, in the highest HT content, with values of 925.6, 902.5 and 903.5 mg/L, respectively. No oleuropein was detected at any sampling time as found by other authors [5,26]. Although *L. plantarum* is mainly known for its β-glucosidase activity or its probiotic potential [27], in all the tests carried out, there was a slight decrease of LAB count in sample with *L. plantarum* that showed an increase only after t14 of fermentation (Figure 1). This suggests that these strains are able to utilize certain metabolic pathways to survive in difficult matrices, which is why there is an increased activity in the last sampling time. An interesting study that may explain the adaptation of *L. plantarum* is proposed by Reveròn et al. [28], who propose a study of transcriptomics and the mechanism of action of *L. plantarum* in response to treatment with pure HT. *C. boidinii* strain used as a potentially resistant strain to several hurdles present in the matrix. Recently, De Melo Pereira et al. [29] reported that the genus *Candida* is commonly found in many fermented foods and beverages obtained by the main types of fermentation (alkaline, alcoholic, acetic, lactic, and mixed processes). In addition to its ubiquitous trait, the *Candida* genus also possesses a complex metabolic mechanism that allows it to survive, compete, and sometimes dominate fermentation processes [30]. Furthermore, it is known that a selected culture, besides the ability to control the fermentation process, should show the ability to survive in the fermentation environment and to exert acidifying activity through the production of organic acids. In the present study, results highlighted that *L. plantarum* inoculated samples exhibited the highest values of all detected acids. In a functional beverage, organic acids can play an additional role in protecting phenols, such as HT and TYR, from oxidation. In addition, different studies revealed that a lactic acid concentration of 0.5% (v/v) produced by LAB prevents pathogens' growth, such as *Salmonella species, Escherichia coli*, and *Listeria monocytogenes* [31,32]. This result confirmed results previously reported, namely, that the fermentation driven by LAB leads to the production of mono–, di–, and tri–carboxylic acids, i.e., acetic, lactic, and propionic acids as intermediaries of biosynthetic metabolic pathways and amino acid metabolism. In detail, Okoye et al. [33] demonstrated through genome study that LAB contain unique and shared secondary metabolite biosynthetic gene clusters with bio preservative potential and a transcription factor, namely CRP (cyclic AMP receptor protein) endowed with novel binding sites involved in organic acid metabolism.

Zooming in on biological activity, results obtained from tested microfiltered fermented OMWW and from pure HT, when tested at a 1:25 dilution, were found to be safe on chosen cell lines. In the present study, the choice of cell lines was based on taking into consideration that the HepG2 is one of the most reliable experimental models for prediction human liver toxicity. Indeed, the liver is responsible for most of the orally administered xenobiotic metabolism, for its anatomical proximity to the gastrointestinal tract and for its histological structure [34], whereas the Caco-2 cell line has been chosen as the most suitable in vitro model to rapidly assess the intestinal permeability and for xenobiotic transport studies [35,36]. Caco-2 cells exhibit a well–differentiated brush border on the apical surface and tight junctions, and express typical small–intestinal microvillus hydrolases and nutrient transporters. The crossing of biological membranes must be taken into account because it correlates with the ability of a pharmacologically active compound to reach the target site where performing the biological function. The intestinal transport of polyphenols seems to be strongly influenced by several factors such as food matrix, biotransformation and conjugation that occur during absorption [37,38]. Many studies have focused on the uptake of individual phenols, such as HT and TYR, which have shown good absorption across the cell membrane, while the uptake of a phytocomplex and how its different composition may affect the transport mechanism has been less explored [39]. In a recent study, Bartolomei et al. [40], demonstrated that a phenolic pool, extracted from extra virgin olive oil (EVOO), induced a protective effect against $H_2 O_2$-induced oxidative stress on Caco-2 and HepG2 cell lines. This observation demonstrated a selective transepithelial transport of certain oleuropein derivatives by Caco-2 cells, confirming that the phytocomplex could be transported with different mechanisms than those involved for single phenolic compounds, separately tested. According to results previously reported both phytocomplex composition and used starter cultures can significantly influence cell membrane crossing. In the present study, microbial cultures differently modulated the response of anti–inflammatory and antioxidant activity. It has been widely reported that phenols contained in EVOO reduce the reactive oxygen species (ROS) and malondialdehyde production, the nitric oxide release and the expression of inducible nitric oxide synthase (iNOS) and cyclooxygenase 2 (COX-2) [41]. Results obtained in the present study confirmed that OMWW samples affected the inhibitory activity towards COX-1 and COX-2, by a modulation of COX-2, according to previous in vitro reports on human monocytes [42]. The same authors demonstrated that HT attenuated ROS–mediated COX-2 transcription induced by bacterial lipopolysaccharide (LPS). COX catalyzes the first step in the biosynthesis of prostaglandins (PG), prostacyclin and thromboxane starting from free arachidonic acid (AA) [43]. Among prostaglandins, PGE_2 is involved in inflammation, angiogenesis and in promoting the growth of several solid tumors, such as breast, ovarian, head and neck cancer, renal cell carcinoma and hematological cancers [44–46]. COX-1 and COX-2 are of great interest because they are targets of non–steroidal anti–inflammatory drugs (NSAIDs), which, when binding to the active site of COX, prevent the AA from reaching the catalytic pocket and, thus, the biosynthesis of prostaglandins. COX inhibition is therefore important in reducing the inflammatory response, tumorigenesis and cancer progression.

Many of the recognized anti–cancer properties of HT are related to other activities, such as ability to modulate the antioxidant system and ROS scavenge [47,48]. Ramirez–Tortosa et al. [49] demonstrated that a supplementation with HT (15 mg/day) is effective into downregulate several transcriptional factors, as described for other antioxidant agents, able to induce, at plasma level, a decrease of metalloproteinase in women with breast cancer. HT, as reported by the European Food Safety Authority but in general the phytocomplex present both in olive oil and in by–products, has a beneficial effect on human health. The interaction between phenols and microorganisms used as starters plays a key role in understanding the mechanism of action and how they can modulate the anti–inflammatory and antioxidant response in the development of degenerative diseases [50].

4. Materials and Methods

4.1. OMWW Sampling

The OMWW samples used in the present study were obtained by a three–phase olive oil extraction system at the Consoli oil company (Adrano, Italy) and collected during a two–year period.

In detail, for Trial I OMWW was collected in the 2019–2020 season and for Trials II and III OMWW samples were collected in the 2020–2021 season. All the Trials are described in Figure 3. For Trial I the fresh produced OMWW was immediately stored at −20 °C at the Di3 A, University of Catania.

For Trial II, the OMWWs were stored at room temperature in the company facilities, until further treatments. To obtain a clear matrix, OMWW samples were subjected to filtration using Oenopad® XF1, XF7 and XFSS filters (OENO S.R.L., Erbusco, Italy), suitable for food matrices and consisting of cellulose, diatomaceous earth and perlite. Different fractions were obtained: the as–is sample (prefiltered or PF sample); and the three fractions (F1, F2, F3) obtained by sequential filtration with filters at different porosity, as: the "XF1" filter (8.0–20 μm) to eliminate solid particulates; the"XF7" filter (2.0–4.0 μm) for clarifying step; the "XFSS" filter (0.20–0.40 μm) for final sterilization. All fractions were collected and stored at −20 °C.

In addition, the Trial III was obtained from the F3 sample, in turn obtained from Trial II, by microfiltration using the Sartoclear Dynamics® kit (Sartorius, Varedo, Italy), connected to a vacuum pump. The latest process allows both the clarification/filtration and cold sterilization in a single step, as the used bottle presented a 0.22-μm polyethersulfone (PES) filter membrane. After processing samples needed for subsequent tests were frozen at −20 °C.

4.2. Set–Up of Fermentation Process

In order to set up the fermentation process some components, such as yeast extract at 1% (*w*/*v*), peptone and glucose at 2% (*w*/*v*), were added into the fresh OMWW samples and to the F3 samples right before fermentation. All components were purchased from Liofilchem (Roseto degli Abruzzi, Italy). The fermentation process was started through the inoculum of microbial pools, consisting of yeast and lactic acid bacteria strains, belonging to the microbial culture collection of the Department of Agricultural, Food and Environmental Sciences (Di3 A) and to ProBioEtna srl, Spin off of University of Catania. In details, the *Candida boidinii* F3 30.1, *Wickerhamomyces anomalus* F5 60.5 and *Lactiplantibacillus plantarum* F 3.5 (DSM 34190) strains were used. All the strains were previously isolated from naturally fermented table olives [51]. One hundred microliters of each yeast inoculum and *L. plantarum* were spotted in Yeast Peptone Dextrose broth (YPD, Sigma-Aldrich, Milano, Italy) and de Man, Rogosa, and Sharpe broth (MRS, Oxoid, UK) and allowed to incubate overnight at a selective temperature of 25 °C and 32 °C, respectively. Then, the strains were inoculated at 0.5%, which corresponded to an initial cell density of 10^7 CFU/mL for yeasts and 10^8 CFU/mL for *L. plantarum* (Figure 4). Seven experimental samples were set up for each Trial: the un–inoculated samples (controls); three single culture inoculated samples; two samples inoculated with each yeast strain in mixed culture with the *L. plantarum* strain;

one three–strain mixed culture sample. All tests were conducted in triplicate in an OMWW total volume of 400 mL. For Trial I and II, the fermentation process was monitored at regular intervals: at T0 (after about 7 h of microbial inoculation); T8 (after 8 days of fermentation); T30 (after 30 days of fermentation). For Trial III, fermentation parameters were monitored at T0, T8, T14 (after 14 days of fermentation) and T21 (end of fermentation). All fermentations were carried out at room temperature (20 ± 4 °C).

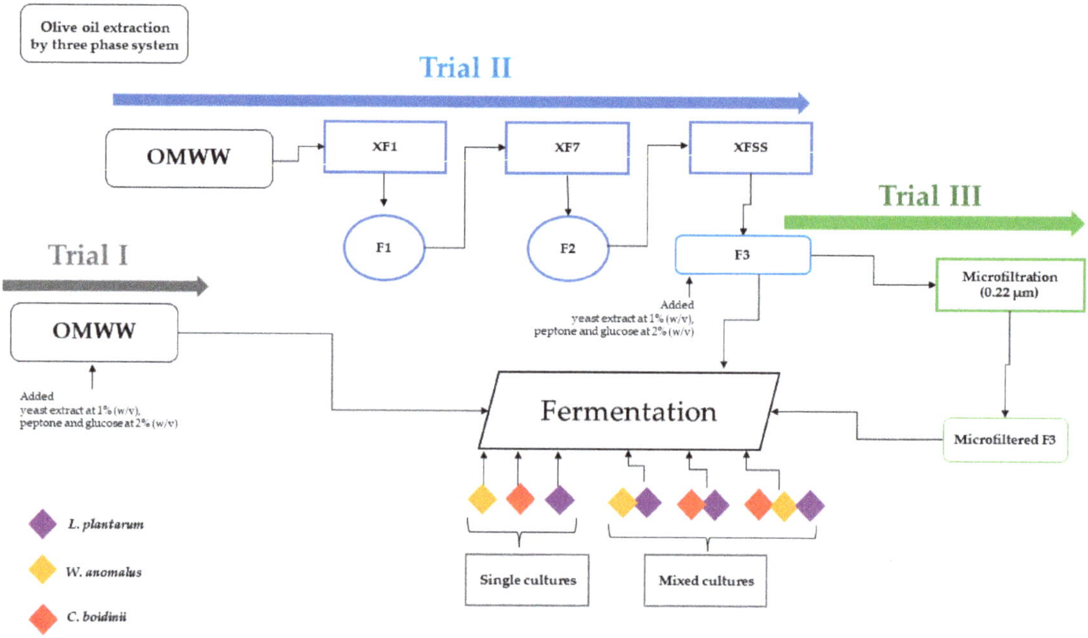

Figure 4. Fermentation process and OMWW obtained Trials.

4.3. Chemical Analyses

The pH, the TSS and the total phenol content were monitored for all samples during fermentation. The pH was measured with a Mettler DL25 pH meter (Mettler–Toledo International Inc., Columbus, OH, USA) and the total soluble solids (TSS), expressed as °Brix, were measured using a refractometer (Atago, RX-5000, Milano, Italy). In addition, the total phenolic content was determined according to the Folin–Ciocalteu's colorimetric method (FC). The tested samples were mixed with 5 mL of commercial FC reagent (Labochimica, Campodarsego, Italy) diluted with water (1:10 *v*/*v*) and added with 4 mL of a 7.5% sodium carbonate solution. Subsequently, samples were left in the dark at room temperature. After 2 h, the absorbance was measured spectrophotometrically at 765 nm (Cary 100 Scan UV-Visible, Agilent, CA, USA). The total phenolic content was expressed as mg gallic acid equivalent (GAE)/L of sample).

4.4. Microbiological Analyses

Samples of Trial I, II and III were serially diluted and poured into agar plates containing specific media and incubated under specific conditions: de Man, Rogosa, and Sharpe Agar (MRSA, Oxoid, Milano, Italy) for lactic acid bacteria counts, incubated at 32 °C for 48 h under anaerobic conditions; Plate Count Agar (PCA, Oxoid, Milano, Italy) for mesophilic aerobic bacteria counts, incubated at 25 °C for 48 h; Violet Red Bile Glucose Agar (VRBGA, Liofilchem, Roseto degli Abruzzi, Italy IT) for the determination of Enterobacteriaceae, incubated aerobically at 30–35 °C for 18–24 h; Sabouraud Dextrose Agar (SAB, Bio-Rad, Hercules, CA, USA) for yeast counts, incubated at 25 °C for 48 h. At the end of fermentation, the presence/absence of *Clostridium perfringens* was also determined in Sulphite-Polymyxin-Sulphadiazine Agar (SPS, Oxoid, UK), incubated at 35–37 °C for 18–48 h, under anaerobic conditions.

Moreover, for starter cultures monitoring, samples of Trial III were subjected to additional counting, in MRS agar and in SAB agar media, for *L. plantarum* and yeast determination, respectively.

All microbiological analyses were performed in triplicate and the results were expressed as Log CFU/mL.

4.5. HPLC Analysis

4.5.1. Detection of Phenols

The HPLC analyses of fermented OMWW samples were performed by directly injecting the filtered samples (0.45 µm PTFE filters, Merck, Darmstadt, Germany) into the HPLC chromatographic system, i.e., Waters Alliance 2695 HPLC liquid chromatography equipped with a Waters 996 photodiode array (PDA) detector set at 280 nm and managed through the Waters Empower software (Waters Corporation, Milford, MA, USA). The column used was a Luna C18 (250 mm × 4.6 mm i.d., 5 m, 100 Å; Phenomenex, Torrance, CA, USA) maintained in an oven at 40 °C. A flow rate of 1 mL/min was used. Chromatographic separation was performed according to Romeo et al. [11]. The internal standard (I.S.), 50 mM pure gallic acid (Fluka, Buchs Switzerland), was used to quantify the phenolic compounds. The identification of phenolic compounds was obtained by comparing the peak retention time with those of pure standards of tyrosol (TYR), oleuropein (OLE), hydroxytyrosol (HT) chlorogenic acid, vanillic acid, caffeic acid, syringic acid, p-coumaric acid, ferulic acid, verbascoside, luteolin-7-o-glucoside, o-coumaric acid, rutin, oleuropein, apigenin-7-o-glucoside, luteolin-4-glucoside, quercetin, luteolin, apigenin (Extrasynthese, Genay, France). All analyses were performed in triplicate for each sample.

4.5.2. Detection of Organic Acids

The determination of organic acids was carried out at the end of fermentation in Trial III. Each sample was filtered through a 0.45 µm PTFE syringe filter (Merck, Germany) before being injected into HPLC (HPLC instruments were described in the previous section) with a DAD detector set at 210 nm (and with spectrum acquisition from 200 to 400 nm). Isocratic elution with 5 mM sulfur acid was performed on a Rezex ROA Organic Acid H+ column (Phenomenex, CA, USA). The run time was set to 50 min at 0.6 mL/min. For calibration, pure standards of lactic, citric, acetic, propionic, isobutyric and butyric acids (all purchased from Sigma–Aldrich, Italy) were injected at different concentrations. All analyses were performed in triplicate for each sample.

4.6. Biological Assays

4.6.1. Cell Culture and Cytotoxicity

Caco-2 cells were grown in Dulbecco's Modified Eagle Medium high glucose (DMEM high glucose, Euroclone S.p.A., Pero, Italy) supplemented with 10% Fetal Bovine Serum (FBS, Euroclone S.p.A., Pero, Italy), 2 mM glutamine (Euroclone S.p.A., Pero, Italy), 100 U/mL of penicillin and 0.1 mg/mL of streptomycin (Euroclone). Caco-2 cells were kindly supplied from Dr. Aldo Cavallini and Dr. Caterina Messa from the Laboratory of Biochemistry National Institute for Digestive Diseases. "S. de Bellis", Bari (Italy).

Human hepatocellular liver carcinoma (HepG2) cell line was purchased from American Type Culture Collection (ATCC). HepG2 cells were cultured in Eagle's Minimum Essential Medium (MEM, Euroclone), supplemented with 10% FBS, 2 mM glutamine (Euroclone), 100 U/mL penicillin and 0.1 mg/mL streptomycin (Euroclone S), 1% Non–Essential Amino Acids (NEAA, Euroclone). Cultured cells were maintained at 37 °C in atmosphere containing 95% of air and 5% of CO_2. Cells were sub-cultivated every 48 h by trypsine–EDTA solution.

Determination of cell growth was performed using the 3-(4.5-dimethylthiazol-2-yl)-2.5-diphenyltetrazolium bromide (MTT) assay (Sigma-Aldrich, Milan, Italy), 10.000 cells/well were seeded into 96-well plates at a volume of 100 μL. After 24 h, 100 μL of microfiltered fermented OMWW samples were added at the appropriate dilution: as such, 1:10, 1:25, 1:50 and 1:100 in triplicate. After 72 h incubation time with extracts, the plates containing the cells were incubated with MTT for 3–4 h at 37 °C and 5% of CO_2. At the end of incubation time, MTT was aspirated, and the formazan crystals were solubilized by using 100 μL of dimethyl sulfoxide/ethanol (1:1) (Sigma–Aldrich). The absorbance values at λ = 570 nm were determined on the Victor Microplate Reader (PerkinElmer, Roma, Italy). Pure HT (Phytolab, Vastenbergsgreuth, Germany) was used as a positive control.

4.6.2. Transport Caco-2 Monolayer

Caco-2 cells were seeded onto a Millicell-96 assay system (Millipore, Burlington, MA, USA) in which a cell monolayer was set in between a filter cell and a receiver plate at a density of 20.000 cells/well. The culture medium was replaced every 48 h and the cells were kept for 21 days in culture. The trans epithelial electrical resistance (TEER) of the monolayers was measured daily before and after the experiment by using an epithelial voltohmmeter (Millicell–ERS). Generally, TEER values greater than 1000 Ω for a 21-day culture are considered optimal. After 21 days of Caco-2 cell growth, the medium was removed from the filter wells and the receiver plate, and they were filled with fresh Hank's balanced salt solution (HBSS) buffer (Invitrogen, Waltham, MA, USA). This procedure was repeated twice, and the plates were incubated at 37 °C for 30 min. After the incubation time, the HBSS buffer was removed and OMWW samples (dilution 1:100) were added to the filter well whereas fresh HBSS was added to the receiver plate. The plates were incubated at 37 °C for 120 min. Afterward, samples were removed from the apical (filter well) and basolateral (receiver plate) side of the monolayer to measure the permeability. The apparent permeability (P_{app}) referred to HT in units of nm/second was calculated using the following Equation (1):

$$P_{app} = \left(\frac{V_A}{\text{Area} \times \text{time}}\right) \times \left(\frac{[\text{sample}]\text{acceptor}}{[\text{sample}]\text{initials}}\right) \quad (1)$$

V_A = the volume (in mL) in the acceptor well; Area = the surface area of the membrane (0.11 cm^2 of the well); time = the total transport time in seconds (7200 s); [sample]acceptor = the concentration of the sample measured by U.V. spectroscopy; [sample]initial = the initial sample concentration (1×10^{-4} M) in the apical or basolateral wells.

4.6.3. Cyclooxygenase Activity Inhibition

Preliminarily, the fermented OMWW samples obtained from Trial III were evaluated for their ability to inhibit *ovine*COX-1 or *human*COX-2 enzymes, measuring the extent (%) of enzyme activity inhibition at 50 μM, at dilution 1:25. The inhibition of the enzyme was evaluated by using a colorimetric COX inhibitor screening assay kit (Catalog No. 7601050, Cayman Chemicals, Ann Arbor, MI, USA) following the manufacturer's instructions. COX is a bifunctional enzyme exhibiting both cyclooxygenase and peroxidase activities. The cyclooxygenase component catalyzes the conversion of arachidonic acid into the hydroperoxide PGG_2 and then peroxidase component catalyzes PGG_2 reduction into the corresponding alcohol PGH_2, the precursor of PGs, thromboxane, and prostacyclin. The c COX inhibitor screening assay colorimetrically measures the peroxidase activity of the cyclooxygenases monitoring the appearance of oxidized N,N,N′,N′-tetramethyl-p-phenylenediamine (TMPD) at λ = 590 nm on the Victor Microplate Reader (PerkinElmer, Italy). Stock solutions of tested samples were dissolved in deionized distillated water.

4.6.4. Antioxidant Activity

Radical scavenging activity was determined as percentage of RSA (radical scavenging activity), according to Palmeri et al. [52]. The values were expressed using the following Equation (2):

$$\text{RSA}~\% = \left(\frac{\text{Blank Absorbance} - \text{Sample Absorbance}}{\text{Blank Absorbance}}\right) \times 100 \qquad (2)$$

Different dilutions of samples were added to the mixture of methanolic solution and 2,2-diphenyl-1-picrylhydrazyl radical 10^{-4} M. The DPPH absorbance values were evaluated at λ = 517 nm by monitoring the kinetics for 5 min with spectrophotometer (Shimadzu UV-1800, Denmark). Pure HT (Phytolab, Germany) was used as a positive control.

4.7. Statistical Analysis

Statistical analysis of the obtained results was performed by means of one–way analysis of variance (ANOVA) and Tukey's HSD post hoc test for separation of means at a significance level of $p \leq 0.05$. For data processing, SPSS software (version 21.0, IBM Statistics, NY, USA) was used for data processing.

5. Conclusions

The microfiltration process resulted in a suitable strategy to obtain a OMWW matrix able to be fermented. The use of selected microbial pools in single and co–cultures showed an increase in HT and TYR contents at the end of fermentation, compared with the control sample. Biological analyses showed that fermentation increases the antioxidant and inflammatory activity of OMWW that resulted to be safe in HepG2 and Caco-2 cell lines. In detail, the phenolic pattern associated to starter microorganisms exhibited an increase of active permeability on Caco-2 monolayer, and a moderate inhibition towards *o*COX-1 and *h*COX-2 was observed. The results confirm that fermented OMWW can be proposed as a new beverage and/or functional ingredient that could include the addition of compounds as flavorings and probiotic microorganisms. Despite the interesting results obtained at lab scale, perspective studies should aim to replay the process at the industrial scale to standardize phenol concentration at each obtained new formulation.

Supplementary Materials: The following supporting information can be downloaded at: https://www.mdpi.com/article/10.3390/molecules28020646/s1, Table S1. Chemical parameters detected in samples of Trial I and Trial II. Table S2. Main microbial groups counted in Trial I and Trial II samples during fermentation. Table S3. Evaluation of tested samples inhibition (as %) on COXs enzymes. Figure S1. Evaluation of inhibition (as %) of different HT concentrations on oCOX-1 and hCOX-2. Figure S2. (a) Trial I at the end of fermentation; (b) Cartons filters after the spinning process and samples from Trial II; (c) Sample of Trial II after the fermentation process. Figure S3. OMWW samples microfiltered at 0.22 μm.

Author Contributions: P.F., P.S.O., N.R., M.M.: Investigation, Methodology, Writing—original draft; P.F., F.V.R.: Software, Methodology; C.C., F.V.R., M.G.P., A.S.: Conceptualization, review & editing; C.C., C.L.R.: Conceptualization, Writing—review & editing; C.C., F.V.R., M.G.P.: Conceptualization, Supervision, Funding acquisition, Review & editing. All authors have read and agreed to the published version of the manuscript.

Funding: This research received no external funding.

Institutional Review Board Statement: This study did not require ethical approval.

Informed Consent Statement: Not applicable.

Acknowledgments: This study was conducted within a Ph.D. research program in Biotecnologie (XXXV cycle) by Paola Foti who received a grant 'Dottorato innovativo con caratterizzazione industriale, PON RI 2014–2020', titled 'Olive oil by-products as a new functional food and source of nutritional food ingredients' from the Department of Agriculture, Food and Environment (Scientific Tutors: C.C. Cinzia Caggia; Flora V. Romeo and Cinzia L. Randazzo). Authors thank the Azienda Olearia Consoli Pasquale & F.lli s.n.c (Adrano, CT), partner of the doctoral program, to kindly supply the OMWW samples. Authors thank Aldo Cavallini and Caterina Messa, from the Laboratory of Biochemistry National Institute for Digestive Diseases, "S. de Bellis", Bari (Italy), to kindly supply the Caco-2 cells.

Conflicts of Interest: The authors declare that they have no known competing financial interest or personal relationship that could have appeared to influence the work reported in this paper.

Sample Availability: Not applicable.

References

1. Gullon, P.; Gullon, B.; Astray, G.; Carpena, M.; Fraga-Corral, M.; Prieto, M.A.; Simal-Gandara, J. Valorization of by-products from olive oil industry and added-value applications for innovative functional foods. *Food Res. Int.* **2020**, *137*, 109683. [CrossRef] [PubMed]
2. Flamminii, F.; Di Mattia, C.D.; Difonzo, G.; Neri, L.; Faieta, M.; Caponio, F.; Pittia, P. From by-product to food ingredient: Evaluation of compositional and technological properties of olive-leaf phenolic extracts. *J. Sci. Food Agric.* **2019**, *99*, 6620–6627. [CrossRef] [PubMed]
3. Lama-Muñoz, A.; del Mar Contreras, M.; Espínola, F.; Moya, M.; Romero, I.; Castro, E. Extraction of oleuropein and luteolin-7-O-glucoside from olive leaves: Optimization of technique and operating conditions. *Food Chem.* **2019**, *293*, 161–168. [CrossRef]
4. Robles-Almazan, M.; Pulido-Moran, M.; Moreno-Fernandez, J.; Ramirez-Tortosa, C.; Rodriguez-Garcia, C.; Quiles, J.L.; Ramirez Tortosa, M. Hydroxytyrosol: Bioavailability, toxicity, and clinical applications. *Food Res. Int.* **2018**, *105*, 654–667. [CrossRef] [PubMed]
5. Dermeche, S.; Nadour, M.; Larroche, C.; Moulti-Mati, F.; Michaud, P. Olive mill wastes: Biochemical characterizations and valorization strategies. *Process. Biochem.* **2013**, *48*, 1532–1552. [CrossRef]
6. Servili, M.; Esposto, S.; Sordini, B.; Veneziani, G.; Urbani, S. *L'acqua di Vegetazione dei Frantoi Oleari: Una Risorsa da Valorizzare*; Accademia dei Georgofili: Florence, Italy, 2021; pp. 317–337.
7. Foti, P.; Occhipinti, P.S.; Romeo, F.V.; Timpanaro, N.; Musumeci, T.; Randazzo, C.L.; Caggia, C. Phenols recovered from olive mill wastewater as natural booster to fortify blood orange juice. *Food Chem.* **2022**, *393*, 133428. [CrossRef]
8. Cedola, A.; Cardinali, A.; D'Antuono, I.; Conte, A.; Del Nobile, M.A. Cereal foods fortified with by-products from the olive oil industry. *Food Biosci.* **2019**, *33*, 100490. [CrossRef]
9. Hadj Saadoun, J.; Bertani, G.; Levante, A.; Vezzosi, F.; Ricci, A.; Bernini, V.; Lazzi, C. Fermentation of Agri-Food Waste: A Promising Route for the Production of Aroma Compounds. *Foods* **2021**, *10*, 707. [CrossRef]
10. Aponte, M.; Ungaro, F.; d'Angelo, I.; De Caro, C.; Russo, R.; Blaiotta, G.; Dal Piaz, F.; Calignano, A.; Miro, A. Improving in vivo conversion of oleuropein into hydroxytyrosol by oral granules containing probiotic *Lactobacillus plantarum* 299v and an Olea europaea standardized extract. *Int. J. Pharmaceut.* **2018**, *543*, 73–82. [CrossRef]
11. Romeo, F.V.; Granuzzo, G.; Foti, P.; Ballistreri, G.; Caggia, C.; Rapisarda, P. Microbial application to improve olive mill wastewater phenolic extracts. *Molecules* **2021**, *26*, 1944. [CrossRef]
12. Ruiz Rodríguez, L.G.; Zamora Gasga, V.M.; Pescuma, M.; Van Nieuwenhove, C.; Mozzi, F.; Sánchez Burgos, J.A. Fruits and fruit by-products as sources of bioactive compounds. Benefits and trends of lactic acid fermentation in the development of novel fruit-based functional beverages. *Food Res. Int.* **2021**, *140*, 109854. [CrossRef] [PubMed]
13. Miao, Z.; Hao, H.; Yan, R.; Wang, X.; Wang, B.; Sun, J.; Li, Z.; Zhang, Y.; Sun, B. Individualization of Chinese alcoholic beverages: Feasibility towards a regulation of organic acids. *LWT* **2022**, *172*, 114168. [CrossRef]
14. Starzyńska-Janiszewska, A.; Fernández-Fernández, C.; Martín-García, B.; Verardo, V.; Gómez-Caravaca, A.M. Solid state fermentation of olive leaves as a promising technology to obtain hydroxytyrosol and elenolic acid derivatives enriched extracts. *Antioxidants* **2022**, *11*, 1693. [CrossRef]
15. Benítez-Cabello, A.; Calero-Delgado, B.; Rodríguez-Gómez, F.; Garrido-Fernández, A.; Jiménez-Díaz, R.; Arroyo-López, F.N. Biodiversity and multifunctional features of lactic acid bacteria isolated from table olive biofilms. *Front. Microbiol.* **2019**, *10*, 836. [CrossRef]
16. Lanza, B.; Di Marco, S.; Bacceli, M.; Di Serio, M.G.; Di Loreto, G.; Cellini, M.; Simone, N. *Lactiplantibacillus plantarum* used as single, multiple, and mixed starter combined with *Candida boidinii* for table olive fermentations: Chemical, textural, and sensorial characterization of final products. *Fermentation* **2021**, *7*, 239. [CrossRef]
17. Zullo, B.A.; Ciafardini, G. Evaluation of physiological properties of yeast strains isolated from olive oil and their in vitro probiotic trait. *Food Microbiol.* **2019**, *78*, 179–187. [CrossRef] [PubMed]
18. Nazhand, A.; Durazzo, A.; Lucarini, M.; Guerra, F.; Souto, S.B.; Souto, E.B.; Santini, A.I. Nutraceuticals and functional beverages: Focus on Prebiotics and Probiotics active beverages. In *Future Foods*; Academic Press: Cambridge, MA, USA, 2022; pp. 251–258.

19. Parra-Perez, A.M.; Pérez-Jiménez, A.; Gris-Cárdenas, I.; Bonel-Pérez, G.C.; Carrasco-Díaz, L.M.; Mokhtari, K.; García-Salguero, L.; Lupiáñez, J.A.; Rufino-Palomares, E.E. Involvement of the PI3K/AKT intracellular signaling pathway in the anticancer activity of hydroxytyrosol, a polyphenol from *Olea europaea*, in hematological cells and implication of HSP60 levels in its anti-inflammatory activity. *Int. J. Mol. Sci.* **2022**, *23*, 7053. [CrossRef]
20. Perrone, M.G.; Vitale, P.; Ferorelli, S.; Boccarelli, A.; Coluccia, M.; Pannunzio, A.; Campanella, F.; Di Mauro, G.; Bonaccorso, C.; Fortuna, C.G.; et al. Effect of mofezolac-galactose distance in conjugates targeting cyclooxygenase (COX)-1 and CNS GLUT-1 carrier. *Eur. J. Med. Chem.* **2017**, *141*, 404–416. [CrossRef]
21. Di Mauro, M.D.; Fava, G.; Spampinato, M.; Aleo, D.; Melilli, B.; Saita, M.G.; Centonze, G.; Maggiore, R.; D'Antona, N. Polyphenolic fraction from olive mill wastewater: Scale-up and in vitro studies for ophthalmic nutraceutical applications. *Antioxidants* **2019**, *8*, 462. [CrossRef]
22. Ricci, A.; Bernini, V.; Maoloni, A.; Cirlini, M.; Galaverna, G.; Neviani, E.; Lazzi, C. Vegetable by-product lacto-fermentation as a new source of antimicrobial compounds. *Microorganisms* **2019**, *7*, 607. [CrossRef]
23. Alfano, A.; Corsuto, L.; Finamore, R.; Savarese, M.; Ferrara, F.; Falco, S.; Santabarbara, G.; De Rosa, M.; Schiraldi, C. Valorization of olive mill wastewater by membrane processes to recover natural antioxidant compounds for cosmeceutical and nutraceutical applications or functional foods. *Antioxidants* **2018**, *7*, 72. [CrossRef]
24. Arienzo, A.; Murgia, L.; Fraudentali, I.; Gallo, V.; Angelini, R.; Antonini, G. Microbiological quality of ready-to-eat leafy green salads during shelf-life and home-refrigeration. *Foods* **2020**, *9*, 1421. [CrossRef]
25. France, T.C.; Kelly, A.L.; Crowley, S.V.; O'Mahony, J.A. Cold microfiltration as an enabler of sustainable dairy protein ingredient innovation. *Foods* **2021**, *10*, 2091. [CrossRef]
26. Allouche, N.; Fki, I.; Sayadi, S. Toward a high yield recovery of antioxidants and purified hydroxytyrosol from olive mill wastewaters. *J. Agric. Food Chem.* **2004**, *52*, 267–273. [CrossRef]
27. Vaccalluzzo, A.; Pino, A.; De Angelis, M.; Bautista-Gallego, J.; Romeo, F.V.; Foti, P.; Caggia, C.; Randazzo, C.L. Effects of different stress parameters on growth and on oleuropein-degrading abilities of *Lactiplantibacillus plantarum* strains selected as tailored starter cultures for naturally table olives. *Microorganisms* **2020**, *8*, 1607. [CrossRef] [PubMed]
28. Reverón, I.; Plaza-Vinuesa, L.; Santamaría, L.; Oliveros, J.C.; de Las Rivas, B.; Muñoz, R.; López de Felipe, F. Transcriptomic evidence of molecular mechanisms underlying the response of Lactobacillus plantarum WCFS1 to hydroxytyrosol. *Antioxidants* **2020**, *9*, 442. [CrossRef] [PubMed]
29. De Melo Pereira, G.V.; Maske, B.L.; de Carvalho Neto, D.P.; Karp, S.G.; De Dea Lindner, J.; Martin, J.G.P.; de Oliveira Hosken, B.; Soccol, C.R. What is *Candida* doing in my food? A review and safety alert on its use as starter cultures in fermented foods. *Microorganisms* **2022**, *10*, 1855. [CrossRef] [PubMed]
30. Palareti, G.; Legnani, C.; Cosmi, B.; Antonucci, E.; Erba, N.; Poli, D.; Testa, S.; Tosetto, A. Comparison between different D-dime cutoff values to assess the individual risk of recurrent venous thromboembolism: Analysis of results obtained in the DULCIS study. *Int. J. Lab. Hematol.* **2016**, *38*, 42–49. [CrossRef]
31. Wang, C.; Chang, T.; Yang, H.; Cui, M. Antibacterial mechanism of lactic acid on physiological and morphological properties of *Salmonella* Enteritidis, *Escherichia coli* and *Listeria monocytogenes*. *Food Control.* **2015**, *47*, 231–236. [CrossRef]
32. Bangar, S.P.; Suri, S.; Trif, M.; Ozogul, F. Organic acids production from lactic acid bacteria: A preservation approach. *Food Biosci.* **2022**, *46*, 101615. [CrossRef]
33. Okoye, C.O.; Dong, K.; Wang, Y.; Gao, L.; Li, X.; Wu, Y.; Jiang, J. Comparative genomics reveals the organic acid biosynthesis metabolic pathways among five lactic acid bacterial species isolated from fermented vegetables. *New Biotechnol.* **2022**, *70*, 73–83. [CrossRef] [PubMed]
34. Ramirez, T.; Strigun, A.; Verlohner, A.; Huener, H.A.; Peter, E.; Herold, M.; Bordag, N.; Mellert, W.; Walk, T.; Spitzer, M.; et al. Prediction of liver toxicity and mode of action using metabolomics in vitro in HepG2 cells. *Arch. Toxicol.* **2018**, *92*, 893–906. [CrossRef] [PubMed]
35. Colabufo, N.A.; Contino, M.; Cantore, M.; Capparelli, E.; Grazia, M.; Cassano, G.; Gasparre, G.; Leopoldo, M.; Berardi, F.; Perrone, R. Naphthalenyl derivatives for hitting P-gp/MRP1/BCRP transporters. *Bioorg. Med. Chem.* **2013**, *21*, 1324–1332. [CrossRef] [PubMed]
36. Riganti, C.; Contino, M.; Guglielmo, S.; Perrone, M.G.; Salaroglio, I.C.; Milosevic, V.; Giampietro, R.; Leonetti, F.; Rolando, B.; Lazzarato, L.; et al. Design, biological evaluation, and molecular modeling of tetrahydroisoquinoline derivatives: Discovery of a potent P-glycoprotein ligand overcoming multidrug resistance in cancer stem cells. *J. Med. Chem.* **2019**, *62*, 974–986. [CrossRef] [PubMed]
37. D'Archivio, M.; Filesi, C.; Varì, R.; Scazzocchio, B.; Masella, R. Bioavailability of the polyphenols: Status and controversies. *Int. J. Mol. Sci.* **2010**, *11*, 1321–1342. [CrossRef] [PubMed]
38. Soler, A.; Romero, M.P.; Macia, A.; Saha, S.; Furniss, C.S.M.; Kroon, P.A.; Motilva, M.J. Digestion stability and evaluation of the metabolism and transport of olive oil phenols in the human small-intestinal epithelial Caco 2/TC7 cell line. *Food Chem.* **2010**, *119*, 703–714. [CrossRef]
39. Serreli, G.; Deiana, M. Extra virgin olive oil polyphenols: Modulation of cellular pathways related to oxidant species and inflammation in aging. *Cells* **2020**, *9*, 478. [CrossRef]

40. Bartolomei, M.; Bollati, C.; Bellumori, M.; Cecchi, L.; Cruz-Chamorro, I.; Santos-Sánchez, G.; Ranaldi, G.; Ferruzza, S.; Sambuy, Y.; Arnoldi, A.; et al. Extra virgin olive oil phenolic extract on human hepatic HepG2 and intestinal Caco-2 Cells: Assessment of the antioxidant activity and intestinal trans-epithelial transport. *Antioxidants* **2021**, *10*, 118. [CrossRef]
41. Abdallah, M.; Marzocco, S.; Adesso, S.; Zarrouk, M.; Guerfel, M. Olive oil polyphenols extracts inhibit inflammatory markers in J774A.1 murine macrophages and scavenge free radicals. *Acta Histochem.* **2018**, *120*, 1–10. [CrossRef]
42. Scoditti, E.; Nestola, A.; Massaro, M.; Calabriso, N.; Storelli, C.; De Caterina, R.; Carluccio, M.A. Hydroxytyrosol suppresses MMP-9 and COX-2 activity and expression in activated human monocytes via PKCα and PKCβ1 inhibition. *Atherosclerosis* **2014**, *232*, 17–24. [CrossRef]
43. Perrone, M.G.; Scilimati, A.; Simone, L.; Vitale, P. Selective COX-1 inhibition: A therapeutic target to be reconsidered. *Curr. Med. Chem.* **2010**, *17*, 3769–3805. [CrossRef] [PubMed]
44. Calvello, R.; Lofrumento, D.D.; Perrone, M.G.; Cianciulli, A.; Salvatore, R.; Vitale, P.; De Nuccio, F.; Giannotti, L.; Nicolardi, G.; Panaro, M.A.; et al. Highly selective cyclooxygenase-1 inhibitors P6 and mofezolac counteract inflammatory state both in vitro and in vivo models of neuroinflammation. *Front Neurol.* **2017**, *9*, 251. [CrossRef] [PubMed]
45. Pati, M.L.; Vitale, P.; Ferorelli, S.; Iaselli, M.; Miciaccia, M.; Boccarelli, A.; Di Mauro, G.D.; Fortuna, C.G.; Souza Domingos, T.F.; Rodrigues Pereira da Silva, L.C.; et al. Translational impact of novel widely pharmacological characterized mofezolac-derived COX-1 inhibitors combined with bortezomib on human multiple myeloma cell lines viability. *Eur. J. Med. Chem.* **2019**, *164*, 59–76. [CrossRef] [PubMed]
46. Perrone, M.G.; Vitale, P.; Miciaccia, M.; Ferorelli, S.; Centonze, A.; Solidoro, R.; Munzone, C.; Bonaccorso, C.; Fortuna, C.G.; Kleinmanns, K.; et al. Fluorochrome selection for imaging intraoperative ovarian cancer probes. *Pharmaceuticals* **2022**, *15*, 668. [CrossRef]
47. Rosignoli, P.; Fuccelli, R.; Sepporta, M.V.; Fabiani, R. In vitro chemo-preventive activities of hydroxytyrosol: The main phenolic compound present in extra-virgin olive oil. *Food Funct.* **2016**, *7*, 301–307. [CrossRef]
48. El-Azem, N.; Pulido-Moran, M.; Ramirez-Tortosa, C.L.; Quiles, J.L.; Cara, F.E.; Sanchez-Rovira, P.; Granados-Principal, S.; Ramirez-Tortosa, M.C. Modulation by hydroxytyrosol of oxidative stress and antitumor activities of paclitaxel in breast cancer. *Eur. J. Nutr.* **2019**, *58*, 1203–1211. [CrossRef]
49. Ramirez-Tortosa, C.; Sanchez, A.; Perez-Ramirez, C.; Quiles, J.L.; Robles-Almazan, M.; Pulido-Moran, M.; Sanchez-Rovira, P.; Ramirez-Tortosa, M. Hydroxytyrosol supplementation modifies plasma levels of tissue inhibitor of metallopeptidase 1 in women with breast cancer. *Antioxidants* **2019**, *8*, 393. [CrossRef]
50. Sanlier, N.; Gokcen, B.B.; Sezgin, A.C. Health benefits of fermented foods. *Crit. Rev. Food Sci. Nutr.* **2019**, *59*, 506–527. [CrossRef]
51. Pino, A.; Vaccalluzzo, A.; Solieri, L.; Romeo, F.V.; Todaro, A.; Caggia, C.; Arroyo-López, F.N.; Bautista-Gallego, J.; Randazzo, C.L. Effect of sequential inoculum of beta-glucosidase positive and probiotic strains on brine fermentation to obtain low salt Sicilian table olives. *Front. Microbiol.* **2019**, *10*, 174. [CrossRef]
52. Palmeri, R.; Siracusa, L.; Carrubba, M.; Parafati, L.; Proetto, I.; Pesce, F.; Fallico, B. Olive leaves, a promising byproduct of olive oil industry: Assessment of metabolic profiles and antioxidant capacity as a function of cultivar and seasonal change. *Agronomy* **2022**, *12*, 2007. [CrossRef]

molecules

Article

Fluorescence Monitoring Oxidation of Extra Virgin Olive Oil Packed in Different Containers

Elísabet Martín-Tornero [1,*], Antonio Fernández [2], Isabel Durán-Merás [1] and Daniel Martín-Vertedor [2,*]

1 Departamento de Química Analítica, Universidad de Extremadura, 06006 Badajoz, Spain
2 Technological Institute of Food and Agriculture, CICYTEX-INTAEX, Junta de Extremadura, Avda. Adolfo Suárez s/n, 06007 Badajoz, Spain
* Correspondence: elisabetmt@unex.es (E.M.-T.); daniel.martin@juntaex.es (D.M.-V.); Tel.: +34-924-012-664 (E.M.-T.)

Abstract: 'Picual' olive oil was stored in different types of containers for 10 months and monitored via quality parameters. In combination with the mentioned analysis, non-destructive fluorescence spectroscopy was performed combined with multivariate analysis to monitor and quantify oil quality levels. Excitation emission matrices (EMMs) were analyzed using parallel factor analysis (PARAFAC). According to the quality parameters, it was observed that Transparent Crystal (TC) and Opaque Crystal (OC) samples were the ones that deteriorated faster due to their higher exposure to light in comparison with Plastic (P) and Canned (C) samples. In a fast and non-destructive manner, the fluorescence spectroscopy-based prototype successfully monitored the oxidation changes in the EVOOs. Unfolded partial least squares (U-PLS) was used to generate a regression model to quantify quality parameters. Good correlation coefficients were found for the peroxide index, K_{232} and the oxidative stability index (r^2 between 0.90 and 0.94 for cross-validation and validation). For all of that, the results obtained confirmed the ability of fluorescence spectroscopy to monitor the quality of olive oil and EEMs combined with U-PLS can be used to analyze these parameters, eluding the classical methods.

Keywords: olive oil; fluorescence spectroscopy; olive oil conservation; packaging; quality parameters

Citation: Martín-Tornero, E.; Fernández, A.; Durán-Merás, I.; Martín-Vertedor, D. Fluorescence Monitoring Oxidation of Extra Virgin Olive Oil Packed in Different Containers. *Molecules* **2022**, *27*, 7254. https://doi.org/10.3390/molecules27217254

Academic Editor: Smaoui Slim

Received: 30 September 2022
Accepted: 23 October 2022
Published: 26 October 2022

1. Introduction

Due to its healthy nutritional properties, it is well known that olive oil is one of the most important products in the Mediterranean diet. Producing countries like Spain or Italy take advantage of this producer role in the sector's economy by ensuring its quality with exhaustive controls, a functional elaboration process, olive oil categorizations (Olive oil, OO; virgin olive oil, VOO; and extra virgin olive oil, EVOO) and packaging. These initiatives help to maintain beneficial substances like unsaturated fatty acids, polyphenols, vitamin E, carotenoids, sterols, etc. At the present time, dietary care and healthy alimentary habits, like consuming EVOO, are growing among consumers, especially when it comes to antioxidant intake due to its long and short terms benefits [1,2] such as preventing and reducing certain gastrointestinal diseases [3].

Having that in mind, the concentration of these bioactive compounds is not only influenced by the cultivar agronomic conditions (such as the cultivar type, maturation stage, agroclimatic conditions and agronomical practices), raw material, harvesting, fruit storage and extraction technology, but also by each factor which can affect it during its commercial life [4–6]. Therefore, some studies have proven that oxygen, light and temperature are variables responsible for increasing deteriorative processes in EVOO as a consequence of oxidative and hydrolytic reactions [7]. When controlling these parameters, EVOO shelf-life can be between 12 and 18 months [8], reaching the second year when good storage is achieved together with well-sealed packaging [9]. Several articles have been published where the influence of the storage time on quality was evaluated [10].

However, EVOO quality and shelf-life is reduced by oxidation, which represents one of the greatest EVOO quality degradation factors during storage [11,12] and which can be counteracted by the antioxidant activity of polyphenolic compounds and tocopherols. The most susceptible fraction to oxidation is the lipid one, which produces an increase of carbonyl and aldehyde compounds that prompt off-flavors and 'oxidative rancidity' at the end, being unsuitable for human consumption [13].

Auto-oxidation contributes to the degradative processes of the olive oil, happening even with lack of light. This is due to a free radical mechanism where, at the beginning, the absorption of O_2 results in the formation of hydroperoxides. On the other hand, when EVOO is exposed to light, photo-oxidation occurs through the action of natural photosensitizers such as chlorophyll. As an outcome, storage and packing conditions of EVOO become crucially important [5,14]. Stefanoudaki et al. [10] studied the evolution of VOO during 15 months of storage inside and outside warehouse conditions, but they did not maintain a controlled temperature.

In order to avoid oxidation by different sources, dark containers, optimal containers that can avoid oxygen penetration, and low temperatures during storage can be helpful in preserving EVOO quality [12,14,15]. Previous studies have evaluated different packaging materials, light transmission influence, temperature effects and time of storage concerning quality physicochemical parameters (such as acidity, peroxide index, K_{232} and K_{270}), sensory attributes and shelf-life of EVOO, finding how organoleptic and quality properties are lost as time goes by [12,15,16].

Fluorescence spectroscopy has been proposed as an alternative technique to analyze and monitor olives and olive oil. It has the advantages over the other conventional methods of its speed of analysis and the minimum sample preparation with the absence of solvents and reagents. Excitation emission matrices (EEMs) have been applied with the aim of determining phenolic compounds in olives [17] or of discriminating between and classifying virgin, pure and olive pomace oil [18], olive oils coming from different regions [19] or olive oils coming from two varieties and submitted to two different irrigation treatments [4]. Moreover, fluorescence spectroscopy has been proposed to monitor olive oil during storage in different conditions. In this sense, the influence of factors such as UV irradiation, sunlight exposure and temperature up to 80 °C were studied [20]. Mishra et al. [21] monitored oils from three different olive varieties exposed and not exposed to light and Lobo Prieto et al. [22] studied the changes in the quality parameters and the relationship with the EEMs of four different cultivars. Fluorescence spectroscopy was also used to monitor oxidation level of four edible vegetable oils during storage at 60 °C [23]. The characteristics of some types of containers were also shown to influence the quality parameters of olive oil, as explained above, and the influence of storage in clear and green glass bottles exposed to light and in darkness was monitored using fluorescence spectroscopy [24]. However, the use of other different commonly used containers has not been studied to date by fluorescence. The aim of this study was to monitor the evolution of the fluorescence fingerprint of the virgin olive oil, during storage at room temperature, in four common different containers. At the same time, excitation emission fluorescence matrices in combination with chemometric algorithms was proposed as an alternative to conventional methods to determine the physicochemical and quality parameters of the olive oil.

2. Results and Discussion

2.1. *Influence of Packaging on the Quality Parameters during Storage at Room Temperature*

Quality indexes were performed for the EVOO samples stored in different containers for 40 weeks (Figure 1). An increase in peroxide index during the first eleven weeks of the study can be observed, followed by a period with values practically constant from week 15 to 29. Around week 34 a slight increase is detected, and a drastic decrease at week 40. In Figure 2A, two clearly different groups can be identified. In the first one, with crystal containers, it can be detected that independently of the crystal nature (TC and

OC), the values of peroxide indexes are higher along the studied period, surpassing the EVOO legal limit at week 34. The second is composed of non-crystal containers (Pl and C) and in this, the values of the peroxide index are lower, which indicates that samples conserved in these containers were less affected by the oxidation. This observation can be explained by how daylight strikes in the sample through the container and how the container protects the oil matrix from degradation. Samples in TC containers were the ones which presented the higher peroxide index, finding a maximum of 21.1 $mEqO_2 \cdot kg^{-1}$. The transparency of the material and its lack of light covering cause different light frequencies to pass through the material, affecting the EVOO, and promoting oxidation and matrix degradation. Similar results were found by Dabbou et al. [24]. However, OC receptacles slightly shield the samples from light irradiation, not surpassing the extra virgin category threshold in any of the studies. With respect to plastic and can vessels, they were able to filter light with such efficiency that the peroxide index did not become higher than 12.0 $mEqO_2 \cdot kg^{-1}$. Other researchers observed maximum values of the peroxide index from month 9 [25]. In addition, Alvarruiz et al. [26] underlined that long-time storage of EVOO increases its peroxide concentration, observing a significant drop after its maximum as in the present research.

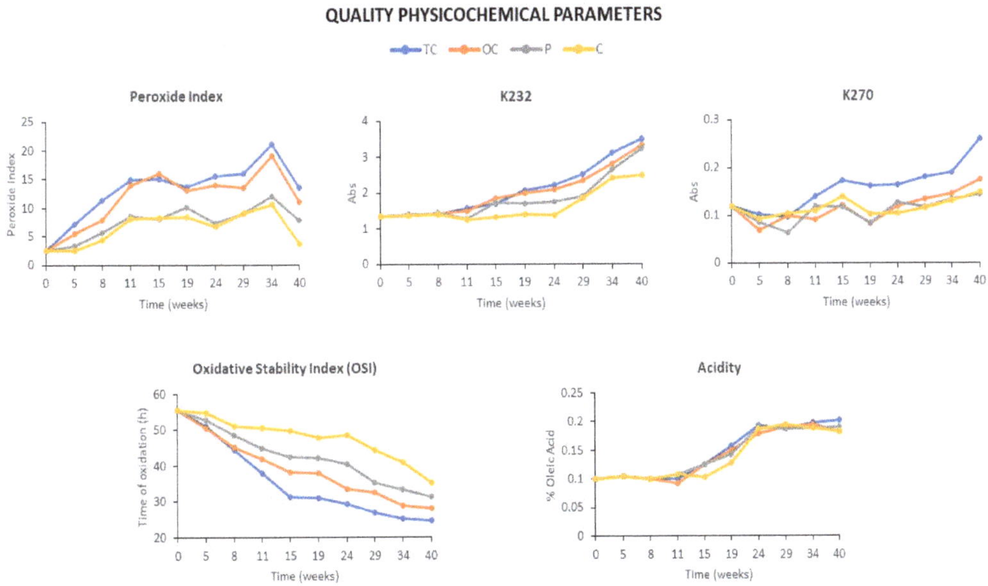

Figure 1. Evolution of quality parameters of olive oil with the storage time in different containers.

Some types of plastic do not correctly protect olive oil from oxidation, letting light and oxygen pass through the material and triggering the oxidation mechanism. It was observed that the peroxide index increases during a 12 month storage period [25,27]. Significant differences were also found by Sanmartin et al. [5] in containers like tinplate, with an increase of peroxide values of 77% after the storage period.

Regarding acidity, initially the percentage of acidity is 0.1% and remains practically constant during the first 11 weeks (Figure 1). From week 15 to 24, an increase up to 0.2% is observed in all packages, keeping this value practically constant until week 40. From week 15 to 24, olive samples in C containers present the lowest values of % oleic acid. The opaque character of metallic cans prevents light penetration through the material and its corresponding effect in EVOO. From week 24 to the end of the study, there are no significant differences between the containers used until week 40. It can be said that samples conserved in TC containers can be differentiated from the rest and also they presented the highest acidity. However, all the containers studied to store the EVOO are effective so that the oil remains in the extra virgin olive oil category (EU Regulation 2568/1991 modified). Previous researchers have found the same trend [26,27].

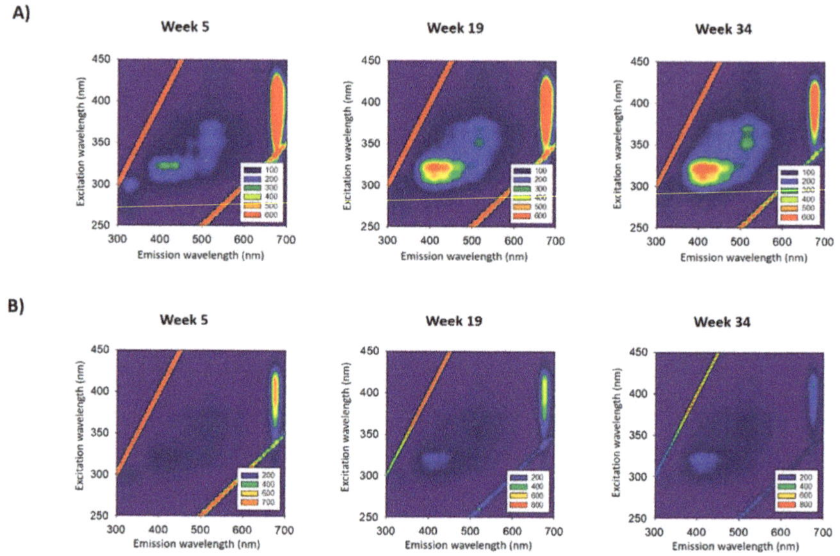

Figure 2. Contour plot of the EEMs of the samples stored in transparent crystal at three different storage times. (**A**) 700 V and (**B**) 630 V.

Data reveal that K_{232}, which is the corresponding parameter to evaluate primary oil oxidation, shows a similar behavior in terms of the percentage of acidity (Figure 2). It remains constant during the first 11 weeks in all the containers studied and increases progressively from week 11 to 40 (Figure 1). From week 19 onwards, differences between containers started to become apparent, reaching a maximum value of 3.5 for oil samples conserved in TC containers. Oil samples in can containers present the lowest values with regards to this parameter in the study, reaching values below the set limit ($\leq$2.5) on week 40. This is an interesting observation because these containers are able to increase the EVOO shelf-life by at least 2.5 months. Results concerning K_{270} did not show clear behavior until week 19. From that week on, K_{270} increased, reaching its highest value in week 40 (Figure 1). EVOO resisted secondary oxidation thanks to the container's light tolerance, with Pl and C being the receptacles that better protected olive oil samples. Olive oils in TC containers show the highest values, exceeding the set limit for the EVOO category ($\leq$0.22). In PT and C packaging, the evolution of this parameter is softer, showing a slight increase up to week 40 with maximum values of 0.15. Similar results were found in other studies [26,27]. In general, glass containers are preferable to plastic vessels for EVOO packaging in order to avoid oxygen permeation [25].

Oxidative stability index (OSI) data showed that, for all packaging, a decrease is observed in the study (Figure 1). During the first 15 weeks of the trial, a considerable decrease in this parameter was observed, from 59 to 50 h for C containers and from 59 to 30 h for TC containers. From that point on, the decrease slows down until week 40. The least significant changes were found with type C containers with a reduction in the OSI of 36%. However, the most significant reduction—of 55.5%—was found with TC containers. Alvarruiz et al. [26] also found a stability loss in the 'Picual' variety of 38% after a long-time storage. A similar tendency was indicated by Iqdiam et al. [27] in different cultivars and oxygen concentrations.

As was expected, oil samples in TC containers were the samples that suffered the deepest oxidation process. In contrast, the C container was the most effective in terms of preserving chemical characteristics to prevent EVOO oxidation due to degradation from exposure to light. This could be justified due to the phenolic compound content. Metallic materials did not let light affect the olive oil samples, preventing phenol degradation and, therefore, maintaining antioxidant resistance. In previous studies, other researchers confirmed the hypothesis that olive oil stability is primarily related to the level of unsaturated fatty acids and antioxidant molecules such as phenols that have a linear correlation with OSI in EVOO [25,27–29].

2.2. Fluorescence Monitoring

2.2.1. EEM Description

In order to obtain the fluorescence fingerprints from the different oil samples packed in different containers, EEMs were registered at the same time that the quality parameters were determined. The conditions employed to collect the EEMs are described in Section 3.6. Figure 2 shows the EEMs from EVOO samples packaged in transparent crystal containers at three different storage times and recorded at two different voltages (630 and 700 V) of the photomultiplier tube. In general terms, the EEMs exhibited physiognomy similar to previous reports [4,22].

In Figure 2, four different fluorescence regions can be observed. In accordance with the literature, the emission spectral region between 320 and 350 nm (excitation between 280 and 310 nm) corresponds to polyphenols and tocopherols present in the olive oil [17,22]. As can be seen, the fluorescence of this region decreases with the storage time, and this behavior was similar for all tested containers. As stated in the bibliography, the fluorescence of this area decreases with storage time due to the action of light, which has been corroborated in the different containers used.

The opposite occurs with the other two more representative spectral regions, one of them with an emission range of 380–460 nm and an excitation range between 320 and 340 nm, and the other with an emission range of 480–550 nm and an excitation range of 350–390 nm, whose fluorescence intensities increase with storage time. These spectral regions have been associated with degradation processes [30].

Moreover, these regions have been related to the primary and secondary oxidation products, respectively. These primary and secondary oxidation products often result from the auto-oxidation and photo-oxidation of oils occurring during the storage, in which triplet oxygen ($^3O_2$) and singlet oxygen ($^1O_2$) react with the oil, respectively [21]. Although several factors can be involved in the formation of these oxidation products, in this study, and given that the only variable is the type of container, our goal is to relate the fluorescence fingerprint with these variables. Finally, the fourth spectral region with emission wavelengths higher than 600 nm, corresponding to chlorophylls and their derivatives, has been widely described in the literature [20,22]. As can be seen in Figure 2, the fluorescence decreases with storage time, as occurs in the first region.

2.2.2. Application of PARAFAC to EEMs in the Different Containers

Since the evolution of the fluorescence of the regions corresponding to polyphenols and chlorophylls has been extensively studied [20,22,24], we have focused the study on the spectral region of the oxidation products, with the objective of figuring out the evolution of the fluorescence fingerprint in this region when different containers for olive oil samples are used. Therefore, the spectral region with excitation wavelengths between 295 and 375 nm and emission wavelengths between 384 and 580 nm was used in the study. To resolve the fluorescence profiles, multivariate exploratory algorithms, such as PARAFAC, were used and two different analyses were carried out, one with all the samples, independently of the container, and the second with the samples of each container.

In the first study, the EEMs were arranged in a three-dimensional structure with dimensions of $17 \times 99 \times 48$ (excitation $\times$ emission $\times$ samples) and then decomposed with

PARAFAC. PARAFAC was performed with one, two, three and four components in orderto select the optimal number. In all the cases, non-negative constraints were applied to all modes, given that all concentration and spectral values are always positive.

Two components were selected as the optimal number, taking into account the core consistency diagnostic (CORCONDIA) [31], the residual analysis [32] and the physiognomy of the loading.

Figure 3 shows the excitation and emission PARAFAC loadings of the two principal components. The excitation profile of the first component shows two maxima at 350 and 370 nm, and the emission profile presents a wide band with a maximum at 525 nm. The presence of fluorescent components with these wavelengths has already been previously described [33–36]. The assignment of this component has been widely discussed by several authors, ruling out that it can be assigned to vitamin E. The most likely assignment is the one proposed by Sikorska et al. [24], who consider that the fluorescent compounds responsible for this signal are formed during the oil storage. In addition, as indicated by Botosoa and Karoui [23], the width of its emission profile and its physiognomy, which is similar to that of a large number of compounds with short excitation wavelengths, indicate that this component corresponds to more than one analyte.

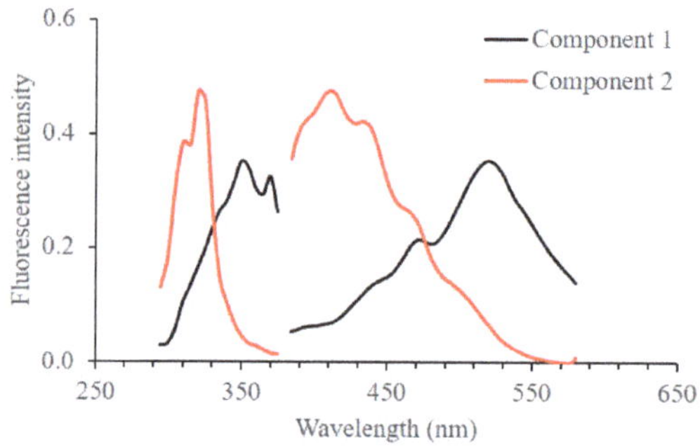

Figure 3. Excitation and emission PARAFAC loadings obtained with the complete set of samples.

With respect to the second component, the excitation profile presents a maximum at 320 nm and a shoulder at 310 nm, and the emission profile has a maximum placed at 414 nm and two shoulders at 434 and 470 nm. This component has also been found in previous studies and it has been ascribed to primary and secondary oxidation products [4,22,33,37]. The same PARAFAC decomposition was performed independently with the samples of each container (12 samples for each type), obtaining the same number of components and similar loading profiles for all of them as when the complete data set was used.

2.2.3. Fluorescence Evolution with the Storage Time in Different Containers

With the aim of studying the evolution of the fluorescence fingerprint of the EVOO in each container type, the evolution of the PARAFAC scores of the two principal components in the spectral region selected were studied. Score values for both components over the 40 weeks for each container are plotted against each other in order to distinguish among the container types (Figure 4).

As can be seen, the score values for the samples in CT are very different from the samples stored in CO, Pl and C. Those samples present higher values with regard to the first and second component than the other containers and they can be easily distinguished. This suggests that the oxidation in those samples was faster than in the other samples due to the photo-oxidation, as was described above. Moreover, it can also be observed that the CO samples stored longer can also be distinguished from samples packed in CO and C containers because they have higher scores with regard to components 1 and 2. However, samples in Pl and C containers are not different if we consider these two components.

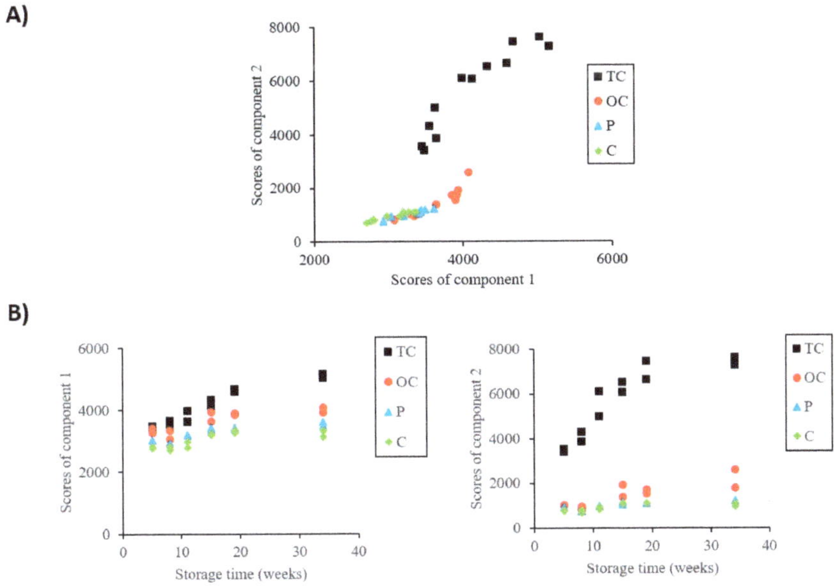

Figure 4. (**A**) PARAFAC scores of all the samples in different containers throughout the entire storage time. (**B**) Evolution of scores 1 and 2 across storage time.

Moreover, it can be seen that all the containers show a sequential shift in the samples measured every 4 weeks along the component 1 and 2 axis. For a better visualization, the PARAFAC score values for the first and second components were plotted as a function of the storage time (Figure 4). The same tendency is observed for both components and all the containers. Score values increased as storage progressed until week 20, and, after that, they remained almost constant. It can also be appreciated that CT samples are those which present higher values for both components, followed by CO, Pl and C containers. CT samples present much higher values for component 2 scores, suggesting that the oxidation that has occurred in these samples is much greater than in the other containers. These results are in agreement with those obtained in the quality parameters, which were described above, and verify the ability of fluorescence spectroscopy to monitor the changes in virgin olive oil during storage.

2.3. Quantification of Quality Parameters by Using the EEMs

After the qualitative analysis of the evolution of the fluorescence of the olive oil across storage time in different containers, the relationship between EEMs and the quality parameters was evaluated. The physico-chemical data indicated the presence of oxidation products, so it is logical to investigate whether it is possible to quantify them by means of the fluorescent region selected in the previous studies.

U-PLS regression was selected to develop regression models between the quality parameters and the EEMs. Samples were divided into two data sets: the calibration set, with 70% of the samples used to optimize and build the model, and the validation set, with the remaining 30% of the samples, that will be used to validate the previous model. First, it was necessary to select the optimum number of components. This was achieved by using the calibration set and the Haaland and Thomas criterion [38]. The optimal number of components is given when the PRESS value is statistically not different to the minimum PRESS value. Four latent variables were selected in all the models. After a model was built, it was validated using the other samples. The results obtained for the different quality parameters (peroxide index, K_{232}, K_{270} and OSI) in the cross-validation and the validation model are summarised in Table 1. The correlation coefficients show that the results obtained with U-PLS and the conventional methods are highly correlated for peroxide index, K_{232} and OSI (r^2 higher than 0.90 in cross-validation and validation). For K_{270}, good results were also found with a correlation coefficient higher than 0.80. The statistical parameters were evaluated through the root mean square error of cross-validation (RMSECV) and prediction (RMSEP), and the relative error of prediction (% REP). These values were considered low for peroxide index, K_{232}, K_{270} and OSI.

Table 1. Statistical parameters with the U-PLS model for the quantification of the quality parameters.

		Peroxide Index	K_{232}	K_{270}	Oxidative Stability Index (OSI)
Cross-validation	r^2_{CV}	0.91	0.90	0.84	0.92
	RMSECV	1.94 $mEqO_2 \cdot kg^{-1}$	0.22	0.017	3.21 h
	REP (%)	19.07	12.94	14.92	7.51
Validation	r^2_V	0.94	0.94	0.86	0.90
	RMSEP	1.53 $mEqO_2 \cdot kg^{-1}$	0.16	0.019	3.60
	REP (%)	14.98	9.22	12.25	8.12

The obtained results prove the accuracy and precision of the developed model for the physico-chemical parameters studied, and suggest that the EEMs combined with U-PLS are a good method to determine these parameters, avoiding the traditional methods.

3. Materials and Methods

3.1. Chemical Reagents

For the analysis of the physico-chemical regulated parameters, ethanol, diethyl ether, sodium thiosulphate, potassium iodine, phenolphthalein, starch, acetic acid, chloroform and cyclohexane, all of analytical grade, were purchased from Fisher Scientific (Fisher Scientific, Waltham, Massachusetts, MO, USA).

3.2. Raw Material

The study was carried out with olive oils obtained from olives of 'Picual' variety cultivated as an experimental olive variety (*Olea europaea* L.) located within the limits of the olive-growing 'Tierra de Barros' area, in the southwest of Badajoz (Spain). The olive orchard was composed of fifteen-year-old olive trees (plantation frame 6 × 7 m^2). The orchard was managed without irrigation, controlling the weeds with post-emergence herbicides. Olives were collected in the morning at the end of October at the veraison stage of maturation, and they were immediately transported to the mill to avoid compositional changes. The oil was extracted within 24 h.

3.3. EVOO Samples

The EVOO was obtained from a local industrial oil mill (Pieralisi, Italy) in Badajoz (Spain). Olives were crushed with a hammer mill and malaxation was carried out at 28 °C for 40 min. A two-phase decanter was used to extract the EVOO. The filtration procedure was carried out immediately after oil extraction. The oil obtained was stored in a stainless-steel silo before being packaged for the experimental tests. After 60 days of storage, the EVOO was placed into four different containers of 250 mL with the same headspace in triplicate: (i) transparent crystal (TC); (ii) opaque crystal (OC); (iii) opaque plastic (Pl); and (iv) metallic can (C). The containers were stored at 20 °C in the laboratory for 10 months. Figure 5 shows the different containers.

Figure 5. Different containers employed in the study.

3.4. Determination of Oil Quality Indexes

At the beginning of the storage period and every 2–3 weeks, measurements of the oil samples stored in each container were taken in order to analyze the quality. To determine the general quality index of the virgin olive oils, free fatty acids, peroxide index, and extinction coefficients (K_{232} and K_{270}) of EVOO samples were determined following the methods described in Regulations EEC 2568/91 and its subsequent amendments. All parameters were determined in duplicate for all samples.

The free acidity was determined by titration of oil dissolved in ethanol-ethyl ether 96° (1:1 v/v) with an ethanolic solution of potassium hydroxide 0.1 N, using phenolphthalein as an indicator. The results were expressed as g of oleic acid per 100 g of oil. The peroxide index was determined diluting the oil samples in acetic acid-chloroform, 3:2 v/v. The dissolved oil was mixed with a solution of acetic acid and potassium iodide in the dark and the released iodine was titrated with a solution of sodium thiosulfate using starch as an indicator. The results were expressed as milliequivalents of active O_2 per kg of oil. For the analysis of the extinction coefficients, K_{270} and K_{232}, the oil was mixed with cyclohexane (1%) and the mixture was introduced in 1 cm optical path quartz cell. After, a UV spectrophotometer (Hewlett-Packard, HP 8452 A) at 270 and 232 nm was used. Finally, the oxidative stability index (OSI) was evaluated in Rancimat 679 equipment (Metrohn Co., Basel, Switzerland). Olive oil (3 g) was placed in a standard glass tube which was heated to 120 °C and had an air flow of 10 L/h. The results were expressed as induction period per hour.

3.5. Statistical Analysis

The results were analyzed with variance analysis (ANOVA) and Tukey's multiple range test. The adequacy of the model was evaluated through a standardized remainder study to check the normality of the data and the homogeneity of the variances. Statistical significance was accepted at a level of $p < 0.05$. For ANOVA, SPSS 18.0 software (SPSS Inc., Chicago, IL, USA) was used. The results are expressed as mean values and the standard deviation (SD) was calculated.

3.6. Fluorescence Measurements

Fluorescence excitation emission matrices (EEMs) were obtained on a Cary Eclipse Varian Fluorescence Spectrophotometer, equipped with two Czerny–Turner monochromators, a xenon light source and a photomultiplier tube as detector [17,39]. The Cary Eclipse software 1.2 was used for data acquisition. A 1.0 cm quartz cell was used. Measurements were made with a variable-angle front-face accessory. The angle of incidence, defined as the angle between the excitation beam and the axis perpendicular to the cell surface, was set to 35°. The slits of excitation and emission monochromators were set to 5 nm. The EEMs were registered as a set of emission spectra over a range of excitation wavelengths. The excitation wavelengths ranged from 250 to 450 nm, each with 5 nm increments. At each excitation wavelength, the emission spectra were recorded from 200 to 700 nm, at

2 nm intervals. Each sample was registered twice with two different photomultiplier tube sensitivities: 630 and 700 V. Data were saved in ASCII format and transferred to a PC for the chemometric analysis described in the next section.

3.7. *Multivariate Analysis*

EEM data were exported to ASCII code and processed using Matlab software (Matlab R2016b). The graphical interface MVC2 (http://www.iquirconicet.gov.ar/descargas/mvc2.rar; accessed on 5 May 2021) was used for PARAFAC [32] and U-PLS [40] calculations.

4. Conclusions

The present study aimed to investigate the evolution of EVOO stored in different containers at room temperature across 40 weeks, and the use of fluorescence spectroscopy coupled with chemometric algorithms to monitor the samples. Physicochemical analysis showed that olive oil stored in transparent crystal containers suffered the worst deterioration among the samples tested due to the transparency of the containers, followed by opaque crystal containers. In this sense, the peroxide index increased during the first weeks, and the acidity remained constant but increased during the last months of the study, while data regarding the oxidative stability index showed a decrease across the study. This reaffirms that photo-oxidation is much faster than auto-oxidation. In places exposed to light, cans may be the most suitable containers to protect olive oil samples against photo-oxidation, maintaining the EVOO category for as long as possible. In that sense, the use of non-transparent containers should be recommended to keep photosensitive samples intact. The classic analysis together with the use of fluorescence gives us an accurate, complete and sensitive monitoring of the evolution of the components of olive oil in a domestic environment. The advantage of using fluorescence signals as fingerprint of samples is their high selectivity and sensitivity. A three-dimensional map of samples offers a huge amount of information concerning complex samples, as foods are. The different spectral regions obtained with the fluorescence analysis of the oils during storage were associated with degradation processes, overall, with primary and secondary oxidation products. On the other hand, EEMs combined with U-PLS proved to be an excellent tool that allowed the establishment of regression models between quality parameters and EEMs. This proves the accuracy and precision of the developed model for the physico-chemical parameters studied, and suggests that the EEMs combined with U-PLS are a good method to determine these parameters, avoiding the traditional one. For all of that, the combination of the determination of the quality parameters along with analytical techniques such as non-destructive fluorescence results in an interesting methodology for studying olive oil deterioration over time, accelerating the process with minimal cost.

Author Contributions: Conceptualization, A.F., E.M.-T. and D.M.-V.; methodology, A.F. and E.M.-T.; software, E.M.-T. and I.D.-M.; validation, E.M.-T. and I.D.-M.; formal analysis, A.F., E.M.-T. and D.M.-V.; investigation, A.F., E.M.-T. and D.M.-V.; resources, D.M.-V. and I.D.-M.; data curation, A.F. and E.M.-T.; writing—original draft preparation, A.F., E.M.-T. and D.M.-V.; writing—review and editing, A.F., E.M.-T. and D.M.-V.; project administration, D.M.-V. and I.D.-M.; funding acquisition, D.M.-V. and I.D.-M. All authors have read and agreed to the published version of the manuscript.

Funding: Financial support was provided by the Ministerio de Ciencia e Innovación de España (Project PID2020-112996GB-I00 funded by MCIN/AEI/10.13039/501100011033) and Junta de Extremadura (Ayuda a Grupos GR21048 and Project IB20016) co-financed by European Funds for Regional Development.

Institutional Review Board Statement: Not applicable.

Informed Consent Statement: Not applicable.

Data Availability Statement: The authors confirm that the data supporting the findings of this study are available within the article and the raw data that support the findings are available from the corresponding author, upon reasonable request.

Conflicts of Interest: The authors declare no conflict of interest.

References

1. Difonzo, G.; Russo, A.; Trani, A.; Paradiso, V.M.; Ranieri, M.; Pasqualone, A.; Summo, C.; Tamma, G.; Silletti, R.; Caponio, F. Green extracts from Coratina olive cultivar leaves: Antioxidant characterization and biological activity. *J. Funct. Foods* **2017**, *31*, 63–70. [CrossRef]
2. Franco, M.N.; Galeano-Díaz, T.; López, O.; Fernández-Bolaños, J.G.; Sánchez, J.; De Miguel, C.; Gil, M.V.; Martín-Vertedor, D. Phenolic compounds and antioxidant capacity of virgin olive oil. *Food Chem.* **2014**, *163*, 289–298. [CrossRef] [PubMed]
3. Terzuoli, E.; Giachetti, A.; Ziche, M.; Donnini, S. Hydroxytyrosol, a product from olive oil, reduces colon cancer growth by enhancing epidermal growth factor receptor degradation. *Mol. Nutr. Food Res.* **2016**, *60*, 519–529. [CrossRef] [PubMed]
4. Martín-Tornero, E.; Fernández, A.; Pérez-Rodriguez, J.M.; Durán-Merás, I.; Prieto, M.H.; Martín-Vertedor, D. Non-destructive fluorescence spectroscopy as a tool for discriminating between olive oils according to agronomic practices and for assessing quality parameters. *Food Anal. Methods* **2022**, *15*, 253–265. [CrossRef]
5. Sanmartin, C.; Venturi, F.; Sgherri, C.; Nari, A.; Macaluso, M.; Flamini, G.; Quartacci, M.F.; Taglieri, I.; Andrich, G.; Zinnai, A. The effects of packaging and storage temperature on the shelf-life of extra virgin olive oil. *Heliyon* **2018**, *4*, e00888. [CrossRef] [PubMed]
6. Sena-Moreno, E.; Cabrera-Bañegil, M.; Pérez-Rodríguez, J.M.; De Miguel, C.; Prieto, M.H.; Martín-Vertedor, D. Influence of Water Deficit in Bioactive Compounds of Olive Paste and Oil Content. *JAOCS J. Am. Oil Chem. Soc.* **2018**, *95*, 349–359. [CrossRef]
7. Lanza, B.; Di Serio, M.G.; Giansante, L.; Di Loreto, G.; Di Giacinto, L. Effect of shelf conditions on the phenolic fraction and oxidation indices of monovarietal extra virgin olive oil from cv. "Taggiasca". *Acta Aliment.* **2015**, *44*, 585–592. [CrossRef]
8. Cicerale, S.; Conlan, X.A.; Barnett, N.W.; Keast, R.S.J. Storage of extra virgin olive oil and its effect on the biological activity and concentration of oleocanthal. *Food Res. Int.* **2013**, *50*, 597–602. [CrossRef]
9. Piscopo, A.; Poiana, M. Packaging and storage of olive oil. In *Olive Germplasm—The Olive Cultivation, Table Olive and Olive Oil Industry in Italy*; Mazzalupo, I., Ed.; IntechOpen: London, UK, 2012; pp. 201–222.
10. Stefanoudaki, E.; Williams, M.; Harwood, J. Changes in virgin olive oil characteristics during different storage conditions. *Eur. J. Lipid Sci. Technol.* **2010**, *112*, 906–914. [CrossRef]
11. Frankel, E.N. Chemistry of extra virgin olive oil: Adulteration, oxidative stability and antioxidants. *J. Agric. Food Chem.* **2010**, *58*, 5991–6006. [CrossRef] [PubMed]
12. Pristouri, G.; Badeka, A.; Kontominas, M.G. Effect of packaging material headspace, oxygen and light transmission, temperature and storage time on quality characteristics of extra virgin olive oil. *Food Control* **2010**, *21*, 412–418. [CrossRef]
13. Morales, M.T.; Przybylski, R. Olive oil oxidation. In *Handbook of Olive Oil*; Aparicio, R., Harwood, J., Eds.; Springer: Berlin/Heidelberg, Germany, 2013; p. 479, ISBN 9781461477761.
14. Gargouri, B.; Zribi, A.; Bouaziz, M. Effect of containers on the quality of Chemlali olive oil during storage. *J. Food Sci. Technol.* **2015**, *52*, 1948–1959. [CrossRef]
15. Cecchi, T.; Passamonti, P.; Cecchi, P. Study of the quality of extra virgin olive oil stored in PET bottles with or without an oxygen scavenger. *Food Chem.* **2010**, *120*, 730–735. [CrossRef]
16. Méndez, A.I.; Falqué, E. Effect of storage time and container type on the quality of extra-virgin olive oil. *Food Control* **2007**, *18*, 521–529. [CrossRef]
17. Cabrera-Bañegil, M.; Martín-Vertedor, D.; Boselli, E.; Durán-Merás, I. Control of olive cultivar irrigation by front-face fluorescence excitation-emission matrices in combination with PARAFAC. *J. Food Compos. Anal.* **2018**, *69*, 189–196. [CrossRef]
18. Guimet, F.; Boqué, R.; Ferré, J. Cluster analysis applied to the exploratory analysis of commercial Spanish olive oils by means of excitation-emission fluorescence spectroscopy. *J. Agric. Food Chem.* **2004**, *52*, 6673–6679. [CrossRef] [PubMed]
19. Guimet, F.; Boqué, R.; Ferré, J. Study of oils from the protected denomination of origin "Siurana" using excitation-emission fluorescence spectroscopy and three-way methods of analysis. *Grasas Aceites* **2005**, *56*, 292–297. [CrossRef]
20. Manzano, J.D.; de la Peña, A.M.; Merás, I.D. Front-Face Fluorescence Combined with Second-Order Multiway Classification, Based on Polyphenol and Chlorophyll Compounds, for Virgin Olive Oil Monitoring Under Different Photo- and Thermal-Oxidation Procedures. *Food Anal. Methods* **2019**, *12*, 1399–1411. [CrossRef]
21. Mishra, P.; Lleó, L.; Cuadrado, T.; Ruiz-Altisent, M.; Hernández-Sánchez, N. Monitoring oxidation changes in commercial extra virgin olive oils with fluorescence spectroscopy-based prototype. *Eur. Food Res. Technol.* **2018**, *244*, 565–575. [CrossRef]
22. Lobo-Prieto, A.; Tena, N.; Aparicio-Ruiz, R.; García-González, D.L.; Sikorska, E. Monitoring virgin olive oil shelf-life by fluorescence spectroscopy and sensory characteristics: A multidimensional study carried out under simulated market conditions. *Foods* **2020**, *9*, 1846–1866. [CrossRef]
23. Botosoa, E.P.; Karoui, R. 3D front face fluorescence spectroscopy as a tool for monitoring the oxidation level of edible vegetable oil during storage at 60 °C. *LWT* **2022**, *154*, 112659. [CrossRef]
24. Sikorska, E.; Khmelinskii, I.V.; Sikorski, M.; Caponio, F.; Bilancia, M.T.; Pasqualone, A.; Gomes, T. Fluorescence spectroscopy in monitoring of extra virgin olive oil during storage. *Int. J. Food Sci. Technol.* **2008**, *43*, 52–61. [CrossRef]
25. Dabbou, S.; Gharbi, I.; Dabbou, S.; Brahmi, F.; Nakbi, A.; Hammami, M. Impact of packaging material and storage time on olive oil quality. *Afr. J. Biotechnol.* **2011**, *10*, 16937–16947. [CrossRef]

26. Alvarruiz, A.; Pardo, J.E.; Copete, M.E.; de Miguel, C.; Rabadán, A.; López, E.; Álvarez-Ortí, M. Evolution of virgin olive oil during long-term storage. *J. Oleo Sci.* **2020**, *69*, 809–814. [CrossRef] [PubMed]
27. Iqdiam, B.M.; Welt, B.A.; Goodrich-Schneider, R.; Sims, C.A.; Baker, G.L.; Marshall, M.R. Influence of headspace oxygen on quality and shelf life of extra virgin olive oil during storage. *Food Packag. Shelf Life* **2020**, *23*, 100433. [CrossRef]
28. Iqdiam, B.M.; Abuagela, M.O.; Marshall, S.M.; Yagiz, Y.; Goodrich-Schneider, R.; Baker, G.L.; Welt, B.A.; Marshall, M.R. Combining high power ultrasound pre-treatment with malaxation oxygen control to improve quantity and quality of extra virgin olive oil. *J. Food Eng.* **2019**, *244*, 1–10. [CrossRef]
29. Baldioli, M.; Servili, M.; Perretti, G.; Montedoro, G.F. Antioxidant activity of tocopherols and phenolic compounds of virgin olive oil. *JAOCS J. Am. Oil Chem. Soc.* **1996**, *73*, 1589–1593. [CrossRef]
30. Dupuy, N.; Le Dréau, Y.; Ollivier, D.; Artaud, J.; Pinatel, C.; Kister, J. Origin of French virgin olive oil registered designation of origins predicted by chemometric analysis of synchronous excitation-emission fluorescence spectra. *J. Agric. Food Chem.* **2005**, *53*, 9361–9368. [CrossRef] [PubMed]
31. Bro, R.; Kiers, H.A.L. A new efficient method for determining the number of components in PARAFAC models. *J. Chemom.* **2003**, *17*, 274–286. [CrossRef]
32. Bro, R. PARAFAC. Tutorial and applications. *Chemom. Intell. Lab. Syst.* **1997**, *38*, 149–171. [CrossRef]
33. Guimet, F.; Ferré, J.; Boqué, R.; Rius, F.X. Application of unfold principal component analysis and parallel factor analysis to the exploratory analysis of olive oils by means of excitation-emission matrix fluorescence spectroscopy. *Anal. Chim. Acta* **2004**, *515*, 75–85. [CrossRef]
34. Kyriakidis, N.B.; Skarkalis, P. Fluorescence spectra measurement of olive oil and other vegetable oils. *J. AOAC Int.* **2000**, *83*, 1435–1439. [CrossRef] [PubMed]
35. Guzmán, E.; Baeten, V.; Pierna, J.A.F.; García-Mesa, J.A. Evaluation of the overall quality of olive oil using fluorescence spectroscopy. *Food Chem.* **2015**, *173*, 927–934. [CrossRef] [PubMed]
36. Tena, N.; Aparicio, R.; García-González, D.L. Chemical changes of thermoxidized virgin olive oil determined by excitation-emission fluorescence spectroscopy (EEFS). *Food Res. Int.* **2012**, *45*, 103–108. [CrossRef]
37. Guimet, F.; Ferré, J.; Boqué, R.; Vidal, M.; Garcia, J. Excitation-emission fluorescence spectroscopy combined with three-way methods of analysis as a complementary technique for olive oil characterization. *J. Agric. Food Chem.* **2005**, *53*, 9319–9328. [CrossRef]
38. Haaland, D.M.; Thomas, E.V. Partial least-squares methods for spectral analyses. 1. Relation to other quantitative calibration methods and the extraction of qualitative information. *Anal. Chem.* **1988**, *60*, 1193–1202. [CrossRef]
39. Cabrera-Bañegil, M.; Martín-Vertedor, D.; Lodolini, E.M.; Durán-Merás, I. Fluorescence Study of Four Olive Varieties Paste According to Sampling Dates and the Control in the Elaboration of Table Olives of "Ascolana tenera". *Food Anal. Methods* **2021**, *14*, 307–318. [CrossRef]
40. Wold, S.; Geladi, P.; Esbensen, K.; Öhman, J. Multi-way principal components-and PLS-analysis. *J. Chemom.* **1987**, *1*, 41–56. [CrossRef]

molecules MDPI

Article

The Effect of Nonthermal Pretreatment on the Drying Kinetics and Quality of Black Garlic

Klaudia Masztalerz [1], Tomasz Dróżdż [2], Paulina Nowicka [3], Aneta Wojdyło [3], Paweł Kiełbasa [2] and Krzysztof Lech [1,*]

1 Institute of Agricultural Engineering, The Faculty of Life Sciences and Technology, Wrocław University of Environmental and Life Sciences, Chełmońskiego 37, 51-630 Wroclaw, Poland
2 Faculty of Production and Power Engineering, University of Agriculture in Krakow, Balicka 116 b, 30-149 Krakow, Poland
3 Department of Fruit, Vegetable and Nutraceutical Plant Technology, Wrocław University of Environmental and Life Sciences, 50-375 Wroclaw, Poland
* Correspondence: krzysztof.lech@upwr.edu.pl; Tel.: +48-71-320-59-59

Abstract: Black garlic is obtained from regular garlic (*Allium sativum* L.) through the aging process and consequently gains many health-promoting properties, including antidiabetic and antioxidant. However, the material is still prone to microbiological deterioration and requires a long time to dry due to its properties. Therefore, this study aimed to investigate the effect of various drying methods on the quality of black garlic as well as determine the influence of selected nonthermal pretreatments on the drying kinetics and quality of black garlic, which is especially important in the case of the materials that are difficult to dry. The Weibull model was chosen to describe drying kinetics. Additionally, color, water activity together with antioxidant activity, phenolic compounds, and antidiabetic potential were determined. This study found that the application of a pulsed electric field (PEF), a constant electric field (CEF) as well as a magnetic field (MF) significantly reduced the time of drying (by 32, 40, and 24 min for a PEF, a CEF, and a MF, respectively, compared to combined drying without the pretreatment), and resulted in high antidiabetic potential. However, the highest content of phenolic compounds (1123.54 and 1125.36 mg/100 g dm for VMD125 and CD3h-VMD, respectively) and antioxidant capacity (ABTS = 6.05 and 5.06 mmol Trolox/100 g dm for VMD500 and CD6h-VMD, respectively) were reported for black garlic treated by vacuum-microwave drying and combined convective pre-drying followed by vacuum-microwave drying. Overall, the nonthermal pretreatment decreased the time of drying and showed very good efficiency in maintaining the antidiabetic potential of black garlic, especially in the case of the materials pretreated by a constant electric field (IC_{50} = 99 and 56 mg/mL, for α-amylase and α-glucosidase, respectively).

Keywords: antidiabetic potential; antioxidant capacity; constant electric field; drying; magnetic field; pretreatment; pulsed electric field; total phenolic content

Citation: Masztalerz, K.; Dróżdż, T.; Nowicka, P.; Wojdyło, A.; Kiełbasa, P.; Lech, K. The Effect of Nonthermal Pretreatment on the Drying Kinetics and Quality of Black Garlic. *Molecules* **2023**, *28*, 962. https://doi.org/10.3390/molecules28030962

Academic Editor: Smaoui Slim

Received: 19 December 2022
Revised: 13 January 2023
Accepted: 16 January 2023
Published: 18 January 2023

1. Introduction

Black garlic is produced during the aging of fresh garlic (*Allium sativum* L.). The process is usually performed for an extended period of time (up to several weeks) at increased temperature (60–90 °C) and high relative humidity (50–95%) [1]. Consequently, the material changes its physical and chemical properties, acquiring the typical dark color in the process. The main bioactive compound responsible for the properties of black garlic is S-allyl cysteine (SAC), which is formed as a result of the conversion of unstable alliin present in fresh garlic [2]. Moreover, the strong taste and smell of garlic changes due to the decreased amount of allicin, which is responsible for its strong off-flavor [3]. An increasing amount of Maillard reaction products in black garlic also influences the taste and aroma, which results in a typical sweet and sour taste that some describe as plum [4]. Black garlic can be characterized by many health-promoting properties, including anti-inflammatory [5,6], antidiabetic [7–9], anticarcinogenic [10,11] as well as the ability to reduce blood pressure [12,13]. These properties make black garlic a very attractive functional food ingredient that can be used in the development of various snacks or other food products, especially in Japan, China, Korea, and also in the USA, where it is gaining recognition [14]. Moreover, fresh garlic is already recognized and accepted in society and there is already a big market for garlic-derived products [1]. Hence, black garlic is among the fastest-growing health food [15]. Therefore, application of various treatments, including drying, can enable obtaining powders that could be used as spices or additives to functional food products.

Convective drying is among the most common drying methods. The process is based on the principle of water evaporation as a result of airflow through the material that absorbs the moisture from the surface. This is followed by the internal diffusion of moisture from the inside of the sample to the outer layer to enable water evaporation. Convective drying has previously been used in studies on garlic [16], Thai basil [17], hemp flowers [18], and kiwiberry [19]. On the other hand, vacuum and vacuum-microwave drying enable reaching a lower final moisture content in the final product. Moreover, vacuum-microwave drying can significantly accelerate the drying process due to volumetric heating occurring as a result of microwaves application, which is intensified by the pressure diffusion mechanism of the Darcy type, resulting from the pressure gradient between the center of the material and surrounding vacuum. These methods were previously used, among others, in studies on true lavender [20] and *Cassia alata* [21]. Drying is a very energy-intensive process, which is mainly due to the long processing times. This can be reduced by combining different drying methods, such as combined convective pre-drying to remove easily accessible unbound water followed by vacuum-microwave finishing drying aimed at reducing the drying time and improving the quality of the material. As a result, significant energy savings can be observed, such as in studies on garlic [16] as well as osmotic dehydration and drying of apples [22].

Another way of reducing process duration and improving the quality of dried products is the application of various pretreatment methods. The most common are thermal methods, including freezing, blanching, or osmotic dehydration. However, nonthermal methods are also gaining recognition due to their effect on the shortening of the drying duration and limiting microbiological contamination [23]. Among these methods is a pulsed electric field (PEF). The use of a PEF is a pretreatment method based on the application of a very short, high-intensity electric field that leads to a cell membrane disintegration [24]. Consequently, electroporation occurs and leads to the intensification of internal water diffusion and therefore a significant reduction in drying time as shown in recent studies on parsnips [25], onions [26], carrots [27] and red bell pepper [28]. Moreover, the application of a PEF as a pretreatment did not affect the nutritional value of apple juice but led to an inactivation of microorganisms extending the product shelf life [29] as well as improving the microbiological safety of the product. A PEF was also previously used before osmotic dehydration, where it showed a significant reduction in the time of osmotic dehydration of blueberries [30]. Another type of nonthermal pretreatment is a constant electric field (CEF). In this method, the material is placed between the electrodes, and a generated constant electric field interacts with the material changing its properties. A CEF was previously applied in studies on Camellia [31] and *Cannabis sativa* [32]. A magnetic field (MF) has been used to accelerate the germination and growth of various seeds, including sunflower seeds [33] as well as in the freezing of plant materials due to its positive effect on the formation of small ice crystals and enhanced freezing rate [34,35]. Moreover, this method was used during convective drying in order to change its properties and accelerate water removal during drying [36].

Application of various nonthermal pretreatment methods such as a pulsed electric field, a constant electric field, and a magnetic field before drying has previously been discussed in the literature, but never in terms of black garlic drying. Therefore, this study aims to determine the influence of process parameters and nonthermal pretreatments on the drying kinetics and quality of the final product, including changes in the antioxidant and antidiabetic potential of black garlic after treatment.

2. Results and Discussion

2.1. Physical Properties of Black Garlic

Table 1 shows the physical properties of black garlic after thermal and nonthermal treatments. In the course of nonthermal pretreatment, the moisture content of the material changed according to the used method, namely, after PEF + H_2O, the moisture content was equal to 1.53 kg/kg; during CEF + H_2O, *Mc* = 1.21 kg/kg; and for MF + H_2O, *Mc* = 1.51 kg/kg. On the other hand, while applying a constant electric field and a magnetic field without water, the samples lost some moisture during pretreatment, resulting in *Mc* = 0.49 and 0.50 kg/kg for a CEF and a MF, accordingly, which is significantly lower compared to fresh material (*Mc* = 0.66 kg/kg). This is due to the effect of nonthermal pretreatment, which, when performed with the material immersed in water, increased the *Mc* as a result of water absorption [37]. The studies by Rizvi Alam et al. [25] and Rahaman et al. [38] also showed that the addition of water when using a PEF, a CEF, and a MF is crucial to ensure the uniform distribution of an electric field in the chamber.

Table 1. Water activity (*aw*), moisture content (*Mc*), and color (CIE *L*a*b**, browning index—*BI*) of black garlic powders after nonthermal treatments and drying.

Method	*Mc* (%)	*aw* (-)	Color			
			*L**(D65)	*a**(D65)	*b**(D65)	*BI*
VMD 500W	8.16 ± 0.4 ^{ab*}	0.2172 ± 0.0054 ^{b}	35.35 ± 0.11 ^{f}	6.21 ± 0.28 ^{a}	8.72 ± 0.68 ^{a}	40.8 ± 3.2 ^{d}
VMD 250W	13.06 ± 0.63 ^{c}	0.229 ± 0.0048 ^{b}	38.01 ± 0.08 ^{d}	6.9 ± 0.11 ^{b}	12.2 ± 0.44 ^{f}	51.5 ± 2.0 ^{c}
VMD 125W	13.42 ± 0.65 ^{c}	0.2891 ± 0.0071 ^{a}	37.76 ± 0.17 ^{d}	7.71 ± 0.09 ^{h}	13.61 ± 0.28 ^{g}	59.0 ± 1.6 ^{b}
CD70-3h/125W	12.95 ± 0.63 ^{c}	0.2936 ± 0.003 ^{a}	38.89 ± 0.11 ^{e}	8.34 ± 0.09 ^{i}	14.8 ± 0.18 ^{c}	62.8 ± 1.1 ^{ab}
CD70-6h/125W	13.44 ± 0.65 ^{c}	0.2994 ± 0.0034 ^{a}	39.70 ± 0.30 ^{e}	9.06 ± 0.27 ^{j}	14.93 ± 0.37 ^{c}	63.1 ± 1.5 ^{a}
CD60-9h/VD60	8.89 ± 0.43 ^{a}	0.3957 ± 0.0058 ^{d}	29.08 ± 0.32 ^{b}	2.31 ± 0.1 ^{d}	2.7 ± 0.12 ^{e}	15.3 ± 0.6 ^{g}
CD70-9h/VD60	8.57 ± 0.41 ^{ab}	0.3479 ± 0.0054 ^{c}	29.16 ± 0.48 ^{b}	1.66 ± 0.14 ^{c}	1.84 ± 0.03 ^{d}	10.5 ± 0.5 ^{h}
PEF + H_2O/CD70-3h/125W	10.51 ± 0.51 ^{e}	0.4694 ± 0.0133 ^{g}	31.00 ± 0.52 ^{a}	4.65 ± 0.02 ^{g}	4.52 ± 0.1 ^{a}	26.3 ± 0.7 ^{e}
CEF + H_2O/CD70-3h/125W	8.89 ± 0.43 ^{a}	0.3891 ± 0.0079 ^{d}	31.23 ± 0.45 ^{a}	4.04 ± 0.04 ^{f}	4.53 ± 0.04 ^{a}	24.8 ± 0.6 ^{ef}
CEF/CD70-3h/125W	8.14 ± 0.39 ^{ab}	0.3691 ± 0.0044 ^{f}	30.92 ± 0.11 ^{a}	3.19 ± 0.05 ^{e}	4.26 ± 0.1 ^{a}	22.0 ± 0.6 ^{f}
MF + H_2O/CD70-3h/125W	7.26 ± 0.35 ^{bd}	0.3485 ± 0.0022 ^{c}	33.57 ± 0.18 ^{c}	5.9 ± 0.02 ^{a}	7.99 ± 0.04 ^{b}	39.6 ± 0.4 ^{d}
MF/CD70-3h/125W	6.41 ± 0.31 ^{d}	0.3211 ± 0.0054 ^{e}	34.28 ± 0.1 ^{c}	6.72 ± 0.03 ^{b}	8.57 ± 0.05 ^{b}	42.6 ± 0.3 ^{d}

* Values followed by the same letter, within the same column, were not significantly different ($p > 0.05$), according to Tukey's HSD test.

The changes in the moisture content as a result of nonthermal pretreatment did not negatively affect the final moisture content after drying (Table 1). When considering the final *Mc* of the material after drying, it can be seen that a higher power of magnetrons during VMD facilitated water removal and enabled obtaining among the lowest values of the *Mc*. Similar results were reported in this study on vacuum-microwave drying of garlic [39]. Samples treated by a PEF, a CEF, or a MF and then combined drying reached a lower moisture content in general. The nonthermal treatment changed the properties of the material and destroyed the cell structure, which led to higher water loss during the process. However, when considering the water activity, it can be seen that even though the pretreated samples exhibited a lower *Mc*, the water activity was higher compared to in non-pretreated samples. This can also be explained by the effect of the pretreatment. When a PEF is applied to the material, electroporation occurs and destroys the cell structure, which not only facilitates water removal but also releases water and chemical compounds from the material matrix [38,40,41]. Consequently, water is more accessible, which leads to increased water activity, even though the material has a lower moisture content than non-pretreated samples. This is in line with the findings presented by Nowacka et al. [24]. Nonetheless, the main factor responsible for microbiological safety is water activity, which in each variant is still below 0.6, making it relatively stable and safe as no proliferation occurs at these values of aw [42].

Color is among the most important characteristics from the consumer's point of view. Black garlic is characterized by a very intense dark brown or even black color as a result of the formation of melanoidins in the material during the aging of fresh garlic [15]. Color analysis showed that the darkest samples (with the lowest L^*) were the ones where vacuum drying was used as a finishing drying method (CD60-9h/VD60 and CD70-9h/VD60). Thus is due to the lowest temperature of the sample when this method was applied (below 60 °C and below 70 °C, respectively). While other drying methods were used, an increase in the L^* parameter could be reported which can be explained by the thermal effect in the material. As a result of drying and temperatures reached during the process, the black garlic samples turned brown instead of the very intense dark color of the fresh material, which could be noticed as increased L^* and *BI* parameters (Figure 1). Despite the highest temperature obtained when the material was dried by VMD500, the *BI* is the lowest among all VMD samples. This can be explained by the exposure time of the material to the high temperature which was the shortest when dried at 500 W (VMD500). Comparison of the samples pretreated by nonthermal methods and untreated samples dried by CD70-3h/125W showed that the pretreatment led to a decrease in L^* after drying which can be explained by the shorter time of vacuum-microwave finishing drying and consequently shorter time of thermal treatment. To the best of the authors' knowledge, there are no studies regarding the color changes during the processing of black garlic, which can be affected by thermal treatment as shown in this study. Therefore, future studies on this issue are needed.

(**a**) (**b**) (**c**)

Figure 1. Fresh black garlic sample (**a**), black garlic after drying (**b**), and black garlic powder (**c**).

2.2. Drying Kinetics

Figure 2 shows the drying kinetics of black garlic treated by convective pre-drying followed by vacuum finishing drying, vacuum-microwave drying, and combined method consisting of convective pre-drying and vacuum-microwave finishing drying. As can be seen, the higher temperature of hot air during convective drying resulted in faster water evaporation and lower MR after 540 min of drying [43]. Then, the application of vacuum drying enabled reaching a lower moisture content while limiting the negative effect of temperature and oxygen, which is typical during CD [44]. While considering the drying kinetics during vacuum-microwave drying, it can be seen that application of VMD reduced the time of drying by 91% in the case of VMD125 compared to CD-VD. Moreover, the higher power of magnetrons during drying resulted in more intense evaporation and further reduced the drying time from 128 at 125 W to 28 at 500 W. This is consistent with previous studies on sour cherries [45] and pears [46]. This can be explained by volumetric heating and temperatures generated during drying. As can be seen, the application of higher power increased the surface temperature of the material up to 140 °C, while drying at 125 W maintained temperatures below 100 °C. Similar behavior was reported by Figiel and Callin-Sanchez et al. [16,39]; however, the temperatures obtained were lower than reported here. This is due to the structure changes occurring in the aging of garlic. As a result, black garlic is much softer, with a gelatin-like texture that disrupts water evaporation and leads to heating up of the material during VMD [47].

According to Figure 2c, it can be seen that increasing the time of convective pre-drying from 3 to 6 h did not affect the drying time during vacuum-microwave finishing drying. Another 3 h of the CD only shortened the VMD time by 8 min. Therefore, prolonged CD resulted in only a little lower MR and this effect was only significant during the first two cycles of VMD. Afterwards, a similar MR was reached and maintained throughout the drying process, which shows that 3 h of convective pre-drying is enough and further prolonging the process does not bring any substantial gains. Different results were obtained in a study on garlic, where a longer time of convective pre-drying resulted in a significant reduction in vacuum-microwave finishing drying duration [16]. However, the proposed pre-treatment times were shorter than the ones presented in this study. Similarly, the study by Castillo-Girones et al. showed that the earlier the switch to vacuum-microwave drying the shorter the overall drying time [48]. This is due to the absorption of microwaves by the water dipoles that are located in the whole volume of the material. Consequently, a higher drying rate can be observed and more intense evaporation occurs compared to convective drying. However, the intense evaporation at the beginning of VMD can exceed the capacity of the vacuum pump [49]. Therefore, it is important to carefully select the process parameters and future research is needed to further optimize the combined drying parameters in black garlic drying.

Based on this, convective pre-drying at 70 °C for 3 h followed by vacuum-microwave finishing drying at 125 W was selected as an optimal method for black garlic drying in the first stage of the experiment. These parameters were then selected for the drying of the materials pretreated by nonthermal methods, i.e., a pulsed electric field, a constant electric field, and a magnetic field.

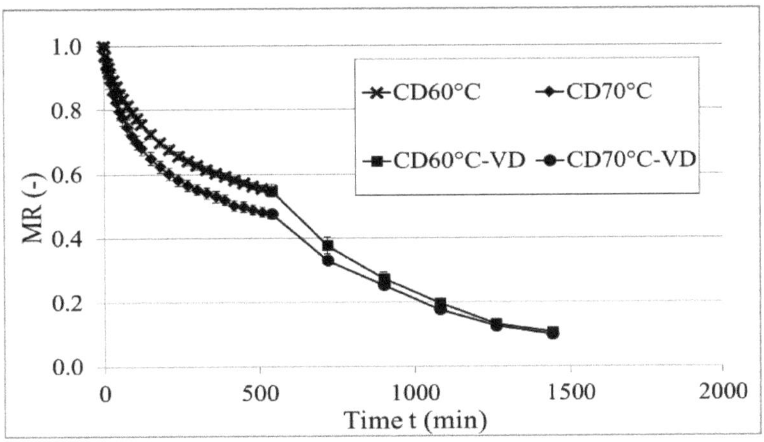

(a)

Figure 2. *Cont.*

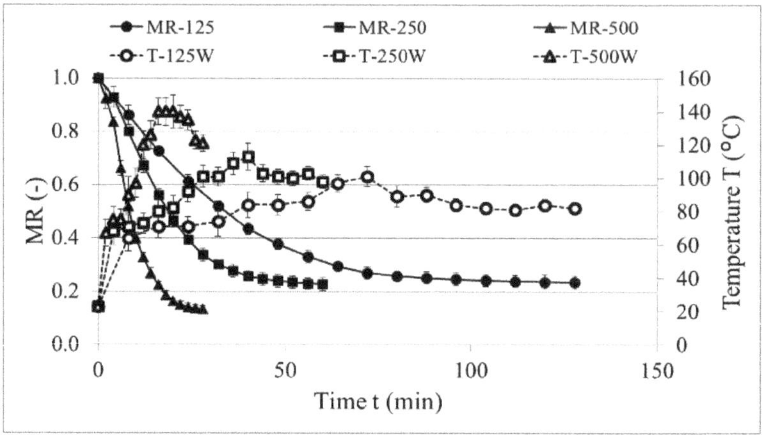

(b)

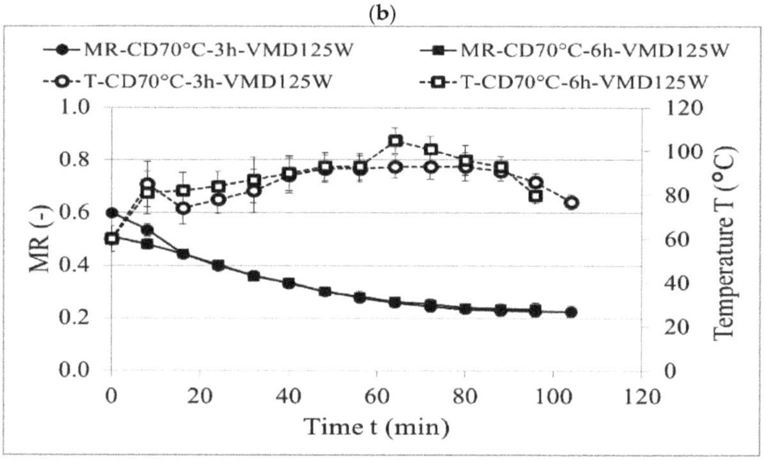

(c)

Figure 2. Drying kinetics of black garlic dried using convective pre-drying and vacuum finishing drying (**a**), vacuum-microwave drying (**b**), and combined convective pre-drying followed by vacuum-microwave finishing drying (**c**).

Figure 3 shows the drying kinetics of the samples pretreated by a pulsed electric field, a constant electric field, and a magnetic field and then dried by convective pre-drying and vacuum-microwave finishing drying. As discussed before, nonthermal pretreatment was performed in water, which changed the initial moisture content in the material. Therefore, the MR was applied to compare the drying kinetics of the materials with a different initial *Mc*. Since the samples pretreated in water had a higher *Mc, a* more accelerated reduction in the MR in the course of drying could be observed. Overall, samples pretreated in water showed a lower final MR than the samples pretreated directly, without water. Water provided the necessary conditions for the uniform distribution of electric fields, which led to a more effective influence of a PEF, a CEF, and a MF [38]. Moreover, PEF pretreatment changed the properties of the material to the highest extent which resulted in the lowest MR after CD and rapid removal of water during drying. This is due to the effect of a PEF, which destroyed the cell structure and consequently allowed for more intense water evaporation during convective drying, which is consistent with the studies on onions [26,37]. Among the lowest MR values was obtained for the samples pretreated by MF + H_2O. This can be explained by the effect of the magnetic field that affects the porosity of the material and as a result, considerably shortens the drying time such as in the studies by Memmedov et al. [36]. A CEF and CEF + H_2O showed the highest MR in those two groups (processed with/without water). This is due to the relatively mild effect of a constant electric field on the material compared to a PEF.

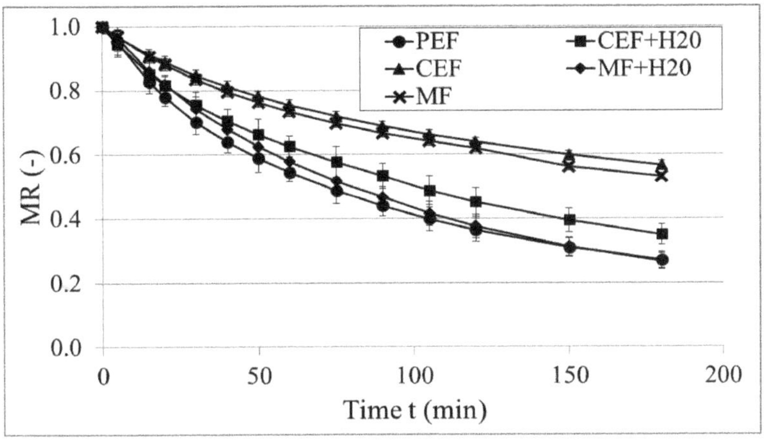

(**a**)

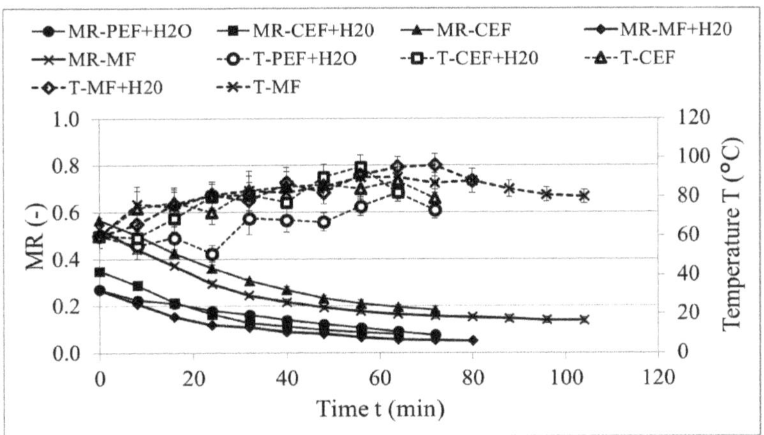

(**b**)

Figure 3. Drying kinetics of pretreated black garlic using a pulsed electric field (PEF), a constant electric field (CEF) and a magnetic field (MF) followed by drying using convective pre-drying (**a**) and vacuum-microwave finishing drying (**b**).

Several mathematical models were used to describe the experimental data for drying kinetics in this study, and among them, the Weibull model was selected to be applied in this study (Equation (1)).

$$MR = a - b \cdot e^{-k \cdot t^n} \tag{1}$$

It can be seen that in all drying variants, R^2 was above 0.99 and RMSE below 0.01 which shows a very good fit (Table 2). This model was previously used to model drying kinetics in figs [50], lemongrass [51], quince [52], and sultana grape fruits [53] and, in each study, showed a very good fit and ability to accurately describe the experimental data. Based on the parameters presented in this study, it can be seen that the drying constant represented as the *k* parameter increased when the power of magnetrons was higher during VMD. Similarly, *k* values were higher when a higher temperature during convective pre-drying was applied.

Table 2. Weibull model parameters (*a, b, k,* and *n*) together with the root mean square error (RMSE) and the coefficient of determination (R^2) according to the used drying and pretreatment method.

Pretreatment	Drying	Constants				Statistics	
		a	*b*	*k*	*n*	RMSE	R^2
-	CD60 °C	0.460	−0.540	0.0180	0.728	0.0014	0.9999
	CD70 °C	0.416	−0.589	0.0290	0.692	0.0033	0.9996
	CD60 °C-VD	−0.001	−0.550	0.0028	0.939	0.0076	0.9959
	CD70 °C-VD	−0.029	−0.506	0.0030	0.901	0.0091	0.9918
	VMD125W	0.230	−0.765	0.0134	1.245	0.0057	0.9994
	VMD250W	0.230	−0.773	0.0136	1.490	0.0023	0.9999
	VMD500W	0.138	−0.864	0.0267	1.630	0.0123	0.9982
	CD70 °C-3h-VMD125W	0.205	−0.395	0.0236	1.060	0.0068	0.9965
	CD70 °C-6h-VMD125W	0.221	−0.287	0.0047	1.450	0.0030	0.9989
PEF + H_2O	CD70 °C-3h	0.171	−0.837	0.0230	0.866	0.0071	0.9990
CEF + H_2O	CD70 °C-3h	0.104	−0.900	0.0220	0.786	0.0044	0.9995
CEF	CD70 °C-3h	0.443	−0.559	0.0170	0.858	0.0014	0.9999
MF + H_2O	CD70 °C-3h	0.139	−0.869	0.0150	0.933	0.0058	0.9993
MF	CD70 °C-3h	0.374	−0.631	0.0180	0.834	0.0052	0.9986
PEF + H_2O	VMD125W	−0.810	−1.080	0.0075	0.764	0.0037	0.9958
CEF + H_2O	VMD125W	0.079	−0.272	0.0180	1.300	0.0036	0.9982
CEF	VMD125W	0.0156	−0.410	0.0120	1.270	0.0015	0.9998
MF + H_2O	VMD125W	0.041	−0.228	0.0460	0.957	0.0054	0.9930
MF	VMD125W	0.139	−0.393	0.0220	1.160	0.0043	0.9986

2.3. *Chemical Analysis*

Bioactive Compounds

Numerous scientific studies have confirmed that phenolic compounds, known as bioactives, have many therapeutic and preventive properties useful for treating chronic diseases, including obesity, diabetes, inflammatory, and neurological diseases [54,55]. Therefore, the phenolic content in the analyzed black garlic samples is shown in Table 3. In the case of all obtained samples, amongthree3 identified groups of phenolics, the dominant ones were flavan-3-ols (monomers and dimers), followed by polymeric procyanidins, and phenolic acids. That consisted of, on average, 96%, 3%, and 1% of all identified compounds, in the case of samples in which the drying process was performed, and, respectively, 90%, 8%, and 2% in fresh black garlic (without additional processing). A significantly higher content of all analyzed phenolics was indicated for black garlic dried by VMD methods (125, 250, and 500 W) and also combined CD-VMD methods (70 °C + 125 W), which were, on average, 2-fold higher when compared with fresh black garlic (Table 3). In turn, the lowest total polyphenols content was determined in samples treated by a PEF. In general, it can be concluded that the drying process resulted in a significant increase in monomeric and dimeric flavan-3-ols, with the simultaneous degradation of polymerized compounds and phenolic acids, while the use of VMD125 W, CD70 °C-3h/125W, and MF + H_2O did not affect the decrease in phenolic acids.

Table 3. The content of phenolic compounds in black garlic after pretreatments and drying.

Method	Phenolic Acids	Flavan-3-ols	Polymeric Proanthocyanidins	Total Polyphenols
	[mg/100 g dm]			
Fresh	11.26 ± 0.01 [a*]	488.24 ± 2.48 [i]	43.41 ± 0.23 [a]	543.92 ± 2.72 [i]
VMD 500 W	2.19 ± 0.04 [k]	1088.86 ± 9.32 [a]	21.25 ± 0.53 [h]	1112.31 ± 9.86 [a]
VMD 250 W	1.98 ± 0.03 [l]	967.67 ± 3.81 [c]	24.94 ± 0.82 [f]	994.59 ± 4.66 [c]
VMD 125 W	10.02 ± 0.38 [c]	1083.67 ± 15.09 [a]	29.86 ± 0.05 [b]	1123.54 ± 15.52 [a]
CD70 °C-3h/125W	10.14 ± 0.21 [c]	1086.46 ± 18.13 [a]	28.76 ± 0.13 [d]	1125.36 ± 18.47 [a]
CD70 °C-6h/125W	2.99 ± 0.02 [i]	1023.53 ± 13.05 [b]	22.37 ± 0.09 [g]	1048.89 ± 13.16 [b]
CD60 °C-9h/VD60	9.46 ± 0.15 [d]	627.13 ± 9.00 [g]	17.61 ± 0.25 [i]	654.20 ± 9.40 [g]
CD70 °C-9h/VD60	9.07 ± 0.24 [e]	709.64 ± 4.04 [f]	24.10 ± 0.03 [f]	742.81 ± 4.31 [f]
PEF + H_2O/CD70-3h/125W	6.55 ± 0.05 [g]	366.55 ± 4.07 [j]	29.30 ± 0.09 [c]	402.40 ± 4.21 [j]
CEF + H_2O/CD70-3h/125W	4.09 ± 0.08 [h]	576.37 ± 2.13 [h]	11.97 ± 0.19 [k]	592.42 ± 2.40 [h]
CEF/CD70-3h/125W	7.54 ± 0.05 [f]	612.78 ± 9.26 [g]	25.57 ± 0.15 [e]	645.89 ± 9.46 [f]
MF + H_2O/CD70-3h/125W	10.57 ± 0.24 [bc]	821.67 ± 10.38 [d]	25.79 ± 0.15 [e]	858.04 ± 10.57 [d]
MF/CD70-3h/125W	2.66 ± 0.01 [j]	776.50 ± 8.82 [e]	21.73 ± 0.43 [h]	800.90 ± 9.26 [e]

* Values followed by the same letter, within the same column, were not significantly different ($p > 0.05$), according to Tukey's HSD test.

So far, many studies have been conducted in which the content of polyphenolic compounds in fresh and black garlic was compared. They showed a several-fold higher concentration of bioactive compounds in the processed product compared to fresh garlic [15,56]. To the best of the authors' knowledge, this is the first time the effect of additional drying on the physicochemical properties of black garlic is being studied. It was shown that it is particularly advantageous in the context of the accumulation of polyphenolic compounds, especially the monomeric and dimeric fractions of flavan-3-ols, to use VMD or combined drying consisting of convective pre-drying and vacuum-microwave finishing drying. This could be due to the heating process improving phenolic content as a result of the cleaving of bound forms (glycosylated, and esterified), thus leading to the increase in free forms of polyphenols. In addition, another reason for an increase in these compounds in the dried sample is the inhibition of enzymatic oxidation involving the antioxidant compounds, including polyphenols. Moreover, an increase in the bioactive compounds could be due to an increase in the levels of polyphenols obtained from the later phase of the browning reaction [56,57].

2.4. *In Vitro Pro-Health Potency and Antioxidant Capacity*

Antioxidant Capacity

The analysis of the antioxidant activity of the examined black garlic varieties showed that the type of pretreatment, drying method, and conditions significantly ($p < 0.05$) affected the antioxidant capacity of the tested samples. The antioxidant properties of black garlic varied widely. The highest antioxidant potential measured by the ORAC method was shown by the black garlic dried by VMD500W (7.90 mmol Trolox/100 g dm), while the lowest one by the black garlic obtained through PEF + H_2O (4.38 mmol Trolox/100 g dm). Similar trends, but slightly lower results, were obtained for the ABTS, and FRAP methods. In these cases, the highest antioxidant potential was recorded for the VMD500W—6.05, and 3.73 mmol Trolox/100 g dm (ABTS, FRAP, respectively), while the lowest value was observed for the PEF + H_2O process—2.03, and 1.09 mmol Trolox/100 g dm (Table 4).

The conducted study showed that the antioxidant activity positively correlated with the content of bioactive compounds, especially monomers and dimmers, flavan-3-ols, and procyanidin polymers. Other authors confirm this relationship. Their studies indicated that the antioxidant activity depends not only on their amount but also on the structure of the compound and the proportion of individual fractions, i.e., anthocyanins, phenolic acids, flavan-3-ols, and flavonols, in the tested material [58,59].

Table 4. Antioxidant, and antidiabetic potential of black garlic dried by different methods.

Method	ABTS	FRAP	ORAC	Inhibition of α-Amylase	Inhibition of α-Glucosidase
	mmol Trolox/100 g dm			IC50 [mg/mL]	
Fresh	2.60 ± 0.24 [e*]	1.35 ± 0.04 [f]	6.67 ± 0.03 [c]	186.56 [h]	211.24 [g]
VMD 500 W	6.05 ± 0.31 [a]	3.73 ± 0.10 [a]	7.90 ± 0.05 [a]	61.73 [a]	63.11 [c]
VMD 250 W	5.18 ± 0.51 [b]	3.08 ± 0.04 [b]	7.85 ± 0.03 [a]	135.04 [e]	208.59 [g]
VMD 125 W	3.50 ± 0.20 [c]	1.92 ± 0.08 [d]	6.82 ± 0.03 [b]	140.87 [f]	186.71 [f]
CD70 °C-3h/125W	3.76 ± 0.31 [c]	2.17 ± 0.06 [c]	5.50 ± 0.02 [g]	111.98 [d]	152.50 [e]
CD70 °C-6h/125W	5.06 ± 0.16 [b]	3.07 ± 0.02 [b]	6.80 ± 0.03 [b]	83.03 [b]	51.21 [a]
CD60 °C-9h/VD60	3.06 ± 0.17 [d]	1.80 ± 0.07 [d]	5.51 ± 0.02 [g]	134.71 [e]	227.79 [hi]
CD70 °C-9h/VD60	2.87 ± 0.12 [de]	1.57 ± 0.02 [e]	5.89 ± 0.03 [d]	229.46 [j]	155.59 [e]
PEF + H_2O/CD70-3h/125W	2.03 ± 0.19 [f]	1.09 ± 0.05 [g]	4.38 ± 0.03 [i]	142.86 [f]	223.77 [h]
CEF + H_2O/CD70-3h/125W	2.71 ± 0.19 [d]	1.18 ± 0.02 [g]	5.60 ± 0.01 [f]	99.00 [c]	56.00 [b]
CEF/CD70-3h/125W	2.49 ± 0.17 [e]	1.41 ± 0.05 [f]	5.18 ± 0.04 [h]	188.39 [h]	149.64 [e]
MF + H_2O/CD70-3h/125W	2.38 ± 0.23 [ef]	1.37 ± 0.06 [f]	5.76 ± 0.02 [e]	153.09 [g]	231.85 [i]
MF/CD70-3h/125W	2.72 ± 0.08 [e]	1.54 ± 0.02 [e]	6.75 ± 0.02 [b]	213.62 [i]	109.37 [d]

* Values followed by the same letter, within the same column, were not significantly different ($p > 0.05$), according to Tukey's HSD test.

2.5. *α-Amylase, and α-Glucosidase Inhibitory Effect*

According to the estimates by the World Health Organization (WHO), chronic noncommunicable diseases, including diabetes are currently the main cause of death worldwide. Type 2 diabetes is characterized by hyperglycemia with impaired carbohydrate, lipid, and protein metabolism resulting from defects in insulin secretion, insulin action, or both. A rapid postprandial increase in glycemia is due to starch degradation by pancreatic amylase, followed by the blocking of the resultant glucose by intestinal α-glucosidase. Therefore, it is suggested that the inhibition of these enzymes is an important strategy for the management of type 2 diabetes. In addition, several studies have investigated the antidiabetic potential of black garlic. It has been shown that black garlic exhibited ameliorative action on glycometabolic biomarkers in diabetic rats, as well as the ability to decrease blood glucose, glycated hemoglobin, and markedly increase serum insulin [15]. Thomson et al. [7] showed garlic extract significantly attenuated the elevation of serum triglyceride, and lowered lipid peroxidation in the kidney, and liver tissues. Therefore, in the present study, black garlic after various treatments was tested for its ability to inhibit α-amylase, and α-glucosidase (Table 4).

The IC_{50} values of the black garlic for inhibiting α-amylase activity ranged from 61.73 mg/mL to 229.46 mg/mL, and the inhibition effect was the highest for black garlic dried by VMD500W, while samples with CD70 °C-9h/VD60 exhibited the lowest inhibition effect (Table 4). In turn, the ability to inhibit α-glucosidase ranged from 51.21 mg/mL (CD70 °C-3h/125W) to 231.85 mg/mL in garlic treated by MF + H_2O. Nevertheless, no recurring pattern was observed in terms of the creation of anti-α-amylase, and anti-α-glucosidase potential by type of fraction, the content of polyphenolic compounds, or drying method. In this case, it is worth emphasizing the high efficiency of black garlic after PEF treatment to inhibit both enzymes, especially since this sample was characterized by both low contents of polyphenolics and low antioxidant activity. This may be due to the higher amount of specific amino acids, which are abundantly described in the literature as molecules with a significant antidiabetic potential [60]. Numerous studies describe the key role of amino acids in shaping the health potential of black garlic [15,57]. The CEF + H_2O treatment may be an effective tool to increase the availability of amino acids in plant materials and thus result in their greater effectiveness in the area of antidiabetic properties, but confirmation of this statement requires detailed studies, which are planned in the future.

3. Materials and Methods

3.1. Material

Reagents for antioxidant and biological activity tests (α-amylase, α-glucosidase, starch, 2,2′-azino-bis(3-ethylbenzothiazoline-6-sulfonic acid) diammonium salt, TPTZ, p-nitrophenyl-α-D-glucopyranoside) were purchased from Sigma-Aldrich (Steinheim, Germany). Standards for polyphenols were acquired from Extrasynthese (Lyon Nord, France). Acetonitrile, methanol, and formic acid for ultra-performance liquid chromatography (UPLC; gradient grade) were purchased from Merck (Darmstadt, Germany). Black garlic was produced from regular garlic (*Allium sativum* L.) as a result of the aging process. The material used in this study was obtained from the local producer ("PASZKÓW" Farma Tadeusz Kaczmarczyk, Świdnica, Poland). Garlic slices were sliced in half and each sample consisted of 100 g of material of the initial moisture content Mc = 0.66 kg/kg db.

3.2. Pretreatment Methods

A pulsed electric field (PEF) treatment was performed in prototype equipment built at the University of Agriculture in Krakow, Poland (model ERTEC-SU1 with Line Parameters Analyzer type AS3 Mini) [29,61]. The material was immersed in water and then placed in a treatment chamber between the electrodes. Each time, the material was treated with 300 impulses with 10 s break between the impulses. The strength of the electric field was fixed at 5 kV/cm.

A constant electric field (CEF) was applied in an apparatus consisting of two flat electrodes placed in a chamber with the material in between [32]. The high voltage pulse generator was set to 9 kV. The treatment consisted of 1272 impulses for 10 s each, with a 5 s break between the impulses. Two variants of CEF treatment were performed: with and without the addition of water to the material.

A magnetic field (MF) was used to treat the material alone and with the addition of water using a magnetic field of 100 mT at the frequency of 50 Hz for 2 h using equipment built at the University of Agriculture in Krakow (Kraków, Poland) (Figure 4).

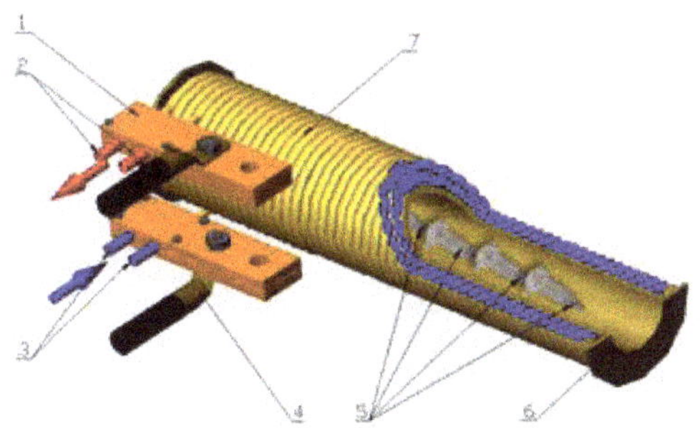

Figure 4. Diagram of the solenoid for magnetic stimulation of black garlic samples: 1—supply chamber; 2—cooling water outlet; 3—cooling water inlet; 4—power cables; 5—sample containers; 6—carcass; 7—coil.

3.3. Drying Methods

In the first stage of the experiment, black garlic cloves were subjected to convective drying (CD) performed using the convective dryer designed and constructed at the Institute of Agricultural Engineering (Wrocław, Poland) [62] at 60 and 70 °C and the air velocity of 0.5 m/s for 9 h (Figure 5). Then, pre-dried black garlic was moved to a vacuum dryer (VD) and dried at 60 °C under 100 Pa (SPT-200, ZEAMiL, Horyzont, Kraków, Poland) until the moisture content of the material was below 13%. Vacuum-microwave drying (VMD) was performed using SM200 dryer (Plazmatronica, Wrocław, Poland) [63] at 125, 250, and 500 W power of magnetrons under the pressure in the range of 50–70 hPa. Combined CD-VMD consisted of 3 h or 6 h of convective pre-drying at 70 °C followed by vacuum-microwave finishing drying at 125 W (CD3h-VMD or CD6h-VMD, respectively).

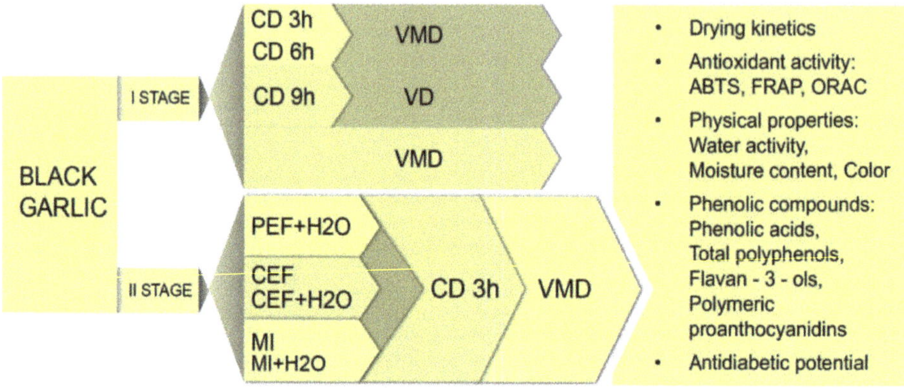

Figure 5. Schematic representation of the two stages of experiments performed in this study and analyses performed in this study. CD—convective drying; VMD—vacuum-microwave drying; VD—vacuum drying; PEF—pulsed electric field; CEF—constant electric field; MF—magnetic field.

In the second stage of this study, black garlic pretreated using a PEF, a CEF, and a MF was subjected to convective pre-drying at 70 °C for 3 h and then vacuum-microwave finishing drying at 125 W.

The surface temperature of all the samples was measured using an infrared camera i50 (Flir Systems AB, Stockholm, Sweden). Experiments were carried out in two repetitions.

3.4. *Physical Properties*

3.4.1. The Moisture Content

The moisture content (Mc) of the samples was measured using vacuum-drying (SPT-200, ZEAMiL, Horyzont, Kraków, Poland) at 100 Pa and 80 °C for 48 h. Measurements were performed in duplicate.

3.4.2. Water Activity

The water activity of the samples both before and after treatments was measured using AquaLab Dew Point 4TE (Decagon Devices Inc., Pullman, WA, USA) water activity meter. Water activity was measured at 25 ± 0.5 °C in triplicate.

3.4.3. Color

Color measurement was performed with the black garlic powder obtained after grinding the material using Profi Cook grinder (PC-KSW 1021) after each drying treatment. Then, the sample was placed in a vessel placed on a Chroma Meter CR-400 colorimeter (Minolta Co., Ltd., Osaka, Japan). The color was expressed within CIE $L^*a^*b^*$ color space meaning lightness (L^*), hues from red to green according to the values of a^* coordinate, and from yellow to blue for b^* coordinate. The browning index (BI) was calculated based on the equations (Equations (2) and (3)) provided by Subhashree et al. [64]:

$$X = \frac{a^* + 1.75 \cdot L^*}{5.645 \cdot L^* + a - 3.012 \cdot b^*} \quad (2)$$

$$BI = \frac{100 \cdot (X - 0.31)}{0.17} \quad (3)$$

3.5. Chemical Analysis

3.5.1. Determination of Phenolic Compounds, including Polymeric Proanthocyanidins by UPLC

Determination of phenolics in black garlic and black garlic after different treatments was performed as described by Nowicka et al. [65] using an Acquity UPLC system (Waters, Milford, MA, USA) with a photodiode and a fluorescence detector with the mass detector G2 Qtof mass spectrometer (Waters, Manchester, UK). The absorbance values of flavan-3-ols and phenolic acids were read at 280 nm, and 320 nm, respectively. The content of polymeric procyanidins was analyzed by the phloroglucinol method [66]. All samples were measured in triplicate, and the results were expressed as mg per 100 g dry mass.

3.5.2. Analysis of Health-Promoting Properties by In Vitro Methods

To analyze the antioxidant activity, the ORAC (oxygen radical absorbance capacity), FRAP (ferric reducing antioxidant power), and ABTS (2,2′-azino-bis(3-ethylbenzothiazoline-6-sulphonic acid)) methods were used as described earlier by Ou et al. [67], Benzie et al. [68], and Re at al. [69], respectively. The obtained results were presented as mmol Trolox per 100 g dry matter (dm).

The α-amylase and α-glucosidase inhibitory effects (antidiabetic activity) of the black garlic were determined according to the procedure described by Nowicka et al. [70]. In the α-amylase activity determination, the primary sample consisted of black garlic extracts, to which the starch solution, as well as α-amylase solution, were added. The reaction of these components was carried out at 37 °C for 15 min, and then it was stopped with 0.4 M HCl followed by the addition of potassium iodide with iodine.

In the case of α-glucosidase activity analysis, the basic sample including back garlic extracts and enzymes, was incubated at 37 °C for 10 min, then the β-D-glucosidase substrate was added, and incubated as before. Acarbose was included as a positive control for α-amylase and α-glucosidase assay. The absorbance measurement was done at 600 nm and 405 nm for α-amylase and α-glucosidase activity, respectively, using a Synergy H1 spectrophotometer (BioTek, Winooski, VT, USA). Both tests were performed in triplicate, and the results were expressed as IC_{50} values.

3.6. Statistical Analysis

Statistica 13.3 software (StatSoft, Krakow, Poland) was used for all statistical analyses. The results were presented as mean ± standard deviation. One-way analysis of variance (ANOVA) was performed in this study. HSD Tukey's least significance test ($p < 0.05$) was used to determine homogenous groups. Table Curve 2D v. 5.0 (Systat Software, Inc., San Jose, CA, USA) was used to fit mathematical models to the experimental data based on the lowest values of the root mean square error (RMSE) and the highest values of the coefficient of determination (R^2).

4. Conclusions

The impact of convective, vacuum-microwave, and combined drying methods was investigated in this study together with the effect of nonthermal pretreatment including a pulsed electric field, a constant electric field, and a magnetic field on the quality of black garlic. This study found that the application of different pretreatment methods significantly reduced the overall time of combined drying as well as leading to the lowest values of the final moisture content among considered treatment variants. Drying kinetics were described by the Weibull model, which presented a very good fit. Moreover, pulsed electric field and magnetic field treatments proved to be effective in maintaining the health-promoting properties of black garlic, especially in terms of antidiabetic potential. However, vacuum-microwave drying positively affected the phenolic content and antioxidant capacity of black garlic due to the thermal effect that led to the cleaving of bound forms of polyphenols and consequently an increase in bioactive compounds. This study showed that nonthermal pretreatments considerably affect the drying process as well as the quality of dried materials; however, future studies are needed to find the optimal process parameters.

Author Contributions: Conceptualization, K.L., K.M. and T.D.; methodology, K.L., K.M., T.D. and P.N.; validation, K.L., K.M. and P.N.; formal analysis, K.L., K.M. and P.N.; investigation, K.L., K.M., P.N. and T.D.; resources, K.L., K.M., T.D., A.W., P.K. and P.N.; writing—original draft preparation, K.M., K.L. and P.N.; writing—review and editing, K.M., K.L., P.N., A.W., P.K. and T.D.; visualization, K.M., K.L. and P.N.; supervision, K.M.; project administration, K.L.; funding acquisition, K.L. All authors have read and agreed to the published version of the manuscript.

Funding: This research was funded by the Program of the Minister MNiSW "Inkubator Innowacyjności 4.0" (Poland) based on the agreement number MNiSW/2020/334/DIR, 28.09.2020.

Institutional Review Board Statement: Not applicable.

Informed Consent Statement: Not applicable.

Data Availability Statement: The data that support the findings of this study are available from the corresponding author upon reasonable request.

Conflicts of Interest: The authors declare no conflict of interest.

Sample Availability: Samples of the compounds are not available from the authors.

References

1. Zhang, X.; Li, N.; Lu, X.; Liu, P.; Qiao, X. Effects of Temperature on the Quality of Black Garlic. *J. Sci. Food Agric.* **2016**, *96*, 2366–2372. [CrossRef] [PubMed]
2. Kimura, S.; Tung, Y.-C.; Pan, M.-H.; Su, N.-W.; Lai, Y.-J.; Cheng, K.-C. Black Garlic: A Critical Review of Its Production, Bioactivity, and Application. *J. Food Drug Anal.* **2017**, *25*, 62–70. [CrossRef] [PubMed]
3. Yuan, H.; Sun, L.; Chen, M.; Wang, J. The Comparison of the Contents of Sugar, Amadori, and Heyns Compounds in Fresh and Black Garlic. *J. Food Sci.* **2016**, *81*, C1662–C1668. [CrossRef] [PubMed]
4. Qiu, Z.; Zheng, Z.; Zhang, B.; Sun-Waterhouse, D.; Qiao, X. Formation, Nutritional Value, and Enhancement of Characteristic Components in Black Garlic: A Review for Maximizing the Goodness to Humans. *Compr. Rev. Food Sci. Food Saf.* **2020**, *19*, 801–834. [CrossRef]
5. Kim, D.; Kang, M.J.; Hong, S.S.; Choi, Y.-H.; Shin, J.H. Antiinflammatory Effects of Functionally Active Compounds Isolated from Aged Black Garlic. *Phytother. Res.* **2017**, *31*, 53–61. [CrossRef]
6. Mirzavandi, F.; Mollahosseini, M.; Salehi-Abargouei, A.; Makiabadi, E.; Mozaffari-Khosravi, H. Effects of Garlic Supplementation on Serum Inflammatory Markers: A Systematic Review and Meta-Analysis of Randomized Controlled Trials. *Diabetes Metab. Syndr. Clin. Res. Rev.* **2020**, *14*, 1153–1161. [CrossRef]
7. Thomson, M.; Al-Qattan, K.K.; JS, D.; Ali, M. Anti-Diabetic and Anti-Oxidant Potential of Aged Garlic Extract (AGE) in Streptozotocin-Induced Diabetic Rats. *BMC Complement. Altern. Med.* **2016**, *16*, 17. [CrossRef]
8. Jung, Y.-M.; Lee, S.-H.; Lee, D.-S.; You, M.-J.; Chung, I.K.; Cheon, W.H.; Kwon, Y.-S.; Lee, Y.-J.; Ku, S.-K. Fermented Garlic Protects Diabetic, Obese Mice When Fed a High-Fat Diet by Antioxidant Effects. *Nutr. Res.* **2011**, *31*, 387–396. [CrossRef]
9. Amor, S.; González-Hedström, D.; Martín-Carro, B.; Inarejos-García, A.M.; Almodóvar, P.; Prodanov, M.; García-Villalón, A.L.; Granado, M. Beneficial Effects of an Aged Black Garlic Extract in the Metabolic and Vascular Alterations Induced by a High Fat/Sucrose Diet in Male Rats. *Nutrients* **2019**, *11*, 153. [CrossRef]
10. Wang, X.; Jiao, F.; Wang, Q.-W.; Wang, J.; Yang, K.; Hu, R.-R.; Liu, H.-C.; Wang, H.-Y.; Wang, Y.-S. Aged Black Garlic Extract Induces Inhibition of Gastric Cancer Cell Growth in Vitro and in Vivo. *Mol. Med. Rep.* **2012**, *5*, 66–72. [CrossRef]
11. Agbana, Y.L.; Ni, Y.; Zhou, M.; Zhang, Q.; Kassegne, K.; Karou, S.D.; Kuang, Y.; Zhu, Y. Garlic-Derived Bioactive Compound S-Allylcysteine Inhibits Cancer Progression through Diverse Molecular Mechanisms. *Nutr. Res.* **2020**, *73*, 1–14. [CrossRef]
12. Asdaq, S.M.; Inamdar, M.N. Potential of Garlic and Its Active Constituent, S-Allyl Cysteine, as Antihypertensive and Cardioprotective in Presence of Captopril. *Phytomedicine* **2010**, *17*, 1016–1026. [CrossRef] [PubMed]
13. Quesada, I.; de Paola, M.; Torres-Palazzolo, C.; Camargo, A.; Ferder, L.; Manucha, W.; Castro, C. Effect of Garlic's Active Constituents in Inflammation, Obesity and Cardiovascular Disease. *Curr. Hypertens. Rep.* **2020**, *22*, 6. [CrossRef]
14. Kim, J.H.; Nam, S.H.; Rico, C.W.; Kang, M.Y. A Comparative Study on the Antioxidative and Anti-Allergic Activities of Fresh and Aged Black Garlic Extracts. *Int. J. Food Sci. Technol.* **2012**, *47*, 1176–1182. [CrossRef]
15. Ahmed, T.; Wang, C.-K. Black Garlic and Its Bioactive Compounds on Human Health Diseases: A Review. *Molecules* **2021**, *26*, 5028. [CrossRef]
16. Calín-Sánchez, Á.; Figiel, A.; Wojdyło, A.; Szarycz, M.; Carbonell-Barrachina, Á.A. Drying of Garlic Slices Using Convective Pre-Drying and Vacuum-Microwave Finishing Drying: Kinetics, Energy Consumption, and Quality Studies. *Food Bioprocess Technol.* **2014**, *7*, 398–408. [CrossRef]

17. Lyczko, J.; Masztalerz, K.; Lipan, L.; Lech, K.; Carbonell-Barrachina, A.; Szumny, A. Chemical Determinants of Dried Thai Basil (O. Basilicum Var. Thyrsiflora) Aroma Quality. *Ind. Crop. Prod.* **2020**, *155*, 112769. [CrossRef]
18. Kwaśnica, A.; Pachura, N.; Masztalerz, K.; Figiel, A.; Zimmer, A.; Kupczyński, R.; Wujcikowska, K.; Carbonell-Barrachina, A.A.; Szumny, A.; Różański, H. Volatile Composition and Sensory Properties as Quality Attributes of Fresh and Dried Hemp Flowers (Cannabis Sativa L.). *Foods* **2020**, *9*, 1118. [CrossRef]
19. Bialik, M.; Wiktor, A.; Rybak, K.; Witrowa-Rajchert, D.; Latocha, P.; Gondek, E. The Impact of Vacuum and Convective Drying Parameters on Kinetics, Total Phenolic Content, Carotenoid Content and Antioxidant Capacity of Kiwiberry (Actinidia Arguta). *Appl. Sci.* **2020**, *10*, 6914. [CrossRef]
20. Łyczko, J.; Jałoszyński, K.; Surma, M.; Masztalerz, K.; Szumny, A. HS-SPME Analysis of True Lavender (Lavandula Angustifolia Mill.) Leaves Treated by Various Drying Methods. *Molecules* **2019**, *24*, 764. [CrossRef]
21. Chua, L.Y.W.; Chua, B.L.; Figiel, A.; Chong, C.H.; Wojdyło, A.; Szumny, A.; Lech, K. Characterisation of the Convective Hot-Air Drying and Vacuum Microwave Drying of Cassia Alata: Antioxidant Activity, Essential Oil Volatile Composition and Quality Studies. *Molecules* **2019**, *24*, 1625. [CrossRef] [PubMed]
22. Masztalerz, K.; Łyczko, J.; Lech, K. Effect of Filtrated Osmotic Solution Based on Concentrated Chokeberry Juice and Mint Extract on the Drying Kinetics, Energy Consumption and Physicochemical Properties of Dried Apples. *Molecules* **2021**, *26*, 3274. [CrossRef] [PubMed]
23. Tylewicz, U.; Castagnini, J.M.; Tappi, S.; Romani, S.; Rocculi, P.; Rosa, M.D. Current Validation of NTP Technologies and Overview of Their Current and Potential Implementation in the Production Chain Including Agri-Food Wastes. In *Nonthermal Processing in Agri-Food-Bio Sciences: Sustainability and Future Goals*; Režek Jambrak, A., Ed.; Food Engineering Series; Springer International Publishing: Cham, Switzerland, 2022; pp. 567–594, ISBN 978-3-030-92415-7.
24. Nowacka, M.; Wiktor, A.; Anuszewska, A.; Dadan, M.; Rybak, K.; Witrowa-Rajchert, D. The Application of Unconventional Technologies as Pulsed Electric Field, Ultrasound and Microwave-Vacuum Drying in the Production of Dried Cranberry Snacks. *Ultrason. Sonochem.* **2019**, *56*, 1–13. [CrossRef] [PubMed]
25. Alam, M.R.; Lyng, J.G.; Frontuto, D.; Marra, F.; Cinquanta, L. Effect of Pulsed Electric Field Pretreatment on Drying Kinetics, Color, and Texture of Parsnip and Carrot. *J. Food Sci.* **2018**, *83*, 2159–2166. [CrossRef]
26. Ostermeier, R.; Giersemehl, P.; Siemer, C.; Töpfl, S.; Jäger, H. Influence of Pulsed Electric Field (PEF) Pre-Treatment on the Convective Drying Kinetics of Onions. *J. Food Eng.* **2018**, *237*, 110–117. [CrossRef]
27. Liu, C.; Pirozzi, A.; Ferrari, G.; Vorobiev, E.; Grimi, N. Impact of Pulsed Electric Fields on Vacuum Drying Kinetics and Physicochemical Properties of Carrot. *Food Res. Int.* **2020**, *137*, 109658. [CrossRef]
28. Rybak, K.; Parniakov, O.; Samborska, K.; Wiktor, A.; Witrowa-Rajchert, D.; Nowacka, M. Energy and Quality Aspects of Freeze-Drying Preceded by Traditional and Novel Pre-Treatment Methods as Exemplified by Red Bell Pepper. *Sustainability* **2021**, *13*, 2035. [CrossRef]
29. Dziadek, K.; Kopeć, A.; Dróżdż, T.; Kiełbasa, P.; Ostafin, M.; Bulski, K.; Oziembłowski, M. Effect of Pulsed Electric Field Treatment on Shelf Life and Nutritional Value of Apple Juice. *J. Food Sci. Technol.* **2019**, *56*, 1184–1191. [CrossRef]
30. Yu, Y.; Yuanshan, Y.; Jin, T.Z.; Fan, X.; Jijun, W.; Jijun, W. Biochemical Degradation and Physical Migration of Polyphenolic Compounds in Osmotic Dehydrated Blueberries with Pulsed Electric Field and Thermal Pretreatments. *Food Chem.* **2018**, *239*, 1219–1225. [CrossRef]
31. Miernik, A.; Kiełbasa, P.; Findura, P.; Byrska, K. Influence of the Constant Electric Field on the Photon Emission Characteristics of Selected Utility Cultivars of the Camellia Plant. *J. Phys. Conf. Ser.* **2021**, *1782*, 12021. [CrossRef]
32. Miernik, A.; Juliszewski, T.; Popardowski, E.; Trzyniec, K.; Kovalyshyn, S.; Wiśniowski, B. Influence of Constant Electric Field Impact on the Intensity and Structure of the Secondary Luminescence of Selected Industrial Cannabis Products. *J. Phys. Conf. Ser.* **2021**, *1782*, 12022. [CrossRef]
33. Vashisth, A.; Nagarajan, S. Effect on Germination and Early Growth Characteristics in Sunflower (Helianthus Annuus) Seeds Exposed to Static Magnetic Field. *J. Plant Physiol.* **2010**, *167*, 149–156. [CrossRef] [PubMed]
34. Kaur, M.; Kumar, M. An Innovation in Magnetic Field Assisted Freezing of Perishable Fruits and Vegetables: A Review. *Food Rev. Int.* **2020**, *36*, 761–780. [CrossRef]
35. Mujumdar, A.S.; Woo, M.W. Effects of Electric and Magnetic Field on Freezing. In *Drying Technologies for Biotechnology and Pharmaceutical Applications*; John Wiley & Sons, Ltd.: Hoboken, NJ, USA, 2020; pp. 283–301, ISBN 978-3-527-80210-4.
36. Memmedov, A.; Kelbaliyev, G.I.; Alisoy, G.T. Solution of an Inverse Problem for Mass Transfer in a Drying Process in a Magnetic Field. *Inverse Probl. Sci. Eng.* **2010**, *18*, 723–736. [CrossRef]
37. Ostermeier, R.; Parniakov, O.; Töpfl, S.; Jäger, H. Applicability of Pulsed Electric Field (PEF) Pre-Treatment for a Convective Two-Step Drying Process. *Foods* **2020**, *9*, 512. [CrossRef]
38. Rahaman, A.; Zeng, X.-A.; Kumari, A.; Farooq, M.A.; Muhammad Adil Farooq; Muhammad Adil Farooq; Siddeeg, A.; Manzoor, M.F. Combined Effect of Pulsed Electric Fields and Ultrasound on Mass Energy Transfer and Diffusion Coefficient of Plum. *Heat Mass Transf.* **2021**, *57*, 1087–1095. [CrossRef]
39. Figiel, A. Drying Kinetics and Quality of Vacuum-Microwave Dehydrated Garlic Cloves and Slices. *J. Food Eng.* **2009**, *94*, 98–104. [CrossRef]
40. Soliva-Fortuny, R.; Balasa, A.; Knorr, D.; Martín-Belloso, O. Effects of Pulsed Electric Fields on Bioactive Compounds in Foods: A Review. *Trends Food Sci. Technol.* **2009**, *20*, 544–556. [CrossRef]

41. Tylewicz, U.; Tappi, S.; Mannozzi, C.; Romani, S.; Dellarosa, N.; Laghi, L.; Ragni, L.; Rocculi, P.; Rosa, M.D. Effect of Pulsed Electric Field (PEF) Pre-Treatment Coupled with Osmotic Dehydration on Physico-Chemical Characteristics of Organic Strawberries. *J. Food Eng.* **2017**, *213*, 2–9. [CrossRef]
42. Tapia, M.S.; Alzamora, S.M.; Chirife, J. Effects of Water Activity (a w) on Microbial Stability as a Hurdle in Food Preservation. In *Water Activity in Foods*; John Wiley & Sons, Ltd.: Hoboken, NJ, USA, 2020; pp. 323–355, ISBN 978-1-118-76598-2.
43. Rasooli Sharabiani, V.; Kaveh, M.; Abdi, R.; Szymanek, M.; Tanaś, W. Estimation of Moisture Ratio for Apple Drying by Convective and Microwave Methods Using Artificial Neural Network Modeling. *Sci. Rep.* **2021**, *11*, 9155. [CrossRef]
44. Calín-Sánchez, Á.; Lipan, L.; Cano-Lamadrid, M.; Kharaghani, A.; Masztalerz, K.; Carbonell-Barrachina Á, A.; Figiel, A. Comparison of Traditional and Novel Drying Techniques and Its Effect on Quality of Fruits, Vegetables and Aromatic Herbs TI2. *Foods* **2020**, *9*, 1261. [CrossRef] [PubMed]
45. Wojdyło, A.; Figiel, A.; Lech, K.; Nowicka, P.; Oszmiański, J. Effect of Convective and Vacuum–Microwave Drying on the Bioactive Compounds, Color, and Antioxidant Capacity of Sour Cherries. *Food Bioprocess Technol.* **2014**, *7*, 829–841. [CrossRef]
46. Taskin, O.; Polat, A.; Izli, N.; Asik, B.B. Intermittent Microwave-Vacuum Drying Effects on Pears. *Pol. J. Food Nutr. Sci.* **2019**, *69*, 101–108. [CrossRef]
47. Ríos-Ríos, K.L.; Montilla, A.; Olano, A.; Villamiel, M. Physicochemical Changes and Sensorial Properties during Black Garlic Elaboration: A Review. *Trends Food Sci. Technol.* **2019**, *88*, 459–467. [CrossRef]
48. Castillo-Gironés, S.; Masztalerz, K.; Lech, K.; Issa-Issa, H.; Figiel, A.; Carbonell-Barrachina, A.A. Impact of Osmotic Dehydration and Different Drying Methods on the Texture and Sensory Characteristic of Sweet Corn Kernels. *J. Food Process. Preserv.* **2021**, *45*, e15383. [CrossRef]
49. Hu, Q.; Zhang, M.; Mujumdar, A.S.; Xiao, G.; Jin-cai, S. Drying of Edamames by Hot Air and Vacuum Microwave Combination. *J. Food Eng.* **2006**, *77*, 977–982. [CrossRef]
50. Babalis, S.J.; Papanicolaou, E.; Kyriakis, N.; Belessiotis, V.G. Evaluation of Thin-Layer Drying Models for Describing Drying Kinetics of Figs (Ficus Carica). *J. Food Eng.* **2006**, *75*, 205–214. [CrossRef]
51. Nguyen, T.V.L.; Nguyen, M.D.; Nguyen, D.C.; Bach, L.G.; Lam, T.D. Model for Thin Layer Drying of Lemongrass (Cymbopogon Citratus) by Hot Air. *Processes* **2019**, *7*, 21. [CrossRef]
52. Tzempelikos, D.A.; Vouros, A.P.; Bardakas, A.V.; Filios, A.E.; Margaris, D.P. Experimental Study on Convective Drying of Quince Slices and Evaluation of Thin-Layer Drying Models. *Eng. Agric. Environ. Food* **2015**, *8*, 169–177. [CrossRef]
53. Karaaslan, S.; Ekinci, K.; Akbolat, D. Drying Characteristics for Sultana Grape Fruit in Microwave Dryer. *Infrastrukt. I Ekol. Teren. Wiej./Infrastruct. Ecol. Rural. Areas* **2017**, *4.1*, 1317–1327. [CrossRef]
54. Noratto, G.; Porter, W.; Byrne, D.; Cisneros-Zevallos, L. Polyphenolics from Peach (Prunus Persica Var. Rich Lady) Inhibit Tumor Growth and Metastasis of MDA-MB-435 Breast Cancer Cells in Vivo. *J. Nutr. Biochem.* **2014**, *25*, 796–800. [CrossRef]
55. Noratto, G.; Martino, H.S.D.; Simbo, S.; Byrne, D.; Mertens-Talcott, S.U. Consumption of Polyphenol-Rich Peach and Plum Juice Prevents Risk Factors for Obesity-Related Metabolic Disorders and Cardiovascular Disease in Zucker Rats. *J. Nutr. Biochem.* **2015**, *26*, 633–641. [CrossRef] [PubMed]
56. Kim, J.-S.; Kang, O.-J.; Gweon, O.-C. Comparison of Phenolic Acids and Flavonoids in Black Garlic at Different Thermal Processing Steps. *J. Funct. Foods* **2013**, *5*, 80–86. [CrossRef]
57. Choi, I.S.; Cha, H.S.; Lee, Y.S. Physicochemical and Antioxidant Properties of Black Garlic. *Molecules* **2014**, *19*, 16811–16823. [CrossRef]
58. Nowicka, P.; Wojdyło, A.; Lech, K.; Figiel, A. Chemical Composition, Antioxidant Capacity, and Sensory Quality of Dried Sour Cherry Fruits Pre-Dehydrated in Fruit Concentrates. *Food Bioprocess Technol.* **2015**, *8*, 2076–2095. [CrossRef]
59. Wojdyło, A.; Figiel, A.; Legua, P.; Lech, K.; Carbonell-Barrachina, Á.A.; Hernández, F. Chemical Composition, Antioxidant Capacity, and Sensory Quality of Dried Jujube Fruits as Affected by Cultivar and Drying Method. *Food Chem.* **2016**, *207*, 170–179. [CrossRef]
60. Poovitha, S.; Parani, M. In Vitro and in Vivo α-Amylase and α-Glucosidase Inhibiting Activities of the Protein Extracts from Two Varieties of Bitter Gourd (Momordica Charantia L.). *BMC Complement. Altern. Med.* **2016**, *16*, 185. [CrossRef]
61. Wesołowski, M.; Nęcka, K.; Dróżdż, T.; Kiełbasa, P. Koncepcja modelowania wyładowania pulsacyjnego pola elektrycznego (PEF) w produktach przemysłu rolno—Spożywczego. *Przegląd Elektrotechniczny* **2018**, *1*, 121–125. [CrossRef]
62. Tulej, W.; Głowacki, S. Modeling of the Drying Process of Apple Pomace. *Appl. Sci.* **2022**, *12*, 1434. [CrossRef]
63. Lech, K.; Figiel, A.; Wojdyło, A.; Korzeniowska, M.; Serowik, M.; Szarycz, M. Drying Kinetics and Bioactivity of Beetroot Slices Pretreated in Concentrated Chokeberry Juice and Dried with Vacuum Microwaves. *Dry. Technol.* **2015**, *33*, 1644–1653. [CrossRef]
64. Subhashree, S.N.; Sunoj, S.; Xue, J.; Bora, G.C. Quantification of Browning in Apples Using Colour and Textural Features by Image Analysis. *Food Qual. Saf.* **2017**, *1*, 221–226. [CrossRef]
65. Nowicka, P.; Wojdyło, A. Content of Bioactive Compounds in the Peach Kernels and Their Antioxidant, Anti-Hyperglycemic, Anti-Aging Properties. *Eur. Food Res. Technol.* **2019**, *245*, 1123–1136. [CrossRef]
66. Kennedy, J.A.; Jones, G.P. Analysis of Proanthocyanidin Cleavage Products Following Acid-Catalysis in the Presence of Excess Phloroglucinol. *J. Agric. Food Chem.* **2001**, *49*, 1740–1746. [CrossRef] [PubMed]
67. Ou, B.; Huang, D.; Hampsch-Woodill, M.; Flanagan, J.A.; Deemer, E.K. Analysis of Antioxidant Activities of Common Vegetables Employing Oxygen Radical Absorbance Capacity (ORAC) and Ferric Reducing Antioxidant Power (FRAP) Assays: A Comparative Study. *J. Agric. Food Chem.* **2002**, *50*, 3122–3128. [CrossRef] [PubMed]

68. Benzie, I.F.F.; Strain, J.J. The Ferric Reducing Ability of Plasma (FRAP) as a Measure of "Antioxidant Power": The FRAP Assay. *Anal. Biochem.* **1996**, *239*, 70–76. [CrossRef]
69. Re, R.; Pellegrini, N.; Proteggente, A.; Pannala, A.; Yang, M.; Rice-Evans, C. Antioxidant Activity Applying an Improved ABTS Radical Cation Decolorization Assay. *Free. Radic. Biol. Med.* **1999**, *26*, 1231–1237. [CrossRef]
70. Nowicka, P.; Wojdyło, A.; Samoticha, J. Evaluation of Phytochemicals, Antioxidant Capacity, and Antidiabetic Activity of Novel Smoothies from Selected Prunus Fruits. *J. Funct. Foods* **2016**, *25*, 397–407. [CrossRef]

molecules MDPI

Article

Antimicrobial Activity of Quercetin, Naringenin and Catechin: Flavonoids Inhibit *Staphylococcus aureus*-Induced Hemolysis and Modify Membranes of Bacteria and Erythrocytes

Artem G. Veiko [1], Ewa Olchowik-Grabarek [2], Szymon Sekowski [2], Anna Roszkowska [2], Elena A. Lapshina [1], Izabela Dobrzynska [3], Maria Zamaraeva [2] and Ilya B. Zavodnik [1,*]

1 Department of Biochemistry, Yanka Kupala State University of Grodno, 230030 Grodno, Belarus
2 Laboratory of Molecular Biophysics, Department of Microbiology and Biotechnology, Faculty of Biology, University of Bialystok, 15-245 Białystok, Poland
3 Laboratory of Bioanalysis, Faculty of Chemistry, University of Bialystok, 15-245 Białystok, Poland
* Correspondence: zavodnik_ib@grsu.by

Abstract: Search for novel antimicrobial agents, including plant-derived flavonoids, and evaluation of the mechanisms of their antibacterial activities are pivotal objectives. The goal of this study was to compare the antihemolytic activity of flavonoids, quercetin, naringenin and catechin against sheep erythrocyte lysis induced by α-hemolysin (αHL) produced by the *Staphylococcus aureus* strain NCTC 5655. We also sought to investigate the membrane-modifying action of the flavonoids. Lipophilic quercetin, but not naringenin or catechin, effectively inhibited the hemolytic activity of αHL at concentrations ($IC_{50} = 65 \pm 5$ μM) below minimal inhibitory concentration values for *S. aureus* growth. Quercetin increased the registered bacterial cell diameter, enhanced the fluidity of the inner and surface regions of bacterial cell membranes and raised the rigidity of the hydrophobic region and the fluidity of the surface region of erythrocyte membranes. Our findings provide evidence that the antibacterial activities of the flavonoids resulted from a disorder in the structural organization of bacterial cell membranes, and the antihemolytic effect of quercetin was related to the effect of the flavonoid on the organization of the erythrocyte membrane, which, in turn, increases the resistance of the target cells (erythrocytes) to αHL and inhibits αHL-induced osmotic hemolysis due to prevention of toxin incorporation into the target membrane. We confirmed that cell membrane disorder could be one of the direct modes of antibacterial action of the flavonoids.

Keywords: flavonoids; antimicrobial agents; *Staphylococcus aureus*; hemolysis; membranes; fluidity

Citation: Veiko, A.G.; Olchowik-Grabarek, E.; Sekowski, S.; Roszkowska, A.; Lapshina, E.A.; Dobrzynska, I.; Zamaraeva, M.; Zavodnik, I.B. Antimicrobial Activity of Quercetin, Naringenin and Catechin: Flavonoids Inhibit *Staphylococcus aureus*-Induced Hemolysis and Modify Membranes of Bacteria and Erythrocytes. *Molecules* **2023**, *28*, 1252. https://doi.org/10.3390/molecules28031252

Academic Editors: Manuel Simões and Smaoui Slim

Received: 27 November 2022
Revised: 19 January 2023
Accepted: 23 January 2023
Published: 27 January 2023

1. Introduction

Increased incidence of infections and considerable growth of resistance of microorganisms to antibiotics, along with production of resistant bacterial strains that substantially complicate treatment, are recognized as a very important global socio-medical problem. Thus, it seems to be of particular relevance to seek compounds that can be used either as an alternative or supplement to antibiotics [1]. The range of such antimicrobial agents also comprises such natural substances as flavonoids [2–4]. Bioactive polyphenols, flavonoids being among them, are essential secondary plant metabolites that are widely present in the human diet. By reducing oxidative stress, modulating cell signaling responses, regulating expression of certain genes or interacting directly with proteins and membranes, flavonoids play multiple important roles in plant defense and prevent numerous human pathologies with different mechanisms such as neurological diseases, cardiovascular diseases, infectious diseases and diabetes [5–10]. The potential of polyphenols (or their synthetic derivatives) as effective antibacterial and antiviral agents with a targeted mode of action, low side effects and the ability to avert bacteria toxin production and biofilm formation have been widely studied and well-documented [11,12].

Specific and nonspecific interactions of flavonoids with prokaryotic proteins and cell wall and membrane components influence bacterial growth, metabolism and pathogenesis, and it also plays an important role in the antibacterial effects of flavonoids [13]. These interactions include covalent bond formation, hydrogen bonding and hydrophobic interactions with bacterial cells. As was shown by proteomic analysis, the action of flavonoids perturbed different groups of bacterial enzymes involved in DNA-related metabolic processes (i.e., bacterial topoisomerases, helicases and DNA gyrases) and influenced ribosomal and membrane proteins, enzymes of cellular transport, bioenergy and lipid metabolisms, cell wall enzymes (i.e., efflux pumps and transporters, ATP synthase, cytochrome c and β-ketoacyl-acyl carrier protein synthases) and other targets [14,15]. These interactions impaired bacterial homeostasis, inhibited toxin and biofilm production and inhibited bacterial motility [14,15]. Recently Yaun et al. showed that the antibacterial activity of plant flavonoids against Gram-positive bacteria correlated with lipophilicity and that the bacteria cell membrane was the main site of flavonoid action [16]. The mechanism of the antimicrobial effect involved damage to phospholipid bilayers and inhibition of the bacterial respiratory chain or ATP synthesis [16].

The flavonol quercetin, one of the most abundant flavonoids, effectively inhibits growth of different drug-resistant Gram-positive and Gram-negative bacteria as well as fungi and viruses due to disruption of bacterial cell walls and membrane damage, intercalation of quercetin with DNA and inhibition of nucleic acid and protein synthesis, reduction of expression of virulence factors, inhibition of the activities of essential virulent enzymes and prevention of biofilm formation. Furthermore, specific structural modifications of quercetin have been shown to enhance its antimicrobial activity [12]. Unique antibacterial properties of quercetin in combination with antibiotics (oxacillin, ampicillin, vancomycin, gentamicin and erythromycin) against methicillin-resistant *S. aureus* were unraveled earlier. Simultaneously, quercetin induced morphological changes and aggregation of *S. aureus* cells, as assessed by electron microscopy [17–19]. In comparison, Wang at al. found that quercetin did not affect *S. aureus* strain (*S. aureus* Newman and its CoA mutant) viability but inhibited bacteria-secreted coagulase, which is essential for the virulence of *S. aureus*. Through this mechanism, quercetin inhibited *S. aureus* virulence. Furthermore, quercetin treatment reduced the retention of bacteria on catheter surfaces, decreased the bacterial load in the kidneys and alleviated kidney abscesses in vivo [20].

It was suggested that the antibacterial activity of another flavonoid, the flavanone naringenin, was dependent on the inhibition of the penicillin-binding protein PBP-2a [21]. The flavonol galangin, when combined with ceftazidime and amoxicillin, caused morphological damage to ceftazidime-resistant *S. aureus* and amoxicillin-resistant *E. coli* [19]. Most likely, the synergic effects of the above compounds were largely due to the variety of mechanisms of flavonoids and antibiotic effects on bacterial cells. In many cases, chalcones (bioprecursors of plant flavonoids) possessed stronger antibacterial activities in comparison with other classes of flavonoids. For example, 3-hydroxychalcone exhibited an approximately six-fold more pronounced effect on *H. influenza* than did the antibiotic azithromycin [22].

The antibacterial activity of flavonoids can be realized directly upon exposure of bacterial cells to the polyphenols, and it can be realized indirectly through effects on virulence factors of microorganisms. During the direct action of polyphenols on microorganisms, the mechanisms of antibacterial activity of flavonoids include [10,23–28]:

1. Influences on the genetic apparatus, repression of genes or inhibition of nucleic acid synthases;
2. Impairments of the membrane bilayer and membrane proteins, changes in membrane structure, fluidity and permeability, membrane pore formation and depolarization and ion leakage;
3. Changes in bacterial metabolism, perturbation in bacterial homeostasis or enzyme inhibition (e.g., DNA gyrase inhibition);
4. Inhibition of adhesions and microbial growth;

5. Reactive oxygen species (ROS) generation;
6. Controlling multidrug resistance, inactivation of bacterial efflux pump transporters (multidrug-resistance pumps) in bacteria and increasing susceptibility to antibiotics;
7. Inhibition of binding to target cells;
8. Metal ion chelation;
9. Perturbation in cell envelope metabolism and envelope synthesis through the inhibition of fatty acids synthesis and;
10. Damage to the bacterial respiratory chain, energy transduction mechanism uncoupling, inhibition of ATP synthase and disruption of bioenergetic status.

Recently, in addition to antibiotic approaches for combating bacteria, topical approaches have become a strategy for affecting virulent factors of microorganisms. This is based on interference with aspects of infectious pathogenesis by its efficient neutralization. It is suggested that such an approach will minimize the risk for drug resistance [29].

Flavonoids can also influence virulence factors, providing an antibacterial effect by:

1. Prevention of toxin secretion, inactivation of toxins and bacterial lipopolysaccharides by interacting with the toxins and changing their conformation and activities [30];
2. Inhibition of biofilm formation and cell adhesion by action on target cells to enhance their toxin resistance [31] and;
3. Perturbation in organization of bacterial quorum and intercellular communication [32].

Recently, we analyzed the antihemolytic activity of four hydrolysable tannins, having different molecular masses and flexibilities, against two *Staphylococcus aureus* strains (8325-4 and NCTC 5655), and we showed a membrane-modifying action of the tannins. We found a correlation between the antihemolytic activity of the tannins studied and their capacity to increase the ordering parameter of the erythrocyte membrane and to change cell zeta-potential [29]. The antihemolytic effect of tannins from sumac leaves (*Rhus typhina* L.) under exposure to *S. aureus* seems to be largely associated with their effects on erythrocyte membrane structure. The sumac tannins incorporated into the erythrocyte membrane and induced transformation of discocytes into echinocytes, which enhanced the rigidity of the erythrocyte lipid bilayer and prevented membrane interaction with bacterial toxins [31].

Despite numerous investigations, the mechanisms of antibiotic and antivirulence activities of flavonoids require further research. For evaluation of antibacterial/antivirulence potential of some flavonoids, we used the clinically important *Staphylococcus aureus* strain NCTC 5655. *S. aureus* is a highly pathogenic Gram-positive, aerobic and toxin-producing microorganism, and it is one of the main microbial food contaminants [33] that causes a wide range of human and animal diseases [34,35]. The pathogenicity of *S. aureus* is largely due to its ability to secrete enzymes and toxins, primarily α-hemolysin (αHL) [29]. This αHL-secreting bacterial strain is widely used to understand the mechanism(s) of pathogenicity of bacteria and evaluation of the targets of antimicrobial drugs [2,3,15,17,20,29,34,36]. At the same time, due to vast genetic plasticity and strain variability, the specificity of quercetin and other flavonoids should be verified using other pathogens. αHL is a small, β-barrel, self-assembling, pore-forming toxin that has long been recognized as an important cause of tissue and cell damage during infection [37]. Bacterial cells secrete αHL, a mature, extracellular, water-soluble, monomeric protein of 293 amino acids (33 kDa). Seven αHL monomers are oligomerized under contact with host cell membranes. The αHL oligomer creates a hemolytic pore in the host cell membrane, which results in a loss of host cell integrity and cell death [38,39]. Recently, it was shown that some flavonoids (e.g., baicalin, kaempferol and quercetin) effectively inhibited the hemolytic activity of *S. aureus* αHL by modulating the expression of virulence factors [40]. In our experiments, we have shown for the first time another mechanism of quercetin's antihemolytic activity, namely the ability to increase the resilience of the erythrocyte membrane to the toxin. The goal of this study was to compare the antihemolytic activity of flavonoids against hemolysis induced by α-hemolysin and the membrane-modifying action of flavonoids.

For analysis of structure–antibacterial activity relationships, we compared three molecules of flavonoids possessing high biochemical activity and representing the different most abundant classes: quercetin (flavonols), catechin (flavanols or flavan-3-ols), and naringenin (flavanones). To characterize the influence of the flavonoids on biophysical properties of erythrocytes and *S. aureus* bacteria, we measured cell diameter, zeta-potential, membrane organization and fluidity using fluorescence probes as well as electrokinetic and lipid bilayer techniques. To determine the susceptibility of the bacterial strain to the applied flavonoids, the lowest concentrations of the flavonoids that prevented bacteria growth (MIC) were assayed. A hemolysis assay was used for evaluation of protective effects of the selected flavonoids under erythrocyte exposure to *S. aureus*.

2. Results

The present work investigated the mechanism(s) of the antibacterial and protective effects against *S. aureus* of the three flavonoids, quercetin, naringenin and catechin (Figure 1), which differ in structure, the number of OH-groups and lipophilicity (lipophilicity refers to the tendency of a compound to partition between a lipophilic organic phase and a polar aqueous phase) (Table 1).

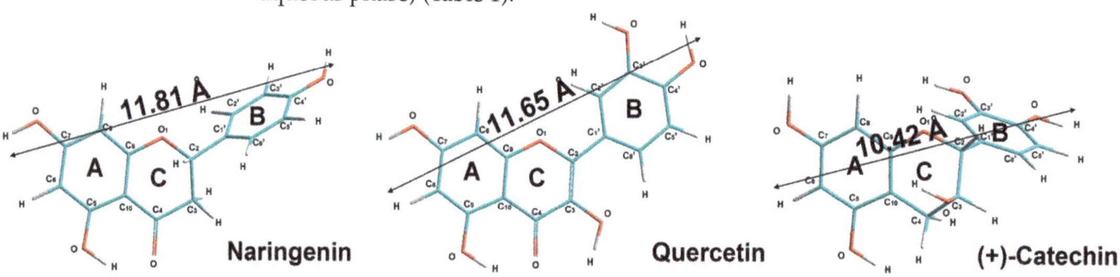

Figure 1. Optimized molecular structures of quercetin, catechin and naringenin and linear sizes of molecules (by ab initio method with 6-31G basis and unrestricted Hartree–Fock (UHF) method with Polak–Ribière gradient algorithm).

Table 1. Antimicrobial activities, interaction with liposomal membrane, water solubility (20–25 °C) [41–43] and calculated molecular parameters of the flavonoids studied.

Parameter	Naringenin	Quercetin	(+)-Catechin
Minimal inhibitory concentrations (MICs), μM	200	100	150
Dipole moments, D	1.602	0.986	2.107
Torsion angles (C3-C2-C1′-C2′)	86	180	118
Stern–Volmer constants of DPH fluorescence quenching in liposomal membranes, μM^{-1}	0.012 ± 0.003	1.66 ± 0.20	0.0012 ± 0.0002
Water solubility, mg/L	4.38	0.51	2260

2.1. Antimicrobial Effects of the Flavonoids

The antimicrobial activity of the flavonoids studied against the *S. aureus* NCTC 5655 strain was evaluated as the minimum inhibitory concentration (MIC) that stops bacterial growth. As Table 1 shows, quercetin had the greatest inhibitory effect, whereas naringenin and catechin demonstrated somewhat weaker effects. Simultaneously, Table 1 presents some flavonoid molecular parameters (e.g., dipole moment and torsion angle) calculated using our quantum-chemical considerations.

2.2. Effects of Flavonoids on Diameter, Zeta-potential and Membrane Structure of the Bacterial Cell

It is known that changes in membrane structure and integrity play a crucial role in bacterial growth and survival [44,45]. To evaluate the interaction of the flavonoids with cell membranes of viable *S. aureus* cells, we carried out a comparative analysis of the effects of the flavonoids on the diameter, zeta-potential (Figure 2) and membrane organization (Figure 3) of bacteria.

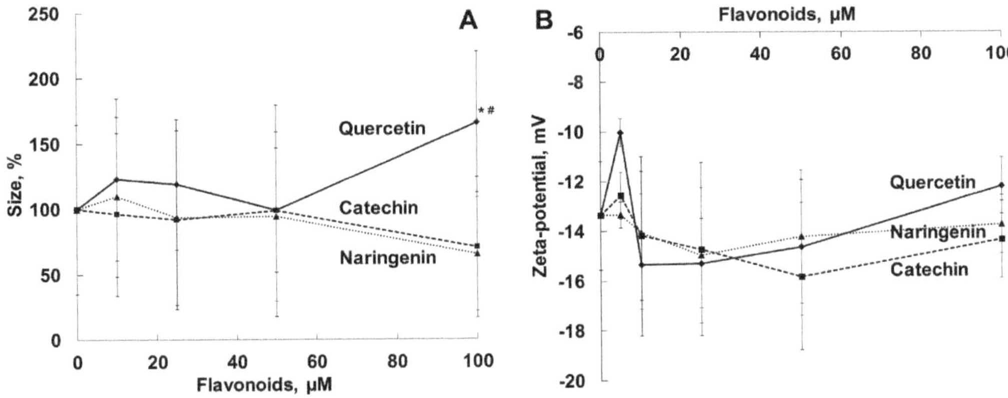

Figure 2. Effects of quercetin, catechin and naringenin on *S. aureus* cell diameter (**A**) and zeta-potential (**B**). Bacterial cells were suspended in PBS, pH 7.4. *—$p < 0.05$ in comparison with the cells in the absence of the flavonoids; #—$p < 0.05$ in comparison with the cells in the presence of other flavonoids.

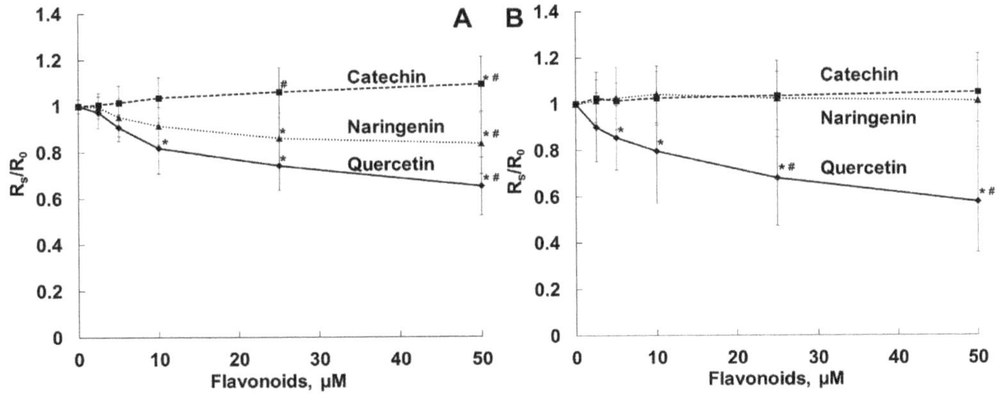

Figure 3. Dependence of fluorescence anisotropy of DPH (**A**) and TMA-DPH probes incorporated in bacterial membranes (**B**) on flavonoid concentrations. Bacterial cells were incubated with TMA-DPH or DPH at a final concentration of 1 µM in PBS, pH 7.4, and anisotropy values were recorded in the absence (R_0) and in the presence (Rs) of the flavonoids. *—$p < 0.05$ in comparison with dye fluorescence anisotropy in the absence of the flavonoids; #—$p < 0.05$ in comparison with dye fluorescence anisotropy in the presence of other flavonoids.

The measurements of the diameters of *S. aureus* cells in the presence of naringenin, quercetin and catechin showed that the interaction of quercetin at a concentration of 100 µM (close to the MIC) with bacteria in the medium resulted in an increase in particle diameter (by 60%). The diameter changes in the presence of catechin and naringenin at concentrations of 10 to 100 µM were not statistically significant (Figure 2A). (We represented the changes in particle diameter in percent due to the high dispersity of the values.) We did not observe any statistically significant changes in bacterial zeta-potential in the presence of the flavonoids studied (Figure 2B).

Figure 3A,B show changes in fluorescence anisotropy (Rs/R_0) of DPH and TMA-DPH probes incorporated into bacterial cell membranes in the presence of the flavonoids studied. This parameter characterizes the state of membranes in the inner hydrophobic and in the aqueous/membrane interface regions, respectively. In our experiment, lipophilic quercetin decreased the Rs/R_0 parameter for both probes, which indicates increased fluidity of bacterial cell membranes at different depths [46,47]. Naringenin also, but to a somewhat lesser extent, reduced the value for DPH fluorescence anisotropy and did not affect the TMA-DPH fluorescence anisotropy value. We did not observe any effects of hydrophilic catechin on the fluorescence anisotropy of either probe.

2.3. Antihemolytic Activities of Flavonoids During Sheep Erythrocyte Hemolysis Caused by S. aureus (NCTC 5655 Strain) and Flavonoid Effects on Erythrocyte and Liposomal Membrane Structures

Antihemolytic activities of the flavonoids were evaluated using the *S. aureus* NCTC 5655 strain, which secretes only αHL toxin [48], and sheep erythrocytes, which manifest high sensitivity to αHL [49]. Preliminary experiments showed that the flavonoids did not induce erythrocyte hemolysis in the range of the concentrations used (1 to 75 μM). The exposure of erythrocytes to NCTC 5655 cells caused effective hemolysis. As Figure 4 shows, quercetin, but not naringenin or catechin, effectively and dose-dependently prevented *S. aureus*-induced hemolysis.

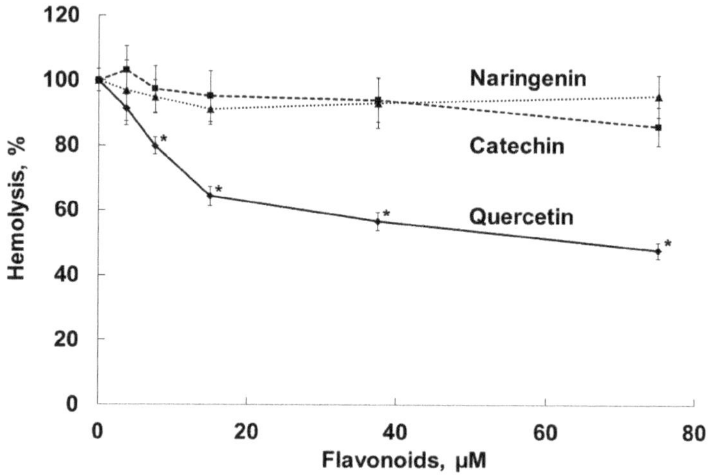

Figure 4. Inhibition by flavonoids (naringenin, quercetin and catechin) of sheep erythrocyte lysis induced by *S. aureus* NCTC 5655 in PBS, pH 7.4, 37 °C, 60 min. Erythrocytes (1% hematocrit) were preincubated with the flavonoids over 30 min, and then 100 μL of bacteria suspension in PBS buffer (OD_{600} = 2.1) was added. The results represent the average values ± standard deviation, n = 6. Hemolysis induced by *S. aureus* in the absence of flavonoids was taken as 100 %. *—$p < 0.05$ compared with erythrocytes in the absence of the flavonoids.

Since the experimental conditions included the presence of quercetin in the medium during hemolysis, the antihemolytic activity of the polyphenol could be associated with neutralization of the αHL toxin itself and/or with modification of the erythrocyte membrane structure, which can limit the effect of αHL on erythrocytes. It is known that αHL oligomerization in the host cell membrane results in pore (transmembrane channels) formation and induced lysis of erythrocytes or other cells [50].

The antihemolytic activity of quercetin was studied using erythrocytes preincubated with polyphenol and subsequently washed with PBS (Figure 5).

The subsequent erythrocyte washing after preliminary exposure to the flavonoid did not remove the tightly bound lipophilic flavonoid and did not change the antihemolytic efficiency of quercetin (naringenin and catechin did not inhibit hemolysis). Quercetin molecules incorporated into host cell membranes, thus changing the membrane properties and preventing the hemolytic activity of αHL.

To gain a deeper insight into the protective effect of the flavonoids on erythrocytes, we studied the effect of these compounds on the structural organization of the erythrocyte membrane by measuring the fluorescence anisotropy values for the DPH and TMA-DPH probes, which differ in their membrane localization (Figure 6).

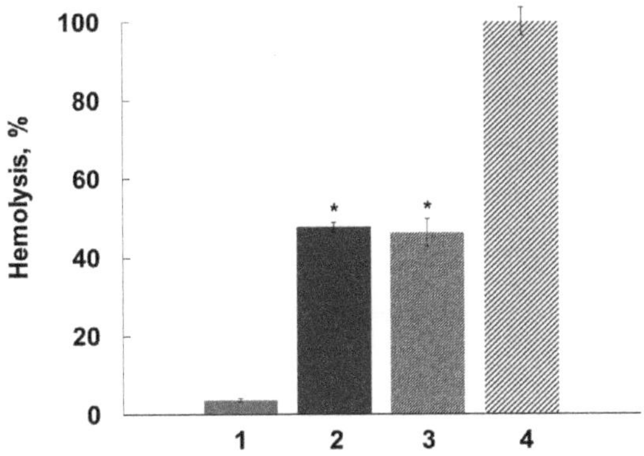

Figure 5. Inhibition by quercetin (75 μM) of sheep erythrocyte lysis induced by *S. aureus* NCTC 5655. 1—hemolysis in the absence of bacterial cells; 2, 3 and 4—hemolysis in the presence of bacterial cells in the medium; 2—hemolysis in the presence of quercetin; 3—hemolysis of erythrocytes washed after preliminary incubation with quercetin; 4—hemolysis induced by *S. aureus* NCTC 5655 in the absence of quercetin. The experimental conditions were similar to those shown in Figure 4. *—$p < 0.05$ in comparison with erythrocytes exposed to bacteria in the absence of flavonoids.

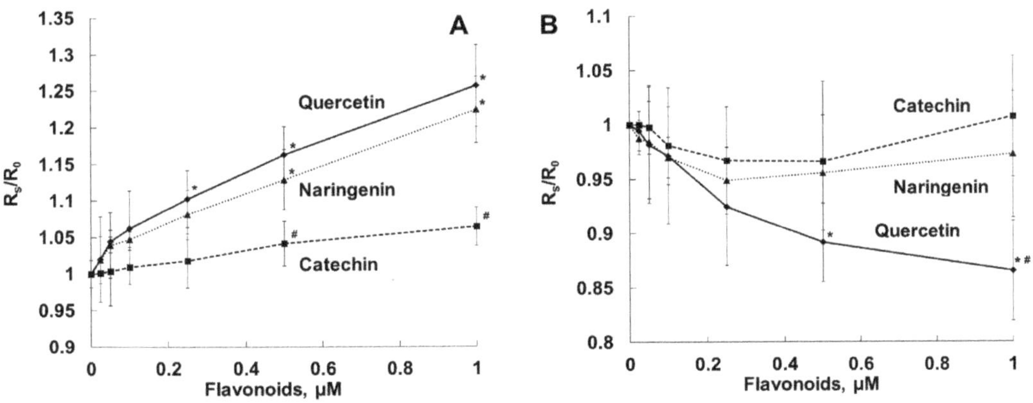

Figure 6. Dependence of fluorescence anisotropy of DPH, located in the hydrophobic region of the erythrocyte membrane (**A**), and TMA-DPH, located in the outer monolayer of the erythrocyte membrane (**B**), on the concentration of the flavonoids. Suspension of erythrocytes (0.01% hematocrit in PBS) was labeled with the fluorescent probes (DPH or TMA-DPH) at a concentration of 1 μM in the dark for 20 min at 37 °C, and then the flavonoids were added and incubated for 30 min at 37 °C. Anisotropy values were registered in the absence (R_0) and in the presence (Rs) of the flavonoids. *—$p < 0.05$ in comparison with dye fluorescence anisotropy in the absence of the flavonoids; #—$p < 0.05$ in comparison with dye fluorescence anisotropy in the presence of other flavonoids.

Using the DPH probe, it was found that quercetin, naringenin and catechin, at low concentrations of up to 1 μM, enhanced fluorescence anisotropy of the probe located in the inner region of the erythrocyte membrane, which is indicative of increased rigidity of the membrane hydrophobic region (Figure 6A). The effect increased in the order of hydrophilic catechin < naringenin < lipophilic quercetin. Applying the TMA-DPH probe, we showed that quercetin, at a concentration of up to 1 μM, diminished probe fluorescence anisotropy, while naringenin and catechin did not significantly influence the membrane fluidity (Figure 6B). In the fluorescence experiments, we used the ratio of the [flavonoid]/[erythrocytes], similar to that in the hemolysis experiments.

The DPH probe, incorporated into the inner part of unilamellar liposomal DMPC vesicles, was also used for assessment of the penetration of the flavonoids into the lipid bilayer. Figure 7 represents the Stern–Volmer plots of the DPH fluorescence quenching by the flavonoids. The calculated Stern–Volmer constants, Ksv, which show the affinity of the flavonoids for the bilayer, are listed in Table 1. Ksv correlated with water solubility of the flavonoids (Table 1).

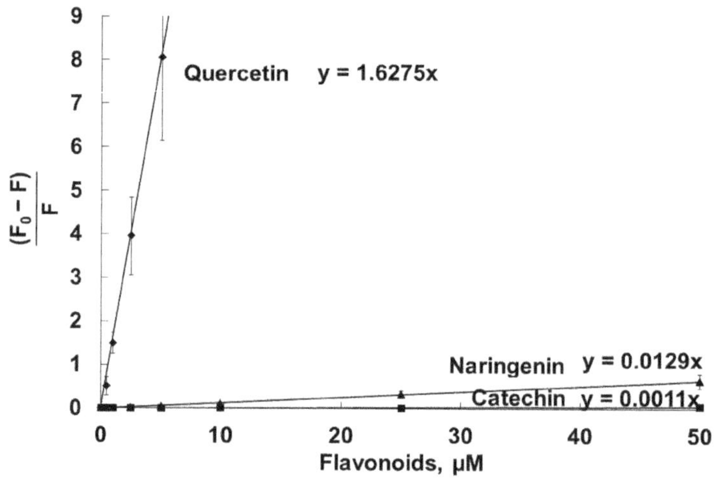

Figure 7. Stern–Volmer plots of quenching by catechin, naringenin and quercetin of the fluorescence of the DPH probe, located in the hydrophobic part of unilamellar liposomal DMPC vesicles. Liposomes (100 µg/mL, PBS, pH 7.4) were labeled with the DPH fluorescent probe at a final concentration of 1 µM, and then the flavonoids were added and incubated at 25 °C for 30 min. Fluorescence intensity values were registered in the absence (F_0) and in the presence (F) of the flavonoids.

3. Discussion

The flavonoids naringenin, quercetin and catechin possess a wide range of biological and pharmacological activities, including their antioxidant, antibacterial, anti-inflammatory, cytoprotective, antitumor and antidiabetic actions [51,52]. At the first step of our work, we studied the effect of naringenin, catechin and quercetin on the NCTC 5655 *S. aureus* strain cell viability and showed that the flavonoids (MIC values were 100–200 µM, Table 1) exerted antibacterial effects. The antibacterial activities of numerous flavonoids were extensively evaluated previously. Table 2 lists the minimal inhibitory concentrations (MICs, the lowest concentrations of substances that prevent visible growth of bacterial strains) of the flavonoids. As can be seen, the MIC values obtained by the authors vary considerably.

Table 2. Minimal inhibitory concentrations of the flavonoids, MICs, for preventing bacteria cell growth.

Flavonoid	MICs	Bacterial Strain	References
Quercetin	120 µM (35.76 µg/mL)	*E. coli*	[52]
	> 3410 µM (> 1024 µg/mL)	*S. aureus* strain Newman	[20]
	830 µM (250 µg/mL)	methicillin-susceptible *S. aureus*	[53]
	1670 µM (500 µg/mL)	methicillin-resistant *S. aureus*	[54]
	> 3330 µM (> 1000 µg/mL)	*S. aureus*	[55]
	1670 µM (500 µg/mL)	*E. coli*	[55]
	170 µM (50 µg/mL)	*S. aureus*	[56]
	200 µM (60 µg/mL)	*P. gingivalis*	[57]
Naringenin	460 µM (125 µg/mL)	methicillin-resistant *S. aureus*	[21]
Glycoside naringin	1720 µM (1000 µg/mL)	*S. aureus*	[56]
Catechin	3550 µM (1000 µg/mL)	*S. aureus*	[56]

Numerous investigations have previously demonstrated that flavonoids effectively reduced bacteria adhesion, changed membrane stability and permeability, reversed antibiotic resistance and improved antibiotic efficacy. As was shown in the case of resistant strains of Staphylococcus spp., quercetin, at sub-inhibitory concentrations, prevented biofilm formation [54]. It was suggested that hydroxylation of the C5, C7, C3′ and C4′ positions of flavonoid molecules as well as geranylation or prenylation at C6 increased the antibacterial effect. On the other hand, the 5′-OH group (as in quercetin) or methoxylation at C3′ and C5 positions has been reported to decrease flavonoids' antibacterial action [27,36,58].

A recent study showed high efficiency of myricetin as an αHL inhibitor [59]. The flavonoid simultaneously reduced the amount of *S. aureus*-produced αHL and neutralized its activity by interfering with αHL polymerization [59]. Lin et al. showed that flavonoids considerably suppressed growth of *S. aureus* or *E. coli* and noticed changes in activities of genes responsible for metabolic control and genetic information processing [24]. Decreased activities of ribosomal proteins [60], impaired synthesis of nucleic acids and inhibition of bacterial topoisomerases by polyphenols were found [61,62]. It was established that flavonoids were capable of enhancing the permeability of *S. aureus*, *E. coli* and *P. aeruginosa* membranes to ions [3,24]. Cushnie and Lamb described leakage of potassium in galangin-induced damage of the *S. aureus* cytoplasmic membrane [3]. In their earlier work, Ikigai et al. evaluated the antibacterial effects of green tea (*Camellia sinensis*) extracts, (−)-epigallocatechin gallate and (−)-epicatechin and demonstrated that the strong bactericidal action of (−)-epigallocatechin gallate caused considerable damage to phosphatidylcholine liposomes. A correlation between the antibacterial activity of catechins against *S. aureus* and *E. coli* and their damaging effect on liposomal membranes was shown [63]. Eumkeb and Chukrathok related apigenin- and naringenin-induced changes in the *Enterobacter cloacae* cytoplasmic membrane to inhibition of energy metabolism and metabolic disorders [19]. Haraguchi et al. showed that chalcones isolated from roots of *Glycyrrhiza inflata*, which demonstated antimicrobial activity, effectively inhibited NADH cytochrome c reductase and oxygen consumption in susceptible bacterial cells. The site for respiratory inhibition was thought to be between CoQ and cytochrome c in the bacterial respiratory chain [64]. Mirzoeva and coauthors found uncoupling of the energy-transducing cytoplasmic membrane and inhibition of the proton-driving force and cell motility of *S. aureus* by an ethanolic extract of propolis and its cinnamic and flavonoid components (quercetin). These effects on the bioenergetic status of the membrane may contribute to the antimicrobial action of propolis [65]. Li and coauthors found that procyanidins elevated the activities of Na^+/K^+-ATPase and Ca^{2+}-ATPase and decreased those of malate dehydrogenase and adenosine triphosphatase in *S. aureus* [66].

The capacity for neutralizing bacterial toxins is an important property of flavonoids. Flavonoids can affect different virulence factors and are capable of neutralizing pore-forming toxins and blocking/modifying the channels formed by them in the membranes of host cells. The antivirulence activities of flavonoids with respect to staphylococcus enterotoxin A depended on the polymerization degree of the flavonoids [67]. High-molecular-mass polyphenols form aggregates with αHL, inhibiting its effect on erythrocytes [68]. An antihemolytic effect was observed when polyphenol-pretreated erythrocytes were washed before addition of *S. aureus*-produced αHL, which agrees with our results and confirms the capacity of plant polyphenols to produce antihemolytic activities at the level of erythrocyte membranes [68].

Polyphenols were demonstrated to inhibit the anion-selective and permeable to urea channel formed by vacuolating cytotoxin (VacA) produced by *Helicobacter pylori* [69] and modify the conductivity of the pore formed by *S. aureus* αHL [70]. It was established that proantocyanidines were capable of detoxifying lipopolysaccharide, the endotoxin of Gram-negative bacteria [71]. The polymeric fraction of black tea thearubigin (polymers of tea catechins) blocked the neuromuscular effects of neurotoxins A, B and E produced by Clostridium botulinum due to catechin chelation of the metalloproteinase moiety of the toxin [33,72]. Black tea caempherol and quercetin glycosides also inhibited the toxic effects of botulotoxins [33]. Morianga et al. reported that polyphenol compounds inactivated the virulence of cholera toxin [73]. The antihemolytic activity of polyphenols may be associated with their effect on bacterial metabolism, which causes a diminution of the amount of toxins released, as was shown earlier for (–)-epigallocatechin [74].

At the next step, we analyzed the influence of the flavonoids on the bacterial cell membrane properties as one of the possible mechanisms of the antimicrobial effects of the flavonoids against αHL-producing NCTC 5655 *S. aureus*. Quercetin (100 μM), but not naringenin or catechin, increased the bacterial cell nanosize measured, probably as a result of cell aggregation (Figure 2A). In our previous experiment, the incorporation of the flavonoids quercetin and naringenin (but not catechin), into the liposomes enlarged the area of the bilayer [39]. He and coworkers have recently shown that quercetin, at sub-MIC concentrations, considerably increased the aggregation rate of *Porphyromas gingivalis* and cell surface hydrophobicity [57]. In our work, *S. aureus* zeta-potential, an electrochemical characteristic of the cell surface, was measured in PBS, pH 7.4, and was found to be -13.4 ± 1.7 mV. The effect of the flavonoids studied on bacterial zeta-potential, as a possible marker of changes in membrane permeability, was not statistically significant (Figure 2B).

In our work, quercetin (10–50 μM) significantly increased the fluidity of the outer layer and the surface area of bacterial cell membranes. In contrast, the antibacterial activity of 2R,3R-dihydromyricetin against *S. aureus* was related to flavonoid-induced morphological changes and a considerable reduction in fluidity of the membrane inner part (using the DPH probe), membrane hyperpolarization and disruption of membrane integrity [75]. Naringenin enhanced the fluidity of the surface area to a lesser extent, and hydrophilic catechin did not affect the fluidity of the bacterial membrane. Earlier we showed that quercetin (10–50 μM), rather than catechin and naringenin, strongly decreased the microfluidity of the liposomal membrane bilayer at different depths and hydration in the region of polar head groups [39]. The different effects of the flavonoids on fluidity of the surface area of cellular and liposomal membranes may be due to flavonoid interactions with the protein component of the biological (bacterial) membrane. A significant correlation between the antimicrobial activity of flavonoids and the liposomal membrane rigidification effect was found earlier, and it was suggested that the flavonoids exerted their antibacterial effect by reducing membrane fluidity [53].

In our experiments, by increasing the fluidity of the bacterial membranes (Figure 3), quercetin, but not naringenin or catechin, influenced *S. aureus* toxicity and inhibited the hemolysis of sheep erythrocytes induced by exposure to *S. aureus* (Figure 4). The antihemolytic effect did not attenuate after washing erythrocytes preliminarily exposed to the flavonoid, as a result of tight binding of quercetin to the erythrocyte membrane (Figure 5). It should be noted that quercetin influenced bacterial membrane organization and inhibited *S. aureus*-induced hemolysis at concentrations lower than the average value for MICs (Table 1). The calculated IC_{50} value, corresponding to the concentration of quercetin that inhibits hemolysis by 50%, was 65 ± 5 μM for the *S. aureus* NCTC 5655 strain.

We demonstrated that the flavonoids (in the order of catechin < naringenin < quercetin) enhanced the stiffness of the erythrocyte membrane hydrophobic region. Simultaneously, quercetin, but not naringenin or catechin, increased the mobility of the erythrocyte membrane surface zone. We suggest that the antihemolytic effect of quercetin was associated with the effect of the flavonoid on the organization of the erythrocyte membrane lipid bilayer, which, in turn, seemed to prevent αHL-induced osmotic hemolysis. The fundamental step of pore formation consists of binding of toxin protomers to a receptor (lipids, glycans or proteins) on the surface of the target cell membrane. Receptors increase the local concentration of the toxin and also promote oligomerization [76]. As was also demonstrated previously, flavonoids could interact with phospholipid and protein components of erythrocyte membranes and protect erythrocytes against hypotonic lysis [77]. It was shown that the liquid disorder phase was preferable for binding of hemolysin to lipids and its oligomerization [78].

The mechanism of flavonoid interactions with membranes and their localization are fundamental to understanding of the pharmacological activities of flavonoids. Using the Stern–Volmer constant, Ksv, of DPH fluorescence quenching by the flavonoids in the liposomal membrane, we estimated the availability of the quencher (flavonoid) to the excited fluorophore (DPH). As Table 1 and Figure 7 demonstrate, quercetin effectively penetrates into the membrane interior. The accessibility of naringenin and catechin to the membrane interior was much lower, in accordance with flavonoid lipophilicity.

For the purposes of demonstration, we provided a molecular model of the structure of αHL consisting of seven polypeptide chains (X-ray structural analysis data were obtained from the Protein Data Bank [https://www.rcsb.org], accessed on 3 March 2021). The diameter of the pore inner cavity formed by αHL was ≈12 Å (Figure 8), whereas the linear sizes of the quercetin, catechin and naringenin molecules were equal to 11.81, 11.65 and 10.42 Å, respectively. (Figure 1). Despite similar linear sizes of the flavonoid molecules and pore diameter, only quercetin had an antihemolytic effect. Earlier, we demonstrated a planar structure of quercetin, but not of catechin or naringenin, and the optimal electron orbital delocalization due to the C2=C3 double bond in the C ring (absent in catechin and naringenin molecules) [79]. The planar geometry of the flavonoid molecule may be crucial for antibacterial activity [80]. Crystallographic data previously showed that *S. aureus* α-hemolysin, a cytolytic endotoxin, formed the mushroom-shaped homo-oligomeric heptamer, which was a solvent-filled transmembrane channel with a hydrophilic interior 100 Å in length and ranging from 14 Å to 46 Å in diameter [81]. Similarly, using penetrating water-soluble polymers, the authors estimated two practically identical openings of the channel with radii of 1.2–1.3 nm and two apparent constrictions with the radii of 0.9 nm and 0.6–0.7 nm, occurring in the channel lumen [82]. These findings correlated with the data indicating that the α-hemolysin channel had a wide vestibule leading into the pore with the diameter of 15 Å [83]. Different organic molecules could block a single αHL-formed pore [70,84]. The dipole moment and the net charge of penetrating molecules play an important role in the transportation through nanopores [83]. The dipole moment of lipophilic quercetin was the lowest, and the largest percentage of quercetin molecules were ionized at the pH of the medium in comparison with the other flavonoids studied. Ostroumova and coauthors showed that the flavonoids hydroxylated in position five of the A-ring and in position 4′ of the B-ring specifically interacted with the voltage sensor of the αHL pore and shifted the αHL-formed channel from a high- to low-conductance state [70]. Olchowik-Grabarek et al. demonstrated that iodogallic acid formed static complexes with α-hemolysin in solutions and inhibited its nanopore conduction in artificial lipid bilayers [85]. It was shown in the case of large tannins that more flexible molecules exerted a more pronounced antihemolytic effect [29].

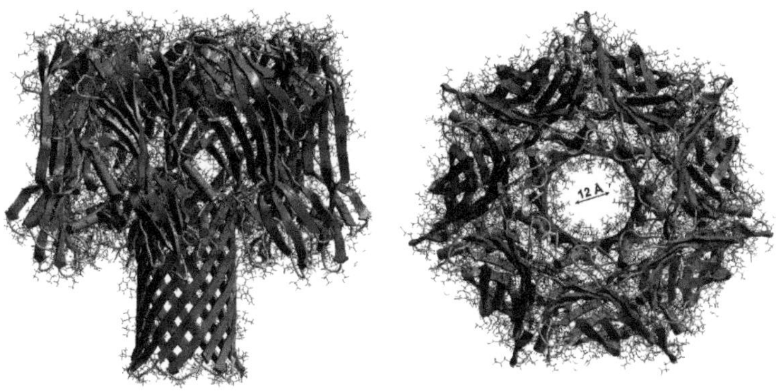

Figure 8. Molecular structure of *S. aureus* α-hemolysin-assembled pore consisting of seven monomers. The diameter of the formed inner cavity is ≈ 12 Å.

Thus, summarizing the results obtained, we can conclude that the flavonoids effectively interacted with biological (bacteria and erythrocytes) and artificial membranes and changed the dynamic characteristics and organization of the membranes. Recently, it was demonstrated that subinhibitory concentrations of antibiotics may not alter membrane fluidity of bacterial cells [86]. In our experiments, quercetin (but not catechin or naringenin) exerted pronounced effects on the membrane structure, fluidity and size of bacterial cells. The high lipophilicity, the C2=C3 double bond and the planarity allowed the quercetin molecule to more effectively inhibit *S. aureus*-induced osmotic hemolysis. Recently, it was concluded that the C2=C3 double bond in flavonoid molecules enhanced Gram-positive bacterial inhibition [58]. It has already been shown that flavonoid molecules (such as quercetin), predominantly located in the hydrophobic region of the membrane bilayer, can initiate the formation of raft-like domains, while the flavonoid molecules located in the polar interface region of the bilayer can fluidize membranes [87]. A specific interaction between flavonoids and some integral membrane proteins is also possible [87]. The antihemolytic activity of quercetin may be due to stiffening of the inner hydrophobic part of the erythrocyte membrane and fluidization of the membrane surface, resulting in a decrease in host cell sensitivity to toxin-induced hemolysis. It is known that the membrane lipid phosphatidylcholine was originally proposed to be an α-hemolysin receptor due to the high-affinity binding observed between αHL and clustered phosphocholine head groups [88]. Previously, it was suggested that the interactions of tannins with the phosphate groups of phospholipids caused stiffening of the erythrocyte membrane outer layer, leading to restriction in the αHL incorporation into the membrane and limitation of the formation of a functional channel [29]. In our experiments, we used sheep erythrocytes that had not been found to possess any specific proteinacious receptors for αHL [29]. One of the possible antihemolytic mechanisms of flavonoids may be physical blocking of the αHL-formed pore. Molecular modeling showed a good agreement of the linear sizes of the polyphenols studied with the diameter of the αHL inner cavity.

4. Conclusions

The emergence of microorganisms with multiple drug resistance and high virulence requires a search for new therapeutic strategies to treat resistant infections. To gain a deeper insight into their antibacterial and protective activities, we carried out a comparative analysis of the effects of flavonoids on the diameter, zeta-potential and membrane organization of viable *S. aureus* cells and on the structure of sheep erythrocyte membranes. The accessibility of naringenin and catechin to the membrane interior was much lower when compared with quercetin, in accordance with flavonoid lipophilicity, as revealed by fluorescence studies. In our experiments, naringenin, catechin and quercetin (100–200 μM) inhibited *S. aureus* cell growth depending on the distorting effect on the bacterial membrane structure. Only lipophilic quercetin at concentrations of 10–80 μM, and not catechin or naringenin, inhibited sheep erythrocyte hemolysis induced by the *S. aureus*-produced αHL toxin. The antihemolytic and membrane-modifying effects were observed at quercetin concentrations well below the minimal inhibitory concentration values for *S. aureus* growth. Using spectroscopic, electrokinetic and lipid bilayer techniques, we report a novel mechanism of antihemolytic activity of quercetin, which consists of enhancing resistance of target cells (erythrocytes) to αHL due to the membrane-modifying effect, restriction of interactions of erythrocytes with the toxin, and prevention of osmotic hemolysis. Previously, it was demonstrated that most antibiotics have intracellular bacterial targets [86]. We confirmed that alteration of bacterial and host cell membranes could be one of the direct modes of antibacterial action of flavonoids.

5. Materials and Methods

5.1. Materials

Naringenin (98%), product no. W530098; quercetin ($\geq$95%), product no. Q4951; (+)-catechin hydrate ($\geq$98%) product no. C1251; 1,6-diphenyl-1,3,5-hexatriene (DPH),

1-(4-trimethylammoniumphenyl)-6-phenyl-1,3,5-hexatriene (TMA-DPH) and 1,2-dimyristoyl-sn-glycero-3-phosphocholine (DMPC) were from Merck/Sigma-Aldrich (St Louis, MO, USA, or Steinheim, Germany). Mueller–Hinton (MH) agar and Mueller–Hinton (MH) broth suitable for microbiology were supplied by Oxoid (Basingstoke, UK). Other reagents and organic solvents were of analytical grade, purchased from POCh (Gliwice, Poland) and used without further purification. Sheep blood was obtained from GrasoBIOTECH (Graso Company, Starogard Gdański, Poland). The flavonoids were used as freshly prepared stock solutions (15 mM) in ethanol. In the preliminary experiments, we showed that ethanol, at the concentrations used (not exceeding 1.3%), did not considerably affect the parameters measured.

5.2. *Bacterial Strain and Growth Condition*

The *S. aureus* strain NCTC 5655, obtained from the National Collection of Type Cultures (Salisbury, England), was used. The bacteria were grown on Mueller Hinton (MH) nutrient agar plates. Before the experiments, the bacteria were grown overnight at 37 °C in Mueller–Hinton (MH) broth with 200 rpm shaking.

5.3. *Antibacterial Activity of Flavonoids*

The minimum inhibitory concentration (MIC), the lowest concentration of a substance that prevents the visible growth of bacteria, was determined by the following experimental procedure. *S. aureus* cells, strain NCTC 5655, were grown overnight at 37 °C in Mueller–Hinton (MH) broth with shaking at 200 rpm, and the cell suspension was adjusted to the absorbance OD_{600} = 0.1 at a wavelength λ = 600 nm. Then 180 μL of the obtained bacterial suspension was added to the wells of a sterile 96-well microtiter plate containing various concentrations of flavonoids in the range of 10–300 μM. Control wells did not contain flavonoids. The plate was shaken on a microplate shaker for 1 min and then was incubated for 24 h at 37 °C. The visible growth of bacteria was evaluated as an increase in OD_{600}. The lowest concentration of flavonoids preventing bacteria growth was taken as its MIC. To evaluate the MIC, the samples were examined using a microplate reader SpectraMax M2 (Molecular Device, San Jose, CA, USA). To avoid the effect of flavonoids on turbidity/absorbance of the medium, negative controls were run for each flavonoid concentration.

5.4. *Inhibition of Sheep Erythrocyte Hemolysis by Flavonoids*

A hemolysis assay was used for analysis of the protective effects of flavonoids. The hemolytic activity of *S. aureus* NCTC 5655 in the presence and in the absence of the flavonoids was measured using sheep erythrocytes as previously described by Olchowik-Grabarek et al. [30]. Erythrocytes were isolated from sheep blood by centrifugation (850× *g*, 15 min) and washed three times with isotonic buffered saline (PBS, 145 mM NaCl, 1.9 mM NaH_2PO_4, 8.1 mM Na_2HPO_4, pH 7.4) at 4 °C. Two milliliters of the erythrocyte suspension with 1% hematocrit were incubated at 37 °C for 30 min in the absence or presence of flavonoids (3.75–75 μM) in PBS buffer. In some experiments, erythrocytes were washed after preliminary exposure to the flavonoids. The *S. aureus* strain NCTC 5655 maintained in MH broth was adjusted to an optical density OD_{600} = 2.1 at a wavelength λ = 600 nm. Then 100 μL of bacteria in MH broth was added to each sample of erythrocytes. After incubation for 60 min at 37 °C, 0.5 mL of the suspension was taken from each sample and mixed with 1 mL of PBS buffer. To obtain 100% hemolysis, 0.5 mL of the suspension was mixed with 1 mL of distilled water. All the samples were centrifuged (850× *g*, 15 min, 4 °C, Hermle Z32 HK, Hermle Labortechnik GmbH, Wehingen, Germany), and the optical density of the supernatants was measured at 540 nm using a Jasco V-770 spectrophotometer (Jasco Corporation, Tokyo, Japan). The results are presented as a percent of hemolysis depending on the concentration of the flavonoids added.

5.5. Measurements of Erythrocyte Membrane Fluidity

Erythrocyte membrane organization was analyzed by fluorescence anisotropy of TMA-DPH and DPH probes (Merck/Sigma-Aldrich (St Louis, MO, USA, or Steinheim, Germany)), which differ in their membrane localization. A suspension of erythrocytes (3 mL of 0.01% hematocrit in PBS) was labeled with a fluorescent probe (DPH or TMA-DPH) at a concentration of 1 μM (20 min, 37 °C) in the dark. DPH was dissolved in tetrahydrofuran, and TMA-DPH was dissolved in methanol. The initial probe concentration was 1 mM. Fluorescence anisotropy values were recorded in the absence and presence of the flavonoids at 37 °C with a Perkin-Elmer LS 55B spectrofluorometer (Perkin-Elmer, Pontyclun, UK) equipped with a fluorescence polarization device. Changes in membrane fluidity after addition of the flavonoids (30 min, 37 °C) in the concentration range from 0.05 to 1.0 μM were determined based on the values for the fluorescence anisotropy of DPH or TMA-DPH probes (R), which were calculated by the fluorescence data manager program according to the standard anisotropy equation:

$$r = \frac{I_{VV} - GI_{VH}}{I_{VV} + 2GI_{VH}} \tag{1}$$

where I_{VV} and I_{VH} are the vertical and horizontal fluorescence intensities, respectively. The factor $G = I_{HV}/I_{HH}$ (grating correction factor) corrects the polarizing effects of the monochromator. The excitation wavelengths were 348 nm for the DPH probe and 340 nm for the TMA-DPH probe, and the fluorescence emission was measured at 426 nm for the DPH probe and 430 nm for the TMA-DPH probe. The results are presented as a ratio (Rs/R_0), where Rs is the fluorescence anisotropy of the probes in the presence of flavonoids and R_0 is the fluorescence anisotropy of the probes in the absence of flavonoids. The DPH probe was localized in the hydrophobic region of the membrane occupied by carbohydrate chains of lipids, whereas the TMA-DPH chain was predominantly localized on the border of the external environment/membrane. The fluorescence anisotropy values of the probes characterized orderliness and motility of the lipid bilayer in the inner and surface membrane regions [29].

5.6. Measurements of *S. aureus* Membrane Fluidity

The structural organization of bacterial membranes was analyzed by the fluorescence anisotropy values of the DPH and TMA-DPH probes. The number of *S. aureus* cells was photometrically standardized. Cells (cell suspension OD_{600} = 0.01 in 10 mM PBS pH 7.4) were incubated with DPH or TMA-DPH at a final concentration of 1 μM for 15 min in the dark at 37 °C. After that, various concentrations of flavonoids (2.5–50 μM) were added to the bacterial cell suspension, and the cells were incubated for 45 min at 37 °C. The values for fluorescence anisotropy of the probes incorporated in the membranes were measured in the absence and in the presence of flavonoids using a Perkin-Elmer LS 55B spectrofluorometer (Perkin-Elmer, Pontyclun, UK).

5.7. *S. aureus* Nanoscale Cell Diameter and Zeta-Potential

Zeta-potential, a tool for studying the alteration in bacterial cell surface permeability, and the average diameter of bacteria were measured using a Malvern Zetasizer Nano ZS (Malvern Instruments Ltd, Malvern, UK). Zeta-potential is an electrokinetic potential associated with the mobility of charged particles, and its changes make it possible to estimate electrostatic forces between interacting particles (bacterial cells). The *S. aureus* cell diameter (nanosize) was analyzed using dynamic light scattering (DLS). The bacterial cells (OD_{600} = 0.01) were suspended in PBS buffer, pH 7.4. The measurements were taken in disposable folded capillary cells (DTS 1070), and the data were analyzed using Malvern software (Malvern Instruments Ltd, Malvern, UK).

5.8. Interaction of Flavonoids with Liposomal Membranes

Using the Stern–Volmer constant, Ksv, of DPH fluorescence quenching by flavonoids in the liposomal membrane, we estimated the availability of the quencher (flavonoid) to the excited fluorophore (DPH) and, therefore, flavonoid penetration into the membrane interior. Liposomes (unilamellar bilayer vesicles) were prepared by an extrusion technique using an Avanti Polar Lipids Mini-Extruder (Birmingham, AL, USA) and were composed of 1,2-dimyristoyl-sn-glycero-3-phosphocholine (DMPC) (14:0), as described previously [89]. The liposomes (100 μg phospholipid/mL) were incubated with DPH at a final concentration of 1 μM for 20 min at 25 °C in PBS, pH 7.4, and fluorescence intensity values were registered in the absence and presence of flavonoids. The results are presented as a Stern–Volmer plot $[(F_0 - F)/F] = Ksv\ [Q]$, where Fo is the fluorescence intensity of the probe in the absence of flavonoids, and the F is the fluorescence intensity of the probe in the presence of flavonoids. Ksv is the Stern–Volmer constant, and [Q] is the flavonoid (quencher) concentration. The excitation and emission wavelengths were 348 and 426 nm, respectively.

5.9. Calculations of Flavonoid and S. aureus α-Hemolysin Molecular Geometries

Flavonoid molecules were theoretically considered by performing both the semi-empirical molecular orbital theory and ab initio calculations. The Austin Model 1 (AM1) semi-empirical method within unrestricted Hartree–Fock (UHF) formalism in the self-consistent field approximation and the Polak–Ribiere algorithm were considered to fully optimize the geometry of the flavonoid molecules [90]. We performed all the calculations by using the HyperChem-8.0 software package (HyperCube, Inc., Gainesville, FL, USA) for searching the conformations with minimum energy [http://www.hyper.com] (accessed on 17 March 2021).

5.10. Statistics

All data are expressed as means ± SD for four to six experiments. Differences between parameter values measured in groups were analyzed using Student's *t*-tests or non-parametric Kruskal–Wallis tests. The normality of distribution was determined by the Shapiro–Wilk test. Statistical analysis was carried out using the STATISTICA 6.0 software package (StatSoft, Inc. Tulsa, OK, USA). The level of significance was set at $p < 0.05$.

Author Contributions: Conceptualization, M.Z., I.B.Z. and A.G.V.; Methodology, E.O.-G., A.G.V. and S.S.; Software, A.G.V., I.D. and S.S.; Investigation, A.G.V., E.O.-G., A.R. and S.S.; Data Curation, E.O.-G., M.Z. and E.A.L.; Writing—Original Draft Preparation, M.Z. and E.A.L.; Writing—Review and Editing, M.Z. and I.B.Z.; Visualization, A.G.V. and E.O.-G.; Supervision, M.Z. and I.B.Z. All authors approved the final version of the manuscript.

Funding: This work was supported by the SWB-7 fund for covering statutory activities of Laboratory of Molecular Biophysics, University of Bialystok, Poland, and by the Belarusian Republican Foundation for Fundamental Research (Joint scientific project BRFFR—NSFC, M23CH-14).

Institutional Review Board Statement: Not applicable.

Informed Consent Statement: Not applicable.

Data Availability Statement: Not applicable.

Conflicts of Interest: The authors declare no conflict of interest.

References

1. Xie, Y.; Yang, W.; Tang, F.; Chen, X.; Ren, L. Antibacterial Activities of Flavonoids: Structure-Activity Relationship and Mechanism. *Curr. Med. Chem.* **2015**, *22*, 132–149. [CrossRef]
2. Cushnie, T.P.T.; Lamb, A.J. Assessment of the Antibacterial Activity of Galangin against 4-Quinolone Resistant Strains of *Staphylococcus aureus*. *Phytomedicine* **2006**, *13*, 187–191. [CrossRef] [PubMed]
3. Cushnie, T.P.T.; Lamb, A.J. Detection of Galangin-Induced Cytoplasmic Membrane Damage in *Staphylococcus aureus* by Measuring Potassium Loss. *J. Ethnopharmacol.* **2005**, *101*, 243–248. [CrossRef] [PubMed]

4. Meyer, J.J.M.; Afolayan, A.J.; Taylor, M.B.; Erasmus, D. Antiviral Activity of Galangin Isolated from the Aerial Parts of *Helichrysum Aureonitens*. *J. Ethnopharmacol.* **1997**, *56*, 165–169. [CrossRef] [PubMed]
5. Shahidi, F.; Ambigaipalan, P. Phenolics and polyphenolics in foods, beverages and spices: Antioxidant activity and health effects—A review. *J. Funct. Foods* **2015**, *18 (Part B)*, 820–897. [CrossRef]
6. Sindhu, R.K.; Verma, R.; Salgotra, T.; Rahman, M.H.; Shah, M.; Akter, R.; Murad, W.; Mubin, S.; Bibi, P.; Qusti, S.; et al. Impacting the Remedial Potential of Nano Delivery-Based Flavonoids for Breast Cancer Treatment. *Molecules* **2021**, *26*, 5163. [CrossRef]
7. Nakajima, V.M.; Ruviaro, A.R.; Barbosa, P.d.P.M.; da Silva, I.F.; de Ávila, A.R.A. Chapter 7—Hesperetin and Naringenin: Protective Effects against Metabolic Syndrome–Associated Inflammation. In *Discovery and Development of Anti-Inflammatory Agents from Natural Products*; Brahmachari, G., Ed.; Natural Product Drug Discovery; Elsevier: Amsterdam, The Netherlands, 2019; pp. 207–239. [CrossRef]
8. Panche, A.N.; Diwan, A.D.; Chandra, S.R. Flavonoids: An Overview. *J. Nutr. Sci.* **2016**, *5*, e47. [CrossRef]
9. Bae, J.; Kim, N.; Shin, Y.; Kim, S.-Y.; Kim, Y.-J. Activity of Catechins and Their Applications. *Biomed. Dermatol.* **2020**, *4*, 8. [CrossRef]
10. Kumar, S.; Pandey, A.K. Chemistry and Biological Activities of Flavonoids: An Overview. *Sci. World J.* **2013**, *2013*, e162750. [CrossRef]
11. Kumar, H.; Bhardwaj, K.; Cruz-Martins, N.; Nepovimova, E.; Oleksak, P.; Dhanjal, D.S.; Bhardwaj, S.; Singh, R.; Chopra, C.; Verma, R.; et al. Applications of Fruit Polyphenols and Their Functionalized Nanoparticles Against Foodborne Bacteria: A Mini Review. *Molecules* **2021**, *26*, 3447. [CrossRef]
12. Nguyen, T.L.A.; Bhattacharya, D. Antimicrobial Activity of Quercetin: An Approach to Its Mechanistic Principle. *Molecules* **2022**, *27*, 2494. [CrossRef] [PubMed]
13. Donadio, G.; Mensitieri, F.; Santoro, V.; Parisi, V.; Bellone, M.L.; De Tommasi, N.; Izzo, V.; Dal Piaz, F. Interactions with Microbial Proteins Driving the Antibacterial Activity of Flavonoids. *Pharmaceutics* **2021**, *13*, 660. [CrossRef] [PubMed]
14. Biharee, A.; Sharma, A.; Kumar, A.; Jaitak, V. Antimicrobial Flavonoids as a Potential Substitute for Overcoming Antimicrobial Resistance. *Fitoterapia* **2020**, *146*, 104720. [CrossRef] [PubMed]
15. Elmasri, W.A.; Zhu, R.; Peng, W.; Al-Hariri, M.; Kobeissy, F.; Tran, P.; Hamood, A.N.; Hegazy, M.F.; Paré, P.W.; Mechref, Y. Multitargeted Flavonoid Inhibition of the Pathogenic Bacterium *Staphylococcus aureus*: A Proteomic Characterization. *J. Proteome Res.* **2017**, *16*, 2579–2586. [CrossRef]
16. Yuan, G.; Guan, Y.; Yi, H.; Lai, S.; Sun, Y.; Cao, S. Antibacterial Activity and Mechanism of Plant Flavonoids to Gram-Positive Bacteria Predicted from Their Lipophilicities. *Sci. Rep.* **2021**, *11*, 10471. [CrossRef]
17. Hirai, I.; Okuno, M.; Katsuma, R.; Arita, N.; Tachibana, M.; Yamamoto, Y. Characterisation of Anti-*Staphylococcus aureus* Activity of Quercetin. *Int. J. Food Sci. Technol.* **2010**, *45*, 1250–1254. [CrossRef]
18. Abreu, A.C.; Serra, S.C.; Borges, A.; Saavedra, M.J.; Mcbain, A.J.; Salgado, A.J.; Simões, M. Combinatorial Activity of Flavonoids with Antibiotics Against Drug-Resistant *Staphylococcus aureus*. *Microb. Drug Resist.* **2015**, *21*, 600–609. [CrossRef]
19. Eumkeb, G.; Chukrathok, S. Synergistic Activity and Mechanism of Action of Ceftazidime and Apigenin Combination against Ceftazidime-Resistant *Enterobacter Cloacae*. *Phytomedicine* **2013**, *20*, 262–269. [CrossRef]
20. Wang, L.; Li, B.; Si, X.; Liu, X.; Deng, X.; Niu, X.; Jin, Y.; Wang, D.; Wang, J. Quercetin Protects Rats from Catheter-related *Staphylococcus aureus* Infections by Inhibiting Coagulase Activity. *J. Cell. Mol. Med.* **2019**, *23*, 4808–4818. [CrossRef]
21. Tomar, A.; Broor, S.; Kaushik, S.; Bharara, T.; Arya, D.S. Synergistic Effect of Naringenin with Conventional Antibiotics Against Methicillin Resistant *Staphylococcus aureus*. *Eur. J. Mol. Clin. Med.* **2021**, *8*, 1770–1784.
22. Farhadi, F.; Khameneh, B.; Iranshahi, M.; Iranshahy, M. Antibacterial Activity of Flavonoids and Their Structure-Activity Relationship: An Update Review. *Phytother. Res.* **2019**, *33*, 13–40. [CrossRef] [PubMed]
23. Cushnie, T.P.T.; Lamb, A.J. Antimicrobial Activity of Flavonoids. *Int. J. Antimicrob. Agents* **2005**, *26*, 343–356. [CrossRef] [PubMed]
24. Lin, Z.; Lin, Y.; Zhang, Z.; Shen, J.; Yang, C.; Jiang, M.; Hou, Y. Systematic Analysis of Bacteriostatic Mechanism of Flavonoids Using Transcriptome and Its Therapeutic Effect on Vaginitis. *Aging* **2020**, *12*, 6292–6305. [CrossRef] [PubMed]
25. Arakawa, H.; Maeda, M.; Okubo, S.; Shimamura, T. Role of Hydrogen Peroxide in Bactericidal Action of Catechin. *Biol. Pharm. Bull.* **2004**, *27*, 277–281. [CrossRef]
26. Cushnie, T.P.T.; Lamb, A.J. Recent Advances in Understanding the Antibacterial Properties of Flavonoids. *Int. J. Antimicrob. Agents* **2011**, *38*, 99–107. [CrossRef]
27. Górniak, I.; Bartoszewski, R.; Króliczewski, J. Comprehensive Review of Antimicrobial Activities of Plant Flavonoids. *Phytochem. Rev.* **2019**, *18*, 241–272. [CrossRef]
28. Osonga, F.J.; Akgul, A.; Miller, R.M.; Eshun, G.B.; Yazgan, I.; Akgul, A.; Sadik, O.A. Antimicrobial Activity of a New Class of Phosphorylated and Modified Flavonoids. *ACS Omega* **2019**, *4*, 12865–12871. [CrossRef]
29. Olchowik-Grabarek, E.; Sekowski, S.; Bitiucki, M.; Dobrzynska, I.; Shlyonsky, V.; Ionov, M.; Burzynski, P.; Roszkowska, A.; Swiecicka, I.; Abdulladjanova, N.; et al. Inhibition of Interaction between *Staphylococcus aureus* α-Hemolysin and Erythrocytes Membrane by Hydrolysable Tannins: Structure-Related Activity Study. *Sci. Rep.* **2020**, *10*, 11168. [CrossRef]
30. Chang, E.H.; Huang, J.; Lin, Z.; Brown, A.C. Catechin-Mediated Restructuring of a Bacterial Toxin Inhibits Activity. *Biochim. Biophys. Acta BBA Gen. Subj.* **2019**, *1863*, 191–198. [CrossRef]

31. Olchowik-Grabarek, E.; Swiecicka, I.; Andreeva-Kovaleskaya, Z.; Solonin, A.; Bonarska-Kujawa, D.; Kleszczyńska, H.; Mavlyanov, S.; Zamaraeva, M. Role of Structural Changes Induced in Biological Membranes by Hydrolysable Tannins from Sumac Leaves (*Rhus typhina* L.) in Their Antihemolytic and Antibacterial Effects. *J. Membr. Biol.* **2014**, *247*, 533–540. [CrossRef]
32. Paczkowski, J.E.; Mukherjee, S.; McCready, A.R.; Cong, J.-P.; Aquino, C.J.; Kim, H.; Henke, B.R.; Smith, C.D.; Bassler, B.L. Flavonoids Suppress *Pseudomonas aeruginosa* Virulence through Allosteric Inhibition of Quorum-Sensing Receptors *. *J. Biol. Chem.* **2017**, *292*, 4064–4076. [CrossRef] [PubMed]
33. Friedman, M. Overview of Antibacterial, Antitoxin, Antiviral, and Antifungal Activities of Tea Flavonoids and Teas. *Mol. Nutr. Food Res.* **2007**, *51*, 116–134. [CrossRef] [PubMed]
34. Oliveira, D.; Borges, A.; Simões, M. *Staphylococcus aureus* Toxins and Their Molecular Activity in Infectious Diseases. *Toxins* **2018**, *10*, 252. [CrossRef] [PubMed]
35. Tong, S.Y.C.; Davis, J.S.; Eichenberger, E.; Holland, T.L.; Fowler, V.G. *Staphylococcus aureus* Infections: Epidemiology, Pathophysiology, Clinical Manifestations, and Management. *Clin. Microbiol. Rev.* **2015**, *28*, 603–661. [CrossRef]
36. Tang, F.; Li, L.; Meng, X.-M.; Li, B.; Wang, C.-Q.; Wang, S.-Q.; Wang, T.-L.; Tian, Y.-M. Inhibition of Alpha-Hemolysin Expression by Resveratrol Attenuates *Staphylococcus aureus* Virulence. *Microb. Pathog.* **2019**, *127*, 85–90. [CrossRef]
37. Fraunholz, M.; Sinha, B. Intracellular *Staphylococcus aureus*: Live-in and Let Die. *Front. Cell Infect. Microbiol.* **2012**, *2*, 43. [CrossRef]
38. Berube, B.J.; Wardenburg, J.B. *Staphylococcus aureus* α-Toxin: Nearly a Century of Intrigue. *Toxins* **2013**, *5*, 1140–1166. [CrossRef]
39. Gouaux, E. α-Hemolysin from *Staphylococcus aureus*: An Archetype of β-Barrel, Channel-Forming Toxins. *J. Struct. Biol.* **1998**, *121*, 110–122. [CrossRef]
40. He, S.; Deng, Q.; Liang, B.; Yu, F.; Yu, X.; Guo, D.; Liu, X.; Dong, H. Suppressing Alpha-Hemolysin as Potential Target to Screen of Flavonoids to Combat Bacterial Coinfection. *Molecules* **2021**, *26*, 7577. [CrossRef]
41. Wen, J.; Liu, B.; Yuan, E.; Ma, Y.; Zhu, Y. Preparation and Physicochemical Properties of the Complex of Naringenin with Hydroxypropyl-β-Cyclodextrin. *Molecules* **2010**, *15*, 4401–4407. [CrossRef]
42. Strugała, P.; Tronina, T.; Huszcza, E.; Gabrielska, J. Bioactivity In Vitro of Quercetin Glycoside Obtained in Beauveria Bassiana Culture and Its Interaction with Liposome Membranes. *Molecules* **2017**, *22*, 1520. [CrossRef] [PubMed]
43. Srinivas, K.; King, J.W.; Howard, L.R.; Monrad, J.K. Solubility of Gallic Acid, Catechin, and Protocatechuic Acid in Subcritical Water from (298.75 to 415.85) K. *J. Chem. Eng. Data* **2010**, *55*, 3101–3108. [CrossRef]
44. Mykytczuk, N.C.S.; Trevors, J.T.; Leduc, L.G.; Ferroni, G.D. Fluorescence Polarization in Studies of Bacterial Cytoplasmic Membrane Fluidity under Environmental Stress. *Prog. Biophys. Mol. Biol.* **2007**, *95*, 60–82. [CrossRef]
45. El Khoury, M.; Swain, J.; Sautrey, G.; Zimmermann, L.; Van Der Smissen, P.; Décout, J.-L.; Mingeot-Leclercq, M.-P. Targeting Bacterial Cardiolipin Enriched Microdomains: An Antimicrobial Strategy Used by Amphiphilic Aminoglycoside Antibiotics. *Sci. Rep.* **2017**, *7*, 10697. [CrossRef] [PubMed]
46. Jasniewski, J.; Cailliez-Grimal, C.; Younsi, M.; Millière, J.-B.; Revol-Junelles, A.-M. Functional Differences in Leuconostoc Sensitive and Resistant Strains to Mesenterocin 52A, a Class IIa Bacteriocin. *FEMS Microbiol. Lett.* **2008**, *289*, 193–201. [CrossRef]
47. Seel, W.; Baust, D.; Sons, D.; Albers, M.; Etzbach, L.; Fuss, J.; Lipski, A. Carotenoids Are Used as Regulators for Membrane Fluidity by *Staphylococcus xylosus*. *Sci. Rep.* **2020**, *10*, 330. [CrossRef]
48. National Collection of Type Cultures UK: Staphylococcus Aureus NCTC 5655. Available online: https://www.culturecollections.org.uk/products/bacteria/detail.jsp?refId=NCTC+5655&collection=nctc (accessed on 2 September 2022).
49. Bernheimer, A.W.; Schwartz, L.L.Y. Isolation and Composition of Staphylococcal Alpha Toxin. *Microbiology* **1963**, *30*, 455–468. [CrossRef]
50. Gurnev, P.A.; Nestorovich, E.M. Channel-Forming Bacterial Toxins in Biosensing and Macromolecule Delivery. *Toxins* **2014**, *6*, 2483–2540. [CrossRef]
51. Kozłowska, A.; Szostak-Węgierek, D. Flavonoids—Food Sources, Health Benefits, and Mechanisms Involved. In *Bioactive Molecules in Food*; Mérillon, J.-M., Ramawat, K.G., Eds.; Reference Series in Phytochemistry; Springer International Publishing: Cham, Switzerland, 2018; pp. 1–27. [CrossRef]
52. Kopustinskiene, D.M.; Jakstas, V.; Savickas, A.; Bernatoniene, J. Flavonoids as Anticancer Agents. *Nutrients* **2020**, *12*, 457. [CrossRef]
53. Wu, T.; He, M.; Zang, X.; Zhou, Y.; Qiu, T.; Pan, S.; Xu, X. A Structure-Activity Relationship Study of Flavonoids as Inhibitors of *E. Coli* by Membrane Interaction Effect. *Biochim. Biophys. Acta BBA Biomembr.* **2013**, *1828*, 2751–2756. [CrossRef]
54. Júnior, S.D.d.C.; Santos, J.V.d.O.; Campos, L.A.d.A.; Pereira, M.A.; Magalhães, N.S.S.; Cavalcanti, I.M.F. Antibacterial and Antibiofilm Activities of Quercetin against Clinical Isolates of *Staphyloccocus aureus* and *Staphylococcus saprophyticus* with Resistance Profile. *Int. J. Environ. Agric. Biotechnol.* **2018**, *3*, 1948–1958. [CrossRef]
55. Adamczak, A.; Ożarowski, M.; Karpiński, T.M. Antibacterial Activity of Some Flavonoids and Organic Acids Widely Distributed in Plants. *J. Clin. Med.* **2020**, *9*, 109. [CrossRef]
56. Gutiérrez-Venegas, G.; Gómez-Mora, J.A.; Meraz-Rodríguez, M.A.; Flores-Sánchez, M.A.; Ortiz-Miranda, L.F. Effect of Flavonoids on Antimicrobial Activity of Microorganisms Present in Dental Plaque. *Heliyon* **2019**, *5*, e0313. [CrossRef]
57. He, Z.; Zhang, X.; Song, Z.; Li, L.; Chang, H.; Li, S.; Zhou, W. Quercetin Inhibits Virulence Properties of *Porphyromas gingivalis* in Periodontal Disease. *Sci. Rep.* **2020**, *10*, 18313. [CrossRef] [PubMed]

58. Shamsudin, N.F.; Ahmed, Q.U.; Mahmood, S.; Ali Shah, S.A.; Khatib, A.; Mukhtar, S.; Alsharif, M.A.; Parveen, H.; Zakaria, Z.A. Antibacterial Effects of Flavonoids and Their Structure-Activity Relationship Study: A Comparative Interpretation. *Molecules* **2022**, *27*, 1149. [CrossRef] [PubMed]
59. Wang, T.; Zhang, P.; Lv, H.; Deng, X.; Wang, J. A Natural Dietary Flavone Myricetin as an α-Hemolysin Inhibitor for Controlling *Staphylococcus aureus* Infection. *Front. Cell Infect. Microbiol.* **2020**, *10*, 330. [CrossRef] [PubMed]
60. Adnan, S.-N.-A.; Ibrahim, N.; Yaacob, W.A. Disruption of Methicillin-Resistant *Staphylococcus aureus* Protein Synthesis by Tannins. *Germs* **2017**, *7*, 186–192. [CrossRef] [PubMed]
61. Baikar, S.; Malpathak, N.; Malpathak, N.; Malpathak, N. Secondary Metabolites as DNA Topoisomerase Inhibitors: A New Era Towards Designing of Anticancer Drugs. *Pharmacogn. Rev.* **2010**, *4*, 12–26. [CrossRef]
62. Plaper, A.; Golob, M.; Hafner, I.; Oblak, M.; Šolmajer, T.; Jerala, R. Characterization of Quercetin Binding Site on DNA Gyrase. *Biochem. Biophys. Res. Commun.* **2003**, *306*, 530–536. [CrossRef]
63. Ikigai, H.; Nakae, T.; Hara, Y.; Shimamura, T. Bactericidal Catechins Damage the Lipid Bilayer. *Biochim. Biophys. Acta BBA Biomembr.* **1993**, *1147*, 132–136. [CrossRef]
64. Haraguchi, H.; Tanimoto, K.; Tamura, Y.; Mizutani, K.; Kinoshita, T. Mode of Antibacterial Action of Retrochalcones from *Glycyrrhiza inflata*. *Phytochemistry* **1998**, *48*, 125–129. [CrossRef] [PubMed]
65. Mirzoeva, O.K.; Grishanin, R.N.; Calder, P.C. Antimicrobial Action of Propolis and Some of Its Components: The Effects on Growth, Membrane Potential and Motility of Bacteria. *Microbiol. Res.* **1997**, *152*, 239–246. [CrossRef] [PubMed]
66. Li, X.; He, C.; Song, L.; Li, T.; Cui, S.; Zhang, L.; Jia, Y. Antimicrobial Activity and Mechanism of Larch Bark Procyanidins against *Staphylococcus aureus*. *Acta Biochim. Biophys. Sin.* **2017**, *49*, 1058–1066. [CrossRef] [PubMed]
67. Shimamura, Y.; Hirai, C.; Sugiyama, Y.; Utsumi, M.; Yanagida, A.; Murata, M.; Ohashi, N.; Masuda, S. Interaction between Various Apple Procyanidin and Staphylococcal Enterotoxin A and Their Inhibitory Effects on Toxin Activity. *Toxins* **2017**, *9*, 243. [CrossRef]
68. Choi, O.; Yahiro, K.; Morinaga, N.; Miyazaki, M.; Noda, M. Inhibitory Effects of Various Plant Polyphenols on the Toxicity of Staphylococcal α-Toxin. *Microb. Pathog.* **2007**, *42*, 215–224. [CrossRef] [PubMed]
69. Tombola, F.; Campello, S.; De Luca, L.; Ruggiero, P.; Del Giudice, G.; Papini, E.; Zoratti, M. Plant Polyphenols Inhibit Vac A, a Toxin Secreted by the Gastric Pathogen *Helicobacter Pylori*. *FEBS Lett.* **2003**, *543*, 184–189. [CrossRef]
70. Ostroumova, O.S.; Efimova, S.S.; Schagina, L.V. 5- and 4′-Hydroxylated Flavonoids Affect Voltage Gating of Single Alpha-Hemolysin Pore. *Biochim. Biophys. Acta BBA Biomembr.* **2011**, *1808*, 2051–2058. [CrossRef]
71. Delehanty, J.B.; Johnson, B.J.; Hickey, T.E.; Pons, T.; Ligler, F.S. Binding and Neutralization of Lipopolysaccharides by Plant Proanthocyanidins. *J. Nat. Prod.* **2007**, *70*, 1718–1724. [CrossRef]
72. Satoh, E.; Ishii, T.; Shimizu, Y.; Sawamura, S.; Nishimura, M. The Mechanism Underlying the Protective Effect of the Thearubigin Fraction of Black Tea (*Camellia sinensis*) Extract against the Neuromuscular Blocking Action of Botulinum Neurotoxins. *Pharmacol. Toxicol.* **2002**, *90*, 199–202. [CrossRef]
73. Morinaga, N.; Iwamaru, Y.; Yahiro, K.; Tagashira, M.; Moss, J.; Noda, M. Differential Activities of Plant Polyphenols on the Binding and Internalization of Cholera Toxin in Vero Cells. *J. Biol. Chem.* **2005**, *280*, 23303–23309. [CrossRef]
74. Shah, S.; Stapleton, P.D.; Taylor, P.W. The Polyphenol (−)-Epicatechin Gallate Disrupts the Secretion of Virulence-Related Proteins by *Staphylococcus aureus*. *Lett. Appl. Microbiol.* **2008**, *46*, 181–185. [CrossRef] [PubMed]
75. Wu, Y.; Bai, J.; Zhong, K.; Huang, Y.; Gao, H. A Dual Antibacterial Mechanism Involved in Membrane Disruption and DNA Binding of 2R,3R-Dihydromyricetin from Pine Needles of *Cedrus Deodara* against *Staphylococcus aureus*. *Food Chem.* **2017**, *218*, 463–470. [CrossRef] [PubMed]
76. Ulhuq, F.R.; Mariano, G. Bacterial Pore-forming Toxins. *Microbiology* **2022**, *168*, 001154. [CrossRef] [PubMed]
77. Chaudhuri, S.; Banerjee, A.; Basu, K.; Sengupta, B.; Sengupta, P.K. Interaction of Flavonoids with Red Blood Cell Membrane Lipids and Proteins: Antioxidant and Antihemolytic Effects. *Int. J. Biol. Macromol.* **2007**, *41*, 42–48. [CrossRef]
78. Schwiering, M.; Brack, A.; Stork, R.; Hellmann, N. Lipid and Phase Specificity of α-Toxin from *S. Aureus*. *Biochim. Biophys. Acta BBA Biomembr.* **2013**, *1828*, 1962–1972. [CrossRef] [PubMed]
79. Veiko, A.G.; Lapshina, E.A.; Zavodnik, I.B. Comparative Analysis of Molecular Properties and Reactions with Oxidants for Quercetin, Catechin, and Naringenin. *Mol. Cell Biochem.* **2021**, *476*, 4287–4299. [CrossRef] [PubMed]
80. Šmejkal, K.; Chudík, S.; Klouček, P.; Marek, R.; Cvačka, J.; Urbanová, M.; Julínek, O.; Kokoška, L.; Šlapetová, T.; Holubová, P.; et al. Antibacterial C-Geranylflavonoids from *Paulownia Tomentosa* Fruits. *J. Nat. Prod.* **2008**, *71*, 706–709. [CrossRef]
81. Song, L.; Hobaugh, M.R.; Shustak, C.; Cheley, S.; Bayley, H.; Gouaux, J.E. Structure of Staphylococcal α-Hemolysin, a Heptameric Transmembrane Pore. *Science* **1996**, *274*, 1859–1865. [CrossRef]
82. Merzlyak, P.G.; Yuldasheva, L.N.; Rodrigues, C.G.; Carneiro, C.M.; Krasilnikov, O.V.; Bezrukov, S.M. Polymeric Nonelectrolytes to Probe Pore Geometry: Application to the Alpha-Toxin Transmembrane Channel. *Biophys. J.* **1999**, *77*, 3023–3033. [CrossRef]
83. Stefureac, R.; Long, Y.-T.; Kraatz, H.-B.; Howard, P.; Lee, J.S. Transport of Alpha-Helical Peptides through Alpha-Hemolysin and Aerolysin Pores. *Biochemistry* **2006**, *45*, 9172–9179. [CrossRef]
84. DeGuzman, V.S.; Lee, C.C.; Deamer, D.W.; Vercoutere, W.A. Sequence-Dependent Gating of an Ion Channel by DNA Hairpin Molecules. *Nucleic Acids Res.* **2006**, *34*, 6425–6437. [CrossRef] [PubMed]

85. Olchowik-Grabarek, E.; Mies, F.; Sekowski, S.; Dubis, A.; Laurent, P.; Zamaraeva, M.; Swiecicka, I.; Shlyonsky, V. Enzymatic Synthesis and Characterization of Aryl Iodides of Some Phenolic Acids with Enhanced Antibacterial Properties. *Biochim. Biophys. Acta Biomembr.* **2022**, *1864*, 184011. [CrossRef] [PubMed]
86. Bessa, L.J.; Ferreira, M.; Gameiro, P. Evaluation of Membrane Fluidity of Multidrug-Resistant Isolates of *Escherichia coli* and *Staphylococcus aureus* in Presence and Absence of Antibiotics. *J. Photochem. Photobiol. B. Biology* **2018**, *181*, 150–156. [CrossRef] [PubMed]
87. Tarahovsky, Y.S.; Muzafarov, E.N.; Kim, Y.A. Rafts Making and Rafts Braking: How Plant Flavonoids May Control Membrane Heterogeneity. *Mol. Cell Biochem.* **2008**, *314*, 65. [CrossRef] [PubMed]
88. Valeva, A.; Hellmann, N.; Walev, I.; Strand, D.; Plate, M.; Boukhallouk, F.; Brack, A.; Hanada, K.; Decker, H.; Bhakdi, S. Evidence that Clustered Phosphocholine Head Groups Serve as Sites for Binding and Assembly of an Oligomeric Protein Pore. *J. Biol. Chem.* **2006**, *281*, 26014–26021. [CrossRef]
89. Veiko, A.G.; Sekowski, S.; Lapshina, E.A.; Wilczewska, A.Z.; Markiewicz, K.H.; Zamaraeva, M.; Zhao, H.; Zavodnik, I.B. Flavonoids Modulate Liposomal Membrane Structure, Regulate Mitochondrial Membrane Permeability and Prevent Erythrocyte Oxidative Damage. *Biochim. Biophys. Acta BBA Biomembr.* **2020**, *1862*, 183442. [CrossRef]
90. Onishi, T. *Quantum Computational Chemistry*; Springer: Singapore, 2018; ISBN 978-981-10-5933-9.

molecules

MDPI

Review

Clinical Potential of Himalayan Herb *Bergenia ligulata*: An Evidence-Based Study

Shubhadeep Roychoudhury [1,*], Dipika Das [1], Sandipan Das [1], Niraj Kumar Jha [2,3,4], Mahadeb Pal [5], Adriana Kolesarova [6], Kavindra Kumar Kesari [7,8], Jogen C. Kalita [9] and Petr Slama [10]

1 Department of Life Science and Bioinformatics, Assam University, Silchar 788011, India
2 Department of Biotechnology, School of Engineering & Technology (SET), Sharda University, Greater Noida 201310, India
3 Department of Biotechnology Engineering and Food Technology, Chandigarh University, Mohali 140413, India
4 Department of Biotechnology, School of Applied & Life Sciences (SALS), Uttaranchal University, Dehradun 248007, India
5 Division of Molecular Medicine, Bose Institute, Kolkata 700054, India
6 Faculty of Biotechnology and Food Sciences, Slovak University of Agriculture in Nitra, 94976 Nitra, Slovakia
7 Department of Bio-products and Bio-systems, School of Chemical Engineering, Aalto University, 00076 Espoo, Finland
8 Department of Applied Physics, School of Science, Aalto University, 00076 Espoo, Finland
9 Department of Zoology, Gauhati University, Guwahati 781014, India
10 Laboratory of Animal Immunology and Biotechnology, Department of Animal Morphology, Physiology and Genetics, Faculty of AgriSciences, Mendel University in Brno, 61300 Brno, Czech Republic
* Correspondence: shubhadeep1@gmail.com

Abstract: Herbal products have been used in traditional systems of medicine and by ethnic healers for ages to treat various diseases. Currently, it is estimated that about 80% of people worldwide use herbal traditional medicines against various ailments, partly due to easy accessibility and low cost, and the lower side effects they pose. *Bergenia ligulata*, a herb ranging from the Himalayas to the foothills, including the north-eastern states of India, has traditionally been used as a remedy against various diseases, most prominently kidney stones. The medicinal properties of *B. ligulata* have been attributed to bergenin, its most potent bioactive component. Apart from bergenin, the other compounds available in *B. ligulata* are arbutin, gallic acid, protocatechuic acid, chlorogenic acid, syringic acid, catechin, ferulic acid, afzelechin, paashaanolactone, caryophyllene, 1,8-cineole, β-eudesmol, stigmasterol, β-sitosterol, parasorbic acid, 3-methyl-2-buten-1-ol, phytol, terpinen-4-ol, tannic acid, isovalaric acid, avicularin, quercetin, reynoutrin, and sitoinoside I. This review summarizes various medicinal properties of the herb, along with providing deep insight into its bioactive molecules and their potential roles in the amelioration of human ailments. Additionally, the possible mechanism(s) of action of the herb's anti-urolithiatic, antioxidative, antipyretic, anti-diabetic, anti-inflammatory and hepatoprotective properties are discussed. This comprehensive documentation will help researchers to better understand the medicinal uses of the herb. Further studies on *B. ligulata* can lead to the discovery of new drug(s) and therapeutics for various ailments.

Keywords: traditional medicine; *Bergenia ligulata*; bioactive compounds; anti-urolithiatic; antioxidant; anti-pyretic; anti-diabetic; anti-inflammatory; cardiovascular diseases

Citation: Roychoudhury, S.; Das, D.; Das, S.; Jha, N.K.; Pal, M.; Kolesarova, A.; Kesari, K.K.; Kalita, J.C.; Slama, P. Clinical Potential of Himalayan Herb *Bergenia ligulata*: An Evidence-Based Study. *Molecules* **2022**, *27*, 7039. https://doi.org/10.3390/molecules27207039

Academic Editor: Smaoui Slim

Received: 11 August 2022
Accepted: 12 October 2022
Published: 18 October 2022

1. Introduction

Traditional herbal medicines are plant-derived natural products that have been used by rural communities for ages for the management of various diseases [1]. Particularly in the tropics and the sub-tropics, an abundance of medicinal plants offers access to effective prevention and management of diseases through self-medication using plant-based medicines. It is estimated that 80% of people worldwide depend on such plant-based traditional medicines for their treatment, and their usage is predominant in developing countries [2]. In recent years, natural products have received renewed global attention from the clinical point of view due to low toxicity, low side effects, cost-effectiveness, and easy accessibility, as compared to modern synthetic medicines [3,4]. *Bergenia ligulata* is a perennial Himalayan herb belonging to the family Saxifragaceae. Another accepted name for the herb is *Bergenia pacumbis*. In the Indian sub-continent, it is distributed along the high-altitude Himalayan regions ranging from Kashmir to Bhutan, including West Bengal and the northeastern states [5,6]. The plant has been used as a folk medicine since ancient times for dissolving kidney stones and is referred to as "*Paashanbheda*" (Sanskrit: *Paashan* meaning "rockstone" and *bheda* meaning "piercing") in the Indian traditional system of medicine—*Ayurveda* [7]. It has simple leaves of orbicular to obovate shape, stout root stock, and solid barrel-shaped cylindrical rhizome (1.5–3 cm in length and 1–2 cm in diameter) [8]. Himalayan communities consume the roots and rhizomes of the plant to treat wounds, septic, cough and cold, cardiac diseases, asthma, inflammation, gastrointestinal disorders, and different kinds of urinary problems [9]. The most abundant bioactive compound in *B. ligulata* is bergenin [10], which, along with its natural derivatives, mainly contributes to the medicinal properties of the herb [11]. Other important phytochemicals in *B. ligulata* include afzelechin [12], ß-sitosterol [13], catechin, leucocyanidin, gallic acid, and tannic acid [14]. These substances have a number of significant biological activities including anti-bacterial, anti-inflammatory, and free-radical-scavenging properties [9]. An in vivo study on Wistar rats revealed the antilithiatic potential of the plant extract (crude aqueous-methanolic extract) when treated for 21 days at a dose of 5–10 mg/kg body weight, significantly inhibiting calcium oxalate (CaC_2O_4) aggregation in the renal ducts [15]. A recent study revealed the ameliorative property of ethanolic extract of *B. ligulata* rhizome against oxalate-mediated renal injury in renal epithelial cells of normal rat kidney 52E (NRK-52E) which was brought about by downregulation of mitogen-activated protein kinases (MAPK), osteopontin (OPN), nuclear factor kappa B (NF-κB), and caspase-3 and reduction of nucleation aggregation and modulation of the crystal structure [16]. Oral administration of the extract, at a dose of 500 mg/kg body weight, in Wistar rats showed strong antipyretic activity against yeast-induced fever [17]. Significant anti-inflammatory and anti-bacterial activities have also been reported after oral administration of 50% ethanolic extract of *B. ligulata* in male Wistar rats at a dose of 1 g/kg body weight [6]. Methanolic extract of *B. ligulata* rhizomes exhibited inhibitory activity against viral RNA and peptide synthesis [18]. In vitro and in vivo studies on *B. ligulata* have shown strong protective properties against *Leishmania donovani* infection, the parasitic load being reduced by >95% at a high dose of 1000 mg/kg body weight in mouse [19]. Notwithstanding its wide-ranging use in traditional medication, scientific evidence on the clinical application of *B. ligulata* remains inadequate. This evidence-based review summarizes the relevant information available on the role of the herb against various kinds of diseases as well as the responsible potent bioactive compounds. The present study also highlights the possible mechanism(s) of action of *B. ligulata* in modulating mammalian physiology.

2. Methodology

For the preparation of the present manuscript, the literature regarding the potential clinical use of *B. ligulata* was searched and articles extracted from online databases such as Pubmed, SCOPUS, Google Scholar, and Science Direct. Keyword strings such as (traditional medicine) AND (*Bergenia ligulata* OR *Saxifraga ligulata* Wall OR *Saxifraga thysanodes* OR *Bergenia pacumbis*), (bioactive molecules) AND (*Bergenia ligulata*), (bergenin) AND (anti-lithiatic activity), (catechin), (bergenin) AND (antioxidant), (inflammation), (pyretic), (cardiovascular disease), (diabetes), (hepatoprotective) were used for searching the literature in SCOPUS and Pubmed. After that, only relevant book chapters, full-text articles, and abstracts were screened, and unrelated articles, as well as publications on languages other than English, were excluded from the study as those publications do not relate to the specific aim of the review article. Thus selected articles (Figure 1) were critically analyzed, and results were organized into several sections in the manuscript. Finally, a possible mechanism of action of *B. ligulata* was also speculated as an outcome of the study. At the end of the article future perspective is also considered.

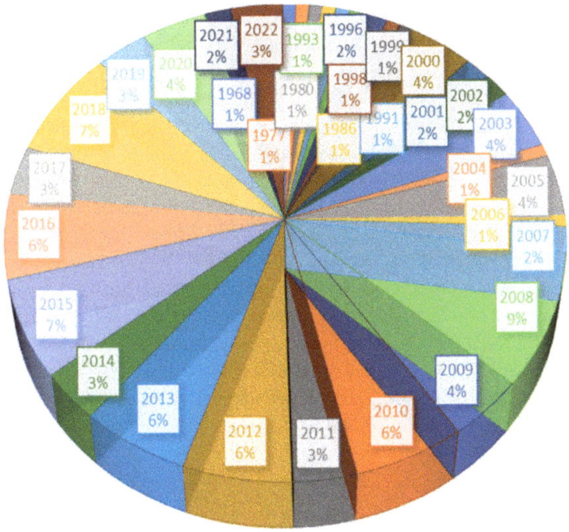

Figure 1. Pie chart illustrating the year-wise distribution of referenced papers.

3. *Bergenia ligulata* as Traditional Herbal Medicine

A huge recovery in the interest and use of medicinal plants has been witnessed in the previous decade. For a considerably long time, the plants belonging to the genus *Bergenia* have received notable attention for their restorative properties against ailments and have been broadly utilized in traditional medicines in various regions, particularly in the Asian continent, including India, Pakistan and Nepal [20]. Rhizomes of the species belonging to the genus *Bergenia* have been used in folk medicine for their antiscorbutic, astringent, diuretic, antipyretic, and ophthalmic properties [21]. The rhizomes are also used in dissolving kidney and gall bladder stones apart from healing cuts, burns, wounds, inflammation, cold, and cough [21]. The first use of the plant, "*Paashanbheda*", for dissolving calculi and treatment of painful urination was narrated in ancient *Charaka Samhita* as early as 600 BC [5]. Ayurveda mentions the use of sap prepared from the leaves of *B. ligulata* for the treatment of urinary diseases, stomach problems, epilepsy, and cold [9]. Similarly, *Ayurveda* and *Unani* medicine systems also mention the utilization of *B. ligulata* roots in the management of vesicular calculi, urinary discharge, exorbitant uterine hemorrhage, ailments of the bladder, diarrhea, menorrhagia, excessive splenic growth, and cardiovascular diseases [7,22]. *Sushruta Samhita* mentions its use in the management of kidney and bladder stones and blood sugar. Some other indigenous Indian literature, including *Bhavaprakash*, *Rajnighantu*, and *Chakradatta*, also prescribed the use of *B. ligulata* in the management of urinary diseases and stones and the purification of the urinary bladder. Various local communities in the central Himalayan region consume different parts of the plant, such as the root rhizome, leaf, or whole plant in the form of sap or liquor or powder against dizziness, headache, vertigo, and kidney stones [9]. For example, the *Bhotia tribes* of central Himalayas consume dried rhizomes powder to treat kidney stones [23]. The tribal communities of Dharchula, Uttar Pradesh, India use roots of *B. ligulata* in healing cuts and wounds, ophthalmic problems, and dissolution of kidney stones and in urinary diseases [9]. The tribal communities in the Chamba district of Himachal Pradesh, India, use leaves of *B. ligulata* in the preparation of tea and consume it to treat common cold [24]. In the eastern Himalayas, leaves of this plant are used to treat cuts, wounds, and boils by the

Monpa tribes of Arunachal Pradesh, India, whereas the Naga tribes of India use the roots in the management of liver diseases and tuberculosis [9]. Similarly, Mizo tribes in India use the decoction prepared from the roots of *B. ligulata* in the management of diarrhea and infection of pulmonary system. Juice prepared from the leaf is used against boils and is also taken orally to dissolve kidney stones [25]. *B. ligulata* is available as an over-the-counter herbal product, particularly as powder [26], as well as stems, roots, and seeds [27].

4. Bioactive Compounds in *Bergenia ligulata*

Ultra-high-performance liquid chromatography coupled with hybrid linear ion trap triple quadrupole mass spectrometry (UHPLC-QqQLIT-MS/MS) has been able to quantify eight major bioactive compounds from the rhizome of the plant: bergenin, arbutin, gallic acid, protocatechuic acid, chlorogenic acid, catechin, syringic acid, and ferulic acid [10]. In comparison with other medicinally important *Bergenia* species such as *B. ciliata, B. purpurascens,* and *B. stracheyi*, the highest total contents of these eight compounds have been noted in *B. ligulata* [10]. Other important phenolic compounds of *B. ligulata* include (+)-afzelechin [28,29], paashaanolactone [30], caryophyllene, 1,8-cineole, β-eudesmol, β-sitosterol, (+)-(6S)-parasorbic acid, 3-methyl-2-buten-1-ol, phytol, and tannic acid [31], isovalaric acid [32], stigmasterol [14], avicularin [7,33] terpinen-4-ol [34], quercetin [35], reynoutrin [36,37], and sitoinoside I [38]. These bioactive compounds, as presented in Table 1, are believed to be responsible for the medicinal properties of the plant [9].

Table 1. Bioactive molecules found in *Bergenia ligulata*.

Chemical Structure	Name of the Molecule	Reference(s)
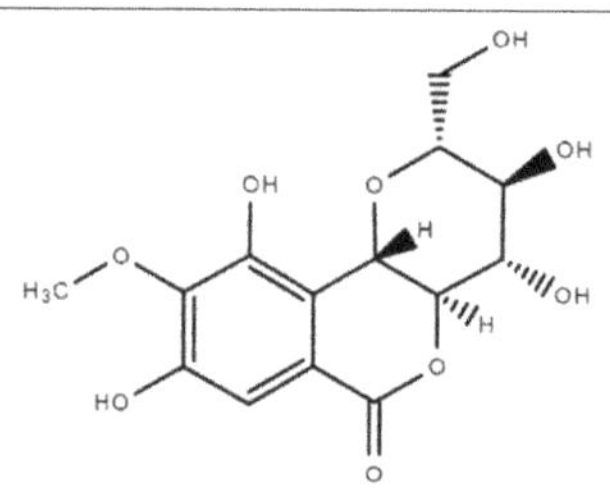	Bergenin	[10]
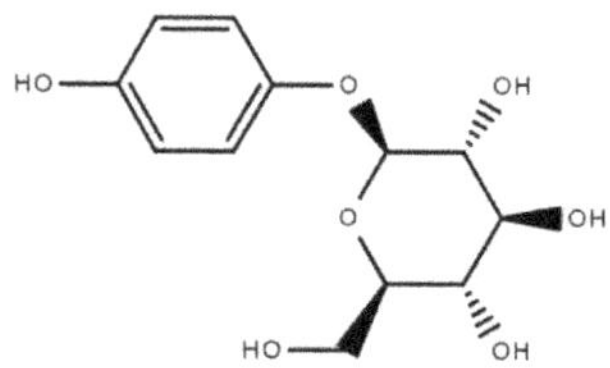	Arbutin	[10]
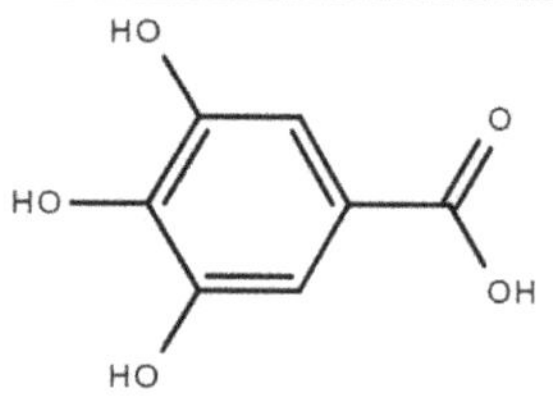	Gallic acid	[10]

Table 1. *Cont.*

Chemical Structure	Name of the Molecule	Reference(s)
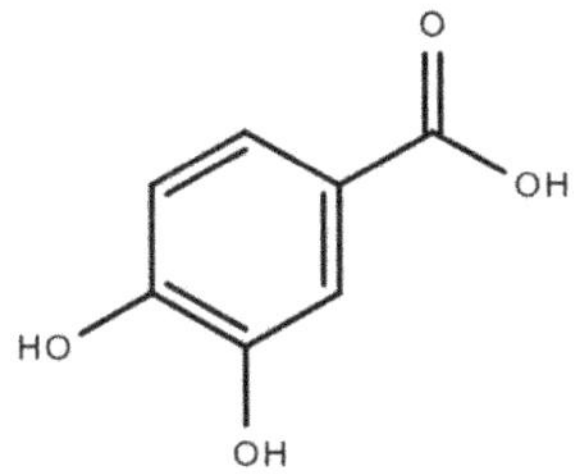	Protocatechuic acid	[10]
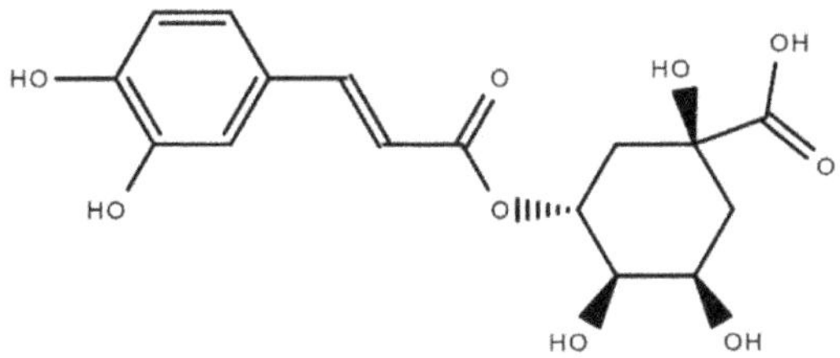	Chlorogenic acid	[10]
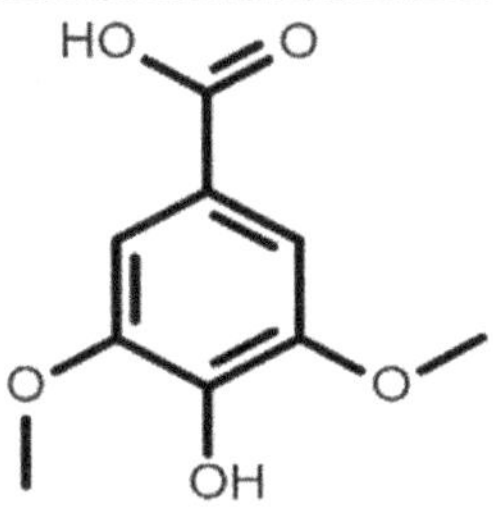	Syringic acid	[10]
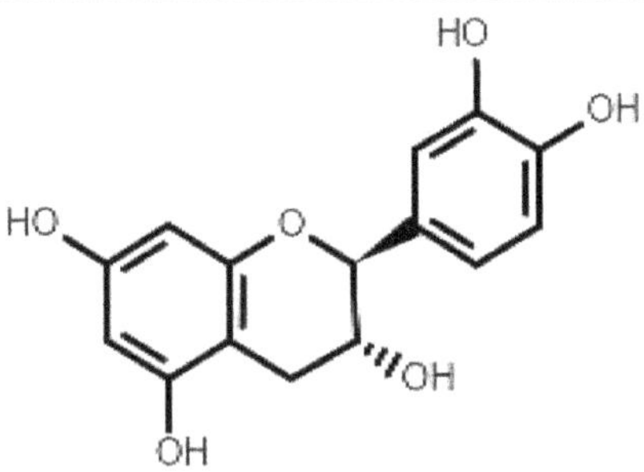	Catechin	[10]
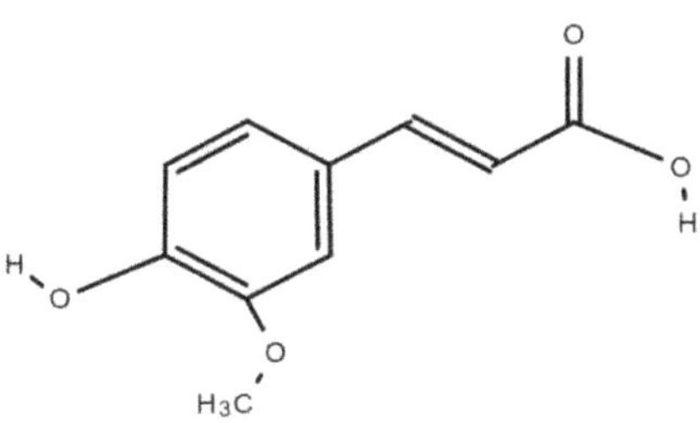	Ferulic acid	[10]

Table 1. *Cont.*

Chemical Structure	Name of the Molecule	Reference(s)
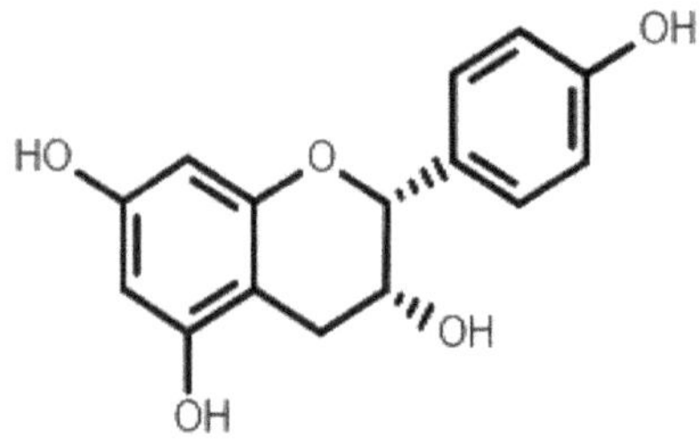	(+)-afzelechin	[28,29]
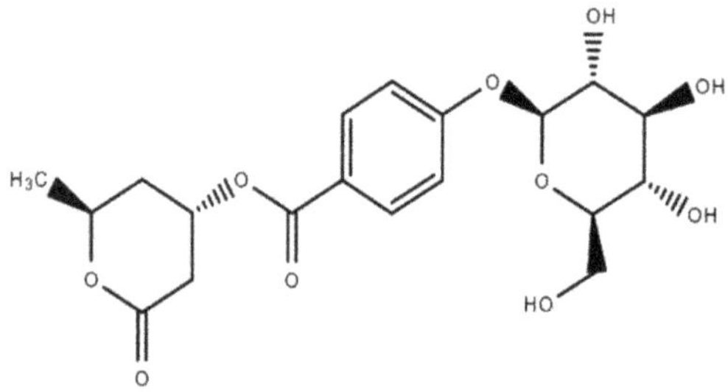	Paashaanolactone	[30]
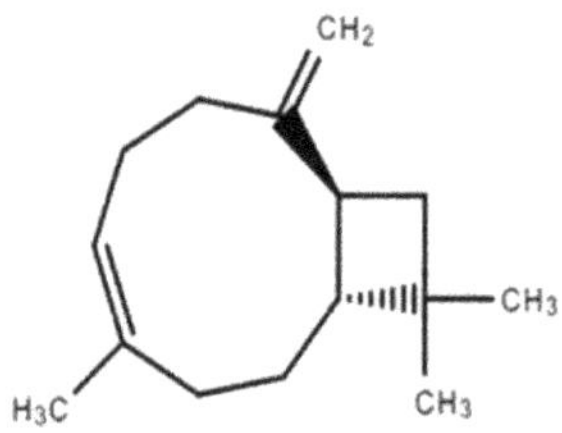	Caryophyllene	[31]
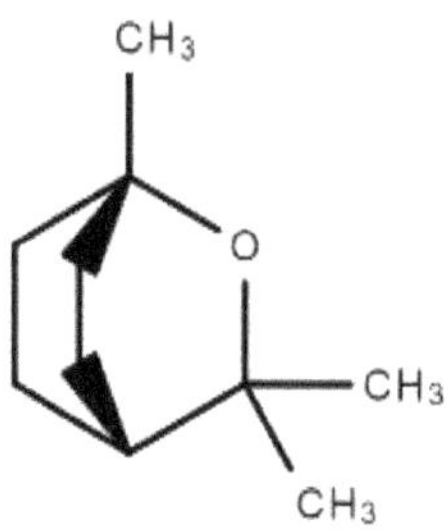	1,8-cineole	[31]
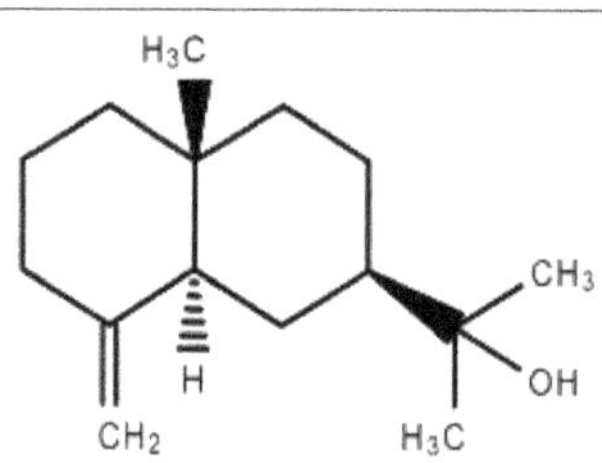	β-eudesmol	[31]

Table 1. *Cont.*

Chemical Structure	Name of the Molecule	Reference(s)
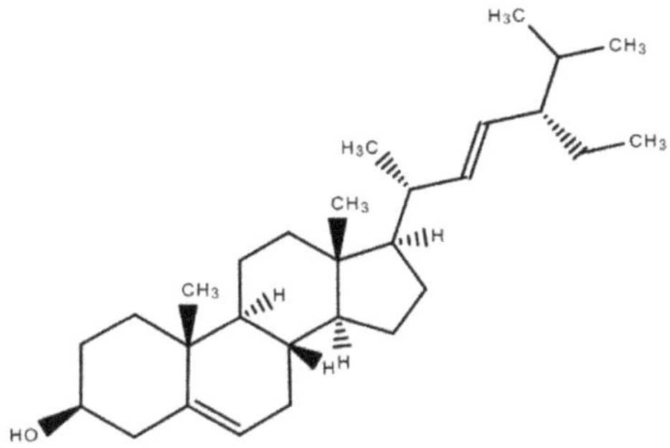	Stigmesterol	[14]
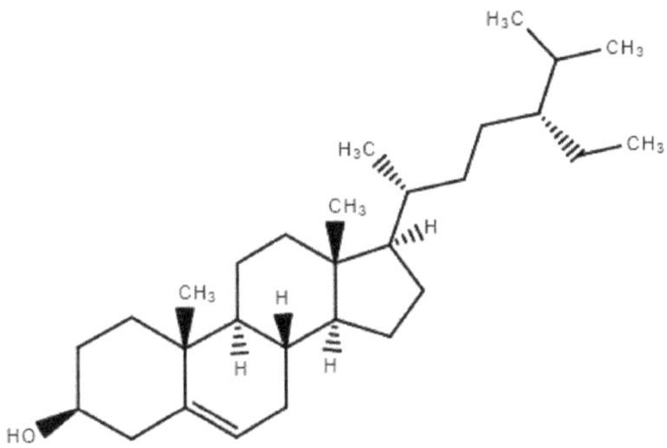	β-sitosterol	[31]
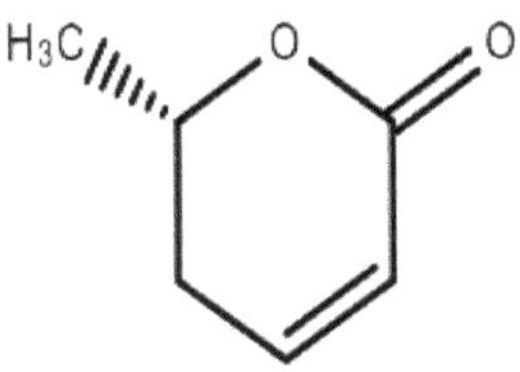	(+)-(6S)-parasorbic acid	[31]
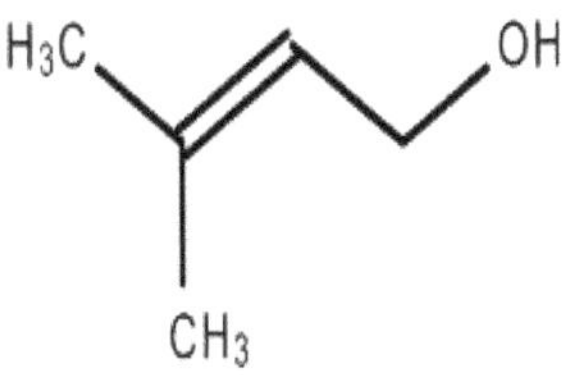	3-methyl-2-buten-1-ol	[31]
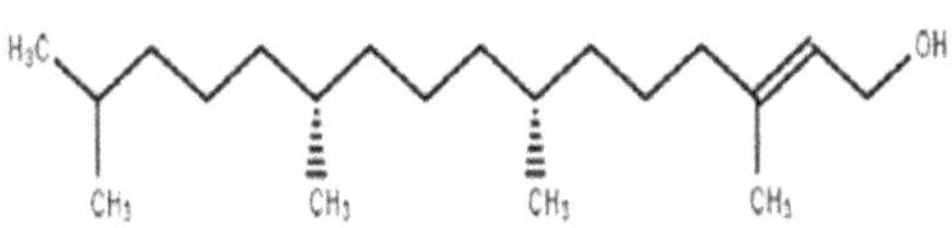	Phytol	[31]

Table 1. *Cont.*

Chemical Structure	Name of the Molecule	Reference(s)
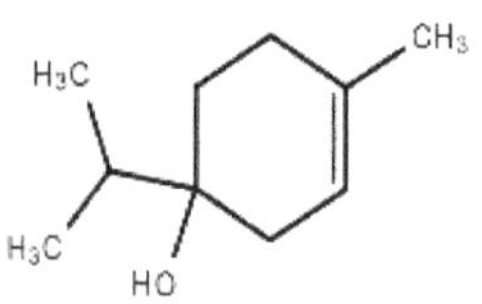	Terpinen-4-ol	[34]
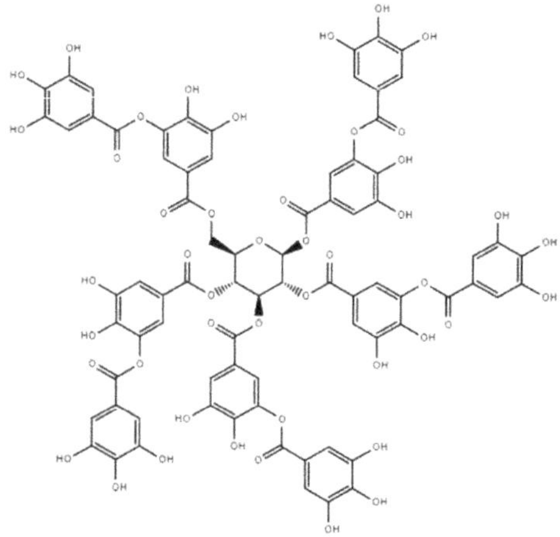	Tannic acid	[31]
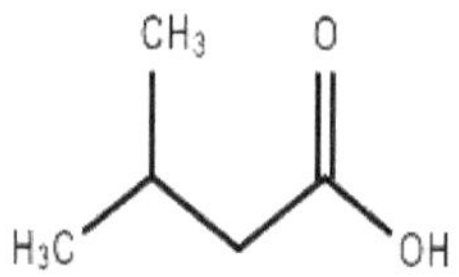	Isovaleric acid	[32]
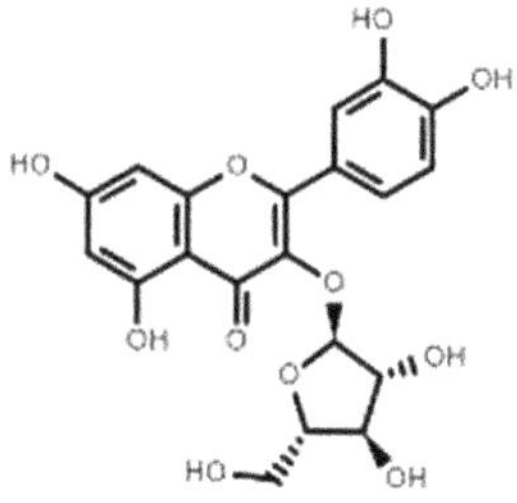	Avicularin	[7,33]
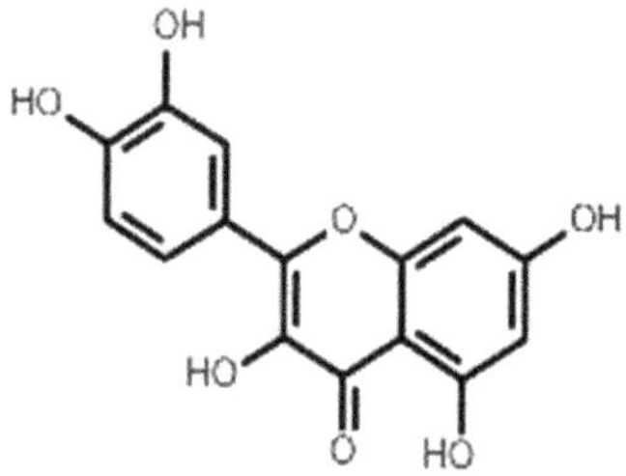	Quercetin	[35]

Table 1. *Cont.*

Chemical Structure	Name of the Molecule	Reference(s)
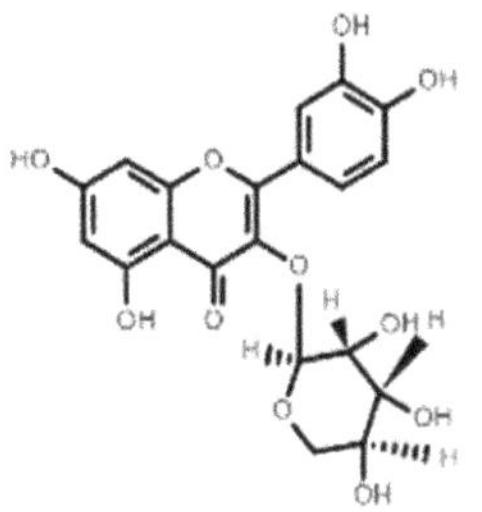	Reynoutrin	[31,37]
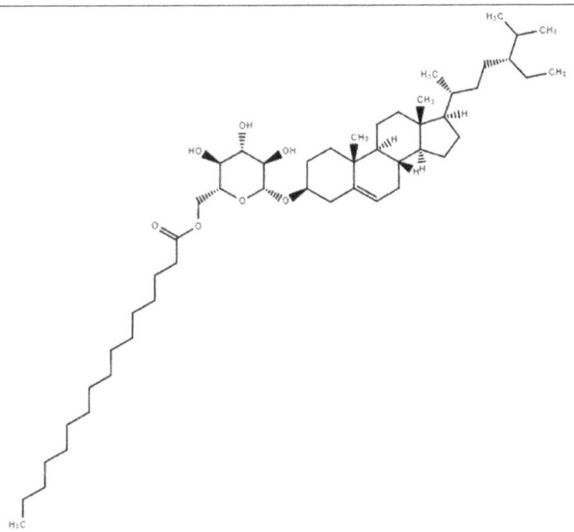	Sitoinoside I	[38]

4.1. Bergenin

Bergenin is a C-glucoside of 4-O-methyl gallic acid, a colorless crystalline polyphenol that comprises an aromatic ring, an annellated δ-lactone ring, and a glucopyranose ring [39]. It shows strong effectiveness in averting stress-induced gastric ulcers in rat models and is widely used as an ingredient of folk medicine for gastritis [40]. Cell line studies revealed that bergenin can induce apoptosis in HeLa (cervical cancer) cells by arresting the cell cycle at G0/G1 phase and can also promote anti-cancer activity by inhibiting the expression of signal transducer and activator of transcription 3 (STAT3) protein and metastasis of cancer cell [41]. Bergenin introduced the highest antioxidant and lipophilic properties which played a vital role in averting neuronal diseases and death [42]. It also accelerates the osteogenesis of bone mesenchymal stem cells through the upregulation of the sirtuin 1 (SIRT1) expression [43]. Strong antimicrobial activity of bergenin has been recorded against *Aspergillus flavus, Aspergillus niger, Escherichia coli, Enterococcus faecalis, Pseudomonas aeruginosa, Staphylococcus aureus,* and *Candida albicans* [44,45].

4.2. Catechins

Catechins are composed of two aromatic rings with hydroxyl group. This colorless crystalline polyphenol [46] has a molecular weight of 290 g/mol [47]. Depending on the distribution of hydroxyl group in aromatic ring, catechins are categorized into two types: (i) free catechins and (ii) esterified catechins [48]. Several studies show strong antiviral properties of catechins against adenovirus [49], enterovirus [50], human immunodeficiency viruses (HIV) [51], influenza virus [52], and tobacco mosaic virus (TMV) [53]. Catechins and their derivatives are effective scavengers of reactive oxygen species (ROS) [54]. Catechins moiety can function as free radical scavengers by requisitioning metal ions, and the B-ring site serves as a major site where the scavenging reaction is carried out [55–57]. Manikandan et al. (2012) reported that catechins possess significant anti-cancer properties and may reduce the proliferation of HCT 116, HCT 15 (human colon adenocarcinoma) and Hep G-2 (human larynx carcinoma) cell lines and are able to induce apoptosis [58].

4.3. Arbutin

Additionally known as *p*-hydroxyphenyl-β-d-glucopyranoside, arbutin is a bioactive hydrophilic polyphenol that has two isomers such as α-arbutin and β-arbutin [59]. Arbutin and its derivatives can directly control the overproduction of melanin by converting tyrosinase into L-DOPA (levodopa) and obstructing the tyrosinase activity without altering the mRNA expression [60]. Arbutin can introduce anti-inflammatory activity by decreasing the production of nitric oxide (NO) and expression of iNOS and cyclooxygenase-2 (COX-2) in lipopolysaccharide-stimulated BV2 cells (murine microglial cells) which also simultaneously suppressed the production of pro-inflammatory cytokines such as interleukin-1β (IL-1β) and tumor necrosis factor-α (TNF-α) and monocyte chemoattractant protein-1 (MCP-1) [61]. Li et al. (2011) reported that arbutin has antitumor activity by inducing TCC-SUP (human bladder cancer) cell proliferation by inhibiting extracellular signal-regulated kinase (ERK) and by accelerating p21 protein expression [62].

4.4. Gallic Acid

Gallic acid is a secondary metabolite present in most plants and is also known as 3,4,5 trihydroxybenzoic acid [63]. Gallic acid alters the integrity of bacterial cell membranes by penetrating the bacterial cell wall and disrupting the cellular respiration and electron transport chain. This compound also negatively alters DNA cleavage by affecting dihydrofolate reductase activity in bacteria [64]. It also shows antiviral activity with respect to hepatitis C virus (HCV) [65] and herpes simplex virus (HSV) [66]. Gallic acid showed anti-cancer activity by arresting the cell cycle and promoted apoptosis by activating the caspases pathway and can also reduce metastasis [64]. Gallic acids also possess gastroprotective activity [67], anti-hypertriglyceridemia activity, and diet-induced anti-hyperglycemic activity [68].

4.5. Protocatechuic Acid

Protocatechuic acid is a secondary metabolite that consists of an aromatic ring and one or more hydroxyl groups and is also known as 3,4-dihydroxybenzoic acid [69]. Protocatechuic acid possesses free radical scavenging property and shows antioxidant activity through reduction of ROS generation in different parts of the body such as the brain, heart, kidney, and liver, and prevents different degenerative diseases [70]. Protocatechuic acid reportedly shows preventive activity against neurodegenerative diseases such as Alzheimer's and Parkinson's diseases by interrupting the aggregation of β-amyloid plaques in brain tissues and preventing hyperphosphorylation of tau protein in neurons [71]. It has the ability to inhibit the growth of Gram-positive and Gram-negative bacteria and also shows antifungal activity [70]. In addition to antioxidant and anti-inflammatory activities, protocatechuic acid produced a significant positive effect in hyperglycemic conditions by increasing plasma insulin level [72]. Lende et al. (2011) suggested that protocatechuic acid can introduce promising anti-inflammatory properties by reducing carrageenan-induced paw edema and Freund's adjuvant arthritis [73].

4.6. Chlorogenic acid

Chlorogenic acid is a phenolic secondary metabolite and is also known as 5-O-caffeoylquinic acid (5-CQA). Chlorogenic acid plays a role in glucose metabolism through the activation of AMP-activated protein kinase (AMPK), which leads to glucose transport from intracellular membrane to plasma membrane by glucose transporter-4 (GLUT4) and increases cardiac glycolysis by activating phosphofructokinase 2 [74]. It can inhibit glucose-6-phosphatase activity and reduce glucose level in circulation, which causes less deposition of fatty acids in the adipose tissue and simultaneously utilizes the stored fat in the body, resulting in a reduction in body weight. It also shows anti-diabetic properties [75]. In addition to antioxidant activity, chlorogenic acid also possesses antibacterial activity against *Escherichia coli, Klebsiella pneumonia, Staphylococcus epidermidis* and antifungal activity against *Candida albicans* and antiviral activities against HIV, HSV-1, and HSV-2 and adenovirus [75].

4.7. Syringic Acid and Ferulic Acid

Syringic acid and ferulic acid are phenolic compounds. Syringic acid contains methoxy groups on the aromatic ring at positions of 3 and 5 [76]. Ferulic acid is also known as 4-hydroxy-3-methoxycinnamic acid [77,78]. Ferulic acid is a potent photoprotective and brightening agent for skin care; it nourishes the skin by balancing collagen and elastin activity [79]. Ferulic acid is also a potent scavenger of free radicals, and it can decrease lipid peroxidation rate in the rat brain by donating electrons from hydroxy and phenoxy groups to neutralize ROS [80]. Ferulic acid has also demonstrated antimicrobial activities against *Enterococcus faecalis, Staphylococcus aureus* [81], and *Cronobacter sakazakii* [82]. Treating *Cronobacter sakazakii* infection with syringic acid deteriorated bacterial cell membrane structure and halted bacterial growth, and this compound can be used as a natural preservative, too [82]. In addition to antioxidant activity, syringic acid appeared effective against acute pancreatitis [83], renal ischemia-reperfusion injury [84], and demyelination and inflammation in sciatic nerves [85]. Syringic acid enhances the working capacity of β-cells in the pancreas and increases the plasma insulin level, which induces more deposition of glycogen in peripheral tissue [85,86].

5. Potential Clinical Use

Bergenia species have been well known since ancient times for their potential curative effects on different human ailments [37]. *B. ligulata* is one of the important members of this genus and, also shows an extensive range of pharmacological activities. It is widely used in Indian traditional medicine as well as other traditional medicine systems of the world. The bioactive compounds found in the plant are diverse and may be responsible for the pharmacological activity of the plant. In traditional medicine systems, the plant is used mainly for its antilithiatic activity; additionally, the whole plant or parts of the plant (root and rhizome) have been used in the management of fever, inflammation, diabetes, microbial infections, wounds, burns [21], amelioration of liver disease, urinary crux, and abdominal and heart diseases [7].

5.1. Antilithiatic Activity

In Indian traditional medicine, the rhizomes of *B. ligulata* have been considered a potential drug in the management of renal stones [87]. Dichloromethane, a bioactive fraction of *B. ligulata* extract showed high efficiency against kidney stone aggregation when administered orally for 21 days at a dose of 7 mg/kg body weight [88]. Spectroscopic analysis revealed bergenin to be the potent antilithiatic bioactive molecule isolated from rhizomes of *B. ligulata* [89]. Methanolic extract of *B. ligulata* and bergenin exhibited marked dissolution of urinary calculi both in kidney and urine constituents [90]. Treatment of rats with aqueous-methanolic extract of *B. ligulata* rhizomes at a dose range of 5–10 mg/kg body weight for 21 days was able to prevent ethylene glycol-induced urolithiasis by inhibiting CaC_2O_4 crystal deposition in the renal tubules and simultaneously improved renal function [15]. An in vitro study confirmed that supplementation of dried leaves aqueous extract of *B. ligulata* has strong inhibitory potential against calcium oxalate monohydrate (COM) and hydrogen phosphate dehydrate crystal formation [91,92]. In ethylene glycol-induced hyperoxaluric rats, the dysfunction of mitochondria during stone crystal formation was manifested by reducing activities of electron transport chain complexes I, II and IV and, also by increasing mitochondrial oxidative stress. Oral administration of bergenin at a dose of 10 mg/kg for 28 days in ethylene glycol-induced hyperoxaluric rats showed amelioration of the damages to mitochondrial complexes as well as the alleviation of oxidative stress showing its potential effectiveness against urolithiasis [93]. A recent study on renal epithelial NRK-52E cells showed that ethanolic extract of *B. ligulata* significantly inhibits the nucleation and aggregation process of calcium oxalate crystals, further modulating the crystal structure by converting COM to the less pernicious form of calcium oxalate dihydrate (COD) [16].

5.2. Antipyretic Activity

Pyrexia or fever may occur due to acute infection or inflammation, or even due to any injury to tissue leading to the release of cytokines that initiate the synthesis of prostaglandin E2 (PgE2) in the hypothalamic area [94]. *B. ligulata* is considered an antipyretic herbal drug, and particularly the dried rhizome is used to prevent such fever [9]. Another study showed that ethanolic extract of roots and rhizomes of *B. ligulata* exhibit antipyretic activity at a dose of 500 mg/kg body weight in albino Wistar rats against yeast-induced fever. The rectal temperatures were recorded at time intervals of 1, 2, 3, 4 and 5 h after administration of *B. ligulata* extract [95].

5.3. Anti-Diabetic Activity

Plants belonging to the genus *Bergenia* play a crucial role in reducing the hyperglycemic condition. Rigorous studies on animal models revealed that *B. ligulata* possesses strong anti-diabetic activity [37]. Another possible mechanism of anti-diabetic action of *B. ligulata* may be attributed to its bioactive compound (+)-afzelechin, which acts as an inhibitor of α-glucosidase enzyme, as ascertained by enzyme inhibition assay [9,96]. The inhibition of α-glucosidase enzyme has been found to be effective in the treatment of hyperglycemia by delaying the absorption of carbohydrates in rat small intestines [96].

5.4. Anti-Inflammatory Activity

Bergenia species have strong anti-inflammatory activity as both the aqueous as well as ethanolic extracts of rhizomes showed effective anti-inflammatory potential in rat models [37]. *B. ligulata* is a well-known herb among folklore medical practitioners due to its anti-inflammatory potential. A study regarding the anti-inflammatory activity ensured its bioactive effect, where the oral application of aqueous, as well 50% ethanolic extracts of *B. ligulata* at a dose of 1 g/kg body weight of male Wistar rats, was able to attenuate the inflammatory response by reducing the level of succinate dehydrogenase (SDH), a key enzyme the level of which has been reported to rise during inflammation [6]. A possible explanation behind the therapeutic effect may be attributed to the bioactive molecule bergenin [97]. A new study on NRK-52 E cells revealed the reduction in inflammatory mediators such as MAPK, OPN, and nuclear factor kappa B (NF-kB) in presence of *B. ligulata* extract [16].

5.5. Hepatoprotective Activity

Bergenia species also have a hepato-protective effect. The administration of ethanolic root extract of *B. ligulata* to albino Wistar rats at a dose range of 25–35 g/kg body weight for 10 days exerted hepatoprotective activity, which was assessed by measuring the levels of serum glutamate pyruvate transaminase (SGPT), serum glutamate oxaloacetate transaminase (SGOT), serum alkaline phosphatase (ALP), and total bilirubin. All these parameters were significantly lower in the *B. ligulata*-treated group as compared to standard drugs [95].

5.6. Cardioprotective Activity

The cardioprotective potential of *B. ligulata* is also attributed to its phytochemical constituents. Several studies have highlighted its effects on different mammalian models. The administration of *B. ligulata* extract at a dose of 50 mg/kg body weight in dogs through an intravenous route showed effective hypotensive activity [7,98]. Alcoholic extract of rhizome also exhibited anti-bradykinin action without altering the action of 5-hydroxytryptamine (5-HT) receptor and acetylcholine as reported from isolated guinea pig ileum [7,9]. The scientific experiments that are carried out in mammalian models shows various pharmacological activities of the *B. ligulata* (Table 2). However, further studies are required for validation of these activities of the bioactive molecules present in the herb.

Table 2. Clinical significance of *Bergenia ligulata* in various experimental models, and their key findings along with calculated human doses.

Plant Part/Chemical Constituent	Mode of Study	Experimental Model	Study Design	Dosage	Calculated Human Dose * [99]	Action	Reference(s)
Ethanolic extract of roots and rhizomes of *B. ligulata*	In vivo	Wistar rat	Yeast induced fever	500 mg/kg	81.08 mg/kg	Rectal temperatures were recorded at a time interval of 1, 2, 3, 4, and 5 h after administration and reduction in temperate was recorded	[17]
Aqueous-methanolic extract of *B. ligulata* rhizomes	In vivo	Rat	Ethylene glycol-induced urolithiasis	5–10 mg/kg	1.62 mg/kg	Inhibition of calcium oxalate (CaC_2O_4) crystal deposition in the renal tubules and simultaneous improvement of renal function	[15]
Bergenin	In vivo	Rat	Ethylene glycol-induced hyperoxaluria	10 mg/kg	1.62 mg/kg	Amelioration of damages to mitochondrial complexes as well as the alleviation of oxidative stress	[93]
Dichloromethane bioactive fraction of *B. ligulata*	In vivo	Rat	Ethylene glycol-induced renal calculi	7 mg/kg	1.13 mg/kg	Inhibition of kidney stone aggregation	[88]
Aqueous and 50% ethanolic extracts of *B. ligulata*	In vivo	Wistar rat	Orally administered	1 g/kg	0.16 gm/kg	Reduction in inflammatory response by lowering the SDH level	[6]
Ethanolic root extract of *B. ligulata*	In vivo	Wistar rats	Orally administered	25–35 gm/kg	5.67 gm/kg	Showed promising hepatoprotective effect by lowering the SGOT, SGPT and ALP levels	[95]
B. ligulata extract	In vivo	Dog	Orally administered	50 mg/kg	27.02 mg/kg	Promising hypotensive activity observed	[7,98]

* Human dose calculation: Reagan-Shaw et al. (2008) formulated conversion of animal doses to human equivalent dose (HED) based on body surface area (BSA). HED (mg/kg) = Animal dose (mg/kg) X animal K_m factor/human K_m factor. K_m is calculated by dividing body weight (kg) by BSA (m^2). Considering 60 kg, 10 kg, and 0.15 kg body weight and 1.6, 0.5 and 0.025 BSA (m^2) for human, dog, and rat, respectively.

6. Possible Mechanism(s) of Action

Bioactive molecules of *B. ligulata* include polyphenols, flavonoids, and quinones that contribute to the pharmacological properties of the plant. The major ingredients of polyphenols include bergenin, arbutin, and catechin, which mostly add value to the medicinal value of the plant [37,100].

6.1. Anti-Urolithiatic Mechanism

The composition of kidney stones mainly depends on the physiologically and chemically altered urine. Depending on the chemical deposition, kidney stones can be classified as calcium stones—i.e., stones formed due to aggregation of calcium oxalate (CaOx) and calcium phosphate ($Ca_3(PO_4)_2$) in renal calculi [101]. Other forms of stone include magnesium ammonium phosphate stones; uric acid stones; cystine stones; and drug-induced stones [101]. The general process of stone formation is common for all types of kidney stones. The most common, i.e., the calcium stone formation process, is mainly attributed to the crystallization of calcium oxalate, and the process involves nucleation, crystal development, aggregation, and the retention of crystal in the renal duct [93,101]. The wide range of action of bergenin can be demonstrated by its inhibitory effect on formation of kidney stones. Bergenin is a C-glycoside of 4-O-methyl gallic acid with an ionizing ability in buffer medium under neutral pH, whereas 4-O-methylglycoside (4-OMG) [102], the hydrolysis product of bergenin has negatively charged free carboxylate group, which enables it to bind with the calcium site of CaOx crystals and inhibit the growth and precipitation of CaOx crystals [89]. Similarly, other phenolic groups of *B. ligulata* such as arbutin and catechin also has the ability to interact with negatively charged oxalate ions by hydrogen bonds, which ultimately modulates the growth of CaOx crystals. Treatment with *B. ligulata* extract directly reduces the activity of lactate dehydrogenase (Figure 2), which is required for the formation of oxalate crystals in kidney [89].

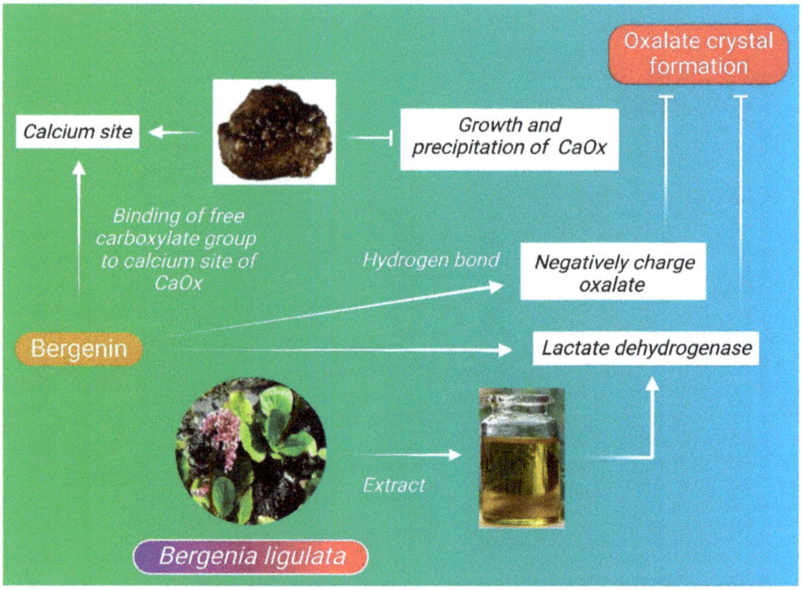

Figure 2. Anti-urolithiatic mechanism of *B. ligulata*. Bergenin, one of the bioactive components of *B. ligulata*, makes it a potent anti-lithiatic herb. Hydrolysis product of bergenin, i.e., 4-O-methylglycoside (4-OMG), has negatively charged free carboxylate group, which enables it to bind with the calcium site of calcium oxalate (CaOx) crystals and inhibit the precipitation of these crystals, thereby impeding renal stone formation.

6.2. Antioxidative Mechanism

Biological reactions in living organisms produce unpaired-electron-containing molecules called free radicals, referred to as ROS and reactive nitrogen species (RNS). Antioxidants are the main defensive mechanism that neutralizes the action of free radicals. A balance between ROS and antioxidants is crucial for maintaining normal biological activity in living organisms. A minimal amount of free radicals are essential for immune responses, phagocytosis, activation of cellular receptors, processing of cellular signaling, and other important biological activities [103], whereas an excessive amount of free radical generation makes cells susceptible to oxidative stress, which ultimately promotes lipid peroxidation, DNA damage, oxidative modification of amino acid, and oxidative-stress-mediated peptide cleavage [103,104]. Bergenin and catechin are two major compounds found in *B. ligulata* that may mostly contribute to the antioxidant property of the plant [14]. Bergenin introduces free-radical-scavenging capacity against hydroxyl radicals by forming aromatic conjugated dienes, and it can also form a complex with Fe (II) that ultimately blocks the generation of hydroxyl radicals in the Fenton reaction [89]. The superoxide anion radical scavenging property of bergenin was assessed through the NADH radical scavenging assay, and the reducing property of the bioactive molecule was attributed to its electron-shifting ability [105]. The 11-O-galloylbergenin of *B. ligulata* has a benzoyl moiety with three hydroxyl groups, i.e., two meta and one para group, which provides 11-O-galloylbergenin a special structural orientation to easily interact with free radicals and scavenge on them [11] (Figure 3).

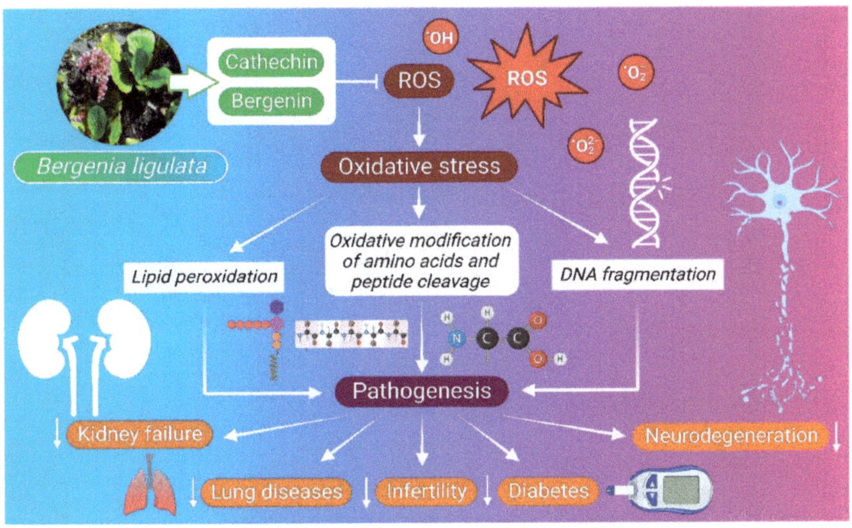

Figure 3. Antioxidative mechanism of *B. ligulata*. Bergenin has been attributed to the free radical scavenging capacity against hydroxyl radicals through the formation of aromatic conjugated dienes and Fe (II) complex, which ultimately block the generation of hydroxyl radicals. One of the compounds of *B. ligulata*, 11-O-galloylbergenin, has a special structural orientation due to the presence of two meta and one para group, which induces free radical scavenging by the herb. ROS—reactive oxygen species.

6.3. Antipyretic Mechanism

Pyrexia is defined as the abnormal elevation of body temperature, where external temperature is more than 38 °C or internal temperature is above 38.4 °C [106]. In the case of pyrogenic pyrexia, the generation of fever involves several mechanisms, most commonly exogenous pyrogens such as microbes or endogenous pyrogens such as interleukin 1 (IL-1) or interleukin 6 (IL-6). TNF-α interacts with organum vasculosum of the lamina terminalis (OVLT), leading to the development of pyrexia. On the contrary, most autoinflammatory conditions are genetic and also related to malfunctioning of pro-inflammatory cytokines such as interlukin-1 or interferon signaling or constitutive NF-kB (nuclear factor kappa-light-chain-enhancer of activated B cells) activation, which are the most potent targets for the management of pyrexia [107]. In pyrexic conditions, there is an increase in prostaglandins (PGE2), which control the thermoregulatory center in the hypothalamus. Some antipyretic drugs introduce their antipyretic activity by blocking prostaglandin biosynthesis [108]. *B. ligulata* has antipyretic properties, particularly ethanolic root and rhizome extracts of the plant contain steroids, alkaloids, flavonoids, and terpenoids, which may block the activity of pyrogens on temperature-sensitive neurons in the preoptic part of the hypothalamus [17]. However, more detailed study is needed to reveal the bioactive molecules that are responsible for the antipyretic property (Figure 4).

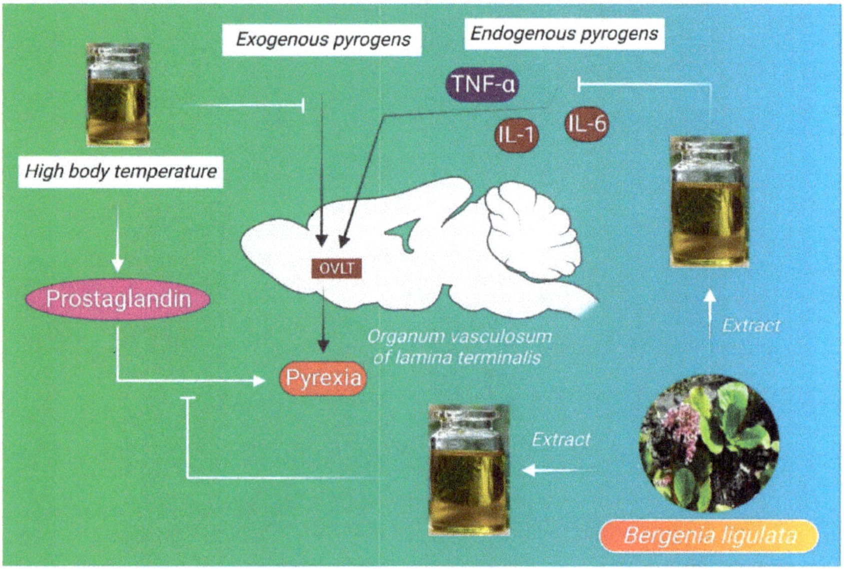

Figure 4. Antipyretic mechanism of *B. ligulata*. The biomolecules present in *B. ligulata* have been suggested to block the activity of pyrogens on temperature-sensitive neurons in the preoptic area of hypothalamus, thereby ameliorating fever. IL-1—interleukin-1, IL-6—interleukin-6, OVLT—organum vasculosum laminae terminalis.

6.4. Anti-Diabetic Mechanism

Diabetes mellitus has several etiologies and is characterized by high blood glucose level resulting from the destruction of pancreatic beta cells, defects in insulin secretion, and abnormalities in insulin receptors [109]. Diabetes can also trigger other diseases such as cardiovascular diseases, neuropathy, retinopathy, and nephropathy [96]. *B. ligulata* extract may stimulate the pancreatic islet cells and increase insulin secretion to maintain normal blood glucose levels [7]. Afzelechin from *B. ligulata* extract exhibited anti-diabetic activity by inhibiting the enzymatic action of α-glucosidase, thus delaying the absorption of dietary carbohydrates in the small intestine and reducing postprandial hyperglycemia and hyper-insulinemia [96,110]. Similarly, (-)-3-O-galloylepicatechin and (-)-3-O-galloylcatechin isolated from *B. ligulata* have demonstrated inhibitory effects against porcine pancreatic α–amylase (Figure 5), which also delays the absorption of glucose in the intestine [9,111].

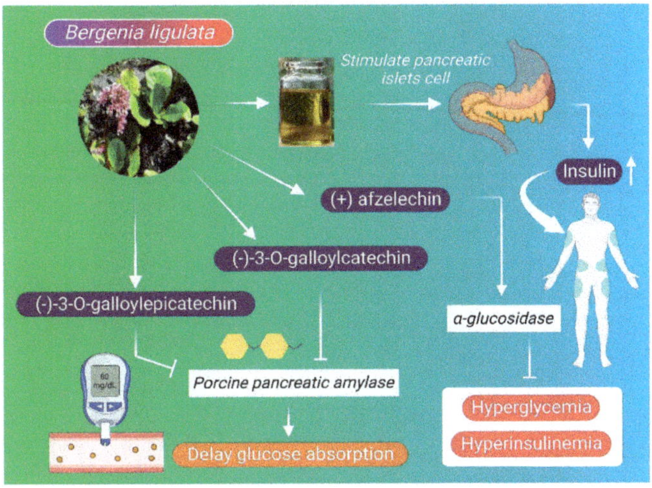

Figure 5. Anti-diabetic mechanism of *B. ligulata*. Plant extract stimulates the pancreatic islet cells and increases insulin secretion for maintaining normal blood glucose levels. (+) afzelechin, a compound of *B. ligulata*, inhibits the enzymatic action of alpha-glucosidase, thus delaying the absorption of dietary carbohydrate in the small intestine and reduces postprandial hyperglycemia and hyperinsulinemia. Additionally, compounds such as (-)-3-O-galloylepicatechin and (-)-3-O-galloylcatechin from *B. ligulata* inhibit pancreatic α-amylase and delay the absorption of glucose in the intestine.

6.5. Anti-Inflammatory Mechanism

Inflammation is a response of the immune system to foreign substances and injuries, which is essential for tissue homeostasis under different noxious conditions [112]. Beta-sitosterol is an important constituent isolated from *B. ligulata* extract that possesses anti-inflammatory properties. Beta-sitosterol directly induces the proliferation rate of T cells and releases interferon and increases natural killer cell activity as well. Anti-inflammatory properties of the *B. ligulata* plant may also be attributed to the synergistic effect of pro-inflammatory enzyme inhibitors that are responsible for the reduction in inflammatory response [6]. Bergenin also inhibits the formation of proinflammatory cytokines such as IL-2, interferon gamma (INF-ɣ), and TNF–α [45]. It also shows anti-microbial properties by inhibiting the growth of microbes in both in vivo and in vitro (Figure 6); however, the exact mechanism of action is yet unknown [37].

6.6. Hepatoprotective Mechanism

Several studies have demonstrated the hepatoprotective activity of *B. ligulata*, although the mechanism of action is poorly understood. The antioxidant properties and cellular restoration capacity of the plant may play significant roles in the recovery of damaged liver tissues. Free radicals such as hydroxyl radicals, hydrogen peroxide, superoxide radicals, and lipid peroxide are predominant in liver diseases. These free radicals are normally generated during the biochemical process of the body or due to exposure to different environmental toxicants or pathological states [113]. Excess amounts of free radicals generate oxidative stress that alters the membrane structure and damage other important components of the cell including lipids, proteins, and nucleic acids [114]. The bioactive compounds of *B. ligulata* may exert hepatoprotective activity (Figure 7) through antioxidant and free-radical scavenging properties. Similarly, it can normalize increased Kuffer cells number and lymphocytic infiltration in infected mice [19]. In Wistar albino rats, *B. ligulata* root extract ameliorated carbon tetrachloride (CCl_4)-induced liver damage along with a reduction in the level of alkaline phosphatase (ALP), total bilirubin level, SGPT, and SGOT [95]. The major bioactive component of *B. ligulata*, i.e., bergenin, might be responsible for the hepatoprotective properties of the plant. Bergenin at doses of 50, 100, and 200 mg/kg body weight showed a strong hepatoprotective effect when administrated orally for 7 successive days in CCl_4-induced liver damage in rats. The administration of bergenin subsequently normalizes the increasing serum enzymatic activities of alanine/aspartate aminotransferase, sorbitol dehydrogenase, and γ-glutamyltransferase in CCl_4-treated rats in a dose-dependent manner. In contrast, the recuperation of the activities of glutathione S-transferase and glutathione reductase was also reported. Additionally, bergenin can prevent the elevation of hepatic malondialdehyde formation and the depletion of reduced glutathione content in the liver of CCl_4-intoxicated rats [115,116].

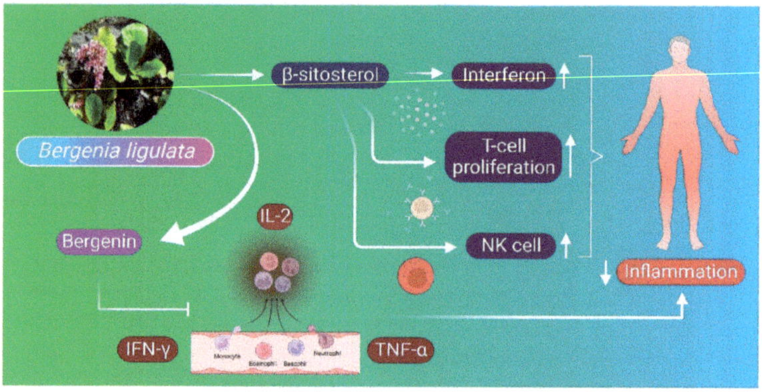

Figure 6. Anti-inflammatory mechanism of *B. ligulata*. Beta-sitosterol is an important constituent of *B. ligulata* plant extract that possesses anti-inflammatory properties. It induces the proliferation of T cells and releases interferon. Anti-inflammatory properties of the plant are attributed to the synergistic effect of pro-inflammatory enzyme inhibitors that are responsible for reducing the inflammatory response. Bergenin also inhibits the formation of proinflammatory cytokines such as interleukin-2 (IL-2), interferon gamma (IFN-Ɣ), and tumor necrosis-alpha (TNF-α), all assisting in the anti-inflammatory response. NK—natural killer cell.

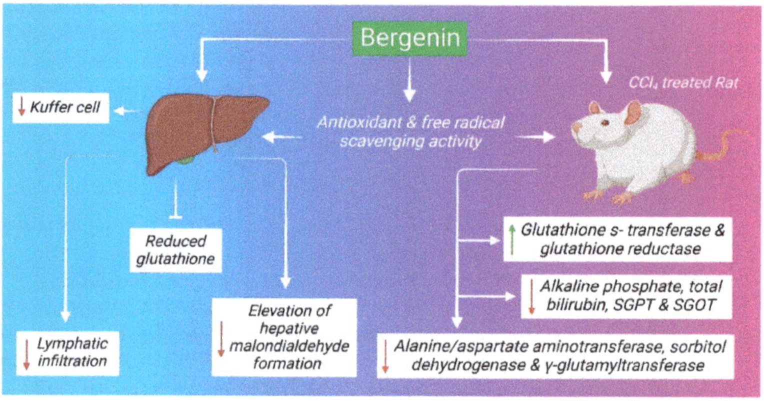

Figure 7. Hepatoprotective mechanism of *B. ligulata*. The plant extract shows hepatoprotective properties through antioxidant activity and free radical scavenging activity. Bioactive compounds reduce the number of Kuffer cells in the liver and inhibit the infiltration of lymphocytes. It also reduces alkaline phosphate, total bilirubin, serum glutamic-pyruvic transaminase (SGPT) and serum glumatic-oxaloacetic transaminase (SGOT), alanine aminotransferase (ALT), sorbitol dehydrogenase, and γ-glutamyl transferase. Further, it helped resist the depletion of reduced glutathione content and inhibit malondialdehyde formation in the liver of CCl_4-treated rats.

7. Future Perspectives

Plant-based natural remedies have been used worldwide for the management of various human ailments for ages. In developing countries, approximately 8 out of 10 people opt for herbal treatment for their primary health care, due to easy accessibility, low toxicity, and inherited traditional knowledge of using plant and their derivatives in the form of either the whole plant or part of the plant, such as root, rhizome, leaf, fruit, or flower. In this modern era, researchers are focusing on herbal medicines by taking reference from indigenous ancient traditional literature such as *Ayurveda*, *Unani*, *Siddha*, traditional Chinese medicine, and others, as a reverse pharmacology approach. Cutting-edge analytical tools and techniques are paving the way for the identification and validation of bioactive molecules from herbs and their use on par with standard modern synthetic drugs. The bioactive molecules of various traditionally used plants have been clinically proven as effective against a range of acute and chronic diseases. Further studies on the traditional herbs may lead to development of several novel drugs against various long-standing human ailments. In this review, we attempted to provide the mechanism(s) of action of *B. ligulata* and/or its phytoconstituents to better understand how the plant executes its anti-urolithiatic, antioxidative, antipyretic, anti-diabetic, anti-inflammatory, and hepatoprotective properties, as well as discuss the important bioactive compounds of clinical importance. The studies conducted on rats and other mammalian models provide a deep insight into the therapeutic potential of *B. ligulata* in the management of various ailments. Despite limited data, *B. ligulata* has been shown to exert antioxidant properties, reduce cellular oxidative stress, and modulate enzyme action (α-glucosidase) under in vivo and in vitro conditions, which renders the herb interesting for further clinical studies.

Author Contributions: Conceptualization, S.R.; data curation, writing—original draft preparation, S.R., D.D., S.D. and N.K.J.; writing—review and editing, M.P., A.K., K.K.K., J.C.K., and P.S. All authors have read and agreed to the published version of the manuscript.

Funding: This research received no external funding.

Institutional Review Board Statement: Not applicable.

Informed Consent Statement: Not applicable.

Data Availability Statement: The study did not report any data.

Conflicts of Interest: The authors declare no conflict of interest.

References

1. Tilburt, J.C.; Kaptchuk, T.J. Herbal Medicine Research and Global Health: An Ethical Analysis. *Bull. World Health Organ.* **2008**, *86*, 594–599. [CrossRef]
2. Akerele, O. Nature's Medicinal Bounty: Don't Throw It Away. In *Proceedings of the World Health Forum*; World Health Organisation: Geneva, Switzerland, 1993; Volume 14, pp. 390–395.
3. Ghatapanadi, S.R.; Johnson, N.; Rajasab, A.H. Medicinal Plants of North Karnataka Used in Treatment of Kidney Stone and Urinary Tract Infections. *Socioscan* **2010**, *2*, 23–24.
4. Arya, P.; Pandey, S.; Verma, V. Kidney Stone Formation and Use of Medicinal Plants as Antiurolithiatic Agents. *Univers. J. Pharm. Res.* **2017**, *2*, 43–48. [CrossRef]
5. Goswami, P.K.; Samant, M.; Srivastava, R.S. Multi Faceted Saxifraga Ligulata. *Int. J. Res. Ayurveda Pharm. IJRAP* **2013**, *4*, 608–611. [CrossRef]
6. Sajad, T.; Zargar, A.; Ahmad, T.; Bader, G.N.; Naime, M.; Ali, S. Antibacterial and Anti-Inflammatory Potential *Bergenia ligulata*. *Am. J. Biomed. Sci.* **2010**, *2*, 313–321. [CrossRef]
7. Gurav, S.S.; Gurav, N.S. A Comprehensive Review: *Bergenia ligulata* Wall—A Controversial Clinical Candidate. *Int. J. Pharm. Sci. Rev. Res.* **2014**, *5*, 1630–1642.
8. Jani, S.; Shukla, V.J.; Harisha, C.R. Comparative Pharmacognostical and Phytochemical Study on *Bergenia ligulata* Wall. and *Ammania buccifera* Linn. *Ayu* **2013**, *34*, 406. [CrossRef]
9. Ruby, K.M.; Dwivedi, J.; Chauhan, R. Pashanbheda a Golden Herb of Himalaya: A Review. *Int. J. Pharm. Rev. Res.* **2012**, *2*, 97–105.
10. Pandey, R.; Kumar, B.; Meena, B.; Srivastava, M.; Mishra, T.; Tiwari, V.; Pal, M.; Nair, N.K.; Upreti, D.K.; Rana, T.S. Major Bioactive Phenolics in *Bergenia* Species from the Indian Himalayan Region: Method Development, Validation and Quantitative Estimation Using UHPLC-QqQLIT-MS/MS. *PLoS ONE* **2017**, *12*, e0180950.

11. Sadat, A.; Uddin, G.; Alam, M.; Ahmad, A.; Siddiqui, B.S. Structure Activity Relationship of Bergenin, p-Hydroxybenzoyl Bergenin, 11-O-Galloylbergenin as Potent Antioxidant and Urease Inhibitor Isolated from *Bergenia ligulata*. *Nat. Prod. Res.* **2015**, *29*, 2291–2294. [CrossRef]
12. Pandey, B.P.; Pradhan, S.P.; Adhikari, K.; Nepal, S. *Bergenia pacumbis* from Nepal, an astonishing enzymes inhibitor. *BMC Complement. Altern. Med.* **2020**, *20*, 1–12. [CrossRef] [PubMed]
13. Das, C.; Kumari, B.; Singh, M.P.; Singh, S. A Literary Review and Therapeutic Action of Pashanbheda (*Bergenia ligulata* Wall) described by Shamhita in Ashmari Roga. *J. Ayurveda Integr. Med.* **2022**, *7*, 105–114.
14. Dix, B.S.; Srivastava, S.N. Tannin Constituents of *Bergenia ligulata* Roots. *Ind. J. Nat. Prod.* **1989**, *5*, 24–25.
15. Bashir, S.; Gilani, A.H. Antiurolithic Effect of *Bergenia ligulata* Rhizome: An Explanation of the Underlying Mechanisms. *J. Ethnopharmacol.* **2009**, *122*, 106–116. [CrossRef]
16. Singh, A.; Tandon, S.; Nandi, S.P.; Kaur, T.; Tandon, C. Downregulation of Inflammatory Mediators by Ethanolic Extract of *Bergenia ligulata* (Wall.) in Oxalate Injured Renal Epithelial Cells. *J. Ethnopharmacol.* **2021**, *275*, 114104. [CrossRef] [PubMed]
17. Nardev, S.; Gupta, A.K.; Vijay, J.; Renu, C. Study on Antipyretic Activity of Extracts of *Bergenia ligulata* Wall. *Int. J. Pharma. Bio Sci.* **2010**, *1*, PS58.
18. Rajbhandari, M.; Wegner, U.; Schoepke, T.; Lindequist, U.; Mentel, R. Inhibitory Effect of *Bergenia ligulata* on Influenza Virus A. *Pharm.-Int. J. Pharm. Sci.* **2003**, *58*, 268–271.
19. Kaur, R.; Kaur, S. Evaluation of In Vitro and In Vivo Antileishmanial Potential of Bergenin Rich *Bergenia ligulata* (Wall.) Engl. Root Extract against Visceral Leishmaniasis in Inbred BALB/c Mice through Immunomodulation. *J. Tradit. Complement. Med.* **2018**, *8*, 251–260. [CrossRef]
20. Gohain, A.; Sharma, A.; Gogoi, H.J.; Cooper, R.; Kaur, R.; Nayik, G.A.; Shaikh, A.M.; Kovács, B.; Areche, F.O.; Ansari, M.J.; et al. *Bergenia pacumbis* (Buch.-Ham. ex D. Don) CY Wu & JT Pan: A Comprehensive Review on Traditional Uses, Phytochemistry and Pharmacology. *Plants* **2022**, *11*, 1129.
21. Kumar, V.; Tyagi, D. Review on Phytochemical, Ethnomedical and Biological Studies of Medically Useful Genus *Bergenia*. *Int. J. Curr. Microbiol. App. Sci.* **2013**, *2*, 328–334.
22. Kirtikar, K.R.; Basu, B.D. *Text Book of Indian Medicinal Plants*; International Book Distributors: Dehradun, India, 2005; Volume 2, pp. 993–994.
23. Samal, P.K.; Dhyani, P.P.; Dollo, M. Indigenous Medicinal Practices of Bhotia Tribal Community in Indian Central Himalaya. *Indian J. Tradit. Knowl.* **2010**, *9*, 140–144.
24. Savita, R. Ethnomedicinal Plants of Chamba District, Himachal Pradesh, India. *J. Med. Plants Res.* **2013**, *7*, 3147–3157.
25. Sharma, H.K.; Chhangte, L.; Dolui, A.K. Traditional Medicinal Plants in Mizoram, India. *Fitoterapia* **2001**, *72*, 146–161. [CrossRef]
26. Amazon. Available online: https://www.Amazon.in/Leayur-Pashan-Bergenia-Ligulata-Powder/Dp/B011UZD8YS (accessed on 29 July 2022).
27. Flipkart. Available online: https://www.Flipkart.Com/Nutrixia-Food-Pashanbhed-Lakdi-Pashan-Bhed-Root-Pakhanved-Jadd-Bergenia-Ligulata-Seed/p/Itma7904d784a9db (accessed on 29 July 2022).
28. Reddy, U.D.C.; Chawla, A.S.; Deepak, M.; Singh, D.; Handa, S.S. High Pressure Liquid Chromatographic Determination of Bergenin and (+)-Afzelechin from Different Parts of Paashaanbhed (*Bergenia ligulata* Yeo). *Phytochem. Anal. Int. J. Plant Chem. Biochem. Tech.* **1999**, *10*, 44–47. [CrossRef]
29. Liu, Y.; Chen, Y.; Liao, B.; Luo, D.; Wang, K.; Li, H.; Zeng, G. Epidemiology of Urolithiasis in Asia. *Asian J. Urol.* **2018**, *5*, 205–214. [CrossRef]
30. Chandrareddy, U.D.; Chawla, A.S.; Mundkinajeddu, D.; Maurya, R.; Handa, S.S. Paashaanolactone from *Bergenia ligulata*. *Phytochemistry* **1998**, *47*, 907–909. [CrossRef]
31. Chauhan, R.; Ruby, K.M.; Dwivedi, J. Secondary Metabolites Found in *Bergenia* Species: A Compendious Review. *Int. J. Pharm. Pharm. Sci.* **2013**, *5*, 9–16.
32. Chauhan, S.K.; Singh, B.; Agrawal, S. Simultaneous Determination of Bergenin and Gallic Acid in *Bergenia ligulata* Wall by High-Performance Thin-Layer Chromatography. *J. AOAC Int.* **2000**, *83*, 1480–1483. [CrossRef]
33. Fujii, M.; Miyaichi, Y.; Tomimori, T. Studies on Nepalese Crude Drugs. XXII: On the Phenolic Constituents of the Rhizome of *Bergenia ciliata* (Haw.) Sternb. *Nat. Med.* **1996**, *50*, 404–407.
34. Kumar, S. Herbaceous Flora of Jaunsar-Bawar (Uttarkhand), India: Enumerations. *Phytotaxonomy* **2012**, *12*, 33–56.
35. Bohm, B.A.; Donevan, L.S.; Bhat, U.G. Flavonoids of Some Species of *Bergenia*, *Francoa*, *Parnassia* and *Lepuropetalon*. *Biochem. Syst. Ecol.* **1986**, *14*, 75–77. [CrossRef]
36. Bajracharya, G.B.; Maharjan, R.; Maharjan, B.L. Potential Antibacterial Activity of *Bergenia purpurascens*. *Nepal J. Sci. Technol.* **2011**, *12*, 157–162. [CrossRef]
37. Koul, B.; Kumar, A.; Yadav, D.; Jin, J.-O. *Bergenia* Genus: Traditional Uses, Phytochemistry and Pharmacology. *Molecules* **2020**, *25*, 5555. [CrossRef]
38. Messaoudi, D.; Bouriche, H.; Demirtas, I.; Senator, A. Phytochemical Analysis and Hepatoprotective Activity of Algerian *Santolina chamaecyparissus* L. Extracts. *Annu. Res. Rev. Biol.* **2018**, *25*, 1–12. [CrossRef]
39. RAStOGi, S.; RAwAt, A.K.S. A Comprehensive Review on Bergenin, a Potential Hepatoprotective and Antioxidative Phytoconstituent. *Herba Pol.* **2008**, *54*, 66–79.

40. Abe, K.; Sakai, K.; Uchida, M. Effects of Bergenin on Experimental Ulcers—Prevention of Stress Induced Ulcers in Rats. *Gen. Pharmacol. Vasc. Syst.* **1980**, *11*, 361–368. [CrossRef]
41. Shi, X.; Xu, M.; Luo, K.; Huang, W.; Yu, H.; Zhou, T. Anticancer Activity of Bergenin against Cervical Cancer Cells Involves Apoptosis, Cell Cycle Arrest, Inhibition of Cell Migration and the STAT3 Signalling Pathway. *Exp. Ther. Med.* **2019**, *17*, 3525–3529. [CrossRef]
42. Takahashi, H.; Kosaka, M.; Watanabe, Y.; Nakade, K.; Fukuyama, Y. Synthesis and Neuroprotective Activity of Bergenin Derivatives with Antioxidant Activity. *Bioorg. Med. Chem.* **2003**, *11*, 1781–1788. [CrossRef]
43. Hou, W.; Ye, C.; Chen, M.; Li, W.; Gao, X.; He, R.; Zheng, Q.; Zhang, W. Bergenin Activates SIRT1 as a Novel Therapeutic Agent for Osteogenesis of Bone Mesenchymal Stem Cells. *Front. Pharmacol.* **2019**, *10*, 618. [CrossRef]
44. da Silva, S.L.; de Oliveira, V.G.; Yano, T.; Nunomura, R.D.C.S. Antimicrobial Activity of Bergenin from *Endopleura uchi* (Huber) Cuatrec. *Acta Amaz.* **2009**, *39*, 187–191. [CrossRef]
45. Patel, D.K.; Patel, K.; Kumar, R.; Gadewar, M.; Tahilyani, V. Pharmacological and Analytical Aspects of Bergenin: A Concise Report. *Asian Pac. J. Trop. Dis.* **2012**, *2*, 163–167. [CrossRef]
46. Bernatoniene, J.; Kopustinskiene, D.M. The Role of Catechins in Cellular Responses to Oxidative Stress. *Molecules* **2018**, *23*, 965. [CrossRef] [PubMed]
47. Shimamura, T.; Zhao, W.-H.; Hu, Z.-Q. Mechanism of Action and Potential for Use of Tea Catechin as an Antiinfective Agent. *Anti-Infect. Agents Med. Chem. (Former. Curr. Med. Chem.-Anti-Infect. Agents)* **2007**, *6*, 57–62. [CrossRef]
48. Gadkari, P.V.; Balaraman, M. Catechins: Sources, Extraction and Encapsulation: A Review. *Food Bioprod. Process.* **2015**, *93*, 122–138. [CrossRef]
49. Weber, J.M.; Ruzindana-Umunyana, A.; Imbeault, L.; Sircar, S. Inhibition of Adenovirus Infection and Adenain by Green Tea Catechins. *Antivir. Res.* **2003**, *58*, 167–173. [CrossRef]
50. Mukoyama, A.; Ushijima, H.; Nishimura, S.; Koike, H.; Toda, M.; Hara, Y.; Shimamura, T. Inhibition of Rotavirus and Enterovirus Infections by Tea Extracts. *Jpn. J. Med. Sci. Biol.* **1991**, *44*, 181–186. [CrossRef]
51. Liu, S.; Lu, H.; Zhao, Q.; He, Y.; Niu, J.; Debnath, A.K.; Wu, S.; Jiang, S. Theaflavin Derivatives in Black Tea and Catechin Derivatives in Green Tea Inhibit HIV-1 Entry by Targeting Gp41. *Biochim. Biophys. Acta BBA-Gen. Subj.* **2005**, *1723*, 270–281. [CrossRef]
52. Ide, K.; Kawasaki, Y.; Kawakami, K.; Yamada, H. Anti-influenza virus effects of catechins: A molecular and clinical review. *Curr. Med. Chem.* **2016**, *23*, 4773–4783. [CrossRef]
53. Okada, F.; Takeo, T.; Okada, S.; Tamemasa, O. Antiviral Effect of Theaflavins on Tobacco Mosaic Virus. *Agric. Biol. Chem.* **1977**, *41*, 791–794.
54. Higdon, J.V.; Frei, B. Tea Catechins and Polyphenols: Health Effects, Metabolism, and Antioxidant Functions. *Crit. Rev. Food Sci. Nutr.* **2003**, *43*, 89–143. [CrossRef]
55. Perron, N.R.; Brumaghim, J.L. A Review of the Iron-Binding Mechanism for Polyphenol Antioxidant Activity. *Cell Biochem. Biophys.* **2009**, *53*, 75–100. [CrossRef] [PubMed]
56. Valcic, S.; Burr, J.A.; Timmermann, B.N.; Liebler, D.C. Antioxidant Chemistry of Green Tea Catechins. New Oxidation Products of (−)-Epigallocatechin Gallate and (−)-Epigallocatechin from Their Reactions with Peroxyl Radicals. *Chem. Res. Toxicol.* **2000**, *13*, 801–810. [CrossRef] [PubMed]
57. Roychoudhury, S.; Agarwal, A.; Virk, G.; Cho, C.-L. Potential Role of Green Tea Catechins in the Management of Oxidative Stress-Associated Infertility. *Reprod. Biomed. Online* **2017**, *34*, 487–498. [CrossRef]
58. Manikandan, R.; Beulaja, M.; Arulvasu, C.; Sellamuthu, S.; Dinesh, D.; Prabhu, D.; Babu, G.; Vaseeharan, B.; Prabhu, N.M. Synergistic Anticancer Activity of Curcumin and Catechin: An in Vitro Study Using Human Cancer Cell Lines. *Microsc. Res. Tech.* **2012**, *75*, 112–116. [CrossRef] [PubMed]
59. Saeedi, M.; Khezri, K.; Seyed Zakaryaei, A.; Mohammadamini, H. A Comprehensive Review of the Therapeutic Potential of A-arbutin. *Phytother. Res.* **2021**, *35*, 4136–4154. [CrossRef]
60. Migas, P.; Krauze-Baranowska, M. The Significance of Arbutin and Its Derivatives in Therapy and Cosmetics. *Phytochem. Lett.* **2015**, *13*, 35–40. [CrossRef]
61. Lee, H.-J.; Kim, K.-W. Anti-Inflammatory Effects of Arbutin in Lipopolysaccharide-Stimulated BV2 Microglial Cells. *Inflamm. Res.* **2012**, *61*, 817–825. [CrossRef]
62. Li, H.; Jeong, Y.-M.; Kim, S.Y.; Kim, M.-K.; Kim, D.-S. Arbutin Inhibits TCCSUP Human Bladder Cancer Cell Proliferation via Up-Regulation of P21. *Pharm.-Int. J. Pharm. Sci.* **2011**, *66*, 306–309.
63. Fernandes, F.H.A.; Salgado, H.R.N. Gallic Acid: Review of the Methods of Determination and Quantification. *Crit. Rev. Anal. Chem.* **2016**, *46*, 257–265. [CrossRef]
64. Kahkeshani, N.; Farzaei, F.; Fotouhi, M.; Alavi, S.S.; Bahramsoltani, R.; Naseri, R.; Momtaz, S.; Abbasabadi, Z.; Rahimi, R.; Farzaei, M.H. Pharmacological Effects of Gallic Acid in Health and Diseases: A Mechanistic Review. *Iran J. Basic. Med. Sci.* **2019**, *22*, 225.
65. Govea-Salas, M.; Rivas-Estilla, A.M.; Rodríguez-Herrera, R.; Lozano-Sepúlveda, S.A.; Aguilar-Gonzalez, C.N.; Zugasti-Cruz, A.; Salas-Villalobos, T.B.; Morlett-Chávez, J.A. Gallic Acid Decreases Hepatitis C Virus Expression through Its Antioxidant Capacity. *Exp. Ther. Med.* **2016**, *11*, 619–624. [CrossRef] [PubMed]

66. Kratz, J.M.; Andrighetti-Fröhner, C.R.; Leal, P.C.; Nunes, R.J.; Yunes, R.A.; Trybala, E.; Bergström, T.; Barardi, C.R.M.; Simões, C.M.O. Evaluation of Anti-HSV-2 Activity of Gallic Acid and Pentyl Gallate. *Biol. Pharm. Bull.* **2008**, *31*, 903–907. [CrossRef] [PubMed]
67. Chatterjee, A.; Chatterjee, S.; Biswas, A.; Bhattacharya, S.; Chattopadhyay, S.; Bandyopadhyay, S.K. Gallic Acid Enriched Fraction of Phyllanthus Emblica Potentiates Indomethacin-Induced Gastric Ulcer Healing via e-NOS-Dependent Pathway. *Evid.-Based Complement. Altern. Med.* **2012**, *2012*, 487380. [CrossRef] [PubMed]
68. Gandhi, G.R.; Jothi, G.; Antony, P.J.; Balakrishna, K.; Paulraj, M.G.; Ignacimuthu, S.; Stalin, A.; Al-Dhabi, N.A. Gallic Acid Attenuates High-Fat Diet Fed-Streptozotocin-Induced Insulin Resistance via Partial Agonism of PPARγ in Experimental Type 2 Diabetic Rats and Enhances Glucose Uptake through Translocation and Activation of GLUT4 in PI3K/p-Akt Signaling Pathway. *Eur. J. Pharmacol.* **2014**, *745*, 201–216. [CrossRef] [PubMed]
69. Kakkar, S.; Bais, S. A Review on Protocatechuic Acid and Its Pharmacological Potential. *Int. Sch. Res. Not.* **2014**, *2014*, 952943. [CrossRef]
70. Semaming, Y.; Pannengpetch, P.; Chattipakorn, S.C.; Chattipakorn, N. Pharmacological Properties of Protocatechuic Acid and Its Potential Roles as Complementary Medicine. *Evid.-Based Complement. Altern. Med.* **2015**, *2015*, 593902. [CrossRef]
71. Krzysztoforska, K.; Mirowska-Guzel, D.; Widy-Tyszkiewicz, E. Pharmacological Effects of Protocatechuic Acid and Its Therapeutic Potential in Neurodegenerative Diseases: Review on the Basis of in Vitro and in Vivo Studies in Rodents and Humans. *Nutr. Neurosci.* **2019**, *22*, 72–82. [CrossRef]
72. Harini, R.; Pugalendi, K.V. Antihyperglycemic Effect of Protocatechuic Acid on Streptozotocin-Diabetic Rats. *J. Basic Clin. Physiol. Pharmacol.* **2010**, *21*, 79–92. [CrossRef]
73. Lende, A.B.; Kshirsagar, A.D.; Deshpande, A.D.; Muley, M.M.; Patil, R.R.; Bafna, P.A.; Naik, S.R. Anti-Inflammatory and Analgesic Activity of Protocatechuic Acid in Rats and Mice. *Inflammopharmacology* **2011**, *19*, 255–263. [CrossRef]
74. Meng, S.; Cao, J.; Feng, Q.; Peng, J.; Hu, Y. Roles of Chlorogenic Acid on Regulating Glucose and Lipids Metabolism: A Review. *Evid.-Based Complement. Altern. Med. ECAM* **2013**, *2013*, 801457. [CrossRef]
75. Naveed, M.; Hejazi, V.; Abbas, M.; Kamboh, A.A.; Khan, G.J.; Shumzaid, M.; Ahmad, F.; Babazadeh, D.; FangFang, X.; Modarresi-Ghazani, F. Chlorogenic Acid (CGA): A Pharmacological Review and Call for Further Research. *Biomed. Pharmacother.* **2018**, *97*, 67–74. [CrossRef] [PubMed]
76. Srinivasulu, C.; Ramgopal, M.; Ramanjaneyulu, G.; Anuradha, C.M.; Kumar, C.S. Syringic Acid (SA)—A Review of Its Occurrence, Biosynthesis, Pharmacological and Industrial Importance. *Biomed. Pharmacother.* **2018**, *108*, 547–557. [CrossRef] [PubMed]
77. Mathew, S.; Abraham, T.E. Ferulic Acid: An Antioxidant Found Naturally in Plant Cell Walls and Feruloyl Esterases Involved in Its Release and Their Applications. *Crit. Rev. Biotechnol.* **2004**, *24*, 59–83. [CrossRef] [PubMed]
78. Boz, H. Ferulic Acid in Cereals—A Review. *Czech J. Food Sci.* **2015**, *33*, 1–7. [CrossRef]
79. Zduńska, K.; Dana, A.; Kolodziejczak, A.; Rotsztejn, H. Antioxidant Properties of Ferulic Acid and Its Possible Application. *Ski. Pharmacol. Physiol.* **2018**, *31*, 332–336. [CrossRef]
80. Srinivasan, M.; Sudheer, A.R.; Menon, V.P. Ferulic Acid: Therapeutic Potential through Its Antioxidant Property. *J. Clin. Biochem. Nutr.* **2007**, *40*, 92–100. [CrossRef]
81. Ergün, B.Ç.; Çoban, T.; Onurdag, F.K.; Banoglu, E. Synthesis, Antioxidant and Antimicrobial Evaluation of Simple Aromatic Esters of Ferulic Acid. *Arch. Pharm. Res.* **2011**, *34*, 1251–1261. [CrossRef]
82. Shi, C.; Song, K.; Zhang, X.; Sun, Y.; Sui, Y.; Chen, Y.; Jia, Z.; Sun, H.; Sun, Z.; Xia, X. Antimicrobial Activity and Possible Mechanism of Action of Citral against Cronobacter Sakazakii. *PLoS ONE* **2016**, *11*, e0159006. [CrossRef]
83. Cikman, O.; Soylemez, O.; Ozkan, O.F.; Kiraz, H.A.; Sayar, I.; Ademoglu, S.; Taysi, S.; Karaayvaz, M. Antioxidant Activity of Syringic Acid Prevents Oxidative Stress in L-Arginine–Induced Acute Pancreatitis: An Experimental Study on Rats. *Int. Surg.* **2015**, *100*, 891–896. [CrossRef]
84. Sancak, E.B.; Akbas, A.; Silan, C.; Cakir, D.U.; Turkon, H.; Ozkanli, S.S. Protective Effect of Syringic Acid on Kidney Ischemia-Reperfusion Injury. *Ren. Fail.* **2016**, *38*, 629–635. [CrossRef]
85. Rashedinia, M.; Alimohammadi, M.; Shalfroushan, N.; Khoshnoud, M.J.; Mansourian, M.; Azarpira, N.; Sabahi, Z. Neuroprotective Effect of Syringic Acid by Modulation of Oxidative Stress and Mitochondrial Mass in Diabetic Rats. *BioMed Res. Int.* **2020**, *2020*, 8297984. [CrossRef] [PubMed]
86. Muthukumaran, J.; Srinivasan, S.; Venkatesan, R.S.; Ramachandran, V.; Muruganathan, U. Syringic Acid, a Novel Natural Phenolic Acid, Normalizes Hyperglycemia with Special Reference to Glycoprotein Components in Experimental Diabetic Rats. *J. Acute Dis.* **2013**, *2*, 304–309. [CrossRef]
87. Garimella, T.S.; Jolly, C.I.; Narayanan, S. In Vitro Studies on Antilithiatic Activity of Seeds of *Dolichos biflorus* Linn. and Rhizomes of *Bergenia ligulata* Wall. *Phytother. Res.* **2001**, *15*, 351–355. [CrossRef] [PubMed]
88. Sharma, I.; Khan, W.; Parveen, R.; Alam, M.; Ahmad, I.; Ansari, M.H.R.; Ahmad, S. Antiurolithiasis Activity of Bioactivity Guided Fraction of *Bergenia ligulata* against Ethylene Glycol Induced Renal Calculi in Rat. *BioMed Res. Int.* **2017**, *2017*, 1969525. [CrossRef] [PubMed]
89. Aggarwal, D.; Kaushal, R.; Kaur, T.; Bijarnia, R.K.; Puri, S.; Singla, S.K. The Most Potent Antilithiatic Agent Ameliorating Renal Dysfunction and Oxidative Stress from *Bergenia ligulata* Rhizome. *J. Ethnopharmacol.* **2014**, *158*, 85–93. [CrossRef]
90. Satish, H.; Umashankar, D. Comparative Study of Methanolic Extract of *Bergenia ligulata* Yeo., with Isolated Constituent Bergenin in Urolithiatic Rats. *BioMed* **2006**, *1*, 80–86.

91. Joshi, V.S.; Parekh, B.B.; Joshi, M.J.; Vaidya, A.B. Herbal Extracts of Tribulus Terrestris and *Bergenia ligulata* Inhibit Growth of Calcium Oxalate Monohydrate Crystals in Vitro. *J. Cryst. Growth* **2005**, *275*, e1403–e1408. [CrossRef]
92. Joshi, V.S.; Parekh, B.B.; Joshi, M.J.; Vaidya, A.D. Inhibition of the Growth of Urinary Calcium Hydrogen Phosphate Dihydrate Crystals with Aqueous Extracts of Tribulus Terrestris and *Bergenia ligulata*. *Urol. Res.* **2005**, *33*, 80–86. [CrossRef]
93. Aggarwal, D.; Gautam, D.; Sharma, M.; Singla, S.K. Bergenin Attenuates Renal Injury by Reversing Mitochondrial Dysfunction in Ethylene Glycol Induced Hyperoxaluric Rat Model. *Eur. J. Pharmacol.* **2016**, *791*, 611–621. [CrossRef]
94. Garg, V.; Dutt, R. Antipyretic Plants: An Updated Review. *Curr. Bioact. Compd.* **2020**, *16*, 4–12. [CrossRef]
95. Nardev, S.; Vijay, J.; Gupta, A.K.; Manoj, G. Evaluation of Ethanolic Extract of Root of *Bergenia ligulata* for Hepatoprotective, Diuretic and Antipyretic Activities. *J. Pharm. Res.* **2009**, *2*, 958–960.
96. Saijyo, J.; Suzuki, Y.; Okuno, Y.; Yamaki, H.; Suzuki, T.; Miyazawa, M. α-Glucosidase Inhibitor from *Bergenia ligulata*. *J. Oleo Sci.* **2008**, *57*, 431–435. [CrossRef] [PubMed]
97. Chen, M.; Ye, C.; Zhu, J.; Zhang, P.; Jiang, Y.; Lu, X.; Wu, H. Bergenin as a Novel Urate-Lowering Therapeutic Strategy for Hyperuricemia. *Front. Cell Dev. Biol.* **2020**, *8*, 703. [CrossRef] [PubMed]
98. Dhar, M.L.; Dhar, M.M.; Dhawan, B.N.; Mehrotra, B.N.; Ray, C. Screening of Indian Plants for Biological Activity: Part I. *Indian J. Exp. Biol.* **1968**, *6*, 232–247. [PubMed]
99. Reagan-Shaw, S.; Nihal, M.; Ahmad, N. Dose Translation from Animal to Human Studies Revisited. *FASEB J.* **2008**, *22*, 659–661. [CrossRef]
100. Árok, R.; Végh, K.; Alberti, Á.; Kéry, Á. Phytochemical Comparison and Analysis of *Bergenia crassifolia* L. (Fritsch.) and *Bergenia cordifolia* Sternb. *Eur. Chem. Bull.* **2012**, *1*, 31–34.
101. Alelign, T.; Petros, B. Kidney Stone Disease: An Update on Current Concepts. *Adv. Urol.* **2018**, *2018*, 3068365. [CrossRef]
102. Kobayashi, H.; De Mejía, E. The Genus Ardisia: A Novel Source of Health-Promoting Compounds and Phytopharmaceuticals. *J. Ethnopharmacol.* **2005**, *96*, 347–354. [CrossRef]
103. Kumar, S.; Pandey, A.K. Free Radicals: Health Implications and Their Mitigation by Herbals. *Br. J. Med. Med. Res.* **2015**, *7*, 438–457. [CrossRef]
104. Benov, L.; Beema, A.F. Superoxide-Dependence of the Short Chain Sugars-Induced Mutagenesis. *Free. Radic. Biol. Med.* **2003**, *34*, 429–433. [CrossRef]
105. Hendrychová, H.; Martin, J.; Tůmová, L.; Kočevar-Glavač, N. Bergenin Content and Free Radical Scavenging Activity of *Bergenia* Extracts. *Nat. Prod. Commun.* **2015**, *10*, 1934578X1501000734. [CrossRef]
106. Stocchetti, N.; Rossi, S.; Zanier, E.; Colombo, A.; Beretta, L.; Citerio, G. Pyrexia in Head-Injured Patients Admitted to Intensive Care. *Intensive Care Med.* **2002**, *28*, 1555–1562. [CrossRef] [PubMed]
107. Walter, E.J.; Hanna-Jumma, S.; Carraretto, M.; Forni, L. The Pathophysiological Basis and Consequences of Fever. *Crit. Care* **2016**, *20*, 200. [CrossRef] [PubMed]
108. Deepa, P.K.; Usha, P.T.A.; Nair, A.C.; Prasannakumari, K.T. Antipyretic Activity of Seeds from Red and White Type of Lotus (Nelumbo Nucifera) in Albino Rat. *Vet. World* **2009**, *2*, 213.
109. Association, A.D. Diagnosis and Classification of Diabetes Mellitus. *Diabetes Care* **2010**, *33*, S62–S69. [CrossRef]
110. Matsuura, H.; Asakawa, C.; Kurimoto, M.; Mizutani, J. α-Glucosidase Inhibitor from the Seeds of Balsam Pear (*Momordica charantia*) and the Fruit Bodies of *Grifola frondosa*. *Biosci. Biotechnol. Biochem.* **2002**, *66*, 1576–1578. [CrossRef]
111. Bhandari, M.R.; Jong-Anurakkun, N.; Hong, G.; Kawabata, J. α-Glucosidase and α-Amylase Inhibitory Activities of Nepalese Medicinal Herb Pakhanbhed (*Bergenia ciliata*, Haw.). *Food Chem.* **2008**, *106*, 247–252. [CrossRef]
112. Chen, L.; Deng, H.; Cui, H.; Fang, J.; Zuo, Z.; Deng, J.; Li, Y.; Wang, X.; Zhao, L. Inflammatory Responses and Inflammation-Associated Diseases in Organs. *Oncotarget* **2018**, *9*, 7204–7218. [CrossRef]
113. Lobo, V.C.; Anita, P.; Naresh, C. Antioxidant and Free Radical Scavenging Activity of *Hygrophila schulli* (Buch.-Ham.) Almeida and Almeida. Seeds. *Adv. Bio Res.* **2010**, *1*, 72–78.
114. Pham-Huy, L.A.; He, H.; Pham-Huy, C. Free Radicals, Antioxidants in Disease and Health. *Int. J. Biomed. Sci. IJBS* **2008**, *4*, 89.
115. Lim, H.-K.; Kim, H.-S.; Choi, H.-S.; Oh, S.; Jang, C.-G.; Choi, J.; Kim, S.-H.; Chang, M.-J. Effects of Acetylbergenin against D-Galactosamine-Induced Hepatotoxicity in Rats. *Pharmacol. Res.* **2000**, *42*, 471–474. [CrossRef] [PubMed]
116. Kim, H.-S.; Lim, H.-K.; Chung, M.-W.; Kim, Y.C. Antihepatotoxic Activity of Bergenin, the Major Constituent of *Mallotus japonicus*, on Carbon Tetrachloride-Intoxicated Hepatocytes. *J. Ethnopharmacol.* **2000**, *69*, 79–83. [CrossRef]

molecules

MDPI

Article

Heterologous Expression and Catalytic Properties of Codon-Optimized Small-Sized Bromelain from MD2 Pineapple

Rafida Razali [1], Fikran Aranda Fahrudin [1], Vijay Kumar Subbiah [1], Kazufumi Takano [2] and Cahyo Budiman [1,*]

[1] Biotechnology Research Institute, Universiti Malaysia Sabah, Kota Kinabalu 88400, Sabah, Malaysia
[2] Department of Biomolecular Chemistry, Kyoto Prefectural University, Hangi-cho, Shimogamo, Sakyo-ku, Kyoto 606-8522, Japan
* Correspondence: cahyo@ums.edu.my

Abstract: Bromelain is a unique enzyme-based bioactive complex containing a mixture of cysteine proteases specifically found in the stems and fruits of pineapple (*Ananas comosus*) with a wide range of applications. MD2 pineapple harbors a gene encoding a small bromelain cysteine protease with the size of about 19 kDa, which might possess unique properties compared to the other cysteine protease bromelain. This study aims to determine the expressibility and catalytic properties of small-sized (19 kDa) bromelain from MD2 pineapple (MD2-SBro). Accordingly, the gene encoding MD2-SBro was firstly optimized in its codon profile, synthesized, and inserted into the pGS-21a vector. The insolubly expressed MD2-SBro was then resolubilized and refolded using urea treatment, followed by purification by glutathione S-transferase (GST) affinity chromatography, yielding 14 mg of pure MD2-SBro from 1 L of culture. The specific activity and catalytic efficiency (k_{cat}/K_m) of MD2-SBro were 3.56 ± 0.08 U mg^{-1} and 4.75 ± 0.23 × 10^{-3} μM^{-1} s^{-1}, respectively, where optimally active at 50 °C and pH 8.0, and modulated by divalent ions. The MD2-SBro also exhibited the ability to scavenge the 2,2-diphenyl-1-picryl-hydrazyl-hydrate (DPPH) with an IC_{50} of 0.022 mg mL^{-1}. Altogether, this study provides the production feasibility of active and functional MD2-Bro as a bioactive compound.

Keywords: bromelain; expression; purification; catalytic activity; metal ion; antioxidant

Citation: Razali, R.; Fahrudin, F.A.; Subbiah, V.K.; Takano, K.; Budiman, C. Heterologous Expression and Catalytic Properties of Codon-Optimized Small-Sized Bromelain from MD2 Pineapple. *Molecules* **2022**, 27, 6031. https://doi.org/10.3390/molecules27186031

Academic Editor: Smaoui Slim

Received: 14 June 2022
Accepted: 8 September 2022
Published: 16 September 2022

Publisher's Note: MDPI stays neutral with regard to jurisdictional claims in published maps and institutional affiliations.

1. Introduction

Bromelain is a member of the papain family that contains a complex and diverse natural mixture of proteases. It belongs to the family of sulfhydryl proteolytic enzymes and has a catalytic mechanism that involves the triad Cys-His-Asn/Glu [1–3]. This enzyme can be found in the pineapple plant (*Ananas comosus*). Depending on the source, it is usually classified as either fruit bromelain or stem bromelain. Apart from the stem and fruits, bromelain was also reported to be present in pineapple peel, core, crown, and leaves [4]. While bromelain is a unique cysteine protease name for pineapple, other cysteine proteases are widely distributed in plants and animals, including papain from papaya (*Carica papaya*) [5] and ficin from *Ficus insipida* [6]. In addition, cysteine protease was also found in viruses [7].

Bromelain possesses significant and notable therapeutic properties such as anti-inflammatory, anti-thrombotic and fibrinolytic effects, inhibition of platelet aggregation, anti-cancer activity, immunomodulatory effects, enhanced wound healing and adsorption of drugs, particularly antibiotics, and cardiovascular and circulatory improvement [8,9]. It is also widely used in the food industry and considered a food supplement approved by the Food and Drug Administration of the United States of America, and is now freely available in the market [1,10]. It can be absorbed into the human intestines without degradation or losing biological activity [11–13]. Moreover, bromelain is also known to have the ability to hydrolyze meat proteins, particularly myofibril and connective fractions, and is considered a meat tenderizer that can be used traditionally [14].

Earlier, the whole-genome sequence of MD2 pineapple revealed the presence of 14 genes encoding cysteine proteases under the bromelain protease group [15]. Interestingly, these 14 genes encode various molecular weights of cysteine proteases ranging from 19 kDa to more than 200 kDa [15,16]. However, most of the studies on cysteine proteases of bromelain deal with medium-sized bromelain with a size of about 20–40 kDa [1,17–19]. So far, no study has reported on small-sized bromelain (20 kDa or less). Production of single cysteine protease bromelain for further applications as an enzyme-based bioactive compound is challenging due to a lengthy purification process. The use of the recombinant approach to producing active bromelain from a single gene is feasible yet challenging due to its solubility issue [1].

Our previous in silico study showed that the small-sized bromelain of MD2 pineapple (MD2-SBro) exhibited different structural features than the medium-sized bromelain of MD2 pineapple (MD2-MBro). Structurally, the Cys catalytic site of MD2-SBro is found to be located at the flexible loop, which is quite mobile and affects its proximity to the substrate [16]. In addition, both MD2-SBro and MD2-MBro also displayed differences in the hydrophobicity of the substrate-binding cavity. Earlier, we demonstrated that recombinant MD2-MBro produced under *Escherichia coli* (*E. coli*) was catalytically active with the specific activity and catalytic efficiency of 6.13 ± 0.01 U mg^{-1} and $5.64 \pm 0.02 \times 10^{-2}$ μM^{-1} s^{-1}, respectively [1].

This report provides the first experimental evidence on catalytic properties of recombinant MD2-SBro produced from *E. coli* host cells. We demonstrated that MD2-SBro was catalytically active with the activity modulated by pH, temperature, and metal ions. In addition, the antioxidant activity of this protein was also detectable.

2. Materials and Methods

2.1. Gene Optimization, Synthesis and Expression System Construction

The MD2-Sbro gene sequence was retrieved from NCBI with accession number OAY85828.1. The gene sequence was then optimized using OptimumGene^{TM} (Piscataway, NJ, USA) according to the codon usage preference of *E. coli*, and chemically synthesized under the GenScript outsource service (Piscataway, NJ, USA). The gene was provided in the pUC18 plasmid, designated as pUC18-SBro. Further, to construct the expression system for the gene, the SBro gene was amplified from the pUC18 plasmid using polymerase chain reaction (PCR) with a pair of specific primers. The sequences of the PCR primers used are as follows: 5′-CGAAGCTTATGGCGGAGTACGGTCGTGTG-3′ (forward, with *Hind*III site) and 5′-GGCTCGAGGCCCCACCAGGAACCCCAGC-3′ (reverse, with *Xho*I site). PCR was performed using KOD polymerase (Toyobo Co., Ltd., Kyoto, Japan) with the GeneAmp PCR system 2400 (Applied Biosystems, Tokyo, Japan). The amplicon was then digested with restriction enzymes of *Hind*III and *Xho*I and ligated into the pGS-21a expression vector using the DNA Ligation Kit, Mighty Mix (Takara, Japan). The success of ligation was confirmed using an insert check with restriction enzymes and nucleotides using the Prism 310 DNA sequencer (Applied Biosystems). The recombinant DNA of the MD2-SBro gene and pGS-21a is designated as an expression system of pGS21-SBro. This expression system allows MD2-SBro to be expressed in a fusion form to a glutathione S-transferase (GST) tag at its N-terminal. This expression system was then transformed into *E. coli* BL21(DE3) using the heat shock method based on Froger and Hall [20].

2.2. Protein Expression

Expression of recombinant MD2-SBro was performed based on Razali et al. [1] with some modifications. Briefly, the transformed cells were cultured in Luria Bertani (LB) media supplemented with 100 μg mL^{-1} ampicillin and incubated at 37 °C at 180 rpm. The enzyme was induced by 1 mM IPTG (isopropyl β-D-1-thiogalactosidase (IPTG) once the OD_{600} reached 0.7, followed by a prolonged incubation at 37 °C, 180 rpm for 5 h. The culture was harvested by centrifugation at 8000× g for 10 min at 4 °C. The cell pellet was then washed twice and resuspended in 20 mM phosphate buffer, pH 8.0, containing 100 mM NaCl, followed by cell lysis by a sonication in ice. The soluble fraction was then separated from the cell debris (pellet) by centrifugation at 35,000× g for 30 min at 4 °C. Soluble and pellet fractions were aliquoted for protein expression and solubility checking under 15% SDS-PAGE (sodium dodecyl sulfate–polyacrylamide gel electrophoresis). The whole fractions were also kept for further steps.

2.3. Solubilization and Refolding of Insoluble Protein

For the protein expressed in an insoluble form, solubilization and refolding steps were performed using urea treatment according to the method described by Kannan et al. [21] and Yamaguchi and Miyazaki [22], with some modifications. Briefly, the inclusion bodies (pellets) obtained after sonication were resuspended in 20 mM phosphate buffer (pH 8.0) containing 8 M urea, 2 mM DTT, and 100 mM NaCl. The sample solution was incubated at 4 °C overnight, followed by centrifugation at 35,000× g, 4 °C for 30 min. The supernatant was then collected for dialysis to refold the protein by removing the urea against 20 mM phosphate buffer (pH 8.0) at 4 °C overnight. Following the dialysis, the sample was centrifuged at 20,000× g for 15 min. The supernatant was collected and considered as solubilized MD2-SBro in a crude form, which was then used for solubility checking under 15% SDS-PAGE and further purification steps.

2.4. Protein Purification

Purification of a crude form of solubilized MD2-Sbro was performed according to [23]. Briefly, the GSTrap HP 5 mL (GE Healthcare; Chicago, IL, USA) column was firstly equilibrated with the binding buffer (20 mM phosphate buffer, pH 8.0 containing 100 mM NaCl). The crude form of solubilized MD2-Sbro, which was filtered previously using a syringe filter 0.22 µM membrane (Pall Life Sciences; Port Washington, NY, USA), was then loaded into the column. The sample was then eluted by linear gradient at gradual increments from 0% to 100% of an elution buffer containing 50 mM Tris-HCl (pH 8.0) with 10 mM reduced glutathione. The presence and purity of eluted MD2-Sbro were then checked from the fractions across the peak of interest in the chromatogram using 15% SDS-PAGE. Finally, the fractions containing MD2-Sbro in acceptable purity were pooled and dialyzed against 50 mM Tris-HCl, pH 8.0.

The purified protein concentration was determined by the NanoDrop™ 2000 (Thermo Fisher Scientific; Waltham, MA, USA) on the basis that the absorbance at 280 nm of 0.1% (1 mg mL^{-1}) solution is 1.69, as calculated based on Goodwin and Morton [24].

2.5. SDS-PAGE

The expression, solubility, and purity of the MD2-SBro protein were confirmed using 15% SDS-PAGE [25]. In addition, the gel was stained with Coomassie Brilliant Blue (CBB) dyes and visualized using a Gel Doc™ XR+ imager (Biorad; Hercules, CA, USA).

2.6. Enzymatic Activity and Kinetic Parameters

The enzymatic activity of the purified protein was determined at 37 °C and pH 8.0 based on the method described by Razali et al. [1], with some modifications. It was determined using N-carbobenzoxyglycine *p*-nitrophenyl ester (N-CBZ-Gly-*p*NP) substrate [26]. Prior to adding the substrate, concentrations of the enzyme with different ranges (0.5–10 µg/mL) were incubated in 20 mM Tris-HCl buffer pH 8.0 at 37 °C for 5 min, with shaking to homogenize the solution. Then, the substrate with 50 µM of final concentration was added to the enzyme solution and incubated for 5 min. The product release was monitored at 340 nm using a Lambda 35 Perkin-Elmer UV-Vis spectrophotometer (Waltham, MA, USA). The amount of p-nitrophenyl (*p*NP) released was calculated based on an extinction coefficient for *p*NP of 6320 M^{-1} cm^{-1}. One unit of the enzyme was defined as the amount of enzyme which produced 1 µmol of the product per minute. To determine the effect of GST on the specific activity of MD2-SBro, the specific activity of the protein was also observed in the absence or presence of free GST protein (Sigma Aldrich; St. Lois, MI, USA). The MD2-SBro was 10 µg/mL, while the concentration of free GST protein was prepared in 1:1, 1:5, and 1:50. The changes in the activity were then observed qualitatively based on the color changes due to the release of the *p*NP moiety.

Meanwhile, the kinetic parameters (V_{max} and K_m) of MD2-SBro were calculated mathematically according to the Michaelis–Menten equation [27], using linear regression analysis through a double reciprocal (Lineweaver–Burk) plot [17]. The enzyme mixture in 20 mM Tris-HCl buffer pH 8.0 was first incubated at 37 °C for 5 min with shaking. The kinetic parameters were then determined against the substrate. Data were obtained by measuring the initial rate of hydrolysis by incubating the enzyme with different range concentrations of the substrate in 20 mM Tris-HCl buffer pH 8.0 at 37 °C. The maximal velocity (V_{max}) and Michaelis constant (K_m) were computed by plotting the data in GraphPad Prism 6 software. The assays were performed in triplicates, and data were shown as the mean ± standard deviation.

2.7. Optimum Temperature and pH

The optimum temperature for the catalytic activity of MD2-SBro was determined at different temperatures from 25–80 °C in 20 mM Tris-HCl buffer pH 8.0 with the detailed conditions described above. Meanwhile, the optimum pH of the protein activity was observed over the pH range of 4.0–10.0. For pH 4.0–6.0, it was measured in 20 mM citrate buffer, 20 mM Tris-HCl buffer for pH 6.0–8.0, and 20 mM Gly-NaOH for pH 8.0–10.0 [1]. The highest activity obtained during the measurement was set as 100% of the specific activity. The assays were performed in triplicates, and data were shown as the mean ± standard deviation.

2.8. Determination of EDTA and Metal Ions Effect

The influence of ethylenediaminetetraacetic acid (EDTA) and different metal ions (Ca^{2+}, Cu^{2+}, Mg^{2+}, Zn^{2+}, Ni^{2+}) on the catalytic activity of MD2-SBro was performed according to the previous reports [1,28–30], with some modifications. The catalytic activity was examined by adding EDTA or different metal ions with a final concentration of 10 mM at the optimum temperature and pH [1].

2.9. Antioxidant Activity

The antioxidant activity was determined using a modified version of the free radical 2,2-diphenyl-1-picrylhydrazyl (DPPH) assay [31]. Various concentrations of proteins (0.05 mg/mL, 0.1 mg/mL, 0.2 mg/mL, and 0.5 mg/mL) were prepared in 20 mM Tris-HCl, pH 8.0. Then, 2.5 mL of MD2-SBro was mixed with 1 mL of DPPH and placed on ice for 30 min. The DPPH test for radical scavenging activity was monitored using a Lambda 35 Perkin-Elmer UV-Vis spectrophotometer (Waltham, MA, USA) at 517 nm. Ascorbic acid, which is known to exhibit antioxidant activity, was used as a positive control. The percentage values of radical scavenging activity by DPPH (A) were calculated using the formula below:

$$A\% = \frac{\text{Abs control} - \text{Abs sample}}{\text{Abs control}} \times 100$$

3. Results

3.1. Gene Optimization

In this study, the gene encoding bromelain MD2-SBro was optimized by increasing the GC content of MD2-SBro from 44.51% to 53.34%. In addition to the changes in GC content, the codon adaptation index (CAI) of the newly synthesized MD2-SBro was also adjusted from 0.40 to 0.95 (Table 1). All the changes were based on the preferences of *E. coli* as the host cells, according to Nussinov [32] and Akhtar et al. [33]. Further, the modification was only performed on the DNA level, where the amino acid sequence was not changed. Accordingly, the translated polypeptide from the optimized MD2-SBro gene is expected to have the same primary structure and fold into the same three-dimensional structure as the polypeptide from the original gene.

Table 1. Gene optimization of the MD2-SBro gene.

Parameters	Original MD2-SBro Gene	Optimized MD2-SBro Gene
GC content	44.51%	53.34%
Codon adaptation index (CAI)	0.40	0.95

3.2. *Protein Expression and Solubilization*

The optimized MD2-SBro was expressed in *E. coli* BL21(DE3), as indicated by the thick band with an apparent size of 50 kDa (Figure 1). This size is comparable to the theoretical size (calculated from the amino acid sequence) of MD2-SBro in a fusion form with a GST-tag, which is 50798.17 Da (GST-tag and linker = 30.32 kDa; MD2-SBro = 19.42 kDa). The 50 kDa band appeared only when IPTG induced the culture. Nevertheless, Figure 1 also showed that the 50 kDa band appeared in a pellet fraction after sonication, which indicated that MD2-SBro was expressed in an insoluble form (inclusion body). As the protein was expressed in an insoluble form, the solubilization and refolding were done using urea treatment, along with a reducing agent of DTT. The solubilized and refolded MD2-SBro was found to be in a soluble form, as shown by the appearance of a 50 kDa band in the soluble fraction after the treatment.

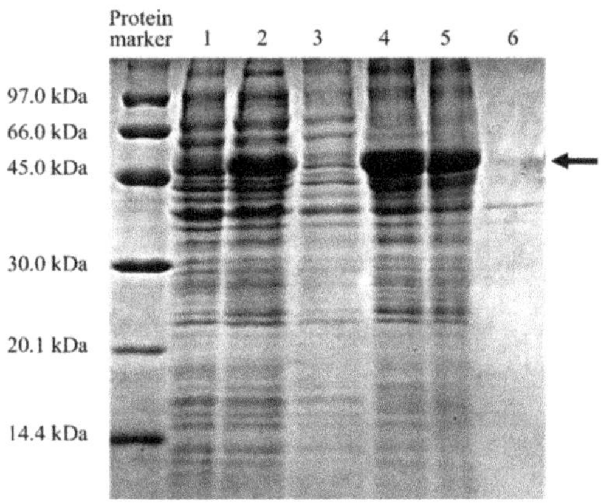

Figure 1. Expression and solubilization check of MD2-SBro protein. Lane 1: Before IPTG induction; Lane 2: After IPTG induction; Lane 3: Soluble fraction obtained after the sonication; Lane 4: Insoluble fraction obtained after the sonication; Lane 5: Soluble fraction obtained after the solubilization; Lane 6: Insoluble fraction obtained after the solubilization. The band corresponding to MD2-SBro is indicated by the arrow.

3.3. *Protein Purification*

The solubilized and refolded MD2-SBro was then purified by GST-affinity chromatography, resulting in a single 50 kDa band in 15% SDS-PAGE (Figure 2). The presence of the contaminants was undetectable under the gel, which showed that the MD2-SBro protein was successfully produced in high purity under single-step chromatography.

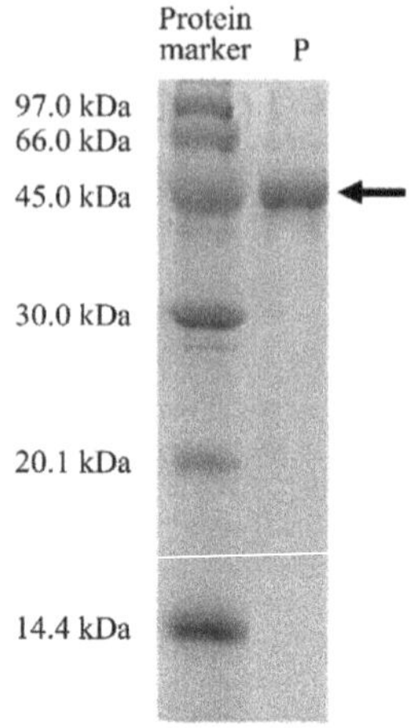

Figure 2. Purification check of MD2-SBro protein. Lane P: Purified protein.

The detail of the purification profiles is shown in Table 2. The amount of purified MD2-SBro obtained from 1 L culture was 14 mg. Meanwhile, the enzymatic activity was calculated based on the amount of *p*-nitrophenol (*p*NP) released upon the digestion of the N-CBZ-Gly-*p*NP substrate. The release of the *p*-nitrophenol moiety is detectable as yellow color and quantitatively measurable by a UV-Vis spectrophotometer. Table 2 also showed that the specific activity of MD2-SBro was 3.56 ± 0.08 U/mg. Interestingly, the purification fold of MD2-SBro was found to be more than 40-fold.

Table 2. Purification profile of MD2-SBro indicating its activity and yield.

Steps	Volume (mL)	Total Protein (mg)	Total Activity (U) *	Specific Activity (U/mg) *	Yield (%)	Purification (Fold)
Cell lysate	70 ± 2.80	510 ± 12	62.30 ± 5.64	0.12 ± 0.002	100	1.0
Glutathione S-transferase (GST) affinity chromatography	22 ± 1.38	14 ± 1.32	49.84 ± 3.17	3.56 ± 0.08	80	42.72

* measured at 37 °C, pH 8.0.

Notably, when the activity of MD2-SBro was observed in the presence of free GST protein, the yellow color of the cocktail reaction was not changed by the addition of free GST protein. Meanwhile, no yellow color was detected when free GST protein was mixed with the substrate without MD2-SBro (Figure 3). To note, the connection of MD2-SBro and the GST-tag is a linker Asp-Asp-Asp-Asp-Lys fragment, which is a cleavage site for enterokinase. The presence of this site allows the production of MD2-SBro free from the tag via digestion by the enterokinase. While it is unlikely that GST modulates or diminishes the activity, it is unclear if the linker participated in the activity. Given the linker is quite short (< 10 amino acids) and located far from the active sites, it is unlikely that the linker affects the activity; nevertheless, this remains to be experimentally confirmed.

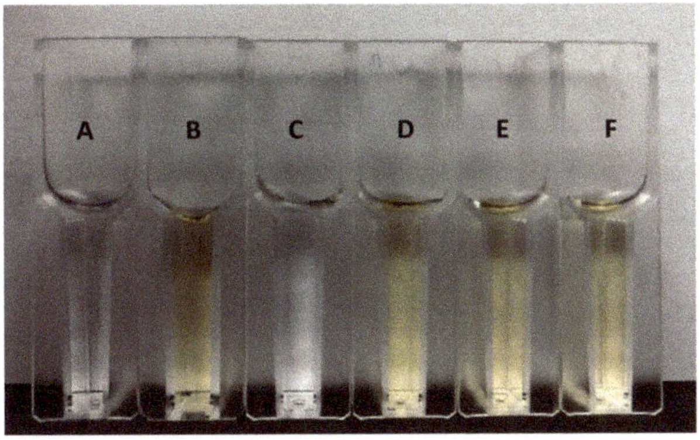

Figure 3. The proteolytic assay mixture of the N-CBZ-Gly-*p*NP substrate. Tube A is a blank (only substrate, without MD2-SBro and free GST protein). Tubes B and C are the mixture containing the substrate with MD2-Sbro and substrate with free GST protein, respectively. Tubes D, E, and F refer to the mixtures containing substrate, MD2-SBro, and free GST protein. The ratio of MD2-SBro and free GST protein were 1:1 (reaction D), 1:10 (reaction E), and 1:50 (reaction F).

3.4. Enzymatic Activity and Kinetic Parameters

Further, to estimate the kinetic parameters V_{max}, K_m, k_{cat}, and catalytic efficiency (k_{cat}/K_m), additional tests were carried out at 37 °C and pH 5.0 by varying substrate concentration. Figure 4 shows the Michaelis–Menten curve and Lineweaver–Burk plot used for the basis of the kinetic parameter's calculation. Accordingly, calculated kinetic parameters of MD2-SBro were shown in Table 3, with the catalytic efficiency of $4.75 \pm 0.23 \times 10^{-3}$ μM^{-1} s^{-1}.

Table 3. Kinetic parameters of MD2-SBro in comparison to MD2-MBro.

Proteins	V_{max} (10^{-3} μM s^{-1})	K_m (μM)	k_{cat} (s^{-1})	k_{cat}/K_m (10^{-3} μM^{-1} s^{-1})	Ref
MD2-SBro	7.20 ± 0.52	42.1 ± 3.81	0.20 ± 0.008	4.75 ± 0.23	This study
MD2-MBro	15 ± 0.5	34.24 ± 1.02	1.93 ± 0.05	56.37 ± 2.08	[1]

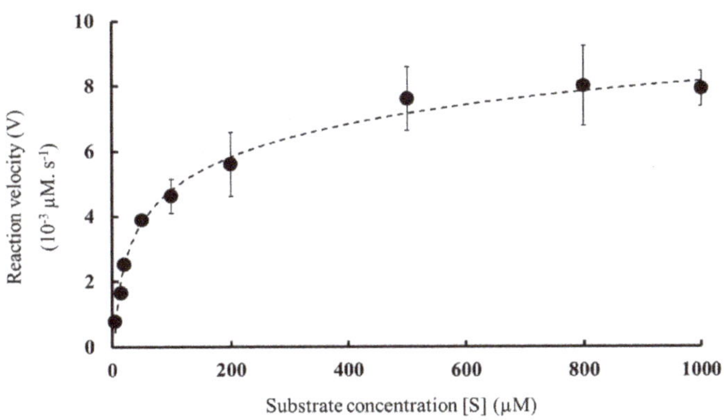

(a)

Figure 4. *Cont.*

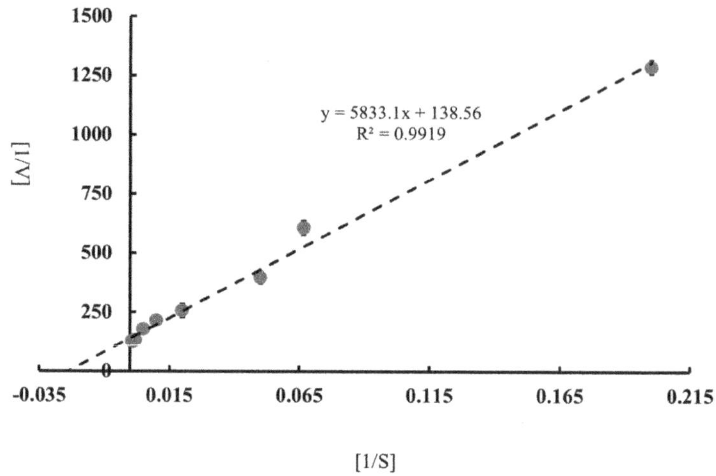

(**b**)

Figure 4. (**a**) Michaelis–Menten curve and (**b**) Lineweaver–Burk double reciprocal plot of MD2-SBro.

3.5. *Optimum Temperature and pH*

The optimum temperature of enzymatic activity was identified by carrying out activity assays at pH 8.0 and varying temperatures from 25–80 °C. The only initial values of enzyme activity were considered to minimize the influence of activity loss due to irreversible denaturation of protein [34]. As demonstrated in Figure 5, the optimum temperature of MD2-SBro was 50 °C.

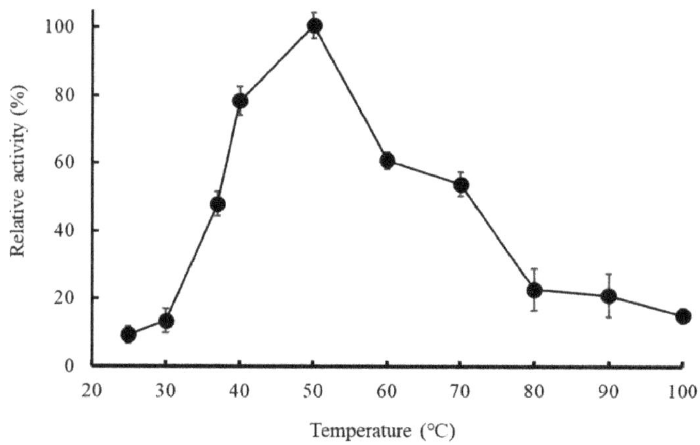

Figure 5. Temperature-dependent activities of the purified MD2-SBro. The highest activity at 50 °C (19.77×10^{-3} U/mg) was adjusted as 100%.

Meanwhile, Figure 6 showed the pH-dependent activity of MD2-SBro, which was measured with pH 4.0–10.0. The data demonstrated that the MD2-SBro was active over a relatively wide pH range, and the highest activity towards the substrate was observed at pH 8.0.

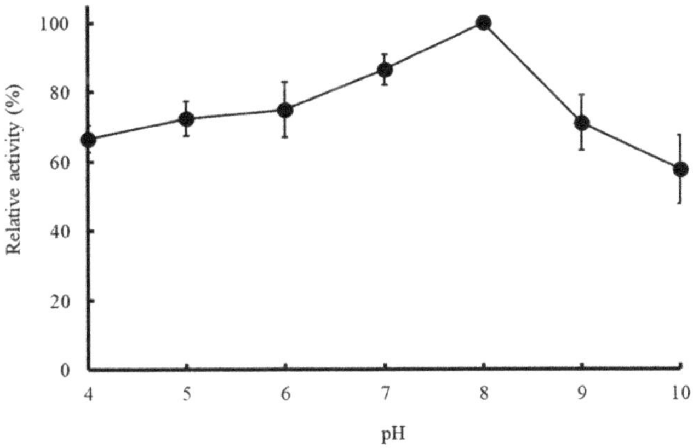

Figure 6. The pH-dependent activities of the purified MD2-SBro. The highest activity at pH 8.0 (3.56×10^{-3} U/mg) was adjusted as 100%.

3.6. *Effect of EDTA and Metal Ions*

As demonstrated in Table 4, the activity of MD2-SBro was decreased to about 17% in the presence of EDTA. In addition, the effect of different types of divalent ions on the catalytic activity of MD2-SBro was found to be varied. The addition of Mg^{2+}, Ni^{2+}, and Ca^{2+} ions increased the catalytic activity of MD2-SBro. Meanwhile, the addition of Zn^{2+} or Cu^{2+} metal ions decreased the activity of MD2-SBro. The reduction by Zn^{2+} was 41% for MD2-SBro. Meanwhile, Cu^{2+} decreased the activity of MD2-SBro by 22%.

Table 4. Relative activity of MD2-SBro in the presence of various metal ions and EDTA.

Metal Ions	Relative Activity (%)
Control	100 ± 5.01
$MgCl_2$	178.32 ± 7.54
$CaCl_2$	121.40 ± 8.71
$NiCl_2$	114.31 ± 10.32
$CuCl_2$	78.09 ± 3.21
$ZnCl_2$	47.36 ± 3.98
EDTA	17.58 ± 1.07

Note: Control refers to the activity with no metal ions or EDTA.

3.7. *Antioxidant Activity*

The antioxidant activity of tested samples was conducted by DPPH assay, which is one of the most stable free radicals and is frequently used to evaluate radical scavengers in many types of samples [35]. The antioxidant activity of MD2-SBro was determined through their ability to scavenge the DPPH radical and therefore inhibit the formation of a radical form of DPPH. Figure 7 shows that the ability of MD2-SBro to inhibit the formation of a radical form of DPPH was in a concentration-dependent fashion. This is similar to the ability of ascorbic acid as a positive control. This indicated that MD2-SBro exhibited antioxidant activity through scavenging DPPH radicals. The calculated IC_{50} value for MD2-SBro to inhibit the DPPH radical formation was 0.022 mg mL^{-1}, slightly higher than the IC_{50} value for ascorbic acid (0.018 mg mL^{-1}).

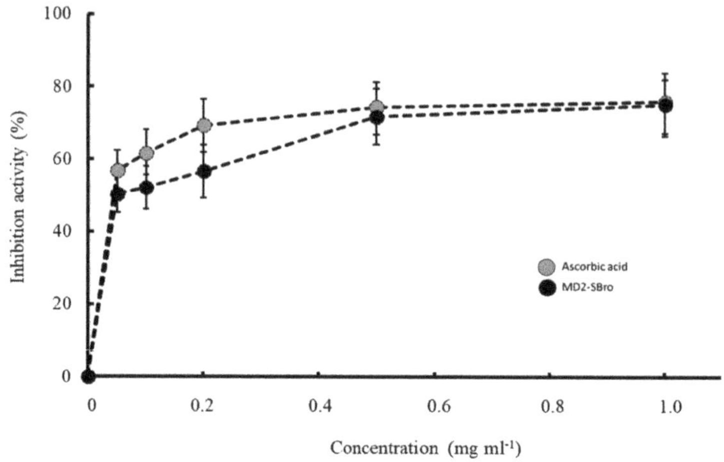

Figure 7. Inhibition of DPPH radical in the presence of different concentrations of ascorbic acid (control) and MD2-SBro.

4. Discussion

The common challenge in the production of recombinant bromelain is dealing with expressibility issues upon heterologous expression. One factor that might account for this issue is the variations and incompatibility of the codon profile of the target gene with the host cells to express [36,37]. Our previous approach using a codon-optimized gene successfully produced recombinant medium-sized bromelain from MD2-pineapple (MD2-MBro, size of 38 kDa) [1]. Accordingly, a similar approach might also work for the heterologous expression of MD2-SBro. In the current study, the gene encoding MD2-SBro was optimized to meet the requirement for expression under *E. coli* host cells. The value of GC content for the optimized MD2-SBro gene was in the range of favorable GC content for *E. coli* host cells [38]. In addition, the final CAI of the MD2-SBro gene was also re-adjusted to be compatible with the codon preferences of the host. Meanwhile, AT-rich regions were removed in the new sequence to avoid premature translational termination [39]. To note, the optimization did not affect the translated amino acid sequence as it dealt only with the changes in the DNA sequences.

Figure 1 confirmed that MD2-SBro was successfully over-expressed in *E. coli* cells. This indicated that the optimized gene of MD2-SBro enables this gene to be compatible with the *E. coli* system for expression. However, MD2-SBro was expressed in an insoluble form (inclusion body), despite originating from the modified gene. Notably, several expression conditions were attempted for MD2-SBro, particularly by varying the incubation temperature and period. Nevertheless, all these conditions resulted in insoluble expressed protein (data not shown). Recently, Bhatwa et al. [40] implied that the formation of an inclusion body is associated with the genetic regulation upon transcription and translation. Codon optimization is essentially related to gene expression through transcription regulation [41]. Nevertheless, as MD2-SBro is expressed as an inclusion body, this indicated that the codon optimization did not sufficiently contribute to the solubility of MD2-SBro upon heterologous expression. This is in agreement with an earlier report that implied the use of codon-optimized sequences did not affect the quality of the inclusion bodies obtained [36]. The formation of the inclusion body of MD2-SBro might be due to post-translational events, particularly the misfolding of protein. Obeng et al. [42] and Razali et al. [43] implied that the production of recombinants in *E. coli* is often challenged by its insolubility due to folding issues. In addition, the high expression level of the expressed protein might also contribute to the formation of the inclusion body due to the high concentrations of folding intermediates, which are prone to clump and aggregated [44–46]. To note, the standard for the heterologous protein to be considered as highly expressed is varied. In this study, 14 mg of MD2-SBro was expressed from 1 L of culture, which sounds to be not a high-level expression. Nevertheless, this expression level is much higher than other recombinant bromelain expressions reported by Iffah et al. [47], Amid et al. [19], and George et al. [48]. To note, all these bromelains were expressed from the nonoptimized gene. Interestingly, MD2-MBro was expressed at a higher level (20 mg/L culture) than MD2-SBro [1] in a fully soluble form. The discrepancy might be due to the differences in the physicochemical properties between both proteins. Bhatwa et al. [40] reported that the formation of the inclusion body of expressed protein is also governed by the structural and physicochemical properties of the proteins themselves. These features include the molecular weight, the number of adjacent hydrophobic residues, and the regions of low complexity.

Of note, MD2-SBro was expressed in a fusion form with a GST-tag. The tag was known not only to assist the purification process, but was also able to enhance solubility. Nevertheless, MD2-SBro remains expressed as an insoluble form or inclusion body. Inclusion bodies were classically considered amorphous types of protein aggregates devoid of any structural regularity [49]. Costa et al. [50] reported that the GST-tag theoretically acts for affinity and solubility enhancer purposes. Nevertheless, Boisselier et al. [51] reported that despite the high solubility of GST, not all GST-tagged fusion proteins are solubilizable. This is possibly due to the uniqueness of each protein, particularly in its hydrophobicity degree. Young et al. [52] reported that hydrophobic regions of the proteins might lead to unspecific interaction, which further caused the aggregation and became an inclusion body.

Following the solubilization of the insoluble protein of MD2-SBro using urea, the protein was successfully refolded. The use of urea in this study is considered the most common and conventional way to solubilize recombinant proteins from the inclusion body upon expression from *E. coli* cells [53]. The refolded protein was found to be in a soluble form (Figure 1), and this showed that the solubilization and refolding process of MD2-SBro inclusion body proteins succeeded. In addition, it also proved that the efficiency of these two steps is high. However, in some cases, the renaturation yields may be limited by the accumulation of inactive misfolded species and aggregates [54,55].

However, the purification yield of MD2-SBro protein after the single-step purification in this study is considered lower than other recombinant bromelains [1,19]. This is probably due to the low recovery of the protein solubilization or refolding, although the expression level was high. Nevertheless, this value is higher than purified recombinant bromelain in the study of Arshad et al. [47] and Bala et al. [56]. Unfortunately, previous studies that involved recombinant bromelain by Muntari et al. [18] and George et al. [48] did not report the purification yield for comparison.

In addition, the specific activity of MD2-SBro was only 3.56 $\pm$ 0.08 U/mg, which was considerably lower than MD2-MBro [1]. Nevertheless, as shown in Table 2, the purification yield of MD2-SBro was 80%, comparable to that of MD2-MBro reported earlier [1]. A high purification fold for MD2-SBro was speculated due to the use of GST affinity chromatography, which was reported to be very specific. MD2-MBro, in contrast, was expressed in a His-tag form, which has less specificity during affinity chromatography. Robichon et al. [57] reported that many indigenous *E. coli* proteins display high affinity to divalent nickel or cobalt ions, mainly due to the presence of clustered histidine residues or biologically relevant metal-binding sites. These indigenous proteins lead to low specificity of Ni-NTA chromatography compared to GST affinity chromatography. In this study, the GST-tag was not removed from MD2-SBro as there were no reports that GST exhibited proteolytic activity that would interfere with the MD2-SBro activity. Earlier, we also demonstrated that the large-sized tag of thioredoxin did not interfere with the activity of recombinant MD2-SBro [1]. In addition, the use of protease to cleave the linker between GST-tag and MD2-SBro is concerning due to the possibility of unspecific cleavage of MD2-SBro by the protease. As shown in Figure 3, it is evident that GST has no effect on the cleavage of *p*NP by MD2-SBro, as indicated by no changes in the yellow color of the cocktail upon the addition of free GST protein at different concentrations. Free GST protease has also demonstrated no proteolytic activity against the substrate due to no yellow color formation. Accordingly, it is suggested that MD2-SBro mainly generates the activity observed in this study.

The calculated kinetic parameters of MD2-SBro (Table 3) clearly demonstrated that catalytic efficiency (k_{cat}/K_m) of MD2-SBro was more than 11-fold lower than that of MD2-MBro, as reported earlier [1]. The differences might be due to their structural discrepancies or the presence of a GST-tag. Earlier, the model structure of MD2-SBro revealed that the Cys-His active site position of MD2-SBro was found to be inappropriate for catalysis. In addition, the substrate-binding pocket of MD2-SBro was found to be less hydrophobic than that of MD2-MBro [16]. This structural feature might account for the low activity of MD2-SBro. Notably, the catalytic efficiency of MD2-SBro was also much lower than the other bromelains, ranging from 17.86–52.53 $\mu M^{-1}\ s^{-1}$, depending on the type of bromelain and substrate used in the assay [19,29].

Further, the optimum temperature of MD2-SBro, which was observed at 50 °C, suggested that the MD2-SBro behaves as a mesophilic protein, where it optimally worked at moderate temperature. This optimum temperature is similar to that of MD2-MBro [1]. Bala et al. [58] reported that non-recombinant bromelain from fruit or stem pineapple generally exhibited optimum activity at a temperature ranging from 40–70 °C. Nevertheless, Corzo et al. [59] discovered that the optimum temperature for the catalytic activity of bromelain was different depending on the substrate. Of note, the optimum temperature of MD2-SBro was higher than favorable growth temperatures in pineapple farms (18–32 °C) [60]. Nevertheless, it remains to be investigated if the optimum temperature of this protein is associated with its biological roles in the pineapple fruit. Meanwhile, the optimum pH of MD2-SBro was found to be 8.0, which is higher than that of MD2-MBro (pH 6.0), as reported earlier [1]. Nevertheless, the optimum pH of MD2-SBro remains in the range of the common optimum pH for stem and fruit bromelain, which was reported to be between 6.0–8.5 [14,61–65]. Interestingly, in the range of pH 4.0–7.0 and pH 9.0–10.0, MD2-SBro remains active with residual activity of > 60% (Figure 5). Of note, the pH optimum of bromelains was reportedly different by many authors due to the use of different substrates [59,66]. Nevertheless, Vernet et al. [67] previously proposed that an acidic pH is more favorable for bromelain if it triggers the pro-domain from the active site. Consequently, this makes the cleavage site within the pro-domain loop accessible to the active site, and the enzyme becomes activated [68]. Notably, MD2-SBro is a small protein with no pro-domain segment [16]. Accordingly, acidic pH is not necessarily required for activating MD2-SBro through the detachment of the pro-domain segment. The absolute specific activity values of MD2-SBro at its optimum temperature (50 °C) and pH (8.0) were 19.77×10^{-3} and 3.56×10^{-3} U/mg, respectively. These values are lower than the specific activity of MD2-MBro at 50 °C (10.22×10^{-2} U/mg) or at pH 8.0 (6.13×10^{-2} U/mg).

It is interesting to find that the activity of MD2-SBro was decreased by EDTA, which is in good agreement with the previous study reported by Hidayani et al. [28]. The negative effect of EDTA on the activity of MD2-SBro is due to the chelating of metal ions in the catalytic site of the enzyme by EDTA and altering of the structure, as was also proposed by Hidayani et al. [28]. Nevertheless, the identities of the metal ions required for the enzyme activity are yet to be investigated. A reasonable way to identify the metal binding site of this protein is through co-crystallization with the metal ions. The putative residues for metal ion coordination are further confirmed through the mutagenesis approach.

Meanwhile, the increasing activity of MD2-SBro due to the presence of different types of divalent ions (Mg^{2+}, Ni^{2+}, and Ca^{2+}) is similar to previous studies of bromelain [1,30,69–72]. The increase of catalytic activity of MD2-SBro in the presence of Mg^{2+}, Ni^{2+}, and Ca^{2+} were 203%, 118%, and 134%, respectively, higher than that in the absence of any metal ions. The observed effects of Ca^{2+} on bromelain activity are in good agreement with the earlier reports [73–75] that calcium ions promote bromelain activity by stabilizing the secondary structure of an enzyme. According to Fadhilah et al. [69], the addition of Mg^{2+} also aids in maintaining the conformation of bromelain, which is important in the occurrence of catalytic activity. Unfortunately, so far, no detailed study on the effect of Ni^{2+} on bromelain activity is available.

Figure 7 also showed a reduction of MD2-SBro activity by adding Zn^{2+} or Cu^{2+} metal ions. Similar results were also reported for the effect of these two ions on bromelain [30,70,72,76]. The effects of Cu^{2+} on bromelain activity corroborate the earlier observations [1,72,74,76,77], which implied that copper ions inhibit the bromelain activity by forming a coordination bond with a catalytic sulfhydryl group.

The interesting bioactivity of bromelain is its antioxidant activity, which remains unknown as to whether it is associated with its catalytic activity. Some reports have clearly demonstrated the antioxidant activity of non-recombinant bromelain against DPPH radical or lipid peroxidation inhibition [78–81]. Nevertheless, there has been no report so far on the antioxidant activity of bromelain produced through the recombinant approach. Figure 7 clearly shows that MD2-SBro could scavenge DPPH as one of the phenotypical antioxidant activities. It is unclear how bromelain scavenges the DPPH radical. However, it might be due to the antioxidant properties of individual amino acids of MD2-SBro. Udenigwe et al. [82] reported that sulfur-containing (SCAA), acidic, and hydrophobic amino acids had strong positive effects on scavenging of 2,2-diphenyl-1-picrylhydrazyl (DPPH). All these residues are found in the MD2-SBro sequence. Chakraborty et al. [83] highlighted that the antioxidant activity of bromelain put this enzyme as a potential food-based bioactive compound for various pharmaceutical applications. Ataide et al. [81] reported that bromelain's antioxidant activity might be associated with bromelain's activity in the modulation of the inflammatory system and skin debridement properties. Of note, most of the antioxidant activity on bromelain used an unpurified (crude) form of bromelain, which leads to a possibility of bias by the antioxidant activity generated by the contamination. The ability of MD2-SBro to scavenge DPPH radically indicated that a single cysteine protease indeed exhibited antioxidant activity. The IC_{50} values of MD2-SBro to scavenge DPPH radical were found to be higher than that of crude bromelain reported by Abbas et al. [80], but lower than that reported by Saptarini et al. [78] and Huang et al. [79]. This indicated that each cysteine protease bromelain possesses unique antioxidant properties.

Of note, the bromelain gene studied in this study originated from MD2 pineapple. This is due to the availability of its whole genome sequence. In addition, MD2 pineapple is currently also the major pineapple variant planted in Malaysia [84]. Therefore, the current study should provide insight into the promising bioactivity of bromelain from MD2 pineapple for further studies, and scale-up productions using a heterologous expression approach.

5. Conclusions

This study demonstrated the success of the production of MD2-SBro, one of the enzyme-based bioactive compounds from MD2-pineapple, using a heterologous expression system using *E. coli* host cells. While recombinant MD2-SBro was produced in the inclusion body, this protein could be solubilized and refolded to form active bromelain. Intriguingly, MD2-SBro is proven to be active, albeit with little specific activity and low catalytic efficiency. The pH and temperature optimum and the metal-ion dependency of this protein were found to behave uniquely compared to other bromelains. Interestingly, the antioxidant activity of MD2-SBro is remarkably higher and close to the well-known antioxidant of ascorbic acid. Future studies in these characterizations may lead to the expansion of small-sized bromelain applications. The expressibility of MD2-SBro in *E. coli* host cells is an important milestone for the production of this protein for further studies and applications as a promising bioactive compound.

Author Contributions: Conceptualization, C.B.; methodology, C.B., R.R. and V.K.S.; investigation, R.R. and F.A.F.; resources, C.B., V.K.S. and K.T.; data curation, C.B., V.K.S. and K.T.; writing—original draft preparation, R.R. and F.A.F.; writing—review and editing, C.B., K.T. and V.K.S.; supervision, C.B. and V.K.S.; project administration, C.B.; funding acquisition, C.B. and V.K.S. All authors have read and agreed to the published version of the manuscript.

Funding: This research was funded by Skim Dana Khas of Universiti Malaysia Sabah, SDK0124-2020.

Institutional Review Board Statement: Not applicable.

Informed Consent Statement: Not applicable.

Data Availability Statement: Not applicable.

Acknowledgments: We thank Nurliana Bt Md Noor for her technical assistance. This research is associated to the Anugerah Penyelidik Muda UMS 2019.

Conflicts of Interest: The authors declare no conflict of interest.

References

1. Razali, R.; Budiman, C.; Kamaruzaman, K.A.; Subbiah, V.K. Soluble expression and catalytic properties of codon-optimized recombinant bromelain form MD2 pineapple in *Escherichia coli*. *Protein J.* **2021**, *40*, 406–418. [CrossRef] [PubMed]
2. Menard, R.; Carrière, J.; Laflamme, P.; Plouffe, C.; Khouri, H.E.; Vernet, T.; Tessier, D.C.; Thomas, D.Y.; Storer, D.C. Contribution of the glutamine 19 side chain to transition-state stabilization in the oxyanion hole of papain. *Biochemistry* **1991**, *30*, 8924–8928. [CrossRef] [PubMed]
3. Otto, H.H.; Schirmeister, T. Cysteine proteases and their inhibitors. *Chem. Rev.* **1997**, *97*, 133–171. [CrossRef] [PubMed]
4. Ketnawa, S.; Chaiwut, P.; Rawdkuen, S. Pineapple wastes: A potential source for bromelain extraction. *Food Bioprod. Process.* **2012**, *90*, 385–391. [CrossRef]
5. Rawlings, N.D.; Barrett, A.J. Families of cysteine peptidases. *Methods Enzymol.* **1994**, *244*, 461–486.
6. Arribére, M.P.; Caffin, O.; Priolo, S. Proteolytic enzymes from the latex of *Ficus pumila* L. (Moraceae). *Acta Farm. Bonaer.* **2000**, *19*, 257–262.
7. Razali, R.; Asis, H.; Budiman, C. Structure-function characteristics of SARS-CoV-2 proteases and their potential inhibitors from microbial sources. *Microorganisms* **2021**, *9*, 2481. [CrossRef]
8. Rathnavelu, V.; Alitheen, N.B.; Sohila, S.; Kanagesan, S.; Ramesh, R. Potential role of bromelain in clinical and therapeutic applications (Review). *Biomed. Rep.* **2016**, *5*, 283–288. [CrossRef]
9. Maurer, H.R. Bromelain: Biochemistry, pharmacology, and medical use. *Cell. Mol. Life Sci.* **2001**, *58*, 1234–1245. [CrossRef]
10. Gomes, H.A.R.; Moreira, L.R.S.; Filho, E.X.F. Chapter 3—Enzymes and food industry: A consolidated marriage. In *Advances in Biotechnology for Food Industry*; Academic Press: Cambridge, MA, USA, 2018; pp. 55–89.
11. Pavan, R.; Jain, S.; Shraddha; Kumar, A. Properties and therapeutic application of bromelain: A review. *Biotechnol. Res. Int.* **2012**, *2012*, 976203. [CrossRef]
12. Chobotova, K.; Vernallis, A.B.; Majid, F.A.A. Bromelain's activity and potential as an anti-cancer agent: Current evidence and perspectives. *Cancer Lett.* **2010**, *290*, 148–156. [CrossRef] [PubMed]
13. Castell, J.V.; Friedrich, G.; Kuhn, C.S.; Poppe, G.E. Intestinal absorption of undegraded proteins in men: Presence of bromelain in plasma after oral intake. *Am. J. Physiol. Cell Physiol.* **1997**, *273*, G139–G146. [CrossRef]
14. Ketnawa, S.; Rawdkuen, S. Application of bromelain extract for muscle foods tenderization. *Food Nutr. Sci.* **2011**, *2*, 393–401. [CrossRef]
15. Redwan, R.M.; Saidin, A.; Kumar, S.V. The draft genome of MD-2 pineapple using hybrid error correction of long reads. *DNA Res.* **2016**, *23*, 427–439. [CrossRef]
16. Razali, R.; Kumar, V.; Budiman, C. Structural insights into the enzymatic activity of cysteine protease bromelain of MD2 pineapple. *Pak. J. Biol. Sci.* **2020**, *23*, 829–838. [CrossRef]
17. Bala, M.; Mel, M.; Jami, M.S.; Amid, A.; Salleh, H.M. Kinetic studies on recombinant stem bromelain. *Adv. Enzym. Res.* **2013**, *1*, 52–60. [CrossRef]
18. Muntari, B.; Amid, A.; Mel, M.; Jami, M.S.; Salleh, H.M. Recombinant bromelain production in *Escherichia coli*: Process optimization in shake flask culture by response surface methodology. *AMB Expr.* **2012**, *2*, 12. [CrossRef]
19. Amid, A.; Ismail, N.A.; Yusof, F.; Salleh, H.M. Expression, purification, and characterization of a recombinant stem bromelain from *Ananas comosus*. *Process Biochem.* **2011**, *46*, 2232–2239. [CrossRef]
20. Froger, A.; Hall, J.E. Transformation of plasmid DNA into *E. coli* using the heat shock method. *J. Vis. Exp.* **2007**, *6*, 253. [CrossRef]
21. Kannan, Y.; Koga, Y.; Inoue, Y.; Haruki, M.; Takagi, M.; Imanaka, T.; Morikawa, M.; Kanaya, S. Active subtilisin-like protease from a hyperthermophilic archaeon in a form with a putative prosequence. *Appl. Environ. Microbiol.* **2001**, *67*, 2445–2452. [CrossRef]
22. Yamaguchi, H.; Miyazaki, M. Refolding techniques for recovering biologically active recombinant proteins from inclusion bodies. *Biomolecules* **2014**, *4*, 235–251. [CrossRef] [PubMed]
23. Mustafa, M.; Ali, L.; Islam, W.; Noman, A.; Zhou, C.; Shen, L.; Zhu, T.; Can, L.; Nasif, O.; Gasparovic, K.; et al. Heterologous expression and characterization of glycoside hydrolase with its potential applications in hyperthermic environment. *Saudi J. Biol. Sci.* **2022**, *29*, 751–757. [CrossRef] [PubMed]
24. Goodwin, T.W.; Morton, R.A. The spectrophotometric determination of tyrosine and tryptophan in proteins. *Biochem. J.* **1956**, *40*, 628–632. [CrossRef] [PubMed]

25. LaemmLi, U.K. Cleavage of structural proteins during the assembly of the head of bacteriophage T4. *Nature* **1970**, *227*, 680–685. [CrossRef]
26. Silverstein, R.M. The assay of the bromelains using N alpha-CBZ-L-lysine *p*-nitrophenyl ester and N-CBZ-glycine *p*-nitrophenyl ester as substrates. *Anal. Biochem.* **1975**, *62*, 478–484. [CrossRef]
27. Nelson, D.L.; Cox, M.M. *Lehnigher Principles of Biochemistry*, 5th ed.; Freeman and Company: New York, NY, USA, 2008.
28. Hidayani, W.A.; Setiasih, S.; Hudiyono, S. Determination of the effect of EDTA and PCMB on purified bromelain activity from pineapple core and in vitro antiplatelet activity. *IOP Conf. Ser. Mater. Sci. Eng.* **2020**, *763*, 012054. [CrossRef]
29. Singh, A.N.; Shukla, A.K.; Jagannadham, M.V.; Dubey, V.K. Purification of a novel cysteine protease, procerain B, from *Calotropis procera* with distinct characteristics compared to Procerain. *Process Biochem.* **2010**, *45*, 399–406. [CrossRef]
30. Lestari, P.; Suyata. Antibacterial activity of hydrolysate protein from Etawa goat milk hydrolysed by crude extract bromelain. *IOP Conf. Ser. Mater. Sci. Eng.* **2019**, *509*, 012111. [CrossRef]
31. Brand-Williams, W.; Cuvelier, E.; Berset, C. Use of a free radical method to evaluate antioxidant activity. *LWT Food Sci. Technol.* **1995**, *28*, 25–30. [CrossRef]
32. Nussinov, R. Eukaryotic dinucleotide preference rules and their implications for degenerate codon usage. *J. Mol. Biol.* **1981**, *149*, 125–131. [CrossRef]
33. Aktar, H.; Aktar, S.; Jan, S.U.; Khan, A.; Zaidi, N.S.S.; Qadri, I. Over expression of a synthetic gene encoding interferon lambda using relative synonymous codon usage bias in *Escherichia coli*. *Pak. J. Pharm Sci.* **2013**, *26*, 1181–1188.
34. Souza, P.M.; Aliakbarian, B.; Filho, E.X.F.; Magalhães, P.O.; Junior, A.P.; Converti, A.; Perego, P. Kinetic and thermodynamic studies of a novel acid protease from *Aspergillus foetidus*. *Macromolecules* **2015**, *81*, 17–21. [CrossRef] [PubMed]
35. Dżugan, M.; Tomczyk, M.; Sowa, P.; Grabek-Lejko, D. Antioxidant activity as biomarker of honey variety. *Molecules* **2018**, *23*, 2069. [CrossRef] [PubMed]
36. Menzella, H.G. Comparison of two codon optimization strategies to enhance recombinant protein production in *Escherichia coli*. *Microb. Cell Fac.* **2011**, *10*, 15. [CrossRef] [PubMed]
37. Wang, Q.; Mei, C.; Zhen, H.; Zhu, J. Codon preference optimization increases prokaryotic Cystatin C expression. *J. Biomed. Biotechnol.* **2012**, *2012*, 732017. [CrossRef]
38. Goh, C.G.K.; Silvester, J.; Mahadi, W.N.S.W.; Chin, L.P.; Ying, L.T.; Leow, T.C.; Kurahashi, R.; Takano, K.; Budiman, C. Expression and characterization of functional domains of FK506-binding protein 35 from *Plasmodium knowlesi*. *Protein Eng. Des. Sel.* **2018**, *31*, 489–498. [CrossRef]
39. Gustafsson, C.; Govindarajan, S.; Minshull, J. Codon bias and heterologous protein expression. *Trends Biotechnol.* **2004**, *22*, 346–353. [CrossRef]
40. Bhatwa, A.; Wang, W.; Hassan, Y.I.; Abraham, N.; Li, X.-Z.; Zhou, T. Challenges associated with the formation of recombinant protein inclusion bodies in *Escherichia coli* and strategies to address them for industrial applications. *Front. Bioeng. Biotechnol.* **2021**, *9*, 630551. [CrossRef]
41. Zhou, Z.; Dang, Y.; Zhou, M.; Li, L.; Yu, C.-H.; Fu, J.; Chen, S.; Liu, Y. Codon usage is an important determinant of gene expression levels largely through its effects on transcription. *Proc. Natl. Acad. Sci. USA* **2016**, *113*, E6117–E6125. [CrossRef]
42. Obeng, E.M.; Brossette, T.; Ongkudon, C.M.; Budiman, C.; Maas, R.; Jose, J. The workability of *Escherichia coli* BL21 (DE3) and *Pseudomonas putida* KT2440 expression platforms with autodisplayed cellulases: A comparison. *Appl. Microbiol. Biotechnol.* **2018**, *102*, 4829–4841. [CrossRef]
43. Razali, R.; Budiman, C.; Kumar, V. Technical data of heterologous expression and purification of SARS-CoV-2 proteases using *Escherichia coli* system. *Data* **2021**, *6*, 99. [CrossRef]
44. Kane, J.F.; Hartley, D.L. Formation of recombinant protein inclusion bodies in *Escherichia coli*. *Trends Biotechnol.* **1988**, *6*, 95–101. [CrossRef]
45. Fahnert, B.; Lile, H.; Neubauer, P. Inclusion bodies: Formation and utilization. *Adv. Biochem. Eng. Biotechnol.* **2004**, *89*, 93–142. [PubMed]
46. Sørensen, H.P.; Mortensen, K.K. Soluble expression of recombinant proteins in the cytoplasm of *Escherichia coli*. *Microb. Cell Fact.* **2005**, *4*, 1. [CrossRef] [PubMed]
47. Arshad, Z.I.M.; Amid, A.; Yusof, F.; Sulaiman, S.Z.; Mudalip, S.K.A.; Man, R.C.; Shaarani, S.M. Comparison of purification methods to purify recombinant bromelain from *Escherichia coli* BL21-A1. *Malays. J. Anim. Sci.* **2017**, *21*, 958–971.
48. George, S.; Bhasker, S.; Madhav, H.; Nair, A.; Chinnamma, M. Functional characterization of recombinant bromelain of Ananas comosus expressed in a prokaryotic system. *Mol. Biotechnol.* **2014**, *56*, 166–174. [CrossRef]
49. Singh, A.; Upadhyay, V.; Upadhyay, A.K.; Singh, S.M.; Panda, A.K. Protein recovery from inclusion bodies of *Escherichia coli* using mild solubilization process. *Microb. Cell Fact* **2015**, *14*, 41. [CrossRef]
50. Costa, S.; Almeida, A.; Castro, A.; Domingues, L. Fusion tags for protein solubility, purification, and immunogenicity in *Escherichia coli*: The novel Fh8 system. *Front. Microbiol.* **2014**, *5*, 63. [CrossRef]
51. Boisselier, E.; Audet, M.L.; Cantin, L.; Salesse, C. A strategy for purifying glutathione S-transferase in the presence of sodium dodecyl sulfate. *BioTechniques* **2018**, *51*, 193–194. [CrossRef]
52. Young, C.L.; Britton, Z.T.; Robinson, A.S. Recombinant protein expression and purification: A comprehensive review of affinity tags and microbial applications. *Biotechnol. J.* **2012**, *7*, 620–634. [CrossRef]

53. Yang, Z.; Zhang, L.; Zhang, Y.; Feng, Y.; Lu, X.; Lan, W.; Wang, J.; Wu, H.; Cao, C.; Wang, X. Highly efficient production of soluble proteins from insoluble inclusion bodies by a two-step-denaturing and refolding method. *PLoS ONE* **2011**, *6*, e22981. [CrossRef] [PubMed]
54. Clark, E.D.B. Protein refolding for industrial processes. *Curr. Opin. Biotechnol.* **2001**, *12*, 202–207. [CrossRef]
55. Lilie, H.; Schwarz, E.; Rudolph, R. Advances in refolding of proteins produced in *E. coli*. *Curr. Opin. Biotechnol.* **1998**, *9*, 497–501. [CrossRef]
56. Bala, M.; Salleh, H.M.; Amid, A.; Mel, M.; Jami, M.S. Recovery of recombinant bromelain from *Escherichia coli* BL21-A1. *Afr. J. Biotechnol.* **2011**, *10*, 18829–18832. [CrossRef]
57. Robichon, C.; Luo, J.; Casuey, T.B.; Benner, J.S.; Samuelson, J.C. Engineering *Escherichia coli* BL21(DE3) derivative strains to minimize *E. coli* protein contamination after purification by immobilized metal affinity chromatography. *Appl. Environ. Microbiol.* **2011**, *7*, 4634–4646. [CrossRef]
58. Bala, M.; Ismail, N.A.; Mel, M.; Jami, M.S.; Salleh, H.M.; Amid, A. Bromelain production: Current trends and perspective. *Arch. Des. Sci.* **2012**, *65*, 360–399.
59. Corzo, C.A.; Waliszewski, K.N.; Welti-Chanes, J. Pineapple fruit bromelain affinity to different protein substrates. *Food Chem.* **2012**, *133*, 631–635. [CrossRef]
60. Bartholomew, D.P.; Paull, R.E.; Rohrbach, K.G. *The Pineapple: Botany, Production and Uses*; CABI Publishing: Wallingford, UK, 2003; pp. 1–301.
61. Harrach, T.; Eckert, K.; Maurer, H.R.; Machleidt, I.; Machleidt, W.; Nuck, R. Isolation and characterization of two forms of an acidic bromelain stem proteinase. *J. Prot. Chem.* **1998**, *17*, 351–361. [CrossRef]
62. Suh, H.J.; Yang, H.C.; Lee, H.; Suwon; Cho, H.Y. Purification and characterization of bromelain isolated from pineapple. *J. Korean Agric. Chem. Soc.* **1992**, *35*, 300–307.
63. Ketnawa, S.; Rawdkuen, S.; Chaiwut, P. Two phase partitioning and collagen hydrolysis of bromelain from pineapple peel Nang Lae cultivar. *J. Biochem. Eng.* **2010**, *52*, 205–211. [CrossRef]
64. Silvestre, M.P.C.; Carreira, R.L.; Silva, M.R.; Corgosinho, F.C.; Monteiro, M.R.P.; Moais, H.A. Effect of pH and temperature on the activity of enzymatic extracts from pineapple peel. *Food Bioproc. Technol.* **2012**, *5*, 1824–1831. [CrossRef]
65. Grzonka, Z.; Kasprzykowski, F.; Wiczk, W. Cysteine Proteases. In *Industrial Enzymes: Structure, Function and Applications*; Springer: Dordrecht, The Netherlands, 2007; pp. 181–195.
66. Ketnawa, S.; Chaiwut, P.; Rawdkuen, S. Extraction of bromelain from pineapple peels. *Food Sci. Technol. Int.* **2011**, *4*, 395–402. [CrossRef] [PubMed]
67. Vernet, T.; Berti, P.J.; de Montigny, C.; Musil, R.; Tessier, D.C.; Ménard, R.; Magny, M.C.; Storer, A.C.; Thomas, D.Y. Processing of the papain precursor. The ionization state of a conserved amino acid motif within the Pro region participates in the regulation of intramolecular processing. *J. Biol. Chem.* **1995**, *270*, 10838–10846. [CrossRef]
68. Verma, S.; Dixit, R.; Pandey, K.C. Cysteine proteases: Modes of activation and future prospects as pharmacological targets. *Front. Pharmacol.* **2016**, *7*, 107. [CrossRef]
69. Fadhilah, Y.; Shoobihah, A.; Setiasih, S.; Handayani, S.; Hudiyono, S. The effect of Ca^{2+}, Mg^{2+} ions, cysteine, and benzoic acid on the activity of purified bromelain from pineapple core extract (*Ananas comosus* [L]. Merr). *AIP Conf. Proc.* **2018**, *2049*, 020029-1–020029-5.
70. Liang, H.Y.; Li, M.; Shi, M.; Liao, A.P.; Wu, R.C. Study on the stability of fruit bromelain. *Adv. Mater. Res.* **2011**, *421*, 19–22.
71. Haq, S.K.; Rasheedi, S.; Sharma, P.; Ahmad, B.; Khan, R.H. Influence of salts and alcohols on the conformation of partially folded intermediate of stem bromelain at low pH. *Int. J. Biochem. Cell Biol.* **2005**, *37*, 361–374. [CrossRef] [PubMed]
72. Shukor, M.Y.; Masdor, N.; Baharom, N.A.; Jamal, J.A.; Abdullah, M.P.A.; Shamaan, N.A.; Syed, M.A. An inhibitive determination method for heavy metals using bromelain, a cysteine protease. *Appl. Biochem. Biotechnol.* **2008**, *144*, 283–291. [CrossRef]
73. Kaul, P.; Sathish, H.A.; Prakash, V. Effect of metal ions on structure and activity of papain from *Carica papaya*. *Nahrung* **2002**, *46*, 2–6. [CrossRef]
74. Kaur, T.; Kaur, A.; Grewal, R.K. Kinetics studies with fruit bromelain (*Ananas comosus*) in the presence of cysteine and divalent ions. *J. Food Sci. Technol.* **2015**, *52*, 5954–5960. [CrossRef]
75. Wang, X.; Liu, Z.; Hu, X.; Huanh, H. Effects of Ca~(2+) on thermo stability and secondary structure of bromelain. *Chin. Food Addit.* **2009**, *30*, 153–155.
76. Masdor, N.A.; Said, N.A.M. Partial purification of crude stem bromelain improves its sensitivity as a protease inhibitive assay for heavy metals. *Aust. J. Basic Appl. Sci.* **2011**, *5*, 1295–1298.
77. Marshall, S.; Golden, J. Characterization of bromelain from *Morinda citrifolia* (Noni). *J. Sci. Res.* **2012**, *4*, 445–456.
78. Saptarini, N.M.; Rahayu, D.; Herawati, I.E. Antioxidant activity of crude bromelain of pineapple (*Ananas comosus* (L.) Merr) Crown from Subang District, Indonesia. *J. Pharm. Bioallied Sci.* **2019**, *11*, S551–S555. [CrossRef]
79. Huang, C.W.; Lin, I.J.; Liu, Y.M.; Mau, J.L. Composition, enzyme and antioxidant activities of pineapple. *Int. J. Food. Prop.* **2021**, *23*, 1244–1251. [CrossRef]
80. Abbas, S.; Shanbhag, T.; Kothare, A. Applications of bromelain from pineapple waste towards acne. Saudi. *J. Biol. Sci.* **2021**, *28*, 1001–2009.

81. Ataide, J.A.; de Carvalho, N.M.; Rebelo, M.d.; Chaud, M.V.; Grotto, D.; Gerenutti, M.; Rai, M.; Mazzola, P.G.; Jozala, A.F. Bacterial nanocellulose loaded with bromelain: Assessment of antimicrobial, antioxidant and physical-chemical properties. *Sci. Rep.* **2017**, *7*, 18031. [CrossRef]
82. Udenigwe, C.C.; Aluko, R.E. Chemometric analysis of the amino acid requirements of antioxidant food protein hydrolysates. *Int. J. Mol. Sci.* **2011**, *12*, 3148–3161. [CrossRef]
83. Chakraborty, A.J.; Mitra, S.; Tallei, T.E.; Tareq, A.M.; Nainu, F.; Cicia, D.; Dhama, K.; Emran, T.B.; Simal-Gandara, J.; Capasso, R. Bromelain a potential bioactive compound: A comprehensive overview from a pharmacological perspective. *Life* **2021**, *11*, 317. [CrossRef]
84. Lasekan, O.; Hussein, F.K. Classification of different pineapple varieties grown in Malaysia based on volatile fingerprinting and sensory analysis. *Chem. Cent. J.* **2018**, *12*, 140. [CrossRef]

molecules MDPI

Review

Specialized Pro-Resolving Mediators in Neuroinflammation: Overview of Studies and Perspectives of Clinical Applications

Mariarosaria Valente [1,2,*], Marta Dentoni [1], Fabrizio Bellizzi [1], Fedra Kuris [1] and Gian Luigi Gigli [1,2,*]

1 Neurology Unit, Dipartimento di Area Medica (DAME), University of Udine, 33100 Udine, Italy; dentoni.marta@spes.uniud.it (M.D.); bellizzi.fabrizio@spes.uniud.it (F.B.); kuris.fedra@spes.uniud.it (F.K.)
2 Clinical Neurology Unit, Department of Neurosciences, S. Maria della Misericordia University Hospital, 33100 Udine, Italy
* Correspondence: mariarosaria.valente@uniud.it (M.V.); gianluigi.gigli@uniud.it (G.L.G.)

Abstract: Specialized pro-resolving mediators (SPMs) are lipid mediators derived from poly-unsaturated fatty acids (PUFAs) which have been demonstrated to have an important role in the inflammation environment, preventing an overreaction of the organism and promoting the resolution of inflammation. Our purpose was to point out the current evidence for specialized pro-resolving mediators, focusing on their role in neuroinflammation and in major neurological diseases.

Keywords: specialized pro-resolving mediators; resolvins; maresins; annexins; lipoxins; protectins; glial cells; neuroinflammation; neurodegeneration; cerebrovascular disorders; multiple sclerosis; dementia

Citation: Valente, M.; Dentoni, M.; Bellizzi, F.; Kuris, F.; Gigli, G.L. Specialized Pro-Resolving Mediators in Neuroinflammation: Overview of Studies and Perspectives of Clinical Applications. *Molecules* **2022**, *27*, 4836. https://doi.org/10.3390/molecules27154836

Academic Editor: Smaoui Slim

Received: 7 July 2022
Accepted: 26 July 2022
Published: 28 July 2022

1. Specialized Pro-Resolving Mediators: Metabolism, Receptors, Pathways

1.1. Overview on Specialized Pro-Resolving Mediators

Inflammation is a cascade event preserved along the evolution from the first multicellular precursor organisms to humans. Its main role is to defend tissues from an insulting agent, such as microbes or direct damage, enabling in most cases a natural return to homeostasis. If inflammation is not someway stopped, it can lead to serious consequences, such as uncontrolled edema [1].

For many years, it was assumed that inflammation was a self-limiting process [1]. However, recent discoveries have shown the presence of an active de-escalation process, promoted by a class of molecules, namely specialized pro-resolving mediators (SPMs). From the beginning of the inflammation process, SPMs reach the site of edema, either transported by blood flow or produced within the inflammatory tissue [1]. Since chronic and/or uncontrolled inflammation plays a key role in a variety of diseases (such as cardiovascular diseases, metabolic syndrome, and neurological diseases), SPMs have a potential therapeutic role. In particular, SPMs are lipid mediators (LMs) derived from PUFAs (polyunsaturated fatty acids), such as AA (Arachidonic Acid), EPA (eicosapentaenoic acid), DHA (docosahexaenoic acid) and n-3 DPA (n-3 docosapentaenoic acid). The properties of ω-3 fish oil fatty acids in human disease and physiology may in part be explained by the formation of autacoids derived from PUFAs [1]. SPMs include lipoxins, resolvins, protectins and maresins, as well as newly identified cysteinyl-conjugated SPMs (cys-SPMs) and n-3 DPA-derived SPMs [2]. In the following paragraphs, each group of SPMs will be analyzed.

1.1.1. Lipoxins

Lipoxins (LX) LXA4 and LXB4 [1,3] were the first discovered SPMs. Lipoxins derive from eicosanoids thanks to a mechanism of lipo-oxygenation. Eicosanoids in turn derive from AA, an ω-6 fatty acid implied in inflammation. AA is converted into LXA4 and LXB4 via 5- lipoxygenase and 15- lipoxygenase. Lipoxins are produced by leukocytes in transcellular biosynthesis steps during interactions between leukocytes and mucosal cells or platelets [1]. Initially, Lipoxins were believed to be agents of anti-inflammation, and their pro-resolution role has only recently been discovered. Aspirin can trigger their biosynthesis thanks to its capacity to promote the formation of LMs via lipo-oxygenation [4].

1.1.2. Resolvins

Resolvins are generated in inflammatory exudates during the phase of resolution. They derive from ω-3 fatty acids EPA and DHA, forming E series (RvE) and D series (RvD) resolvins, respectively. Their synthesis is also promoted by aspirin (likewise Lipoxins) [1]. Resolvins act in several ways in order to interrupt the inflammation cascade. In particular, they [1]:

- inhibit the production of pro-inflammatory mediators, such as chemokines and cytokines
- enhance scavenging of pro-inflammatory chemokines
- promote the recruitment of monocytes and phagocytes' clearance via the lymphatic system
- limit PMN (polymorphonuclear cells) migration and infiltration
- Focusing on the subclass of Resolvins, we can find [2]:
- E-series Resolvins: RvE1, RvE2, RvE3 and the recent RvE4;
- D-series Resolvins: RvD1, 17R-ResolvinD1, RvD2, RvD3 and 17R-Resolvin D3, RvD4, RvD5.

1.1.3. Protectins

Protectins consist of Protectin D1/Neuroprotectin D1 (PD1/NPD1). They are biosynthesized from DHA via the 15-LOX mechanism. It can be found in human cell types, murine exudates and brain tissue; in this last case, it is called "NeuroprotectinD1" (NPD1), whereas PD1 operates in peripheral tissue. PD1/NPD1 has neuroprotective properties in the brain, retina and Central Nervous System (CNS). Its aspirin-triggered epimer, 17R-NPD1, has the same actions as NPD1 in controlling PMN, enhancing macrophage functions and attenuating experimental stroke [2].

1.1.4. Maresins

Maresins were first identified in human macrophages, in a pathway initiated by 12-LOX. Their name derives from an acronym: Macrophage Mediators in Resolving Inflammation. Maresin1 (MaR1) is able to promote the regeneration of tissues in an experimental model of simple organisms (planaria) with a strong capability of regeneration. In human cells, it is produced by platelets and PMN interactions. MaR1 promotes tissue regeneration and repair and has a neuroprotective role [2].

1.1.5. Recently Discovered SPMs

Recently discovered peptide-lipid conjugated SPMs include cysteinyl-SPMs (cys-SPMs). They consist of three series of SPMs, each one with three bioactive members: maresin conjugates in tissue regeneration (MCTR), protectin conjugates in tissue regeneration (PCTR) and resolving conjugates in tissue regeneration (RCTR). They show pro-repair and pro-regenerative actions [2]. For the sake of completeness, it is important to mention n3-DPA-derived SMPs: RvDn-3 DPA, MaRn-3 DPA and PDn-3 DPA, 13-series resolvins (RvTs). They share the potent actions of DPA and EHA-derived SPMs in the resolution of systemic inflammation and neuro-inflammation. RvTs' biosynthesis is promoted by atorvastatin via S-nitrosylation of cyclooxygenase-2 (COX-2) [2].

1.2. Receptors and Pathways

It is important to emphasize that these endogenous mediators of resolution do not act thanks to an "inhibition" of inflammation pathways: instead, they actively promote specific pathways in order to obtain a return to homeostasis. There are specific G-protein-coupled seven-transmembrane receptors (GPCR) activated by SPMs [1]. Every single class of SPM demonstrates stereoselective activation of its own GPCR. SPMs show affinities for ligand-receptors in the nano-picomolar range, thus demonstrating a potent action in vitro and in vivo [2].

Resolvin E1 (RvE1) acts via ChemR23 (GPCR for RvE1). It is also a partial agonist on the LTB_4 (leukotriene B4) receptor (BLT1), activated by LTB_4 as well. Nevertheless, RvE1 has a different mechanism of action, which is a time and dose-dependent phosphorylation of Akt and p70S6K (ribosomal protein S6 kinase) via ChemR23 [1].

Resolvin D1 (RvD1) binds two separate GPCR on human leukocytes: ALX/FPR2 (LXA_4 receptor) and GPR32 (GPCR for RvD1) [1]. ALX/FPR2 receptors can also be activated by Annexin-1 and Chemerin [5]. Deficits in ALX/FPR2 in experimental models (mice) amplify cardiomyopathy, age-related obesity, and leukocyte-directed endothelial dysfunction [6].

MaR1 can activate two classes of receptors:

- leucine-rich repeat-containing G protein-coupled receptor 6 (LGR6), a phagocyte's receptor
- retinoic acid-related orphan receptor α (ROR-α), a liver macrophages' nuclear receptor

Stimulating the LGR6 receptor, MaR1 can promote phagocytosis, efferocytosis, and the phosphorylation of select proteins [7]. NPD1/PD1's receptor, GPR37, increases intracellular Ca^{2+} in macrophages and promotes phagocytosis [8]. RvD5n-3 DPA binds an orphan receptor, GPR101, with high stereospecificity [2]; in experimental KO models of GPR101, there is a lack of protective action of RvD5n-3 in inflammatory arthritis [9].

These receptors demonstrate overlapping actions (for example, ALX, GPR18, LGR6 and GPR101 can promote calcium mobilization via cAMP signal) and distinct actions too; thus, they could act in tandem to promote defense from injury, inflammation, and infection [2].

1.3. *Mechanism of Action*

The first signs of inflammation response are vasodilation and changes in vessel permeability. These factors not only permit the recruitment of cells implied in the inflammatory response but also give substrates for the biosynthesis of important molecules, such as SPMs [1]. Apparently, ω-3 PUFAs, AA, EPA and DHA can be found within inflammatory exudates during very early phases, as demonstrated in various works [10,11]. Therefore, the inflammation response is counterbalanced early by pro-resolution mediators. This avoids an excess of an inflammatory response that can be disruptive for the organism and for the tissue itself [1].

2. Specialized Pro-Resolving Mediators and Neuroinflammation

2.1. *Neuroinflammation and Its Resolution*

While inflammation is usually a self-limiting physiological process, when persistent or dysregulated it can become harmful to human tissues; if this happens within the CNS it is referred to as neuroinflammation and many studies proved that chronic neuroinflammation could ultimately lead to neurodegeneration [12–14]. In this picture, an emerging concept is the resolution of neuroinflammation which contributes to brain homeostasis; a great deal of attention has been paid to the topic in the last few years. The main actors of this specular process are the so-called SPMs, whose characteristics have been explained in the previous chapter. In the last decade, several research groups started to investigate the role of SPMs in the nervous tissue as regulators of the inflammation process that may contribute to the crosstalk between glial cells and neurons in several neurological pathologies [15].

2.2. *The Role of Glial Cells in Neuroinflammation and the Contribution of SPMs*

Nervous tissue is composed of about 100 billion neurons and 80 to 100 billion glial cells, namely ectoderm-derived astrocytes and oligodendrocytes, and mesoderm-derived microglial cells. Astrocytes play a key role in the metabolism and metabolic support of nervous parenchyma and specifically neurons, i.e., lactate shuttle, the glutamate–glutamine cycle, and ketone bodies supply. Neuroinflammation has lately been interpreted as a condition of metabolic imbalance and energetic depletion, both in the acute and chronic settings. It hence derives that glial cells play a crucial role in the control of neuroinflammation, by regulating nervous tissue metabolism.

As demonstrated, brain tissue contains high levels of PUFAs, mainly DHA and AA, which are the principal precursors of SPMs. The main PUFA source is unesterified plasma fatty acid pool rather than endogenous synthesis; such a source is severely impacted by dietary supply according to studies conducted on rodents [16,17]. Interestingly, the hippocampus and prefrontal cortex contain the highest DHA content while the hypothalamus has the lowest [15]. As for their proportion of representation in the human brain, astrocytes contain 10–12% of DHA, oligodendrocytes 5%, and microglial cells up to 2% [18]. Astrocytes, the most abundant glial cells present in the nervous tissue, take part in many vital processes, such as the migration of developing axons and certain neuroblasts, the regulation of blood flow, electrolyte homeostasis, blood–brain barrier (BBB), and synapse function. Moreover, they seem to be the main glial cells involved in neuroinflammation, although they show significant diversity in this process. For instance, they express high levels of the ALX/FPR2 receptor, which has a central role in the regulation of astrogliosis, an active inflammatory path that leads to neural protection, repair and ultimately to glial scarring [15]. LXA_4 and RvD1, the two SPMs that bind this receptor, promote the inhibition of astrocytes' pro-inflammatory activities [19]. Moreover, it has been observed that peripheral RvD1 administration in brain injury models improved its functional recovery through an ALX/FPR2-regulated pathway probably induced by astrocytes [15]. Another important receptor expressed by astrocytes and playing an important role in neuroprotection is ChemR23/ERV1, expressed in the human hippocampus, which binds RvE1: animal studies demonstrated that peripheral administration of RvE1 in Alzheimer's disease (AD), in combination with LXA_4, reduced astrocyte activation [20]. Other receptors involved in the neuroprotection and resolution of inflammation are GPR37, GPR18 and LgR6, whose expression in astrocytes is challenged, and further studies both in vivo and in vitro are needed on this subset. Besides their main function of myelin synthesis, oligodendrocytes, the second most represented cell population in the CNS, may play a role in the resolution of neuroinflammation thanks to the latest evidence on their active production of immune-regulatory factors or their receptors [21]. Comparing oligodendrocytes with astrocytes, ALX/FPR2 is not expressed by these cells; the only SPMs receptor identified seems to be GPR37 [22]. On the other hand, microglia, the immune cells of the CNS, thanks to their very physiological role, seem to express all the known SPM receptors and are susceptible to the effects of different SPMs categories (lipoxins, RvE, RvD, protectins and maresins) [15,23]. Nonetheless, the cellular origin of SPMs in these cells, as in astrocytes and oligodendrocytes, has not yet been demonstrated and only a few in vitro studies have tried to investigate it [16].

3. Specialized Pro-Resolving Mediators and Potential Applications in Neuroinflammatory Conditions

3.1. Specialized Pro-Resolving Mediators in Ischemic Stroke and Cerebrovascular Events

The concept of ischemic stroke has been expanded to include not only what happens inside the vessel, but also in the surrounding environment, the so-called "neurovascular unit", which includes the interaction between glia, neurons, vascular cells, and matrix components; after the acute event, secondary neuroinflammation takes place, bringing about detrimental effects producing further injury and neuronal death, and promotion of recovery [24]. Several studies have investigated the possible role of pro-resolving mediators in improving post-stroke prognosis; however, they have mostly been conducted on rodents, and applications in humans remain speculative and in need of further research. Table 1 provides a summary of in vivo studies on SPMs in ischemic stroke and cerebrovascular events.

3.1.1. Resolvins in Ischemic Stroke and Cerebrovascular Events

Though preliminary studies indicate a decrease in the risk of cardiovascular diseases thanks to *n*-3 PUFAs supplementation, large double-blind studies did not show clear beneficial effects; however, PUFAs may play a role both through the reduction of pro-inflammatory factors, as well as through the stimulation of the resolution of inflammation [25]. The human body metabolizes *n*-3 PUFAs into RvD2 via the lipoxygenase pathway. Exogenous supply of RvD2 via intraperitoneal injection in a middle cerebral artery occlusion (MCAO) mouse model was able to reverse induced brain injury, including infarction, inflammatory response, brain edema, and neurological dysfunction [26]. Apparently, the capacity of *n*-3 PUFAs to generate RvD2 was reduced by middle cerebral artery occlusion/reperfusion (MCAO/R), making their supplementation less effective than direct RvD2 injection: in the early brain ischemia/reperfusion (I/R) injury process, the metabolic processing of fish oil (especially DHA) may be blocked. In the same animal model, neutrophil membrane-derived nanovesicles loaded with RvD2 were shown to alleviate inflammation and protect the mouse brain from ischemic stroke injury, thus providing a possible therapeutic strategy [27]. Despite what has just been said about PUFAs, acute post-ischemic administration of triglyceride emulsions containing only DHA (tri-DHA) conferred neuroprotection against hypoxic-ischemic injury in neonatal mice [28,29]. RvD1 may play a role in modulating stroke risk factors by preventing atherosclerosis: supplementation with exogenous RvD1 improved plaque stability in fat-fed Ldlr-/- via an increase in fibrous cap thickness and decreased lesional oxidative stress and necrosis [30]. Further studies demonstrate how RvD2 and RvE1 supply may significantly decrease or slow down atherosclerotic changes as well [25]. Post-stroke blood levels of RvD1 have also been found to correlate with cognitive performance: in a prospective study assessing the impact of eicosanoids on cognitive function in stroke survivors, prostaglandin E2, 9S-, 13S-HODE and RvD1 were all strongly associated with the post-stroke cognitive impairment, while RvD1 only correlated with better cognitive performance [31].

3.1.2. Maresins in Ischemic Stroke and Cerebrovascular Events

The intracerebroventricular injection of MaR1 may play a protective role against I/R injury by inhibiting pro-inflammatory reactions and NF-kB p65 activation and nuclear translocation: in a MCAO mice model, MaR1 significantly reduced the infarct volume and neurological defects, protecting the brain tissue and neurons from injury [32]. MaR1 treatment also attenuated cerebral I/R injury by reducing inflammatory responses and mitochondrial damage via the activation of SIRT1 signaling [33].

3.1.3. Annexins in Ischemic Stroke and Cerebrovascular Events

The pro-resolving protein Annexin A1 (AnxA1) has been studied too; targeting the AnxA1/Fpr2/ALX pathway may represent another novel treatment strategy for resolving thrombo-inflammation [34,35]. Macrophages can differentiate into two subtypes, depending on cytokines and chemokines production during inflammation response. In particular, different chemokines can attract Th1 and Th2 or T regulatory (Tr) cells, and this response is integrated by M1 and M2 macrophages in circuits of amplification and regulation of T-cell responses. M1 macrophages are effector cells that kill microorganisms and tumor cells, producing a multitude of proinflammatory cytokines. On the other hand, M2 cells tune inflammatory responses and adaptive Th1 immunity, scavenge debris, and promote angiogenesis, tissue remodeling and repair [36]. Ac2-26 (annexin/lipocortin 1-mimetic peptide) administered to a transient MCAO/R mouse model was shown to modulate microglial/macrophage polarization towards M2 anti-inflammatory phenotype and alleviate subsequent cerebral inflammation; this was achieved by regulating the FPR2/ALX-dependent AMPK-mTOR pathway [37]. The same study pointed to plasma AnxA1 as a potential biomarker for the outcomes of acute ischemic stroke patients receiving endovascular thrombectomy. AnxA1 administration was shown to be beneficial in intracerebral hemorrhage, attenuating neuroinflammation via the AnxA1/FPR2/p38 signaling pathway [38]. In addition, in cerebral I/R injury AnxA1 may shift the platelet phenotype from pro-pathogenic to regulatory: it was able to reduce the propensity of platelets to aggregate and cause thrombosis by affecting integrin ($\alpha IIb\beta 3$) activation [39].

3.1.4. Lipoxins in Ischemic Stroke and Cerebrovascular Events

LXA4 is another potent anti-inflammatory mediator exerting a neuroprotective effect following a cerebrovascular event. It has been postulated to regulate microglial M1/M2 polarization after cerebral I/R injury via the Notch signaling pathway, and to downregulate the expression of the proinflammatory cytokines IL-1β and TNF-α [40]. The intracerebroventricular injection of LXA4 or its synthetic analogs was shown to decrease infarct volumes and improve neurological function in mice. One study pointed to Nrf2 upregulation being involved in the neuroprotective effects of LXA4; such effects were partially blocked by Boc2, a specific antagonist of the LXA4 receptor (ALXR) [41]. However, LXA4 induced Nrf2 expression and its nuclear translocation, as well as HO-1 expression and GSH synthesis; the latter two effects were not blocked by Boc2, indicating that Nrf2 upregulation may be ALXR independent. In addition, the PPAR agonist rosiglitazone has been shown to be neuroprotective by increasing LXA4 and reducing leukotriene B4 (LTB4) in experimental stroke [42].

As LXA4 is rapidly inactivated, potent analogs have been synthesized, including BML-111. Post-ischemic treatment with BML-111 significantly reduced infarct size, decreased vasogenic edema, protected against BBB disruption, and reduced hemorrhagic transformation in rats [43]. Similarly, post-ischemic, intravenous treatment with BML-111 for 1 week was shown to induce early protective effects, reducing infarct volume, and improving sensorimotor function at 1 week; however, it did not reduce infarct size or improve behavioral deficits 4 weeks after ischemic stroke [44]. Another stable synthetic analog of LXA4 is lipoxin A4 methyl ester (LXA4 ME). Intracerebroventricular injection in I/R injury mice ameliorated neurological dysfunctions, reduced infarction volume, attenuated neuronal apoptosis and had overall an anti-inflammatory effect [45,46]. It was also shown to reduce BBB dysfunction and MMP-9 expression while increasing TIMP-1 expression [47]. One study suggested that intracerebroventricular injection for two consecutive weeks in mice after the acute event could alleviate spatial learning and memory impairments, thus exerting beneficial effects on the cognitive impairment induced by chronic cerebral hypoperfusion. This was likely achieved through attenuating oxidative injury and reducing neuronal apoptosis in the hippocampus with the activation of the ERK/Nrf2 signaling pathway [48]. Interestingly, one of the few studies on LXA4 conducted in humans showed that the levels of LXA4 in the acute phase of ischemic stroke were significantly reduced in post-stroke cognitive impairment patients compared with those with no cognitive impairment upon Mini-Mental State Examination (MMSE) testing [49]. LXA4 has also been reported to reduce neuroinflammation by activating FPR2 and inhibiting p38 in a rat model of subarachnoid hemorrhage (SAH); intracerebroventricular injection of exogenous LXA4 reduced brain water content and BBB leakage, and improved neurological function, memory and learning after the event [50]. An amelioration of endothelial dysfunction, microflow recovery, and suppression of neutrophil infiltration was also shown, possibly involving the LXA4/FPR2/ERK1/2 pathway [51].

3.1.5. Protectins in Ischemic Stroke and Cerebrovascular Events

Protectins, also called neuroprotectins, may play a role in stroke too. NPD1 was demonstrated to reduce infarct volume, inhibit the activation of NF-kB, and reduce the expression of COX-2 and infiltration polymorphonuclear leukocytes [52]. The intracerebroventricular injection of NPD1 in the rat I/R injury model has been shown to significantly reduce infarct volume and improve neurological scores, through the inhibition of calpain-mediated TRPC6 proteolysis and the subsequent activation of CREB via the Ras/MEK/ERK pathway [53]. Infarct size reduction in aged rats via the activation of the Akt and p70S6K pathways has also been demonstrated [54]. NPD1 seems to counter uncompensated oxidative stress by upregulating ring finger protein 146 (Iduna) in neurons and astrocytes, which facilitates DNA repair and protects against cell death; in fact, Iduna is usually downregulated in the penumbra after cerebral ischemia [55]. NPD1 may also work on mitochondria-related cell death pathways, which play a major role in ischemic brain injury. Following NPD1 acute intraperitoneal injection in mice, ischemic core expansion was prevented by about 40%; brain mitochondria showed a preserved membrane structure, together with a reduction of mitochondrial BAX translocation and activation [56]. NPD1 administration has also been demonstrated to promote neurogenesis and angiogenesis, BBB integrity, penumbra protection and subsequent long-term neurobehavioral recovery after experimental ischemic stroke [57]. By administering aspirin plus DHA, the synthesis of aspirin-triggered NPD1 (AT-NPD1) in the brain was discovered; the total chemical synthesis of this molecule and usage in mice MCAO model was shown to promote sustained neurobehavioral recovery, reduce infarct volume and brain edema, and protect white matter [58].

Table 1. Summary of in vivo studies on SPMs in ischemic stroke and cerebrovascular events.

Reference	Type of Study	Animal Model	Pro-Resolving Mediator	Delivery (Or Measurement If the Study Was Non-Interventional)	Outcome
Zuo et al., 2018 [26]	Animal study	MCAO mouse model	RvD2	intraperitoneal	↓ infarction, inflammation, edema, and neurological dysfunction; compared with ω-3 fatty acid oral supplements, better rescue effect on cerebral infarction
Dong et al., 2019 [27]	Animal study	MCAO mouse model	RvD2	Intravenous infusion of RvD2-loaded nanovesicles	↓ inflammation; ↑ neurological function
Fredman et al., 2016 [30]	Animal study	fat-fed Ldlr-/- mice	RvD1	Immunoprecipitation injection	↓ atherosclerosis
Kotlęga et al., 2021 [31]	Human study	-	RvD1	blood levels of endogenous pro-resolving mediators	Post-stroke blood levels of RvD1 correlated with a better cognitive performance
Xian et al., 2016 [32]	Animal study	MCAO mouse model	MaR1	Intracerebroventricular	↓infarct volume and neurological defects by inhibiting NF-kB p65 function
Xian et al., 2019 [33]	Animal study	MCAO mouse model	MaR1	Intracerebroventricular	↓ inflammation and mitochondrial damage via activation of SIRT1 signaling
Vital et al., 2020 [34]	Animal study	Lipopolysaccharide and sickle transgenic mice models of thrombo-inflammation	AnxA1 mimetic peptide Ac2-26	Intravenous	↓ thrombo-inflammation via Fpr2/ALX receptor and ↓ platelet aggregation
Gavins et al., 2007 [35]	Animal study	MCAO in wild-type or AnxA1−/− mice	AnxA1 mimetic peptide Ac2-26	Intravenous	↓ inflammation via receptors of the FPR family
Xu et al., 2021 [37]	Animal study	MCAO mouse model	AnxA1 mimetic peptide Ac2-26	Intravenous	↓ inflammation by regulating the FPR2/ALX-dependent AMPK-mTOR pathway
Ding et al., 2020 [38]	Animal study	Collagenase-induced ICH mouse model	Recombinant human AnXA1	Intracerebroventricular	↓ inflammation via the FPR2/p38/COX-2 pathway
Senchenkova et al., 2019 [39]	Animal study	MCAO in wild-type or AnxA1−/− mice	Whole protein AnXA1	Intravenous	↓ platelet aggregation by affecting integrin (αIIbβ3) activation
Li et al., 2021 [40]	Animal study	MCAO mouse model	LXA4	Intracerebroventricular	↓ proinflammatory cytokines and regulate microglial M1/M2 polarization via the Notch signaling pathway
Wu et al., 2013 [41]	Animal study	MCAO mouse model	LXA4	Intracerebroventricular	↓infarct volume and ↑ neurological function through Nrf2 upregulation
Hawkins et al., 2014 [43]	Animal study	MCAO mouse model	LXA4 analog BML-111	Intravenous	↓ infarct size, edema, BBB disruption, and hemorrhagic transformation

Table 1. *Cont.*

Reference	Type of Study	Animal Model	Pro-Resolving Mediator	Delivery (Or Measurement If the Study Was Non-Interventional)	Outcome
Hawkins et al., 2017 [44]	Animal study	MCAO mouse model	LXA4 analog BML-111	Intravenous	↓ infarct volume; and ↑ neurological function at 1 week. No reduction of infarct size or improvement of behavioral deficits 4 weeks after ischemic stroke
Wu et al., 2010 [45]	Animal study	MCAO mouse model	LXA4 ME	Intracerebroventricular	↓ proinflammatory cytokines, neurological dysfunctions, infarction volume, and neuronal apoptosis
Ye et al., 2010 [46]	Animal study	MCAO mouse model	LXA4 ME	Intracerebroventricular	↓ proinflammatory cytokines, neurological dysfunctions, infarction volume, and neuronal apoptosis
Wu et al., 2012 [47]	Animal study	MCAO mouse model	LXA4 ME	Intracerebroventricular	↓ BBB dysfunction and MMP-9 expression; ↑ TIMP-1 expression
Jin et al., 2014 [48]	Animal study	BCCAO	LXA4 ME	Intracerebroventricular	Amelioration of cognitive impairment via ↓oxidative injury and ↓neuronal apoptosis in the hippocampus with the activation of the ERK/Nrf2 signaling pathway
Wang et al., 2021 [49]	Human study	-	LXA4, RvD1, RvD2, RvE1, MaR1	blood levels of endogenous pro-resolving mediators	↓ LXA4 in patients with post-stroke cognitive impairment
Guo et al., 2016 [50]	Animal study	endovascular perforation model of SAH	Exogenous LXA4	Intracerebroventricular	↓ neuroinflammation by activating FPR2 and inhibiting p38
Liu et al., 2019 [51]	Animal study	endovascular perforation model of SAH	Recombinant LXA4	Intracerebroventricular	↓ endothelial dysfunction and neutrophil infiltration, possibly involving the LXA4/FPR2/ERK1/2 pathway
Yao et al., 2013 [53]	Animal study	MCAO mouse model	NPD1	Intracerebroventricular	↓ infarct volume and ↑ neurological scores through inhibition of calpain-mediated TRPC6 proteolysis and activation of CREB via the Ras/MEK/ERK pathway
Eady et al., 2012 [54]	Animal study	MCAO mouse model	NPD1	Intravenous	↓ infarct size in aged rats via activation of Akt and p70S6K pathways
Belayev et al., 2017 [55]	Animal study	MCAO mouse model	DHA (NPD1 precursor)	Intravenous	↓ oxidative stress by upregulating ring finger protein 146 (Iduna) in neurons and astrocyte
Zirpoli et al., 2021 [56]	Animal study	Unilateral cerebral hypoxia-ischemia injury mouse model	NPD1	Intraperitoneal	↓ ischemic core expansion, preserved mitochondrial structure and ↓ BAX translocation and activation
Belayev et al., 2018 [57]	Animal study	MCAO mouse model	NPD1	Intracerebroventricular	↑ neurogenesis and angiogenesis, BBB integrity, and long-term neurobehavioral recovery
Bazan et al., 2012 [58]	Animal study	MCAO mouse model	AT-NPD1	Intravenous	↓ infarct volume and brain edema; ↑ neurobehavioral recovery

AnXA1: Annexin A1; AT-NPD1: aspirin-triggered NPD1; BBB: Blood–Brain Barrier; BCCAO: bilateral common carotid artery occlusion; DHA: docosahexaenoic acid; FPR: formyl-peptide receptor; LXA4: Lipoxin A4; LXA4 ME: Lipoxin A4 Methyl Ester; MaR1: Maresin1; MCAO: middle cerebral artery occlusion; NPD1: Neuroprotectin D1; RvD1: Resolvin D1; RvD2: Resolvin D2; RvE1: Resolvin E1; SAH: Sub Arachnoid Hemorrhage.

3.2. Specialized Pro-Resolving Mediators in Neurological Immune-Mediated Disorders

Multiple sclerosis (MS) is a neuroinflammatory disease in which unresolved and uncontrolled inflammation leads to a pathological disease state, thus representing a classical model of chronic inflammation; in this context, SPMs could be instrumental in resolving the pathologic inflammation. However, there are minimal data available on the functional status of SPMs in MS; it seems that SPMs have neuroprotective action in MS by exerting pro-resolving effects in the pre-clinical model; however, little is known about the direct effect of SPMs on oligodendrocytic or neuronal cells [59]. Table 2 provides a summary of in vivo studies on SPMs in neurological immune-mediated disorders.

3.2.1. Annexins in Neurological Immune-Mediated Disorders

The potential role of SPMs has been studied in experimental autoimmune encephalomyelitis (EAE), a mouse model of MS. Across the literature, the role of AnxA1 appears to be contentious. However, it is worth underlying that AnxA1 likely exerts a dual function on the innate and adaptive immune systems; in the innate immune system, endogenous AnxA1 plays an anti-inflammatory role that controls events occurring early in the inflammatory process; on the contrary, in the adaptive immune system its role is controversial, and it may as well be pro-inflammatory [60,61]. Dated works demonstrating annexin-1 immunoreactivity in plaque lesions in both experimental mice and MS patients led to hypothesizing a possible contribution to anti-inflammatory processes [62–64]. Another study pointed to the potential therapeutic benefit of annexin-1 administration in MS, as intracerebroventricular administration in EAE rats significantly reduced neurological severity, and immunoneutralization of endogenous brain annexin-1 failed to exacerbate the clinical features of EAE [65]. In another work, the potential modulatory role of AnxA1 in the development of EAE was investigated, and a direct correlation between AnxA1 T cells expression and severity of disease was shown. MOG35-55-induced EAE development was impaired in AnxA1 null mice, which showed decreased signs of the disease compared to wild type mice at the peak and reduced infiltration of T cells in the spinal cord [60]. Moreover, reduced in vitro recall proliferative response to MOG35-55 in Annexin A1 null T cells was demonstrated, with a significantly reduced Th1/Th17 phenotype, as compared to wild type cells [60]. Authors thus concluded that the identification and generation of neutralizing antibodies against AnxA1 could play a therapeutic role in MS. A more recent study tried to shed some light on the ability of AnxA1 to influence T cell effector function in relapsing/remitting MS (RRMS); by measuring circulating expression levels of AnxA1 in RRMS patients, it was found that they are inversely correlated with disease score and progression [61]. In addition, at the cellular level, there was impaired AnxA1 production in CD4+CD252 conventional T and CD4+RORgt+ T (Th17) cells from RRMS subjects that were associated with an increased migratory capacity in an in vitro BBB model. Authors associated AnxA1 anti-inflammatory action with the STAT3 signaling pathway.

3.2.2. Resolvins in Neurological Immune-Mediated Disorders

As far as concerns the therapeutic potential of RvD1 in EAE mice, oral administration was very effective in attenuating disease progression by suppressing autoreactive T cells and inducing an M2 phenotype of monocytes/macrophages and resident brain microglial cells, though not affecting the number of infiltrating cells [66].

3.2.3. Lipoxins in Neurological Immune-Mediated Disorders

LXA4 may play a role in MS as well. Intraperitoneal injection of LXA4 was shown to ameliorate EAE clinical symptoms and inhibit CD4+ and CD8+ T cell infiltration into the CNS; in addition, LXA4 potently reduced encephalitogenic Th1 and Th17 effector functions, both in vivo and in isolated human T cells from healthy donors and patients with RRMS [67]. The same study demonstrated that LXA4 affects the spinal cord lipidome by significantly reducing the levels of pro-inflammatory LMs during EAE.

3.2.4. Maresins in Neurological Immune-Mediated Disorders

A recent paper analyzed SPMs in active brain lesions, serum, and peripheral blood mononuclear cells (PBMCs) in MS patients and in the spinal cord of EAE mice, showing that levels of MaR1 and other SPMs were below the limit of detection or not increased in mice [68]. Similarly, they were undetected in serum and active brain lesion samples of MS patients, which may be linked to impaired expression of the enzymes involved in the biosynthetic pathways of SPMs. When exogenous MaR1 was administered to mice, various pro-inflammatory cytokines were suppressed, the number of Th1 cells was reduced and the number of Tregs increased, while macrophages underwent polarization towards an anti-inflammatory phenotype [68].

3.2.5. Differential Expression of SPMs in MS

The possibility of non-exhaustive or possibly 'delayed' resolution pathways in MS was also suggested by a study observing that LM pathways are regulated differentially in the cerebrospinal fluid (CSF) of MS patients, depending on disease severity [69]. Specifically, in patients with highly active MS, RvD1 was significantly upregulated and NPD1 was detected in this group only. In line with these results, another study group showed distinct LM profiles that significantly correlated with disease severity in MS patients' peripheral blood [70]. In particular, relapsing and progressive MS patients were associated with high eicosanoid levels, whereas the majority of pro-resolving LM were either significantly reduced or below limits of detection and correlated with disease progression. Furthermore, the expression of several enzymes and associated receptors involved in SPM biosynthesis was found to be reduced in the blood-derived leukocytes of MS patients. These findings support the idea that differentially expressed mediators, such as LXA4, LXB4, RvD1 and NPD1 reduced MS-derived monocyte activation and cytokine production and inhibited inflammation-induced BBB dysfunction and monocyte trans-endothelial migration. The same study group recently presented data at the 35th Annual Congress of the European Committee for Treatment and Research in Multiple Sclerosis (ECTRIMS, 2019) suggesting impaired production of SPMs in MS patients. Comprehensive metabolipidomics profiling was used to identify the spectrum of LM signatures in the CSF of patients across different clinical courses, including relapsing, remitting, and progressive modes of MS; CSF analysis revealed lower levels of LXB4 and RvD3 in different clinical courses of the disease (unpublished) [59].

3.2.6. SPMs in Demyelinating Disorders Other Than MS

Other than the previously discussed findings in MS, few studies on neurological demyelinating disorders and pro-resolving mediators have been published. A recent paper pointed to inflammation resolution impairment in neuromyelitis optica spectrum disorders (NMOSD), showing that RvD1 levels were significantly decreased, whereas leukotrienes B4 (LTB4) levels were significantly increased in the CSF of NMOSD patients [71]. Furthermore, AQP4-IgG titer was negatively correlated with RvD1 levels in the CSF of NMOSD patients, indicating such antibodies may contribute to increased and unresolved inflammation. SPMs have been investigated in experimental autoimmune neuritis (EAN) too, a model of acute inflammatory demyelinating polyradiculoneuropathy. Annexin-1 expression was found to be increased in the inflamed sciatic, which may indicate immunoregulatory functions in-situ and contribute to the termination of the autoimmune response [72]. In addition, in the EAN model RvD1, its synthetic enzyme and receptor were found to be increased in the peripheral nervous system (PNS) during the recovery stage of EAN; intraperitoneal RvD1 injection led to macrophage phagocytosis of apoptotic T cells in PNS, thereby upregulating TGFβ by macrophages, increasing local Treg cell counts, and finally promoting inflammation resolution and disease recovery [73].

Table 2. Summary of in vivo studies on SPMs in neurological immune-mediated disorders.

Reference	Type of Study	Model	Pro-Resolving Mediator	Delivery (Or Measurement If the Study Was Non-Interventional)	Outcome
Paschalidis N et al., 2009 [60]	Animal study	MOG34-55 -induced EAE in AnxA1 null mice compared to MOG34-55 -induced EAE in control mice	Absence of AnxA1 expression	Measurement of disease activity in spinal cord; lymph-node cells (respectively, by isolation of T-cells and/or fixation with haematoxylin and eosin; and by test ELISA for Th1/Th17 cytokine profile)	↓ signs of the disease in AnxA1 null mice compared to wild type mice ↓ infiltration of T cells in the spinal cord of AnxA1 null mice compared to wild type
Huitinga I et al., 1998 [65]	Animal study	EAE rats (MS mouse model)	AnxA1	Intracerebroventricular administration	↓ neurological severity
Poisson LM, 2015 [66]	Animal study	EAE rats (MS mouse model)	RvD1	Oral administration	Attenuation of disease progression by suppressing autoreactive T cells and inducing an M2 phenotype of monocytes/macrophages and resident brain microglial cells
Derada Troletti C et al., 2021 [67]	Animal study	EAE rats (MS mouse model)	LXA4	Intraperitoneal injection	Improvement of EAE clinical symptoms and inhibit CD4+ and CD8+ T cell infiltration into the CNS
Derada Troletti C et al., 2021 [67]	In vivo and in vitro study	Human T cells from healthy donors and patients with relapsing-remitting MS	LXA4	Measurement of T-cell functions	↓ encephalitogenic Th1 and Th17 effector functions
Sánchez-Fernández A et al., 2022 [68]	Animal study	EAE rats (MS mouse model)	MaR1	Intraperitoneal injection	Suppression of various pro-inflammatory cytokines, ↓ number of Th1 cells ↑ of Tregs polarization of macrophages towards an anti-inflammatory phenotype
Prüss H et al., 2013 [69]	Human study	MS patients	RvD1 NDP1	CSF levels	↑ of RvD1 Only detection of NDP1
Kooij G et al., 2020 [70]	Human study	NMOSD patients	RvD1 LTB4	CSF levels	RvD1 ↓ LTB4 ↑
Luo B et al., 2016 [73]	Animal study	EAN (experimental autoimmune neuritis) model	RvD1	Intraperitoneal injection	Macrophage phagocytosis of apoptotic T cells in PNS, ↑ TGFβ by macrophages, ↑ local Treg cell counts, and promotion of inflammation resolution and disease recovery

AnXA1: Annexin A1; AT-NPD1: aspirin-triggered NPD1; BCCAO: bilateral common carotid artery occlusion; DHA: docosahexaenoic acid; EAE: Experimental Autoimmune Encephalitis; EAN: Experimental Autoimmune Neuritis; FPR: formyl-peptide receptor; LXA4: Lipoxin A4; LXA4 ME: Lipoxin A4 Methyl Ester; LTB4: Leukotriene B4; MaR1: Maresin1; NPD1: Neuroprotectin D1; PNS: peripheral nervous system; RvD1: Resolvin D1.

3.3. Specialized Pro-Resolving Mediators in Neurodegenerative Diseases

AD is the most common type of dementia; a growing body of evidence suggests that inflammation is involved in its pathogenesis. Epidemiological studies suggest that the use of anti-inflammatory drugs is associated with a lower incidence of AD; however, clinical trials with anti-inflammatory drugs have not been successful [74]. Given these premises, the possibility of promoting resolution rather than inhibiting inflammation looks appealing.

The potential benefit of working on inflammation resolution is supported by several observations. First of all, a shift in the LM profile in the CSF from pro-resolving to pro-inflammatory occurs as AD progresses: in a recent study, liquid chromatography–tandem mass spectrometry was used to analyze pro-resolving and pro-inflammatory LMs in the CSF of patients with cognitive impairment ranging from subjective impairment to a diagnosis of AD; LMs profile correlated to cognition, CSF tau, and β-amyloid. RvD4, RvD1, NPD1, MaR1, and RvE4 were lower in AD and/or mild cognitive impairment (MCI) compared to subjective cognitive impairment (SCI); on the other hand, pro-inflammatory mediators were higher in AD and MCI [75]. Similarly, it was found that the levels of the MaR1, NPD1 and RvD5, were lower in the entorhinal cortex of AD patients as compared to age-matched controls, while levels of the pro-inflammatory prostaglandin D2 (PGD2) were higher in AD [76]. In addition, RvD4 showed a negative correlation to AD tangle biomarkers and positive correlations to cognitive test scores [75]. Similar findings have been reported in mice, where SPMs in the brain cortex were substantially lower in mice with an APOE4 genotype [77]. The finding that SPMs receptors are increased in the AD brain in post-mortem studies and correlate to Braak stages, suggests a prominent role of resolution pathways; the increase in these receptors may either represent a primary factor in the pathogenesis of the disease or a consequence of failed resolution [78]. The same study group investigated age-related changes in the LM profile in the APP knock-in (APP KI) mouse model of AD, concluding that the brain lipidome appeared to be modified preferentially during aging as compared to amyloid pathology, as the oldest age group was the one with the greatest increase in LMs, despite an early onset of Aβ pathology [79]. In this case, the SPMs biosynthetic enzymes were found to be increased, while their receptor expression decreased in the aged App KI mice, in disagreement with their previous work [78] on AD patients. The discrepancy may be explained by the fact that the stage of AD pathology in 18-month-old App KI mice is likely less advanced compared to that seen in human post-mortem brains [79].

3.3.1. SPMs Administration in AD Models

Several in vivo mouse studies support the potential benefit deriving from SPM use in AD. Table 3 provides a summary of in vivo studies on SPMs in neurodegenerative diseases. When a mixture of the SPMs including RvE1, RvD1, RvD2, MaR1 and NPD1 was administered to mice via intranasal delivery, an amelioration of memory deficits occurred, together with a restoration of gamma oscillation deficits, and a prominent decrease in microglial activation [80]. Intraperitoneal injection of RvE1 and LXA4, alone or in combination, increased the concentration of RvE1, LXA4, and RvD2 in the hippocampus of a murine model, reversed the inflammatory process and decreased the neuroinflammation associated with Aβ pathology; the levels of SPMs in the hippocampus of 5xFAD mice were in fact shown to be significantly lower than in wild-type mice [20]. Similarly, intracerebroventricular administration of LXA4 was able to inhibit the inflammatory response induced by β-amyloid in the cortex and hippocampus of experimental mice, in particular, the production of IL-1b and TNFa [81]. Other than LXA4, the effect of aspirin-triggered LXA4 (ATL) has been investigated too; ATL is generated after the acetylation of COX-2, and displays the same anti-inflammatory activity as the native lipoxins and is more resistant to metabolic inactivation [82]. Subcutaneous injection of ATL was able to reduce NF-kB activation and levels of proinflammatory cytokines and chemokines, as well as create an anti-inflammatory cerebral milieu, resulting in the recruitment of microglia in an alternative phenotype. Such microglia showed improved phagocytic function towards Aβ, ultimately leading to a reduction in synaptotoxicity and improvement in cognition [83]. Not only was ATL demonstrated to enhance the cognitive performance of 3xTg-AD mice and reduce Aβ load, but also to decrease the levels of phosphorylated-tau (p-tau) [84]. Furthermore, intracerebroventricular supply of MaR1 improved the cognitive decline of experimental mice; MaR1 was able to attenuate microglial activation, reduce pro-inflammatory cytokines in favor of anti-inflammatory ones, and up-regulate the levels of proteins related to survival pathways including PI3K/AKT, ERK and down-regulate the levels of proteins associated with inflammation, autophagy, and apoptosis pathways, such as p38, mTOR and caspase 3 [85]. NPD1 seems to play a role in decreasing inflammatory signaling in AD [86–88]; however, to our knowledge, NPD1 alone has never been administered to mice.

The potential role of AnxA1 in the AD murine model has been investigated too. AnxA1 is a pro-resolving mediator that helps to restore the integrity of the BBB and inhibit microglial activation in the brain; interestingly, these functions depend on AnxA1 integrity, and enzymatic cleavage generates pro-inflammatory fragments [89]. When AnxA1 level was measured in the blood and CSF of patients with AD and behavioral variant of frontotemporal dementia (bvFTD), reduced plasma levels of AnxA1 were observed in bvFTD compared to AD and controls, while no difference was shown in the CSF; moreover, a significant cleavage of AnxA1 in PBMCs in both dementia groups was shown [89]. A link between AnxA1, neuroinflammation and amyloid pathology is further suggested by the identification of elevated cleaved AnxA1 in the brains of patients with neurodegenerative dementias including AD, positively correlating with amyloidogenic brain Aβ, inflammatory and pro-apoptotic markers [90]. However, intact AnxA1 protein was found to be increased in the brain of both AD patients and animal models and induce the clearance and degradation of the amyloid-β peptide in vitro by acting on formyl peptide receptor-like 1 (FPRL1) [91]. The increases in ANXA1 observed in AD brains suggest that upregulation of AnxA1 could represent an adaptive response of microglia during inflammatory conditions and an attempt to turn down inflammation at the early disease stage; in later stages with chronic production of Aβ and pro-inflammatory cytokines, microglia change their neuroprotective phenotype in favor of a more pro-inflammatory activation state [91]. Still, surprisingly, when APP/PS1 double-transgenic AD mice were treated for 20 weeks with the anti-inflammatory FPR2 agonist Ac2-26, Ac2-26-treatment did not show any beneficial effect [92]. As previously mentioned, AnxA1 protects against BBB breakdown in AD: treatment with human recombinant ANXA1 (hrAnxA1) in the murine brain endothelial cell line bEnd.3 was able to rescue β-amyloid 1–42 -induced BBB disruption via inhibition of RhoA-ROCK signaling pathway [93]. Similarly, intravenous injection of hrAnxA1 was able to decrease BBB permeability, and reduce β-amyloid load and p-tau build-up in 5xFAD mice and Tau-P301L mice; in addition, the prolonged treatment with hrAnxA1 reduced the memory deficits and increased synaptic density in young 5xFAD mice [94].

Few studies have examined the neuroinflammation-modulating effects of *n*-3 PUFA feeding in the AD murine model; one study reported fish oil feeding managed to attenuate neuroinflammatory gene expression; however, no alteration in the levels of SPMs, brain eicosanoids or docosanoids was detected [95]. When investigating the ability of PUFAs to influence the production of SPMs in AD patients, a randomized, placebo-controlled trial found unchanged levels of the SPMs LXA4 and RvD1 in the group supplemented with *n*-3 FAs, whereas a decrease was documented in the placebo group, indicating that PUFA supplementation managed to prevent reduction in SPMs released from PBMCs [96]. Another work pointed to the ability of n3-PUFA to increase amyloid-β phagocytosis and RvD1 in patients with MCI [97].

3.3.2. SPMs and Sphingosine Kinase

As a conclusive remark on AD and pro-resolving mediators, SPMs have been found to be regulated by sphingosine kinases (especially SphK1) that act by monitoring COX-2, a potent inhibitor of SPMs production [98]. SphK1 generates *N*-acetyl sphingosine (*N*-AS) from acetyl-CoA and sphingosine; *N*-AS then acetylates serine 565 (S565) of COX-2, and the *N*-AS-acetylated COX-2 induces the production of SPMs [99]. In a mouse model of AD, microglia showed a reduction in *N*-AS generation, leading to decreased acetyl-S565 COX2 and SPMs production; mouse treatment with *N*-AS increases acetylated COX-2 and *N*-AS-triggered SPMs in microglia, leading to resolution of neuroinflammation, an increase in microglial phagocytosis, and improved memory [99].

3.3.3. SPMs in Neurodegenerative Disorders Other Than AD

While most literature on SPMs and neurodegenerative disorders focuses on AD, neuroinflammation is also one of the hallmarks of Parkinson's disease (PD) and may play a role in midbrain dopamine (DA) neuron degeneration. Still, the effects of stimulating the resolution of inflammation in PD remain largely unexplored. Both in vitro [100,101] and in vivo models seem to point to a possible role of SPMs. In a lipopolysaccharide (LPS)-induced rat model of PD the effects of intrathecal injection of RvD2 on substantia nigra pars compacta (SNpc) were studied; RvD2 was shown to recover neural injury by suppressing inflammatory mediator expression [101]. In fact, LPS-induced inflammation in SNpc increased the expression of NO, iNOS, TNF-a, IL-1, IL-18, IL-6, IL-1b, ROS production, the translocation of NF-kB p65, IkBa, and IKKb expression in glial cells; after injection of RvD2, the treatment prevented development of behavioral defects and TLR4/NF-kB pathway activation. Another study on rats overexpressing human α-synuclein (Syn) demonstrated that prior to nigral degeneration they display altered DA neuron properties, and reduced striatal DA outflow and motor deficits; these early alterations are coupled with microglia activation and perturbations in inflammatory and pro-resolving mediators, namely IFN-γ and RvD1 [102]. When early and chronic intraperitoneal injection of RvD1 was provided, central and peripheral inflammation, as well as neuronal dysfunction and motor deficits were prevented. Interestingly, the same work demonstrated that endogenous RvD1 is decreased in human patients with early PD [102]. Supporting the role of SPMs in PD, homozygous missense variants in the AnxA1 were recently suggested to cause parkinsonism by leading to extracellular *SNCA* accumulation, neuroinflammation, as well as defects in intracellular signaling pathways and synaptic plasticity; however, such mutations seem to be exceedingly rare, and pathogenicity could not be further explored [103].

Other than AD and PD, a third relevant neurodegenerative pathology is amyotrophic lateral sclerosis (ALS). To our knowledge, no in vivo study on the role of SPMs in ALS has been performed. Although the cause of neuronal degeneration in ALS has not been fully elucidated, there is evidence of macrophage and T cell infiltration into the spinal cord, which may be responsible for motor neuron death. In ALS macrophages, aggregated superoxide dismutase-1 (SOD-1) stimulated the expression of inflammatory cytokines, including IL-1β, IL-6, and TNF-α; it was shown that RvD1 was able to inhibit macrophage IL-6 and TNF-α production, thus suppressing inflammation [104]. Another study investigated the effects of MaR1 on motor neuron cell death, finding it protected motor neuron-like NSC-34 cells against serum-free and $SOD1^{G93A}$ or $TDP\text{-}43^{A315T}$-induced cell death, as well as H_2O_2- or tunicamycin-induced cell death [105].

Table 3. Summary of in vivo studies on SPMs in neurodegenerative diseases.

Reference	Type of Study	Model	Pro-Resolving Mediator	Delivery (Or Measurement If the Study Was Non-Interventional)	Outcome
Do K V et al., 2022 [75]	Human, non-interventional	Patients with AD, MCI, SCI	RvD4	CSF levels of RvD4	Negative correlation to AD tangle biomarkers, and positive correlations to cognitive test scores
Zhu M. et al., 2016 [76]	Human study	Patients with AD	MaR1, NPD1, RvD5	Postmortem tissue samples from the entorhinal cortex	↓ concentration of pro-resolving mediators in the entorhinal cortex of AD patients as compared to age-matched controls, while levels of the pro-inflammatory prostaglandin D2 were higher in AD
Martinsen A. et al., 2019 [77]	Animal study	APOE4 Female mice	Various SPMs	Brain postmortem tissue samples	↓ SPMs in mice with the APOE4 genotype
Emre C. et al., 2020 [78]	Human study	Patients with AD	SPMs receptors	Brain postmortem tissue samples	↑ SPMs receptors
Emre C, Do K V. et al., 2021 [79]	Animal study	APP KI mouse model of AD	LMs profile	Brain postmortem tissue samples	↑ microglia proliferation starting from a young age in the App KI mice, while ↓ astrocyte numbers in older ages Brain lipidome appears to be modified preferentially during aging as compared to amyloid pathology, as the oldest age group was the one with the greatest increase in LMs, despite an early onset of Aβ pathology
Emre C, Arroyo-García et al., 2022 [80]	Animal study	Murine model of AD	RvE1, RvD1, RvD2, MaR1 and NPD1	Intranasal	Amelioration of memory deficits; restoration of Gamma oscillation deficits; ↓ microglial activation
Kantarci A. et al., 2017 [20]	Animal study	Murine model of AD	RvE1 and LXA4	Intraperitoneal	↑ RvE1, LXA4, and RvD2 in the hippocampus; reversing of the inflammatory process, ↓ neuroinflammation

Table 3. *Cont.*

Reference	Type of Study	Model	Pro-Resolving Mediator	Delivery (Or Measurement If the Study Was Non-Interventional)	Outcome
Wu J. et al., 2011 [81]	Animal study	Murine model of AD	LXA4	Intracerebroventricular	Inhibiting the inflammatory response induced by β-amyloid in the cortex and hippocampus (in particular, production of IL-1b and TNFa)
Serhan CN., 2005 [82]	Animal study	Murine model of AD	ATL	Subcutaneous	↓ NF-kB activation and levels of proinflammatory cytokines and chemokines; creating an anti-inflammatory cerebral milieu, resulting in the recruitment of microglia in an alternative phenotype
Medeiros R. et al., 2013 [83]	Animal study	Murine model of AD	ATL	Subcutaneous	↓ phosphorylated-tau (p-tau)
Yin P. et al., 2019 [85]	Animal study	Murine model of AD	MaR1	Intracerebroventricular	Improving cognitive decline of experimental mice: attenuating microglial activation, ↓ the pro-inflammatory cytokines in favor of anti-inflammatory ones, and ↑ the levels of proteins related to survival pathway including PI3K/AKT, ERK; ↓ levels of proteins associated with inflammation, autophagy, and apoptosis pathways, such as p38, mTOR and caspase 3
Schröder N et al., 2020 [92]	Animal study	Murine model of AD	Ac2-26	Intraperitoneal injection	No beneficial effect
Park JC et al., 2017 [93]	In vitro and in vivo (animal study)	Aβ-42 treated murine brain endothelial cell line bEnd.3; Murine model of AD	Human recombinant ANXA1; ANXA1	Administration of human recombinant ANXA1 in Aβ-42 treated murine brain endothelial cell line bEnd.3; *ANXA1 levels on blood of murine model of AD*	rescuing β-amyloid 1–42 -induced BBB disruption via inhibition of RhoA-ROCK signaling pathway in brain endothelial cell line bEnd.3; ↓ ANXA1 in a murine model of AD

Table 3. *Cont.*

Reference	Type of Study	Model	Pro-Resolving Mediator	Delivery (Or Measurement If the Study Was Non-Interventional)	Outcome
Ries M. et al., 2021 [94]	Animal study	Murine model of AD	Human recombinant AnxA1	Intravenous injection	↓ β-amyloid load and p-tau build-up in 5xFAD mice and Tau-P301L mice; prolonged treatment reduced the memory deficits and increased synaptic density in young 5xFAD mice
Tian Y. et al., 2015 [101]	Animal study	Rat model of PD	RvD2	Intrathecal injection on substantia nigra pars compacta	recovering neural injury by suppressing inflammatory mediator expression
Krashia P. et al., 2019 [102]	Animal study	Rats overexpressing human α-synuclein (Syn)	RvD1	Chronic intraperitoneal injection	preventing central and peripheral inflammation, as well as neuronal dysfunction and motor deficits

AD: Alzheimer Disease; Ac2-26: annexin/lipocortin 1-mimetic peptide; ATL: Aspirin-triggered LXA4; LXA4: Lipoxin A4; LMs: Lipid Mediators; MaR1: Maresin1; MCI: Mild cognitive impairment; *N*-AS: *N*-acetyl sphingosine; n3-PUFAs: omega-3 polyunsaturated fatty acids; NPD1: Neuroprotectin D1; PD: Parkinson Disease; RvD1: Resolvin D1; RvD4: Resolvin D4; RvD5: Resolvin D5; RvE1: Resolvin E1; SCI: Subjective cognitive impairment.

4. Materials and Methods

The PubMed library was searched for journal articles published in English up to 4 May 2022; we used the entry words "pro-resolving mediators", "resolvins", "maresins", "annexins", "lipoxins", "protectins", "neuroinflammation", "central nervous system", "glial cells", and "Alzheimer", "Parkinson", "neurodegenerative", "stroke", "cerebrovascular". We purposefully focused on findings in human studies and on pre-clinical studies which have been implemented in the animal model, as pre-clinical animal studies more closely resemble possible future applications in clinical practice. In vitro studies have not been discussed in this work. However, a concise overview of the main in vitro models currently used in the field of neurological diseases and SPMs research has been provided in Table 4.

Table 4. Overview of the main in vitro models currently used in the field of neurological disease and SPMs research.

In Vitro Model	Brief Model Description
Cerebrovascular diseases	
OGD of rat cortical neurons	Primary cortical neurons are subjected to OGD mimicking ischemic injury.
OGD/R of BV2 murine microglial cell	BV2 murine microglial cells are subjected to OGD mimicking ischemic injury, with subsequent reoxygenation and exposure to a glucose-containing medium.
OGD/R of rat astrocytes	Primary astrocytes are exposed to OGD mimicking ischemic injury, with subsequent reoxygenation and exposure to a glucose-containing medium.
Immune-mediated demyelinating disorders	
Imiquimod and ssRNA40- stimulated PBMCs	Freshly isolated PBMCs are stimulated with Imiquimod (TLR 7 agonist) and ssRNA40 (TLR 8 agonist) to induce inflammatory changes.
Mixed glial cell model	This model is meant to study the expression of inflammatory mediators and myelin genes under inflammation; mixed glial cell cultures are treated with a combination of pro-inflammatory cytokines to create an inflammatory environment.
Co-culture studies	Mouse brain microglial cells are co-cultured with rat oligodendrocyte progenitor cells and then processed for expression of myelin genes. Co-culture systems allow studying the interactions between cell populations.
Neurodegenerative diseases	
AD models	
AD patients PBMCs	PBMCs drawn from the venous blood of AD patients.
Aβ40- or Aβ42-exposed PBMCs	PBMCs are isolated from patients' peripheral venous blood and incubated with Aβ40 or Aβ42 to mimic the AD environment.
Aβ42-exposed human CHME3 microglial cells	Human microglial cell line CHME3 has also been employed and incubated with Aβ42.
Aβ42-treated HNG	Co-cultures of neurons derived from mice and mouse brain mixed glial cells, subsequently stimulated by Aβ42. Co-culture of human cells has been employed too (primary human neuronal-glial co-culture).
Aβ-stimulated BV2 microglial cells	Mouse microglial cell line BV2 gets incubated with Aβ42.

Table 4. *Cont.*

In Vitro Model	Brief Model Description
HNG transfected with βAPPsw	HNG cells may either be challenged with Aβ42 oligomeric peptide as described above or transfected with beta amyloid precursor protein (βAPP)sw to mimic AD in vitro.
STS-induced apoptosis in neuroblastoma cells	Neuroblastoma cell line SH-SY5Y represents a model of human neuronlike cells. To study neuronal survival, apoptosis can be induced by incubating the differentiated SH-SY5Y cells with STS.
PD models	
LPS-induced murine microglial cells	Rat microglial cells incubated with the addition of LPS, which induces inflammatory changes.
MPP+ -treated PC12 pheochromocytoma cells	MPTP is an environmental toxin that specifically damages DA neurons; the same applies to its metabolite MPP+, explaining why they are commonly used to obtain in vitro PD models. In this case, PC12 rat pheochromocytoma cells are treated with MPP+.
ALS models	
Fibrillar wild type SOD-1-stimulated PBMCs	In ALS PBMCs, in vitro aggregated SOD-1 is used to stimulate the expression of inflammatory cytokines.
SOD1G93A or TDP-43A315T- transfected motor neuron-like NSC-34	NCS-34 cells are transfected with SOD1G93A plasmid or TDP-43A315T plasmid, which induces cell death thus providing a model of motor neuron degeneration.

Aβ40: Amyloid Beta 1–40; Aβ42: Amyloid Beta 1–42; AD: Alzheimer's Disease; ALS: Amyotrophic Lateral Sclerosis; $(\beta APP)_{sw}$: Swedish double mutation $APP695_{sw}$, K595N-M596L; DA: dopamine; HNG: Human Neuronal-Glial co-culture; LPS: Lipopolysaccharide; MPP+: 1-methyl-4-phenylpyridium; MPTP: 1-methyl-4-phenyl-1,2,3,6-tetrahydropyridine; NSC-34: Neuroblastoma spinal cord 34; OGD: oxygen-glucose deprivation; OGD/R: oxygen-glucose deprivation/reoxygenation; PBMCs: Peripheral Blood Mononuclear Cells; PD: Parkinson's disease; SOD-1: Superoxide Dismutase-1; STS: Staurosporine; TDP-43: TAR DNA-binding protein 43; TLR: Toll-like receptor.

5. Conclusions

Inflammation is a reaction to a harmful agent, physiologically self-contained thanks to the intervention of endogenous molecules which promote its resolution. If persistent or dysregulated, inflammation itself becomes noxious for human tissues.

Chronic, low-grade inflammation of the CNS is considered the pathophysiological foundation of many neurological disorders and of the neurodegenerative processes themselves.

In this picture, pro-resolution is a spontaneous collateral biochemical mechanism led by SPMs in the inflamed tissues and the identification of these molecules contributed to the understanding of the inflammation processes; their use in the clinical setting could potentially be an important tool for clinicians.

We have analyzed the bulk of the evidence in this article on the role of SPMs in the control of inflammation processes in several models of the most important neurological disorders. This amount of evidence has moved an interest among patients and physicians for the clinical use of SPMs.

Unfortunately, at the moment, this interest cannot be certainly defined as evidence-based.

A preliminary problem to be explored for future clinical studies relates to the route of administration of SPMs. In fact, regarding the administration of SPMs, there is at present no evidence on whether they can actually cross the BBB, although their characteristic of small lipophilic molecules makes this possibility plausible, similarly to what is known for their precursors DHA and EPA [74].

A second aspect concerns the definition of precise and clinically meaningful outcome measures, specific for the diseases to be treated.

Finally, little is known about the toxicology of SPMs, although no side effects have been reported so far. However, since research on SPMs is relatively new, an effort should be made to conduct future studies on safety, in order to rule out possible harmful effects. This is even more important if we consider that a few, not recent, studies on ω-3 PUFAs, which are SPMs precursors, have pointed out that they may impact platelet aggregation and reduce the immune response to infections [106–109].

Having in mind these considerations, we believe that the information coming from animal studies should prompt investigators and industry to fill the scientific gap with robust clinical studies on SPMs, which are tremendously needed.

It is likely that the clinical use of SPMs will not be as potent as that of anti-inflammatory drugs, but their action is likely more physiological, and it could probably be better tolerated by patients. In addition, their effects could be potentiated by the synergic action of other "natural" approaches to the control of chronic low-grade inflammation, such as those based on nutrition and lifestyle.

Author Contributions: Conceptualization, M.V. and M.D.; methodology, M.D.; validation, G.L.G. and M.V.; resources, M.V. and G.L.G.; writing—original draft preparation, M.D., F.K., F.B.; writing—review and editing, M.D., F.K., F.B., M.V. and G.L.G.; supervision, G.L.G. and M.V.; project administration: M.V. All authors have read and agreed to the published version of the manuscript.

Funding: This review received no external funding.

Informed Consent Statement: Not applicable.

Data Availability Statement: Not applicable.

Conflicts of Interest: The authors declare no conflict of interest.

References

1. Fredman, G.; Serhan, C.N. Specialized proresolving mediator targets for RvE1 and RvD1 in peripheral blood and mechanisms of resolution. *Biochem. J.* **2011**, *437*, 185–197. [CrossRef]
2. Chiang, N.; Serha, C.N. Specialized pro-resolving mediator network: An update on production and actions. *Essays Biochem.* **2020**, *64*, 443–462. [CrossRef]
3. Petasis, N.A.; Akritopoulou-Zanze, I.; Fokin, V.; Bernasconi, G.; Keledjian, R.; Yang, R.; Uddin, J.; Nagulapalli, K.C.; Serhan, C.N. Design, synthesis and bioactions of novel stable mimetics of lipoxins and aspirin-triggered lipoxins. *Prostaglandins Leukot. Essent. Fat. Acids* **2005**, *73*, 301–321. [CrossRef] [PubMed]
4. Clària, J.; Serhan, C.N. Aspirin triggers previously undescribed bioactive eicosanoids by human endothelial cell-leukocyte interactions. *Proc. Natl. Acad. Sci. USA* **1995**, *92*, 9475–9479. [CrossRef] [PubMed]
5. Perretti, M.; D'Acquisto, F. Annexin A1 and glucocorticoids as effectors of the resolution of inflammation. *Nat. Rev. Immunol.* **2009**, *9*, 62–70. [CrossRef] [PubMed]
6. Tourki, B.; Kain, V.; Shaikh, S.R.; Leroy, X.; Serhan, C.N.; Halade, G.V. Deficit of resolution receptor magnifies inflammatory leukocyte directed cardiorenal and endothelial dysfunction with signs of cardiomyopathy of obesity. *FASEB J.* **2020**, *34*, 10560–10573. [CrossRef]
7. Chiang, N.; Libreros, S.; Norris, P.C.; de la Rosa, X.; Serhan, C.N. Maresin 1 activates LGR6 receptor promoting phagocyte immunoresolvent functions. *J. Clin. Investig.* **2019**, *129*, 5294–5311. [CrossRef]
8. Bang, S.; Xie, Y.-K.; Zhang, Z.-J.; Wang, Z.; Xu, Z.-Z.; Ji, R.-R. GPR37 regulates macrophage phagocytosis and resolution of inflammatory pain. *J. Clin. Investig.* **2018**, *128*, 3568–3582. [CrossRef]
9. Flak, M.B.; Koenis, D.S.; Sobrino, A.; Smith, J.; Pistorius, K.; Palmas, F.; Dalli, J. GPR101 mediates the pro-resolving actions of RvD5n-3 DPA in arthritis and infections. *J. Clin. Investig.* **2020**, *130*, 359–373. [CrossRef]
10. Bannenberg, G.L.; Chiang, N.; Ariel, A.; Arita, M.; Tjonahen, E.; Gotlinger, K.H.; Hong, S.; Serhan, C.N. Molecular Circuits of Resolution: Formation and Actions of Resolvins and Protectins. *J. Immunol.* **2005**, *174*, 4345–4355. [CrossRef]
11. Kasuga, K.; Yang, R.; Porter, T.F.; Agrawal, N.; Petasis, N.A.; Irimia, D.; Toner, M.; Serhan, C.N. Rapid Appearance of Resolvin Precursors in Inflammatory Exudates: Novel Mechanisms in Resolution. *J. Immunol.* **2008**, *181*, 8677–8687. [CrossRef] [PubMed]
12. Avital, A.; Goshen, I.; Kamsler, A.; Segal, M.; Iverfeldt, K.; Richter-Levin, G.; Yirmiya, R. Impaired interleukin-1 signaling is associated with deficits in hippocampal memory processes and neural plasticity. *Hippocampus* **2003**, *13*, 826–834. [CrossRef]
13. Baruch, K.; Deczkowska, A.; David, E.; Castellano, J.M.; Miller, O.; Kertser, A.; Berkutzki, T.; Barnett-Itzhaki, Z.; Bezalel, D.; Wyss-Coray, T.; et al. Aging-induced type I interferon response at the choroid plexus negatively affects brain function. *Science* **2014**, *346*, 89–93. [CrossRef]
14. Deleidi, M.; Jäggle, M.; Rubino, G. Immune aging, dysmetabolism, and inflammation in neurological diseases. *Front. Neurosci.* **2015**, *9*, 172. [CrossRef]
15. Tiberi, M.; Chiurchiù, V. Specialized Pro-resolving Lipid Mediators and Glial Cells: Emerging Candidates for Brain Homeostasis and Repair. *Front. Cell. Neurosci.* **2021**, *15*, 673549. [CrossRef]
16. Shang, P.; Zhang, Y.; Ma, D.; Hao, Y.; Wang, X.; Xin, M.; Zhang, Y.; Zhu, M.; Feng, J. Inflammation resolution and specialized pro-resolving lipid mediators in CNS diseases. *Expert Opin. Ther. Targets* **2019**, *23*, 967–986. [CrossRef]

17. Joffre, C.; Rey, C.; Layé, S. N-3 polyunsaturated fatty acids and the resolution of neuroinflammation. *Front. Pharmacol.* **2019**, *10*, 1022. [CrossRef] [PubMed]
18. Rey, C.; Nadjar, A.; Joffre, F.; Amadieu, C.; Aubert, A.; Vaysse, C.; Pallet, V.; Layé, S.; Joffre, C. Maternal n-3 polyunsaturated fatty acid dietary supply modulates microglia lipid content in the offspring. *Prostaglandins Leukot. Essent. Fat. Acids* **2018**, *133*, 1–7. [CrossRef]
19. Svensson, C.I.; Zattoni, M.; Serhan, C.N. Lipoxins and aspirin-triggered lipoxin inhibit inflammatory pain processing. *J. Exp. Med.* **2007**, *204*, 245–252. [CrossRef]
20. Kantarci, A.; Aytan, N.; Palaska, I.; Stephens, D.; Crabtree, L.; Benincasa, C.; Jenkins, B.G.; Carreras, I.; Dedeoglu, A. Combined administration of resolvin E1 and lipoxin A4 resolves inflammation in a murine model of Alzheimer's disease. *Exp. Neurol.* **2018**, *300*, 111–120. [CrossRef]
21. Peferoen, L.; Kipp, M.; van der Valk, P.; van Noort, J.M.; Amor, S. Oligodendrocyte-microglia cross-talk in the central nervous system. *Immunology* **2014**, *141*, 302–313. [CrossRef] [PubMed]
22. Yang, H.J.; Vainshtein, A.; Maik-Rachline, G.; Peles, E. G protein-coupled receptor 37 is a negative regulator of oligodendrocyte differentiation and myelination. *Nat. Commun.* **2016**, *7*, 10884. [CrossRef] [PubMed]
23. Gutiérrez, I.L.; Novellino, F.; Caso, J.R.; García-Bueno, B.; Leza, J.C.; Madrigal, J.L.M. CCL2 Inhibition of Pro-Resolving Mediators Potentiates Neuroinflammation in Astrocytes. *Int. J. Mol. Sci.* **2022**, *23*, 3307. [CrossRef] [PubMed]
24. Jayaraj, R.L.; Azimullah, S.; Beiram, R.; Jalal, F.Y.; Rosenberg, G.A. Neuroinflammation: Friend and foe for ischemic stroke. *J. Neuroinflamm.* **2019**, *16*, 142. [CrossRef]
25. Tułowiecka, N.; Kotlęga, D.; Prowans, P.; Szczuko, M. The role of resolvins: EPA and DHA derivatives can be useful in the prevention and treatment of ischemic stroke. *Int. J. Mol. Sci.* **2020**, *21*, 7628. [CrossRef] [PubMed]
26. Zuo, G.; Zhang, D.; Mu, R.; Shen, H.; Li, X.; Wang, Z.; Li, H.; Chen, G. Resolvin D2 protects against cerebral ischemia/reperfusion injury in rats. *Mol. Brain* **2018**, *11*, 9. [CrossRef] [PubMed]
27. Dong, X.; Gao, J.; Zhang, C.Y.; Hayworth, C.; Frank, M.; Wang, Z. Neutrophil Membrane-Derived Nanovesicles Alleviate Inflammation To Protect Mouse Brain Injury from Ischemic Stroke. *ACS Nano* **2019**, *13*, 1272–1283. [CrossRef]
28. Mayurasakorn, K.; Niatsetskaya, Z.V.; Sosunov, S.A.; Williams, J.J.; Zirpoli, H.; Vlasakov, I.; Deckelbaum, R.J.; Ten, V.S. DHA but Not EPA Emulsions Preserve Neurological and Mitochondrial Function after Brain Hypoxia-Ischemia in Neonatal Mice. *PLoS ONE* **2016**, *11*, e0160870. [CrossRef]
29. Williams, J.J.; Mayurasakorn, K.; Vannucci, S.J.; Mastropietro, C.; Bazan, N.G.; Ten, V.S.; Deckelbaum, R.J. N-3 Fatty Acid Rich Triglyceride Emulsions Are Neuroprotective after Cerebral Hypoxic-Ischemic Injury in Neonatal Mice. *PLoS ONE* **2013**, *8*, e56233. [CrossRef]
30. Fredman, G.; Hellmann, J.; Proto, J.D.; Kuriakose, G.; Colas, R.A.; Dorweiler, B.; Connolly, E.S.; Solomon, R.; Jones, D.M.; Heyer, E.J.; et al. An imbalance between specialized pro-resolving lipid mediators and pro-inflammatory leukotrienes promotes instability of atherosclerotic plaques. *Nat. Commun.* **2016**, *7*, 12859. [CrossRef]
31. Kotlęga, D.; Peda, B.; Drozd, A.; Zembroń-Łacny, A.; Stachowska, E.; Gramacki, J.; Szczuko, M. Prostaglandin E2, 9S-, 13S-HODE and resolvin D1 are strongly associated with the post-stroke cognitive impairment. *Prostaglandins Other Lipid Mediat.* **2021**, *156*, 106576. [CrossRef]
32. Xian, W.; Wu, Y.; Xiong, W.; Li, L.; Li, T.; Pan, S.; Song, L.; Hu, L.; Pei, L.; Yao, S.; et al. The pro-resolving lipid mediator Maresin 1 protects against cerebral ischemia/reperfusion injury by attenuating the pro-inflammatory response. *Biochem. Biophys. Res. Commun.* **2016**, *472*, 175–181. [CrossRef] [PubMed]
33. Xian, W.; Li, T.; Li, L.; Hu, L.; Cao, J. Maresin 1 attenuates the inflammatory response and mitochondrial damage in mice with cerebral ischemia/reperfusion in a SIRT1-dependent manner. *Brain Res.* **2019**, *1711*, 83–90. [CrossRef] [PubMed]
34. Vital, S.A.; Senchenkova, E.Y.; Ansari, J.; Gavins, F.N.E. Targeting AnxA1/Formyl Peptide Receptor 2 Pathway Affords Protection against Pathological Thrombo-Inflammation. *Cells* **2020**, *9*, 2473. [CrossRef] [PubMed]
35. Gavins, F.N.E.; Dalli, J.; Flower, R.J.; Granger, D.N.; Perretti, M. Activation of the annexin 1 counter-regulatory circuit affords protection in the mouse brain microcirculation. *FASEB J.* **2007**, *21*, 1751–1758. [CrossRef]
36. Mantovani, A.; Sozzani, S.; Locati, M.; Allavena, P.; Sica, A. Macrophage Polarization: Tumor-Associated Macrophages as a Paradigm for Polarized M2 Mononuclear Phagocytes. *Trends Immunol.* **2002**, *23*, 549–555. [CrossRef]
37. Xu, X.; Gao, W.; Li, L.; Hao, J.; Yang, B.; Wang, T.; Li, L.; Bai, X.; Li, F.; Ren, H.; et al. Annexin A1 protects against cerebral ischemia–reperfusion injury by modulating microglia/macrophage polarization via FPR2/ALX-dependent AMPK-mTOR pathway. *J. Neuroinflamm.* **2021**, *18*, 119. [CrossRef] [PubMed]
38. Ding, Y.; Flores, J.; Klebe, D.; Li, P.; McBride, D.W.; Tang, J.; Zhang, J.H. Annexin A1 attenuates neuroinflammation through FPR2/p38/COX-2 pathway after intracerebral hemorrhage in male mice. *J. Neurosci. Res.* **2020**, *98*, 168–178. [CrossRef]
39. Senchenkova, E.Y.; Ansari, J.; Becker, F.; Vital, S.A.; Al-Yafeai, Z.; Sparkenbaugh, E.M.; Pawlinski, R.; Stokes, K.Y.; Carroll, J.L.; Dragoi, A.; et al. Novel Role for the AnxA1-Fpr2/ALX Signaling Axis as a Key Regulator of Platelet Function to Promote Resolution of Inflammation. *Circulation* **2019**, *140*, 319–335. [CrossRef] [PubMed]
40. Li, Q.Q.; Ding, D.H.; Wang, X.Y.; Sun, Y.Y.; Wu, J. Lipoxin A4 regulates microglial M1/M2 polarization after cerebral ischemia-reperfusion injury via the Notch signaling pathway. *Exp. Neurol.* **2021**, *339*, 113645. [CrossRef]

41. Wu, L.; Liu, Z.J.; Miao, S.; Zhou, L.B.; Cai, L.; Wu, P.; Ye, D.Y.; Wu, Q.; Li, H.H. Lipoxin A4 ameliorates cerebral ischaemia/reperfusion injury through upregulation of nuclear factor erythroid 2-related factor 2. *Neurol. Res.* **2013**, *35*, 968–975. [CrossRef] [PubMed]
42. Sobrado, M.; Pereira, M.P.; Ballesteros, I.; Hurtado, O.; Fernández-López, D.; Pradillo, J.M.; Caso, J.R.; Vivancos, J.; Nombela, F.; Serena, J.; et al. Synthesis of Lipoxin A4 by 5-Lipoxygenase Mediates PPAR -Dependent, Neuroprotective Effects of Rosiglitazone in Experimental Stroke. *J. Neurosci.* **2009**, *29*, 3875–3884. [CrossRef] [PubMed]
43. Hawkins, K.E.; DeMars, K.M.; Singh, J.; Yang, C.; Cho, H.S.; Frankowski, J.C.; Doré, S.; Candelario-Jalil, E. Neurovascular protection by post-ischemic intravenous injections of the lipoxin A 4 receptor agonist, BML-111, in a rat model of ischemic stroke. *J. Neurochem.* **2014**, *129*, 130–142. [CrossRef]
44. Hawkins, K.E.; DeMars, K.M.; Alexander, J.C.; de Leon, L.G.; Pacheco, S.C.; Graves, C.; Yang, C.; McCrea, A.O.; Frankowski, J.C.; Garrett, T.J.; et al. Targeting resolution of neuroinflammation after ischemic stroke with a lipoxin A4 analog: Protective mechanisms and long-term effects on neurological recovery. *Brain Behav.* **2017**, *7*, e00688. [CrossRef]
45. Wu, Y.; Ye, X.H.; Guo, P.P.; Xu, S.P.; Wang, J.; Yuan, S.Y.; Yao, S.L.; Shang, Y. Neuroprotective effect of lipoxin a4 methyl ester in a rat model of permanent focal cerebral ischemia. *J. Mol. Neurosci.* **2010**, *42*, 226–234. [CrossRef]
46. Ye, X.H.; Wu, Y.; Guo, P.P.; Wang, J.; Yuan, S.Y.; Shang, Y.; Yao, S.L. Lipoxin A4 analogue protects brain and reduces inflammation in a rat model of focal cerebral ischemia reperfusion. *Brain Res.* **2010**, *1323*, 174–183. [CrossRef]
47. Wu, Y.; Wang, Y.P.; Guo, P.; Ye, X.H.; Wang, J.; Yuan, S.Y.; Yao, S.L.; Shang, Y. A lipoxin A 4 analog ameliorates blood-brain barrier dysfunction and reduces MMP-9 expression in a rat model of focal cerebral ischemia-reperfusion injury. *J. Mol. Neurosci.* **2012**, *46*, 483–491. [CrossRef]
48. Jin, W.; Jia, Y.; Huang, L.; Wang, T.; Wang, H.; Dong, Y.; Zhang, H.; Fan, M.; Lv, P. Lipoxin A4 methyl ester ameliorates cognitive deficits induced by chronic cerebral hypoperfusion through activating ERK/Nrf2 signaling pathway in rats. *Pharmacol. Biochem. Behav.* **2014**, *124*, 145–152. [CrossRef]
49. Wang, X.; Miao, Z.; Xu, X.; Schultzberg, M.; Zhao, Y. Reduced Levels of Plasma Lipoxin A4 Are Associated with Post-Stroke Cognitive Impairment. *J. Alzheimer's Dis.* **2021**, *79*, 607–613. [CrossRef]
50. Guo, Z.; Hu, Q.; Xu, L.; Guo, Z.; Ou, Y.; He, Y.; Yin, C.; Sun, X.; Tang, J.; Zhang, J.H. Lipoxin A4 Reduces Inflammation Through Formyl Peptide Receptor 2/p38 MAPK Signaling Pathway in Subarachnoid Hemorrhage Rats. *Stroke* **2016**, *47*, 490–497. [CrossRef] [PubMed]
51. Liu, L.; Zhang, P.; Zhang, Z.; Hu, Q.; He, J.; Liu, H.; Zhao, J.; Liang, Y.; He, Z.; Li, X.; et al. LXA4 ameliorates cerebrovascular endothelial dysfunction by reducing acute inflammation after subarachnoid hemorrhage in rats. *Neuroscience* **2019**, *408*, 105–114. [CrossRef]
52. Miao, Z.; Schultzberg, M.; Wang, X.; Zhao, Y. Role of polyunsaturated fatty acids in ischemic stroke—A perspective of specialized pro-resolving mediators. *Clin. Nutr.* **2021**, *40*, 2974–2987. [CrossRef] [PubMed]
53. Yao, C.; Zhang, J.; Chen, F.; Lin, Y. Neuroprotectin D1 attenuates brain damage induced by transient middle cerebral artery occlusion in rats through TRPC6/CREB pathways. *Mol. Med. Rep.* **2013**, *8*, 543–550. [CrossRef] [PubMed]
54. Eady, T.N.; Belayev, L.; Khoutorova, L.; Atkins, K.D.; Zhang, C.; Bazan, N.G. Docosahexaenoic Acid Signaling Modulates Cell Survival in Experimental Ischemic Stroke Penumbra and Initiates Long-Term Repair in Young and Aged Rats. *PLoS ONE* **2012**, *7*, e46151. [CrossRef] [PubMed]
55. Belayev, L.; Mukherjee, P.K.; Balaszczuk, V.; Calandria, J.M.; Obenaus, A.; Khoutorova, L.; Hong, S.; Bazan, N.G. Neuroprotectin D1 upregulates Iduna expression and provides protection in cellular uncompensated oxidative stress and in experimental ischemic stroke. *Cell Death Differ.* **2017**, *24*, 1091–1099. [CrossRef] [PubMed]
56. Zirpoli, H.; Sosunov, S.A.; Niatsetskaya, Z.V.; Mayurasakorn, K.; Kollareth, D.J.M.; Serhan, C.N.; Ten, V.S.; Deckelbaum, R.J. NPD1 rapidly targets mitochondria-mediated apoptosis after acute injection protecting brain against ischemic injury. *Exp. Neurol.* **2021**, *335*, 113495. [CrossRef] [PubMed]
57. Belayev, L.; Hong, S.H.; Menghani, H.; Marcell, S.J.; Obenaus, A.; Freitas, R.S.; Khoutorova, L.; Balaszczuk, V.; Jun, B.; Oriá, R.B.; et al. Docosanoids Promote Neurogenesis and Angiogenesis, Blood-Brain Barrier Integrity, Penumbra Protection, and Neurobehavioral Recovery After Experimental Ischemic Stroke. *Mol. Neurobiol.* **2018**, *55*, 7090–7106. [CrossRef] [PubMed]
58. Bazan, N.G.; Eady, T.N.; Khoutorova, L.; Atkins, K.D.; Hong, S.; Lu, Y.; Zhang, C.; Jun, B.; Obenaus, A.; Fredman, G.; et al. Novel aspirin-triggered neuroprotectin D1 attenuates cerebral ischemic injury after experimental stroke. *Exp. Neurol.* **2012**, *236*, 122–130. [CrossRef] [PubMed]
59. Zahoor, I.; Giri, S. Specialized Pro-Resolving Lipid Mediators: Emerging Therapeutic Candidates for Multiple Sclerosis. *Clin. Rev. Allergy Immunol.* **2021**, *60*, 147–163. [CrossRef]
60. Paschalidis, N.; Iqbal, A.J.; Maione, F.; Wood, E.G.; Perretti, M.; Flower, R.J.; D'Acquisto, F. Modulation of experimental autoimmune encephalomyelitis by endogenous Annexin A1. *J. Neuroinflamm.* **2009**, *6*, 33. [CrossRef]
61. Colamatteo, A.; Maggioli, E.; Azevedo Loiola, R.; Sheikh, M.H.; Calì, G.; Bruzzese, D.; Maniscalco, G.T.; Centonze, D.; Buttari, F.; Lanzillo, R.; et al. Reduced Annexin A1 Expression Associates with Disease Severity and Inflammation in Multiple Sclerosis Patients. *J. Immunol.* **2019**, *203*, 1753–1765. [CrossRef] [PubMed]
62. Elderfield, A.J.; Newcombe, J.; Bolton, C.; Flower, R.J. Lipocortins (annexins) 1, 2, 4 and 5 are increased in the central nervous system in multiple sclerosis. *J. Neuroimmunol.* **1992**, *39*, 91–100. [CrossRef]

63. Elderfield, A.J.; Bolton, C.; Flower, R.J. Lipocortin 1 (annexin 1) immunoreactivity in the cervical spinal cord of Lewis rats with acute experimental allergic encephalomyelitis. *J. Neurol. Sci.* **1993**, *119*, 146–153. [CrossRef]
64. Probst-Cousin, S.; Kowolik, D.; Kuchelmeister, K.; Kayser, C.; Neundörfer, B.; Heuss, D. Expression of annexin-1 in multiple sclerosis plaques. *Neuropathol. Appl. Neurobiol.* **2002**, *28*, 292–300. [CrossRef]
65. Huitinga, I.; Bauer, J.; Strijbos, P.J.L.M.; Rothwell, N.J.; Dijkstra, C.D.; Tilders, F.J.H. Effect of annexin-1 on experimental autoimmune encephalomyelitis (EAE) in the rat. *Clin. Exp. Immunol.* **1998**, *111*, 198–204. [CrossRef]
66. Poisson, L.M.; Suhail, H.; Singh, J.; Datta, I.; Denic, A.; Labuzek, K.; Hoda, M.N.; Shankar, A.; Kumar, A.; Cerghet, M.; et al. Untargeted plasma metabolomics identifies endogenous metabolite with drug-like properties in chronic animal model of multiple sclerosis. *J. Biol. Chem.* **2015**, *290*, 30697–30712. [CrossRef]
67. Derada Troletti, C.; Enzmann, G.; Chiurchiù, V.; Kamermans, A.; Tietz, S.M.; Norris, P.C.; Jahromi, N.H.; Leuti, A.; van der Pol, S.M.A.; Schouten, M.; et al. Pro-resolving lipid mediator lipoxin A4 attenuates neuro-inflammation by modulating T cell responses and modifies the spinal cord lipidome. *Cell Rep.* **2021**, *35*, 109201. [CrossRef] [PubMed]
68. Sánchez-Fernández, A.; Zandee, S.; Mastrogiovanni, M.; Charabati, M.; Rubbo, H.; Prat, A.; López-Vales, R. Administration of Maresin-1 ameliorates the physiopathology of experimental autoimmune encephalomyelitis. *J. Neuroinflamm.* **2022**, *19*, 27. [CrossRef] [PubMed]
69. Prüss, H.; Rosche, B.; Sullivan, A.B.; Brommer, B.; Wengert, O.; Gronert, K.; Schwab, J.M. Proresolution Lipid Mediators in Multiple Sclerosis—Differential, Disease Severity-Dependent Synthesis—A Clinical Pilot Trial. *PLoS ONE* **2013**, *8*, e55895. [CrossRef] [PubMed]
70. Kooij, G.; Troletti, C.D.; Leuti, A.; Norris, P.C.; Riley, I.; Albanese, M.; Ruggieri, S.; Libreros, S.; van der Pol, S.M.A.; van het Hof, B.; et al. Specialized pro-resolving lipid mediators are differentially altered in peripheral blood of patients with multiple sclerosis and attenuate monocyte and blood-brain barrier dysfunction. *Haematologica* **2020**, *105*, 2056–2070. [CrossRef] [PubMed]
71. Wang, X.; Jiao, W.; Lin, M.; Chao, L.; Caiyun, L.; Ying, W.; Di, M.; Xiuzhe, W.; Ping, Y.; Jiachun, F.; et al. Resolution of inflammation in neuromyelitis optica spectrum disorders. *Mult. Scler. Relat. Disord.* **2019**, *27*, 34–41. [CrossRef] [PubMed]
72. Gold, R.; Oelschläger, M.; Pepinsky, R.B.; Sommer, C.; Hartung, H.P.; Toyka, K.V. Increased lipocortin-1 (annexin-1) expression in the sciatic nerve of Lewis rats with experimental autoimmune neuritis. *Acta Neuropathol.* **1999**, *98*, 583–589. [CrossRef]
73. Luo, B.; Han, F.; Xu, K.; Wang, J.; Liu, Z.; Shen, Z.; Li, J.; Liu, Y.; Jiang, M.; Zhang, Z.; et al. Resolvin D1 programs inflammation resolution by increasing TGF-β expression induced by dying cell clearance in experimental autoimmune neuritis. *J. Neurosci.* **2016**, *36*, 9590–9603. [CrossRef]
74. Zhu, M.; Wang, X.; Sun, L.; Schultzberg, M.; Hjorth, E. Can inflammation be resolved in Alzheimer's disease? *Ther. Adv. Neurol. Disord.* **2018**, *11*, 17–19. [CrossRef] [PubMed]
75. Do, K.V.; Hjorth, E.; Wang, Y.; Jun, B.; Kautzmann, M.A.I.; Ohshima, M.; Eriksdotter, M.; Schultzberg, M.; Bazan, N.G. Cerebrospinal Fluid Profile of Lipid Mediators in Alzheimer's Disease. *Cell. Mol. Neurobiol.* **2022**, 1–15. [CrossRef]
76. Zhu, M.; Wang, X.; Hjorth, E.; Colas, R.A.; Schroeder, L.; Granholm, A.; Serhan, C.N.; Schultzberg, M. Pro-Resolving Lipid Mediators Improve Neuronal Survival and Increase Aβ42 Phagocytosis. *Mol. Neurobiol.* **2016**, *53*, 2733–2749. [CrossRef]
77. Martinsen, A.; Tejera, N.; Vauzour, D.; Harden, G.; Dick, J.; Shinde, S.; Barden, A.; Mori, T.A.; Minihane, A.M. Altered SPMs and age-associated decrease in brain DHA in APOE4 female mice. *FASEB J.* **2019**, *33*, 10315–10326. [CrossRef] [PubMed]
78. Emre, C.; Hjorth, E.; Bharani, K.; Carroll, S.; Granholm, A.C.; Schultzberg, M. Receptors for pro-resolving mediators are increased in Alzheimer's disease brain. *Brain Pathol.* **2020**, *30*, 614–640. [CrossRef]
79. Emre, C.; Do, K.V.; Jun, B.; Hjorth, E.; Gómez Alcalde, S.; Kautzmann, M.I.; Gordon, W.C.; Nilsson, P.; Bazan, N.G.; Schultzberg, M. Age-related changes in brain phospholipids and bioactive lipids in the APP knock-in mouse model of Alzheimer's disease. *Acta Neuropathol. Commun.* **2021**, *9*, 116. [CrossRef]
80. Emre, C.; Arroyo-García, L.E.; Do, K.V.; Jun, B.; Ohshima, M.; Gómez Alcalde, S.; Cothern, M.L.; Maioli, S.; Nilsson, P.; Hjorth, E.; et al. Intranasal delivery of pro-resolving lipid mediators rescues memory and gamma oscillation impairment in App NL-G-F/NL-G-F mice. *Commun. Biol.* **2022**, *5*, 245. [CrossRef]
81. Wu, J.; Wang, A.; Min, Z.; Xiong, Y.; Yan, Q.; Zhanga, J.; Xu, J.; Zhang, S. Lipoxin A4 inhibits the production of proinflammatory cytokines induced by β-amyloid in vitro and in vivo. *Biochem. Biophys. Res. Commun.* **2011**, *408*, 382–387. [CrossRef] [PubMed]
82. Serhan, C.N. Lipoxins and aspirin-triggered 15-epi-lipoxins are the first lipid mediators of endogenous anti-inflammation and resolution. *Prostaglandins Leukot. Essent. Fat. Acids* **2005**, *73*, 141–162. [CrossRef]
83. Medeiros, R.; Kitazawa, M.; Passos, G.F.; Baglietto-Vargas, D.; Cheng, D.; Cribbs, D.H.; LaFerla, F.M. Aspirin-triggered lipoxin A4 stimulates alternative activation of microglia and reduces alzheimer disease-like pathology in mice. *Am. J. Pathol.* **2013**, *182*, 1780–1789. [CrossRef] [PubMed]
84. Dunn, H.C.; Ager, R.R.; Baglietto-Vargas, D.; Cheng, D.; Kitazawa, M.; Cribbs, D.H.; Medeiros, R. Restoration of Lipoxin A4 Signaling Reduces Alzheimer's Disease-Like Pathology in the 3xTg-AD Mouse Model. *J. Alzheimer's Dis.* **2015**, *43*, 893–903. [CrossRef] [PubMed]
85. Yin, P.; Wang, X.; Wang, S.; Wei, Y.; Feng, J.; Zhu, M. Maresin 1 Improves Cognitive Decline and Ameliorates Inflammation in a Mouse Model of Alzheimer's Disease. *Front. Cell. Neurosci.* **2019**, *13*, 466. [CrossRef]

86. Zhao, Y.; Calon, F.; Julien, C.; Winkler, J.W.; Petasis, N.A.; Lukiw, W.J. Docosahexaenoic acid-derived neuroprotectin D1 induces neuronal survival via secretase- and PPARγ-mediated mechanisms in Alzheimer's disease models. *PLoS ONE* **2011**, *6*, e15816. [CrossRef]
87. Stark, D.T.; Bazan, N.G. Neuroprotectin D1 induces neuronal survival and downregulation of amyloidogenic processing in Alzheimer's disease cellular models. *Mol. Neurobiol.* **2011**, *43*, 131–138. [CrossRef]
88. Dai, S.; Zhou, F.; Sun, J.; Li, Y. NPD1 Enhances Autophagy and Reduces Hyperphosphorylated Tau and Amyloid-β42by Inhibiting GSK3β Activation in N2a/APP695swe Cells. *J. Alzheimer's Dis.* **2021**, *84*, 869–881. [CrossRef]
89. Fraga, V.G.; Magalhães, C.A.; Loures, C.D.M.G.; Cruz de Souza, L.; Cerqueira Guimarães, H.; Alves Gomes Zauli, D.; das Graças Carvalho, M.; Ferreira, C.N.; Caramelli, P.; Pires de Sousaa, L.; et al. Inflammatory and Pro-resolving Mediators in Frontotemporal Dementia and Alzheimer's Disease. *Neuroscience* **2019**, *421*, 123–135. [CrossRef] [PubMed]
90. Chua, X.Y.; Chong, J.R.; Cheng, A.L.; Lee, J.H.; Ballard, C.; Aarsland, D.; Francis, P.T.; Lai, M.K.P. Elevation of inactive cleaved annexin A1 in the neocortex is associated with amyloid, inflammatory and apoptotic markers in neurodegenerative dementias. *Neurochem. Int.* **2022**, *152*, 105251. [CrossRef] [PubMed]
91. Ries, M.; Loiola, R.; Shah, U.N.; Gentleman, S.M.; Solito, E.; Sastre, M. The anti-inflammatory Annexin A1 induces the clearance and degradation of the amyloid-β peptide. *J. Neuroinflamm.* **2016**, *13*, 234. [CrossRef]
92. Schröder, N.; Schaffrath, A.; Welter, J.A.; Putzka, T.; Griep, A.; Ziegler, P.; Brandt, E.; Samer, S.; Heneka, M.T.; Kaddatz, H.; et al. Inhibition of formyl peptide receptors improves the outcome in a mouse model of Alzheimer disease. *J. Neuroinflamm.* **2020**, *17*, 131. [CrossRef] [PubMed]
93. Park, J.C.; Baik, S.H.; Han, S.H.; Cho, H.J.; Choi, H.; Kim, H.J.; Choi, H.; Lee, W.; Kim, D.K.; Mook-Jung, I. Annexin A1 restores Aβ1-42-induced blood–brain barrier disruption through the inhibition of RhoA-ROCK signaling pathway. *Aging Cell* **2017**, *16*, 149–161. [CrossRef] [PubMed]
94. Ries, M.; Watts, H.; Mota, B.C.; Yanez Lopez, M.; Donat, C.K.; Baxan, N.; Pickering, J.A.; Chau, T.W.; Semmler, A.; Gurung, B.; et al. Annexin A1 restores cerebrovascular integrity concomitant with reduced amyloid-β and tau pathology. *Brain* **2021**, *144*, 1526–1541. [CrossRef] [PubMed]
95. Hopperton, K.E.; Trépanier, M.O.; James, N.C.E.; Chouinard-Watkins, R.; Bazinet, R.P. Fish oil feeding attenuates neuroinflammatory gene expression without concomitant changes in brain eicosanoids and docosanoids in a mouse model of Alzheimer's disease. *Brain Behav. Immun.* **2018**, *69*, 74–90. [CrossRef]
96. Wang, X.; Hjorth, E.; Vedin, I.; Eriksdotter, M.; Freund-Levi, Y.; Wahlund, L.; Cederholm, T.; Palmblad, J.; Schultzberg, M. Effects of n-3 FA supplementation on the release of proresolving lipid mediators by blood mononuclear cells: The omegAD study. *J. Lipid Res.* **2015**, *56*, 674–681. [CrossRef]
97. Fiala, M.; Halder, R.C.; Sagong, B.; Ross, O.; Sayre, J.; Porter, V.; Bredesen, D.E. ω-3 supplementation increases amyloid-β phagocytosis and resolvin D1 in patients with minor cognitive impairment. *FASEB J.* **2015**, *29*, 2681–2689. [CrossRef] [PubMed]
98. Ayub, M.; Jin, H.K.; Bae, J.S. Sphingosine kinase-dependent regulation of pro-resolving lipid mediators in Alzheimer's disease. *Biochim. Biophys. Acta Mol. Cell Biol. Lipids* **2022**, *1867*, 159126. [CrossRef]
99. Lee, J.Y.; Han, S.H.; Park, M.H.; Song, I.; Choi, M.; Yu, E.; Park, C.; Kim, H.; Kim, S.H.; Schuchman, E.H.; et al. N-AS-triggered SPMs are direct regulators of microglia in a model of Alzheimer's disease. *Nat. Commun.* **2020**, *11*, 2358. [CrossRef] [PubMed]
100. Xu, J.; Gao, X.; Yang, C.; Chen, L.; Chen, Z. Resolvin D1 attenuates Mpp+-induced parkinson disease via inhibiting inflammation in PC12 cells. *Med. Sci. Monit.* **2017**, *23*, 2684–2691. [CrossRef]
101. Tian, Y.; Zhang, Y.; Zhang, R.; Qiao, S.; Fan, J. Resolvin D2 recovers neural injury by suppressing inflammatory mediators expression in lipopolysaccharide-induced Parkinson's disease rat model. *Biochem. Biophys. Res. Commun.* **2015**, *460*, 799–805. [CrossRef]
102. Krashia, P.; Cordella, A.; Nobili, A.; La Barbera, L.; Federici, M.; Leuti, A.; Campanelli, F.; Natale, G.; Marino, G.; Calabrese, V.; et al. Blunting neuroinflammation with resolvin D1 prevents early pathology in a rat model of Parkinson's disease. *Nat. Commun.* **2019**, *10*, 3945. [CrossRef] [PubMed]
103. Li, C.; Ou, R.; Gu, X.; Hou, Y.; Chen, Y.; Wei, Q.; Zhang, L.; Lin, J.; Liu, K.; Huang, J.; et al. ANXA1 and the risk for early-onset Parkinson's disease. *Neurobiol. Aging* **2022**, *112*, 212–214. [CrossRef] [PubMed]
104. Liu, G.; Fiala, M.; Mizwicki, M.T.; Sayre, J.; Magpantay, L.; Siani, A.; Mahanian, M.; Chattopadhyay, M.; La Cava, A.; Wiedau-Pazos, M. Neuronal phagocytosis by inflammatory macrophages in ALS spinal cord: Inhibition of inflammation by resolvin D1. *Am. J. Neurodegener. Dis.* **2012**, *1*, 60–74. [PubMed]
105. Ohuchi, K.; Ono, Y.; Joho, M.; Tsuruma, K.; Ogami, S.; Yamane, S.; Funato, M.; Kaneko, H.; Nakamura, S.; Hara, H.; et al. A Docosahexaenoic Acid-Derived Pro-resolving Agent, Maresin 1, Protects Motor Neuron Cells Death. *Neurochem. Res.* **2018**, *43*, 1413–1423. [CrossRef]
106. Bright, J.M.; Sullivan, P.S.; Melton, S.L.; Schneider, J.F.; McDonald, T.P. The effects of n-3 fatty acid supplementation on bleeding time, plasma fatty acid composition, and in vitro platelet aggregation in cats. *J. Vet. Intern. Med.* **1994**, *8*, 247–252. [CrossRef]
107. Kelley, D.S.; Taylor, P.C.; Nelson, G.J.; Mackey, B.E. Dietary docosahexaenoic acid and immunocompetence in young healthy men. *Lipids* **1998**, *33*, 559–566. [CrossRef]

108. Cooper, A.L.; Gibbons, L.; Horan, M.A.; Little, R.A.; Rothwell, N.J. Effect of dietary fish oil supplementation on fever and cytokine production in human volunteers. *Clin. Nutr.* **1993**, *12*, 321–328. [CrossRef]
109. Lee, T.H.; Hoover, R.L.; Williams, J.D.; Sperling, R.I.; Ravalese, J., 3rd; Spur, B.W.; Robinson, D.R.; Corey, E.J.; Lewis, R.A.; Austen, K.F. Effect of dietary enrichment with eicosapentaenoic and docosahexaenoic acids on in vitro neutrophil and monocyte leukotriene generation and neutrophil function. *N. Engl. J. Med.* **1985**, *312*, 1217–1224. [CrossRef] [PubMed]

molecules MDPI

Article

Bioactive Compounds in Rosehip (*Rosa canina*) Powder with Encapsulating Agents

Marta Igual [1], Patricia García-Herrera [2], Rosa M. Cámara [2], Javier Martínez-Monzó [1], Purificación García-Segovia [1,*] and Montaña Cámara [2,*]

1 Food Investigation and Innovation Group, Food Technology Department, Universitat Politècnica de València, Camino de Vera s/n, 46022 Valencia, Spain; marigra@upvnet.upv.es (M.I.); xmartine@tal.upv.es (J.M.-M.)
2 Departamento Nutrición y Ciencia de los Alimentos, Facultad de Farmacia, Universidad Complutense de Madrid (UCM), Pza. Ramón y Cajal, s/n, 28040 Madrid, Spain; patrigar@pdi.ucm.es (P.G.-H.); rosacama@ucm.es (R.M.C.)
* Correspondence: pugarse@tal.upv.es (P.G.-S.); mcamara@ucm.es (M.C.); Tel.: +34-963879694 (P.G.-S.); +34-913941802 (M.C.); Fax: +34-913941802 (M.C.)

Abstract: *Rosa canina* pseudo-fruits contain interesting bioactive compounds. This work aims to evaluate the use of different biopolymers as encapsulating agents on the content of organic acids, minerals, fibers, phenols, carotenoids, and the antioxidant activity of the powdered product. Fruits were ground and freeze-dried with or without biopolymers (maltodextrin, resistant maltodextrin, cyclodextrin, and pea protein). Rosehip formulated purees with encapsulating agents are an interesting food ingredient rich in fiber and minerals that could be used in the food industry in order to obtain different functional foods. Results obtained in this study show that all formulated samples are a good source of potassium, calcium, magnesium, and manganese. Both rosehip without biopolymers and rosehip with pea protein formulations are also a good source of Zn. Formulation with pea protein can be claimed as a good source of Fe. All formulations are food ingredients with a very high content of ascorbic acid. Comparing the encapsulating agents, depending on the studied bioactive compound samples behaved differently. In conclusion, it can be indicated that pea protein is recommended as an encapsulating agent since the rosehip with pea protein sample has the highest content of fiber, minerals, organic acids, and carotenoids among the encapsulating agents studied.

Keywords: rosehip; bioactive compounds; antioxidant capacity; encapsulation

Citation: Igual, M.; García-Herrera, P.; Cámara, R.M.; Martínez-Monzó, J.; García-Segovia, P.; Cámara, M. Bioactive Compounds in Rosehip (*Rosa canina*) Powder with Encapsulating Agents. *Molecules* **2022**, *27*, 4737. https://doi.org/10.3390/molecules27154737

Academic Editor: Smaoui Slim

Received: 1 July 2022
Accepted: 21 July 2022
Published: 25 July 2022

1. Introduction

Wild rose (*Rosa canina* L.) is a native shrub that belongs to the *Rosaceae* family and is widespread in northern Europe, Asia, the Middle East, and North America. For centuries, the pseudo-fruits of *Rosa canina* (rosehip) have been recognized as valuable food and medicine constituents due to their notable content of pro-health compounds [1]. The beneficial health effects are related to their rich content in flavonoids, carotenoids, fatty acids, and vitamins [2]. However, like other fruits, it is perishable, its production is seasonal, and its consumption is made difficult by the physiology of the pseudo-fruit. The pseudo-fruits, rosehips, are aggregate fruits comprising several achenes enclosed by an enlarged, red, fleshy floral cup (hypanthium) (Figure 1). The medium weight of the fruit is 2.8 to 2.9 g, distributed between the pericarp (65–70%) and the hard and hairy seeds (30–35%) [3,4].

The commercialization of dried fruit-based products can offer solutions to problems related to the short shelf life of fruits and their seasonality and a means of providing microbiologically stable products because of their low water activity. Furthermore, dried fruit both facilitates the shipping operations and makes them more profitable due to its lower volume and weight and easier handling. In the case of rosehips, drying [5] and specifically obtaining powder represents an alternative for its consumption due to the complexity of the use of the fruit, which is limited to infusions or jams. Obtaining a powdered product would encourage its consumption as rehydrated as a juice or infusion, or to be added to desserts, dairy products, salads, ice cream, snacks, among other things, and even for enriching almost any food in bioactive compounds [6–8].

Figure 1. Rosehips of *Rosa canina* and lengthwise section fruit.

Freeze-drying is one of the drying methods that provide the highest retention of chemical profile and antioxidant activity in foods, attributed to its less intense heating [9]. In fact, several studies have successfully obtained a vegetable/fruit snack with good physical, chemical, and functional properties by using freeze-drying [10–14]. Moreover, to improve the quality and stability of the freeze-dried fruit, the addition of high molecular weight additives to the product before drying as a carrier and anticaking agents is a widely used alternative to stabilize hygroscopic powders [15,16]. The use of biopolymers improves and maintains the characteristics of powder products but also allows the microencapsulation of bioactive compounds from the matrix. Maltodextrins are usually added during the production of food powders in order to act as encapsulating or wall materials, contributing to keeping the desired functional properties in the finished product [16]. However, there are other encapsulate agents with prebiotic effects as resistant maltodextrins [17] or proteins biopolymers from plants as pea protein [18] as an alternative. In addition, the typical use of cyclodextrins in cosmetics is being extended to food and, therefore, could also be a possibility to be explored [19].

Microencapsulation provides a physical barrier around the microencapsulated compounds, reducing the contact and reactivity of the encapsulated material with the environment [20]. Consequently, microencapsulation has been proved to be an excellent tool for the stabilization of bioactive compounds [21,22] and the inclusion of compounds in food matrices as food ingredients [23].

Depending on the final composition, formulated food ingredients rich in bioactive compounds could be of great interest for industry purposes both in functional foods formulation as well as to be used as an ingredient in food supplements [24].

In order to obtain a powdered rosehip product that facilitates its commercialization, this work aims to evaluate the use of different biopolymers as encapsulating agents on the content of organic acids, minerals, fiber, phenols, carotenoids, and the antioxidant activity of the powdered product.

2. Results and Discussion

2.1. General Parameters

Table 1 shows the values of Xw, Brix degrees, and pH of the purees obtained according to Figure 2 and as explained in Section 2.2. The water content of the powdered products obtained from each puree is also shown (Table 1). The addition of biopolymers to the rosehip puree reduced significantly ($p < 0.05$) the Xw of the mixtures. Furthermore, the addition of MD, RMD, and PP increased significantly ($p < 0.05$) the Brix degrees of the mixtures and, therefore, the soluble solids, as was observed by other works [17,25]. The pH increased significantly ($p < 0.05$) in the mixtures with CD and PP. The Xw of the powder products obtained showed significant differences ($p < 0.05$). The addition of any biopolymer reduced significantly ($p < 0.05$) the Xw of the powder, but this decrease was more drastic in the samples with PP. Probably PPR presented a greater facility for the water exit from the matrix. Generally, powder samples that come from purees with higher water and/or solutes content presented higher water content after freeze-drying. Water content is related to the drying efficiency, powder flowability, stickiness, and storage stability due to its effect on glass transition and crystallization behavior [26]. In this sense, R would be more susceptible to undesirable physical changes. The use of the biopolymers studied, especially PP, could protect against these changes. This fact was observed by other authors in orange pulp [27].

Table 1. Mean values (and standard deviations) of water content in g water/100 g (Xw puree), Brix degree and pH of puree, and water content in g water/100 g (Xw powder) of formulated rose hip purees and powders.

	Puree Samples			Freeze-Dried Samples
Sample	**Xw Puree**	**Brix Degrees**	**pH**	**Xw Powder**
R	88.00 (0.06) ^{a}	11.87 (0.20) ^{c}	3.76 (0.02) ^{c}	2.48 (0.01) ^{a}
MDR	80.08 (0.02) ^{c}	21.27 (0.22) ^{a}	3.78 (0.02) ^{c}	2.16 (0.08) ^{b}
RMDR	79.73 (0.03) ^{d}	21.17 (0.21) ^{a}	3.76 (0.02) ^{c}	2.15 (0.04) ^{b}
CDR	80.34 (0.12) ^{b}	12.00 (0.20) ^{c}	3.84 (0.02) ^{b}	1.80 (0.07) ^{c}
PPR	80.38 (0.26) ^{b}	16.10 (0.20) ^{b}	4.57 (0.02) ^{a}	0.86 (0.03) ^{d}

The same letter in superscript within column indicates homogeneous groups established by ANOVA ($p < 0.05$). R: rosehip; MDR: maltodextrin rosehip; RMDR: resistant maltodextrin rosehip; CDR: cyclodextrin rosehip; PPR: pea protein rosehip.

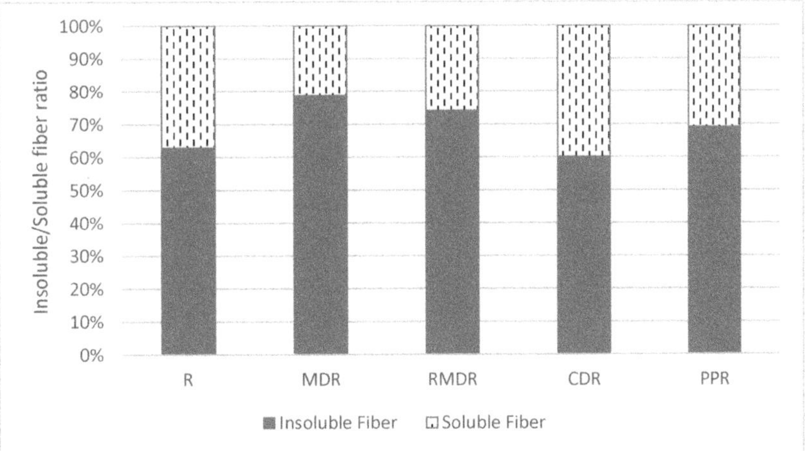

Figure 2. Insoluble/soluble fiber ratio in samples analyzed. R: rosehip; MDR: maltodextrin rosehip; RMDR: resistant maltodextrin rosehip; CDR: cyclodextrin rosehip; PPR: pea protein rosehip.

2.2. Insoluble, Soluble, and Total Dietary Fiber

Results from insoluble (IF), soluble (SF), and total dietary fiber (TF) are shown in Table 2. Rosehip powdered puree is characterized by its high IF content, accounting for more than 60% of the TF (Figure 2). Formulation of rosehip with starch derivates results in a decrease in TF as well in IF and SF fractions. Even with lower content, CDR formulation keeps the same ratio IF:SF as rosehip R.

Table 2. Mean values (and standard deviations) of insoluble, soluble, and total fiber content of rosehip formulated samples (g/100 g).

Sample	Insoluble Fiber	Soluble Fiber	Total Fiber
R	17.12 (0.26) ^{d}	10.11 (0.00) ^{e}	27.13 (0.09) ^{d}
MDR	12.61 (0.53) ^{c}	3.42 (0.34) ^{b}	16.03 (0.55) ^{b}
RMDR	8.03 (0.65) ^{a}	2.81 (0.06) ^{a}	10.35 (0.75) ^{a}
CDR	10.02 (0.46) ^{b}	6.68 (0.19) ^{c}	16.70 (0.57) ^{b}
PPR	17.86 (0.66) ^{d}	8.01 (0.06) ^{d}	25.87 (0.66) ^{c}

The same letter in superscript within column indicates homogeneous groups established by ANOVA ($p < 0.05$). R: rosehip; MDR: maltodextrin rosehip; RMDR: resistant maltodextrin rosehip; CDR: cyclodextrin rosehip; PPR: pea protein rosehip.

It is important to highlight the significant decrease in SF on MDR and RMDR samples in which the IF represents almost 80% of total fiber. In the case of the rosehip puree formulated with pea protein, the PPR sample, fiber content, and profile are more similar to the original R puree. That means that the most significant changes, a reduction in fiber content, occur in starch derivates formulated samples.

Fiber consumption in Western societies is insufficient, and its deficiency is directly linked to certain diseases. The required daily fiber intake can be obtained from foods such as fruits and vegetables, whole grains, legumes, nuts, and others, or by eating foods enriched with fiber as a functional ingredient [28]. Results from our study show that all the formulated samples analyzed can be considered valuable ingredients with to be used in the food industry for food products fiber enrichment as all formulations can be classified as "high fiber content" according to Regulation 1169/2011 and Regulation 1924/2006 [29,30].

2.3. Mineral Content

Respecting the mineral content, potassium is the main microelement, and manganese is the main microelement found in all formulated samples, as expected for vegetable samples (Table 3). It is important to note that Na was not quantified in any of the analyzed samples (Na limit of detection (LOD) = 0.394 ppm and quantification (LOQ) = 1.314 ppm). The rosehip sample (R) showed higher macro and microelements content than other rosehip formulations with starch derivates as encapsulating agents. Iron content in R is lower (0.24 mg/100 g) than the content reported by other authors [31–33], who found 5.69, 1.22, and 2.34 mg/100 g, respectively. Manganese values for the rosehip sample are, according to other authors [32,34] although higher than values (1.46–3.20 mg/100 g, respectively) reported by [33,35]. Within the encapsulating agents studied, the PPR sample presented a higher potassium and magnesium content than the rest of the formulations due to its plant origin and pea minerals contribution. Potassium content in all samples is higher than expected, as values reported by [34,35], who studied the potassium content in rose canine, were in the range of 914–944 mg/100 g. Looking for the relations between minerals, the calcium/magnesium ratio was favorable to calcium in all samples studied, being the lower calcium content for PPR. Ratio Fe/Cu always is higher than 1 in all samples. Further, all samples presented a ratio of Mn/Zn favorable for Mn as it would be expected for wild fruits [36].

According to labeling regulation 1924/2006, it can be said that a food is a "source of [mineral]" or "high content of [mineral]" when it covers 15% and 30%, respectively, of the nutrient reference values. All formulated samples could be marketed as a good source of potassium, calcium, magnesium, and manganese. Both R and PPR formulations are also good sources of Zn, and PPR is the only one that can be claimed as a good source of Fe (6.05 mg/100 g).

Table 3. Mean values (and standard deviations) of main macro and microelements (mg/100 g) in formulated rosehip samples.

Macroelements				
Sample	**Na**	**K**	**Ca**	**Mg**
R	nd	1733.5 (19.1) ^{d}	534.5 (15.8) ^{b}	257.0 (3.2) ^{c}
MDR	nd	1167.8 (8.0) ^{a}	331.0 (10.1) ^{a}	184.7 (1.6) ^{a}
RMDR	nd	1307.7 (72.3) ^{bc}	329.2 (2.8) ^{a}	183.5 (2.0) ^{a}
CDR	nd	1241.9 (69.8) ^{ab}	317.2 (8.8) ^{a}	187.9 (2.9) ^{a}
PPR	nd	1356.8 (11.2) ^{c}	316.7 (1.7) ^{a}	217.2 (6.1) ^{b}
Microelements				
Sample	**Mn**	**Cu**	**Fe**	**Zn**
R	7.39 (0.36) ^{c}	0.74 (0.00) ^{b}	0.24 (0.02) ^{a}	2.16 (0.12) ^{c}
MDR	4.37 (0.64) ^{a}	0.59 (0.07) ^{ab}	1.78 (0.27) ^{c}	0.54 (0.03) ^{a}
RMDR	5.33 (0.16) ^{b}	0.39 (0.02) ^{a}	1.02 (0.03) ^{b}	1.43 (0.09) ^{b}
CDR	4.96 (0.07) ^{ab}	0.51 (0.01) ^{ab}	0.62 (0.02) ^{ab}	1.23 (0.14) ^{b}
PPR	5.04 (0.42) ^{ab}	1.52 (0.17) ^{c}	6.05 (0.43) ^{d}	3.68 (0.26) ^{d}

The same letter in superscript within column indicates homogeneous groups established by ANOVA ($p < 0.05$). R: rosehip; MDR: maltodextrin rosehip; RMDR: resistant maltodextrin rosehip; CDR: cyclodextrin rosehip; PPR: pea protein rosehip.

2.4. *Bioactive Compounds and Antioxidant Capacity*

Results from bioactive compounds analyzed in formulated samples, such as organic acids, total polyphenols, and carotenoids, are shown in Tables 4 and 5.

Table 4. Mean values (and standard deviations) of main organic acids (g/100 g).

Sample	Quinic	Malic	Ascorbic	Citric	Fumaric
R	0.68 (0.01) ^{d}	1.55 (0.07) ^{d}	0.44 (0.02) ^{d}	6.05 (0.24) ^{c}	0.175 (0.001) ^{b}
MDR	0.50 (0.00) ^{b}	1.07 (0.01) ^{ab}	0.25 (0.02) ^{a}	4.08 (0.06) ^{ab}	0.171 (0.008) ^{ab}
RMDR	0.49 (0.01) ^{b}	1.10 (0.04) ^{b}	0.37 (0.02) ^{c}	4.08 (0.13) ^{ab}	0.172 (0.003) ^{ab}
CDR	0.45 (0.04) ^{a}	0.98 (0.07) ^{a}	0.37 (0.02) ^{c}	3.74 (0.41) ^{a}	0.166 (0.003) ^{ab}
PPR	0.58 (0.02) ^{c}	1.26 (0.05) ^{c}	0.31 (0.03) ^{b}	4.24 (0.08) ^{b}	0.165 (0.003) ^{a}

The same letter in the superscript within column indicates homogeneous groups established by ANOVA ($p < 0.05$). R: rosehip; MDR: maltodextrin rosehip; RMDR: resistant maltodextrin rosehip; CDR: cyclodextrin rosehip; PPR: pea protein rosehip.

Table 5. Mean values (and standard deviations) of total carotenoids (TC), total phenols (TP) content, and antioxidant capacity (AC).

Sample	TC ($mg_{\beta\text{-carotene}}$/100 g)	TP (mg_{GAE}/100 g)	AC (mg_{TE}/100 g)
R	74.4 (0.2) ^{a}	2482 (8) ^{a}	1793 (9) ^{a}
MDR	24.43 (0.03) ^{d}	1275 (4) ^{b}	955 (7) ^{b}
RMDR	24.32 (0.12) ^{d}	1220 (4) ^{c}	928 (4) ^{c}
CDR	28.4 (0.2) ^{c}	628 (8) ^{e}	607 (13) ^{e}
PPR	45.9 (0.2) ^{b}	799 (4) ^{d}	712 (10) ^{d}

The same letter in superscript within column indicates homogeneous groups established by ANOVA ($p < 0.05$). R: rosehip; MDR: maltodextrin rosehip; RMDR: resistant maltodextrin rosehip; CDR: cyclodextrin rosehip; PPR: pea protein rosehip.

Regarding the organic acids, Figure 3 and Table 4 show the organic acids profile and content on analyzed samples. As expected, the rosehip R sample is the one with a higher organic acid content, being citric acid the predominant. This disagrees with other authors who found malic or ascorbic acid as the majority organic acid for rosehip [37,38], but it agrees with [39], who reported malic and citric acid as the main organic acids in fruits of rosehip fruits. Nevertheless, the ascorbic acid content in our samples is inside the range (0.2–0.85 g/100 g) reported by [40].

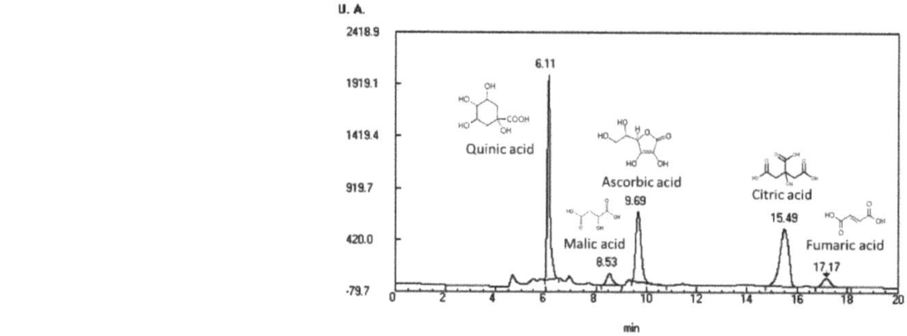

Figure 3. –HPLC-UV organic acid profile of rosehip sample.

Rosehip fruit is characterized by its high ascorbic acid content. In our study, the ascorbic acid decrease could be due to the obtaining process of the powder by the high temperatures or light exposition. With the independence of that, all formulated samples analyzed can be considered as very valuable food ingredients and claimed as "high ascorbic acid content", according to Regulation 1169/2011; thus, these ingredients could be considered as and be of great interest for the food industry. As expected, PPR organic acid profile is different from starch derivate formulations due to the pea protein contribution to the organic acid content of the final formulation.

Table 5 shows the total content of carotenoids and phenols mean values and standard deviations in brackets, as well as the antioxidant capacity of the powdered products. It can be observed that the bioactive compounds content studied or antioxidant capacity is higher in R compared to the rest samples. This fact is due to all R content being rosehips and providing these bioactive compounds and the rest of the samples presenting 48% of biopolymers in their composition.

Depending on the studied bioactive compound, samples behaved differently. In the case of carotenoids, the use of PP allowed obtaining a rosehip powder product with a higher content of them; however, the content of phenolic compounds is higher in the rosehip sample encapsulated with MD. This behavior probably will be related to the affinity of encapsulation agents with the bioactive compounds. Maltodextrins are usually employed to encapsulate hydrosoluble compounds such as ascorbic acid, phenolic compounds, or fruit juices, and proteins are used to encapsulate liposoluble substances such as lycopene or polyunsaturated fatty acids [41]. In the case of CDR, the TP value was significantly ($p < 0.05$) the lowest, mainly due to cyclodextrin conformation. They are shaped like a conical toroid with the secondary OH groups (corresponding to carbons 2 and 3 of glucose) on the widest face, the primary OH groups (corresponding to carbon 6) on the opposite face, and the Hs oriented inward of the cavity. Therefore, they have a hydrophobic cavity and a hydrophilic outer part. This fact makes them capable of forming inclusion complexes with essentially apolar molecules and of a suitable size for the host–guest interaction [19]. In this way, the cyclodextrin would be easier to encapsulate and protect TC than TP. As can be seen in Table 5, the TC values for CDR are the next highest, after PPR, in the samples containing biopolymers. AC values showed the same trend as TP values; in addition, the Pearson coefficient between AC-TP was 0.9973 ($p < 0.05$); therefore, they are highly correlated, and TP presents an important role in the AC of studied products.

Figure 4 shows TP and AC EE % in rosehip freeze-dried samples. EE refers to the potential of the wall material to encapsulate or hold the core material inside the microcapsule [42]. EE is also related to the shelf life of the phenolic compounds content and AC in the powder. In the first place, the practically null EE% values of R (TP and AC) stand out since it does not contain biopolymers in its composition, and there is no encapsulation. Low EE% CDR values can also be observed, reaching only 14% for TP and 5% for AC. As indicated above, the conformation of the cyclodextrin generates a cavity in which to stay the molecules to be encapsulated, but these molecules must be hydrophobic to be linked and carry out the host–guest interaction [19]. Figure 5 evaluates the EE% for TP that are hydrophilic compounds. MDR, RMDR, and PPR presented EE for TP from 50% to 60% and for AC from 45% to 55%, values similar to those found in references for other vegetable products [18,43]. The sample with the highest EE % values was MDR.

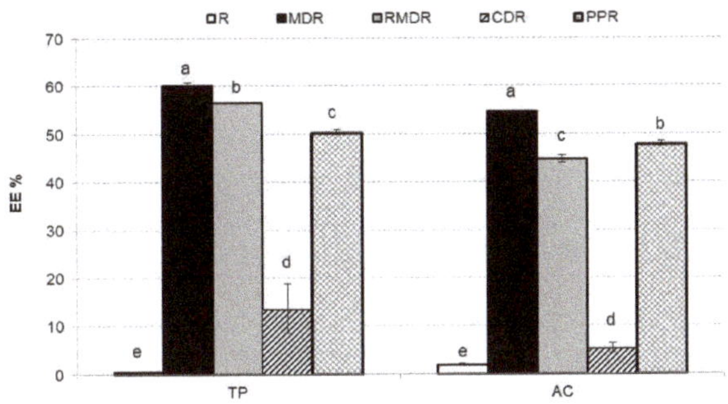

Figure 4. Mean values and standard deviation of encapsulation efficiencies percentage in rosehip freeze-dried formulated samples for total phenols and antioxidant capacity. Letters indicate homogeneous groups established by the ANOVA ($p < 0.05$) for total phenols and antioxidant capacity. R: rosehip; MDR: maltodextrin rosehip; RMDR: resistant maltodextrin rosehip; CDR: cyclodextrin rosehip; PPR: pea protein rosehip.

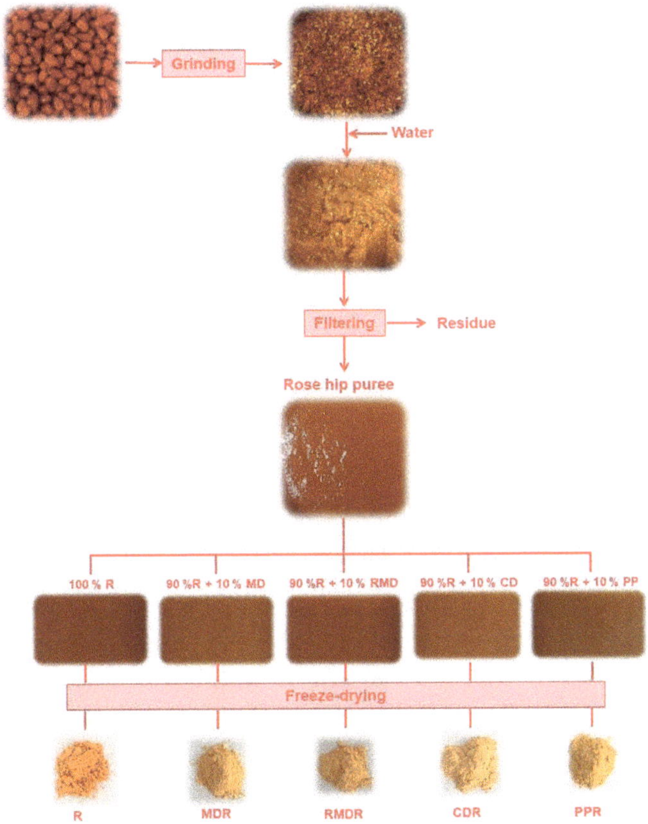

Figure 5. Scheme of rosehip powders obtaining.

3. Materials and Methods

3.1. *Raw Materials*

Rosehip (*R. canina*) fruits were manually harvested in Aldehuela (Teruel, Spain) in October 2020. Maltodextrin (GLUCIDEX® 12) (MD), pea protein powder (Nutralys® S85F) (PP), and beta cyclodextrin (KLEPTOSE®) (CD) were supplied by Roquette S.L. (Valencia, Spain). Resistant maltodextrin (Fibersol-2®) (RMD) was purchased from ADM/Matsutani, LLC (Decatur, IL, USA).

3.2. *Sample Preparation*

Figure 5 shows a scheme of powder processing. Rosehips (1000 g) were washed with distilled water and homogenized within a Thermomix (TM 21, Vorwerk, Valencia, Spain) for 1 min at 5200 rpm. Then, distilled water (1000 g) was added and newly re-homogenized for 5 min at 5200 rpm. The mixture was filtered using a sieve (light of mesh diameter 1 mm, Cisa 029077, 1 series). Four different formulations were prepared by adding 10 g of MD, RMD, CD, or PP to 90 g of the filtered mixture. Moreover, a control sample (R) without biopolymers was prepared. Five purees were as follows: R (100% rosehip), MDR (10% maltodextrin: 90% rosehip), RMDR (10% resistant maltodextrin: 90% rosehip), CDR (10% cyclodextrin: 90% rosehip), and PPR (10% pea protein: 90% rosehip). Then, the control rosehip and formulated purees were freeze-dried. A puree layer (0.5 cm thickness) was placed in a standardized aluminum plate (15 cm diameter and 5 cm height). Consecutively, samples were stored at −45 °C (Vertical Freezer, CVF450/45, Ing. Climas, Barcelona, Spain) for 24 h before being dried in a Lioalfa-6 Lyophyliser (Telstar, Spain) at 2600 Pa and −56.6 °C for 48 h. The freeze-dried samples were ground in a grinder (Minimoka, Taurus, Lleida, Spain) to obtain a free-flowing powder.

3.3. *Analytical Determinations*

3.3.1. Water Content, Degree Brix, and pH

Water content (Xw, grams water per 100 g of product), degree Brix (grams soluble solids per 100 g liquid phase), and pH were determined for the control and formulated purees. Water content was also determined in powder samples. The Xw was determined by drying the sample to a constant weight at 70 °C in a vacuum oven [44]. Degree Brix was measured in previously homogenized samples with a refractometer at 20 °C (Abbemat 200, Anton Paar, Austria). pH determination of purees was made using a Basic 20 pH meter (Crison, Spain). All determinations were performed in triplicate.

3.3.2. Fiber

Total, soluble, and insoluble fiber were determined by AOAC 991.43 enzymatic gravimetric method [45]. The first samples were under enzymatic digestion with α-amylase, protease, and amyloglucosidase (Sigma-Aldrich, St. Louis, MO, USA) in order to eliminate the protein and starch present in the samples. To obtain the insoluble fiber, we proceeded to the filtrate of the liquid obtained on crucibles with a Gooch Pyrex filter plate, and later, they were dried in an oven at 100 °C and then weighed. The reserved liquid of the insoluble fiber was stored in a 500 mL flask with the addition of 400 mL of 96% *v*/*v* ethanol and precipitated from one day to the next. Next, it was filtered in new Gooch Pyrex crucibles with the same insoluble fiber conditions. In both cases, the content of protein and ash was determined, and the content corresponding to insoluble and soluble fiber was calculated, respectively.

3.3.3. Ash Content and Mineral Composition

Method 930.05 of AOAC for ash determination was used [45]. Incineration was performed in an oven (Muffle MR 170, W.C. Heraeus Hanau, Hanau, Germany) for 24 h at 550 C, and ashes were gravimetrically quantified. The residue was extracted with HCl (50% *v*/*v*) and HNO_3 (50% *v*/*v*) to measure Fe, Cu, Mn, and Zn and was directly quantified. To avoid interferences between different elements, a dilution with 1.16 La_2O_3/100 HCl (resulting $LaCl_2$) was performed to analyze Ca and Mg and with CsCl (0.2 g/100 g solution) to analyze Na and K. All measurements were performed in atomic absorption spectroscopy (AAS) in Analyst 200 Perkin Elmer equipment (Perkin Elmer, Waltham, MA, USA), comparing absorbance responses with analytical standard solutions for AAS.

3.3.4. Organic Acids

Organic acids were determined based on protocols described by Sánchez-Mata et al. [46]. Extraction was performed with 0.5 g of sample in 25 mL of 3% m-phosphoric acid and analyzed using an HPLC-UV methodology. The HPLC equipment used was a liquid chromatograph (Micron Analítica, Madrid, Spain) equipped with a Sphereclone ODS (2) 250 * 4.60 mm, 5 μm Phenomenex column, isocratic pump (model PU-II), an AS-1555 automatic injector (Jasco, Tokyo, Japan), and a UV-visible detector (Thermo Separation Spectra Series UV100, Waltham, MO, USA), 215 nm for organic acids. The mobile phase was 1.8 mM H_2SO_4 (pH = 2.6), with a flow rate of 0.4 mL/min for organic acids, and injection volume was 100 μL for samples and serial volumes for the standard curve (20, 30, 40, 50, 60, 70, 80, 90, and 100 μL). The compounds were identified by chromatographic comparisons with authentic standards (quinic (0.152 mg/mL), ascorbic (0.155 mg/mL), malic (0.403 mg/mL), fumaric (0.254 mg/mL) and citric acids (0.307 mg/mL), all from Sigma, St. Louis, MO, USA), using linear calibration curves of all compounds for quantification purposes. All data were analyzed using Biocrom 2000 3.0 software (Biocrom, Madrid, Spain).

3.3.5. Total Phenols (TP)

Determining TP was based on the Folin–Ciocalteu method. Briefly, 1 g of sample was mixed with 5 mL methanol, 0.5 mL HCl 5N, NaF 2 mM and centrifugated at 12,857× *g*, 4 °C, 10 min using an Eppendorf centrifuge (Eppendorf, Hamburg, Germany). From the supernatant, 250 μL were mixed in a 25 mL volumetric flask with 1.25 mL Folin–Ciocalteu reagent and stored in a darker place for 8 min. Afterward, 3.75 mL Na_2CO_3 with a concentration of 7.5% was added and further stored for 120 min [47]. Absorbance was measured at 765 mm in a UV-3100PC spectrophotometer (VWR, Leuven, Belgium). The total phenolic content was expressed as mg of gallic acid (Sigma-Aldrich, Steinheim, Germany) equivalents (GAE) per 100 g.

3.3.6. Total Carotenoids (TC)

The TC in the samples (1 g) was extracted with a solvent hexane/acetone/ethanol mixture (50:25:25, *v*/*v*/*v*) for 30 min following the Olives et al. [48] method in triplicate. The absorbance of the hexane layer of the sample extracts was measured at 446 nm in a UV-visible spectrophotometer (Thermo Electron Corporation). The TC content was expressed as mg of β-carotene (Fluka-Biochemika, Buchs, Switzerland) per 100 g.

3.3.7. Antioxidant Capacity (AC)

AC was assessed using the free radical scavenging activity of the samples evaluated with the stable radical 2,2-diphenyl-1-picryl-hydrazyl-hydrate (DPPH) following Igual et al. [14] methodology in triplicate. Samples were mixed with methanol. The homogenate was centrifuged (12,857× *g*, 10 min, 4 °C) to obtain the supernatant. A total of 0.1 mL of supernatant was added to 3.9 mL of DPPH (0.030 g/L, Sigma-Aldrich, Steinheim, Germany) in methanol. A UV-visible spectrophotometer (Thermo Electron Corporation) was used at the absorbance at 515 nm. The results were expressed as milligram Trolox equivalents (TE) per 100 g.

3.3.8. Encapsulation Efficiencies (EE)

To evaluate the EE, analyzed total phenols or antioxidant capacity in each case (TP or AC) content, represented as TB, and surface analyzed bioactive compounds (SB) content of the samples were determined after freeze-drying [18]. For TB determination, samples were treated according to TP or AC. For SB determination, samples were not ground to destroy microcapsules. Only samples were extracted with the solvents in a vortex for 30 s and filtered through a 0.45 µm-size filter following the procedure of [42]. The % EE was calculated using the following Equation (1), where EE refers to encapsulation efficiencies, TB refers to total phenols or antioxidant capacity content, and SB to surface analyzed bioactive compounds.

$$\%\ \text{EE} = \frac{(\text{TB} - \text{SB})}{\text{TB}} \times 100 \quad (1)$$

3.4. Statistical Analysis

Analysis of variance (ANOVA) was applied with a confidence level of 95% ($p < 0.05$) to evaluate the differences among samples. Statgraphics Centurion XVII Software, version 17.2.04 (Statgraphics Technologies, Inc., The Plains, VA, USA) was used. The method used to discriminate between means is Fisher's least significant difference procedure.

4. Conclusions

Rosehip formulated purees with encapsulating agents are an interesting food ingredient rich in fiber and minerals that could be used in the food industry in order to obtain different functional foods. All formulated samples could be marketed as good sources of potassium, calcium, magnesium, and manganese. Both R and PPR formulations are also good sources of Zn, and PPR is the only one that can be claimed as a good source of Fe. In addition, all formulations can be considered food ingredients with a very high content of ascorbic acid. Comparing the encapsulating agents, depending on the studied bioactive compound samples behaved differently. In the case of carotenoids, the use of PP allowed obtaining a rosehip powder product with a higher content of them; however, the content of phenolic compounds is higher in the rosehip sample encapsulated with MD. In general, we can conclude that pea protein is recommended as an encapsulating agent since the PPR formulations have the highest content of bioactive compounds: fiber, minerals, organic acids, and carotenoids, among the encapsulating agents studied.

Author Contributions: Conceptualization, M.I., J.M.-M. and P.G.-S.; methodology, M.I., P.G.-H., R.M.C., J.M.-M., P.G.-S. and M.C.; writing—original draft preparation, M.I., P.G.-H. and M.C.; writing—review and editing, M.I., P.G.-H., R.M.C., J.M.-M., P.G.-S. and M.C.; supervision, J.M.-M., P.G.-S. and M.C.; funding acquisition, J.M.-M., P.G.-S. and M.C. All authors have read and agreed to the published version of the manuscript.

Funding: This research was funded by UPV-CUINA, ref: 20170573 ("Alimentación, Salud y Gartronomía" research line) and UCM-ALIMNOVA Research Group, ref: 951505 (grants: GR29-20 and GRFN 19-21).

Institutional Review Board Statement: Not applicable.

Informed Consent Statement: Not applicable.

Data Availability Statement: Not applicable.

Conflicts of Interest: The authors declare no conflict of interest.

Sample Availability: Samples of the compounds are not available from the authors.

Abbreviations

R	Rose hips
MD	Maltodextin
RMD	Resistant maltodextin
CD	Ciclodextrin
PP	Pea protein powder
MDR	90% MD + 10% R
RMDR	90% RMP + 10% R
CDR	90% CD + 10% R
PPR	90% PP + 10% R
TP	Total phenols
GAE	Gallic acid equivalent
TC	Total carotenoids
AC	Antioxidant capacity
DPPH	2,2-diphenyl-1-picryl-hydrazyl-hydrate
TE	Trolox equivalent
EE	Encapsulation efficiencies
TB	Total phenols or Antioxidant Capacity content
SB	Surface analyzed bioactive compounds

References

1. Roman, I.; Stănilă, A.; Stănilă, S. Bioactive compounds and antioxidant activity of *Rosa canina* L. biotypes from spontaneous flora of Transylvania. *Chem. Cent. J.* **2013**, *7*, 73. [CrossRef] [PubMed]
2. Mármol, I.; Sánchez-De-Diego, C.; Jiménez-Moreno, N.; Ancín-Azpilicueta, C.; Rodríguez-Yoldi, M. Therapeutic applications of rose hips from different *Rosa* species. *Int. J. Mol. Sci.* **2017**, *18*, 1137. [CrossRef] [PubMed]
3. Patel, S. Rose hip as an underutilized functional food: Evidence-based review. *Trends Food Sci. Technol.* **2017**, *63*, 29–38. [CrossRef]
4. Dashbaldan, S.; Rogowska, A.; Pączkowski, C.; Szakiel, A. Distribution of triterpenoids and steroids in developing rugosa rose (*Rosa rugosa* thunb.) accessory fruit. *Molecules* **2021**, *26*, 5158. [CrossRef]
5. Moldovan, C.; Babotă, M.; Mocan, A.; Menghini, L.; Cesa, S.; Gavan, A.; Barros, L. Optimization of the drying process of autumn fruits rich in antioxidants: A study focusing on rosehip (*Rosa canina* L.) and sea buckthorn (*Elaeagnus rhamnoides* (L.) A. Nelson) and their bioactive properties. *Food Funct.* **2021**, *12*, 3939–3953. [CrossRef]
6. Hui, Y.H.; Clary, C.; Farid, M.M.; Fasina, O.O.; Noomhorm, A.; Welti-Chanes, J. *Food Drying, Science and Technology*; DEStech Publications Inc.: Lancaster, PA, USA, 2008.
7. Liapis, A.I.; Bruttini, R. Freeze Drying. In *Handbook of Industrial Drying*; Mujumdar, A.S., Ed.; CRC Press: Boca Raton, FL, USA, 2007; pp. 257–283.
8. Igual, M.; Chiş, M.S.; Păucean, A.; Vodnar, D.C.; Ranga, F.; Mihăiescu, T.; Martínez-Monzó, J.; García-Segovia, P. Effect on Nutritional and Functional Characteristics by Encapsulating *Rose canina* Powder in Enriched Corn Extrudates. *Foods* **2021**, *10*, 2401. [CrossRef]
9. An, K.; Zhao, D.; Wang, Z.; Wu, J.; Xu, Y.; Xiao, G. Comparison of different drying methods on Chinese ginger (*Zingiber officinale* Roscoe): Changes in volatiles, chemical profile, antioxidant properties, and microstructure. *Food Chem.* **2016**, *197*, 1292–1300. [CrossRef]
10. Ciurzynska, A.; Marczak, W.; Lenart, A.; Janowicz, M. Production of innovative freeze-dried vegetable snack with hydrocolloids in terms of technological process and carbon footprint calculation. *Food Hydrocoll.* **2020**, *108*, 105993. [CrossRef]
11. Egas-Astudillo, L.A.; Martínez-Navarrete, N.; Camacho, M.M. Impact of biopolymers added to a grapefruit puree and freezedrying shelf temperature on process time reduction and product quality. *Food Bioprod. Process.* **2020**, *120*, 143–150. [CrossRef]
12. Leiton-Ramírez, Y.M.; Ayala-Aponte, A.; Ochoa-Martínez, C.I. Physicochemical properties of guava snacks as affected by drying technology. *Processes* **2020**, *8*, 106. [CrossRef]
13. Silva-Espinoza, M.A.; Ayed, C.; Foster, T.; Camacho, M.M.; Martínez-Navarrete, N. The impact of freeze-drying conditions on the physico-chemical properties and bioactive compounds of a freeze-dried orange puree. *Foods* **2020**, *9*, 32. [CrossRef] [PubMed]
14. Igual, M.; Cebadera, L.; Cámara, R.M.; Agudelo, C.; Martínez-Navarrete, N.; Cámara, M. Novel ingredients based on grapefruit freeze-dried formulations: Nutritional and bioactive value. *Foods* **2019**, *8*, 506. [CrossRef] [PubMed]
15. Otálora, M.C.; Carriazo, J.G.; Iturriaga, L.; Nazareno, M.A.; Osorio, C. Microencapsulation of betalains obtained from cactus fruit (*Opuntia ficus-indica*) by spray drying using cactus cladode mucilage and maltodextrin as encapsulating agents. *Food Chem.* **2015**, *187*, 174–181. [CrossRef] [PubMed]
16. Telis, V.R.N.; Martínez-Navarrete, N. Collapse and color changes in grapefruit juice powder as affected by water activity, glass transition and addition of carbohydrate polymers. *Food Biophys.* **2009**, *4*, 83–93. [CrossRef]
17. Igual, M.; García-Segovia, P.; Martínez-Monzó, J. Resistant maltodextrin's effect on the physicochemical and structure properties of spray dried orange juice powders. *Eur. Food Res. Technol.* **2021**, *247*, 1125–1132. [CrossRef]

18. García-Segovia, P.; Igual, M.; Martínez-Monzó, J. Beetroot Microencapsulation with Pea Protein Using Spray Drying: Physicochemical, Structural and Functional Properties. *Appl. Sci.* **2021**, *11*, 6658. [CrossRef]
19. Hernández, M.G. Ciclodextrinas: Complejos de inclusión con polímeros. *Rev. Iberoame. Polim.* **2007**, *8*, 301–313.
20. Gharsallaoui, A.; Roudaut, G.; Chambin, O.; Voilley, A.; Saurel, R. Applications of spray-drying in microencapsulation of food ingredients: An overview. *Food Res. Int.* **2007**, *40*, 1107–1121. [CrossRef]
21. Bajaj, S.R.; Marathe, S.J.; Singhal, R.S. Co-encapsulation of vitamins B12 and D3 using spray-drying: Wall material optimization, product characterization, and release kinetics. *Food Chem.* **2021**, *335*, 127642. [CrossRef]
22. Solomando, J.C.; Antequera, T.; Ruiz-Carrascal, J.; Perez-Palacios, T. Improvement of encapsulation and stability of EPA and DHA from monolayered and multilayered emulsions by high-pressure homogenization. *J. Food Process Preserv.* **2019**, *44*, e14290. [CrossRef]
23. Chindapan, N.; Niamnuy, C.; Davahastin, S. Physical properties, morphology and saltiness of salt particles as affected by spray drying conditions and potassium chloride substitution. *Powder Technol.* **2018**, *326*, 265–271. [CrossRef]
24. Domínguez Díaz, L.; Fernández-Ruiz, V.; Cámara, M. The frontier between nutrition and pharma: The international regulatory framework of functional foods, food supplements and nutraceuticals. *Crit. Rev. Food Sci. Nutr.* **2020**, *60*, 1738–1746. [CrossRef] [PubMed]
25. Silva-Espinoza, M.A.; Camacho, M.M.; Martínez-Monzó, J.; Martínez-Navarrete, N. Impact of the Freeze-Drying Conditions Applied to Obtain an Orange Snack on Energy Consumption. *Foods* **2021**, *10*, 2756. [CrossRef] [PubMed]
26. Mahdavi, S.A.; Jafari, S.M.; Assadpoor, E.; Dehnad, D. Microencapsulation optimization of natural anthocyanins with maltodextrin, gum Arabic and gelatin. *Int. J. Biol. Macromol.* **2016**, *85*, 379–385. [CrossRef]
27. Pacheco, C.; García-Martínez, E.; Moraga, G.; Piña, J.; Nazareno, M.A.; Martínez-Navarrete, N. Development of dried functional foods: Stabilization of orange pulp powder by addition of biopolymers. *Powder Technol.* **2020**, *362*, 11–16. [CrossRef]
28. Cámara, M.; Fernández-Ruiz, V.; Morales, P.; Sánchez-Mata, M.C. Fibre Compounds and Human Health. *Curr. Pharm. Des.* **2017**, *23*, 2835–2849. [CrossRef]
29. European Parliament and Council of the European Union. Regulation (EU) No 1169/2011 of the European Parliament and of the Council of 25 October 2011 on the Provision of Food Information to Consumers. 2011. Available online: https://eur-lex.europa.eu/legal-content/EN/TXT/?uri=CELEX%3A32011R1169&qid=1604656217441 (accessed on 22 June 2022).
30. European Parliament and Council of the European Union. Regulation (EC) No 1925/2006 of the European Parliament and of the Council of 20 December 2006 on the Addition of Vitamins and Minerals and of Certain Other Substances to Foods. 2006. Available online: https://eur-lex.europa.eu/legal-content/EN/ALL/?uri=CELEX%3A32006R1925 (accessed on 22 June 2022).
31. Singh, K.; Singh, D.; Sheetal, B.H.A.T.; Sharma, Y.P.; Gairola, S. Nutraceutical potential of rose hips of three wild *Rosa* species from Western Himalaya, India. *Not. Bot. Horti Agrobot. Cluj-Napoca* **2021**, *49*, 12471. [CrossRef]
32. Smanalieva, J.; Iskakova, J.; Oskonbaeva, Z.; Wichern, F.; Darr, D. Investigation of nutritional characteristics and free radical scavenging activity of wild apple, pear, rosehip, and barberry from the walnut-fruit forests of Kyrgyzstan. *Eur. Food Res. Technol.* **2020**, *246*, 1095–1104. [CrossRef]
33. Paunović, D.; Kalušević, A.; Petrović, T.; Urošević, T.; Djinović, D.; Nedović, V.; Popović-Djordjević, J. Assessment of chemical and antioxidant properties of fresh and dried rosehip (*Rosa canina* L.). *Not. Bot. Horti Agrobot. Cluj-Napoca* **2019**, *47*, 108–113. [CrossRef]
34. Ercisli, S. Chemical composition of fruits in some rose (*Rosa* spp.) species. *Food Chem.* **2007**, *104*, 1379–1384. [CrossRef]
35. Kazaz, S.; Baydar, H.; Erbas, S. Variations in Chemical Compositions of *Rosa damascena* Mill. and *Rosa canina* L. Fruits. *Czech J. Food Sci.* **2009**, *27*, 178–184. [CrossRef]
36. Ruiz Rodríguez, B.M. Frutos Silvestres de Uso Tradicional en la Alimentación: Evaluación de su Valor Nutricional, Compuestos Bioactivos y Capacidad Antioxidante. Ph.D. Thesis, Facultad de Farmacia, Universidad Complutense de Madrid, Madrid, Spanish, 2014.
37. Akagić, A.; Oras, A.V.; Oručević Žuljević, S.; Spaho, N.; Drkenda, P.; Bijedić, A.; Memić, S.; Hudina, M. Geographic variability of sugars and organic acids in selected wild fruit species. *Foods* **2020**, *9*, 462. [CrossRef] [PubMed]
38. Cunja, V.; Mikulic-Petkovsek, M.; Weber, N.; Jakopic, J.; Zupan, A.; Veberic, R.; Schmitzer, V. Fresh from the ornamental garden: Hips of selected rose cultivars rich in phytonutrients. *J. Food Sci.* **2016**, *81*, C369–C379. [CrossRef] [PubMed]
39. Okatan, V.; Çolak, A.M.; Güçlü, S.F.; Korkmaz, N.; Sękara, A. Local genotypes of dog rose from Interior Aegean region of Turkey as a unique source of pro-health compounds. *Bragantia* **2019**, *78*, 397–408. [CrossRef]
40. Adamczak, A.; Buchwald, W.; Zieliński, J.; Mielcarek, S. Flavonoid and organic acid content in rose hips (*Rosa* L., sect. *Caninae dc. Em. Christ.*). *Acta Biol. Crac. Ser. Bot.* **2012**, *54*, 105–112. [CrossRef]
41. Comunian, T.A.; Silva, M.P.; Souza, C.J. The use of food by-products as a novel for functional foods: Their use as ingredients and for the encapsulation process. *Trends Food Sci. Technol.* **2021**, *108*, 269–280. [CrossRef]
42. Idham, Z.; Muhamad, I.I.; Sarmidi, M.R. Degradation kinetics and color stability of spray-dried encapsulated anthocyanins from *Hibiscus sabdariffa* L. *J. Food Process. Eng.* **2012**, *35*, 522–542. [CrossRef]
43. Čakarević, J.; Šeregelj, V.; Šaponjac, V.T.; Ćetković, G.; Brunet, J.Č.; Popović, S.; Kostić, M.H.; Popović, L. Encapsulation of beetroot juice: A study on the application of pumpkin oil cake protein as new carrier agent. *J. Microencapsul.* **2020**, *37*, 121–133. [CrossRef]
44. AOAC, Association of Official Analytical Chemists. *Official Methods of Analysis*, 17th ed.; no. 934.06; AOAC: Gaithersburg, MD, USA, 2000.

45. Horwitz, W.; Latimer, G.W., Jr. (Eds.) *Official Methods of Analysis of the AOAC International*, 19th ed.; AOAC: Gaithersburg, MD, USA, 2012.
46. Sánchez-Mata, M.C.; Cabrera-Loera, R.D.; Morales, P.; Fernández-Ruiz, V.; Cámara, M.; Díez-Marqués, C.; Pardo-de-Santayana, M.; Tardío, J. Wild vegetables of the Mediterranean area as valuable sources of bioactive compounds. *Genet. Resour. Crop Evol.* **2012**, *59*, 431–443. [CrossRef]
47. Igual, M.; García-Martínez, E.; Camacho, M.M.; Martínez-Navarrete, N. Stability of micronutrients and phytochemicals of grapefruit jam as affected by the obtention process. *Food Sci. Technol. Int.* **2016**, *22*, 203–212. [CrossRef]
48. Olives Barba, A.I.; Cámara Hurtado, M.; Sánchez Mata, M.C.; Fernández Ruiz, V.; López Sáenz De Tejada, M. Application of a UV-vis detection-HPLC method for a rapid determination of lycopene and β-carotene in vegetables. *Food Chem.* **2006**, *95*, 328–336. [CrossRef]

molecules

MDPI

Article

Acrylamide-Derived Ionome, Metabolic, and Cell Cycle Alterations Are Alleviated by Ascorbic Acid in the Fission Yeast

Marek Kovár [1], Alica Navrátilová [2], Renata Kolláthová [3], Anna Trakovická [2] and Miroslava Požgajová [4,*]

[1] Institute of Plant and Environmental Science, Faculty of Agrobiology and Food Resources, Slovak University of Agriculture in Nitra, Tr. A. Hlinku 2, 94976 Nitra, Slovakia; marek.kovar@uniag.sk

[2] Institute of Nutrition and Genomics, Faculty of Agrobiology and Food Resources, Slovak University of Agriculture in Nitra, Tr. A. Hlinku 2, 94976 Nitra, Slovakia; alica.navratilova@uniag.sk (A.N.); anna.trakovicka@uniag.sk (A.T.)

[3] Institute of Animal Husbandry, Faculty of Agrobiology and Food Resources, Slovak University of Agriculture in Nitra, Tr. A. Hlinku 2, 94976 Nitra, Slovakia; renata.kollathova@uniag.sk

[4] AgroBioTech Research Center, Slovak University of Agriculture in Nitra, Tr. A. Hlinku 2, 94976 Nitra, Slovakia

* Correspondence: miroslava.pozgajova@uniag.sk; Tel.: +421-37-641-4919

Abstract: Acrylamide (AA), is a chemical with multiple industrial applications, however, it can be found in foods that are rich in carbohydrates. Due to its genotoxic and cytotoxic effects, AA has been classified as a potential carcinogen. With the use of spectrophotometry, ICP-OES, fluorescence spectroscopy, and microscopy cell growth, metabolic activity, apoptosis, ROS production, MDA formation, CAT and SOD activity, ionome balance, and chromosome segregation were determined in *Schizosaccharomyces pombe*. AA caused growth and metabolic activity retardation, enhanced ROS and MDA production, and modulated antioxidant enzyme activity. This led to damage to the cell homeostasis due to ionome balance disruption. Moreover, AA-induced oxidative stress caused alterations in the cell cycle regulation resulting in chromosome segregation errors, as 4.07% of cells displayed sister chromatid non-disjunction during mitosis. Ascorbic acid (AsA, Vitamin C), a strong natural antioxidant, was used to alleviate the negative impact of AA. Cell pre-treatment with AsA significantly improved AA impaired growth, and antioxidant capacity, and supported ionome balance maintenance mainly due to the promotion of calcium uptake. Chromosome missegregation was reduced to 1.79% (44% improvement) by AsA pre-incubation. Results of our multiapproach analyses suggest that AA-induced oxidative stress is the major cause of alteration to cell homeostasis and cell cycle regulation.

Keywords: acrylamide; antioxidant capacity; ascorbic acid; cell cycle; ionome; morphology; oxidative stress; *Schizosaccharomyces pombe*

Citation: Kovár, M.; Navrátilová, A.; Kolláthová, R.; Trakovická, A.; Požgajová, M. Acrylamide-Derived Ionome, Metabolic, and Cell Cycle Alterations Are Alleviated by Ascorbic Acid in the Fission Yeast. *Molecules* **2022**, *27*, 4307. https://doi.org/10.3390/molecules27134307

Academic Editor: Smaoui Slim

Received: 16 June 2022
Accepted: 1 July 2022
Published: 5 July 2022

1. Introduction

Acrylamide (AA) is a monomeric substrate with the chemical formula C_3H_5NO and a molecular weight of 71.08. It is a synthetic, water-soluble, colorless, and odorless substance [1] widely used as an industrial chemical in the textile, plastic, and paper industries [2,3]. Acrylamide was discovered in foods in 2002 as a product of the so-called Maillard reaction between asparagine residues and reducing sugars at high temperature (above 120 °C), hence fried and baked starch-rich foods such as potato chips, French fries, crackers, cookies, coffee, and bread are the main sources of the accidental AA ingestion [4]. Intake of food-derived AA is estimated to be 0.3–2.0 μg kg^{-1} b.w. (body weight) in the human population [5]. The toxicology of AA has gained attention as its toxic spectrum was found to be wider than expected. AA was classified as a potential carcinogen by the International Agency for Research on Cancer (IARC) as its major metabolite glycidamide triggers neurotoxicity, genotoxicity, and reproductive toxicity in animals and humans due to its ability to bind DNA [6–9]. Intracellularly, AA non-enzymatically or by glutathione-S-transferases conjugates with glutathione (GSH) resulting in the formation of N-acetyl-S-(3-amino-3-oxypropyl)-cysteine. This in turn leads to depletion of the cellular GSH, thus triggering elevation of reactive oxygen species (ROS) formation leading to oxidative stress. Additionally, enhanced sensitivity to AA of cells depleted from Cu, Zn-superoxide dismutase (Sod1p), and AA-mediated decrease in catalase (CAT) expression and increase in malondialdehyde (MDA) formation, confirmed the oxidative activity of AA [10–12]. Due to increasing evidence of AA toxicity, the need to understand the mechanism of AA activity and the prophylactic/protective therapy's capability to reduce or minimize AA-induced alterations has risen. It has been shown that compounds with antioxidant properties such as curcumin, β-carotene, vanillin, caffeic acid, rosmarinic acid, Vitamin E, or 5-amino salicylic acid have the ability to inhibit AA-triggered cytotoxicity through their direct effect on ROS scavenging activity [5,13,14]. In addition, the amount of AA in the food colorant ammonia caramel that is produced by Maillard reaction and caramelization has been reduced by ascorbic acid supplementation to a mixture of glucose and ammonia, thus preserving the colorant from AA contamination [15].

Abiotic stress caused by variable substances that leads to oxidative stress of the organism has been associated with cellular ionome destabilization resulting in physiological alterations in a variety of model systems [16,17]. The term ionome determines the collection of all ions in a given biological system, as non-metal and metal elements in nutrition play an irreplaceable role in multiple biological processes. Hence, the ion composition of the cell might represent a reliable tool to study physiological responses of the organism to environmental variations, or nutritional status.

A simple eukaryotic model organism, yeast *Schizosaccharomyces pombe*, was used to investigate the role of AA in cell growth, cell morphology, cell antioxidant capacity, ROS generation, ion balance, and cell cycle progression. *S. pombe* is a rod-shaped, non-pathogenic, widely used, convenient model system that shares many functional similarities of biological processes with higher eukaryotes [18,19]. As the underlying mechanism of AA-induced toxicity is often linked to the overproduction of ROS that triggers oxidative stress, in the presented study we have investigated the possible role of a strong antioxidant, and free radical neutralizer, ascorbic acid (AsA, Vitamin C) [20], to prevent the cytostatic and cytotoxic effect of AA. Although the impact of AA on different model systems has been intensively studied, to our knowledge, its influence on the cell cycle progression and ionome homeostasis in *S. pombe* has yet not been elucidated. Thus, in our study, we provided a newly exerted approach to stress response studies through the determination of the ionome balance sustainment.

2. Results

2.1. The time and Dose-Dependent Effect of Ascorbic Acid on the Cell Growth Intensity and the Impact of Acryl Amide Addition

S. pombe cells were grown overnight (o/n) to the exponential growth phase, and OD_{600} was adjusted to 0.3. Cells were divided into four groups, one group was left untreated the other three groups were incubated with 1, 10, and 100 mM AsA for 30 min. Afterward, indicated AA concentrations were subjected to the cells and OD_{600} was measured every 3 h of incubation. As expected, increasing AA concentrations affected cell growth substantially. Cell treatment with 1 and 10 mM AsA improved the cell growth both, with and without (w/o) AA addition. However, cell treatment with 100 mM AsA either did not affect or reduced the growth of cells, indicating the prooxidative effect of such a high AsA concentration (Figures 1 and S1).

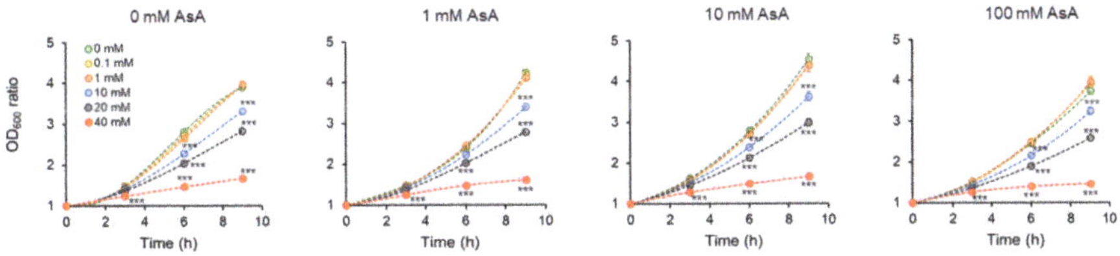

Figure 1. Determination of the cell growth intensity upon AA addition with and w/o AsA pre-treatment. Evaluated was the OD_{600} ratio determined every three hours of incubation (3 h, 6 h, and 9 h) and calculated as the difference to the time point 0 h. Cells un- and pre-treated with AsA were subjected to indicated AA concentrations (0 mM green, 0.1 mM yellow, 1 mM orange, 10 mM blue, 20 mM grey, and 40 mM red). Increasing AA concentration led to a marked reduction of the cell growth ability. AsA treatment improved growth ability at 1 mM and 10 mM concentration, while 100 mM did not improve or even reduce the growth of cells. Each point represents the mean value ± SD of 4 individual samples. Statistical differences are indicated as $p < 0.001$ ***.

Similarly, the relative cell growth ratio (*RGR*) representing the cell mass gain after every third hour of incubation revealed the positive effect of 1 mM AsA and an even more pronounced positive effect of 10 mM AsA on the cell growth under AA stress, while 100 mM AsA affected the *RGR* negatively (Figure S2). Generation time (gt) that represents the time required for the cell doubling, revealed the positive effect of 10 mM AsA pretreatment, as it shortened the generation time of AA exposed cells, although without reaching statistical significance. Similar to previous observations, 100 mM AsA pre-treatment led to prolongation of the cell growth (Figure S3). For further analyses, the 10 mM AsA concentration was used, as this concentration showed the best protective effect against AA.

To determine the toxic concentration of AA for *S. pombe* cells, the IC_{50} value representing the concentration of AA that leads to 50% cell growth inhibition was calculated (Table 1, Figure S4). Notably, the IC_{50} value is slightly higher under conditions of 10 mM AsA pre-treatment, meaning that a higher AA concentration is required to inhibit the growth of 50% of cells compared to AsA untreated cells, thus confirming the significant role of AsA to protect cells against AA-induced growth alterations.

Table 1. IC_{50} value indicating inhibitory concentration of AA to cell growth.

AsA	IC_{50} (mM of AA)	R^2
0 mM	30.8337	0.932
10 mM	37.8238	0.916

R^2—coefficient of determination.

Determination of cell growth on solid media confirmed the detrimental effect of increasing AA concentration. However, in this particular experiment, AsA pre-incubation was not sufficient to alleviate the negative impact of AA. Cells were allowed to grow on the solid media for 2–3 days, hence as they spent all added AsA they were no longer protected from AA (Figure S5).

2.2. Cell Morphology

As the shape of the single-celled living organism reflects its health condition, we determined changes in the length (μm), width (μm), volume (μm^3), and surface (μm^2) of *S. pombe* cells subjected to AA with and w/o AsA pre-treatment. Increasing AA concentration led to cell prolongation, while cell width remained almost constant. Accordingly, cell volume and surface increased with rising AA concentration. Interestingly, pre-treatment of AsA did not significantly improve cell shape compared to untreated cells (Figure 2).

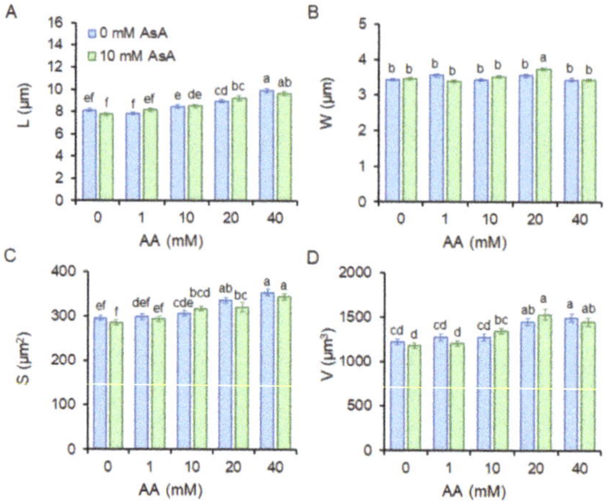

Figure 2. Cell morphology. Length (**A**), width (**B**), surface (**C**), and volume (**D**) were determined in *S. pombe* cells upon addition of indicated AA concentrations, with (green bars) and without (blue bars) AsA pre-treatment. Different letters above bars indicate statistical significance of four independent experiments. Each bar with tick represents mean value ± standard deviation (SD) (n = 100). Statistical significance is determined by Duncan's post-hoc test, different letters above bars indicate statistical difference at a 0.05 level of significance.

2.3. Cell Viability and Metabolic Activity

Metabolic activity serves as an indicator of mitochondrial functionality. Mitochondria of viable and metabolically active cells reduce TTC to formazan detectable at 485 nm. Enhancement of formazan generation upon stimuli caused by a low concentration of AA (1 mM) indicates excitement of cellular metabolism aiming to protect internal homeostasis. However, higher AA concentrations markedly reduced metabolic activity, suggesting AA-induced deterioration of mitochondrial metabolism and function. Cell pre-treatment with AsA significantly protected mitochondrial function resulting in improvement of metabolic activity (Figure 3). Methylene blue staining is a very useful method to distinguish between viable and dead cells. As it permeates only dead cells, blue-stained cells are considered dead. The percentual proportion of living to dead cells is calculated. Strikingly, although the addition of AA markedly hinders cell growth, cells persist alive trying to survive and overcome the negative effect of AA. Even the highest AA concentration (40 mM) leads to only approximately 4–5% cell death despite almost complete growth retraction. AsA pre-treatment did not cause any significant differences (Figure S6).

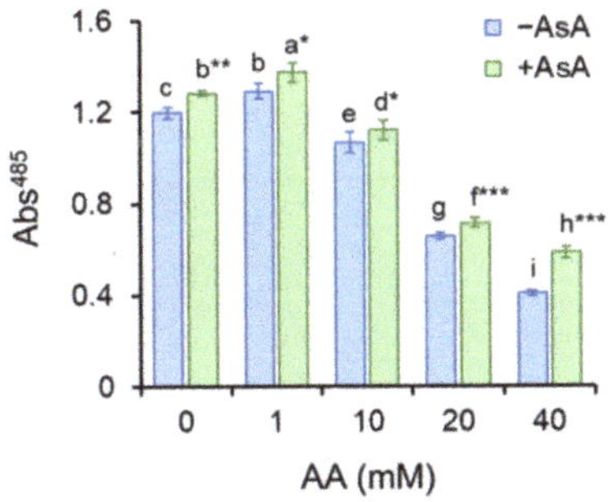

Figure 3. Metabolic activity. Metabolic activity determined as formazan production, detectable at 485 nm and normalized to protein unit, from yellowish tetrazolium chloride (TTC), indicates functionality of mitochondria, thereby indirectly representing cell vitality. Individual bars represent mean value of 8 samples from two independent experiments ± SD. Statistical significance, determined by Duncan's post-hoc test, was set up as $p < 0.05$ *, 0.01 **, 0.001 *** different letters above bars indicate statistical difference.

2.4. *Ascorbic Acid Mitigates Aa-Induced Enhancement of ROS Production and Oxidative Stress*

Generation of ROS was determined by the use of fluorescence spectroscopy. H_2DCFDA converts to fluorescent DCF in the presence of ROS, thus serving as an undirect ROS indicator. The addition of AA led to an increase in ROS production. Cell pre-treatment with AsA (10 mM) reduced AA-derived production of ROS significantly (Figure 4). Enhanced ROS formation may lead to oxidative stress. To test this, the formation of malondialdehyde (MDA), the end product of lipid peroxidation, was investigated. As the MDA content is calculated in relation to the protein content, the amount of the protein from the whole-cell extract was determined (Figure S7A). Interestingly, OD_{600} values and protein content (μg mL^{-1}) showed a positive mutual correlation upon AA and AsA treatment of cells, meaning that the increase in the protein content is related to the cell amount (Figure S7B). As assumed, MDA formation increased upon AA addition, in particular, the addition of 10 mM AA enhanced the generation of MDA dramatically. Cell pre-treatment with AsA reduced MDA formation significantly, thereby protecting cells from AA-mediated oxidative stress (Figure 5).

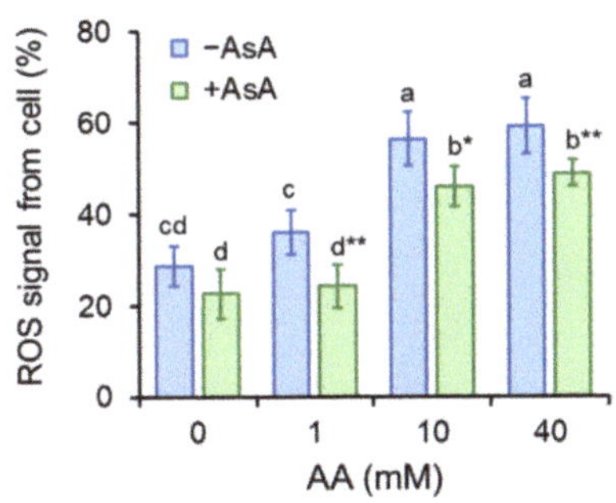

Figure 4. ROS generation. AA-induced enhancement of ROS formation detected as enhancement of fluorescence signal was compared to the control group w/o AA addition and to the group of cells pre-treated with AsA. Individual bars represent mean ± SD of 8 individual samples and two independent experiments. Statistical significance is determined by Duncan's post-hoc test, different letters above bars indicate statistical difference as $p < 0.05$ *, 0.01 **.

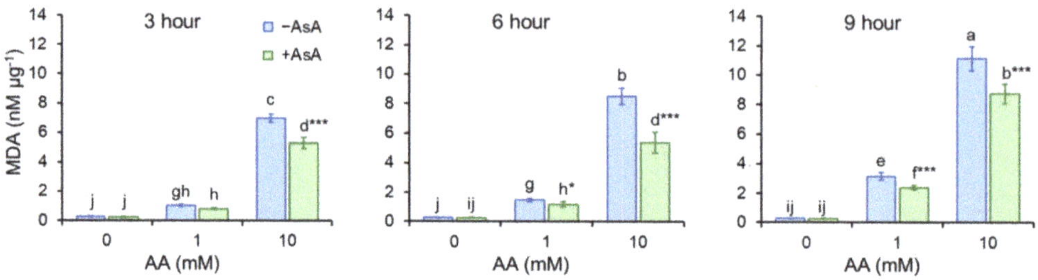

Figure 5. MDA formation. AA triggers oxidative stress resulting in lipid peroxidation. As indicated by enhanced MDA production, AA causes peroxidation of membrane lipids in a dose dependent manner. AsA pre-treatment protect cells from AA-triggered oxidative stress, as it significantly reduces MDA production. Bars represent mean ± SD of 4 individual samples. Statistical significance is determined by Duncan's post-hoc test, different letters above bars indicate statistical difference as $p < 0.05$ *, 0.001 ***.

2.5. Acrylamide-Reduced Antioxidant Cell Capacity Is Significantly Restored by AsA

Cells undergoing oxidative stress activate protective machineries to eliminate the detrimental consequences of stress. Catalase (CAT) is the enzyme that catalyzes the decomposition of hydrogen peroxide to water and molecular oxygen. Presence of AA in the growth media increases CAT activity in the time and dose dependent manner. AsA pre-treatment protects cells from the negative impact of AA thus CAT activity is significantly lower, compared to the untreated control (Figure 6A). It is worth to mention that CAT activity affects MDA content logarithmically, as the increasing CAT activity reduces MDA content, with the more pronounced positive effect in the samples supplemented with AsA (Figure S8).

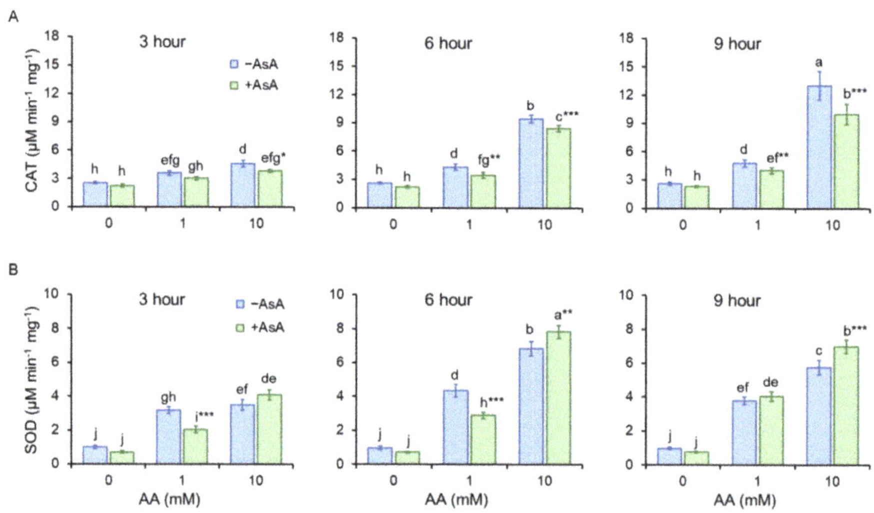

Figure 6. Activity of the antioxidant enzyme system catalase (CAT) and superoxide dismutase (SOD). The antioxidant capacity of cells was determined via detection of antioxidant enzymes activity CAT (**A**) and SOD (**B**). AsA pre-treatment significantly improved AA-altered antioxidant capacity of cells. Bars represent mean ± SD of 4 individual samples. Statistical significance is determined by Duncan's post-hoc test, different letters above bars indicate statistical difference as $p < 0.05$ *, 0.01 **, 0.001 ***.

Superoxide dismutase (SOD) is the enzymatic antioxidant that helps to maintain redox balance by converting the highly reactive superoxide anion to oxygen and less reactive hydrogen peroxide. Acrylamide at lower concentration (1 mM) induces SOD activation that protects cells from oxidative stress, however, at the same AA concentration, AsA treatment serves as protective antioxidant and defends cells with the need of significantly less SOD activation. Cell incubation with high AA concentration (10 mM) increased SOD activation substantially, AsA pretreatment contributed to the cell protection by additional enhancement of SOD activation (Figure 6B).

2.6. *Acrylamide Toxicity Triggers Mild Apoptotic Events in S. pombe*

As yeast cells exposed to AA displayed an intracellular overload of ROS which is often connected to cell death by apoptosis, double staining with Annexin V-FITC and PI was performed. Annexin V-FITC detects the exposure of PS at the outer layer of the plasma membrane, an early sign of apoptosis. The uptake of PI requires detrimental alteration of the cell membrane referring to its irreversible damage with permanent loss of barrier function, resulting in cell death. Hence, cells that were Annexin V positive and PI negative are considered apoptotic, and PI-positive cells are considered necrotic.

Although AA exposure to mammalian cells has been associated with events resulting in programmed cell death, *S. pombe* cells showed only mild signs of apoptosis from AA exposure as only the highest AA concentration (40 mM) enhanced cell apoptosis (Figure 7). Additionally, PI staining revealed only a moderate increase in necrotic cells compared to the untreated control (Figure 7) which is consistent with the cell viability determination by methylene blue staining (Figure S6). Despite the compelling endurance of *S. pombe* cells against AA-triggered death, AsA pre-treatment enhanced cell ability to survive, thus, preserving their vitality.

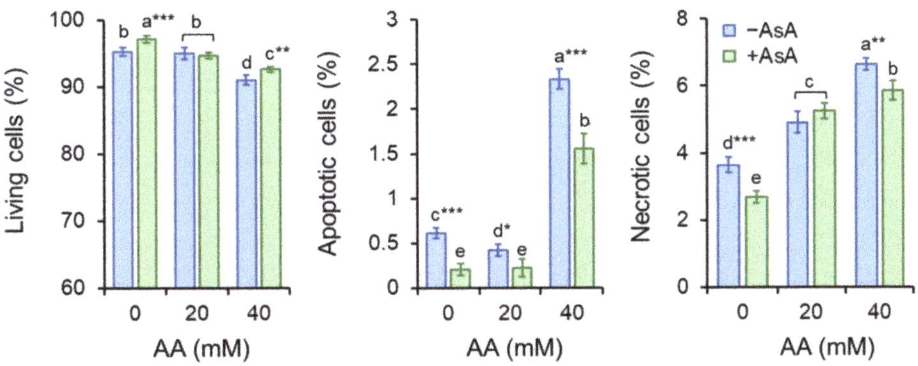

Figure 7. Yeast cell apoptosis determination. To evaluate cell death by apoptosis, Annexin V-FITC signal was determined upon AA exposure of cells un- and pre-treated with AsA. Cells were co-stained with PI to distinguish between apoptosis and necrosis. PI positive cells were considered necrotic. Bars represent mean ± SD of at least 150 counted cells. Statistical significance is determined by Duncan's post-hoc test, different letters above bars indicate statistical difference. Significance of statistical differences are indicated as $p < 0.05$ *, 0.01 **, 0.001 ***.

Altogether, the results show that the loss of cultivability particularly in the presence of lower AA concentration is not connected to cell death as the cell undergoes apoptosis to a similar extent as the control group unexposed to AA.

2.7. *Acrylamide Triggers Errors in Chromosome Segregation during Cell Cycle*

The cell cycle ensures the growth and development of living organisms, and its error-free progression is thus fundamental for normal life. Segregation of chromosomes during mitosis results in the production of two identical daughter cells from one mother cell. *S. pombe* strain JG 15,457 carries chromosome II marked with GFP (GFP-tagged LacI molecules bind to lacO repeats inserted within the centromere II, cen2-GFP). Segregation of sister chromatids of chromosome II was determined during anaphase in PFA fixed cells 6 h after AA addition. Sister chromatids of the control and AsA pre-treated cells segregated to opposite poles, while AA addition led to sister chromatid non-disjunction in 4% of anaphase cells. However, AsA pre-treatment significantly reduced the non-disjunction of sister chromatids to 1.79% in AA incubated cells. This suggests that AA mediates alterations in the regulation of chromosome segregation most likely through enhancement of ROS production, therefore AsA pre-treatment and its antioxidant property is able, at least to some extent, to protect cells from errors in chromosome segregation (Figure 8).

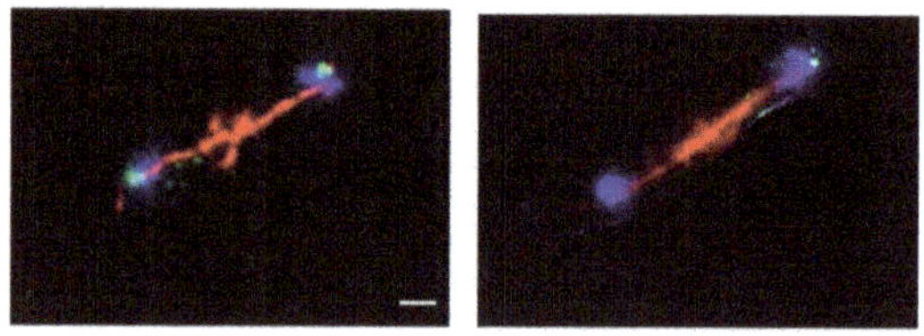

	Normal segregation (%)	Missegregation (%)
control	100.00 ± 0.00	0.00 ± 0.00 ^{C}
AsA	99.88 ± 0.05	0.12 ± 0.05 ^{C}
AA	95.93 ± 0.40	4.07 ± 0.40 ^{A}
AA +AsA	98.21 ± 0.55	1.79 ± 0.55 ^{B***}

Figure 8. Chromosome segregation during mitosis. Segregation of chromosome II during mitosis was analyzed in cells exposed to 20 mM AA for 6 h with and without 10 mM AsA pre-treatment. Control cells were grown in YES medium without AA and AsA addition. At least 200 cells from 4 inde-pendent experiments were scored for normal or missegregated chromosome II and expressed in percent. Statistical significance is determined by Duncan's post-hoc test, different letters above bars indicate statistical difference. Significance of statistical differences are indicated as $p < 0.001$ ***.

2.8. Ionome Balance

Maintaining intracellular ionome homeostasis is an important mechanism by which cells adapt to the stress caused by cell disrupting agents such as AA. Accordingly, AA-mediated cell homeostasis alterations triggered an imbalance in the intracellular mineral element content. Relative to the untreated control, 1 mM AA exposure for 3 h caused a decrease in the content of most of the evaluated elements, except for Cu, while cells exposed to 10 mM AA displayed increased (Ca, Cu, Mg, Mn, Sr) or slightly decreased (Fe, K, Zn) ion concentrations. Marked enhancement of all determined element levels was detected upon cell pre-incubation with AsA as compared to the untreated control. This, in turn, resulted in the increase in the ion content of cells co-treated with AA and AsA compared to AA-exposed cells revealing the positive impact of AsA treatment in the protection of ion homeostasis maintenance (Figure 9A). The second messenger Ca was the dominant component that accumulated in the cell to prevent the ionome balance disruption. Strikingly, Sr content increased in the same fashion as the Ca level possibly due to utilization of the open Ca channels. Changes in levels of ions related to the enzymatic antioxidant system (Cu, Mn, Zn) might be connected to AA-triggered ROS overproduction and oxidative stress. Variables standardization by *Z-scores* reveals that in general, AsA treatment led to enhancement of the concentration of most of the analyzed elements compared to AsA untreated cells (Figure 9B). Moreover, correlation analyses show positive mutual interactions among all tested elements after 3 h of AA exposure (Figure 9C). A different situation was observed after 9 h of incubation, as most of the evaluated mineral elements showed reduced concentration relative to the untreated control (Figure 9D). The most prominent drop in the concentration was observed with K of AsA preincubated cells, which is probably related to the form of ascorbate used in our study, the Na salt. After 9 h of incubation, all AsA was spent, therefore K was no longer required to compensate for the boost of Na. Interestingly, AsA treatment led to enhancement of Fe levels resulting in Cu regulation. Standardized *Z-scores* clearly created clusters of cell responses dependent on the treatment revealing that the treatment is the variable responsible for changes in the ion content (Figure 9E). Ion interactions display positive mutual correlations, except for K due to its marked drop in concentration. The most prominent positive correlation of all tested ions emerged toward Ca (Figure 9F). Altogether our results clearly show that AA exposure causes marked ion balance disruption in the cell that leads to cell homeostasis alterations. However, to our surprise, yeast cells subjected to long-term AA exposure show signs of adaptation to the toxicant.

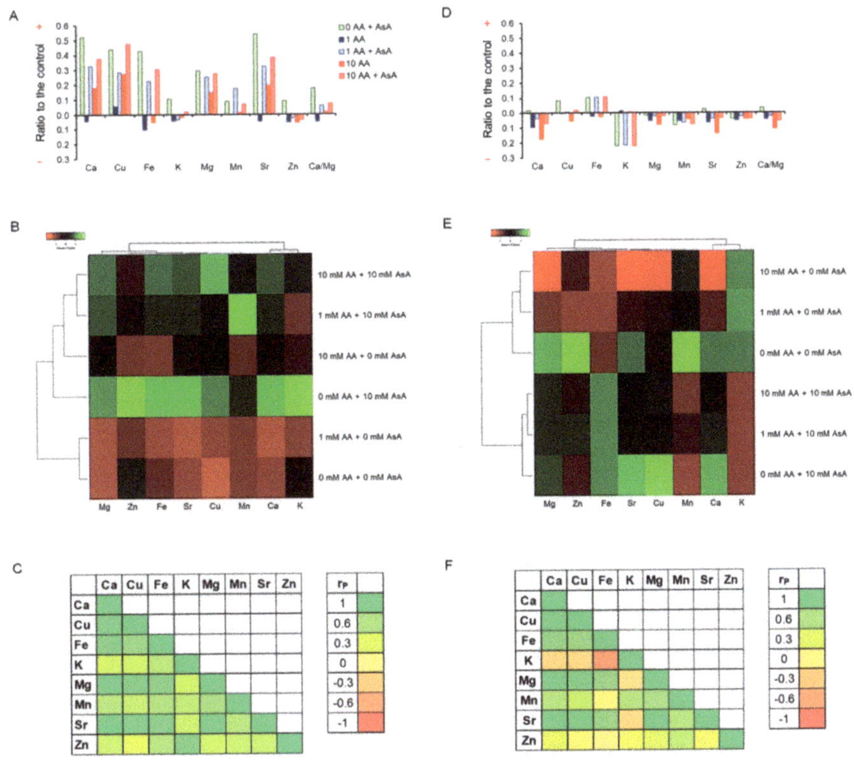

Figure 9. AA mediates disruption of the ionome balance. Differences of the ion content upon treatment are compared to the untreated control after 3 h (**A**) and 9 h (**D**) of incubation and expressed as increase (+) or decrease (−) relative to the control. Control is set up as 0.0 on the x axis. *Z-score* clusters of variables to adjust for the differences in magnitude between different ions after 3 h (**B**) and 9 h (**E**) of incubation. Correlation analyses of mutual ion interactions by Pearson test (r_P) after 3 h (**C**) and 9 h (**F**) of incubation.

3. Discussion

Acrylamide is a chemical substance often used in industry, however, with a potential unintended human exposure. Due to the reported risks linked with its use, the effect of its exposure to eukaryotic cells requires adequate assessment. In the present study, we used fission yeast *Schizosaccharomyces pombe* as a eukaryotic model organism to investigate the effect of AA-induced oxidative stress on chromosome segregation during mitosis. Yeast *S. pombe* represents a convenient model system for elucidating a variety of biological processes including growth, division, metabolic activity, and antioxidant capacity of the organism. This nonpathogenic single-celled eukaryote is easy to grow and manipulate. Additionally, its genome contains protein-coding genes responsible for cell division and cellular organization, which are also found in the genome of higher eukaryotes. Hence, it became a popular model system to study the basic principles of a cell that can be used to understand more complex organisms [21,22]. The eukaryotic mitotic cell cycle is a highly complex event that guarantees duplication and faithful segregation of all components required for cell survival to daughter cells. To ensure genomic integrity, the error-free DNA synthesis phase (S-phase), and the mitosis phase (M-phase) are required to occur exactly once per cell cycle [23]. Progression of the cell cycle is regulated by a variety of not only internal but also external factors. Within the last decades, attention has been paid to reactive oxygen species (ROS) that seem to play an indisputable role in the regulation of cell cycle progression [24,25]. Although ROS are important mediators of natural physiological processes, their overproduction contributes to oxidative stress that might cause cell damage through oxidization of physiologically important biomolecules and alteration of cell signaling pathways [26,27]. Despite the relative resistance of *S. pombe* cells toward AA toxicity determined by the IC_{50} value (Table 1), the addition of AA to growth media resulted in the dose and time-dependent growth retardation (Figure 1) and AA-mediated enhancement of ROS production (Figure 4). This is related to the cellular metabolic activity, which was at first slightly stimulated upon 1 mM AA concentration, whereas higher concentrations of AA dramatically reduced the mitochondrial functionality (Figure 3). Similarly, many studies on a variety of model systems showed that AA decreases cell viability, induces protein modifications, increases the production of ROS, and alters mitochondrial function [26,28–30]. Therefore, to counteract the toxic impact of AA, we investigated the possible protective effect of a strong natural antioxidant, ascorbic acid (AsA, Vitamin C). Growth of AA exposed cells upon pre-treatment of AsA significantly improved, however only 1 mM and even more pronounced 10 mM concentrations were able to improve cell growth, 100 mM AsA caused growth alterations, suggesting pro-oxidative properties of such high concentration. Hence, 10 mM AsA was used throughout the experiments. The significant positive effect of AsA pre-treatment resulted in the reduction of ROS production, and improved metabolic activity. Other authors tested various compounds with antioxidant properties to eliminate AA toxicity. Jackfruit flake digest and a naturally occurring flavonoid myricitrin reduced excessive ROS production resulting in alleviation of AA-triggered mitochondrial disorders in Caco-2 cells [31,32], in HepG2 cells AA-induced ROS overproduction was reduced by curcumin or mulberry digest [33,34]. As ROS-induced oxidative stress often triggers membrane lipid peroxidation, we determined the level of MDA, the end product of lipid peroxidation, upon AA addition. MDA content increased significantly even after cell exposure to 1 mM AA, while 10 mM AA led to marked enhancement of MDA levels. Notably, cell pre-incubation with AsA for 30 min led to significant protection against oxidative stress resulting in the reduction of MDA content. These results are supported by a previous study showing that AsA reduces enhanced ROS production or MDA content triggered by the AA metabolite, glycidamide, and exposure to Sertoli cells [35]. The enzymatic antioxidant defense system of the cell is activated upon exposure of the organism to oxidative stress. The activity of the superoxide scavenging enzyme SOD which converts superoxide radicals to less toxic H_2O_2 and CAT which catalyzes the direct decomposition of H_2O_2 to H_2O was determined every three hours after cells were exposed to AA. Our results reveal that 1 mM AA enhances the formation of superoxide already after 3 h of exposure, as the SOD activity increases significantly. Interestingly, AsA pre-treatment reduced the formation of superoxide, as the activity of SOD is lower as compared to AsA untreated cells. Cell treatment with 10 mM AA for 3 h led to increased SOD activity of both pre-treated and untreated cells with AsA suggesting the supportive role of AsA in cell protection against AA toxicity. Six hours of cell incubation with AA led to a noticeable enhancement of SOD activity with AsA protective effect similar to that observed at three hours of incubation. Apparently, after 9 h of AA exposure, the formed superoxide was neutralized as the SOD activity decreased as compared to its activity after 6 h of incubation (Figure 6B). In accordance with our observations, the SOD activity of human erythrocytes exposed to AA increased in a dose-dependent manner [36]. We assume that due to SOD-triggered superoxide radical decomposition to H_2O_2, CAT activity increased upon AA exposure. Its activity depended on the time of exposure and the dose of AA. AsA pre-treatment reduced the CAT activity of AA exposed cells, as its own antioxidant activity protected cells, and the CAT activity was therefore not required at such a high level (Figure 6A). Similarly, Celik et al. [37] detected increased CAT activity upon AA addition to HEK293 cells, although the activity of the enzyme was dependent on the dose of AA exposed to cells. Several scientific publications present decreased SOD and CAT activity of various model systems after AA exposure. This might result from the different behavior and sensitivity of the used model organism toward AA, different doses, and exposition times of AA, and/or due to manipulation of the model organism that already caused stress. However, the remaining common characteristic is the protection of the model organism against AA-induced oxidative stress via substances with antioxidant properties [38–40]. Taken together, the presented results showed that AsA pre-treatment significantly reduces ROS levels and promotes SOD and CAT activity to protect cells exposed to AA. Rod-shaped *S. pombe* cells are normally 6–7 μm in length and 2–3 μm in width. Before undergoing the G2–M transition, they grow until their length reaches approximately 14 μm. However, the size at which cells enter mitosis largely depends on environmental stimuli such as nutrient availability or stress. Poor nutrient conditions force cells to divide at a smaller size than in nutrient-rich media. Cell exposure to challenging conditions initiates signaling through the regulatory network, so the shape of cells is often the result of the prior signaling. The size modulation typically occurs through the MAP (mitogen-activated protein) kinase and TOR (target of rapamycin) signaling pathways acting on Cdc25 [19,41]. AA exposure significantly modulated the shape of *S. pombe* cells in a dose-dependent manner, the higher the concentration, the bigger the cell (Figure 2). Interestingly, AsA pre-treatment had only minor effects on the cell shape regulation of AA-exposed cells.

Strikingly, AA exposure in a dose close to IC_{50} value (40 mM) hindered cell growth substantially, however, the cell's vital functions and its metabolism were preserved as only approximately 3–4% of cells were unviable (Figure S4). Moreover, *S. pombe* cells show fairly high resistance against AA-triggered cell death by apoptosis (Figure 7). This suggests that despite attenuated cell vitality, most of the cells are able to survive. A negative condition of such survival is the accumulation of cell cycle regulation alterations as AA exposure leads to marked cell cycle delay and errors in chromosome segregation (Figure 8). The entry of mitosis is a strictly regulated process that controls the attachments between kinetochores and microtubules through sophisticated pathways including error correction (EC) and spindle assembly checkpoint (SAC) pathways. The Aurora B kinase, Ark1 in *S. pombe*, is the major kinase of these pathways that facilitates the establishment of accurate kinetochore–microtubule attachments. Alterations of its function have been associated with frequent occurrence of sister chromatids non-disjunction during mitosis [42,43] Hence, we assume that AA might negatively influence Ark1 function that in turn leads to erroneous segregation of chromosomes. However, Sickles et al. [44] also presented a potential mechanism of acrylamide-induced disruption of cell cycle mitotic activity that results in genomic instability. According to their study, the microtubule-depolymerizing kinesins may be inhibited by AA, which results in the failure of the migration of chromosomes from the metaphase plate. Similarly, Adler et al. [45] already in 1993 in their report suggested that AA might induce aneuploidy in mammalian cells through concentration-dependent spindle disturbances resulting in improper functioning of the spindle. Many studies showing AA altered cell cycle [46,47] are consistent with our findings that AA induces improper segregation of chromosomes and thus reduces mitotic activity. However, the underlying mechanism of its impact on cell cycle regulation remains unclear. One of the most important mechanisms of AA-induced toxicity is oxidative stress caused by ROS overproduction. As AsA pre-treatment reduced chromosome missegregation events we suggest that ROS-derived alterations of the cell cycle regulators are responsible for the alteration of mitotic chromosome segregation.

Ionome is critical for a variety of biological processes including growth or defense responses however, conditions for the regulation of the ionome homeostasis in the fission yeast under stress are only partially known. Despite the conserved role of each particular element, the yeast ion composition varies according to changes in the surrounding environment. As we have shown previously, acute contamination by Cd or Ni causes not only dramatic changes in the ion content of the cell but also affects ion trafficking through mutual mineral element interactions [48]. In line with this, AA-triggered stress results in ion balance disruption in *S. pombe* cells, leading to changes in mineral element concentrations of cells exposed to AA (Figure 9A,D). Cell response to AA toxicity resulted in a reduction in Fe levels which leads to an increase in Cu content. In addition, levels of Cu, Mn, and Zn ions that are involved in the antioxidant defense system of the cell were modulated by AA exposure, hence supporting AA-induced oxidative stress. Moreover, AA exposure caused accumulation of the second messenger and enzyme co-factor Ca, in order to reduce the negative impact of AA on the cell. Similarly, an influx of Ca from extracellular sources was reported by Popa et al. [49] in *Saccharomyces cerevisiae* cells exposed to high levels of oxidative stress. The authors concluded that Ca overload upon oxidative stress mediates the cytotoxic effect of the stressor rather than serving as an adaptation mediator for the cell. However, we believe that Ca protects cell viability as cells exposed to AA concentrations that almost completely hinder cell growth; only a small percent of such cells underwent apoptosis or lost vitality. In accordance, cell co-treatment with the strong antioxidant, AsA, caused a marked boost in Ca levels. This implies that Ca uptake is involved in the reduction of oxidative stress, thus protecting cells from AA toxicity. In line with this, in our previous study, we described the positive effect of AsA against another toxicant causing oxidative stress that disrupts the ionome, cadmium (Cd) [50]. AA-induced perturbations in the ionome led to changes in mineral elements' content and their mutual interactions in the dose in a time-dependent manner. Long-term exposure to AA led to downregulation of previously upregulated ion contents indicating initialization of cell adaptation (Figure 9A–F). Cell homeostasis, however, largely depends on the ionome balance referring to the complexity of the regulation of its sustainability. Over 600 genes in the yeast genome have been described to significantly impact the ionome. Clustering based on the ionome phenotype identified genes that target physical and genetic interaction networks between genes within a particular mineral nutrient regulatory function [51]. Thus, a possible explanation of the AA's ability to alter ion balance might arise from the AA-mediated changes in the expression of transporter genes resulting in ion accumulation in the cell although the exact mechanism still needs to be elucidated.

4. Materials and Methods

4.1. Yeast Cultivation and Growth Conditions

The wild-type *Schizosaccharomyces pombe* strain *SP72 h+ ade6-M210 ura4-D18 leu 1–32* was used for all studies except for the immunostaining experiment, for which the *JG 15,457* strain (cen2(D107)::KanR-ura4 + -lacO his7 + ::lacI-GFP) carrying chromosome II labeled with GFP was used. Yeast cells were cultured in the complete YES liquid medium containing 0.5% yeast extract, and 3% glucose, supplemented with 225 mg L^{-1} of amino acids (adenine, L-histidine, L-leucine, L-lysine, and uracil) at 30 °C, under aerobic conditions with vigorous shaking (150 rpm). The fission yeast grows at temperatures ranging from 18 to 37 °C with the optimum at 30 °C [52]. To avoid the collateral effect of the restrictive growth conditions and the AA-mediated effect on the cell behavior, cells were cultured at the optimum temperature and on the complete nutrient media.

4.2. Growth Intensity Determination

Cells from the overnight (o/n) culture were divided into four equal portions, three of which were supplemented with either 1, 10, or 100 mM of L-ascorbic acid (AsA) (Sigma–Aldrich, St. Louis, MO, USA), respectively, diluted in distilled water, for 30 min, while the third was left untreated, and incubated at 30 °C and 150 rpm. The density of cultured cells was then adjusted to an OD_{600} = 0.3, transferred to 24 well plates, and treated with various concentrations of AA (0, 1, 10, 20, 30, and 40 mM). Optical density at 600 nm of the culture incubated at 30 °C and 150 rpm for 9 h was measured every 3 h by Glomax Multi Detection System (Promega Corporation, Madison, WI, USA). Hence, as one life cycle of *S. pombe* takes approximately 3 h, the cell growth intensity upon AA addition was analyzed within three life cycles. The total growth intensity (OD_{600} ratio) is calculated as a ratio in the cell density at 3 h, 6 h, or 9 h, respectively, to the 0 h time point [41].

4.3. Relative Growth Rate (RGR) and Generation Time (gt)

RGR represents the cell mass gain determined as the increase in the OD_{600} value after each third hour of incubation and is calculated by Equation (1):

$$RGR = \frac{log(\Delta OD_{600})}{\Delta t} \tag{1}$$

where ΔOD_{600} represents the difference between optical cell density measured at time $t + 1$ and previous time t, and Δt is the time interval (3 h).

Generation time (*gt*) represents cell doubling time and is calculated by Equation (2):

$$gt = \frac{log\ (2)}{m} \tag{2}$$

where m is the gradient of the regression line.

4.4. IC_{50} Value

Represents concentration of AA that leads to 50% cell growth reduction. To calculate the IC_{50} value, cells were treated with serially diluted AA using log 2 dilution starting at 400 mM. Cells without (w/o) AA treatment were used as a control. Cells were incubated in 96 well plates for 9 h at 30 °C w/o shaking, and OD_{600} was determined every hour. For the calculation, the freely available online calculator was used, Quest Graph™ IC50 Calculator (https://www.aatbio.com/tools/ic50-calculator/, last accessed on 22 November 2021; AAT Bioquest, Sunnyvale, CA, USA).

4.5. Characterization of the Yeast Morphology

Yeast morphology analysis was performed as previously described [50]. Briefly, cells incubated with AA for 9 h with and w/o ASA pre-treatment were visualized under bright-field microscopy at 40× magnification (Leica DMI 6000, Leica microsystems, Wetzlar, Germany). Microscopy images were captured, and cell morphometric analyses were determined by the ImageJ software v. 1.52r (National Institutes of Health, CA, USA).

Cell volume (V; μm^3) was calculated according to Equation (3):

$$V = \frac{4}{3}\pi L W^2 \tag{3}$$

where L represents the cell length, and W is the cell width.

Cell surface (S; μm^2) was calculated according to Equation (4):

$$S = 2\pi\left(W^2 + LW\frac{arcsine}{\varepsilon}\right) \quad (4)$$

where, factor ε is calculated as $\varepsilon = \frac{\sqrt{L^2 - W^2}}{L}$.

4.6. Spot Test

To solid YES media 0, 0.1, 1, 10, and 20 mM of AA was added. Cells from the overnight culture were divided into two groups, one was left untreated, and the second was pre-treated with 10 mM AsA. Serially diluted cells resulting approximately in 10,000, 1000, 100, and 10 cells/spot, were placed on plates. After 2–3 days of incubation at 30 °C, the size and density of spots were compared.

4.7. Preparation of the Cell Extract for Biochemical Analyses

Control and AA-treated cells with and w/o AsA preincubation were collected by centrifugation at 8500 rpm for 90 s, washed 3 times with sterile H_2O, and resuspended in PBS (pH 7.0). Cells were either directly used for further analyses or stored at −80 °C. Cell homogenization was achieved by sonication (Digital Sonifier 450, Branson Ultrasonics Corp, Danbury, CT, USA) at 3 × 30 s intervals (repeated 1 s pulses followed by a 1 s pause giving 15 pulses within 30 s, and power 80 W representing 20% of the full power capacity) on ice. Cell debris was removed after 15 min centrifugation at 14,000× *g* and 4 °C. In the collected supernatant the protein level, metabolic activity, catalase (CAT) activity, superoxide dismutase (SOD) activity, and malondialdehyde (MDA) content were determined.

4.8. Metabolic Activity

Metabolic activity of yeast was calculated according to [53] with a few modifications. Briefly, yeast suspensions were centrifugated for 90 s at 10,000× *g* and washed with PBS (pH 7.0). Pellets were resuspended in 1 mL 0.5% 2,3,5-triphenyltertrazolium chloride (TTC) diluted in PBS and incubated for 20 h in the dark at 30 °C. After incubation, the pellet was washed twice with PBS, and generated red formazan was extracted by addition of 1 mL ethanol:acetone (2:1) mixture prior to cell lysis by sonication. Absorbance was measured at 485 nm and metabolic activity was calculated as relative units (r.u.) of absorbance per mg protein.

4.9. Cell Viability

Cells un- and pre-treated with AsA exposed to AA for 6 h were washed in PBS and 0.05% of methylene blue (Sigma–Aldrich) with 0.1% natrium citrate (Sigma–Aldrich) was added for 5 min. Methylene blue is able to penetrate only dead cells, hence blue-stained cells are considered as dead. Microscopic slides were prepared and a percentual portion of the dead to living cells was determined.

4.10. Biochemical Analysis

Agilent Cary 60 UV/VIS spectrophotometer (Agilent Technologies, Santa Clara, CA, USA) was used to determine CAT activity represented as the stepwise decrease in the absorbance at 240 nm for 90 s which determines the H_2O_2 decomposition. Addition of 50 µL of 30 mM H_2O_2 to 100 µL of the solution containing sample initialized reaction; the final volume of the reaction was 600 µL. The molar absorption coefficient of 36 mM^{-1} cm^{-1} was used to calculate specific catalase activity.

Total superoxide dismutase (SOD) activity was assayed according to [54] with slight modifications. Briefly, 100 µL of homogenized sample solution was added into 880 µL of a reactive mixture of 50 mM phosphate buffer (pH 7.8) containing 1 mM EDTA, 13 mM L-methionine, and 75 µM NBT (nitroblue tetrazolium). Finally, 20 µL of 2 mM riboflavin was added and the reaction was started by light irradiation (5000 lux) for 10 min at 20 °C. Absorbance of the samples was measured spectrophotometrically at 560 nm.

Malondialdehyde (MDA) content that represents lipid peroxidation was evaluated as previously described by [50]. Briefly, the TBA solution (15% trichloroacetate (TCA) containing 0.375% (w/v) thiobarbituric acid (TBA) was added to the supernatant of each sample and incubated at 95 °C for 30 min. The sample was rapidly cooled on ice and centrifuged at 8500 rpm for 60 s, the absorbance of the supernatant was measured at 532 and 600 nm at the Agilent Cary 60 UV/VIS spectrophotometer. The molar absorption coefficient 153 mM^{-1} cm^{-1} was used to calculate MDA content in nmol μg^{-1} protein.

The Bradford assay [55] was used to determine protein concentration at 600 nm using bovine serum albumin (Sigma–Aldrich, St. Louis, MO, USA) as a standard.

4.11. Determination of ROS Generation

Generation of the total ROS was performed as previously described by [56] with slight modifications. Control and AA-treated cell cultures with and w/o AsA (10 mM) preincubation were adjusted to $OD_{600} = 1$, washed with PBS, and incubated with 10 μM H_2DCFDA (Sigma–Aldrich) at 30 °C in the dark for 1 h without shaking. H_2DCFDA is a compound that is oxidized by ROS to a highly fluorescent DCF, which is fluorescently detectable at 498 nm wavelength, thus providing an estimate of ROS levels in the cell. The excessive fluorescent dye was washed off and the cells were resuspended in PBS. Fluorescence filter with 490 nm excitation wavelength at the Glomax Multi Detection System (Promega Corporation, Madison, WI, USA) was used for the DCF fluorescence detection. The measured values were normalized to the untreated cells.

4.12. Detection of Apoptosis and Necrosis

Evaluation of the yeast cells' apoptosis was performed according to [57] with some modifications. Annexin V- FITC (fluorescein isothiocyanate) -conjugated that specifically binds to phosphatidylserine (PS) residues, was used to detect the externalization of PS which is an apoptosis marker. Propidium iodide (PI) (Sigma Aldrich) penetrates dead cells and serves as a marker to differentiate between apoptosis and necrosis. After 1 h AA exposure with and without AsA treatment, yeast cells were collected by centrifugation and washed twice with phosphate-buffered saline (PBS), pH 6.8. Washed cells were resuspended in sorbitol buffer (1.2 M sorbitol, 0.5 mM $MgCl_2$, and 35 mM K_2HPO_4), pH 6.8 to a final concentration of 1×10^7 cells mL^{-1}. For cell wall digestion, cells were incubated with 10 μg mL^{-1} zymolyase (Roche) in sorbitol buffer for 1 h at 37 °C. Afterward, 1 mL of spheroplasts were centrifuged at 500 rpm for 5 min and resuspended in 60 μL of incubation buffer (containing the Annexin-V-FITC and PI) and incubated for 10 min at room temperature (RT). Cells were visualized by the fluorescence microscope (Leica DMI 6000, Leica microsystems, Wetzlar, Germany). Two independent experiments were evaluated, each containing at least 150 cells. Cells were observed in the stepwise-selected microscope fields to avoid counting the same cells.

4.13. Determination of Ion Composition

Ion composition was evaluated by the use of ICP-OES (ICP-OES 720, Agilent Technologies Australia (M) Pty Ltd., Santa Clara, CA, USA) as previously described [48]. Briefly, yeast cells exposed to 0-, 1-, and 10-mM AA for 3 and 9 h with and without AsA preincubation were washed three times with deionized water and incubated at 55 °C for 12 h. Weighted yeast pellets were placed into PTFE digestion tubes and 3 mL of extra pure HNO_3 was added for pressure microwave digestion by the ETHOS-One (Milestone, Srl., Sorisole (BG), Italy) microwave digestion system. Mineralized samples were filtered through a quantitative Munktell filter paper No. 390 (Munktell & Filtrak, Bärenstein, Germany) into 25 mL volumetric flasks and filled with deionized H_2O in 4 biological replicates. Afterwards, the ion content detected by ICP-OES was calculated to μg g^{-1} of the yeast dry matter and expressed as the content ratio to the untreated control.

4.14. Immunostaining and Fluorescence Microscopy

Chromosome segregation was analyzed by the use of fluorescence microscopy as previously described [58]. Shortly, yeast strain *JG15457* with the chromosome II marked with GFP was grown in YES medium at 30 °C and 150 rpm to achieve exponential growth. Cells were collected after 6 h of incubation, fixed by 2% PFA, and stained with primary TAT1 mouse monoclonal anti-tubulin and rabbit polyclonal anti-GFP antibodies, DNA was visualized using DAPI. Analyses were performed at the fluorescence microscope (Leica DMI 6000, Leica microsystems, Wetzlar, Germany) equipped with a digital camera. At least 200 cells in the anaphase stage of the mitotic cell cycle were evaluated for correct or incorrect segregation of chromosome II.

4.15. Statistical Analysis

Data are expressed as the mean ± standard deviation (SD). Statistical significance of obtained differences was analyzed by the ANOVA Duncan's and Fisher LSD post-hoc test using the Statistica 10 software (StatSoft Inc., Tulsa, OK, USA). Lavene's and Cochran's tests were used to evaluate data homogeneity and normality distribution of the results. Limits of the statistical significance were set up to $p < 0.05$ *, 0.01 **, 0.001 ***. Time-dependent ionome changes were calculated for each ion as a ratio to the untreated control. Absolute ion concentrations were then *Z-score*-transformed to adjust for the differences in magnitude between different ions according to Formula (5):

$$Z - score^{Individual} = \frac{\left(Concentration^{individual} - Mean\ concentration\right)}{SD} \quad (5)$$

Z-scores were used for cluster analysis by the average linkage clustering method with Pearson distance measurement. Pearson correlations between individual ions were calculated. *Z-score* and ion correlations were visualized by Heatmapper [59].

5. Conclusions

In this study, we investigated the underlining toxic effect of AA on cell vital functions by multiapproach analyses covering large-scale biological processes and stress responses. AA-induced enhancement of ROS production led to oxidative stress in *S. pombe* resulting in cell cycle arrest and alterations in chromosome segregation. Additionally, the cell disturbing activity of AA was confirmed by its negative effect on ion balance maintenance. Supplementation of ascorbic acid significantly protected cells against AA-mediated cytotoxicity due to its direct ROS scavenging activity and intracellular Ca uptake support. Our complex study extended the knowledge of AA-induced toxicity from the known metabolic disorders to disrupted ionome balance and altered chromosome segregation resulting in cell cycle arrest without marked signs of apoptosis. Additionally, we show that AsA treatment not only reduced ROS production but also improved ionome homeostasis and prevented errors in chromosome segregation during mitosis. Though, our results suggest that the use of AsA, a natural antioxidant supplementation, could significantly attenuate AA-induced cytotoxicity which might have health-protective implications in situations of dietary AA exposure. However, as AA is known to accumulate in the cell, further investigations of low-dose AA exposure for a longer period are required.

Supplementary Materials: The following are available online at https://www.mdpi.com/article/10.3390/molecules27134307/s1, Figure S1: Cell growth intensity upon AA addition with and w/o AsA pre-treatment, Figure S2: Relative growth rate of cells ex-posed to AA with and w/o AsA pre-treatment, Figure S3: Generation time (gt), Figure S4: Determination of IC50 upon AA exposure with and w/o AsA pre-treatment, Figure S5: Spot test, Figure S6: Cell viability upon AA addition with and w/o AsA pre-treatment, Figure S7: Protein content, Figure S8: CAT/MDA ratio.

Author Contributions: Conceptualization, M.K., M.P. and A.N.; methodology, M.K., A.N.; validation, M.P., M.K. and A.T.; formal analysis, M.K., A.N., R.K.; investigation, M.P.; resources, A.T.; data curation, M.K.; writing—original draft preparation, M.P.; writing—review and editing, M.K., A.N., A.T.; visualization, M.P., M.K.; project administration, A.T.; funding acquisition, A.N., M.P. All authors have read and agreed to the published version of the manuscript.

Funding: This research was supported by the SUA grant agency projects under grants 12-GASPU-2021, and GA FAPZ 2/2021; and the project "Drive4SIFood" of the European Union Operational Program Integrated Infrastructure Managing Authority Ministry of Transport and Construction of the Slovak Republic under grant number ITMS2014+ 313011V336.

Institutional Review Board Statement: Not applicable.

Informed Consent Statement: Not applicable.

Data Availability Statement: The authors confirm that the data supporting the findings of this study are available within the article and its Supplementary Materials.

Conflicts of Interest: The authors declare no conflict of interest.

Sample Availability: Samples of the compounds are available from the authors or from commercial sources.

References

1. Jakobsen, L.S.; Granby, K.; Knudsen, V.K.; Nauta, M.; Pires, S.M.; Poulsen, M. Burden of disease of dietary exposure to acrylamide in Denmark. *Food Chem. Toxicol.* **2016**, *90*, 151–159. [CrossRef] [PubMed]
2. Parzefall, W. Minireview on the toxicity of dietary acrylamide. *Food Chem. Toxicol.* **2008**, *46*, 1360–1364. [CrossRef] [PubMed]
3. Friedman, M. Chemistry, biochemistry, and safety of acrylamide. *A review. J. Agric. Food Chem.* **2003**, *51*, 4504–4526. [CrossRef] [PubMed]
4. Adani, G.; Filippini, T.; Wise, L.A.; Halldorsson, T.I.; Blaha, L.; Vinceti, M. Dietary Intake of Acrylamide and Risk of Breast, Endometrial, and Ovarian Cancers: A Systematic Review and Dose–Response Meta-analysis. *Cancer Epidemiol. Biomark. Prev.* **2020**, *29*, 1095–1106. [CrossRef] [PubMed]
5. Hong, Z.; Minghua, W.; Bo, N.; Chaoyue, Y.; Haiyang, Y.; Haiqing, Y.; Chunyu, X.; Yan, Z.; Yuan, Y. Rosmarinic acid attenuates acrylamide induced apoptosis of BRL-3A cells by inhibiting oxidative stress and endoplasmic reticulum stress. *Food Chem. Toxicol.* **2021**, *151*, 112156. [CrossRef]
6. IARC. Monographs on the evaluation of carcinogenic risks to humans. *Int. Agency Res. Cancer IARC* **1994**, *60*, 389–433. Available online: https://publications.iarc.fr/Book-And-Report-Series/Iarc-Monographs-On-The-Identification-Of-Carcinogenic-Hazards-To-Humans/Some-Industrial-Chemicals-1994 (accessed on 2 May 2022).
7. Shipp, A.; Lawrence, G.; Gentry, R.; McDonald, T.; Bartow, H.; Bounds, J.; Macdonald, N.; Clewell, H.; Allen, B.; Van Landingham, C. Acrylamide: Review of toxicity data and dose-response analyses for cancer and noncancer effects. *Crit. Rev. Toxicol.* **2006**, *36*, 481–608. [CrossRef]
8. Kacar, S.; Sahinturk, V.; Kutlu, H.M. Effect of acrylamide on BEAS-2B normal human lung cells: Cytotoxic, oxidative, apoptotic and morphometric analysis. *Acta Histochem.* **2019**, *121*, 595–603. [CrossRef]
9. Song, D.; Xu, C.; Holck, A.L.; Liu, R. Acrylamide inhibits autophagy, induces apoptosis and alters cellular metabolic profiles. *Ecotoxicol. Environ. Saf.* **2021**, *208*, 111543. [CrossRef]
10. Fuhr, U.; Boettcher, M.I.; Kinzig-Schippers, M.; Weyer, A.; Jetter, A.; Lazar, A.; Taubert, D.; Tomalik-Scharte, D.; Pournara, P.; Jakob, V.; et al. Toxicokinetics of Acrylamide in Humans after Ingestion of a Defined Dose in a Test Meal to Improve Risk Assessment for Acrylamide Carcinogenicity. *Cancer Epidemiol. Biomark. Prev.* **2006**, *15*, 266–271. [CrossRef]
11. Kwolek-Mirek, M.; Zadrąg-Tęcza, R.; Bednarska, S.; Bartosz, G. Yeast Saccharomyces cerevisiae devoid of Cu,Zn-superoxide dismutase as a cellular model to study acrylamide toxicity. *Toxicol. Vitr.* **2011**, *25*, 573–579. [CrossRef]
12. Albalawi, A.; Alhasani, R.H.A.; Biswas, L.; Reilly, J.; Shu, X. Protective effect of carnosic acid against acrylamide-induced toxicity in RPE cells. *Food Chem. Toxicol.* **2017**, *108*, 543–553. [CrossRef]
13. Kommuguri, U.N.; Pallem, P.V.S.; Bodiga, S.; Bodiga, V.L. Effect of dietary antioxidants on the cytostatic effect of acrylamide during copper-deficiency in Saccharomyces cerevisiae. *Food Funct.* **2014**, *5*, 705–715. [CrossRef]
14. Ahmad Bainmahfouz, F.R.; Ali, S.S.; Al-Shali, R.A.; El-Shitany, N.A.E.-A. Vitamin E and 5-amino salicylic acid ameliorates acrylamide-induced peripheral neuropathy by inhibiting caspase-3 and inducible nitric oxide synthase immunoexpression. *J. Chem. Neuroanat.* **2021**, *113*, 101935. [CrossRef]
15. Chen, H.; Gu, Z. Effect of Ascorbic Acid on the Properties of Ammonia Caramel Colorant Additives and Acrylamide Formation. *J. Food Sci.* **2014**, *79*, C1678–C1682. [CrossRef]
16. Zheng, L.; Zhu, H.-Z.; Wang, B.-T.; Zhao, Q.-H.; Du, X.-B.; Zheng, Y.; Jiang, L.; Ni, J.-Z.; Zhang, Y.; Liu, Q. Sodium selenate regulates the brain ionome in a transgenic mouse model of Alzheimer's disease. *Sci. Rep.* **2016**, *6*, 39290. [CrossRef]
17. López-Orenes, A.; Bueso, M.C.; Conesa, H.; Calderón, A.A.; Ferrer, M.A. Seasonal ionomic and metabolic changes in Aleppo pines growing on mine tailings under Mediterranean semi-arid climate. *Sci. Total Environ.* **2018**, *637–638*, 625–635. [CrossRef]

18. Hoffman, C.S.; Wood, V.; Fantes, P.A. An Ancient Yeast for Young Geneticists: A Primer on the Schizosaccharomyces pombe Model System. *Genetics* **2015**, *201*, 403–423. [CrossRef]
19. Hayles, J.; Nurse, P. Introduction to Fission Yeast as a Model System. *Cold Spring Harb. Protoc.* **2018**, *2018*, pdb.top079749. [CrossRef]
20. Villagran, M.; Ferreira, J.; Martorell, M.; Mardones, L. The Role of Vitamin C in Cancer Prevention and Therapy: A Literature Review. *Antioxidants* **2021**, *10*, 1894. [CrossRef]
21. Vo, T.V.; Das, J.; Meyer, M.J.; Cordero, N.A.; Akturk, N.; Wei, X.; Fair, B.J.; Degatano, A.G.; Fragoza, R.; Liu, L.G.; et al. A Proteome-wide Fission Yeast Interactome Reveals Network Evolution Principles from Yeasts to Human. *Cell* **2016**, *164*, 310–323. [CrossRef] [PubMed]
22. Vyas, A.; Freitas, A.V.; Ralston, Z.A.; Tang, Z. Fission Yeast Schizosaccharomyces pombe: A Unicellular "Micromammal" Model Organism. *Curr. Protoc.* **2021**, *1*, e151. [CrossRef] [PubMed]
23. Moser, B.A.; Russell, P. Cell cycle regulation in Schizosaccharomyces pombe. *Curr. Opin. Microbiol.* **2000**, *3*, 631–636. [CrossRef]
24. Boonstra, J.; Post, J.A. Molecular events associated with reactive oxygen species and cell cycle progression in mammalian cells. *Gene* **2004**, *337*, 1–13. [CrossRef]
25. Sundaram, G.; Palchaudhuri, S.; Dixit, S.; Chattopadhyay, D. MAPK mediated cell cycle regulation is associated with Cdc25 turnover in S. pombe after exposure to genotoxic stress. *Cell Cycle* **2008**, *7*, 365–372. [CrossRef]
26. Yang, L.; Dong, L.; Zhang, L.; Bai, J.; Chen, F.; Luo, Y. Acrylamide Induces Abnormal mtDNA Expression by Causing Mitochondrial ROS Accumulation, Biogenesis, and Dynamics Disorders. *J. Agric. Food Chem.* **2021**, *69*, 7765–7776. [CrossRef]
27. Agus, H.H.; Kok, G.; Derinoz, E.; Oncel, D.; Yilmaz, S. Involvement of Pca1 in ROS-mediated apoptotic cell death induced by alpha-thujone in the fission yeast (Schizosaccharomyces pombe). *FEMS Yeast Res.* **2020**, *20*, foaa022. [CrossRef]
28. Huang, Y.-S.; Hsieh, T.-J.; Lu, C.-Y. Simple analytical strategy for MALDI-TOF-MS and nanoUPLC—MS/MS: Quantitating curcumin in food condiments and dietary supplements and screening of acrylamide-induced ROS protein indicators reduced by curcumin. *Food Chem.* **2015**, *174*, 571–576. [CrossRef]
29. Hung, C.H.; Lin, Y.C.; Tsai, Y.G.; Lin, Y.C.; Kuo, C.H.; Tsai, M.L.; Kuo, C.H.; Liao, W.T. Acrylamide Induces Mitophagy and Alters Macrophage Phenotype via Reactive Oxygen Species Generation. *Int. J. Mol. Sci.* **2021**, *22*, 1683. [CrossRef]
30. Nowak, A.; Zakłos-Szyda, M.; Żyżelewicz, D.; Koszucka, A.; Motyl, I. Acrylamide Decreases Cell Viability, and Provides Oxidative Stress, DNA Damage, and Apoptosis in Human Colon Adenocarcinoma Cell Line Caco-2. *Molecules* **2020**, *25*, 368. [CrossRef]
31. Qu, D.; Liu, C.; Jiang, M.; Feng, L.; Chen, Y.; Han, J. After In Vitro Digestion, Jackfruit Flake Affords Protection against Acrylamide-Induced Oxidative Damage. *Molecules* **2019**, *24*, 3322. [CrossRef]
32. Chen, W.; Feng, L.; Shen, Y.; Su, H.; Li, Y.; Zhuang, J.; Zhang, L.; Zheng, X. Myricitrin Inhibits Acrylamide-Mediated Cytotoxicity in Human Caco-2 Cells by Preventing Oxidative Stress. *BioMed Res. Int.* **2013**, *2013*, e724183. [CrossRef]
33. Cao, J.; Liu, Y.; Jia, L.; Jiang, L.-P.; Geng, C.-Y.; Yao, X.-F.; Kong, Y.; Jiang, B.-N.; Zhong, L.-F. Curcumin Attenuates Acrylamide-Induced Cytotoxicity and Genotoxicity in HepG2 Cells by ROS Scavenging. *J. Agric. Food Chem.* **2008**, *56*, 12059–12063. [CrossRef]
34. Zhang, L.; Xu, Y.; Li, Y.; Bao, T.; Gowd, V.; Chen, W. Protective property of mulberry digest against oxidative stress—A potential approach to ameliorate dietary acrylamide-induced cytotoxicity. *Food Chem.* **2017**, *230*, 306–315. [CrossRef]
35. Orta Yilmaz, B.; Yildizbayrak, N.; Aydin, Y. Vitamin C inhibits glycidamide-induced genotoxicity and apoptosis in Sertoli cells. *J. Biochem. Mol. Toxicol.* **2020**, *34*, e22545. [CrossRef]
36. Catalgol, B.; Ozhan, G.; Alpertunga, B. Acrylamide-induced oxidative stress in human erythrocytes. *Hum. Exp. Toxicol.* **2009**, *28*, 611–617. [CrossRef]
37. Celik, F.S.; Cora, T.; Yigin, A.K. Investigation of Genotoxic and Cytotoxic Effects of Acrylamide in HEK293 Cell Line. *J. Cancer Prev. Curr. Res.* **2018**, *9*, 260–264. [CrossRef]
38. Gül, M.; Kayhan Kuştepe, E.; Erdemli, M.E.; Altınöz, E.; Gözükara Bağ, H.G.; Gül, S.; Göktürk, N. Protective effects of crocin on acrylamide-induced testis damage. *Andrologia* **2021**, *53*, e14176. [CrossRef]
39. Reshmitha, T.R.; Nisha, P. Lycopene mitigates acrylamide and glycidamide induced cellular toxicity via oxidative stress modulation in HepG2 cells. *J. Funct. Foods* **2021**, *80*, 104390. [CrossRef]
40. Abdel-Moneim, A.M.; Elsawy, H.; Alzahrani, A.M.; Ali, A.; Mahmoud, O. Silymarin Ameliorates Acrylamide-Induced Hyperlipidemic Cardiomyopathy in Male Rats. *BioMed Res. Int.* **2019**, *2019*, e4825075. [CrossRef]
41. Požgajová, M.; Navrátilová, A.; Šebová, E.; Kovár, M.; Kačániová, M. Cadmium-Induced Cell Homeostasis Impairment is Suppressed by the Tor1 Deficiency in Fission Yeast. *Int. J. Mol. Sci.* **2020**, *21*, 7847. [CrossRef]
42. Zhou, X.; Zheng, F.; Wang, C.; Wu, M.; Zhang, X.; Wang, Q.; Yao, X.; Fu, C.; Zhang, X.; Zang, J. Phosphorylation of CENP-C by Aurora B facilitates kinetochore attachment error correction in mitosis. *Proc. Natl. Acad. Sci. USA* **2017**, *114*, E10667–E10676. [CrossRef]
43. Hálová, L.; Petersen, J. Aurora promotes cell division during recovery from TOR-mediated cell cycle arrest by driving spindle pole body recruitment of Polo. *J. Cell Sci.* **2011**, *124*, 3441–3449. [CrossRef]
44. Sickles, D.W.; Sperry, A.O.; Testino, A.; Friedman, M. Acrylamide effects on kinesin-related proteins of the mitotic/meiotic spindle. *Toxicol. Appl. Pharmacol.* **2007**, *222*, 111–121. [CrossRef]
45. Adler, I.-D.; Zouh, R.; Schmid, E. Perturbation of cell division by acrylamide in vitro and in vivo. *Mutat. Res. Lett.* **1993**, *301*, 249–254. [CrossRef]

46. Chen, J.-H.; Tsou, T.-C.; Chiu, I.-M.; Chou, C.-C. Proliferation Inhibition, DNA Damage, and Cell-Cycle Arrest of Human Astrocytoma Cells after Acrylamide Exposure. *Chem. Res. Toxicol.* **2010**, *23*, 1449–1458. [CrossRef]
47. Hassan, H.A.; EL-Kholy, W.M.; EL-Sawi, M.R.F.; Galal, N.A.; Ramadan, M.F. Myrtle (Myrtus communis) leaf extract suppresses hepatotoxicity induced by monosodium glutamate and acrylamide through obstructing apoptosis, DNA fragmentation, and cell cycle arrest. *Environ. Sci. Pollut. Res.* **2020**, *27*, 23188–23198. [CrossRef]
48. Pozgajova, M.; Navratilova, A.; Arvay, J.; Duranova, H.; Trakovicka, A. Impact of cadmium and nickel on ion homeostasis in the yeast Schizosaccharomyces pombe. *J. Environ. Sci. Health B* **2020**, *55*, 166–173. [CrossRef] [PubMed]
49. Popa, C.V.; Dumitru, I.; Ruta, L.L.; Danet, A.F.; Farcasanu, I.C. Exogenous Oxidative Stress Induces Ca2+ Release in the Yeast Saccharomyces Cerevisiae. *FEBS J.* **2010**, *277*, 4027–4038. [CrossRef] [PubMed]
50. Navrátilová, A.; Kovár, M.; Požgajová, M. Ascorbic acid mitigates cadmium-induced stress, and contributes to ionome stabilization in fission yeast. *Environ. Sci. Pollut. Res.* **2021**, *28*, 15380–15393. [CrossRef] [PubMed]
51. Yu, D.; Danku, J.; Baxter, I.; Kim, S.; Vatamaniuk, O.K.; Vitek, O.; Ouzzani, M.; Salt, D.E. High-resolution genome-wide scan of genes, gene-networks and cellular systems impacting the yeast ionome. *BMC Genom.* **2012**, *13*, 623. [CrossRef]
52. Petersen, J.; Russell, P. Growth and the Environment of Schizosaccharomyces pombe. *Cold Spring Harb. Protoc.* **2016**, *2016*, pdb.top079764. [CrossRef]
53. Bayliak, M.M.; Burdylyuk, N.I.; Lushchak, V.I. Quercetin increases stress resistance in the yeast Saccharomyces cerevisiae not only as an antioxidant. *Ann. Microbiol.* **2016**, *66*, 569–576. [CrossRef]
54. Beauchamp, C.; Fridovich, I. Superoxide dismutase: Improved assays and an assay applicable to acrylamide gels. *Anal. Biochem.* **1971**, *44*, 276–287. [CrossRef]
55. Bradford, M.M. A rapid and sensitive method for the quantitation of microgram quantities of protein utilizing the principle of protein-dye binding. *Anal. Biochem.* **1976**, *72*, 248–254. [CrossRef]
56. Ďurovcová, I.; Goffa, E.; Šestáková, Z.; Mániková, D.; Gaplovská-Kyselá, K.; Chovanec, M.; Ševčovičová, A. Acute Exposure to Bisphenol A Causes Oxidative Stress Induction with Mitochondrial Origin in Saccharomyces cerevisiae Cells. *J. Fungi* **2021**, *7*, 543. [CrossRef]
57. Sousa, C.A.; Soares, H.M.V.M.; Soares, E.V. Nickel Oxide Nanoparticles Trigger Caspase- and Mitochondria-Dependent Apoptosis in the Yeast Saccharomyces cerevisiae. *Chem. Res. Toxicol.* **2019**, *32*, 245–254. [CrossRef]
58. Pozgajova, M.; Cipak, L.; Trakovicka, A. Prp4 kinase is required for proper segregation of chromosomes during meiosis in Schizosaccharomyces pombe. *Acta Biochim. Pol.* **2013**, *60*, 871–873. [CrossRef]
59. Babicki, S.; Arndt, D.; Marcu, A.; Liang, Y.; Grant, J.R.; Maciejewski, A.; Wishart, D.S. Heatmapper: Web-enabled heat mapping for all. *Nucleic Acids Res.* **2016**, *44*, W147–W153. [CrossRef]

molecules

MDPI

Article

Fatty Acid Profile and Antioxidant Capacity of Dabai (*Canarium odontophyllum* L.): Effect of Origin and Fruit Component

Shanti Faridah Salleh [1], Olaide Olawunmi Ajibola [2,3,4,*], Crilio Nolasco-Hipolito [5], Ahmad Husaini [2], Carvajal Zarrabal-Octavio [6], Samuel Lihan [3], Gbadebo Clement Adeyinka [7], Firdaus R. Rosli [2], Idris Adewale Ahmed [8,9], Mohamed Zaky Zayed [10] and Rosmawati Saat [2]

1 Institute of Sustainable and Renewable Energy (ISuRE), University Malaysia Sarawak, Kota Samarahan 94300, Malaysia; sshanti@unimas.my
2 Faculty of Resource Science and Technology, University Malaysia Sarawak, Kota Samarahan 94300, Malaysia; haahmad@unimas.my (A.H.); firdaus.rizwan@gmail.com (F.R.R.); srosma@unimas.my (R.S.)
3 Institute of Biodiversity and Environmental Conservation, University Malaysia Sarawak, Kota Samarahan 94300, Malaysia; lsamuel@unimas.my
4 Department of Biochemistry, Memorial University of Newfoundland, St. John's, NL A1C 5S7, Canada
5 Institute of Biotechnology, University del Papaloapan, Circuito Central 200, Col. Parque Industrial, San Juan Bautista Tuxtepec 68301, Mexico; cnolasco@unpa.edu.mx
6 Biochemistry and Nutrition Chemistry Area, University of Veracruz, Juan Pablo II s/n, Boca del Rio 94294, Mexico; ocarvajal@uv.mx
7 Department of Chemical Engineering, Mangosuthu University of Technology, Durban 4031, South Africa; adeyinkagbadebo78.ga@gmail.com
8 Center for Natural Products Research and Drug Discovery, University Malaysia, Kuala Lumpur 50603, Malaysia; idrisahmed@um.edu.my
9 Department of Biotechnology, Faculty of Applied Science, Lincoln University College, Kelana Jaya, Petaling Jaya 47301, Malaysia
10 Forestry and Wood Technology Department, Faculty of Agriculture (EL-Shatby), Alexandria University, Alexandria 21527, Egypt; mzmohamedzaky86@gmail.com
* Correspondence: 20010164@siswa.unimas.my or olaideajibola@gmail.com; Tel.: +60-1-6651-3450

Citation: Salleh, S.F.; Ajibola, O.O.; Nolasco-Hipolito, C.; Husaini, A.; Zarrabal-Octavio, C.; Lihan, S.; Adeyinka, G.C.; Rosli, F.R.; Ahmed, I.A.; Zayed, M.Z.; et al. Fatty Acid Profile and Antioxidant Capacity of Dabai (*Canarium odontophyllum* L.): Effect of Origin and Fruit Component. *Molecules* **2022**, *27*, 3840. https://doi.org/10.3390/molecules27123840

Academic Editor: Smaoui Slim

Received: 29 April 2022
Accepted: 3 June 2022
Published: 15 June 2022

Abstract: In the present work, the influence of geographical location on the fatty acid profiles, antioxidant potential, as well as cytotoxicity of edible dabai fruit fractions (kernel, skin, and pulp) were analyzed. The fatty acid profiles were determined by Gas Chromatography (GC), and the antioxidant activity was quantified with free 2,2-diphenyl-1-picr/ylhdrazyl, while the cytotoxicity was assessed by the brine shrimp lethality test. The results showed that the samples from Sibu, Serian, and Kapit geographical locations had a high content of the saturated fatty acids, ranging from 46.63% to 53.31% in the three fractions. The highest mono-saturated fatty acids (MUFA) content was found in Sibu. Serian and Kapit kernel fractions MUFA, however, ranged from 21.2% to 45.91%. No fatty acid composition was detected in Bentong and Kanowit. The fatty acid composition and DPPH free radical scavenging antioxidant activity of dabai were statistically independent using a multivariate analysis in different localities in Malaysia. The skin fraction had a more appreciable antioxidant potential and toxicity level than the pulp and kernel fractions. The highest antioxidant activity (EC_{50} 198.76 ± 1.06 μg/mL) with an LC_{50} value of 1387.22 μg/mL was obtained from the Sibu skin fraction. Therefore, the fatty acid composition, antioxidant, as well as cytotoxicity analyses of the extracts from different localities indicated that "geographical location" remarkably influenced fatty acid composition, antioxidant activity, and toxicity.

Keywords: *Canarium odontophyllum*; cytotoxicity; antioxidant; fatty acid profiles; geographical location

1. Introduction

Canarium odontophyllum L. (Sibu olive) also known as dabai in Sarawak, Malaysia, is an evergreen tree from the Burseraceae family. It is one of the sources of local commercial fruits and local timber trees. The fruit is known as Kembayau in Brunei and Sabah [1]. It is a seasonal fruit with two peak production seasons: mainly July–August and November–December, depending on the weather pattern. In Southeast Asia Borneo Island, it is propagated from seeds [2].

Recently, Sibu olive has been reported to have the potential to be tapped for commercial purposes due to its medicinal and nutritional values. Hence, it is one of the most important sources of employment opportunities for most of the unemployed youths and women in the rural population [3]. During the past two decades, its cultivation has attracted considerable attention because of its general acceptability by consumers and its export potential [4]. The kernel and pulp of this fruit are rich in fat. The fat of the dabai kernel is highly saturated, while the fat of the dabai pulp is moderately saturated [5]. The highly saturated fat of the dabai kernel is the substitute for palm kernel fat and cocoa butter in making chocolates [6]. The fat derived from the dabai kernel and pulp has been biologically proven to enhance the lipid profile of laboratory rats treated with dabai fat [7]. The defatted dabai peel extract has also been reported to demonstrate protective properties, like lipid peroxidation and inhibition of oxidative stress [8].

The skin, flesh, and kernel of dabai have been reported to possess antioxidant properties [9]. According to Prasad et al. [10], the ethyl acetate fraction of dabai peel had stronger antioxidant capacities than other fractions. Besides antioxidants, the study also reported that the peel fraction had the highest phenolic and flavonoid contents compared to other fractions [10]. Stem bark and leaves of dabai have been shown to have weaker cytotoxicity against the human colorectal carcinoma (HCT) 116-cancer cell lines [11]. However, the previous study has shown that differences in climate, cultivar, and maturity exert an impact on the nutritional and physical properties of dabai fruits; a study had shown that dabai fruits have been proven to be an excellent source of lipids, protein, and energy, as well as certain essential minerals, such as calcium, magnesium, and phosphorus [12]. In addition, this fruit has high antioxidant properties and a creamy taste. Anthocyanin is the primary phenolic compound in dabai fruit. For example, the antioxidant abilities of different fruit parts (peel, pulp, and seeds) depend on the extraction method and the solvent used.

To the best of our knowledge, no published study is available on the influence of geographical location on the fatty composition, antioxidant activity, and toxicity of different edible fractions of dabai fruits. Furthermore, the origin effect has not been thoroughly studied, despite the reported effects of genotypes. Thus, this study aims to assess the effect of geographical location on the different edible fractions of dabai fruit (kernel, skin, and pulp) and to identify the best locality with the highest fatty composition, antioxidant properties, and non-toxicity potential among the different localities in Sarawak, Malaysia.

2. Materials and Methods

Matured dabai fruits from five localities in Sarawak, Malaysia, namely Sibu, Serian, Kapit, Kanowit, and Betong, were collected from the farm within these geographical locations. Fruits were cleaned under running tap water without any physical harm to remove any impurities or dirt. Fruits were allowed to dry overnight in the shade to avoid degradation of phytochemicals, before being stored in a freezer at a temperature of −20 °C. Frozen fruits were removed from the freezer and thawed at room temperature. Fruit parts, such as skin, pulp, and kernel, were separated manually. A knife was used to peel the skin fraction; the fraction of the pulp was scraped using a spoon, while the kernel fraction was collected using a sledgehammer to break the hard-shelled seeds. Prior to the analysis, the skin and pulp fractions were stored in the freezer at a temperature of about −20 °C, whereas the kernel fraction was dried in an oven for 6 h at a temperature of 30 °C before it was blended to form a powder [13].

2.1. Sample Extraction

A 100 g sample of each fraction of *C. odontophyllum* L. fruits (skin, pulp, and kernel) was extracted with n-hexane in a 2 L Erlenmeyer flask. The mixture was homogenized and soaked at an ambient temperature for 24 h before it was filtered through filter paper (TISH Scientific, Cleves, OH, USA). A rotary evaporator (Heidolph Laborota 4000, Heidolph Instruments GmbH & Co KG, Schwabach, Germany) was used to concentrate the extracted oil samples at 40 °C [14]. The extracted oil was flushed with nitrogen gas to remove the solvent's residual before it was stored in a dark room. The extraction was done in triplicate.

2.2. Evaluation of the Fatty Acids Composition in the Extract Samples

The percentage of fatty acids from extracted oils in the skin, pulp, and kernel fractions of dabai from different geographical locations was qualitatively evaluated in this study. Briefly, the fatty acid standard was run using Gas Chromatography Mass Spectrometry (GC-MS) to know the retention time for individual fatty acid components (based on chromatogram peak), and this served as a reference for possible identification. The sample extract was then injected into GC-MS following the same procedure as done for the standard, a sample chromatogram was obtained, and each peak was identified using standard retention time. Each peak in the samples was further confirmed with the NIST mass spectral library, where 95 to 98% marge was considered. The percentage yield of individual fatty acid was thereby obtained.

A GC-MC Shimadzu brand (Model QP-2010 Plus) with an HP-5 fused capillary column (5% phenyl methylpolysiloxane stationary phase), with 30.0 m × 0.25 mm × 0.25 μm capacity, was used. The GC-MS configuration was programmed with the injector and detector temperature set at 200 °C and 210 °C, respectively. A split mode injection (1:10) was conducted. and helium was employed as a carrier gas at a controlled constant flow rate of 1.0 mL/min. The initial column oven temperature was 40 °C for 4 min, raised to 210 °C at the rate of 4 °C/min, and kept for a further 15 min. A 1.0 μL injection volume of each oil sample (dissolved with 200-μL n-hexane) was injected into the GC column.

2.3. DPPH Free Radical Scavenging Antioxidant Activity

The method by Ajibola et al. [15,16] was adapted to determine 1, 1-diphenyl-2-picryl-hydrazyl (DPPH) free radical scavenging activities (RSA) of skin, pulp, and kernel fractions of dabai extracts. The extracts were first kept at 50 °C for 2 h, employing an orbital shaker incubator (New Brunswick C2, Edison, NJ, USA) with an agitation rate of 180 rpm. The homogenates were then filtered through filter papers employing Buchner funnels, before being freeze-dried and kept at −30 °C until further tests were determined. All the freeze-dried extracts were completely dissolved in 10.0 mL of methanol (CH_3OH) and sonicated for 6 min to form a stock solution. The working standard solutions (10, 50, 100, 500, 1000 μg/mL) were further prepared from the stock solution. For each concentration, 1.0 mL of the solution was pipetted into a screw-top dark vial before it was thoroughly mixed with 3.0 mL of 0.1 mM solution of DPPH in CH_3OH solution. The mixture was incubated in the dark at 25 °C for 30 min and analyzed at 517 nm employing a UV/vis spectrophotometer (Shimadzu Co., Kyoto, Japan), with the absorbance of 4 mL of 0.1 mM of DPPH in CH_3OH solution, without the sample extract counted as a negative control and ascorbic acid as a positive control. Each solution was read in triplicate. The DPPH free radical scavenging activity (RSA) (%) was calculated according to Equation (1):

$$\text{DPPH free radical scavenging activity } (\%) = \left(\frac{Ab_0 - Ab_1}{Ab_0}\right) \times 100 \quad (1)$$

where

Ab_0 = the absorbance of the control (without sample) at 517 nm;
Ab_1 = the absorbance of the sample at 517 nm

The RSA of the extracts was expressed in terms of EC_{50} value, which is the effective concentration by which 50% of DPPH free radical was scavenged. The EC_{50} value was determined from the sigmoid curve of antioxidant activity (%) versus the sample concentration; the lower the value of EC_{50}, the stronger the extract has to act as a DPPH scavenger [17]. In comparison to both the extracts and ascorbic acid (positive standard), a standard antioxidant was prepared to find the extract with the strongest scavenging activity.

2.4. Brine Shrimp Lethality Bioassay

Artemia salina (brine shrimp) larvae were used to perform toxicity tests. About 0.7 g of *A. salina* egg was added into a conical beaker containing 1.0 L of seawater, with pH 7.6, temperature 26–28 °C, salinity 22 ppt, and continuous airflow for the hatching process to occur for 48 h [18]. For the brine shrimp lethality test, exactly 5.0 mg of lipid extracts was weighed in the calibrated sample bottle before it was dissolved in 5.0 mL of methanol. Exactly 10 ppm, 50 ppm, 100 ppm, 250 ppm, and 500 ppm of the solution were transferred into the test plates in triplicate. The solvent was evaporated to dryness in the fume hood before 0.2 mL of dimethyl sulfoxide (DMSO) and 4.8 mL of seawater were added to each of the six-well plates. Exactly 0.2 mL of the diluted sample was transferred into every well plate, and 10 nauplii (larvae) of *A. salina* cyst were added and incubated for 24 h under direct light at 25 °C. Three replicates per concentration were conducted in this biological activity. The solution containing 5.0 mg of thymol in 5.0 mL of seawater with 10 nauplii without the sample was used as a positive control (standard), whereas the solution containing 4.50 mL of seawater and 0.5 mL of DMSO with 10 nauplii without the sample was used as a negative control. During the study, no feed or air was needed since the feeding of brine shrimps with dry yeast suspension during the toxicity assessment was considered insignificant [19]. Populations of dead and surviving nauplii were counted in each well, and LC_{50} was calculated. Based on Thangapandi Veni calculations [19], toxicity characteristics of the extracts were estimated as follows (Equation (2)) [20]:

$$Mortality\ (\%) = \frac{F_{test} - D_{control}}{B_{control}} \times 100 \qquad (2)$$

where F_{test} was the population of dead larvae in each test plate, $B_{control}$ was the population of live larvae in control plates, and $D_{control}$ was the population of dead larvae in each control plate.

2.5. Statistical Analysis

All results were expressed as group mean ± standard deviation (SD), analyzed using SPSS v. 21.0 (IBM statistical software 21.0). The statistical test was investigated using Tukey's post hoc test, and the p-value < 0.05 was considered to be a significant difference. The determination of the mean death of the *A. salina* nauplii at different concentrations of the samples and the concentration that kills 50% (LC_{50}) of the brine shrimp were made using Microsoft Excel, and the probit analysis test was conducted on statistical software SPSS ($p < 0.05$).

2.6. Multivariate Analysis

A Principal Component Analysis (PCA) based on a correlation matrix was performed to elucidate the correlated fatty acid composition and DPPH free radical scavenging antioxidant activity in different localities of dabai fruit. The results were further analyzed by a multivariate statistical approach employing a hierarchical cluster analysis (HCA, Pvclust function in R) associated with a proximity score matrix represented as a proximity heat-map.

3. Results and Discussion

3.1. Determination of Fatty Acids

The kernel and pulp of *C. odontophyllum* fruits, locally called Sibu olive and dabai, are rich in fat [5]. The highly saturated fat of the dabai kernel is not only used as a substitute for palm kernel fat and cocoa butter in making chocolates [6], but it has also been biologically employed to enhance the lipid profile of laboratory animals [7]. The antioxidant properties of the defatted dabai peel extract have also been reported [8]. Despite the potential of the dabai fruit, the origin effect has not been thoroughly studied, despite the reported effects of genotypes.

Figure 1 presents the GC-MS chromatogram for the mixture of FAME standards, whereby a total of 14 individual FAMEs were eluted. Table 1 presents the retention times for the identified components of the FAMEs standards, analyzed by GC-MS.

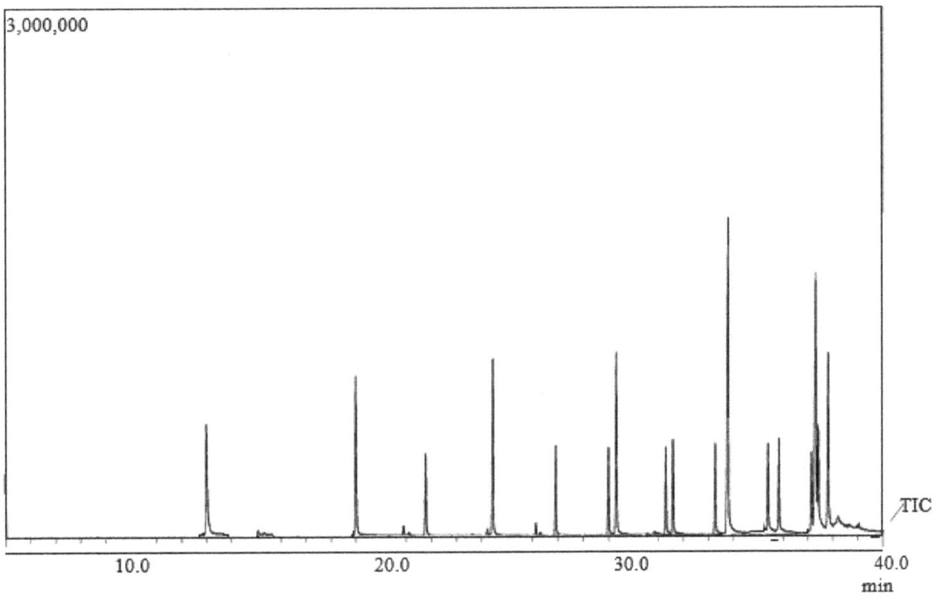

Figure 1. GC-MS chromatogram for the mixture of FAME standards.

Table 1. Retention times for identified components of FAME standards analyzed by GC-MS.

No.	Fatty Acid (as Methyl Esther)	IUPAC	Retention Times, t_{Rx} (min)
1	Undecanoic acid methyl ester	C11:0	13.14
2	Lactic acid methyl ester	C12:0	18.12
3	Tridecanoic acid methyl ester	C13:0	20.62
4	Myristic acid methyl ester	C14:0	23.65
5	Pentadecanoic acid methyl ester	C15:0	26.71
6	Palmitoleic acid methyl ester	C16:1	29.01
7	Palmitic acid methyl ester	C16:0	29.43
8	Heptanoic acid methyl ester	C17:0	31.76
9	Linolenic acid methyl ester	C18:3n3	33.17
10	Linoleic acid methyl ester	C18:2n6c	33.64
11	Oleic acid methyl ester	C18:1n9c	35.69
12	Stearic acid methyl ester	C18:0	36.26
13	Arachidonic acid methyl ester	C20:4n6	37.21
14	Eicosadienoic acid methyl ester	C20:2	37.86

Tables 2–4 present the total amount of fatty acids in the skin, pulp, and kernel fractions of dabai fruit obtained from different geographical locations, respectively. The dominant fatty acid was palmitic acid, ranging from 26.40% and 39.66%; followed by linoleic acid, ranging from 7% to 47.04%; oleic acid, ranging from 7% to 45.91%; and stearic acid, ranging from 9.01% to 23.42% (Table 2). Heptanoic acid was present in Kapit (skin) and Serian (pulp) in trace quantities, and eicosadienoic was present in Serian (skin) and Sibu (pulp).

Furthermore, only the Serian sample contained a trace quantity of elaidic acid, whereas the Sibu sample was the only one with a small amount of arachidic acid (0.81%).

Table 2. Fatty acid profiles (%) found in the skin fractions of *Canarium odontophyllum* L. fruit from different localities.

	Sibu	Serian	Kapit	Betong	Kanowit
SFA					
C16:0	39.66 ± 0.14 ^{bc}	27.90 ± 0.67 ^{a}	36.49 ± 0.33 ^{b}	ND	ND
C17:0	ND	ND	1.81 ± 0.21 ^{a}	ND	ND
C18:0	9.01 ± 0.32 ^{a}	23.42 ± 0.98 ^{c}	11.50 ± 0.11 ^{b}	ND	ND
C20:0	ND	2.73 ± 0.83 ^{a}	ND	ND	ND
Total:	48.67 ± 0.21 ^{a}	54.05 ± 0.77 ^{a}	49.80 ± 0.59 ^{a}	ND	ND
MUFA					
C18:1n9c	ND	ND	43.20 ± 0.32 ^{a}	ND	ND
C18:1n9t	ND	ND	ND	ND	ND
Total:	ND	ND	43.20 ± 0.32 ^{a}	ND	ND
PUFA					
C18:2n6c	47.04 ± 0.09 ^{c}	35.65 ± 0.25 ^{b}	7.00 ± 0.50 ^{a}	ND	ND
C20:2	ND	0.81 ± 0.11 ^{a}	ND	ND	ND
Total:	47.04 ± 0.09 ^{c}	36.46 ± 0.35 ^{b}	7.00 ± 0.50 ^{a}	ND	ND

The triplicate measurements were performed with the same 100 g of each fraction sample. Values are presented as mean ± SD (n = 3). Non-identical superscripts in the same row represent significant differences at $p < 0.05$. SFA = saturated fatty acid; PUFA = polyunsaturated fatty acids; MUFA = mono-unsaturated fatty acids; ND = not detected.

Table 3. Total amount of fatty acids (%) found in the pulp fractions of dabai fruit oil from different localities.

	Sibu	Serian	Kapit	Betong	Kanowit
SFA					
C16:0	29.86 ± 0.99 ^{ab}	36.49 ± 0.76 ^{b}	27.90 ± 0.88 ^{a}	ND	ND
C17:0	ND	1.81 ± 0.77 ^{a}	ND	ND	ND
C18:0	16.77 ± 0.11 ^{ab}	11.50 ± 0.25 ^{a}	23.42 ± 0.55 ^{b}	ND	ND
Total:	46.63 ± 0.15 ^{ab}	49.80 ± 0.33 ^{b}	41.87 ± 0.45 ^{a}	ND	ND
MUFA					
C18:1n9c	13.38 ± 0.97 ^{bc}	7.00 ± 0.86 ^{a}	10.30 ± 0.75 ^{b}	ND	ND
Total:	13.38 ± 0.15 ^{bc}	7.00 ± 0.35 ^{a}	10.30 ± 0.55 ^{b}	ND	ND
PUFA					
C18:2n6c	38.58 ± 0.66 ^{ab}	43.20 ± 0.73 ^{b}	35.65 ± 0.79 ^{a}	ND	ND
C20:2	1.40 ± 0.12 ^{a}	ND	ND	ND	ND
Total:	39.98 ± 0.06 ^{ab}	43.20 ± 0.58 ^{b}	35.65 ± 0.13 ^{a}	ND	ND

The triplicate measurements were performed with the same 100 g of each fraction sample. Results are expressed as mean value ± SD (n = 3). Non-identical letters in the same row indicate significantly different values at $p < 0.05$. SFA = saturated fatty acid; PUFA = polyunsaturated fatty acids; MUFA = mono-unsaturated fatty acids; ND = not detected.

It was observed that the Betong and Kanowit samples had no fatty acid compositions compared to the other locations.

In addition, the values obtained in this study indicated that saturated fatty acids (SFAs) and polyunsaturated fatty acids (PUFAs) are the major groups of skin oil as a result of palmitic acid and linoleic acid, while the monounsaturated fatty acids (MUFAs) were in minute quantities. The SFAs ranged from 48.67% to 54.05%, PUFAs ranged from 7% to 47%, and MUFAs ranged from 0% to 43.20% in all of the five localities (Table 2). In this study, no fatty acid content was detected in the dabai skin samples from Betong and Kanowit. The non-detectability of the fatty acids at these two locations does not necessarily mean that they were not present in the samples, but the values may have been below the detection limit of the instrument. More importantly, other overriding factors for the availability of fatty acids in the samples are post-harvesting factors, such as cultural practices; soil type; and climatic conditions, which could play a crucial role in the quantity and quality of fatty acids in dabai fruit. These factors play an important role in the availability of bioactive phytochemicals, nutritional qualities, and antioxidant properties of seed oil. Another possible factor could be differences in genetic characteristics of the seed-bearing plants, which could be the case for the observation in this study.

Table 4. Total amount of fatty acids (%) found in the kernel fractions of dabai fruit oil from different localities.

	Sibu	Serian	Kapit	Betong	Kanowit
SFA					
C16:0	35.82 ± 0.33 [b]	38.32 ± 0.44 [bc]	26.24 ± 0.55 [a]	ND	ND
C18:0	16.68 ± 0.48 [b]	10.64 ± 0.37 [a]	15.63 ± 0.26 [b]	ND	ND
C20:0	0.81 ± 0.59 [a]	ND	ND	ND	ND
Total:	53.31 ± 0.80 [c]	48.96 ± 0.91 [b]	41.87 ± 0.39 [a]	ND	ND
MUFA					
C18:1n9c	45.91 ± 0.28 [bc]	43.65 ± 0.19 [b]	21.22 ± 0.37 [a]	ND	ND
Total:	45.91 ± 0.28 [bc]	43.65 ± 0.19 [b]	21.22 ± 0.37 [a]	ND	ND
PUFA					
C18:2n6c	3.04 ± 0.13 [a]	3.83 ± 0.26 [a]	35.59 ± 0.39 [b]	ND	ND
Total:	3.04 ± 0.13 [a]	3.83 ± 0.26 [a]	35.59 ± 0.39 [b]	ND	ND

The triplicate measurements were performed with the same 100 g of each fraction sample. Results are expressed as mean $\pm$ SD (n = 3). Non-identical letters in the same row indicate significantly different values at $p < 0.05$. SFA = saturated fatty acid; PUFA = polyunsaturated fatty acids; MUFA = mono-unsaturated fatty acids; ND = not detected.

The production of these fatty acid contents was significantly influenced by the location and geographical factors, which is consistent with the findings of Parcerisa et al. [21]. Studies have shown that geographical factors such as ecology, location, growing conditions, species, altitude, climate, season, soil type, maturity, and harvest period have a significant impact on the fatty acid composition [22]. Other studies like Liu et al. [23], Górnaś et al. [24], He et al. [25], and Sicari et al. [26] have reported a similar trend. Non-detectability of fatty acids could as well be due to the presence of interferences, which could have a high impact in suppressing the fatty acid peaks.

It is also important to observe that the dabai from the Serian geographical location had a higher SFAs (54.05%) content. On the other hand, the dabai skin from Kapit had the highest MUFAs (43.20%) and the lowest PUFAs (7%) content.

Chua et al. [5] reported that the total PUFAs, MUFAs, and SFAs contents of six common dabai skin genotypes from Sibu and Kuching markets, Sarawak, Malaysia, ranged from 13.12% to 41.75%, 43.10% to 48.01%, and 10.24% to 42.62%, respectively. These findings were in agreement with our PUFAs and SFAs values, except for MUFA values.

Similarly, the result showed that the SFA and PUFA are the major groups of dabai pulp fatty acid due to linoleic acid, palmitic acid, and stearic acid, while MUFAs were in minute quantities. The SFAs ranged from 0% to 49.80%, PUFAs ranged from 0% to 43.20%, and MUFAs ranged from 0% to 13% in all of the five localities (Table 3). No fatty acid content was detected in the dabai pulp in Betong and Kanowit

Samples from Serian had higher SFAs (49.80%) and PUFAs (43.20%) contents, while the dabai pulps from Kapit had the lowest SFAs (41.87%) and PUFAs (35.65%) contents.

Azlan et al. [27] reported that the total PUFAs, MUFAs, and SFAs contents of dabai pulp from the Sarawak Agriculture Department using Soxhlet extraction were 12.76%, 42.82%, and 44.43%, respectively. Jelani et al. [28] also reported that the total MUFAs, PUFAs, and SFAs contents of two-mixture clones from the Semongok Agriculture Research

Centre were 25.52%, 15.94%, and 59.54%, respectively. The results obtained are in agreement with Azlan et al. [27] and Jelani et al. [28] for SFAs values but not for PUFA and MUFA values. Research on other plants has indicated that the oil of the fruits grown in low-temperature localities contains more unsaturated fatty acids than the fat of fruits grown in hot localities [22].

Except for PUFA values, our findings are in line with Shakirin et al. [7], who observed that the total MUFAs of the dabai pulp oil were lower compared to high-fat fruits, such as avocado (65% to 68%) and olive (56% to 86%). The fatty acid profile of dabai pulp contains an oil that is comparable to palm oil [5].

The results showed that the SFA and MUFA are the major groups of dabai kernel fatty acid, containing oleic acid, palmitic acid, and stearic acid, while PUFAs were in minute quantities, except for the dabai kernel samples from the Kapit region. The SFAs ranged from 41.87% to 53.31%, PUFAs ranged from 3.04% to 35.59%, and MUFAs ranged from 21.22% to 47.21% for the three localities (Sibu, Serian, and Kapit), as shown in Table 4. The fatty acid compositions contained in samples have multiple functions, and their bioactivity and mechanism of action highly depends on their composition and environmental factors.

Interestingly, Sibu samples had the highest SFAs (53.31%) and PUFAs (45.91%), while Kapit kernels had the lowest SFAs (41.87%) and PUFAs (35.65%) contents.

An earlier study by Shakirin et al. [7] reported that the dabai kernel has slightly higher SFAs than MUFAs and PUFAs using Soxhlet extraction. Azlan et al. [27] obtained similar results from their study of the determination of fatty acid content of dabai kernel oil extracted by employing Soxhlet extraction for SFAs and MUFAs values, except for the PUFAs values. They reported that the total of SFAs, MUFAs, and PUFAs content were 60.84%, 35.11%, and 3.78%, respectively. This disagreement for the PUFA values may be due to geographical conditions of the environment, climate, and varieties [22,28,29].

In this study, oleic and palmitic acids accounted for more than 80% of the total fatty acids from Sibu and Serian. In comparison to Ibrahim et al. [30], the oleic and palmitic acids in the tested kernel oil were higher. The kernel oil had a similar fatty acid profile to *C. ovatum*, with oleic and palmitic acids accounting for 32.6–38.2% and 44.4–59.6%, respectively [31]. The SFAs values in the two studied geographical locations (Sibu and Serian) were higher in comparison to *C. album* L., although kernel oil can be characterized as a saturated fatty acid. However, it was less saturated compared to coconut oil [7]. In addition, oil from the dabai kernel contained some saturated fatty acids (lauric and myristic acid), similar to coconut oil.

Multivariate Statistical Analysis of the total amount of fatty acids (%) found in the kernel, pulp, and skin fractions of dabai fruit oil from different localities.

The multivariate statistical analysis revealed interconnected correlation patterns among the fatty acid composition (%) found in the kernel, pulp, and skin fractions of dabai fruit oil from different localities. (Figure 2).

The spatial arrangement of the fatty acid composition found in the kernel, pulp, and skin fractions of dabai fruit oil from Serian and Kapit in the double-positive quadrant signified the presence of interrelated fatty acids (Figure 2a). Moreover, clustering of the fatty acid composition exhibited a mixed pattern in describing the variation of the three localities. Furthermore, the presence of a fairly distinct fatty acid composition in Sibu, Serian, and Kapit was highlighted by their isolated spatial location as well as the correlation matrix (Table 5). It was interesting that the clustering analysis using a hierarchical method (Figure 2b) showed a concordant pattern with the PCA analysis. The fatty acid composition of dabai in the three localities was grouped into various clusters, according to their proximities. The proximity heat map (Figure 2c) corroborated the correlation matrix, demonstrating a similar fatty acid composition in the three localities of dabai.

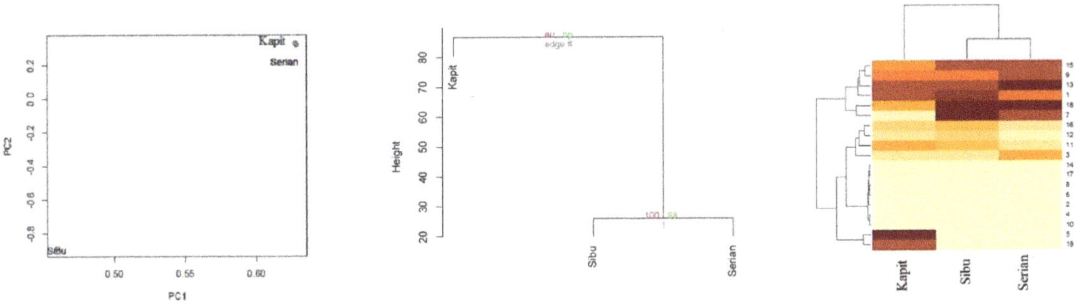

(a) (b) (c)

Figure 2. The multivariate statistical analysis revealed interconnected correlation patterns among the fatty acid composition (%) found in the kernel, pulp, and skin fractions of dabai fruit oil from different localities. (**a**) PCA based on the correlation matrix of fatty acids (%); (**b**) A consensus tree of the relationships among Sibu, Serian, and Kapit in the fatty acids (%). The Pvclust package in R was used to cluster these traits according to the Euclidean distance matrix. The numbers at the forks were the percentages of approximately unbiased (AU; in red) p-values and bootstrap probabilities (BP, in green), estimated from 1000 bootstrapping samples; (**c**) A heatmap showing cluster groups in both the fatty acids (%) and different localities (Sibu, Serian, and Kapit). The numbers on the right side represent the codes for 19 fatty acids (%) found in the kernel, pulp, and skin fractions within three localities that showed a mixed pattern.

Table 5. Correlation Coefficient Matrix of the fatty acids (%) found in the kernel, pulp, and skin. Fractions of dabai fruit oil from different localities.

	Sibu	**Serian**	**Kapit**
Sibu	1.0000		
Serian	0.9387524	1.0000	
Kapit	0.4684037	0.4738977	1.0000

Note: *p*-values for statistical tests are in parentheses.

3.2. Antioxidant Activity Measured Using DPPH

The values of DPPH free radical scavenging activity at 1000 µg/mL for skin, pulp, and kernel fractions of dabai fruit oils from the different geographical locations are shown in Table 6. At a concentration of 1000 µg/mL, the skin fraction of dabai from the Sibu locality exhibited the highest percentage (69.84 ± 0.14%) compared to the other fruit lipids.

The results of the test of variance of the antioxidant bioactivity in terms of the EC_{50} of the skin, pulp, and kernel fractions of dabai extracts showed that the effect of geographical location was significant ($p < 0.05$) (Table 6).

These data agreed with Rashid et al. [32], who reported that dabai skin is the principal antioxidant source because of it being rich in phenolic contents. In addition, the EC_{50} value for a skin fraction of dabai from Sibu was 198.76 ± 1.06 µg/mL; thus, the strength for this extract to act as a DPPH scavenger was higher compared to the skin fractions of dabai from Serian, Betong, Kapit, and Kanowit (Table 7). However, the EC_{50} values for other dabai lipid extracts could not be obtained due to their very low percentage of antioxidant activity (Table 8). These findings were in line with the works of Jelani et al. [28] and Rashid et al. [32]. It may be due to the low total phenolic, flavonoid, and anthocyanin contents.

Table 6. DPPH free RSA at 1000 μg/mL for three different fractions of extracted dabai fruit lipids of different localities with standard ascorbic acid.

Fruit Lipids	Percentages of DPPH Scavenging Activity at 1000 μg/mL (±S.D.%)
C. odontophyllum L. skin fraction	
Sibu	69.84 ± 0.01
Serian	65.08 ± 0.30
Kapit	57.38 ± 0.51
Betong	60.09 ± 0.12
Kanowit	53.61 ± 0.14
C. odontophyllum L. pulp fraction	
Sibu	29.68 ± 0.52
Serian	29.49 ± 0.17
Kapit	30.08 ± 0.58
Betong	30.08 ± 0.38
Kanowit	31.41 ± 0.57
C. odontophyllum L. kernel fraction	
Sibu	12.00 ± 0.18
Serian	11.37 ± 1.76
Kapit	13.36 ± 0.59
Betong	14.97 ± 0.59
Kanowit	15.60 ± 0.14
Ascorbic acid (Standard reference)	96.92 ± 0.00

Table 7. Correlation Coefficient Matrix of the values of DPPH free radical scavenging activity at 1000 μg/mL for the skin, pulp, and kernel fractions of dabai fruit oils from the different geographical locations.

	Sibu	Serian	Kapit	Betong	Kanowit
Sibu	1.0000				
Serian	0.9991089	1.0000			
Kapit	0.9981758	0.9967917	1.0000		
Betong	0.9953717	0.9940903	0.9988797	1.0000	
Kanowit	0.9956498	0.9938128	0.9992088	0.9997455	1.0000

Note: *p*-values for statistical tests are in parentheses.

The result confirmed that the EC_{50} values varied from 320.64 ± 1.09 to 198.76 ± 1.06 μg/mL. The EC_{50} value of the standard ascorbic acid solution was lower at 18.59 ± 1.06 μg/mL compared to EC_{50} values for all dabai lipid extracts, indicating that ascorbic acid had the strongest power to scavenge DPPH radicals. Since EC_{50} is inversely associated with the anti-radical ability of the compounds, the lower the EC_{50}, the higher the antioxidant activity [33]. Based on the outcomes of this study, the effect of the geographical location on the antioxidant ability of the skin fraction was very profound, which is consistent with the findings of Dastoor et al. [34], Koohsari et al. [35], and Zargoosh et al. [36]. The skin fraction of the dabai from Sibu, with the lowest $EC_{50,}$ had the highest antioxidant bioactivity. The natural antioxidant components contained in the fruit extract fractions have multiple functions, and their bioactivity and mechanism of action greatly depend on their geographical and cultivation conditions, since these conditions affect the synthesis of the formation of secondary bioactive compounds in the plant. Plant geographical location, as a result of environmental differences, weather differences, and soil conditions, can influence the formation of secondary bioactive compounds in the plant. In addition, it might also be due to salinity induced by the metabolic systems [36]. The relative geographical habitat of this locality is directly associated with the increase in antioxidant activity [37].

Table 8. EC_{50} values of different fractions of *C. odontophyllum* L. lipids from different localities.

Lipid Sample	EC_{50} Value (±S.D. μg/mL)
Skin fraction	
Sibu	198.76 ± 1.06 [c]
Serian	214.88 ± 1.22 [bc]
Kapit	298.24 ± 1.08 [ab]
Betong	289.08 ± 1.07 [b]
Kanowit	320.64 ± 1.09 [a]
Pulp fraction	
Sibu	N.D.
Serian	N.D.
Kapit	N.D.
Betong	N.D.
Kanowit	N.D.
Kernel fraction	
Sibu	N.D.
Serian	N.D.
Kapit	N.D.
Betong	N.D.
Kanowit	N.D.
Ascorbic acid (control)	18.59 ± 1.06 [d]

The triplicate measurements were performed with the same 100 g of each fraction sample. Values are presented as mean ± SD (n = 3). Non-identical superscripts in the same row represent significant differences at $p < 0.05$. ND: not detected.

The topography could have also contributed to the antioxidant property of the collected samples from Serian and Sibu. This may be due to reduced temperature and increased exposure to ultraviolet radiation of the organ responsible for the synthesis of antioxidant elements.

Multivariate Statistical Analysis of the values of DPPH free radical scavenging activity at 1000 μg/mL for skin, pulp, and kernel fractions of dabai fruit oils from the different geographical locations.

The multivariate statistical analysis revealed interconnected correlation patterns among the values of DPPH free radical scavenging activity at 1000 μg/mL for skin, pulp, and kernel fractions of dabai fruit oils from the different geographical locations (Figure 3).

The spatial arrangement of the DPPH (%) found in the kernel, pulp, and skin fractions of dabai fruit oil from Sibu, Serian, Kapit, Betong, and Kanowit in the double-positive quadrant signified the presence of interrelated DPPH (%) (Figure 3a). Moreover, clustering of DPPH (%) exhibited a mixed pattern in describing the variation of the five localities. Furthermore, the presence of fairly distinct DPPH (%) in Sibu, Serian, Kapit, Betong, and Kanowit was highlighted by their isolated spatial location as well as the correlation matrix (Table 7). It was interesting that the clustering analysis using a hierarchical method (Figure 3b) showed a concordant pattern with the PCA analysis. The values of DPPH free radical scavenging activity at 1000 μg/mL for the skin, pulp, and kernel fractions of dabai fruit oils from the five localities were grouped into various clusters according to their proximities. The proximity heat map (Figure 3c) corroborated the correlation matrix, demonstrating similar DPPH (%) in the five localities of dabai.

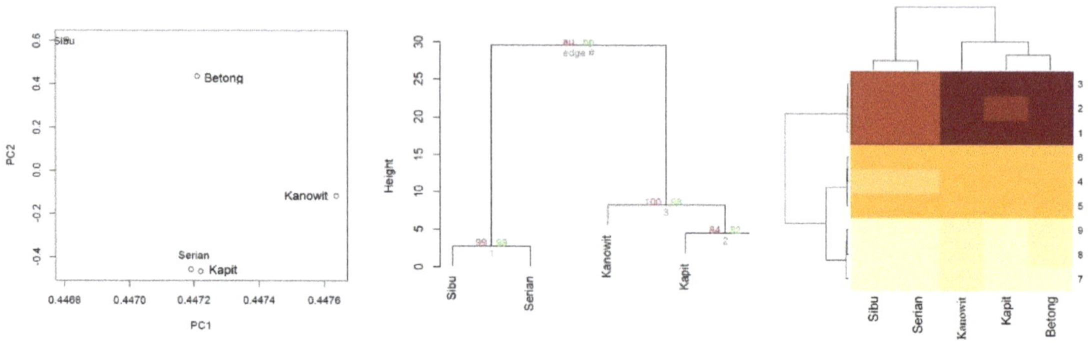

(**a**) (**b**) (**c**)

Figure 3. Multivariate statistical analysis revealed interconnected correlation patterns among the values of DPPH free radical scavenging activity at 1000 µg/mL for skin, pulp, and kernel fractions of dabai fruit oils from the different geographical locations. (**a**) PCA based on the correlation matrix of DPPH (%); (**b**) A consensus tree of the relationships among Sibu, Serian, Kapit, Betong, and Kanowit in the DPPH (%). The Pvclust package in R was used to cluster these traits according to the Euclidean distance matrix. The numbers at the forks were the percentages of approximately unbiased (AU; in red) p-values and bootstrap probabilities (BP, in green) estimated from 1000 bootstrapping samples; (**c**) A heatmap showing cluster groups in both the DPPH (%) and different localities (Sibu, Serian, Kapit, Betong, and Kanowit). The numbers on the right side represent the codes for 9 DPPH free radical scavenging activity at 1000 µg/mL for skin, pulp, and kernel fractions of dabai fruit oils from the different geographical locations, which showed a mixed pattern.

3.3. Brine Shrimp Lethality Bioassay (BSLA)

The results of the brine shrimp lethality (BSL) test at various concentrations for *Canarium* spp. oil extracts, as well as their plots of percentage mortality versus concentrations, are shown in Table 9, respectively. An LC_{50} value of less than 1000 µg/mL is toxic, whereas an LC_{50} value of higher than 1000 µg/mL is non-toxic [38,39].

The *Artemia* species are very useful and suitable for toxicity evaluation of bioactive substances in crude lipid fruit fractions. The BSL procedure is one of the most useful, reliable, and routine assays in the laboratory. This assay has been used for toxicity screening of some pesticides, heavy metals, food additives, and pharmaceutical compounds [16–19]. Due to its low cost, simplicity, and high sensitivity, the BSL assay has been receiving great attention from many researchers [16–19]

Natural deaths (mortalities), which were evaluated in blank seawater and wells treated with the positive control only, usually did not exceed 25%. This seemed to be a result of a lack of oxygen because most of the *A. salina* nauplii did not survive beyond 48 h of the assay [16–19]. In this regard, factors like age, composition, pH, the salinity of the matrix, and the temperature of larvae are effective factors in natural mortality [16].

During the study, no feed or air was needed because feeding brine shrimps with dry yeast suspension during the toxicity assessment is considered insignificant [16–19].

The results showed that the skin fraction had higher lethality concentrations (LC_{50}) than other fractions at all geographical locations. There was a significant difference ($p< 0.05$) in the effect of the geographical location on the toxicity of the three fractions against *A. salina* (Table 9). Basri et al. [11] suggest that the contents of saponins, terpenoids, tannins, and flavonoids influence toxicity. This is in line with Khoo et al. [40,41], who reported that the polyphenol components found in the dabai skin contain saponin, flavonoid, carotenoids, and anthocyanins; the fruit pulp contains anthocyanin and carotenoids; and the kernel contain α tocopherols, γ- tocopherols, and flavonoid [42]. Therefore, it was suspected that the cytotoxicity level against *A. salina* in the skin fraction extracts was induced by saponin, flavonoid, and anthocyanins contents, since the concentration of these three molecules might be richer in the skin compared to those in the kernel and pulp. Anthocyanin is known to possess antimicrobial, anti-obesity, anti-inflammatory, antidiabetic, and anticancer effects, as well as the prevention of chronic diseases [43]. The anticancer and other mechanisms of anthocyanins are based on their antioxidant potential, which is linked to their ability to scavenge free radicals, inhibit the enzymes involved in ROS formation, and prevent the oxidation of extracellular and cellular biomolecules [44].

Table 9. Average death of *A. salina* nauplii at different concentrations of dabai extracts of different localities.

Sample Extracts	Average Death of *A. salina* Nauplii						LC_{50} (µg/mL)
	Concentration (ppm)						
	1	10	50	100	250	500	
Skin							
Sibu	0.00 ± 0.00	0.00 ± 0.00	1.00 ± 1.00	0.67 ± 1.15	1.00 ± 0.73	1.67 ± 2.08	1387.22
Serian	0.00 ± 0.00	1.00 ± 1.73	1.33 ± 1.53	1.33 ± 1.15	1.67 ± 0.58	1.67 ± 1.15	1441.56
Kapit	0.00 ± 0.00	1.00 ± 1.00	1.33 ± 2.31	1.00 ± 1.73	1.00 ± 0.00	2.00 ± 1.00	1754.74
Betong	0.00 ± 0.00	1.67 ± 2.08	0.33 ± 0.58	1.00 ± 1.00	1.33 ± 0.58	1.67 ± 1.53	1528.52
Kanowit	0.00 ± 0.00	0.00 ± 0.00	1.00 ± 1.00	0.67 ± 1.15	0.33 ± 0.58	1.33 ± 0.58	1842.65
Pulp							
Sibu	0.67 ± 1.15	1.00 ± 1.00	1.33 ± 1.15	0.67 ± 0.58	1.33 ± 1.53	2.00 ± 1.00	7182.10
Serian	0.00 ± 0.00	0.00 ± 0.00	1.00 ± 1.00	0.67 ± 1.15	1.33 ± 1.53	2.00 ± 1.00	6249.25
Kapit	0.00 ± 0.00	0.00 ± 0.00	0.33 ± 0.58	1.00 ± 1.00	1.33 ± 1.53	2.00 ± 1.00	5061.03
Betong	0.67 ± 1.15	0.67 ± 0.58	1.00 ± 0.00	2.00 ± 1.00	1.67 ± 1.53	2.67 ± 2.08	7196.53
Kanowit	0.33 ± 0.58	0.67 ± 1.15	1.00 ± 1.00	0.67 ± 0.58	2.00 ± 0.00	2.67 ± 1.53	4192.32
Kernel							
Sibu	0.33 ± 0.58	1.00 ± 0.00	1.33 ± 0.58	2.33 ± 0.58	3.00 ± 0.00	3.67 ± 0.58	2742.65
Serian	1.00 ± 1.73	0.67 ± 0.58	2.00 ± 1.00	3.00 ± 0.00	2.67 ± 1.53	4.00 ± 1.00	2578.46
Kapit	0.67 ± 1.15	1.33 ± 1.53	2.00 ± 1.73	2.33 ± 0.58	3.00 ± 2.65	4.33 ± 1.15	2928.31
Betong	1.00 ± 1.00	1.33 ± 0.58	2.00 ± 1.00	3.33 ± 2.08	3.00 ± 1.00	4.67 ± 2.08	2261.04
Kanowit	0.00 ± 0.00	0.33 ± 0.58	1.33 ± 0.58	1.67 ± 0.58	2.67 ± 0.58	3.00 ± 0.00	1954.74
Thymol (positive control)	5.00 ± 1.00	7.33 ± 0.58	10.00 ± 0.00	10.00 ± 0.00	10.00 ± 0.00	10.00 ± 0.00	10.32
DMSO + seawater (negative control)	0	0	0	0	0	0	-

Flavonoids are found in all growing parts of the plant, being reported as the most abundant plant pigment, along with carotenoids and chlorophyll, also providing taste and fragrance to seeds, fruits, and flowers, which make them attractive to other organisms [45,46]. These bioactive compounds are also among the largest groups of secondary metabolites [47]. Besides their significance in plants, flavonoids are crucial for human health due to their significant pharmacological bioactivities. However, the production of these compounds in the pulp and kernel of dabai fraction extracts, although under the control of genetic factors, is remarkably affected by the geographical location. Investigations have revealed that environmental factors. such as geographical location. can change the level of phytochemical components [35,48]. Climatic factors, like lower temperatures, high relative humidity. as well as high rainfall. might be a reason for the higher compound contents in different geographical locations [49–51].

Higher LC_{50} values (>1000 mg/L) were observed in the kernel, pulp. and skin fraction extracts from the different localities. This indicates that the oil extract might be safe for pharmaceutical uses. More evidence (e.g., clinical trials) might be needed to substantiate its safety status. Hence, dabai cultivation in the Sarawak region, specifically in Sibu and Serian, could serve as a major contributor to the economy. Dabai has good fatty acid profiles, remarkable antioxidant activity, and is non-toxic to human life.

4. Conclusions

The extracts of the edible fractions from dabai fruits (kernel, oil, and pulp) collected at five different geographical locations in Sarawak showed different fatty acid compositions, antioxidants, and cytotoxicity. Among the locations, the Sibu and Serian samples' extracts showed a higher fatty acid composition and antioxidant potential. They also exhibited higher mortality of the brine shrimp than the samples from other locations (Betong, Kapit, and Kanowit). A significant correlation between antioxidant properties and cytotoxicity showed that certain phenolic bioactive compounds were the main contributors to the cytotoxicity properties of dabai fruits (kernel, oil, and pulp). Regarding the differences between the fraction extracts, we concluded that the geographical locations of samples influenced the fatty acid composition and phytochemicals of the plant. Furthermore, this study contributes to the origin effect, which has not been thoroughly studied, despite the reported effects of genotypes. Climate change might have also affected the reported values in this study compared to previous reports.

For any future application of the extract of the edible fraction of *C. odontophyllum* L. fruits as an antifungal biopesticide in organic agriculture, the location of the sample's collection needs to be included in the standardization of the formulations. The management of the geographical differences is essential to any successful agricultural and bioengineering industrial application.

Author Contributions: O.O.A., C.N.-H., S.F.S. and C.Z.-O.: sampling of dabai from different locations and the schematic approach of the experiments; Material and Method: O.O.A., F.R.R., S.F.S., G.C.A., A.H. and S.L.; Formal analysis and Writing; O.O.A., F.R.R., I.A.A. and M.Z.Z.; Revising: C.N.-H., G.C.A., O.O.A., R.S. and C.Z.-O. All authors have read and agreed to the published version of the manuscript.

Funding: The research received no external funding.

Institutional Review Board Statement: Not applicable.

Informed Consent Statement: Not applicable.

Data Availability Statement: Not available.

Acknowledgments: The authors thank Universiti Malaysia Sarawak for the support of this research.

Conflicts of Interest: The authors declare no conflict of interest.

References

1. Ding, P.; Tee, Y.K. Physicochemical characteristics of dabai (*Canarium odontophyllum* Miq.) fruit. *Fruits* **2011**, *66*, 47–52. [CrossRef]
2. Hanim, S.F.; Azrina, A.; Khoo, H.E.; Amin, I. Protective effects of pulp and kernel oils from *Canarium odontophyllum* fruit in normal and hypercholesterolemic rabbits. *Int. Food Res. J.* **2015**, 2, 1318–1326.
3. Azlan, A.; Nasir, N.N.M.; Amom, Z.; Ismail, A. Physical properties of skin, flesh, and kernel of Canarium odontophyllum fruit. *J. Food Agric. Environ.* **2009**, *7*, 55–57.
4. Khoo, H.E.; Azlan, A.; Abd Kadir, N.A.A. Fatty Acid Profile, Phytochemicals, and Other Substances in Canarium odontophyllum Fat Extracted Using Supercritical Carbon Dioxide. *Front. Chem.* **2019**, *7*, 5. [CrossRef]
5. Chua, H.P.; Nicholas, D.; Adros Yahy, M.N. Physical properties and nutritional values of dabai fruit (Canarium odontophyllum) of different genotypes. *J. Trop. Agric. Food Sci.* **2015**, *43*, 1–10.
6. Azlan, A.; Khoo, H.E.; Shapie, W.K.W.; Kadir, N.A.A.; Sultana, S. Nutritional quality and sensory evaluation of dabai-fortified cocoa bar. *Int. J. Food Prop.* **2020**, *23*, 1324–1336. [CrossRef]

7. Shakirin, F.H.; Azlan, A.; Ismail, A.; Amom, Z.; Yuon, L.C. Protective Effect of Pulp Oil Extracted from *Canarium odontophyllum* Miq. Fruit on Blood Lipids, Lipid Peroxidation, and Antioxidant Status in Healthy Rabbits. *Oxidative Med. Cell. Longev.* **2012**, *2012*, 1318–1326. [CrossRef]
8. Khoo, H.E.; Azlan, A.; Ismail, A.; Abas, F.; Hamid, M. Inhibition of Oxidative Stress and Lipid Peroxidation by Anthocyanins from Defatted Canarium odontophyllum Pericarp and Peel Using In Vitro Bioassays. *PLoS ONE* **2014**, *9*, e81447. [CrossRef]
9. Azrina, A.; Nadiah, M.N.; Amin, I. Antioxidant properties of methanolic extract of Canarium odontophyllum fruit. *Int. J. Food Res.* **2010**, *17*, 319–326.
10. Prasad, K.N.; Chew, L.Y.; Khoo, H.; Yang, B.; Azlan, A.; Ismail, A. Carotenoids and antioxidant capacities from Canarium odontophyllum Miq. fruit. *Food Chem.* **2011**, *124*, 1549–1555. [CrossRef]
11. Basri, D.F.; Alamin, Z.A.Z.; Chan, K.M. Assessment of cytotoxicity and genotoxicity of stem bark extracts from Canarium odontophyllum Miq. (dabai) against HCT 116 human colorectal cancer cell line. *BMC Complement. Altern. Med.* **2015**, *16*, 36. [CrossRef] [PubMed]
12. Chew, L.Y.; Khoo, H.E.; Amin, I.; Azrina, A.; Lau, C.Y. Analysis of Phenolic Compounds of Dabai (*Canarium odontophyllum* Miq.) Fruits by High-Performance Liquid Chromatography. *Food Anal. Methods* **2012**, *5*, 126–137. [CrossRef]
13. Chua, H.P.; Nicholas, D.; Suzalyna, M. Phenolic and flavonoid contents and antioxidant activities of selected dabai (Canarium odontophyllum) genotypes. *J. Trop.Agric. Food Sci.* **2014**, *42*, 105–114.
14. Djarkasi, G.S.; Raharjo, S.; Noor, Z.; Sudarmadji, S. Sifat fisik dan kimia minyak kenari. *agriTECH* **2007**, *27*, 165–170. [CrossRef]
15. Ajibola, O.O.; Lihan, S.; Hussaini, A.; Hipolito, C.N.; Octavio, C.Z.; Sarbini, S.R.; Mohamad, S.A.S. Cell Viability, Physicochemical and Sensory Characteristics of Probiotic Coconut Juice During Cold Storage. *J. Sustain. Sci. Manag.* **2021**, *16*, 1–13. [CrossRef]
16. Olaide, A.O.; Lihan, S.; Husaini, A.; Saat, R.; Mohammad, F.S.; Adewale, I.A.; Abideen, W.A. Use of the Lactococcus lactis IO-1 for developing a novel functional beverage from coconut water. *Ann. Univ. Dunarea de Jos Galati Fascicle VI Food Technol.* **2020**, *44*, 118–131. [CrossRef]
17. Gülçin, I. Antioxidant Activity of Eugenol: A Structure–Activity Relationship Study. *J. Med. Food* **2011**, *14*, 975–985. [CrossRef]
18. Ajibola, O.O.; Lihan, S.; Hussaini, A.; Saat, R.; Ahmed, I.A.; Abideen, W.; Sinang, F.M.; Sing, N.N.; Adeyinka, G.C. Toxicity Assessment of Lactococcus lactis IO-1 Used in Coconut Beverages against Artemia salina using Brine Shrimp Lethality Test. *Appl. Food Biotechnol.* **2020**, *7*, 127–134. [CrossRef]
19. Ajibola, O.O. Potential Use of Lactococcus Lactis Subsp. Lactis IO- 1 in Fermented Coconut Juice. Master's Thesis, Faculty of Resource Science and Technology, University Malaysia Sarawak, Kota Samarahan, Malaysia, 2020. Available online: https://ir.unimas.my/id/eprint/31400 (accessed on 20 April 2022).
20. Kuspradini, H.; Rosiarto, A.M.; Putri, A.S.; Kusuma, I.W. Antioxidant and toxicity properties of anthocyanin extracted from red flower of four tropical shrubs. *Nusant. Biosci.* **2016**, *8*, 135–140. [CrossRef]
21. Parcerisa, J.; Richardson, D.G.; Rafecas, M.; Codony, R.; Boatella, J. Fatty acid, tocopherol and sterol content of some hazelnut varieties (Corylus avellana L.) harvested in Oregon (USA). *J. Chromatogr. A* **1998**, *805*, 259–268. [CrossRef]
22. Tüfekci, F.; Karataş, S. Determination of geographical origin Turkish hazelnuts according to fatty acid composition. *Food Sci. Nutr.* **2018**, *6*, 557–562. [CrossRef] [PubMed]
23. Liu, L.; Guan, L.-L.; Yang, Y.-X. A Review of Fatty Acids and Genetic Characterization of Safflower (*Carthamus tinctorius* L.) Seed Oil. *World J. Tradit. Chin. Med.* **2016**, *2*, 48–52. [CrossRef]
24. Górnaś, P.; Rudzińska, M.; Segliņa, D. Lipophilic composition of eleven apple seed oils: A promising source of unconventional oil from industry by-products. *Ind. Crop. Prod.* **2014**, *60*, 86–91. [CrossRef]
25. He, Z.; Zhu, H.; Li, W.; Zeng, M.; Wu, S.; Chen, S.; Qin, F.; Chen, J. Chemical components of cold pressed kernel oils from different Torreya grandis cultivars. *Food Chem.* **2016**, *209*, 196–202. [CrossRef] [PubMed]
26. Sicari, V.; Pellicano, T.M.; Messina, F. Comparison of Physicochemical Characteristics and Composition of Bergamot Oil Seed Extracted from Three Different Cultivars. *Emir. J. Food Agric.* **2017**, *29*, 470–475. [CrossRef]
27. Azlan, A.; Prasad, K.N.; Khoo, H.E.; Abdul-Aziz, N.; Mohamad, A.; Ismail, A.; Amom, Z. Comparison of fatty acids, vitamin E and physicochemical properties of Canarium odontophyllum Miq. (dabai), olive and palm oils. *J. Food Compos. Anal.* **2010**, *23*, 772–776. [CrossRef]
28. Tuntiprapas, P.; Shimada, S.; Pongparadon, S.; Prathep, A.; Saensouk, P.; Theerakulpisut, P.; Kerdphol, R. Is Halophila major (Zoll.) Miquel a big H. ovalis (R. Brown) JD Hooker? An evaluation based on age, morphology, and ITS sequence. *Sci. Asia* **2015**, *43*, 347–353. [CrossRef]
29. Esteki, M.; Farajmand, B.; Kolahderazi, Y.; Simal-Gandara, J. Chromatographic Fingerprinting with Multivariate Data Analysis for Detection and Quantification of Apricot Kernel in Almond Powder. *Food Anal. Methods* **2017**, *10*, 3312–3320. [CrossRef]
30. Ibrahim, H.A.R. The noble oil of brunei. In Proceedings of the 10th Asian Food Conference (Theme: Food for mankind—Contribution of Science and Technology), Kuala Lumpur, Malaysia, 21—23 August 2007.
31. Kakuda, Y.; Jahaniaval, F.; Marcone, M.F.; Montevirgen, L.; Montevirgen, Q.; Umali, J. Characterization of pili nut (*Canarium ovatum*) oil: Fatty acid and triacylglycerol composition and physicochemical properties. *J. Am. Oil Chem. Soc.* **2000**, *77*, 991–997. [CrossRef]
32. Rashid, N.A.H.A.; Shamsudin, R.; Arifin, S.H.; Abdullah, W.N.Z.Z. Morphological and quality characteristics of genus of *Canarium* L.: A review. *IOP Conf. Ser. Earth Environ. Sci.* **2021**, *733*, 012015. [CrossRef]

33. Noroozi, S.A.; Yousefzadeh, K.; Kamal Sadat Aslan, S.; Mansourifar, S. Evaluation of Changes in Essential Oil, Chlorophyll, Carotenoid, Anthocyanins, and Flavonoids of *Mentha longifolia* (L.) Huds. subsp. *longifolia* in different habitats of Marand. *J. Ecoph. Med. Plants* **2017**, *5*, 52–64.
34. Dastoor, R.; Bakhshi, D.; Ali Akbar, A.R. Evaluation and comparison of total phenol, total flavonoid, resveratrol and antioxidant capacity in fruits of the species Vitis vinifera, Pistacia vera, Sambucus nigra and Ilex spinigera. *J. Ecoph. Med. Plants* **2017**, *5*, 37–48.
35. Koohsari, H.; Khormali, H.; Khormali, A. Evaluation of Flavonoids and Phenolic Compounds, Antioxidant and Antibacterial Activity of *Hypericum Perforatum* L. Collected from Two Sites in North Country. *J. Ecoph. Med. Plants.* **2017**, *5*, 78–90.
36. Zargoosh, Z.; Ghavam, M.; Bacchetta, G.; Tavili, A. Effects of ecological factors on the antioxidant potential and total phenol content of Scrophularia striata Boiss. *Sci. Rep.* **2019**, *9*, 16021. [CrossRef] [PubMed]
37. Ayaz, F.A.; Kadioglu, A.; Turgut, R. Water stress effects on the content of low molecular weight carbohydrates and phenolic acids in *Ctenanthe setosa* (Rosc.) Eichler. *Can. J. Plant Sci.* **2000**, *80*, 373–378. [CrossRef]
38. Mehrpour, M.; Kashefi, B.; Moghaddam, M. Investigation of phytochemical and antioxidant compounds of different organs of the Angushei medicinal plant in two natural habitats of Semnan and Khorasan provinces. *J. Ecoph. Med. Plants* **2016**, *4*, 56–68.
39. Meyer, B.N.; Ferrigni, N.R.; Putnam, J.E.; Jacobsen, L.B.; Nichols, D.E.; McLaughlin, J.L. Brine Shrimp: A Convenient General Bioassay for Active Plant Constituents. *Planta Med.* **1982**, *45*, 31–34. [CrossRef]
40. Barradas-Dermitz, D.M.; Cirilo, N.H.; Hayward-Jones, P.M.; Aurelio, R.H.; Calderon-Garcidueñas, A.L.; Ajibola, O.O.; Octavio, C.Z. Contribution to the Phytomedicinal Study of the Solid and Aqueous Extract of *Anthurium Schlechtendalii* Kunth Root against Liver Damage in a Rat Model. *Pharmacologyonline* **2021**, *3*, 1283–1291.
41. Khoo, H.E.; Azlan, A.; Ismail, A.; Abas, F. Antioxidative Properties of Defatted Dabai Pulp and Peel Prepared by Solid Phase Extraction. *Molecules* **2012**, *17*, 9754–9773. [CrossRef]
42. Kadir, N.A.A.A.; Azlan, A.; Abas, F.; Ismail, I.S. Hepatoprotective Effect of Supercritical Carbon Dioxide Extracted Dabai Pulp Oil and Its Defatted Pulp. *Molecules* **2021**, *26*, 671. [CrossRef]
43. Shakirin, F.H.; Prasad, N.; Ismail, A.; Yuon, L.C.; Azlan, A. Antioxidant capacity of underutilized Malaysian *Canarium odontophyllum* (dabai) Miq. fruit. *J. Food Compos. Anal.* **2010**, *23*, 777–781. [CrossRef]
44. He, K.; Li, X.; Chen, X.; Ye, X.; Huang, J.; Jin, Y.; Li, P.; Deng, Y.; Jin, Q.; Shi, Q.; et al. Evaluation of antidiabetic potential of selected traditional Chinese medicines in STZ-induced diabetic mice. *J. Ethnopharmacol.* **2011**, *137*, 1135–1142. [CrossRef] [PubMed]
45. Khoo, H.E.; Azlan, A.; Tang, S.T.; Lim, S.M. Anthocyanidins and anthocyanins: Colored pigments as food, pharmaceutical ingredients, and the potential health benefits. *Food Nutr. Res.* **2017**, *61*, 1361779. [CrossRef] [PubMed]
46. Stalikas, C.D. Extraction, separation, and detection methods for phenolic acids and flavonoids. *J. Sep. Sci.* **2007**, *30*, 3268–3295. [CrossRef] [PubMed]
47. Koes, R.E.; Quattrocchio, F.; Mol, J.N.M. The flavonoid biosynthetic pathway in plants: Function and evolution. *BioEssays* **1994**, *16*, 123–132. [CrossRef]
48. De Rijke, E.; Out, P.; Niessen, W.M.A.; Ariese, F.; Gooijer, C.; Brinkman, U.A.T. Analytical separation and detection methods for flavonoids. *J. Chromatogr. A* **2006**, *1112*, 31–63. [CrossRef]
49. Rezende, W.P.; Borges, L.L.; Santos, D.L.; Paula, N.M.A.A.J.R. Effect of Environmental Factors on Phenolic Compounds in Leaves of *Syzygium jambos* (L.) Alston (Myrtaceae). *Mod. Chem. Appl.* **2015**, *3*, 1–6. [CrossRef]
50. Gholizadeh Moghaddam, N.; Hosseini, B.; Alirezalou, A. Evaluation of variation of some phytochemical indices of leaf extract of genotypes of different species of Barberry. *J. Ecoph. Med. Plants* **2017**, *3*, 1–12.
51. Afolabi, F.; Adeyinka, G.C.; Ajibola, O.O.; Bakare, B.F. Comparative assessment of heavy metal contamination of abandoned and active dumpsite of Osun waste management, Ejigbo Road, Osogbo, Osun State, Nigeria. *Int. J. Environ. Anal. Chem.* **2021**, *101*, 1–17. [CrossRef]

molecules

Article

Neurotoxin (N-Oxalyl-L-α,β-Diamino Propionic Acid) Content in Different Plant Parts of Grass Pea (*Lathyrus sativus* L.) Spanning Seedling to Maturity Stage: Does It Increase over Time?

Surendra Barpete [1,2,*], Priyanka Gupta [1], Debjyoti Sen Gupta [3], Jitendra Kumar [3], Arpan Bhowmik [4] and Shiv Kumar [1,2,*]

1 International Center for Agricultural Research in the Dry Areas (ICARDA), Rabat Institute, Rabat 6299, Morocco; vidhiguptaniwari@gmail.com
2 ICARDA—Food Legume Research Platform, Amlaha 466113, India
3 ICAR—Indian Institute of Pulses Research, Kanpur 208024, India; debgpb@gmail.com (D.S.G.); jitendra73@gmail.com (J.K.)
4 ICAR-Indian Agricultural Statistics Research Institute, Library Avenue, New Delhi 110012, India; arpan.stat@gmail.com
* Correspondence: surendrabarpete@gmail.com (S.B.); sk.agrawal@cgiar.org (S.K.)

Citation: Barpete, S.; Gupta, P.; Sen Gupta, D.; Kumar, J.; Bhowmik, A.; Kumar, S. Neurotoxin (N-Oxalyl-L-α, β-Diamino Propionic Acid) Content in Different Plant Parts of Grass Pea (*Lathyrus sativus* L.) Spanning Seedling to Maturity Stage: Does It Increase over Time? *Molecules* **2022**, *27*, 3683. https://doi.org/10.3390/molecules27123683

Academic Editor: Smaoui Slim

Received: 20 February 2022
Accepted: 20 May 2022
Published: 8 June 2022

Abstract: ODAP (N-oxalyl-L-2,3-diaminopropionic acid) is present in the seeds of grass pea. In this study, variation of total ODAP accumulation in leaves throughout the crop growth starting from 40 days after sowing to maturity, and the distribution pattern of ODAP in different plant parts including the seeds at the mature stage was analyzed. Five grass pea accessions were evaluated for two subsequent growing seasons in one location of ICARDA, Aleppo (Syria). The results found that the rate of accumulation of total ODAP varied during plant development. Increased rates of synthesis were noticed in young leaves of grass pea. The highest total ODAP content in leaves was noted in the early growth stage (40–50 days after sowing). Mean total ODAP content in leaves ranged from 0.17 to 0.96 percent during 2010–2011 and from 0.19 to 1.28 percent during 2011–2012. During maturity, the total ODAP content was lowest in the seeds than in leaves, stems, pod cover, seed coat, and cotyledons. The ranges of total ODAP content were 0.13 (seed)–0.34 (stem), 0.20 (seed)–1.01 (leaf), 0.22 (seed)–0.62 (leaf), 0.21 (seed)–0.66 (leaf), and 0.21 (seed)–0.78 (leaf) percent in B387, B222, B390, Bio-520, and B587 accessions, respectively, during maturity. The results indicated that the rate of accumulation and synthesis of total ODAP varied during the plant lifespan. The lowest total ODAP content of leaves was observed after 130 days of sowing. The lower total ODAP content after the early vegetative stage of grass pea plants makes them suitable as a feed.

Keywords: ODAP; neurotoxin; grass pea; *Lathyrus sativus*; growth stage; neurolathyrism; non-protein amino acids

1. Introduction

Grass pea (*Lathyrus sativus* L.) is considered an easy-to-adapt crop in a wide range of climatic conditions due to its demonstrated ability to survive under drought, intense precipitation events, elevated temperatures, and other edaphic stresses [1]. Its protein-rich seeds (>30% protein) are highly valued as human food and animal feeds. Grass pea assumes the role of survival food for poor masses and is the only source of feed and fodder for animals in rural areas of South Asia and Sub-Saharan Africa during drought years [2,3]. Continuous consumption of grass pea over long periods (>3 months) makes poor people vulnerable to 'neurolathyrism' due to a plant toxin called β-N-oxalyl-L-α, β-diaminopropionic acid (β-ODAP) present in its seeds and vegetative parts [4]. Lathyrism causes muscle spasms, cramps, and weakness of the lower limbs. Spastic paraparesis, sensory, and bladder dysfunction also may occur. Coarse tremor of upper parts was also observed. This medical condition is irreversible, however, mostly, life expectancy is not reduced [5]; Lambein et al. [6] postulated that since ODAP is an amino acid that accumulates in all tissues at all growth stages, it may have a role in drought tolerance that has made the grass pea such a useful species over long period of its domestication. However, homoarginine found in its seeds is an alternative substrate for the synthesis of Nitric Oxide (NO) in the human body that has a crucial role in the maintenance of cardiovascular and cerebral metabolism [7].

This crop remains popular among the resource-poor farmers in marginal areas due to the ease with which it can be grown successfully under adverse agro-climatic conditions without much production inputs [3,8,9], despite the ban on its sale in some countries [10]. Presently, it is grown on more than 1.5 million ha with 1.20 million tons production [1] and efforts are underway (in Australia, southern Europe, and America) to expand it as a break crop between cereals and as a bonus crop on fallow lands for feed and fodder [11,12]. However, the presence of ODAP in different plant parts of grass pea is the major hindrance in its further expansion as a food, feed, and fodder crop. Accumulation of ODAP content in grass pea cultivars is highly influenced by environmental factors [13–15]. An array of physiological and developmental processes influences ODAP content during ontogenetic changes in the life cycle of the plant under drought and heat stresses [1]. Therefore, due to the increasing interest in *Lathyrus* species as alternative pulses, grass pea breeding has a focus to reduce ODAP and increase methionine, homoarginine, and other nutritional factors [16–18]. Since grass pea plant is used as green fodder for animals and dry seeds for human food and animal feed purposes, the present study was undertaken to assess total ODAP concentration in developing plant parts at various growth stages to understand the pattern of total ODAP accumulation in grass pea. The objectives of this study were to find out (a) the total ODAP accumulation in leaf throughout the crop growth staring from 40 days after sowing to maturity, and (b) the distribution pattern of total ODAP in different plant parts, including seeds at the mature stage.

2. Materials and Methods

2.1. Plant Material

Five grass pea accessions (B222, B387, B390, B587, and Bio520) with similar phenology were selected for the present study in order to avoid the confounding effect of growth stages on the total ODAP content. The experiments were carried out at the International Center for Agricultural Research in the Dry Areas (ICARDA) at its experimental farm located at Tel Hadya (360°56′ E, 360°01′ N, 284 m AMSL) in Syria during 2010–2011 and 2011–2012. The soil at the experimental site is classified as clay with 7.9 pH and mean soil temperature ranging from 11.0 to 30.9 °C at 5 cm depth during the crop season. The climate of the region is typically the Mediterranean with hot dry summers and cold wet winters, with highly variable rainfall. The average precipitation during the crop season was 270 and 240 mm during 2010–2011 and 2011–2012, respectively. The experiments were carried out following the Randomized Complete Block Design (RCBD) with three replications. The individual plot size was 4.8 m^2 with four rows of 4 m length spaced 30 cm apart. Seeds were sown at a 20 cm distance within the row for ease of sampling. No irrigation was given at any growth stage. Care was taken to ensure normal plant growth throughout the crop life cycle.

2.2. Plant Material Sampling and Data Recording

To record leaf total ODAP concentration at different growth stages, plant samples were collected randomly from the middle rows of each plot at 10-day intervals, from 40 days after sowing until maturity. At the maturity stage, ten plants were harvested from each plot to obtain enough materials for the chemical analysis. During 2011–2012, reaching the maturity stage in addition to leaf samples different other plant parts were collected separately (shoot, pod cover, seed coat, cotyledons, and seed) for the total ODAP analysis.

All the plant parts except the matured seeds were dried at 40 °C, weighed and carefully mixed to give reasonable representative samples for grinding. The matured seeds were air dried. The testa of seeds was not removed from any genotype during the study.

2.3. Determination of Total ODAP Content

Total ODAP content in different plant parts and seeds was measured following UV-Spectrophotometer (Cadex Model: SB038, St-Jean-sur-Richelieu, QC, Canada) method suggested by Rao (1978) [19] and modified by Briggs et al., (1983) [20]. The modifications involved the two times extraction of 0.5 g of the flour with 60% ethanol followed by hydrolysis with 3M KOH (Sigma, St. Louis, MO, USA) in boiling water bath for 30 min. After centrifugation for 15 min, an aliquot (250 μL) of the hydrolysate was diluted with 750 μL water and reacted with 2 mL o-phthalaldehyde (Sigma, St. Louis, MO, USA) reagent. The mixture was incubated at 40 °C for 2 h before measuring the absorbance at 425 nm. The total ODAP standard curve ($r^2 = 0.99$) was calibrated using DAP-HCl (Sigma, St. Louis, MO, USA).

2.4. Statistical Analysis

Repeated measures (RM)- analysis of variance (ANOVA) was performed on the mean-total ODAP content in leaf samples of five accessions for 2010–2011 and 2011–2012 using SAS (r) Proprietary Software 9.4 (SAS Institute, Cary, NC, USA). For each year, pairwise comparisons between all the level of time were done using *t* test. Here, *p*-value is adjusted based on Bonferroni Corrected Method. The effect size (η^2) for RM-ANOVA which is measure of association between time and response variable was calculated as follows:

$$\text{Effect Size}\left(\eta^2\right) = \frac{\text{SS}_{\text{Time}}}{(\text{SS}_{\text{Time}} + \text{SS}_{\text{Error}})} \tag{1}$$

Two-way analysis of variance was performed for total ODAP content among different plant parts in different lathyrus accessions during mature stage in 2011–2012. Equality of mean value of total ODAP content of tested accessions between growing years was calculated using *t* test. Similarly, equality of mean values of total ODAP content of specific growth stages between two growing years was also tested using *t* test.

3. Results

3.1. Total ODAP Content in Tested Grass Pea Accessions in 2010–2011

The highest total ODAP content was noted in early growth stage (40–50 days after sowing, DAS) (Table 1). The levels of total ODAP decreased apparently during later vegetative phase (>130 DAS), and subsequently remained lowest at mature stage in seeds of different accessions (Table 1). The ranges of distribution of total ODAP from early vegetative stage to maturity were 0.89 (40 DAS)–0.17% (maturity stage), 1.06 (50 DAS)–0.19% (maturity stage), 0.96 (40 DAS)–0.14% (maturity stage), 1.09 (50 DAS)–0.17 (maturity stage), and 1.03 (50 DAS)–0.21 (maturity stage) in accessions B387, B222, B390, Bio520, and B587, respectively (Table 1).

However, the genotype B390 showed lower total ODAP content compared to other four grass pea accessions at maturity stage. Repeated measures (RM)- analysis of variance (ANOVA) was calculated for total ODAP content in leaf samples of five accessions for 2010–2011 and found that time effect is significant at 1% level of significance (Table 2). The effect size (η^2) for repeated measure ANOVA which is measure of association between time and response variable is calculated as follows:

$$\text{Effect Size}\left(\eta^2\right) = \frac{\text{SS}_{\text{Time}}}{(\text{SS}_{\text{Time}} + \text{SS}_{\text{Error}})} = \frac{3.202}{3.429} = 0.934 \tag{2}$$

Table 1. Total ODAP content (%) in different plant development stage in *L. sativus* during the year 2010–2011.

S. No.	Day *	Mean ± SE (%)				
		B387	**B222**	**B390**	**Bio-520**	**B587**
1	40	0.89 ± 0.11	1.02 ± 0.08	0.96 ± 0.10	0.93 ± 0.22	1.00 ± 0.15
2	50	0.75 ± 0.09	1.06 ± 0.10	0.94 ± 0.11	1.09 ± 0.20	1.03 ± 0.06
3	60	0.88 ± 0.20	0.93 ± 0.22	0.95 ± 0.07	0.89 ± 0.30	0.95 ± 0.28
4	70	0.70 ± 0.23	1.00 ± 0.25	0.87 ± 0.08	0.64 ± 0.21	0.99 ± 0.12
5	80	0.64 ± 0.18	0.80 ± 0.20	0.70 ± 0.15	0.59 ± 0.03	0.83 ± 0.07
6	90	0.70 ± 0.05	0.89 ± 0.09	0.59 ± 0.06	0.49 ± 0.20	0.66 ± 0.04
7	100	0.62 ± 0.20	0.66 ± 0.10	0.56 ± 0.22	0.53 ± 0.14	0.61 ± 0.09
8	110	0.59 ± 0.07	0.61 ± 0.05	0.49 ± 0.07	0.42 ± 0.17	0.57 ± 0.18
9	120	0.50 ± 0.03	0.54 ± 0.20	0.43 ± 0.23	0.39 ± 0.11	0.50 ± 0.09
10	130	0.48 ± 0.10	0.43 ± 0.20	0.39 ± 0.15	0.36 ± 0.15	0.43 ± 0.12
11	Mature stage (seeds)	0.17 ± 0.09	0.19 ± 0.10	0.14 ± 0.09	0.17 ± 0.10	0.21 ± 0.04

Day * = sample collection day after sowing.

Table 2. Repeated Measure ANOVA for leaf total ODAP content during 2010–2011.

Source of Variation	Df	Sum Sq	Mean Sq	F Value	Pr (>F)
Time	10	3.202	0.3202	56.43	$<2 \times 10^{16}$ *
Error (Genotype × Time)	40	0.227	0.0057		

* significant at 1% level of significance.

The effect size based on the above repeated measure ANOVA is 0.934, which is quite high enough and, thus, time (Days after sowing) is having significant impact on character (total ODAP content) under study. The trend in the total ODAP content accumulation in 2010–2011 as time increases has a downward movement (Figure 1). The mean neurotoxin content of five *Lathyrus* accessions from 40 DAS to maturity stage ranged from 0.97 to 0.17 percent (Table 3).

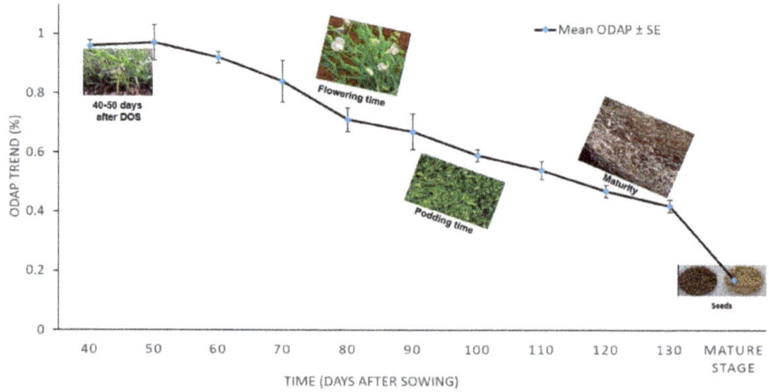

Figure 1. Mean leaf total ODAP content (%) of tested *Lathyrus* accessions grown during 2010–2011 at Tel Hadya. Y axis presents leaf total ODAP content in percent, X axis denotes Time, Time points were: 1 = 40 DAS, 2 = 50 DAS, 3 = 60 DAS, 4 = 70 DAS, 5 = 80 DAS, 6 = 90 DAS, 7 = 100 DAS, 8 = 110 DAS, 9 = 120 DAS, 10 = 130 DAS, and 11 = mature stage (seeds).

Table 3. Mean total ODAP content during different plant development stages in five accessions of grass pea during 2010–11.

* Days	40	50	60	70	80	90	100	110	120	130	Mature Stage (seeds)
** Mean ODAP% ± SE	0.96 ± 0.02	0.97 ± 0.06	0.92 ± 0.02	0.84 ± 0.07	0.71 ± 0.04	0.67 ± 0.06	0.59 ± 0.02	0.54 ± 0.03	0.47 ± 0.02	0.42 ± 0.02	0.17 ± 0.02

Day * = Sample collection day after sowing ** Mean total ODAP content in leaf of five acceccsions of grass pea.

3.2. Total ODAP Content in Tested Grass Pea Accessions in 2011–2012

The highest total ODAP content was noted in the early growth stage (40 DAS) (Table 4). The levels of total ODAP decreased apparently during the later vegetative phase (>130 DAS), and subsequently remained lowest at the mature stage in seeds of all the accessions (Table 3). The ranges of distribution of total ODAP from early vegetative stage to maturity were 0.12 (maturity stage)–1.47 percent, 0.21 (maturity stage)–1.19 percent, 0.21 (maturity stage)–1.27 percent, 0.20 (maturity stage)–1.34 percent, and 0.20 (maturity stage)–1.13 percent in accessions B387, B222, B390, Bio520 and B587, respectively, at 40 DAS (Table 4).

Table 4. Total ODAP content (%) in different plant development stage in *L. sativus* during the year 2011–2012.

Sl. No	Day *	Mean ± SE (%)				
		B387	B222	B390	Bio-520	B587
1	40	1.47 ± 0.2	1.19 ± 0.06	1.27 ± 0.14	1.34 ± 0.19	1.13 ± 0.03
2	50	1.06 ± 0.04	1.02 ± 0.04	1.10 ± 0.08	1.10 ± 0.09	1.06 ± 0.07
3	60	1.04 ± 0.03	1.09 ± 0.04	1.01 ± 0.02	1.08 ± 0.05	1.03 ± 0.02
4	70	0.93 ± 0.06	0.94 ± 0.03	1.02 ± 0.02	1.10 ± 0.06	0.98 ± 0.02
5	80	0.80 ± 0.17	1.02 ± 0.24	0.81 ± 0.15	1.00 ± 0.20	0.91 ± 0.17
6	90	0.92 ± 0.21	1.08 ± 0.05	1.19 ± 0.10	1.37 ± 0.09	1.12 ± 0.08
7	100	1.22 ± 0.08	1.06 ± 0.11	0.86 ± 0.09	0.98 ± 0.16	0.98 ± 0.17
8	110	1.12 ± 0.27	0.92 ± 0.15	0.85 ± 0.12	0.89 ± 0.12	0.84 ± 0.14
9	120	0.74 ± 0.05	0.84 ± 0.13	0.76 ± 0.04	0.67 ± 0.02	0.90 ± 0.10
10	130	0.80 ± 0.20	0.58 ± 0.11	0.76 ± 0.06	0.74 ± 0.03	0.58 ± 0.10
11	Mature stage (seeds)	0.12 ± 0.03	0.21 ± 0.03	0.21 ± 0.04	0.20 ± 0.09	0.21 ± 0.22

Day * = sample collection day after sowing.

However, the genotype B387 showed lower total ODAP content compared to the other four grass pea accessions at maturity stage (Table 4). Repeated measures (RM)—analysis of variance (ANOVA) was calculated for total ODAP content in leaf samples of five accessions for 2011–2012 and found that time effect is significant at 1% level of significance (Table 5). The effect size (η^2) for repeated measure ANOVA which is measure of association between time and response variable is calculated as follows:

$$\text{Effect Size}\left(\eta^2\right) = \frac{SS_{Time}}{(SS_{Time} + SS_{Error})} = \frac{4.152}{3.429} = 0.908 \tag{3}$$

The effect size based on the above repeated measure ANOVA is 0.908, which is quite high enough and, thus, time (Days after sowing) is having significant impact on character (total ODAP content) under study. The trend in the ODAP content accumulation in 2011–2012 as time increases has a downward movement (Figure 2). The mean of neurotoxin content for five *Lathyrus* accessions ranged from 1.28 to 0.19% from 40 days after seed germination to maturity stage, respectively (Table 6).

Table 5. Repeated Measure ANOVA for leaf total ODAP content during 2011–2012.

Source of Variation	Df	Sum Sq	Mean Sq	F Value	Pr (>F)
Time	10	4.152	0.4152	39.59	$<2 \times 10^{16}$ *
Error (Genotype × Time)	40	0.419	0.0105		

* Significant at 1% level of significance.

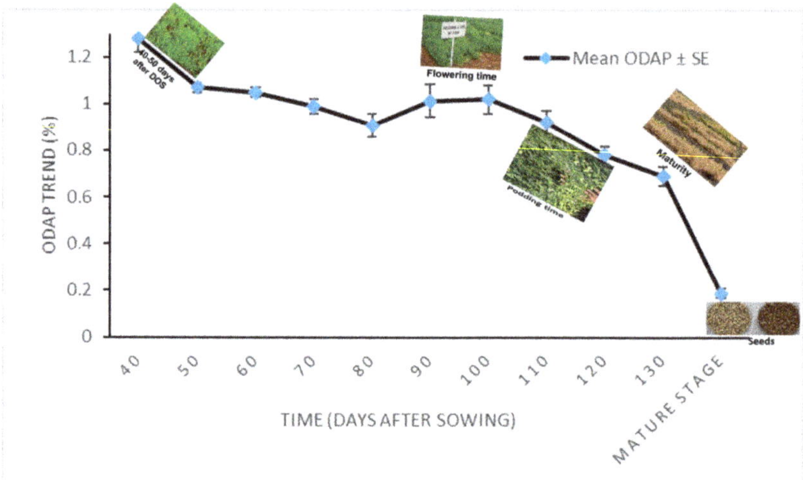

Figure 2. Mean leaf total ODAP content during different time points of sampling of tested *Lathyrus* accessions grown during 2011–2012 at Tel Hyada. Y axis presents leaf total ODAP content in percent, X axis denotes Time. Time points were: 1 = 40 DAS, 2 = 50 DAS, 3 = 60 DAS, 4 = 70 DAS, 5 = 80 DAS, 6 = 90 DAS, 7 = 100 DAS, 8 = 110 DAS, 9 = 120 DAS, 10 = 130 DAS, and 11 = mature stage (seeds).

Table 6. Mean total ODAP content during different plant development stages in five accessions of grass pea during 2011–12.

Day *	40	50	60	70	80	90	100	110	120	130	Mature Stage (Seeds)
Mean ** ODAP% ± SE	1.28 ± 0.06	1.07 ± 0.02	1.05 ± 0.02	0.99 ± 0.03	0.91 ± 0.05	1.13 ± 0.07	1.02 ± 0.06	0.92 ± 0.05	0.78 ± 0.04	0.69 ± 0.04	0.19 ± 0.02

Day * = Sample collection day after sowing ** Mean total ODAP content in leaf of five acceccsions of grass pea.

3.3. *Distribution of Total ODAP in Different Plant Parts at Mature Stage*

At the maturity stage when plants senescence, total ODAP content was determined in the stem, leaf, pod cover, seeds, seed coat, and cotyledons, and in ANOVA analysis, it was found that the tested genotypes significantly differed in total ODAP content (%) in case of leaf and seed coat samples. (Tables 7 and 8, Figure 3). The total ODAP content was lowest in the seeds than in leaves, stems, pod cover, seed coat, and cotyledons (Figure 3). The ranges of total ODAP content were 0.13 (seed)–0.34 (stem), 0.20 (seed)–1.01 (leaf), 0.22 (seed)–0.62 (leaf), 0.21 (seed)–0.66 (leaf), and 0.21 (seed)–0.78 (leaf) percent in B387, B222, B390, Bio-520, and B587 accessions, respectively, during maturity. The leaves and stems seemed to be the major source of total ODAP (Table 8 and Figure 3) compared with different seed parts. The mean seed total ODAP content was lower compared to the mean total ODAP content of seed cover, pod cover, and cotyledons (Table 8 and Figure 3). The mean total ODAP content in pod cover, seed coat, and cotyledons were almost the same. All the accessions were having similar total ODAP content in seeds except for B387, which was having lowest total ODAP content (Table 8).

Table 7. Analysis of Variance of Total ODAP content (%) of different *L. sativus* accessions at mature stage.

Source	DF	Mean Square (Stem)	Mean Square (Leaf)	Mean Square (Pod Cover)	Mean Square (Seed)	Mean Square (Seed Coat)	Mean Square (Cotyledons)
Rep	2	0.060	0.007	0.017	0.003	0.005	0.009
Genotypes	4	0.026	0.157 *	0.026	0.005	0.017 *	0.006
Error	8	0.032	0.019	0.017	0.001	0.004	0.003
Total	14						

* Significant at $p < 0.05$.

Table 8. Mean total ODAP content (%) in different plant parts at the mature stage in five *L. sativus* accessions grown during 2011–2012 at Tel Hyada, Aleppo, Syria.

Plant Part	Total ODAP Content (%) (Mean ± SE)					
	B387	B222	B390	Bio-520	B587	Total
Stem	0.342 ± 0.08	0.395 ± 0.02	0.266 ± 0.03	0.321 ± 0.03	0.514 ± 0.09	0.368 ± 0.05
Leaf	0.387 ± 0.05	1.009 ± 0.02	0.614 ± 0.06	0.663 ± 0.02	0.784 ± 0.01	0.691 ± 0.03
Pod cover	0.163 ± 0.01	0.285 ± 0.02	0.413 ± 0.08	0.219 ± 0.02	0.299 ± 0.02	0.276 ± 0.03
Seed	0.125 ± 0.02	0.200 ± 0.04	0.219 ± 0.04	0.211 ± 0.01	0.213 ± 0.03	0.193 ± 0.02
Seed coat	0.144 ± 0.03	0.260 ± 0.01	0.257 ± 0.06	0.354 ± 0.05	0.283 ± 0.03	0.260 ± 0.03
Cotyledons	0.181 ± 0.07	0.291 ± 0.02	0.249 ± 0.04	0.258 ± 0.02	0.379 ± 0.04	0.272 ± 0.04

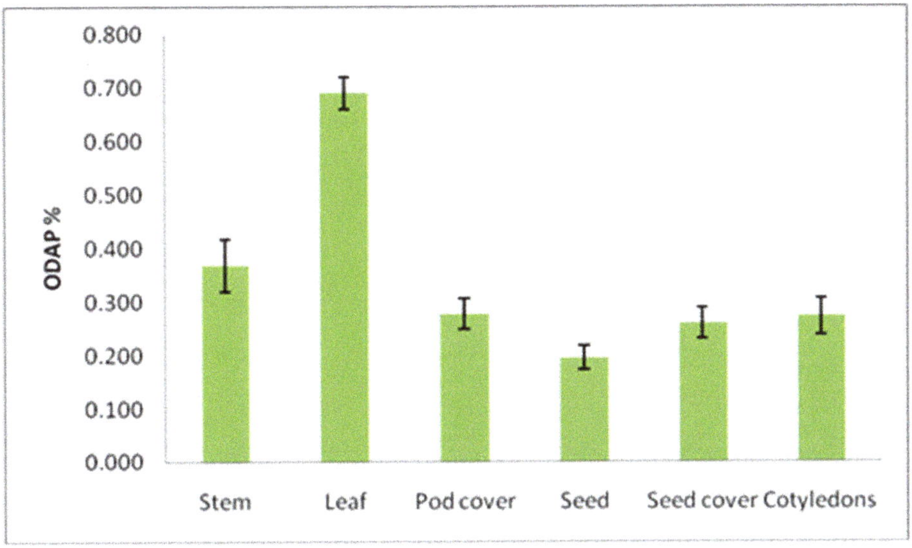

Figure 3. ODAP content variation in diffrent plant parts at mature stage in grass pea. Y axis presents ODAP content in percent, and X axis denotes different plant parts as source of sampling for ODAP estimation.

3.4. Yearly Variation in Total ODAP

The mean total ODAP content of *L. sativus* samples harvested during 2011–2012 was significantly higher (0.91%) (*t* test *p* value 0.042) compared to samples harvested during 2010–2011 (0.66%) (calculated from Tables 3 and 6, respectively). The lowest seed total

ODAP content (0.14%) was recorded in genotype B390 during the first growing season (2010–2011), and genotype B387 was having lowest seed total ODAP content (0.12%) during 2011–2012. While comparing the different plant stages between the two growing seasons (2010–2011 and 2011–2012), significant differences for total ODAP content (%) were observed for 40, 60, 80, 90, 100, 110, 120, and 130 days after sowing. Further, during maturity stage, no significant differences for total ODAP content (%) between the two growing years were observed (Table 9).

Table 9. Mean total ODAP content (%) in different plant development stages in *L. sativus* during the year 2010–2011 and 2011–2012.

	Days of Sampling after Sowing										
Genotypes	**40**	**50**	**60**	**70**	**80**	**90**	**100**	**110**	**120**	**130**	**Mature Seed**
Mean	0.959 (2010) 1.278 (2011)	0.972 (2010) 1.066 (2011)	0.920 (2010) 1.048 (2011)	0.839 (2010) 0.992 (2011)	0.713 (2010) 0.907 (2011)	0.667 (2010) 1.134 (2011)	0.594 (2010) 1.021 (2011)	0.536 (2010) 0.922 (2011)	0.471 (2010) 0.782 (2011)	0.418 (2010) 0.692 (2011)	0.173 (2010) 0.193 (2011)
Variance	0.003 (2010) 0.018 (2011)	0.019 (2010) 0.001 (2011)	0.001 (2010) 0.001 (2011)	0.027 (2010) 0.005 (2011)	0.010 (2010) 0.011 (2011)	0.022 (2010) 0.027 (2011)	0.003 (2010) 0.017 (2011)	0.006 (2010) 0.013 (2011)	0.004 (2010) 0.008 (2011)	0.002 (2010) 0.010 (2011)	0.001 (2010) 0.002 (2011)
t test (*p* value)	0.004	0.173	0.000	0.107	0.017	0.002	0.001	0.000	0.000	0.002	0.364

Note: Number within parenthesis denotes year.

4. Discussion

4.1. Total ODAP Content in Tested Grass Pea Accessions in 2010–2011 and 2011–2012

The amount of total ODAP content at most developmental stages was significantly different except at maturity among all the accessions of grass pea between growth seasons 2010–2011 and 2011–2012 (Tables 1, 4 and 9). During the early vegetative phase, rapid variation in the seed total ODAP content was noticed in all accessions. Moreover, total ODAP content in the seeds grown in field, showed similar trends during both seasons (Figures 1 and 2). The data reported in this study were collected from two years of field experimentation, the pattern of accumulation and variation in the amount of total ODAP during different developmental stages varied during growth seasons as suggested by Addis & Narayan [21] and Jiao et al., [22]. However, it is not known whether the accumulation of total ODAP in tissues results from de novo synthesis in the tissue or translocation from other tissues.

The present results showed that most early developmental stages attained maximum total ODAP content and leaves are the most active site for total ODAP biosynthesis. It might be due to the increased level of free nitrogen-containing compounds in the cells during seed germination and early developmental stage [22]. Xiong et al., [15] reported that ODAP concentration varied significantly during vegetative and reproductive stages among the seven grass pea genotypes. The β-ODAP content decreased in leaves in early reproductive development and in pods as they matured. The net amount of β-ODAP in leaves and pods at early podding was positively associated with seed β-ODAP concentration at maturity. In the present study, a similar trend was also observed where total ODAP content gradually decreased from early vegetative phase to reproductive phase to maturity.

4.2. Changes in the Concentration of Total ODAP in Different Parts of Plants Depending on the Developmental Stage

The plants showed visible effects of ontogeny on total ODAP accumulation. Total ODAP content was low in the mature seeds compared to leaves, stems, pod cover, seed coat, and cotyledons (Table 7). Similarly, Xiong et al., [15] reported that ODAP content decreased in plant parts as they are matured. Kuo et al., [13] suggested that ODAP can be formed in the ripening seeds and is not exclusively transported from the ODAP pools in the pericarp. Jiao et al., [22] reported that total ODAP mainly accumulates in young or fast-growing tissues (including seedlings, ripening seeds, and young leaves), which are all sink tissues, containing a relative abundance of free nitrogen-containing compounds. In the present study, it was notified that during maturity, leaf of B222 exhibited higher (1.009%) total ODAP content compared to other accessions (Table 8). It might be due to the presence of high ODAP extracts, some of the tissues were colored with pigments. The previous report suggested that the presence of certain pigments and amino acids might affect the spectrophotometric measurement of total ODAP concentration [20,21]. On average, seed coat had 0.26% total ODAP (Table 7). However, pod cover on average showed a similar percentage (0.27%) of total ODAP. The mean total ODAP content in seed coat (0.26%) and pod cover (0.27%) agreed with the results of Addis and Narayan [21], who confirmed the rate of synthesis and accumulation of ODAP varied during the developing fruits of *L. sativus*.

4.3. Yearly Variation in Total ODAP

The level of total ODAP in harvested seeds of all accessions was substantially higher in the growing season 2011–2012 than 2010–2011 (Tables 3 and 6). Piergiovanni et al. [23] reported that grass pea had widest year-to-year variation for seed total ODAP content as was recorded in the three *Lathyrus* accessions. Moreover, they also reported interesting considerations on factors affecting the ODAP content between locations as well as growing seasons for grain yield and ODAP content [23]. The ODAP storage in seeds changes from one year to the next, and the impact of growing location, might be relative to soil composition, yield, sowing and harvesting date, environmental conditions of growing location, genotype × environment interaction, etc. Similarly, as a consequence of the large overlap of sowing and harvesting periods between growing seasons, it is predictable that weather conditions (rainfall quantity and/or average temperature) might have a major influence in regulating the ODAP accumulation [24]. This is not surprising because it is well known that the amounts of some antinutritional factors (for example, ODAP and trypsin inhibitor content) stored in legume grains are modulated by high rainfall [23] and/or high temperature during specific crop growth stages [25,26]. The environmental conditions that can affect grass pea seed quality traits have been reported in recent studies dealing with morphological and compositional seed traits [23,27]. In the present study five (B222, 387, 390, Bio520, and 587) grass pea accessions introduced from ICARDA and were grown for December to June during 2010–2011 and 2011–2012 at the International Center for Agricultural Research in the Dry Areas, Aleppo, Syria, which is located at 360°56′ E, 360°01′ N latitude with an altitude of 284 m above sea level. The soil is clay with pH 7.9, with mean soil temperature in December to June ranging from 11.0 to 30.90 °C, respectively, at 5 cm depth. The weather variables during the testing periods for grass pea and average rainfall were 270 mm and 240 mm in 2010–2011 and 2011–2012, respectively. The lower rainfall could have caused higher mean total ODAP content of 2011–2012 grown *L. sativus* samples.

Author Contributions: S.B. and P.G. performed the research work. S.B. and P.G. contributed to phenotypic data collection and biochemical analysis. D.S.G. and J.K. analyzed data and prepared a primary draft of the manuscript. A.B. helped in the statistical analysis of data. S.K. supervised all research activity and wrote the final draft of the manuscript. All authors have read and agreed to the published version of the manuscript.

Funding: This research work was supported by International Center for Agricultural Research in the Dry Areas, Rabat, Morocco.

Institutional Review Board Statement: Not applicable.

Informed Consent Statement: Not applicable.

Data Availability Statement: The original contributions presented in the study are included in the article; further inquiries can be directed to the corresponding author/s.

Conflicts of Interest: The authors declare that the research was conducted in the absence of any commercial or financial relationships that could be construed as potential conflict of interest.

Abbreviations

ODAP	N-oxalyl-L-α, β-diaminopropionic acid
OPA	O-phthalaldehyde
DAS	Days after sowing

References

1. Kumar, S.; Bejiga, G.; Ahmed, S.; Nakkoul, H.; Sarkar, A. Genetic Improvement of grass pea for low neurotoxin (β-ODAP) content. *Food Chem. Toxicol.* **2011**, *49*, 589–600. [CrossRef]
2. Kumar, S.; Gupta, P.; Barpete, S.; Sarker, A.; Amri, A.; Mathur, P.N.; Baum, M. Grass pea. In *Genetic and Genomic Resources for Grain Legume Improvement*; Singh, M., Upadhyaya, H.D., Bisht, I.S., Eds.; Elsevier: London, United Kingdom, 2013; pp. 269–293. [CrossRef]
3. Lambein, F.; Travella, S.; Kuo, Y.H.; Montagu, M.V.; Heijde, M. Grass pea (*Lathyrus sativus* L.): Orphan crop, nutraceutical or just plain food? *Planta* **2019**, *250*, 821–838. [CrossRef] [PubMed]
4. Yan, Z.Y.; Spencer, P.S.; Li, Z.X.; Liang, Y.M.; Wang, Y.F.; Wang, C.Y.; Li, F.M. *Lathyrus sativus* (grass pea) and its neurotoxin ODAP. *Phytochemistry* **2006**, *67*, 107–121. [CrossRef] [PubMed]
5. Mora, G.; Adriano, C. Disorders of upper and lower motor neurons. In *Prognosis of Neurological Diseases*; Springer: Milano, Italy, 2015; pp. 261–272.
6. Lambein, F.; Kuo, Y.H.; Kusama-Eguchi, K.; Ikegami, F. β-N-oxalyll-α,β-diaminopropionic acid, a multifunctional plant metabolite of toxic reputation. *Arkivoc* **2007**, *9*, 45–52. [CrossRef]
7. Singh, S.S.; Rao, S.L.N. Lessons from neurolathyrism: A disease of the past & the future of *Lathyrus sativus* (Khesari dal). *Indian J. Med. Res.* **2013**, *138*, 32–37. Available online: https://www.ncbi.nlm.nih.gov/pmc/articles/PMC3767245/pdf/IJMR-138-32.pdf (accessed on 12 January 2022).
8. Hao, X.; Yang, T.; Liu, R.; Hu, J.; Yao, Y.; Burlyaeva, M.; Wang, Y.; Ren, G.; Zhang, H.; Zong, X.; et al. An RNA sequencing transcriptome analysis of grass pea (*Lathyrus sativus* L.) and development of SSR and KASP markers. *Front. Plant Sci.* **2017**, *8*, 1873. [CrossRef]
9. Kumar, S.; Gupta, P.; Barpete, S.; Choukri, H.; Maalouf, F.; Sarkar, A. Grass Pea. In *The Beans and the Peas*; Pratap, A., Gupta, S., Eds.; Woodhead Publishing; Elsevier: Amsterdam, The Netherlands, 2020; pp. 273–287. [CrossRef]
10. Tadesse, W. Stability of grasspea (*Lathyrus sativus* L.) varieties for ODAP content and grain yield in Ethiopia. *Lathyrus Lathyrism Newslett.* **2003**, *3*, 32–34. Available online: https://www.clima.uwa.edu.au/data/assets/pdf_file/0005/919805/Tadesse.pdf (accessed on 18 January 2022).
11. Siddique, K.H.M.; Loss, S.P.; Herwig, S.P.; Wilson, J.M. Growth, yield and neurotoxin (ODAP) concentration of three Lathyrus species in Mediterranean type environments of Western Australia. *Aust. J. Exp. Agric.* **1996**, *36*, 209–218. [CrossRef]
12. Hanbury, C.D.; White, C.L.; Mullan, B.P.; Siddique, K.H.M. A review of the potential of *Lathyrus sativus* L. and *L. cicera* L. grain for use as animal feed. *Anim. Feed. Sci. Technol.* **2000**, *87*, 1–27. [CrossRef]
13. Kuo, Y.H.; Khan, J.K.; Lambein, F. Biosynthesis of the neurotoxin â-ODAP in developing pods of *Lathyrus sativus*. *Phytochemistry* **1994**, *35*, 911–913. [CrossRef]
14. Tadesse, W.; Bekele, E. Variation and association of morphological and biochemical characters in grass pea (*Lathyrus sativus* L.). *Euphytica* **2003**, *130*, 315–324. [CrossRef]
15. Xiong, J.L.; Xiong, Y.; Bai, X.; Kong, H.; Tan, R.; Zhu, H.; Siddique, K.H.M.; Wang, J.; Turner, N.C. Genotypic variation in the concentration of β-*N*-Oxalyl-L-α,β-diaminopropionic acid (β-ODAP) in grasspea (*Lathyrus sativus* L.) seeds is associated with an accumulation of leaf and pod β-ODAP during vegetative and reproductive stages at three levels of water Stress. *J. Agric. Food Chem.* **2015**, *63*, 6133–6141. [CrossRef] [PubMed]
16. Yusuf HK, M.; Hoque, K.; Uddin, A.; Roy, B.C.; Lambein, F. Homoarginine antagonizes the toxicity of *Lathyrus* toxin in one-day-chiks. *Bangladesh J. Physiol. Pharmacol.* **1995**, *10*, 74–75.
17. Shamin, M.Z.; Hossain, M.S.; Islam, K.; Yusuf, H.K.M.; Lambein, F. Mechanism of ODAP toxicity in one-day-old chicks. *Dhaka Univ. J. Biol. Sci.* **2002**, *11*, 1–7. Available online: http://hdl.handle.net/1854/LU-271535 (accessed on 12 January 2022).
18. Fikre, A.Y.; Kuo, Y.H.; Ahmed, S.; Gheysen, G.; Lambein, F. Effect of Methionine Supplement on Physical Responses and Neurological Symptoms in Broiler Chicks Fed Grass Pea (*Lathyrus sativus*) Based Starter Ration. *Food Chem. Toxicol.* **2010**, *48*, 11–17. [CrossRef]
19. Rao, S.L.N. A sensitive and specific colorimetric method for the determination of α, β- diaminopropionic acid and the *Lathyrus sativus* neurotoxin. *Anal. Biochem.* **1978**, *86*, 386–395. [CrossRef]

20. Briggs, C.J.; Parreno, N.; Campbell, C.G. Phytochemical assessment of *Lathyrus* species for the neurotoxic agent, β-N-oxalyl-L-α, β-diaminopropionic acid. *Planta Med.* **1983**, *47*, 188–190. [CrossRef]
21. Addis, G.; Narayan, R.K.J. Developmental variation of the neurotoxin, b-N-oxalyl-L-a, b-diaminopropionic acid (ODAP), in *Lathyrus sativus*. *Ann. Bot.* **1994**, *74*, 209–215. [CrossRef]
22. Jiao, C.J.; Xu, Q.L.; Wang, C.Y.; Li, F.M.; Li, Z.X.; Wang, Y.F. Accumulation pattern of toxin β-ODAP during lifespan and effect of nutrient elements on β-ODAP content in *Lathyrus sativus* seedlings. *J. Agric. Sci.* **2006**, *144*, 369–375. [CrossRef]
23. Piergiovanni, A.R.; Lupo, F.; Zaccardelli, M. Environmental effect on yield, composition and technological seed traits of some Italian ecotypes of grass pea (*Lathyrus sativus* L.). *J. Sci. Food Agric.* **2011**, *91*, 122–129. [CrossRef]
24. Piergiovanni, A.R.; Damascelli, A. L-Homoarginine accumulation in grass pea (*Lathyrus sativus* L.) dry seeds. A preliminary survey. *Food Nutr. Sci.* **2011**, *2*, 207–213. [CrossRef]
25. Berger, J.D.; Siddique, K.H.M.; Loss, S.P. Cool season grain legumes for Mediterranean environments: Species × environment interaction in seed quality traits and anti-nutritional factors in the genus *Vicia*. *Aust. J. Agric. Res.* **1999**, *50*, 389–402. [CrossRef]
26. Urga, K.; Fufa, H.; Biratu, E.; Husain, A. Evaluation of *Lathyrus sativus* cultivated in Ethiopia for proximate composition, minerals, β-ODAP and anti-nutritional components. *Afr. J. Food Agric. Nutr. Dev.* **2005**, *5*, 1–15.
27. Polignano, G.B.; Bisignano, V.; Tomaselli, V.; Uggenti, P.; Alba, V.; Gatta, C.D. Genotype × Environment interaction in grass pea (*Lathyrus sativus* L.) lines. *Int. J. Agron.* **2009**, *2009*, 1–17. [CrossRef]

molecules

Article

Interactive Effects of Molybdenum, Zinc and Iron on the Grain Yield, Quality, and Nodulation of Cowpea (*Vigna unguiculata* (L.) Walp.) in North-Western India

Salwinder Singh Dhaliwal [1], Vivek Sharma [1], Arvind Kumar Shukla [2], Janpriya Kaur [1], Vibha Verma [1], Manmeet Kaur [1], Prabhjot Singh [1], Marian Brestic [3,*], Ahmed Gaber [4] and Akbar Hossain [5,*]

1 Department of Soil Science, Punjab Agricultural University, Ferozepur Rd, Ludhiana 141027, India; ssdhaliwal@pau.edu (S.S.D.); sharmavivek@pau.edu (V.S.); janpriyakaur89@pau.edu (J.K.); vermavibha@pau.edu (V.V.); manmeetgill885@gmail.com (M.K.); prabh@pau.edu (P.S.)

2 ICAR-Indian Institute of Soil Science, Bhopal 462038, India; arvindshukla2k3@yahoo.co.in

3 Department of Plant Physiology, Slovak University of Agriculture, Tr. A. Hlinku 2, 949 01 Nitra, Slovakia

4 Department of Biology, College of Science, Taif University, P.O. Box 11099, Taif 21944, Saudi Arabia; a.gaber@tu.edu.sa

5 Department of Agronomy, Bangladesh Wheat and Maize Research Institute, Dinajpur 5200, Bangladesh

* Correspondence: marian.brestic@uniag.sk (M.B.); akbar.hossain@bwmri.gov.bd (A.H.)

Citation: Dhaliwal, S.S.; Sharma, V.; Shukla, A.K.; Kaur, J.; Verma, V.; Kaur, M.; Singh, P.; Brestic, M.; Gaber, A.; Hossain, A. Interactive Effects of Molybdenum, Zinc and Iron on the Grain Yield, Quality, and Nodulation of Cowpea (*Vigna unguiculata* (L.) Walp.) in North-Western India. *Molecules* **2022**, *27*, 3622. https://doi.org/10.3390/molecules27113622

Academic Editor: Smaoui Slim

Received: 17 April 2022
Accepted: 3 June 2022
Published: 5 June 2022

Abstract: Micronutrient deficiency is a major constraint for the growth, yield and nutritional quality of cowpea which results in nutritional disorders in humans. Micronutrients including molybdenum (Mo), iron (Fe) and zinc (Zn) play a pivotal role in crop nutrition, and their role in different metabolic processes in crops has been highlighted. In order to increase the nutritional quality of cowpea, a field experiment was conducted for two years in which the effect of Mo along with iron (Fe) and zinc (Zn) on productivity, nitrogen and micronutrient uptake, root length and the number of nodules in cowpea cultivation was investigated. It was found that the foliar application of Fe and Zn and their interaction with Mo application through seed priming as well as soil application displayed increased yield, nutrient concentration, uptake and growth parameters which helped to enhance the nutritional quality of cowpea for consumption by the human population. The results of the above experiments revealed that among all the treatments, the soil application of Mo combined with the foliar application of 0.5% each of $FeSO_4{\cdot}7H_2O$ and $ZnSO_4{\cdot}7H_2O$ (M_2F_3 treatment) enhanced the grain and stover yield of cowpea, exhibiting maximum values of 1402 and 6104.7 kg ha^{-1}, respectively. Again, the M_2F_3 treatment resulted in higher Zn, Fe and Mo concentrations in the grain (17.07, 109.3 and 30.26 mg kg^{-1}, respectively) and stover (17.99, 132.7 and 31.22 mg kg^{-1}, respectively) of cowpea. Uptake of Zn, Fe and Mo by the grain (25.23, 153.3 and 42.46 g ha^{-1}, respectively) as well as the stover (104.2, 809.9 and 190.6 g ha^{-1}, respectively) was found to be maximum for the M_2F_3 treatment. The root length (30.5 cm), number of nodules per plant (73.0) and N uptake in grain and stover (55.39 and 46.15 kg ha^{-1}) were also higher for this treatment. Overall, soil application of Mo along with the foliar application of $FeSO_4{\cdot}7H_2O$ (0.5%) and $ZnSO_4{\cdot}7H_2O$ (0.5%) significantly improved yield outcomes, concentration, uptake, root length, nodules $plant^{-1}$ and N uptake of cowpea to alleviate the micronutrient deficiency.

Keywords: cowpea; Mo soil treatment; Fe and Zn foliar application

1. Introduction

More than half of the world's population consumes micronutrients at concentrations lower than their daily minimal requirements [1]. Recent data on human nutrient deficiency have shown that 'hidden hunger' affects more than two billion people globally [2]. Among different micronutrients, molybdenum (Mo) is also an essential trace nutrient due to its pivotal role in more than 60 enzymes that catalyze various redox reactions [3]. Its crucial role in nitrogen fixation through the enzyme nitrogenase and nitrate reductase is well known and thus affects nitrogen transport in plants [4]. On the other hand, the deficiency of Mo in crops leads to the reduced growth of flowers, smaller sizes and less maturity, consequently resulting in a lower grain yield.

Additionally, iron (Fe) and zinc (Zn) are essential nutrients for plants and their deficiencies in crops are the most common nutrient deficiencies around the globe [5]. Iron is an important structural component of numerous enzymes that are involved in various metabolic processes of plants [6]. Despite the higher abundance in the earth's crust, its poor bioavailability in soil is due to the rapid binding with soil particles and the formation of insoluble complexes under aerobic conditions [7]. On the other hand, Zn is also a trace element that is considered crucial as it possesses antioxidant properties and is required for proper growth, immune system development, enzyme activation and neurobehavioral development [8]. A lack of Zn in the diet may result in serious health-related issues, such as stunted growth in children, increased illness susceptibility, poor birth outcomes and harm to the brain and immune system [9,10]. This has led to a growing interest in how the micronutrient content of crops might be modified in order to benefit human health and nutrition.

To improve the crop's nutritional value, a variety of traditional interventions have been applied, including dietary supplementation, food fortification and dietary diversification [11]. Due to the lack of infrastructure, these strategies have been found to be unsuccessful. In this view, an alternate key to malnutrition named biofortification has been suggested. It is a method for enhancing the concentration of the desired mineral in a crop using specialized techniques such as plant breeding and agronomic procedures [12,13]. Further, agronomic biofortification through foliar sprays, seed priming and soil treatments are considered convenient ways to improve the nutrient content in the crop [14]. Foliar application has led to an improvement in the micronutrient status of crops, as nutrients are rapidly absorbed by the leaves at suitable growth stages [15]. On the other hand, seed priming, a pre-sowing treatment, regulates seed germination through the controlled hydration of seeds that enables the pre-germination activity. Additionally, soil treatment involves the addition of micronutrients directly into the soil which helps to pass on the available nutrients to plants for adequate plant growth.

Cowpea is a lucrative summer season vegetable and is valued for its proteins, minerals and energy [16]. Cowpea has been referred to as the poor man's meat due to its high level of protein and is consumed by more than 200 million people from Africa, Asia and North and South America on a daily basis [17]. Besides being nutritious, it helps to sustain the productivity of the cropping systems through its ability to fix atmospheric nitrogen [18]. Poor management and inadequate cultivation on agriculturally marginal and sub-marginal lands are considered the root causes of the low productivity of pulse crops. To maximize the yield potential of pulses, the adoption of acceptable production technologies is crucial, and this can be accomplished through the use of fertilizers and micronutrients. Since Mo, Zn and Fe play key roles in the plant metabolism [19], the biofortification of these micronutrients in cowpea through the foliar application, seed priming or soil treatment would provide a potential increase in the micronutrient levels in crops, which might improve the nutritional level of crops required by humans.

Several researchers have probed the effect of enhancing bioavailable Mo, Zn and Fe in various crops through the process of biofortification. The iron biofortification of cowpea has been found to escalate the grain yield and bioavailable Fe content in cowpea [20]. Another study revealed that the seed treatment with Mo at 0.5 g/kg seed and biofertilizers in combination with the foliar application of boron significantly increased the yield attributes of cowpea [21]. The seed priming with Mo increased the grain yield and net return in chickpea [22]. However, many studies have reported the sole application of biofertilizers in cowpea [23,24] but a comparative analysis of the effect of foliar application, seed priming and soil treatment using Mo, Zn and Fe along with their interactive effects on cowpea has not been explored so far. This interactive effect could be better explored in terms of micronutrient accumulation in grain and stover along with the improvement in crop quality which could benefit human health by combating hidden hunger. The objective of the present study was to assess the influence of Mo, Zn and Fe molecules interactions on the yield, concentration and uptake of these micronutrients to enhance the food quality of crops to increase the nutritional security of the consumers.

2. Materials and Methods

2.1. Site Specification and Characteristics

The two-year field experiment was conducted at the Farm Research Area, Department of Soil Science, Punjab Agricultural University (PAU), Ludhiana, Punjab in the Indo-Gangetic plains of north-western India during the Kharif season (June–October). The experimental soil possessed a pH of 7.21, an EC of 0.34 dS m^{-1}, an OC of 0.31% and had a sandy loam texture. DTPA-extractable micronutrient levels in the soil were initially 1.16 and 4.86 mg kg^{-1} for Zn and Fe, respectively. The region has a subtropical climate along with hot, rainy summers as well as dry winters. The annual rainfall ranges from 400 to 600 mm and the months of July to September receive the majority of the rainfall, which is around 70%.

2.2. Treatment Details

The present experiment involved three main plot treatments of Mo application, i.e., no molybdenum (M_0), seed priming with 500 mg kg^{-1} Mo solution (M_1), soil application of Mo with 1.25 kg ha^{-1} (M_2) and four subplot treatments of Zn and Fe as foliar application, i.e., no foliar spray (F_0), 0.5% $FeSO_4{\cdot}7H_2O$ (F_1), 0.5% $ZnSO_4{\cdot}7H_2O$ (F_2) and 0.5% each of $FeSO_4{\cdot}7H_2O$ + $ZnSO_4{\cdot}7H_2O$ (F_3), respectively. The details of the treatments are given in Table 1. The experiment was laid out in a split-plot design with three replications.

Table 1. Treatment details of the field experiment.

Treatments	Details
M_0F_0	Control
M_0F_1	0.5% $FeSO_4{\cdot}7H_2O$
M_0F_2	0.5% $ZnSO_4{\cdot}7H_2O$
M_0F_3	0.5% $FeSO_4{\cdot}7H_2O$ + 0.5% $ZnSO_4{\cdot}7H_2O$
M_1F_0	Mo seed priming
M_1F_1	Mo seed priming + 0.5% $FeSO_4{\cdot}7H_2O$
M_1F_2	Mo seed priming + 0.5% $ZnSO_4{\cdot}7H_2O$
M_1F_3	Mo seed priming + 0.5% $FeSO_4{\cdot}7H_2O$ + 0.5% $ZnSO_4{\cdot}7H_2O$
M_2F_0	Mo soil application
M_2F_1	Mo soil application + 0.5% $FeSO_4{\cdot}7H_2O$
M_2F_2	Mo soil application + 0.5% $ZnSO_4{\cdot}7H_2O$
M_2F_3	Mo soil application + 0.5% $FeSO_4{\cdot}7H_2O$ + 0.5% $ZnSO_4{\cdot}7H_2O$

The recommended doses of N (19.0 kg ha^{-1}) and P (55 kg ha^{-1}) were applied as basal through urea and di-ammonium phosphate at the time of sowing. The sowing of the cowpea variety 'CL 857' was performed at 30 cm plant to plant spacing and 45 cm row to row spacing. The foliar application of $FeSO_4{\cdot}7H_2O$ and $ZnSO_4{\cdot}7H_2O$ was applied twice (at 40 days and 50 days of sowing) as per the experimental details.

2.3. Harvesting and Analysis

The plants were manually harvested at the physiological maturity stage and grain as well as stover samples were collected for further analysis. Grain and stover yields were measured from the net plot area leaving the border rows and were later converted to kg ha^{-1}. In order to measure the growth parameters, five plant samples were selected randomly from the central rows to measure the root length and the number of nodules. Root samples were washed with distilled water and the residual water was removed with absorbent paper. The nodules were removed quickly from the roots and counted.

To measure the dry weight, the samples were air-dried before drying in an oven at 65 °C for 48 h. A mechanical grinder was used to grind oven-dried plant samples to a fine powder. On an electric hot plate, the grounded samples of grain and stover weighing 1.0 g each were subjected to the digestion using a mixture of di-acid, i.e., HNO_3 and $HClO_4$ acid in a 3:1 ratio [25]. The micronutrient contents of Zn and Fe in digested extracts of the plant were measured using an atomic absorption spectrophotometer (Model AAS 240 FS, Company Varian, Labexchange - Die Laborgerätebörse GmbH, Bruckstr. 58, D-72393 Burladingen Germany) [26]. The samples were analyzed for N using Kjeldahl's method [27]. Additionally, the Mo content in samples was measured by the method given by Purushottam et al. [28]. A sample (5 g) was weighed into a dry 25 mL graduated cylinder and digested with 5 mL of aqua regia for about an hour on a hot plate. After complete decomposition of the sample, 0.5 mL of phosphoric acid was added and the solution was made up to a suitable volume with demineralized water in order to determine the Mo content using AAS. The micronutrient uptake by the grain and stover of cowpea (g ha^{-1}) was calculated using the following equation:

$$\text{Uptake}\left(\text{g ha}^{-1}\right) = \frac{\text{concentration}\left(\text{mg kg}^{-1}\right) \times \text{Yield}\left(\text{q ha}^{-1}\right)}{10} \tag{1}$$

2.4. Micronutrient Use Efficiency Indices

The mobilization efficiency index (MEI) determines the translocation of nutrients towards the grain as well as the stover of crops. In the present study, the MEI calculation was performed using the following equation:

$$\text{MEI} = \frac{\text{Nutrient cocnentartion in grain}\left(\text{mg kg}^{-1}\right)}{\text{Nutrient concentration in stover}\left(\text{mg kg}^{-1}\right)} \tag{2}$$

The physiological efficiency (PE) indicates the increase in yield per unit of absorbed nutrient by plants and identifies the role of nutrients in increasing the crop yield. In the present study, the determination of PE of Zn, Fe and Mo viz. (PE_{Zn}), (PE_{Fe}), (PE_{Mo}) was completed through the equations given below [29]:

$$\text{PE} = \frac{Y_t - Y_c}{NU_t - NU_c} \tag{3}$$

where, Y_t and Y_c denote the grain yield (kg ha^{-1}) of cowpea in Mo, Zn and Fe-treated plots as well as in the control, respectively; NU_t and NU_c denote the total nutrient (Zn, Fe and Mo) uptake (kg ha^{-1}) of cowpea in Mo, Zn and Fe-treated plots as well as in the control, respectively.

2.5. Statistical Analysis

Data were analyzed statistically using SPSS version 16.0 (SPSS Inc., Chicago, IL, USA) packages. A two-way analysis of variance (ANOVA) and the Duncan Multiple Range test were performed to assess the significant difference between the treatment results on the crop.

3. Results

The results of the study suggest that the Mo seed, as well as soil application along with a combined foliar spray of Fe and Zn, improved the yield, N and micronutrient uptake, root length and the number of nodules. The various parameters analyzed in the present study are described in the following sections.

3.1. Grain and Stover Yield

The two-year mean data for Kharif 2020 and 2021 seasons demonstrated that the application of Mo, Fe and Zn had a significant effect on the grain and stover yield of cowpea (Table 2). In the main plot treatments, the average of the two-year data revealed that Mo application irrespective of the application method (M_1 and M_2) had a significant positive impact on the grain and stover yield (Figure 1a). Moreover, Mo application through soil (M_2) resulted in a significantly higher grain and stover yield (1307.4 and 3886.8 kg ha^{-1}) as compared to the Mo seed priming (M_1). In sub-plot treatments, F_2 showed a significantly higher grain and stover yield (1306.6 and 3966.5 kg ha^{-1}) over the other treatments followed by F_1, F_2 and F_0 (Figure 1b).

Table 2. Effect of different levels of Zn, Fe and Mo molecules on the grain and stover yield of cowpea.

Treatments	Grain Yield (kg ha^{-1})			Stover Yield (kg ha^{-1})		
Interaction	I Year	II Year	Mean	I Year	II Year	Mean
M_0F_0	884.7 ^{f} ± 153.3	848.2 ^{f} ± 49.9	866.5 ^{f} ± 25.8	2662.4 ^{h} ± 112.2	2973.9 ^{e} ± 20.0	2818.2 ^{g} ± 65.2
M_0F_1	1138.8 ^{de} ± 84.1	1175.9 ^{cd} ± 23.6	1157.4 ^{d} ± 26.3	3490.0 ^{cde} ± 123.4	3502.8 ^{d} ± 50.6	3496.4 ^{e} ± 51.5
M_0F_2	1089.9 ^{e} ± 46.5	1022.2 ^{e} ± 90.8	1056.1 ^{e} ± 47.8	3150.25 ^{g} ± 168.6	3280.8 ^{d} ± 154.2	3215.5 ^{f} ± 10.2
M_0F_3	1206.7 ^{bcd} ± 31.6	1204.2 ^{cd} ± 76.5	1205.4 ^{cd} ± 53.3	3676.5 ^{bc} ± 89.8	3969.6 ^{bc} ± 186.7	3823.0 ^{bc} ± 68.5
M_1F_0	943.8 ^{f} ± 40.3	951.7 ^{ef} ± 72.3	947.7 ^{f} ± 55.8	3231.2 ^{fg} ± 258.9	3279.1 ^{d} ± 169.1	3255.1 ^{f} ± 63.5
M_1F_1	1201.9 ^{bcd} ± 49.4	1251.3 ^{bc} ± 120.8	1226.6 ^{bcd} ± 34.8	3574.5 ^{bcde} ± 244.3	4001.9 ^{c} ± 114.9	3788.2 ^{bcd} ± 91.4
M_1F_2	1160.1 ^{de} ± 40.1	1138.7 ^{d} ± 137.7	1149.4 ^{e} ± 15.1	3393.8 ^{ef} ± 506.6	3969.4 ^{bc} ± 94.1	3681.6 ^{cde} ± 291.7
M_1F_3	1299.9 ^{ab} ± 181.9	1323.1 ^{b} ± 50.4	1311.6 ^{ab} ± 16.4	3756.9 ^{ab} ± 203.8	4067.7 ^{bc} ± 198.8	3912.3 ^{b} ± 30.6
M_2F_0	1180.2 ^{cde} ± 70.6	1334.7 ^{b} ± 90.7	1257.5 ^{bc} ± 109.2	3346.2 ^{cf} ± 115.8	3856.4 ^{c} ± 114.2	3601.3 ^{de} ± 42.3
M_2F_1	1286.4 ^{abc} ± 97.5	1358.2 ^{ab} ± 87.3	1322.3 ^{ab} ± 50.8	3709.3 ^{ab} ± 86.1	4186.6 ^{b} ± 339.5	3947.9 ^{ab} ± 179.2
M_2F_2	1222.4 ^{bcd} ± 125.8	1271.5 ^{bc} ± 70.8	1246.9 ^{bcd} ± 34.7	3634.8 ^{bcd} ± 186.7	4032.7 ^{bc} ± 129.9	3833.7 ^{bc} ± 40.2
M_2F_3	1350.2 ^{a} ± 102.2	1455.6 ^{a} ± 93.8	1402.9 ^{a} ± 74.4	3880.7 ^{a} ± 188.4	4447.9 ^{a} ± 411.4	4164.3 ^{a} ± 157.7
LSD (0.05)	112.2	107.4	97.2	187.5	245.2	209.4

M_0: No molybdenum, M_1: Molybdenum seed priming, M_2: Molybdenum soil treatment, F_0: No fertilizer application, F_1: Fe application, F_2: Zn application, F_3: Fe+Zn application. The mean with a similar or dissimilar letter(s) was evaluated with the least significant difference (LSD) multiple range tests using a probability level of $p \leq 0.05$ along with standard deviation.

The effect of the interaction between Mo application and foliar spray further suggests that there was a significant improvement in the grain yield with the maximum value of 1402.9 kg ha^{-1} observed for the M_2F_3 treatment (Mo soil application + Fe + Zn foliar spray) which was statistically on par with the M_2F_1 treatment (Mo soil application + Fe foliar spray) and M_1F_3 (Mo seed application + Fe + Zn foliar spray) with grain yields of 1322.3 and 1311.6kg ha^{-1}, respectively. However, the grain yield was minimum with the M_0F_0 treatment, i.e., the control with the mean value of 866.5 kg ha^{-1} which was statistically on par with M_1F_0 treatment, i.e., Mo seed application with no foliar spray (947.7 kg ha^{-1}).

Likewise, the interactive effect of Mo as well as Fe and Zn foliar application showed that the M_2F_3 treatment resulted in a maximum stover yield of 6104.7 kg ha^{-1} which was statistically on par with the M_2F_1 treatment (3947.9 kg ha^{-1}). However, the minimum value of stover yield was observed with the M_0F_0 treatment with a mean value of 2818.24 kg ha^{-1}. Therefore, the results conclude that Mo soil treatment along with the combined spray of Fe and Zn significantly improved the grain as well as the stover yield of cowpea.

The effect of the interaction between Mo application and foliar spray further showed a significant improvement in grain yield with the maximum value of 1402.9 kg ha^{-1} observed for the M_2F_3 treatment (Mo soil application + Fe + Zn foliar spray) which was not statistically different from the M_2F_1 treatment (Mo soil application + Fe foliar spray) and M_1F_3 (Mo seed application + Fe + Zn foliar spray) with grain yields of 1322.3 and 1311.6kg ha^{-1}, respectively. However, the grain yield was minimum in the M_0F_0 treatment, i.e., the control with the mean value of 866.5 kg ha^{-1} which was statistically on par with the M_1F_0 treatment, i.e., Mo seed application with no foliar spray (947.7 kg ha^{-1}).

Likewise, the interactive effect of Mo as well as Fe and Zn foliar application showed that the M_2F_3 treatment resulted in a maximum stover yield of 6104.7 kg ha^{-1} which was statistically on par with the M_2F_1 treatment (3947.9 kg ha^{-1}). However, the minimum value of stover yield was observed with the M_0F_0 treatment with a mean value of 2818.24 kg ha^{-1}. Therefore, the results conclude that Mo soil treatment along with the combined spray of Fe and Zn significantly improved the grain as well as the stover yield of cowpea.

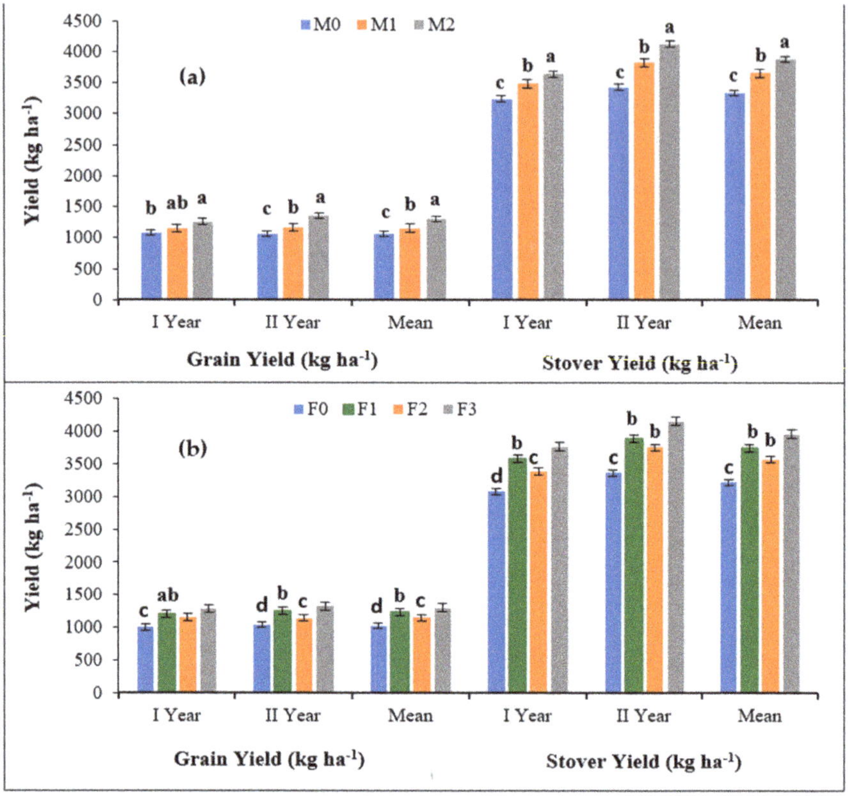

Figure 1. (**a**) Different methods of molybdenum and (**b**) Zn application on the grain and stover yield of cowpea over two years. The column representing the mean with a similar or dissimilar letter(s) was evaluated with the least significant difference (LSD) multiple range tests using a probability level of $p \leq 0.05$ along with standard deviation.

3.2. Micronutrient Concentration in Grain and Stover

The mean of two-year data pertaining to the concentrations of micronutrients in the grain and stover of cowpea is presented in Table 3. The concentration of micronutrients (Zn, Fe and Mo) in grain, as well as the stover of cowpea, increased significantly with the application of Mo either alone or in combination with the foliar application of Zn and Fe over the treatments in which no Mo, Fe and Zn were added. In the main plot treatments, the Mo application (M_1 and M_2) had a significant positive impact on the Zn and Fe concentrations in grain and stover (Figure 2a). Further, Mo application through the soil (M_2) resulted in significantly higher Zn and Fe concentrations in the grain (14.72 and 98.14 mg kg^{-1}) and stover (17.01 and 125.2 mg kg^{-1}) as compared to the Mo seed priming (M_1). For the Mo concentration in grain and stover, the results of M_2 (25.66 and 27.09 mg kg^{-1}) were significantly higher than M_0, whereas the results of M_1 (23.36 and 24.74 mg kg^{-1}) were statistically on par with M_0 (21.34 and 23.54 mg kg^{-1}). In the sub-plot treatments, the F_3 treatment resulted in the maximum Zn, Fe and Mo concentrations in grain (15.49, 98.22 and 26.05 mg kg^{-1}) and stover (17.25, 127.7 and 28.09 mg kg^{-1}) followed by F_1 and F_2, whereas minimum values were recorded in F_0 (Figure 2b). The concentration of micronutrients (Zn, Fe and Mo) in the grain as well as the stover of cowpea, increased significantly with the application of Mo either alone or in combination with the foliar application of Zn and Fe over the treatments in which no Mo, Fe and Zn were added.

Table 3. Effect of the application of Zn, Fe and Mo molecule on the concentration of micronutrients in the grain and stover of cowpea.

Treatments	Grain Concentration (mg kg^{-1})			Stover Concentration (mg kg^{-1})		
Interaction	**Zn**	**Fe**	**Mo**	**Zn**	**Fe**	**Mo**
M_0F_0	9.94 ^{g} ± 6.1	62.55 ^{e} ± 21.7	18.44 ^{d} ± 2.0	12.72 ^{f} ± 0.7	103.9 ^{g} ± 5.3	21.50 ^{d} ± 3.8
M_0F_1	13.65 ^{de} ± 5.3	78.46 ^{cde} ± 37.6	21.98 ^{bcd} ± 6.9	15.74 ^{bcd} ± 2.4	122.8 ^{cd} ± 0.6	24.58 ^{bd} ± 0.7
M_0F_2	12.82 ^{ef} ± 4.7	71.56 ^{de} ± 30.8	22.31 ^{bcd} ± 8.9	13.47 ^{ef} ± 3.9	116.2 ^{e} ± 3.7	22.12 ^{cd} ± 4.7
M_0F_3	14.11 ^{cd} ± 5.5	89.10 ^{bc} ± 54.8	22.64 ^{bcd} ± 1.1	16.57 ^{abc} ± 2.7	124.1 ^{bc} ± 1.1	25.9 ^{abcd} ± 1.9
M_1F_0	10.74 ^{g} ± 7.8	78.87 ^{cd} ± 37.3	20.71 ^{c} ± 2.9	14.79 ^{de} ± 0.2	108.37 ^{f} ± 2.6	22.46 ^{cd} ± 2.3
M_1F_1	14.08 ^{cd} ± 7.4	90.95 ^{bc} ± 46.4	24.07 ^{bc} ± 6.2	16.47 ^{bc} ± 2.6	124.0 ^{bc} ± 8.5	26.20 ^{abcd} ± 0.4
M_1F_2	13.55 ^{de} ± 7.1	88.02 ^{bc} ± 44.2	23.40 ^{bc} ± 12.0	15.98 ^{bcd} ± 2.9	120.9 ^{d} ± 3.3	23.20 ^{bcd} ± 2.1
M_1F_3	15.31 ^{b} ± 6.6	96.22 ^{ab} ± 54.1	25.24 ^{b} ± 5.5	17.20 ^{ab} ± 1.9	126.4 ^{b} ± 3.7	27.09 ^{abc} ± 5.2
M_2F_0	11.84 ^{f} ± 7.2	83.66 ^{bcd} ± 43.6	22.23 ^{bcd} ± 6.8	16.05 ^{cd} ± 0.8	114.2 ^{e} ± 6.8	23.24 ^{bcd} ± 0.2
M_2F_1	15.22 ^{b} ± 7.7	105.6 ^{a} ± 45.7	26.06 ^{ab} ± 10.7	17.19 ^{ab} ± 3.3	130.5 ^{a} ± 2.1	28.40 ^{ab} ± 8.0
M_2F_2	14.77 ^{bc} ± 7.4	93.86 ^{abc} ± 42.1	24.10 ^{b} ± 9.2	16.81 ^{abc} ± 3.5	123.4 ^{cd} ± 0.9	25.50 ^{bcd} ± 1.4
M_2F_3	17.07 ^{a} ± 6.8	109.3 ^{a} ± 64.4	30.26 ^{a} ± 3.7	17.99 ^{a} ± 3.7	132.7 ^{a} ± 1.4	31.22 ^{a} ± 3.4
LSD (0.05)	1.0	16.2	4.2	1.5	2.9	5.5

M_0: No molybdenum, M_1: Molybdenum seed priming, M_2: Molybdenum soil treatment, F_0: No fertilizer application, F_1: Fe application, F_2: Zn application, F_3: Fe+Zn application. The mean with a similar or dissimilar letter(s) was evaluated with the least significant difference (LSD) multiple range tests using a probability level of $p \leq 0.05$ along with standard deviation.

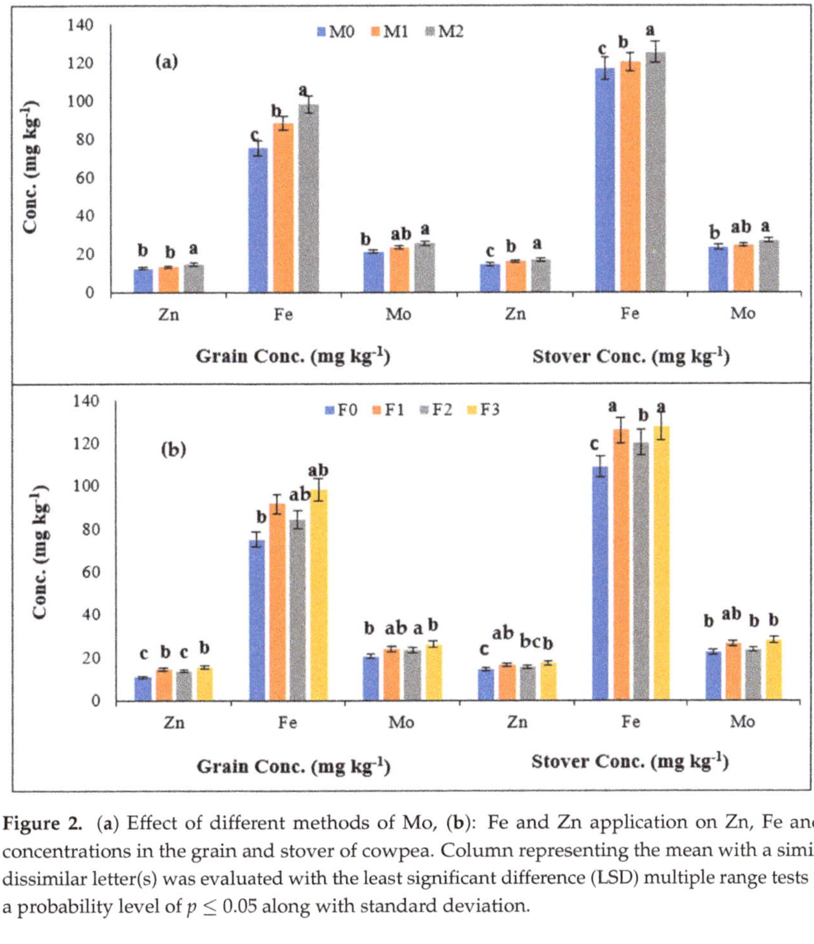

Figure 2. (**a**) Effect of different methods of Mo, (**b**): Fe and Zn application on Zn, Fe and Mo concentrations in the grain and stover of cowpea. Column representing the mean with a similar or dissimilar letter(s) was evaluated with the least significant difference (LSD) multiple range tests using a probability level of $p \leq 0.05$ along with standard deviation.

In the main plot treatments, the Mo application (M_1 and M_2) had a significant positive impact on Zn and Fe concentrations in grain and stover (Figure 2a). Further, Mo application through soil (M_2) resulted in significantly higher Zn and Fe concentrations in the grain (14.72 and 98.14 mg kg^{-1}) and stover (17.01 and 125.2 mg kg^{-1}) as compared to the Mo seed priming (M_1). For the Mo concentration in grain and stover, the results of M_2 (25.66 and 27.09 mg kg^{-1}) were significantly higher than M_0, whereas the results of M_1 (23.36 and 24.74 mg kg^{-1}) were statistically on par with M_0 (21.34 and 23.54 mg kg^{-1}). In the sub-plot treatments, the F_3 treatment resulted in the maximum Zn, Fe and Mo concentrations in the grain (15.49, 98.22 and 26.05 mg kg^{-1}) and stover (17.25, 127.7 and 28.09 mg kg^{-1}) followed by F_1 and F_2, whereas minimum values were recorded in F_0 (Figure 2b). Except for the Zn concentration in the grain, in all other cases, the results of F_3 treatment were statistically on par with F_1. The results of the interaction between Mo, Zn and Fe showed that the concentrations of Zn, Fe and Mo ranged from 9.94 to 17.07, 62.55 to 109.3 and 18.44 to 30.26 mg kg^{-1}, respectively, in the grain of cowpea under different treatments. The interactive effects show that the M_2F_3 treatment involving Mo soil treatment along with combined foliar Fe and Zn application showed maximum concentrations of Zn, Fe and Mo (17.07, 109.3 and 30.26 mg kg^{-1}) as compared to the M_0F_0 treatment with concentration values of 9.94, 62.55 and 18.44 mg kg^{-1}, respectively. Moreover, in the case of Fe, the results of the M_2F_3 treatment were found to be statistically on par with the M_2F_1 and M_2F_2 treatments (105.6 and 93.86 mg kg^{-1}, respectively). Additionally, for Mo, its concentration with the M_2F_3 treatment (30.26 mg kg^{-1}) was not statistically different from the M_2F_1 treatment (26.06 mg kg^{-1}). In the case of stover, the interactive studies between Mo, Fe and Zn applications show that the maximum increase in micronutrient level was observed in the M_2F_3 treatment with concentrations of 17.99, 132.7 and 31.22 mg kg^{-1} for Zn, Fe and Mo, respectively. However, the M_0F_0 treatment revealed the minimum micronutrient concentrations of 12.72, 103.9, and 21.50 mg kg^{-1}, respectively. In case of Zn, the M_2F_3 treatment was statistically on par with the M_2F_1, M_2F_2, M_1F_3 and M_0F_3 treatments (17.19, 16.81, 17.20 and 16.57 mg kg^{-1}, respectively). For Fe, the results of the M_2F_3 treatment were not statistically different from the M_2F_1 treatment (130.5 mg kg^{-1}). Similarly, for Mo, its concentration in the M_2F_3 treatment (31.22 mg kg^{-1}) was not statistically different from the M_2F_1, M_1F_3, M_1F_1 and M_0F_3 treatments (28.40, 27.09, 26.20 and 25.90 mg kg^{-1}).

3.3. *Micronutrient Uptake by Grain and Stover*

The uptake of micronutrients increased significantly in both grain as well as the stover of cowpea with the combined application of Mo, Fe and Zn over the treatments in which no or sole application of Mo, Fe and Zn was carried out (Table 4). In the main plot treatments, the highest increase in micronutrient uptake was observed in the M_2 treatment among the different Mo applications with uptake values of 22.24, 128.3 and 33.55 g ha^{-1} for Zn, Fe and Mo, respectively, in the grain of cowpea (Figure 3a). The results of Fe uptake in the M_2 treatment were statistically on par with the M_1 treatment (102.6 g ha^{-1}).

In the case of stover uptake in cowpea, M_2 treatment, i.e., soil Mo application led to the highest micronutrient uptakes of 55.52, 430.6 and 84.76 g ha^{-1} for Zn, Fe and Mo, respectively. Moreover, the results of the M_2 treatment were statistically on par with the M_1 treatment with uptake values of 61.76, 471.4 and 99.10 g ha^{-1} for Zn, Fe and Mo, respectively. In the subplot treatments, the F_3 treatment showed a higher improvement in micronutrient uptakes with values of 22.55, 128.3 and 34.04 g ha^{-1} for Zn, Fe and Mo, respectively (Figure 3b). Moreover, the Zn, Fe and Mo uptake in the stover of cowpea were recorded as the highest in F_3 as compared to no or the sole application of Fe and Zn (68.94, 513.6 and 113.29 g ha^{-1}, respectively). The effect of the interaction between Mo, Fe and Zn suggests that the M_2F_3 treatment showed the maximum micronutrient uptake in the grain of cowpea with values of 25.23, 153.3 and 42.46 g ha^{-1} for Zn, Fe and Mo, respectively. In the case of Fe, the results of the M_2F_3 treatment were not statistically different from the M_2F_1 (139.7 g ha^{-1}) and M_1F_3 (126.2 g ha^{-1}) treatments. However, the minimum uptake of

11.02, 54.20 and 15.98 g ha^{-1} for the Zn, Fe and Mo micronutrients was observed in the M_0F_0 treatment, i.e., the control.

Table 4. Effect of the application of Zn, Fe and Mo molecules on the uptake of micronutrients in the grain and stover of cowpea.

Treatments	Uptake in Grain (g ha^{-1})			Uptake in Stover (g ha^{-1})		
	Zn	Fe	Mo	Zn	Fe	Mo
M_0F_0	11.02 ^{f} ± 5.6	54.20 ^{g} ± 17.2	15.98 ^{g} ± 0.4	35.85 ^{g} ± 18.4	292.81 ^{g} ± 158.7	60.59 ^{d} ± 20.1
M_0F_1	18.22 ^{d} ± 5.7	90.81 ^{def} ± 45.6	25.44 ^{ef} ± 4.5	55.03 ^{cde} ± 41.8	429.35 ^{cd} ± 304.1	85.94 ^{bc} ± 56.7
M_0F_2	14.22 ^{e} ± 5.6	75.57 ^{efg} ± 29.1	23.56 ^{f} ± 3.8	43.31 ^{ef} ± 39.5	373.64 ^{ef} ± 236.4	71.12 ^{d} ± 69.5
M_0F_3	19.98 ^{cd} ± 6.7	107.4 ^{cd} ± 65.9	27.29 ^{def} ± 0.7	63.34 ^{bc} ± 43.8	474.43 ^{bc} ± 296.4	99.01 ^{b} ± 73.1
M_1F_0	13.99 ^{e} ± 7.3	74.75 ^{fg} ± 35.8	19.63 ^{g} ± 1.5	48.14 ^{fg} ± 24.7	352.75 ^{f} ± 243.9	73.10 ^{cd} ± 61.7
M_1F_1	20.19 ^{c} ± 8.5	111.6 ^{bc} ± 60.1	29.53 ^{ce} ± 4.6	62.39 ^{bcd} ± 40.7	469.73 ^{bc} ± 306.4	99.25 ^{b} ± 56.0
M_1F_2	18.37 ^{d} ± 8.4	101.1 ^{cdef} ± 49.5	26.89 ^{def} ± 6.6	58.83 ^{bcde} ± 36.9	445.10 ^{cd} ± 229.6	85.41 ^{c} ± 35.2
M_1F_3	22.56 ^{b} ± 8.4	126.2 ^{abc} ± 72.4	33.09 ^{bc} ± 4.0	67.39 ^{ab} ± 44.7	494.51 ^{b} ± 321.9	105.98 ^{ab} ± 96.1
M_2F_0	20.19 ^{c} ± 7.2	105.2 ^{cde} ± 63.9	27.95 ^{de} ± 6.4	57.80 ^{def} ± 21.0	411.26 ^{de} ± 206.2	83.69 ^{cd} ± 47.8
M_2F_1	22.74 ^{b} ± 9.3	139.7 ^{ab} ± 65.9	34.46 ^{b} ± 8.2	67.88 ^{ab} ± 46.9	515.20 ^{ab} ± 298.3	112.12 ^{ab} ± 111.4
M_2F_2	20.97 ^{bc} ± 8.6	117.0 ^{bcd} ± 55.8	30.05 ^{cd} ± 6.5	64.44 ^{bc} ± 46.5	473.07 ^{bc} ± 282.4	97.75 ^{bc} ± 66.4
M_2F_3	25.23 ^{a} ± 8.1	153.3 ^{a} ± 98.6	42.46 ^{a} ± 0.3	74.54 ^{a} ± 54.6	552.60 ^{a} ± 310.8	130.00 ^{a} ± 93.7
LSD (0.05)	1.8	30.0	4.3	11.1	47.0	24.7

M_0: No molybdenum, M_1: Molybdenum seed priming, M_2: Molybdenum soil treatment, F_0: No fertilizer application, F_1: Fe application, F_2: Zn application, F_3: Fe+Zn application. The mean with a similar or dissimilar letter(s) was evaluated with the least significant difference (LSD) multiple range tests using a probability level of $p \leq 0.05$ along with standard deviation.

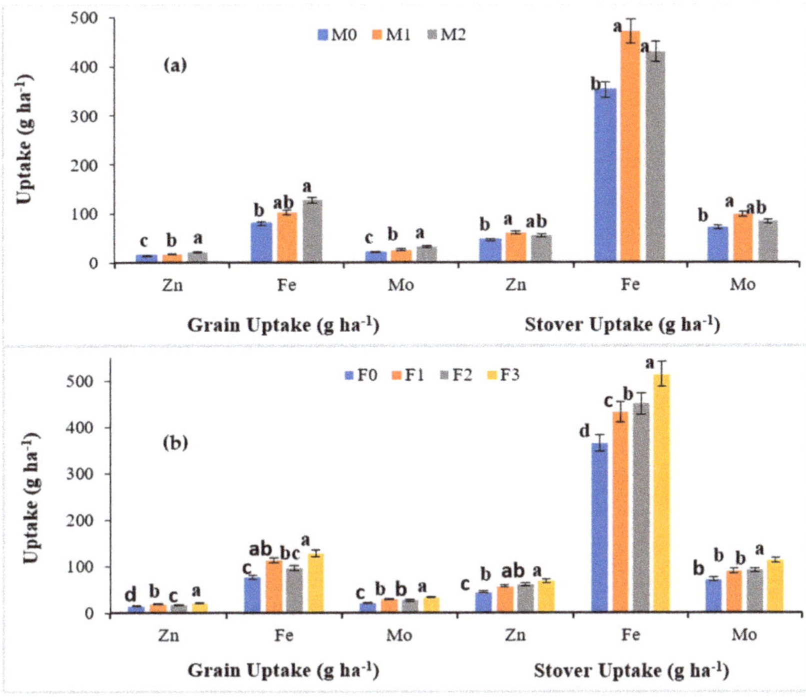

Figure 3. (**a**) Effect of different methods of Mo and (**b**) Fe and Zn application on Zn, Fe and Mo uptake in the grain and stover of cowpea. The column representing the mean with a similar or dissimilar letter(s) was evaluated with the least significant difference (LSD) multiple range tests using a probability level of $p \leq 0.05$ along with standard deviation.

In the case of Zn, Fe and Mo uptake in stover, the combined use of Mo soil treatment along with Fe and Zn foliar application, i.e., M_2F_3 treatment, further enhanced the Zn, Fe and Mo uptake to 104.2, 809.9 and 190.6 g ha^{-1}, respectively. The results of this treatment were statistically on par with the M_2F_1 (88.29 g ha^{-1}) and M_1F_3 (88.66 g ha^{-1}) treatments in the case of Zn. Similarly for Fe, the M_2F_3 treatment was not statistically different from the M_2F_1 treatment with values of 757.4 g ha^{-1}. However, the minimum uptake of 41.24, 431.5 and 89.21 g ha^{-1} for Zn, Fe and Mo micronutrients was observed in the M_0F_0 treatment, i.e., the control.

3.4. Root Length, Nodules and N Uptake

The results of two years' mean data concerning the root length, the number of nodules plant^{-1} and N concentration in grain as well as the stover of cowpea are given in Table 5.

Table 5. Effect of the application of Zn, Fe and Mo molecules on root length, number of nodules and N concentration of cowpea.

Treatments	Root Length (cm)	Nodules Plant^{-1} (no)	N Uptake (kg ha^{-1})	
Molybdenum			Grain	Stover
Interaction				
M_0F_0	20.3 ± 2.5	31.0 ^{f} ± 1.5	29.36 ^{f} ± 2.2	22.07 ^{f} ± 5.4
M_0F_1	23.5 ± 3.2	34.5 ^{ef} ± 2.2	40.83 ^{d} ± 2.1	39.62 ^{de} ± 2.6
M_0F_2	24.9 ± 0.5	44.0 ^{de} ± 4.6	41.69 ^{d} ± 0.5	36.22 ^{e} ± 4.8
M_0F_3	27.7 ± 2.1	47.0 ^{d} ± 3.2	47.93 ^{cd} ± 0.8	44.38 ^{bcd} ± 2.5
M_1F_0	28.3 ± 2.5	62.0 ^{bc} ± 3.7	36.62 ^{e} ± 1.0	34.10 ± 1.9
M_1F_1	28.5 ± 4.1	56.0 ^{cd} ± 1.2	50.83 ^{bc} ± 2.1	56.20 ^{a} ± 1.0
M_1F_2	29.8 ± 2.0	57.0 ^{cd} ± 1.0	46.02 ^{d} ± 3.,2	43.67 ^{bcd} ± 2.0
M_1F_3	31.3 ± 0.4	57.0 ^{cd} ± 1.5	53.25 ^{ab} ± 2.5	42.16 ^{cd} ± 2.2
M_2F_0	29.3 ± 0.9	53.0 ^{cde} ± 0.8	47.19 ^{cd} ± 4.7	47.26 ^{b} ± 3.4
M_2F_1	30.5 ± 1.0	62.0 ^{bc} ± 3.1	55.91 ^{a} ± 3.1	47.12 ^{b} ± 3.6
M_2F_2	27.5 ± 2.5	69.0 ^{ab} ± 2.3	49.93 ^{bc} ± 2.0	44.30 ^{bcd} ± 0.6
M_2F_3	30.5 ± 2.3	73.0 ^{a} ± 1.0	55.39 ^{a} ± 1.0	46.15 ^{bc} ± 2.7
LSD (0.05)	NA	10.5	3.54	4.87

M_0: No molybdenum, M_1: Molybdenum seed priming, M_2: Molybdenum soil treatment, F_0: No fertilizer application, F_1: Fe application, F_2: Zn application, F_3: Fe+Zn application. The mean with a similar or dissimilar letter(s) was evaluated with the least significant difference (LSD) multiple range tests using a probability level of $p \leq 0.05$ along with standard deviation.

In the main plot treatments (Figure 4a), the root length of cowpea increased significantly with Mo application over the control (24.1 cm). The results of seed priming of Mo (29.5 cm) were statistically on par with the soil application of Mo (29.5 cm). Moreover, the number of plant^{-1} nodules increased significantly with Mo application over the control (39.0). Moreover, the results under soil application (64.0) were significantly higher over the seed priming of Mo (58.0). The N uptake increased significantly with the Mo application over the control in the grain and stover of cowpea (39.75 and 35.07 kg ha^{-1}, respectively).

The results of the seed-primed Mo treatment (M_1) were statistically on par with the soil application of Mo treatment (M_2). In the sub-plot treatment (Figure 4b), the root length also increased significantly with the foliar application of Fe+Zn (29.8 cm) over the control (25.9 cm). The number of plant^{-1} nodules was significantly higher with the combined application of Fe+Zn (59.0), i.e., F_3 over the F_0 (49.0), F_1 (51.0) and F2 (57.0). For grain N uptake, the highest values were obtained with the combined application of Fe+Zn (52.20 kg ha^{-1}) under the F_3 treatment, which was significantly higher over the F_0 (37.56 kg ha^{-1}). A similar trend was observed for stover N uptake with the foliar application of Fe and Zn. Under the interactive effects, the root length showed a non-significant variation. The highest number of plant^{-1} nodules in cowpea were recorded in M_2F_3 (73.0), whereas the lowest value was recorded in M_0F_0 (31.0). The results of M_2F_3

were statistically on par with the M_2F_2 treatment (69.0). The interactive effects of Mo, Fe and Zn also showed significant effects on N uptake in the grain and stover. The maximum N uptake in grain was recorded under the M_2F_1 treatment (55.91 kg ha^{-1}), whereas the minimum value was observed under the M_0F_0 treatment (29.36 kg ha^{-1}). The results of M_2F_1 were statistically at par with M_1F_3 and M_2F_3. The data for stover N uptake showed that the highest results were obtained under the M_1F_1 treatment (56.20 kg ha^{-1}) and the lowest value was recorded under the M_0F_0 treatment (22.07 kg ha^{-1}).

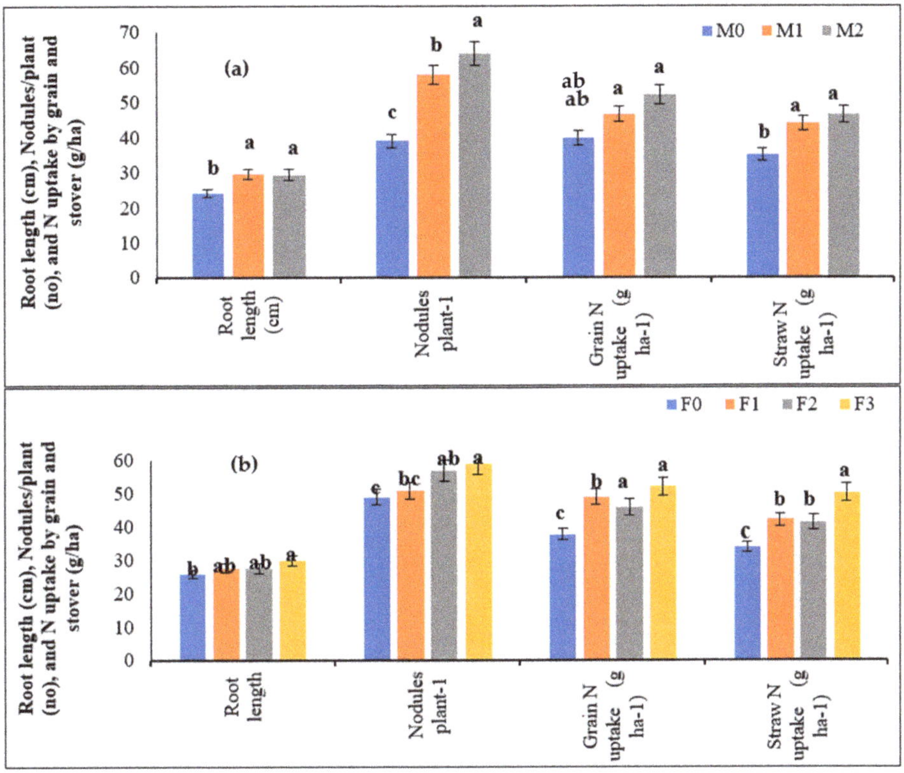

Figure 4. (**a**) Effect of different methods of Mo and (**b**) Fe and Zn application on root length, nodules plant^{-1}, N uptake in grain and stover yield of cowpea. The column representing the mean with a similar or dissimilar letter(s) was evaluated with the least significant difference (LSD) multiple range tests using a probability level of $p \leq 0.05$ along with standard deviation.

3.5. *Efficiency Indices*

The results of Table 6 demonstrate that MEI-Zn was at a maximum with the M_2F_3 treatment (0.949) showing the soil-applied Mo along with Fe and Zn foliar application and was lowest in the M_1F_0 treatment (0.726). Similarly, for Fe and Mo, MEI was highest in the M_2F_3 treatment with values of 0.824 and 0.969, respectively and was lowest in the M_0F_0 treatment for Fe (0.602), and M_1F_2 for Mo (0.708). Additionally, the results of PE-Zn, PE-Fe and PE-Mo were highest in the M_2F_0 treatment with values of 0.572 q/g, 0.069 q/g and 0.350 q/g, respectively. However, the lowest PE values were observed in the M_2F_3 treatment (0.323 q/g) for Zn, M_2F_1 (0.051 q/g) for Fe and M_2F_3 (0.195 q/g) for Mo, respectively.

Table 6. Effect of the application of Zn, Fe and Mo molecules on micronutrient use efficiencies by cowpea.

Treatments	Mobilization Efficiency			Physiological Efficiency (q g^{-1})		
	Zn	Fe	Mo	Zn	Fe	Mo
M_0F_0	0.781 ^{b} ± 0.02	0.602 ^{c} ± 0.06	0.858 ^{c} ± 0.01	-	-	-
M_0F_1	0.867 ^{a} ± 0.03	0.639 ^{bc} ± 0.04	0.894 ^{abc} ± 0.04	0.368 ^{bc} ± 0.02	0.056 ^{b} ± 0.001	0.281 ^{abc} ± 0.02
M_0F_2	0.952 ^{a} ± 0.05	0.616 ^{bc} ± 0.01	0.709 ^{d} ± 0.04	0.355 ^{bc} ± 0.01	0.057 ^{b} ± 0.005	0.337 ^{ab} ± 0.04
M_0F_3	0.852 ^{a} ± 0.09	0.718 ^{abc} ± 0.06	0.874 ^{bc} ± 0.07	0.389 ^{bc} ± 0.01	0.057 ^{b} ± 0.005	0.268 ^{abc} ± 0.03
M_1F_0	0.726 ^{b} ± 0.1	0.728 ^{ab} ± 0.08	0.922 ^{abc} ± 0.11	0.499 ^{ab} ± 0.05	0.068 ^{a} ± 0.002	0.342 ^{a} ± 0.03
M_1F_1	0.855 ^{ab} ± 0.12	0.733 ^{ab} ± 0.11	0.919 ^{abc} ± 0.15	0.385 ^{bc} ± 0.09	0.056 ^{b} ± 0.004	0.254 ^{abc} ± 0.01
M_1F_2	0.848 ^{a} ± 0.07	0.728 ^{ab} ± 0.12	0.708 ^{d} ± 0.17	0.388 ^{bc} ± 0.12	0.057 ^{b} ± 0.001	0.325 ^{ab} ± 0.01
M_1F_3	0.890 ^{a} ± 0.02	0.761 ^{a} ± 0.11	0.932 ^{abc} ± 0.10	0.354 ^{bc} ± 0.11	0.056 ^{b} ± 0.001	0.246 ^{abc} ± 0.02
M_2F_0	0.738 ^{b} ± 0.01	0.733 ^{ab} ± 0.06	0.956 ^{ab} ± 0.07	0.572 ^{a} ± 0.12	0.069 ^{a} ± 0.004	0.350 ^{a} ± 0.09
M_2F_1	0.885 ^{a} ± 0.01	0.809 ^{a} ± 0.02	0.918 ^{abc} ± 0.09	0.359 ^{bc} ± 0.06	0.051 ^{c} ± 0.003	0.224 ^{bc} ± 0.11
M_2F_2	0.879 ^{a} ± 0.03	0.761 ^{a} ± 0.05	0.945 ^{abc} ± 0.05	0.361 ^{bc} ± 0.05	0.057 ^{b} ± 0.001	0.273 ^{abc} ± 0.12
M_2F_3	0.949 ^{a} ± 0.05	0.824 ^{a} ± 0.04	0.969 ^{a} ± 0.01	0.323 ^{c} ± 0.04	0.052 ^{c} ± 0.002	0.195 ^{c} ± 0.11
LSD (0.05)	0.16	0.12	0.09	0.15	0.001	0.11

M_0: No molybdenum, M_1: Molybdenum seed priming, M_2: Molybdenum soil treatment, F_0: No fertilizer application, F_1: Fe application, F_2: Zn application, F_3: Fe+Zn application. The mean with a similar or dissimilar letter(s) was evaluated with the least significant difference (LSD) multiple range tests using a probability level of $p \leq 0.05$ along with standard deviation.

4. Discussion

4.1. Grain and Stover Yield

The results displayed in Table 2 and Figure 1 reveal that the grain as well as the stover yield of cowpea significantly increased for two years and was the maximum for M_2 treatment among the Mo treatments. Thus, the presence of Mo in the M_1 and M_2 treatments improved the grain as well as the stover yield as compared to the control. The trend might be attributed to the vital role of Mo in the synthesis and activity of molybdoenzymes which regulates the N fixation, thus increasing the N content and crop yield [30]. The improved yield with Mo application could also be ascribed to its outstanding role in photosynthesis and respiration processes. In the absence of molybdenum in soil, the plant molybdoenzymes could break and adversely affect the nitrogen fixation by soil bacteria which results in a reduced yield [31]. Additionally, the higher yield observed in the foliar Fe application as compared to Zn was largely related to the crucial role of Fe in the synthesis of growth promoters such as auxins, photosynthesis, seed maturation and nucleic acid metabolism which results in a significantly higher grain and stover yield [32,33]. However, Zn foliar application reduced the yield attributes which may have been due to the lower macronutrient concentrations in the grain.

Additionally, the double and triple micronutrient application exhibited superior grain and stover yields over single micronutrients which might have been due to the synergistic interactions involved among Mo, Fe and Zn. Combined soil treatment with Mo along with the foliar application of Fe had a favorable effect on nitrogenase activity in nodules and nitrate reductase activity in the plant system. Studies in the literature have reported that the application of essential nutrients (N, Fe and Mo) at the optimum level positively influences the metabolic processes and thus leads to a higher yield of cowpea. The application of Mo along with micronutrients and Rhizobium inoculation has also recorded enhanced cowpea growth and nodulation [34]. Another study reported that the seed treatment with Mo solution can overcome the internal Mo deficiencies and thus maintain the activity of molybdoenzymes [35]. Similarly, a significant effect on the root growth and yield of soybean has been observed under the seed inoculation with Rhizobium and Mo [36].

4.2. Micronutrient Concentrations in the Grain and Stover

The data (Table 3 and Figure 2) revealed that the sole and combined application of Mo, Fe and Zn led to a significant improvement in the micronutrient concentrations in cowpea grain and stover as compared to control which might have been due to the immediate absorption of available micronutrients by plant leaves. Among the Mo treatments, the M_2 treatment proved beneficial for enhancing the micronutrient concentration, where Mo played a major role in the functioning of nitrate and nitrite reductase [31]. The present findings are concordant with previous studies, where Mo supplementation increased the Mo concentration in '*Le-Conte*' pear [37], grapes (cv. Merlot) [38], peanut [39] and lettuce [40]. Togay et al. [41] also reported the enhanced concentration of Fe, P, Mn, Cu and Mo in lentil (*Lens culinaris* Medic.) through the combined Fe (20 kg ha^{-1}) and Mo (6 g kg^{-1} seed) application. Our findings are in line with the aforementioned studies, suggesting that the application of Mo increased the concentration of Zn, Fe and Mo in the grain and stover of cowpea.

However, the foliar application of Fe+Zn in the F_3 treatment also resulted in an increased micronutrient concentration in cowpea as compared to other treatments including the control, i.e., F_0. This might have been due to the higher availability of these micronutrients to the crop at the optimum level of application [13]. Another important point is the interaction between micronutrients which affects their uptake, distribution and utilization in plants [42]. Additionally, the combined application of Mo through soil treatment along with the Fe and Zn foliar spray further increased the micronutrient levels in both the grain and stover of cowpea; thus, it could be inferred that Mo, Fe and Zn possessed the appropriate mechanisms for the translocation of micronutrients to the grain and stover in cowpea. The enhancement in the nutrient content might have been due to an increased absorption as well as the assimilation of the micronutrients that resulted in balanced nutritional value in the crop for higher growth and thereby a higher nutrient content. Similar results were observed by Hristozkova et al. [43] where Mo enhanced the accumulation of nutrients in cowpea plant tissues. Gad and Kandil [44] added that the presence of Mo and N significantly increased the composition of minerals such as N, P, K, Fe, Mn, Zn, Cu and Mo in cowpea with all nitrogen levels as compared to the untreated plants.

4.3. Micronutrient Uptake by the Grain and Stover

The results of the present study (Table 4 and Figure 3) demonstrated that micronutrient uptake was found to increase significantly with external supplementation. The trend could be coupled with the joint impact of yield as well as concentration. Moreover, the exogenous supply of nutrients through the treatment with Mo, Fe and Zn molecules increased the availability of these nutrients in the soil to a remarkable extent. The results in the present study are in agreement with previous studies in which the application of Mo resulted in an improved Fe, P, Mn and Mo uptake in rice [45]. Similarly, Ndakidemi et al. [46] suggested a significant increase in Mo uptake (0.644 mg $plant^{-1}$) in the roots of the common bean (*Phaseolus vulgaris* L.) with the Mo application at 12 g kg^{-1} as compared to the control. This could be attributed to the absorption of an increased quantity of Mo from the soil which led to its improved concentration and uptake in the common bean. Overall, the combined application of Mo, Fe and Zn molecules was most effective in increasing the micronutrient uptake in the grain and stover of cowpea.

4.4. Root Length, Nodules and N Concentration of Cowpea

The present findings indicate the beneficial effects of Mo application on root length as with Mo application there was a significant increase in root length of cowpea (Table 5 and Figure 4). The direct effect of Mo on root growth has not been reported yet; however, the results might be attributed to the enhanced activity of various enzymes, such as nitrogenase and nitrate reductase, the growth of root nodules and the promotion of the hormone synthesis to be transported in roots [47,48]. Similar results have been reported by Liu et al. [49] in which Mo application enhanced the root length of soybean. Additionally, Mo also plays a crucial role in N metabolism through Mo enzymes and plays a key role in carrying out redox reactions [8,31]. The trend of the number of nodules and nitrogen uptake can also be explained based on the above reasons. The Mo cofactor 'FeMoCo' increases the activity of enzymes involved in nitrogen fixation that catalyzes the inorganic nitrogen assimilation. After the uptake by roots, the nitrates are directed towards the plant vacuoles. In plants, nitrate is reduced to ammonium (NH_4^+) through an enzymatic reaction. Initially, the transformation of NO_3^- to NO_2^- occurs in the cytoplasm in the presence of nitrate reductase followed by its conversion in NO_4^+ in proplastids or chloroplasts catalyzed by nitrite reductase [50]. The results of the present study have demonstrated the significant association of N accumulation with soil Mo application. The absence of Mo promotes nitrate accumulation and indicates less N assimilation by the plants [40]. Similar results have been reported in alfalfa nodulation with Mo supplementation [51].

4.5. Efficiency Indices of Cowpea

The results of MEI indicated that the MEI of Mo was higher in the presence of the Mo soil treatment along with Fe and Zn foliar application (Table 6). On the other hand, the results of PE indicated an increase in grain production with the absorbed nutrient. The higher values for PE-Zn, PE-Fe and PE-Mo were found in the M_2F_3 treatment involving Mo soil application along with Fe and Zn foliar application as compared to no or sole applications of Mo, Fe and Zn.

5. Conclusions

The findings of a two-year study clarified that the supplementation of Mo, Fe and Zn molecules through ammonium molybdate, $FeSO_4{\cdot}7H_2O$ and $ZnSO_4{\cdot}7H_2O$ influenced the yield, quality and root system of cowpea. The treatment involving Mo soil application along with a foliar spray of $FeSO_4{\cdot}7H_2O$ (0.5%) + $ZnSO_4{\cdot}7H_2O$ (0.5%) resulted in the highest increased yield, micronutrient concentration and uptake in cowpea. The root length and the number of nodules were also enhanced with the Mo application. The presence of Mo and Fe molecules enhanced the N content and helped in nitrogen fixation which in turn improved the nutritional quality of the produce. Additionally, the efficiency indices, i.e., MEI and PE were maximum in the treatment involving Mo application along with the foliar spray of Fe and Zn. Thus, Mo soil treatment along with Fe and Zn application could be considered the most efficient strategy for enhancing the grain and stover yield along with the availability of micronutrients for improved cultivation of cowpea.

Author Contributions: Conceptualization, S.S.D., V.S., A.K.S., J.K., V.V., M.K. and P.S.; methodology, S.S.D., V.S., A.K.S., J.K., V.V., M.K. and P.S.; software, S.S.D. and A.H.; validation, S.S.D., V.S., A.K.S., J.K., V.V., M.K. and P.S.; formal analysis, S.S.D. and A.H.; investigation, S.S.D., V.S., A.K.S., J.K., V.V., M.K. and P.S.; resources, S.S.D.; data curation, S.S.D. and A.H.; writing—original draft preparation, S.S.D., V.S., A.K.S., J.K., V.V., M.K. and P.S.; writing—review and editing, A.G., M.B. and A.H.; visualization, S.S.D., V.S., A.K.S., J.K., V.V., M.K. and P.S.; supervision, S.S.D.; project administration, S.S.D., A.G., A.K.S. and A.H.; funding acquisition, A.G., A.K.S. and A.H. All authors have read and agreed to the published version of the manuscript.

Funding: This research was also partially funded by the Department of Soil Science, Punjab Agricultural University, Ludhiana, India and ICAR- Indian Institute of Soil Science, Bhopal, 462038, Madhya Pradesh, India. This research was also partially funded by the Taif University Researchers for funding this research with Supporting Project number (TURSP-2020/39), Taif University, Taif, Saudi Arabia.

Institutional Review Board Statement: Not applicable.

Informed Consent Statement: Not applicable.

Data Availability Statement: All data are available in the manuscripts.

Conflicts of Interest: The authors would hereby like to declare that there is no conflict of interest in the article.

References

1. Silva, V.M.; Boleta, E.H.M.; Martins, J.T.; Dos Santos, F.L.M.; Silva, A.C.R.; Alcock, T.D.; Wilson, L.; De S'a, M.E.; Young, S.D.; Broadley, M.R.; et al. Agronomic biofortification of cowpea with selenium: Effects of selenate and selenite applications on selenium and phytate concentrations in seeds. *J. Sci. Food Agric.* **2019**, *99*, 5969–5983. [CrossRef] [PubMed]
2. Masuda, H.; Aung, M.S.; Kobayashi, T.; Nishizawa, N.K. Iron biofortification: The gateway to overcoming hidden hunger. In *The Future of Rice Demand: Quality beyond Productivity*; Costa De Oliveira, A., Pegoraro, C., Ebeling Viana, V., Eds.; Springer: Cham, Switzerland, 2020; pp. 149–177.
3. Baker, A.V.; Philbeam, D.J. *Handbook of Plant Nutrition*; Taylor and Francis Group: New York, NY, USA, 2007; pp. 375–394.
4. Liu, P.; Yang, Y. Research on development of molybdenum in soil and its effects on vegetation. *Agri-Environ. Prot.* **2001**, *20*, 280–282.
5. Gödecke, T.; Stein, A.J.; Qaim, M. The global burden of chronic and hidden hunger: Trends and determinants. *Glob. Food Sec.* **2019**, *17*, 21–29. [CrossRef]
6. Briat, J.F. *Iron Nutrition and Implications for Biomass Production and the Nutritional Quality of Plant Products. The Molecular and Physiological Basis of Nutrient Use Efficiency in Crops*; Wiley-Blackwell: Hoboken, NJ, USA, 2011; pp. 311–334.
7. Gómez-Galera, S.; Rojas, E.; Sudhakar, D.; Zhu, C.; Pelacho, A.; Capell, T.; Christou, P. Critical evaluation of strategies for mineral fortification of staple food crops. *Transgenic Res.* **2010**, *19*, 165–180. [CrossRef] [PubMed]
8. Fageria, N.K.; Baligar, V.C.; Clark, R.B. Micronutrients in crop production. *Adv. Agron.* **2002**, *77*, 185–268.
9. Black, R.E.; Lindsay, H.A.; Bhutta, Z.A.; Caulfield, L.E.; Onnis, M.D. Maternal and child under-nutrition: Global and regional exposures and health consequences. *Lancet* **2008**, *371*, 243–260. [CrossRef]
10. Cakmak, I. Enrichment of cereal grains with zinc: Agronomic or genetic biofortification? *Plants Soil* **2008**, *302*, 1–17. [CrossRef]
11. White, P.J.; Broadley, M.R. Biofortifying crops with essential mineral elements. *Trends Plant Sci.* **2005**, *10*, 586–593. [CrossRef]
12. Dhaliwal, S.S.; Sadana, U.S.; Manchanda, J.S.; Dhadli, H.S. Biofortification of wheat grains with zinc and iron in *Typic Ustochrept* soils of Punjab. *Ind. J. Fert.* **2009**, *5*, 13–20.
13. Dhaliwal, S.S.; Sharma, V.; Shukla, A.K.; Verma, V.; Sandhu, P.S.; Behera, S.K.; Singh, P.; Kaur, J.; Singh, H.; Abdel-Hafez, S.H.; et al. Interactive effect of foliar application of nitrogen, zinc and iron on productivity and oil nutritional quality of Indian mustard (*Brassica juncea* L.). *Agronomy* **2021**, *11*, 2333. [CrossRef]
14. Ramos, D.P.; Tavares, T.C.O.; Sousa, S.A.; Nascimento, V.L.; Martinez, R.A.S.; Junior, A.F.C.; Fidelis, R.R. Agronomic biofortification of cowpea with selenium by foliar fertilization: Effect of doses in three cultivars. *J. Plant Nutr.* **2020**, *43*, 538–547. [CrossRef]
15. Izydorczyk, G.; Ligas, B.; Mikula, K.; Witek-Krowiak, A.; Moustakas, K.; Chojnacka, K. Biofortification of edible plants with selenium and iodine–A systematic literature review. *Sci. Total Environ.* **2021**, *754*, 141983. [CrossRef] [PubMed]
16. Manzeke, M.G.; Mtambenengwe, F.; Nezomba, H.; Watts, M.J.; Broadley, M.R.; Mapfumo, P. Zinc fertilization increases productivity and grain nutritional quality of cowpea (*Vigna unguiculata* [L.] Walp.) under integrated soil fertility management. *Field Crops Res.* **2017**, *213*, 231–244. [CrossRef]
17. Teka, T.A.; Retta, N.; Bultosa, G.; Admassu, H.; Astatkie, T. Protein fractions, in vitro protein digestibility and amino acid composition of select cowpea varieties grown in Ethiopia. *Food Biosci.* **2020**. [CrossRef]
18. Dhanasekar, P.; Souframanien, J.; Suprasanna, P. Breeding Cowpea for Quality Traits: A Genetic Biofortification Perspective. In *Breeding for Enhanced Nutrition and Bio-Active Compounds in Food Legumes*; Gupta, D.S., Gupta, S., Kumar, J., Eds.; Springer: Cham, Switzerland, 2021. [CrossRef]
19. Moura, J.O.; Rocha, M.M.; Gomes, R.L.F.; Freire Filho, F.R.; Damasceno, K.J.; Ribeiro, V.Q. Path analysis of iron and zinc contents and others traits in cowpea. *Crop Breed. Appl. Biotechnol.* **2012**, *12*, 245–252. [CrossRef]
20. Marquez-Quiroz, C.; De-la-Cruz-Lazaro, E.; Osorio-Osorio, R.; Sanchez-Chavez, E. Biofortification of cowpea beans with iron: Iron's influence on mineral content and yield. *J. Soil Sci. Plant Nutr.* **2015**, *15*, 839–847. [CrossRef]
21. Chatterjee, R.; Bandyopadhyay, S. Effect of boron, molybdenum and biofertilizers on growth and yield of cowpea (*Vigna unguiculata* L. Walp.) in acid soil of the eastern Himalayan region. *J. Saudi Soc. Agric. Sci.* **2017**, *16*, 332–336. [CrossRef]
22. Singh, S.; Bawa, S.S.; Singh, S.; Sharma, S.C.; Kumar, V. Effect of seed priming with molybdenum on the performance of rainfed chickpea (*Cicer arietinum* L.). *Agric. Res. J.* **2014**, *51*, 124–127.
23. López-Morales, D.; de la Cruz-Lázaro, E.; Sánchez-Chávez, E.; Preciado-Rangel, P.; Márquez-Quiroz, C.; Osorio-Osorio, R. Impact of Agronomic Biofortification with Zinc on the Nutrient Content, Bioactive Compounds, and Antioxidant Capacity of Cowpea Bean (*Vigna unguiculata* L. Walpers). *Agronomy* **2020**, *10*, 1460. [CrossRef]
24. Silva, V.M.; Nardeli, A.J.; Mendes, N.; Rocha, M.M.; Wilson, L.; Young, S.D.; Broadley, M.R.; White, P.J.; Reis, A. Agronomic biofortification of cowpea with zinc: Variation in primary metabolism responses and grain nutritional quality among 29 diverse genotypes. *Plant Physiol. Biochem.* **2021**, *162*, 378–387. [CrossRef]
25. Kumar, B.; Dhaliwal, S.S. Zinc biofortification of dual-purpose cowpea [*Vigna unguiculata* (L.) Walp.] for enhancing the productivity and nutritional quality in a semi-arid region of India. *Arch. Agron. Soil Sci.* **2021**. [CrossRef]
26. Lindsay, W.L.; Norvell, W.A. Development of a DTPA soil test for zinc, iron, manganese, and copper. *Soil Sci. Soc. Am. J.* **1978**, *42*, 421–428. [CrossRef]

27. Jackson, M.L. A manual of methods useful for instruction and research in soil chemistry, physical chemistry, soil fertility and soil genesis. In *Soil Chemical Analysis-Advanced Course*, 2nd ed.; Department of Science, University of Wisconsin Madison: Madison, WI, USA, 1973.
28. Purushottam, A.; Naidu, P.P.; Lal, S.S. Determination of molybdenum by atomic-absorption spectrophotometry. *Talanta* **1972**, *19*, 1193–1198. [CrossRef]
29. Dhaliwal, S.S.; Sharma, V.; Shukla, A.K.; Kaur, J.; Verma, V.; Singh, P.; Singh, H.; Abdel-Hafez, S.H.; Sayed, S.; Gaber, A.; et al. Enrichment of zinc and iron micronutrients in lentil (*Lens culinaris* Medic.) through biofortification. *Molecules* **2021**, *26*, 7671. [CrossRef] [PubMed]
30. Rana, M.S.; Bhantana, P.; Imran, M.; Saleem, M.H.; Moussa, M.G.; Khan, Z.; Khan, I.; Alam, M.; Abbas, M.; Binyamin, R.; et al. Molybdenum potential vital role in plants metabolism for optimizing the growth and development. *Ann. Environ. Sci. Toxicol.* **2020**, *4*, 32–44.
31. Kaiser, B.N.; Gridley, K.L.; Brady, J.N.; Phillips, T.; Tyerman, S.D. The role of molybdenum in agricultural plant production. *Ann. Bot.* **2005**, *96*, 745–754. [CrossRef] [PubMed]
32. Pal, V.; Singh, G.; Dhaliwal, S. Yield enhancement and biofortification of chickpea *Cicer arietinum* L. grain with iron and zinc through foliar application of ferrous sulfate and urea. *J. Plant Nutr.* **2019**, *42*, 1789–1802. [CrossRef]
33. Schmidt, W.; Thomine, S.; Buckhout, T.J. Iron nutrition and interactions in plants. *Front. Plant Sci.* **2020**, *10*, 1670. [CrossRef]
34. Subasinghe, S.; Dayatilake, G.A.; Senaratne, R. Effect of B, Co and Mo on nodulation, growth and yield of cowpea (*Vigna unguiculata*). *Trop. Agric. Res. Ext.* **2003**, *6*, 108–112. [CrossRef]
35. Kothari, M.L. Effect of modes and levels of molybdenum application on grain yield protein content and nodulation of chickpea grown on loamy sand soil. *Commun. Soil Sci. Plant Anal.* **2002**, *33*, 18–23.
36. Sable, S.; Sontakey, P.Y.; Nair, B.; Manapure, P.; Deotale, R.D. Influence of seed inoculation with rhizobium and molybdenum on soybean roots. *J. Soils Crop.* **2000**, *10*, 126–130.
37. Abd-El-Latif, F.M.; Bakry, K.A.; El-Gioushy, S.F.; Hussein, A.M.; Mohamed, M.S. Effect of foliar spray with molybdenum and iron on vegetive growth and nutritional status of pear trees. *J. Plant Prod.* **2020**, *11*, 655–659.
38. Longbottom, M.L.; Dry, P.R.; Sedgley, M. Effects of sodium molybdate foliar sprays on molybdenum concentration in the vegetative and reproductive structures and on yield components of *Vitis vinifera* cv. Merlot. *Aust. J. Grape Wine Res.* **2010**, *16*, 477–490. [CrossRef]
39. Soares Filho, S.I.B.; Lazarini, E.; Orioli Júnior, V.; Bernardes, J.V.S. Sowing dates and molybdenum foliar application for two peanut cultivars. *Rev. Ciênc. Agríc.* **2020**, *18*, 27. [CrossRef]
40. Steiner, F.; Zoz, T.; Zuffo, A.M.; Machado, P.P.; Zoz, J.; Zoz, A. Foliar application of molybdenum enhanced quality and yield of crispleaf lettuce (*Lactuca sativa* L., cv. Grand Rapids). *Acta Agron.* **2018**, *67*, 73–78. [CrossRef]
41. Togay, N.; Togay, Y.; Erman, M.; Çig, F. Effect of Fe (iron) and Mo (molybdenum) application on the yield and yield parameters of lentil (*Lens culinaris* Medic.). *Legume Res.* **2015**, *38*, 389–393.
42. Dhaliwal, S.S.; Sharma, V.; Shukla, A.K.; Verma, V.; Behera, S.K.; Singh, P.; Alotaibi, S.S.; Gaber, A.; Hossain, A. Comparative efficiency of mineral, chelated and nano forms of zinc and iron for improvement of zinc and iron in chickpea (*Cicer arietinum* L.) through biofortification. *Agronomy* **2021**, *11*, 2436. [CrossRef]
43. Hristozkova, M.; Geneva, M.; Stancheva, I. Response of pea plants (*Pisum sativum* L.) to reduced supply with molybdenum and copper. *Int. J. Agric. Biol.* **2006**, *8*, 218–220.
44. Gad, N.; Kandil, H. Evaluate the effect of molybdenum and different nitrogen levels on cowpea (*Vigna unguiculata*). *J. Appl. Sci. Res.* **2013**, *9*, 1490–1497.
45. Zakikhani, H.; Khanif, Y.M.; Anuar, A.R.; Radziah, O.; Soltangheisi, A. Effects of different levels of molybdenum on uptake of nutrients in rice cultivars. *Asian J. Crop Sci.* **2014**, *6*, 236–244. [CrossRef]
46. Ndakidemi, P.A.; Bambara, S.; Makoi, J.H.J.R. Micronutrient uptake in common bean (*Phaseolus vulgaris* L.) as affected by *rhizobium* inoculation, and the supply of molybdenum and lime. *Plant Omics* **2011**, *4*, 40–52.
47. Gupta, U.C.; Lipsett, J. Molybdenum in soils, plants, and animals. *Adv. Agron.* **1981**, *34*, 73–115.
48. Marschner, H. *Mineral Nutrition of Higher Plants*; Academic Press: San Diego, CA, USA, 1995.
49. Liu, P.; Yang, Y.S.; Xu, G.D.; Fang, Y.H.; Yang, Y.A.; Kalin, R.M. The effect of molybdenum and boron in soil on the growth and photosynthesis of three soybean varieties. *Plant Soil Environ.* **2005**, *51*, 197–205. [CrossRef]
50. Rosales, E.P.; Iannone, M.F.; Groppa, M.D.; Benavides, M.P. Nitric oxide inhibits nitrate reductase activity in wheat leaves. *Plant Physiol. Biochem.* **2011**, *49*, 124–130. [CrossRef] [PubMed]
51. Adhikari, L.; Missaoui, A.M. Nodulation response to molybdenum supplementation in alfalfa and its correlation with root and shoot growth in low pH soil. *J. Plant Nutr.* **2017**, *40*, 2290–2302. [CrossRef]

molecules

MDPI

Article

Towards Bioprospection of Commercial Materials of *Mentha spicata* L. Using a Combined Strategy of Metabolomics and Biological Activity Analyses

Juan Camilo Henao-Rojas [1,*], Edison Osorio [2], Stephanie Isaza [3], Inés Amelia Madronero-Solarte [1], Karina Sierra [2], Isabel Cristina Zapata-Vahos [4], Jhon Fredy Betancur-Pérez [5], Jorge W. Arboleda-Valencia [5,6] and Adriana M. Gallego [7,*]

1 Corporación Colombiana de Investigación Agropecuaria-Agrosavia, Centro de Investigación La Selva, Kilómetro 7, Vía a Las Palmas, Vereda Llanogrande, Rionegro 054048, Colombia; imadronero@agrosavia.co

2 Grupo de Investigación en Sustancias Bioactivas GISB, Facultad de Ciencias Farmacéuticas y Alimentarias, Universidad de Antioquia, Cl. 70 No. 52-21, Medellin 0500100, Colombia; edison.osorio@udea.edu.co (E.O.); karina.sierra@udea.edu.co (K.S.)

3 Hierbas y Plantas Tropicales SAS-HIPLANTRO, Cra. 56a No. 72a 101, Itagüí 055410, Colombia; estefania@hiplantro.com

4 Facultad de Ciencias de la Salud, Atención Primaria en Salud, Universidad Católica de Oriente, Rionegro 054040, Colombia; izapata@uco.edu.co

5 Centro de Investigaciones en Medio Ambiente y Desarrollo—CIMAD, Facultad de Ciencias Contables, Económicas y Administrativas, Universidad de Manizales, Cra. 9 No 19-03, Manizales 170001, Colombia; jbetancur@umanizales.edu.co (J.F.B.-P.); jwilliam.arboleda@udea.edu.co (J.W.A.-V.)

6 Grupo de Investigación FITOBIOL, Instituto de Biología, Facultad de Ciencias Exactas y Naturales, Universidad de Antioquia, Cl. 67 No 53-108, Medellin 050010, Colombia

7 Biomasnest, Medellin 050010, Colombia

* Correspondence: jhenao@agrosavia.co (J.C.H.-R.); adriana.gallego.02@gmail.com (A.M.G.)

Citation: Henao-Rojas, J.C.; Osorio, E.; Isaza, S.; Madronero-Solarte, I.A.; Sierra, K.; Zapata-Vahos, I.C.; Betancur-Pérez, J.F.; Arboleda-Valencia, J.W.; Gallego, A.M. Towards Bioprospection of Commercial Materials of *Mentha spicata* L. Using a Combined Strategy of Metabolomics and Biological Activity Analyses. *Molecules* **2022**, *27*, 3559. https://doi.org/10.3390/molecules27113559

Academic Editor: Smaoui Slim

Received: 15 April 2022
Accepted: 27 May 2022
Published: 31 May 2022

Abstract: Spearmint (*Mentha spicata* L.) has been widely studied for its diversity of compounds for product generation. However, studies describing the chemical and biological characteristics of commercial spearmint materials from different origins are scarce. For this reason, this research aimed to bioprospecting spearmint from three origins: Colombia (Col), Mexico (Mex), and Egypt (Eg). We performed a biological activity analysis, such as FRAP, DPPH, and ABTS, inhibition potential of *S. pyogenes*, *K. pneumoniae*, *E. coli*, *P. aeuroginosa*, *S. aureus*, *S aureus* Methicillin-Resistant, and *E. faecalis*. Furthermore, we performed chemical assays, such as total polyphenol and rosmarinic acid, and untargeted metabolomics via HPLC-MS/MS. Finally, we developed a causality analysis to integrate biological activities with chemical analyses. We found significant differences between the samples for the total polyphenol and rosmarinic acid contents, FRAP, and inhibition analyses for Methicillin-Resistant *S. aureus* and *E. faecalis*. Also, clear metabolic differentiation was observed among the three commercial materials evaluated. These results allow us to propose data-driven uses for the three spearmint materials available in current markets.

Keywords: spearmint; commercial materials; antimicrobial activity; antioxidant activity; untargeted metabolomics; bioprospection

1. Introduction

Approximately 25–30 Mentha species have been reported worldwide, including *Mentha piperita* L., *Mentha arvensis* L., *Mentha suaveolens* Ehrh, and *Mentha spicata* L., stand out as some of the most representative of the *Lamiaceae* family. Among them, spearmint (*Mentha spicata* L.) has presented an increase in worldwide cultivation, a higher recognition for its intense aroma, and more reports regarding its stimulant, diaphoretic, antiseptic, gastrointestinal respiratory, and antispasmodic effects [1–4]. The producing regions are concentred in the United States, Egypt, Australia, and some areas of Asia. Recently, Mentha production and demand have increased in Latin American countries, such as Mexico and Colombia [5].

The commercialization of *Mentha spicata* L. as fresh leaves is common in restaurants and haute cuisine. As dried-ground leaves, it is used at the agro-industrial level in the production of extracts for the pharmaceutical, cosmetic, confectionery, functional foods, toothpaste, or herbal infusions industries. Additionally, spearmint extracts have recently been shown to possess antibacterial, antifungal, antiviral, insecticidal, and antioxidant properties [6–8].

Some authors have reported spearmint compounds with multiple applications in the generation of functional foods and personal care products [9], primarily phenolic acids such as protocatechuic acid, hydroxybenzoic acid, 4-hydroxy cinnamic acid, caffeic acid, syringic acid and ferulic acids, gallic acid, vanillic acids, p-coumaric acid, and rosmarinic acids [10]. For similar applications, flavonoids such as thymonin, naringenin, rutin, quercetin, esculetin, nodifloretin, luteolin, and scopoletin have also been reported [10].

A considerable amount of research has been carried out on the identification, biochemical characterization, localization, and health benefits of the metabolites of spearmint [11,12]. However, few studies have focused on evaluating the biological properties, quality, and chemical diversity of spearmint's commercial materials present in the international market [3,7]. Materials from different countries were chosen because they retain the highest uses by food and cosmetic companies in the current Colombian market. Its proper use could generate high rates of productive performance, efficiency, and homogeneity of the products obtained, impacting direct economic costs.

The present work focused on identifying the main chemical properties and biological activities of commercial spearmint extracts grown in three different origins of Colombia (Col), Mexico (Mex), and Egypt (Eg). Our results provide tools for data-driven decision-making in selecting specific raw materials of *Mentha spicata* L. for the future generation of products.

2. Results

2.1. Chemical Measurements of *Mentha spicata* L.

The TPC ranged from 1183 to 1781 mg GA/100 g extract (Figure 1a), and significant differences were presented among the samples. Mexico had the highest TPC value, followed by Eg and then Col. The content of rosmarinic acid (RA) was higher in Eg, followed by Mex, and next by the Col samples. Eg showed a significant difference in RA compared to the other two origins (Figure 1b).

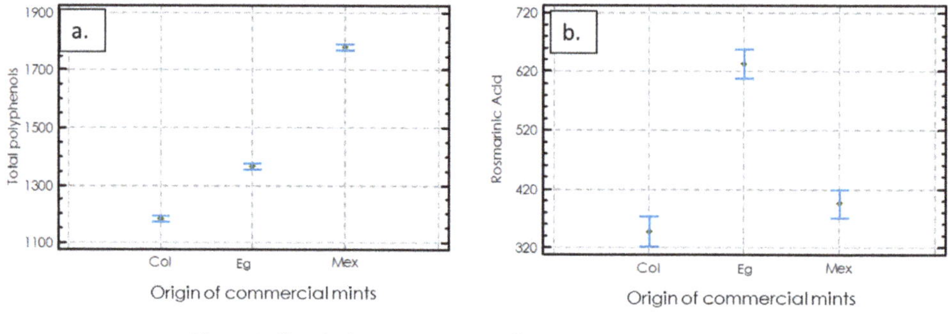

Figure 1. Chemical measurements in three commercial materials of *Mentha spicata L.* (**a**) Total polyphenol content (TPC) (mg of gallic acid (GA) per 100 g of extract); (**b**) rosmarinic acid concentration (mg rosmarinic acid (RA) per 100 g sample). Statistical analysis uses a one-way ANOVA (Tukey test, $p < 0.05$). Equal letters mean that there is no statistically significant difference.

2.2. Biological Activity Assays

2.2.1. Antioxidant Activities

The results of the antioxidant activity for the spearmint extracts are shown in Figure 2. The reductive capacity FRAP of Mex (Figure 2a) presented a significantly higher value of 1963.3 ± 76.5 mg eq of ascorbic acid (AA)/100 g than the other two origins. In addition, although no sample showed significant differences for the ABTS analysis, Mex showed the highest inhibition percentage (27.4 ± 2.0 mmol eq TX/100 g) (Figure 2b). Similarly, we did not find significant differences between the samples; however, Eg showed the highest value of 2,2-diphenyl-1-picrylhydrazyl (DPPH) among the samples (13.1 ± 2.3 mmol eq TX/100 g) (Figure 2c).

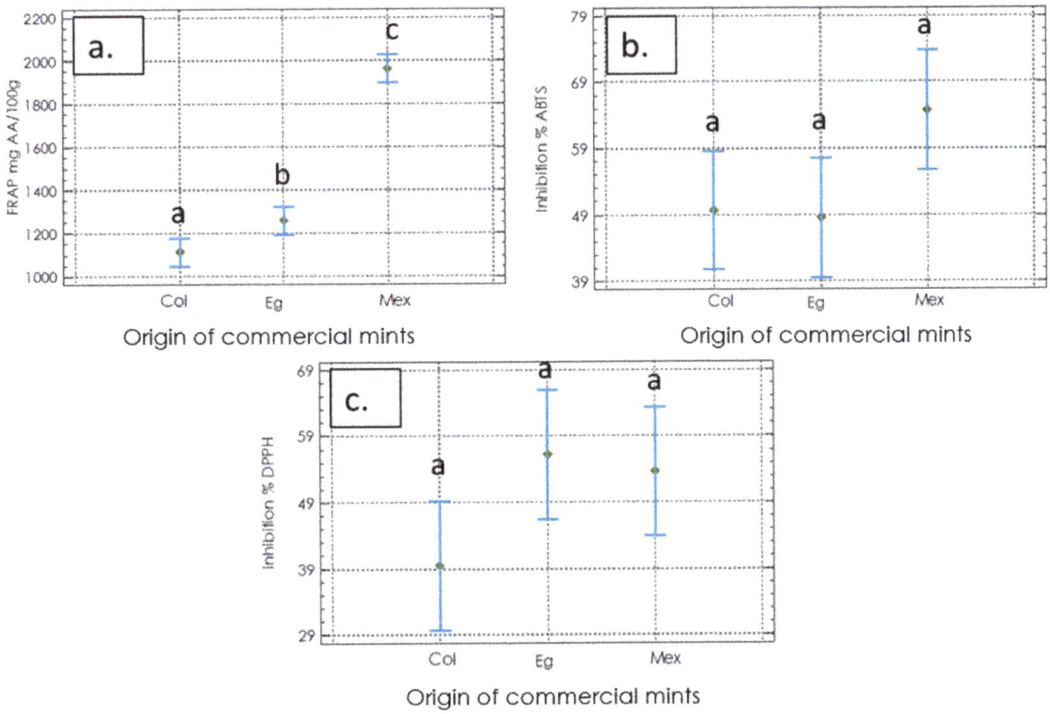

Figure 2. Antioxidant activities of three commercial materials from *Mentha spicata* L. (**a**) Ferric Reducing Antioxidant Power (FRAP activity) (mg Ascorbic Acid/ 100 g sample); (**b**) ABTS radical inhibition capacity (%); (**c**) DPPH radical inhibition capacity (%). Statistical analysis uses a one-way ANOVA (Tukey test, $p < 0.05$). Equal letters mean that there is no statistically significant difference.

2.2.2. Antimicrobial Activities

The antimicrobial activity assay showed that extracts obtained from different commercial materials of Mentha spicata inhibit the growth of Staphylococcus aureus Methicillin-resistant (SARM) (ATCC 43300) and Enterococcus faecalis (ATCC 19433) but did not show effects on the other tested bacteria. After 24 h of incubation, the growth of SARM showed a growth reduction of 53% and 24% with Col and Mex extracts, respectively. The Eg extract did not affect SARM's growth (Figure 3a). In addition, the extract obtained from Colombian plants resulted in a more stable reduction of both bacteria compared with the positive control. Likely, E. faecalis was reduced by 56%, 68%, and 91% for Col, Eg, and Mex extracts, respectively (Figure 3b). Furthermore, we evaluated other gram-negative bacteria (Escherichia coli, Klebsiella pneumoniae, Pseudo-monas aeruginosa, and Streptococcus pyogenes), which did not show inhibition in spearmint extracts. Altogether, the results show the promising antimicrobial activity of M. spicata commercial extracts against gram-positive bacteria, such as SARM and E. faecalis, which are among the most significant human bacterial pathogens in clinical medicine.

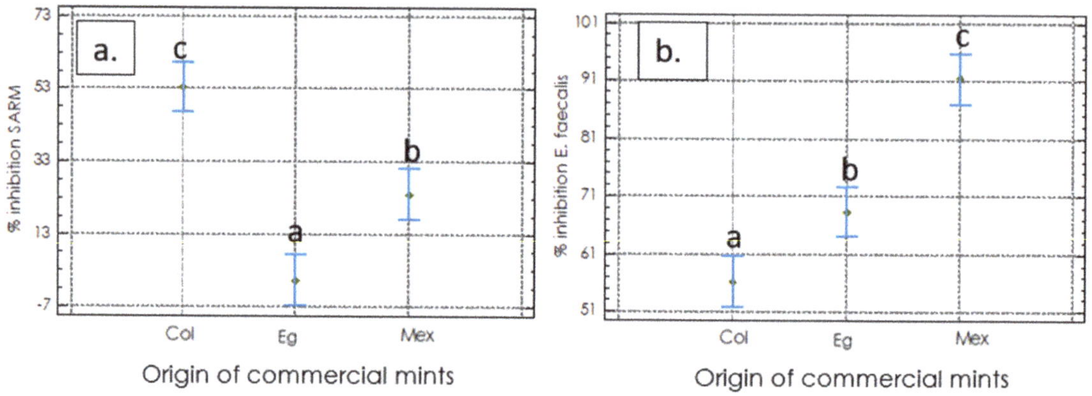

Figure 3. Antimicrobial activity of extracts from spearmint commercial materials against opportunistic bacteria; (**a**) percentage inhibition on Staphylococcus aureus Methicillin-resistant (SARM) (ATCC 43300); (**b**) percentage inhibition on Enterococcus faecalis (ATCC 19433). All data were collected in triplicate. One-way ANOVA using the Tukey test ($p < 0.05$) was used. The same letters mean that there is no statistically significant difference.

2.3. Metabolomic Analysis

The metabolic profiling for the spearmint samples from the three different origins of Col, Eg, and Mex was analyzed using an untargeted approach via HPLC-MS/MS. The Partial Least Squares Discriminant Analysis (PLS-DA) was carried out to assess the discriminatory and predictive ability of the metabolite profiles among the three spearmint origins. Also, PLSD-DA was used to distinguish samples of different geographic origins by chemical composition. The PLS-DA plot showed that all biological replicates for each origin were grouped together, suggesting the consistency of our experimental groups (Figure 4a). Also, the Col samples were further separated from Eg and Mex origins and showed a low metabolic variability within samples compared to Eg and Mex. PC1 and PC2 accounted for 85.3% of the total variation. The PLS-DA was validated using the leave one cross-validation algorithm (LOOCV) to determine the model's performance. A high predictive accuracy ($R^2Y = 0.837$) was found; the exogenous variables (origin) explained the endogenous variation of the dependent variable. Likewise, the model's predictive power ($Q^2 = 0.738$) is considered reliable. Subsequently, we were able to tentatively automatically annotate 932 metabolites on our HPLC/MS data (Supplementary Table S1).

Furthermore, about 72 features were selected for further fragmentation and annotation via MS/MS using the Variable Importance Parameter (VIP) value > 2.0 criteria. Then, we manually annotated six metabolites using an in-house database built for mint, which is represented in the loading plot (Figure 4b). We tentatively annotated one metabolite of each chemical class of cycloparaffin, flavone, carotenoid, hydroxycoumarin, cinnamate ester, and caffeic acid ester (Table 1 and Supplementary Table S2). Esculetin contributed to the separation of Col, the rosmarinic acid separated Eg from the other two origins, and the rest of the metabolites were grouped for the Mex origin.

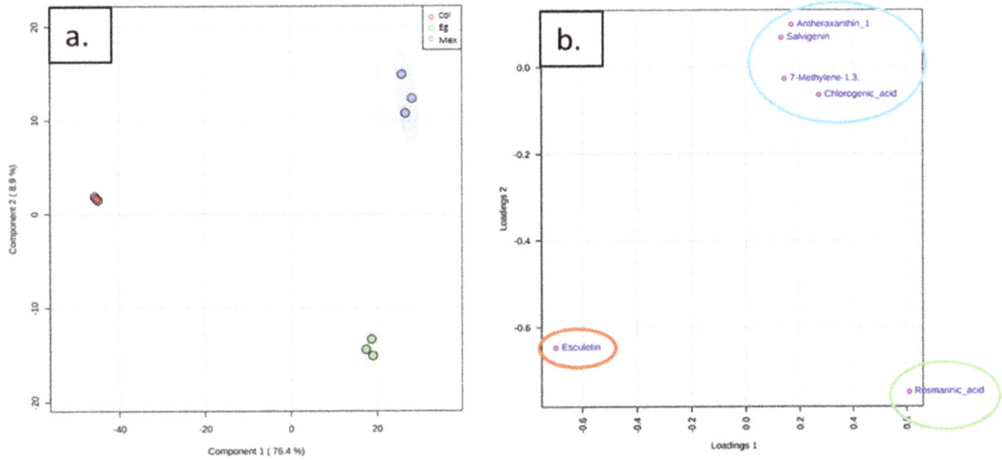

Figure 4. Multivariate analysis of discriminant metabolites in spearmint commercial materials. (**a**) PLS-DA for Mentha of three origins Col, Mex, and Eg. (**b**) Loading plot showing the six annotated metabolites for spearmint.

Table 1. List of annotated metabolites of spearmint (*Mentha spicata* L.).

No.	RT	*m/z*	Adduct	Molecular Formula	Anotación	Class
1	5.835	119.08387	$[M + H]^+$	C_9H_{12}	7-Methylene-1,3,5-cyclooctatriene	Cycloparaffin
2	18.619	585.45056	$[M + H]^+$	$C_{40}H_{56}O_3$	Antheraxanthin	Carotenoid
3	18.609	329.1127	$[M + H]^+$	$C_{18}H_{16}O_6$	Salvigenin	Flavone
4	6.296	361.091	$[M + H]^+$	$C_{18}H_{16}O_8$	Rosmarinic acid	Caffeic acid ester
5	19.208	179.10556	$[M + H]^+$	$C_9H_6O_4$	Esculetin	Hydroxycoumarin
6	18.078	355.09079	$[M + H]^+$	$C_{16}H_{18}O_9$	Chlorogenic acid	Cinnamate ester

Next, we plotted a heatmap for the six annotated metabolites to display differences in their expression among the three spearmint origins (Figure 5). The clustering analysis allowed us to identify two groups agreeing with the PLS-DA. Most metabolites showed a contrasting pattern in Col (C1) compared to the Eg and Mex origins (C2). Most of the metabolites showed a low expression in Col compared to the other two origins. Furthermore, we performed a correlation heatmap for these six metabolites. Interestingly, esculetin was negative and significantly (p-value < 0.05) correlated with the rest of the metabolites (Figure S2 and Table S3).

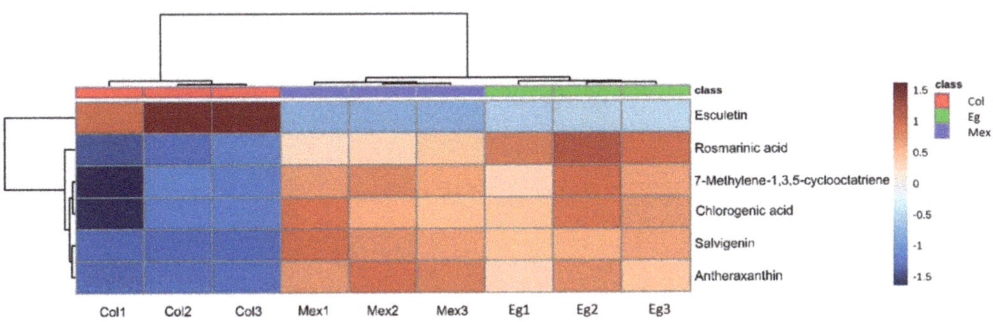

Figure 5. Heatmap showing expression patterns for six clustered metabolites from three commercial mints.

2.4. Causality Analysis

A causality analysis has an approach that allows us to understand, in a multidimensional way, how a set of predictor variables influence a response variable. A causality analysis differs from a correlation analysis in the former, which considers the sum of the variables and is not based on pair-wise relationships. A causality analysis is helpful because, in a multi-component matrix, some can produce synergy between them while others act oppositely. By performing this analysis, we were able to discriminate which metabolites and in what proportion influence the different biological activities studied.

The results show that there are predictors (e.g., chemical analysis or specific metabolites) generating positive or negative causal relationships against the antioxidant activity. Here, a positive causal relationship means a predictor favors the antioxidant activity. In contrast, a negative causal relationship means a predictor disfavor of the antioxidant activity. TPC and the metabolites, chlorogenic acid and 7-methylene-1,3,5-cyclooctatriene, contributed to the antioxidant capacity, while esculetin and salvigenin did not contribute to the antioxidant capacity (Figure 6a). This is in concordance with reports where a positive correlation between antioxidant activities and TPC and chlorogenic acid has been reported [13–15].

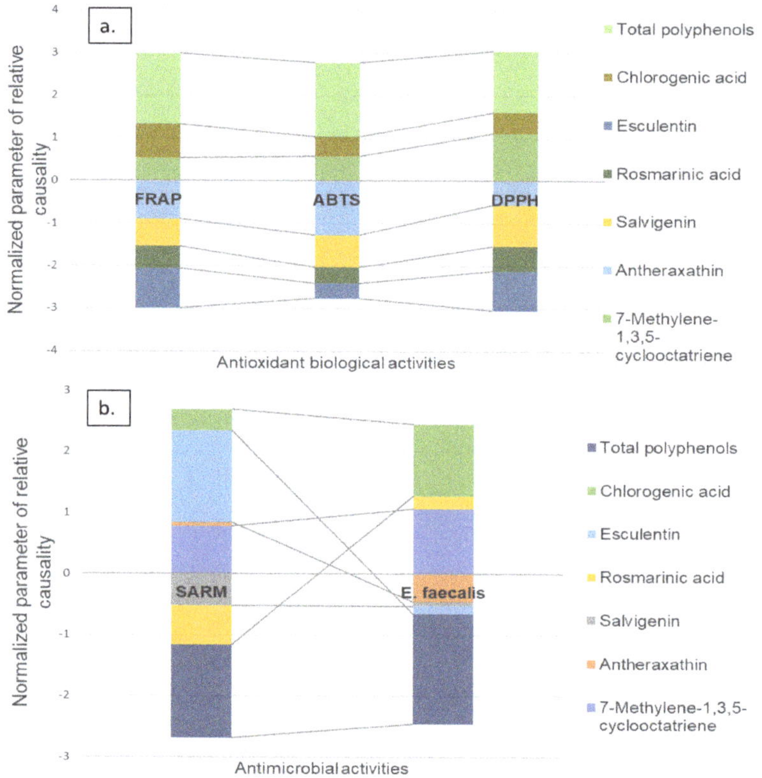

Figure 6. Causal relationships between antioxidant and antimicrobial activities and the main discriminating metabolites in *Mentha spicata* L. commercial materials of (**a**) antioxidant causality analysis; (**b**) antimicrobial causality analysis.

Regarding the antimicrobial activities, the chlorogenic acid, esculetin, and 7-methylene-1,3,5-cyclooctatriene showed a positive contribution to the increase of the SARM inhibitory capacity. Contrary, the metabolites anteraxanthin, rosmarinic acid, salvigenin, and TPC, showed a negative influence on SARM's inhibitory capacity. A different case was presented for the inhibitory activity of *E. faecalis*. Here, the chlorogenic acid, 7-Methylene-1,3,5-cyclooctatriene, and TPC increased the inhibitory capacity for this microbial strain. In contrast, esculetin, rosmarinic acid, and anteraxanthin reduced the effectiveness in the inhibitory activity of *E. faecalis* (Figure 6b).

The causality analysis results indicate that it is generally possible to obtain the predictive capacity of the antioxidant and antimicrobial activities of the samples of Mentha spicata extracts based on their discriminant metabolites. The above modified polynomial models with an adjusted R square greater than 98.5.

We further searched by uses reported in the literature for the annotated metabolites (Table 2). We found uses reported for antheraxanthin, salvigenin, rosmarinic acid, esculetin, and chlorogenic acid, except for 7-Methylene-1,3,5-cyclooctatriene. Whereas antheraxanthin reports cosmetical uses, the rest of the metabolites are reported to have the potential for health-related uses.

Table 2. Bioprospection of annotated metabolites in *Mentha spicata* L.

Compound	Food/Pharmacy Uses	References
7-Methylene-1,3,5-cyclooctatriene	Reports not found.	
Antheraxanthin	Skin care.	[16–18]
Salvigenin	Antiinflammatory and analgesic properties.	[19]
Rosmarinic acid	Antibacterial, antioxidant, anticancer, antiinflammatory, immunomodulatory, and coadjuvant activities in the treatment of cancer, diabetes, neuroprotective, and prevention of cognitive decline in Alzheimer's disease.	[20–23]
Esculetin	Anti-inflamatory, anticoagulant, liver-protection, antidiabetic, antioxidant, antitumor, and UV-filters.	[24,25]
Chlorogenic acid	High potential antioxidant, hepatoprotective, cardioprotective, anti-inflammatory, antipyretic, neuroprotective, anti-obesity, antiviral, antimicrobial, anti-hypertension, and free radicals.	[26]

3. Discussion

This study provides a characterization of the antioxidant and antimicrobial activities and chemical composition (using chromatographical and spectrometry) for commercial spearmint of the three origins of Col, Mex, and Eg. Studying the bioprospection of the currently available commercial spearmint materials is relevant because it allows the mint market companies to adjust their portfolios and make better decisions using data-driven strategies.

With respect to antioxidant activities, the DPPH free radical inhibition capacity is consistent with reports from the Oman region (Arabian Peninsula), which reported 54.68% inhibition. In this study, the authors associated the inhibition mainly with the presence of terpenes in the extracts of spearmint [27]. Furthermore, our three commercial extracts exhibited a higher reducing capacity measured by FRAP than Yousuf et al. [28]. The high capacity to trap free radicals is likely since these mint species' metabolites with antioxidant characteristics have a high hydrogen donating capacity [27]. It is congruent with our results where Mex showed the highest reducing capacity and two discriminant potent antioxidants, such as the chlorogenic acid and the antheraxanthin. These metabolites work as a hydrogen donator and electron donator, respectively. In addition, the botanical extracts of mint showed no significant difference in their ABTS radical inhibition capacity value, reporting activities between 48 and 60%, which is similar to that referenced by [29], who found a value of 40.2 $\pm$ 0.2%.

The higher Rosmarinic Acid (RA) content in the Eg sample could be due to genotype-environment interaction effects. *Mentha spicata* L. materials planted in the regions of Egypt have undergone some breeding process (natural or artificial) that favors the accumulation of rosmarinic acid, as reported by [30]. The RA can be further enhanced by the breeding process used in Central and Latin America (CLATAM), which is primarily clonal. Given the impossibility of the plant to naturally complete its flowering stage in these regions, it substantially reduces its diversity by sexual reproduction [31,32]. Additionally, as an emerging product in CLATAM, the production conditions are not fully optimized, which favors the bioaccumulation of target metabolites in this species. Consequently, abiotic factors, such as agronomic practices, cultivation conditions (open field or protected agriculture), and fertilization may influence the concentration of RA.

Some of the metabolites found in our spearmint extracts have also been previously reported for their antimicrobial activity on *S. aureus* and *S. aureus* methicillin-resistant but in cocoa extracts. Similarly, spearmint-related metabolites also had significant activity against gram-positive bacteria, such as *S. aureus*, in coffee extracts (Martínez-Tomé et al.) [33,34]. Among these metabolites, chlorogenic acids have been reported to have an antimicrobial effect on both gram-positive and gram-negative bacteria, exhibiting bactericidal activity against *P. fluorescens* and *S. aureus* but no sporicidal activity against *B. cereus* and *C. sporogenes* [34,35]. In this case, byproducts containing chlorogenic acids are considered promissory useful sources for pharmaceutical, cosmetic, and food industries [36]. In our results, the positive effect of spearmint extracts on Gram-positive bacteria may be due to their bacterial membrane composition, which is more easily permeable than Gram-negative bacteria membranes. Gram-positive membranes are more likely penetrable, generate complexes with extracellular proteins and soluble proteins quickly, and binding to the bacterial DNA [22].

The chemical and pharmacological analysis of the spearmint extracts (commercial materials from different sources) allows the determination of the potential of the samples for the possible development of ingredients or products of interest in the cosmetic, food, and pharmaceutical industries. In this way, spearmint's chemical properties and pharmacological characteristics must be considered simultaneously. The accumulation of phenolic compounds in plant tissues is a distinctive feature of environmental stress. Polyphenolic compounds help plants cope with multiple biotic and abiotic stresses, such as drought, heavy metals, salinity, temperature, and ultraviolet light. As these conditions are variable in different origins, that could shape the metabolic profile of each commercial material. Furthermore, a specific metabolic profile varies significantly depending on the extract preparation, particularly when it comes to industrial processing for different industries [9].

The causality analysis shows a positive influence on all antioxidant activities, the total polyphenols, and chlorogenic acid, which is not surprising given the multiple reports of these compounds as inhibitors of oxidative radicals [26]. On the contrary, the compounds such as salvigenin and esculetin show a decrease in antioxidant biological activities. This is in agreement with the studies where salvigenin and esculetin present low antioxidant activity. Indeed, salvigening is reported with lipid-lowering and mitochondrial-stimulating activity, while esculentin shows anti-inflammatory, anti-apoptotic, anticancer, antidiabetic, and neuroprotective activities [37,38]. The antimicrobial causality analysis shows that the compounds with the most significant influence on the MRSA and *E. faecalis* inhibition are chlorogenic acid and esculentin, which are metabolites that are reported as compounds with an antibiotic capacity on gram-positive bacteria [38,39].

We found that esculetin contributed to the separation of the Col sample, and this sample also showed the highest SARM percentage inhibition. Therefore, we could propose that spearmint from Colombia could serve in products targeting antioxidant activities, UV protection, and antimicrobial activities. In contrast, the rosmarinic acid contributed to the separation of the Eg sample from the two other origins. It also agreed with the rosmarinic content for the Eg sample. Consequently, based on the health potential given for rosmarinic acid, the Eg spearmint could be used for pharmaceutical uses. Finally, a higher and varied number of metabolites allowed separate Mex samples from the other commercial samples, including the potent antioxidant chlorogenic acid and antheraxhanthin.

Furthermore, the Mex samples showed the highest inhibition percentage for *E. fecalis* and the highest polyphenol and antioxidant activity by FRAP. In this regard, the Mex spearmint samples have a higher potential to serve in both pharmaceutical and cosmetic uses [40]. Moreover, spearmint extracts have great potential in food industries since an increased awareness among consumers has increased the consumption of medicinal plants, mainly as natural antioxidants [41]. The applications of mint in food have been described as flavoring and preservative due to its antimicrobial properties [12]. Some applications have reported mint for fish and seafood, confectionery, chewing gum, and the cheese industry [3]. Altogether, and thanks to the metabolic composition and the biological characterization of commercial materials, the mint market can strengthen the decision making and apply spearmint for uses in cosmetics [42,43], pharmaceutics [44,45], and food [46].

4. Materials and Methods

4.1. *Mentha spicata* L. *Commercial Materials*

For the present investigation, three foliar commercial materials of dry and ground *Mentha spicata* L. were selected, which presented an active agro-industrial market for their commercialization in food companies or natural products in Colombia. For this, three different origins were selected: Eg (mint produced in the El-Fayun region, Egypt), Mex (mint produced in Atlixco, Mexico), and Col (mint produced in El Retiro-Antioquia, Colombia). The average climatic conditions of the production areas were obtained using the CHELSA database (https://chelsa-climate.org/) (accessed on 30 March 2022) and are consolidated in the Supplementary Material Figure S1.

4.2. *Biological Activities and Chemical Asssays*

4.2.1. Preparation of Botanical Extracts of *Mentha spicata* L.

For the preparation of the botanical extracts, the methodology optimized by Sierra et al. [47] was used, for which 4 $\pm$ 0.05 g of each of the commercial mint materials (dry and ground plant material) were weighed, and a 30 mL of 80% HPLC-grade ethanol extraction solution in type I water was added. Subsequently, they were subjected to sonication with a frequency of 37 Hz, 30 $\pm$ 5 °C for 30 min (Elma P60H, Singer, La Vergne, TN, USA, EE. UU.) and then centrifugation at 8000 rpm for 20 min (Sorvall, Thermo Scientific, Waltham, MA, USA, EE. UU.) was applied to recover the supernatant fractions. All extraction assays were performed in triplicate.

4.2.2. Determination of Total Phenolic Content (TPC)

The Folin-Ciocalteu method was carried out to determine the TPC content of each extract [48]. Gallic acid, in a dynamic range of 10 to 100 μg/mL, was used as a reference standard. Then, 25 μL of the extract was mixed with 125 μL of Folin-Ciocalteu's reagent (1:10), both of which were diluted in distilled water. The mixture was shaken and incubated in darkness for 5 min at room temperature, followed by the addition of 100 μL of Na_2CO_3 (7.5% *w*/*v*). After 60 min of incubation at room temperature in the dark, absorbance readings were performed at 765 nm using a Synergy HT multimodal microplate reader (Biotek Instruments, Inc.; Wonooski, VT, USA). The total polyphenol content was calculated using a calibration curve with gallic acid. The results are expressed as milligrams of gallic acid (GA) per 100 g of extract, as appropriate (mg GA/g extract). All measurements were performed in triplicate.

4.2.3. DPPH Radical Scavenging Capacity Assay

According to Brand-Williams et al. [49], the ability of Mentha leaf extracts to scavenge the DPPH (2,2-diphenyl-1-picrylhydrazyl) radical was determined. The reaction was composed of the Aliquots (10 μL) of the leaf extract and 990 μL of the DPPH standard solution. The absorbance was determined at 517 nm after 30 min in the dark (Biotek Instruments, Inc.; Wonooski, VT, USA). The results were expressed as millimoles of the Trolox Equivalent per 100 g of fresh Mentha (mmolTE 100 g^{-1}).

4.2.4. Reducing Power Assay (FRAP)

The reduction ability was measured using the methods of Benzie and Strain [50] with some modifications. Aliquots of 50 μL of the sample extract were mixed with 50 μL of an acetate buffer, pH 3.6, and adjusted to 1000 μL with a FRAP solution ($FeCl_3$, TPTZ (Tripyridyl-s-triazine) in HCl 40 mM). The increase in absorbance was measured at 590 nm (Biotek Instruments, Inc.; Wonooski, VT, USA). The FRAP values were expressed as milligrams of the Ascorbic Acid Equivalent per 100 g of fresh Mentha (mgAAE 100 g^{-1}).

4.2.5. ABTS Radical Scavenging Capacity Assay

Radical scavenger activity against the stable radical, ABTS, was measured according to Mesa-Vanegas et al. [51]. An aliquots of 10 μL of the sample extract and 990 μL of the standard ABTS were mixed. The absorbance was determined at 732 nm after 30 min in the dark (Biotek Instruments, Inc.; Wonooski, VT, USA). The results were expressed as millimoles of the Trolox Equivalents per 100 g of fresh Mentha (mmolTE 100 g^{-1}).

4.2.6. Quantification of Rosmarinic Acid (RA)

Rosmarinic acid was identified and quantified using a methodology previously established in the laboratory [47]. It is based on reverse phase HPLC/DAD chromatographic analysis (Agilent Technologies, Palo Alto, CA, USA). Chromatographic separation was performed with a Zorbax SB RRTT C18 column (50 mm × 4.6 mm with 1.8 μm of particle size) of fast resolution and high performance using the mobile phase previously filtered through 0.45 μm nylon membrane filters and degassed using an ultrasonic bath before the analysis. The isocratic elution curve consisted of water with formic acid 0.5% (A) and acetonitrile (B), and the linear gradient used was as follows: 0 min, 16% B; 4 min, 16% B; 8 min, 20% B; 11 min, 40% B; 12 min, 45% B; 13 min, 50% B; 14 min, 60% B; and 15 min, 16% B, at a flow rate of 1.0 mL/min. The column temperature was maintained at 30 °C, and the injection volume of both the calibration curve of rosmarinic acid (CAS No. 20283-92-5, purity 99%, European Pharmacopoeia reference, Europe) and the sample solutions was 5 μL. The wavelength was set at 329 nm to monitor the chromatographic profile. All measurements were performed in triplicate [47].

4.3. *Antimicrobial Activity in Commercial Materials of Mentha spicata* L.

Hydroalcoholic extracts were prepared as described above to evaluate the antimicrobial activity of three commercial materials of *Mentha spicata* L. These extracts were used for the bacterial inhibition test following the protocol published by Balouiri et al. [52]. The inhibitory effect from the extracts was carried out against opportunistic pathogens, such as *Enterococcus faecalis* (ATCC 19433), *Escherichia coli* (ATCC 25922 and XL1 blue), *Klebsiella pneumoniae* (ATCC 700603), *Pseudomonas aeruginosa* (ATCC 27853), *Staphylococcus aureus* Methicillin-resistant (MRSA) (ATCC 43300), and *Streptococcus pyogenes* (ATCC 19615). The bioassay was realized in a 96-well microtiter; the wells were adjusted to 200 μL as the final volume. In addition, each treatment consisted of a 170 μL LB culture medium (50% Bactotryptone, 25% yeast extract, 35% NaCl, and diluted in distilled water), 20 μL of each extract, and 10 μL of the respective overnight bacterial culture. It was used as a negative control (hydroalcoholic solvent) and as a positive control (Gentamicin 40 μg*L^{-1}). Afterward, the bacteria cultures were incubated at 37 °C, and the growth was monitored by measuring the absorbance at OD 600 until 24 h. Each treatment was carried out in triplicate, and the percentage of the inhibition was calculated by using the following equation (Control OD − (Sample OD/Control OD)) × 100.

4.4. *Metabolic Profiling of Spearmint Based on HPLC-MS and HPLC-MS/MS*

An ACQUITY H-CLASS UPLC-Xevo G2-XS-QTOF (Waters, Herts, UK) kit was used to obtain the undirected metabolic profile. The MS system was operated using electrospray ionization in positive mode, using an ACQUITY UPLC BEH C18 column (100 mm × 2.1 mm, 1.7 μm particle size) as the stationary phase. The working temperature was 55 °C, with a capillary voltage of 5.5 kV and a cone voltage of 30 V. The mobile phase used was a binary gradient of ultrapure water + 1% formic acid (solution A) and acetonitrile + 0.1% formic acid (solution B). The elution began with a linear gradient, starting with A:60%–B:40%

for 17.5 min, followed by B:100% from 17.5 min to 25 min, and ending with the gradient A:60%–B:40% until the end at 30 Minutes. The flow rate was 0.250 mL/min, and the injection volume was 5 μL.

The data was processed with MS-DIAL V. 4.70 (Yokohama, Japan) [53]. For the peak detection, a weighted smoothing average algorithm was used as linear, setting the smoothing level to 1 scan, the minimum peak width at 3 scans, and the minimum peak height at 1000 amplitude. For the alignment for the peaks, we used 0.1 min in the RT tolerance and 0.025 Da in the MS1 tolerance. Subsequently, the peak areas obtained were normalized to the sum of the total ion chromatograms, logarithmically transformed, and scaled from Pareto, and the missing values were imputed using the baseline value and the signals that were not present in 70% of the samples were removed.

Statistical analysis was performed with MetaboAnalyst 4.0 software (Montreal, QC, Canadá) [54] and then by a Principal Component Analysis (PCA). Then, a PLS-DA was carried out to determine the representative characteristics that present the greatest variation among the mint samples. The top six VIPs with values greater than 2.0 were selected to be considered as potential biomarkers of the differences between the samples. For the data annotation, an automatic annotation was first performed using MSDIAL with the Fiehn/Vaniya Natural Products Library. MS and RT were used to match in the compound annotation. Fragmented VIP metabolites were compared on their precursor ion with data obtained from the Metlin, PlantCyc, and MoNA bases, using a threshold of 20 ppm. Additionally, manual curing was performed using a base of in-house data built from the mint reports.

Metabolites Annotation

For the metabolite annotation process, extraction of the experimental spectra was performed using MZmine software (Okinawa, Japan). After that, a first analysis was performed, which consisted of a manual search in the database established for Mentha. For those that could not be annotated, an automatic screening process was performed, in which the experimental spectra were compared with several compounds obtained from the Mass Bank of North America database, specifically a database for *Mentha spicata* built from this database. It should be noted that this comparison takes into account various adducts in order to identify more precisely the metabolite that best fits. To corroborate this information, a literature analysis was carried out to identify those metabolites that have been found in *Mentha spicata* L. in positive ionization and were obtained from liquid chromatography. Finally, with this list of compounds, the search for theoretical spectra was performed using the FOODB database and the simulation of the spectrum with CFM-ID using SMILES. With these spectra, an automatic comparison was performed, taking into account the mass and intensity to establish those metabolites that present the best fit with respect to the experimental one.

4.5. Causality Analysis between Metabolomic Profiles and Biological Activities

The determination of causal relationships between variables associated with functional activities and metabolites' candidates of commercial materials of *Mentha spicata* L. used a multiple regression model adjusted with a linear polynomial of degree one (Equation (1)). In this model, each response variable (φ) was assumed as a determined functional potential. On the other hand, the relative intensities of the discriminant metabolites (VIP) were taken as fixed effect covariates in the proposed model [55]. The matrix was subjected to a Cochrane–Orcutt optimization process, where the classic least-squares procedure is modified to allow autocorrelation between successive residuals. In this case, both the autocorrelation standard (ρ) and the model parameters (β_k) are determined interactively using between 30 and 50 iterations until the change in the derived value of each parameter, compared to the previous step, is less than 0.01, as proposed by Sun, Lang, and Boning (2021) for situations in which the model residuals are not independent.

$$\varphi = \beta_0(1-\rho) + \beta_1 x_1 + \beta_2 x_2 + \ldots . \beta_k x_k \quad (1)$$

where:

φ = Related biological activity factor
β_k = Discriminating adjustment parameters associated with the cause − effect relationship
x_k = Discriminating metabolites associated
ρ = Autocorrelation standard.

5. Conclusions

This work shows the first comparison of the attributes of the functional interest between *Mentha spicata* raw materials available in the international aromatic plant market, demonstrating the high heterogeneity in chemical variability and possible applications of the available raw materials. The results confirm that the characteristics of their chemical quality and antioxidant and antimicrobial activity show marked differences between the materials evaluated. The Col samples have potential as antioxidants and antimicrobials, probably for cosmetics, based on their percentage of inhibition SARM and esculetin. The Eg samples may work for pharma uses based on their rosmarinic content. Also, due to the metabolic diversity in the Mex samples, Mentha from this origin could have potential in varied industries, such as cosmetics and pharma, and is also supported by its antimicrobial activities. Finally, according to consumer awareness, spearmint extracts have great potential in food industries, mainly in which consumers may improve their health by increasing the intake of functional plant-based foods.

Supplementary Materials: The following supporting information can be downloaded at: https://www.mdpi.com/article/10.3390/molecules27113559/s1. Figure S1: average monthly climatic conditions of the cultivation areas of the commercial products of *Mentha spicata* L. evaluated; Figure S2: Pearson correlation analysis of the top 10 discriminant metabolites in commercial Mentha spicata samples; Table S1: list of metabolites automatically annotated via HPLC/MS database; Table S2: top 10 of discriminant metabolites reliably annotated in commercial materials of *Mentha spicata* L.; Table S3: *p*-values and Pearson correlation coefficients between top 10 discriminant metabolites.

Author Contributions: Conceptualization, J.C.H.-R., J.W.A.-V. and A.M.G.; methodology, J.W.A.-V., J.F.B.-P., I.C.Z.-V., I.A.M.-S., K.S., E.O. and S.I.; validation, J.C.H.-R., A.M.G., and J.W.A.-V.; formal analysis, J.C.H.-R., A.M.G. and J.W.A.-V.; investigation, J.W.A.-V., J.F.B.-P., I.C.Z.-V., I.A.M.-S., K.S., E.O. and S.I.; data curation, A.M.G. and J.C.H.-R.; writing—original draft preparation, J.C.H.-R., A.M.G., J.W.A.-V., J.F.B.-P., I.C.Z.-V., I.A.M.-S., K.S., E.O. and S.I.; writing—review and editing, J.C.H.-R. and A.M.G. All authors have read and agreed to the published version of the manuscript.

Funding: This research was funded by the Minciencias and Secretary of Agriculture and Rural Development from Government of Antioquia; grant number 710680562744; contract 80740-006-2019.

Institutional Review Board Statement: Not applicable.

Informed Consent Statement: Not applicable.

Data Availability Statement: Not applicable.

Acknowledgments: The authors thank the science, technology, and innovation fund of the General Royalties System (SGR) of Colombia and the Ministry of Science and Technology (Minciencias) for funding this research. Also, to the Colombian Agricultural Research Corporation—Agrosavia, University Catolica of Oriente, University of Manizales, and the companies, Hiplantro and Biomasnest, for the time dedicated by the researchers and the use of the assets inherent in the preparation of this document.

Conflicts of Interest: The authors declare no conflict of interest.

Sample Availability: Samples of the compounds are not available from the authors.

References

1. Akdoğan, M.; Tamer, M.N.; Cüre, E.; Cüre, M.C.; Köroğlu, B.K.; Delibaş, N. Effect of spearmint (*Mentha spicata Labiatae*) teas on androgen levels in women with hirsutism. *Phytother. Res.* **2007**, *21*, 444–447. [CrossRef]
2. Begaa, S.; Messaoudi, M.; Ouanezar, A.; Hamidatou, L.; Malki, A. Chemical elements of Algerian *Mentha spicata* L. used in the treatment of digestive system disorders by employing instrumental neutron activation analysis technique. *J. Radioanal. Nucl. Chem.* **2018**, *317*, 1107–1112. [CrossRef]
3. Mahendran, G.; Verma, S.K.; Rahman, L.-U. The traditional uses, phytochemistry and pharmacology of spearmint (*Mentha spicata* L.): A review. *J. Ethnopharmacol.* **2021**, *278*, 114266. [CrossRef]
4. Cirlini, M.; Mena, P.; Tassotti, M.; Herrlinger, K.A.; Nieman, K.M.; Dall'Asta, C.; Del Rio, D. Phenolic and Volatile Composition of a Dry Spearmint (*Mentha spicata* L.) Extract. *Molecules* **2016**, *21*, 1007. [CrossRef]
5. Villagran, E.; Henao-Rojas, J.C.; Franco, G. Thermo-Environmental Performance of Four Different Shapes of Solar Greenhouse Dryer with Free Convection Operating Principle and No Load on Product. *Fluids* **2021**, *6*, 183. [CrossRef]
6. Zaidi, S.; Dahiya, P. In vitro antimicrobial activity, phytochemical analysis and total phenolic content of essential oil from Mentha spicata and Mentha piperita. *Int. Food Res. J.* **2015**, *22*, 16.
7. Chauhan, S.S.; Agarwal, R. Evaluation of antibacterial activity of volatile oil from *Mentha spicata* L. *J. Drug Deliv. Ther.* **2013**, *3*, 120–121. [CrossRef]
8. Wu, Z.; Tan, B.; Liu, Y.; Dunn, J.; Martorell Guerola, P.; Tortajada, M.; Cao, Z.; Ji, P. Chemical Composition and Antioxidant Properties of Essential Oils from Peppermint, Native Spearmint and Scotch Spearmint. *Molecules* **2019**, *24*, 2825. [CrossRef]
9. Zhang, L.-L.; Chen, Y.; Li, Z.-J.; Li, X.; Fan, G. Bioactive properties of the aromatic molecules of spearmint (*Mentha spicata* L.) essential oil: A review. *Food Funct.* **2022**, *13*, 3110–3132. [CrossRef]
10. Eftekhari, A.; Khusro, A.; Ahmadian, E.; Dizaj, S.M.; Hasanzadeh, A.; Cucchiarini, M. Phytochemical and nutra-pharmaceutical attributes of Mentha spp.: A comprehensive review. *Arab. J. Chem.* **2021**, *14*, 103106. [CrossRef]
11. Pearson, W.; Fletcher, R.S.; Kott, L.S.; Hurtig, M.B. Protection against LPS-induced cartilage inflammation and degradation provided by a biological extract of *Mentha spicata*. *BMC Complementary Altern. Med.* **2010**, *10*, 19. [CrossRef]
12. Kee, L.A.; Shori, A.B.; Baba, A.S. Bioactivity and health effects of Mentha spicata. *Food Nutr. Metab.* **2017**, *5*, 1–2.
13. Farnad, N.; Heidari, R.; Aslanipour, B. Phenolic composition and comparison of antioxidant activity of alcoholic extracts of Peppermint (*Mentha piperita*). *J. Food Meas. Charact.* **2014**, *8*, 113–121. [CrossRef]
14. Elansary, H.O.; Szopa, A.; Kubica, P.; Ekiert, H.; Klimek-Szczykutowicz, M.; El-Ansary, D.O.; Mahmoud, E.A. Polyphenol Profile and Antimicrobial and Cytotoxic Activities of Natural *Mentha* × *piperita* and *Mentha longifolia* Populations in Northern Saudi Arabia. *Processes* **2020**, *8*, 479. [CrossRef]
15. Uribe, E.; Marín, D.; Vega-Gálvez, A.; Quispe-Fuentes, I.; Rodríguez, A. Assessment of vacuum-dried peppermint (*Mentha piperita* L.) as a source of natural antioxidants. *Food Chem.* **2016**, *190*, 559–565. [CrossRef]
16. Pistelli, L.; Sansone, C.; Smerilli, A.; Festa, M.; Noonan, D.M.; Albini, A.; Brunet, C. MMP-9 and IL-1β as Targets for Diatoxanthin and Related Microalgal Pigments: Potential Chemopreventive and Photoprotective Agents. *Mar. Drugs* **2021**, *19*, 354. [CrossRef]
17. Tominaga, K.; Hongo, N.; Karato, M.; Yamashita, E. Cosmetic benefits of astaxanthin on humans subjects. *Acta Biochim. Pol.* **2012**, *59*, 43–47. [CrossRef]
18. Davinelli, S.; Nielsen, M.E.; Scapagnini, G. Astaxanthin in Skin Health, Repair, and Disease: A Comprehensive Review. *Nutrients* **2018**, *10*, 522. [CrossRef]
19. Mansourabadi, A.H.; Sadeghi, H.M.; Razavi, N.; Rezvani, E. Anti-inflammatory and Analgesic Properties of Salvigenin, Salvia officinalis Flavonoid Extracted. *J. Future Nat. Prod.* **2016**, *2*, 31–41.
20. Alagawany, M.; Abd El-Hack, M.E.; Farag, M.R.; Gopi, M.; Karthik, K.; Malik, Y.S.; Dhama, K. Rosmarinic acid: Modes of action, medicinal values and health benefits. *Anim. Health Res. Rev.* **2017**, *18*, 167–176. [CrossRef]
21. González-Vallinas, M.; Molina, S.; Vicente, G.; de la Cueva, A.; Vargas, T.; Santoyo, S.; García-Risco, M.R.; Fornari, T.; Reglero, G.; Ramírez de Molina, A. Antitumor effect of 5-fluorouracil is enhanced by rosemary extract in both drug sensitive and resistant colon cancer cells. *Pharmacol. Res.* **2013**, *72*, 61–68. [CrossRef]
22. Ngo, Y.L.; Lau, C.H.; Chua, L.S. Review on rosmarinic acid extraction, fractionation and its anti-diabetic potential. *Food Chem. Toxicol.* **2018**, *121*, 687–700. [CrossRef]
23. Caruso, G.; Godos, J.; Privitera, A.; Lanza, G.; Castellano, S.; Chillemi, A.; Bruni, O.; Ferri, R.; Caraci, F.; Grosso, G. Phenolic Acids and Prevention of Cognitive Decline: Polyphenols with a Neuroprotective Role in Cognitive Disorders and Alzheimer's Disease. *Nutrients* **2022**, *14*, 819. [CrossRef]
24. Liang, C.; Ju, W.; Pei, S.; Tang, Y.; Xiao, Y. Pharmacological Activities and Synthesis of Esculetin and Its Derivatives: A Mini-Review. *Molecules* **2017**, *22*, 387. [CrossRef]
25. Casciaro, B.; Moros, M.; Rivera-Fernández, S.; Bellelli, A.; de la Fuente, J.M.; Mangoni, M.L. Gold-nanoparticles coated with the antimicrobial peptide esculentin-1a(1-21)NH2 as a reliable strategy for antipseudomonal drugs. *Acta Biomater.* **2017**, *47*, 170–181. [CrossRef]
26. Naveed, M.; Hejazi, V.; Abbas, M.; Kamboh, A.A.; Khan, G.J.; Shumzaid, M.; Ahmad, F.; Babazadeh, D.; FangFang, X.; Modarresi-Ghazani, F.; et al. Chlorogenic acid (CGA): A pharmacological review and call for further research. *Biomed. Pharmacother.* **2018**, *97*, 67–74. [CrossRef]

27. Alsaraf, S.; Hadi, Z.; Akhtar, M.J.; Khan, S.A. Chemical profiling, cytotoxic and antioxidant activity of volatile oil isolated from the mint (*Mentha spicata* L.,) grown in Oman. *Biocatal. Agric. Biotechnol.* **2021**, *34*, 102034. [CrossRef]
28. Yousuf, T.; Akter, R.; Ahmed, J.; Mazumdar, S.; Talukder, D.; Nandi, N.C.; Nurulamin, M. Evaluation of acute oral toxicity, cytotoxicity, antidepressant and antioxidant activities of Japanese mint (*Mentha arvensis* L.) oil. *Phytomedicine Plus* **2021**, *1*, 100140. [CrossRef]
29. Bardaweel, S.K.; Bakchiche, B.; Alsalamat, H.A.; Rezzoug, M.; Gherib, A.; Flamini, G. Chemical composition, antioxidant, antimicrobial and Antiproliferative activities of essential oil of *Mentha spicata* L. (Lamiaceae) from Algerian Saharan atlas. *BMC Complementary Altern. Med.* **2018**, *18*, 201. [CrossRef]
30. Golparvar, A.R.; Hadipanah, A.; Mehrabi, A.M. Diversity in chemical composition from two ecotypes of (*Mentha longifolia* L.) and (*Mentha spicata* L.) in Iran climatic conditions. *J. Biodivers. Environ. Sci.* **2015**, *6*, 26–33.
31. Devi, A.; Sharma, G. Morphological, phenological and cytological comparison of Mentha longifolia and M. spicata from sub-tropical and temperate regions of Jammu province (J&K). *Vegetos* **2022**, *35*, 179–187. [CrossRef]
32. López-Hernández, F.; Cortés, A.J. Whole Transcriptome Sequencing Unveils the Genomic Determinants of Putative Somaclonal Variation in Mint (*Mentha* L.). *Mol. Sci.* **2022**, *23*, 5291. [CrossRef] [PubMed]
33. Martínez-Tomé, M.; Jiménez-Monreal, A.M.; García-Jiménez, L.; Almela, L.; García-Diz, L.; Mariscal-Arcas, M.; Murcia, M.A. Assessment of antimicrobial activity of coffee brewed in three different ways from different origins. *Eur. Food Res. Technol.* **2011**, *233*, 497. [CrossRef]
34. Farjana, A.; Zerin, N.; Kabir, M.S. Antimicrobial activity of medicinal plant leaf extracts against pathogenic bacteria. *Asian Pac. J. Trop. Dis.* **2014**, *4*, S920–S923. [CrossRef]
35. Suárez-Quiroz, M.L.; Taillefer, W.; López Méndez, E.M.; González-Ríos, O.; Villeneuve, P.; Figueroa-Espinoza, M.-C. Antibacterial Activity and Antifungal and Anti-Mycotoxigenic Activities Against A spergillus flavus and A. ochraceus of Green Coffee Chlorogenic Acids and Dodecyl Chlorogenates. *J. Food Saf.* **2013**, *33*, 360–368. [CrossRef]
36. Chaves-Ulate, E.; Esquivel-Rodríguez, P. Ácidos clorogénicos presentes en el café: Capacidad antimicrobiana y antioxidante. *Agron. Mesoam.* **2019**, *30*, 299–311. [CrossRef]
37. Serino, E.; Chahardoli, A.; Badolati, N.; Sirignano, C.; Jalilian, F.; Mojarrab, M.; Farhangi, Z.; Rigano, D.; Stornaiuolo, M.; Shokoohinia, Y.; et al. Salvigenin, a Trimethoxylated Flavone from Achillea Wilhelmsii C. Koch, Exerts Combined Lipid-Lowering and Mitochondrial Stimulatory Effects. *Antioxidants* **2021**, *10*, 1042. [CrossRef]
38. Zhang, L.; Xie, Q.; Li, X. Esculetin: A review of its pharmacology and pharmacokinetics. *Phytother. Res.* **2022**, *36*, 279–298. [CrossRef]
39. Wang, X.; Du, J.; Zhou, J. Antibiotic activities of extracts from Prunus mume fruit against food-borne bacterial pathogens and its active components. *Ind. Crops Prod.* **2019**, *133*, 409–413. [CrossRef]
40. Jiménez, N.; Carrillo-Hormaza, L.; Pujol, A.; Álzate, F.; Osorio, E.; Lara-Guzman, O. Antioxidant capacity and phenolic content of commonly used anti-inflammatory medicinal plants in Colombia. *Ind. Crops Prod.* **2015**, *70*, 272–279. [CrossRef]
41. Park, Y.J.; Baek, S.-A.; Choi, Y.; Kim, J.K.; Park, S.U. Metabolic Profiling of Nine Mentha Species and Prediction of Their Antioxidant Properties Using Chemometrics. *Molecules* **2019**, *24*, 258. [CrossRef] [PubMed]
42. Scherer, R.; Lemos, M.F.; Lemos, M.F.; Martinelli, G.C.; Martins, J.D.L.; da Silva, A.G. Antioxidant and antibacterial activities and composition of Brazilian spearmint (*Mentha spicata* L.). *Ind. Crops Prod.* **2013**, *50*, 408–413. [CrossRef]
43. Fatiha, B.; Didier, H.; Naima, G.; Khodir, M.; Martin, K.; Léocadie, K.; Caroline, S.; Mohamed, C.; Pierre, D. Phenolic composition, in vitro antioxidant effects and tyrosinase inhibitory activity of three Algerian Mentha species: *M. spicata* (L.), *M. pulegium* (L.) and *M. rotundifolia* (L.) Huds (*Lamiaceae*). *Ind. Crops Prod.* **2015**, *74*, 722–730. [CrossRef]
44. Sõukand, R.; Pieroni, A.; Biró, M.; Dénes, A.; Dogan, Y.; Hajdari, A.; Kalle, R.; Reade, B.; Mustafa, B.; Nedelcheva, A.; et al. An ethnobotanical perspective on traditional fermented plant foods and beverages in Eastern Europe. *J. Ethnopharmacol.* **2015**, *170*, 284–296. [CrossRef]
45. Bhatia, H.; Sharma, Y.P.; Manhas, R.K.; Kumar, K. Ethnomedicinal plants used by the villagers of district Udhampur, J&K, India. *J. Ethnopharmacol.* **2014**, *151*, 1005–1018. [CrossRef]
46. Biswas, A.K.; Chatli, M.K.; Sahoo, J. Antioxidant potential of curry (*Murraya koenigii* L.) and mint (*Mentha spicata*) leaf extracts and their effect on colour and oxidative stability of raw ground pork meat during refrigeration storage. *Food Chem.* **2012**, *133*, 467–472. [CrossRef]
47. Sierra, K.; Naranjo, L.; Carrillo-Hormaza, L.; Franco, G.; Osorio, E. Spearmint (*Mentha spicata* L.) Phytochemical Profile: Impact of Pre/Post-Harvest Processing and Extractive Recovery. *Molecules* **2022**, *27*, 2243. [CrossRef]
48. Agudelo, C.; Bravo, K.; Ramírez-Atehortúa, A.; Torres, D.; Carrillo-Hormaza, L.; Osorio, E. Chemical and Skincare Property Characterization of the Main Cocoa Byproducts: Extraction Optimization by RSM Approach for Development of Sustainable Ingredients. *Molecules* **2021**, *26*, 7429. [CrossRef]
49. Brand-Williams, W.; Cuvelier, M.E.; Berset, C. Use of a free radical method to evaluate antioxidant activity. *LWT Food Sci. Technol.* **1995**, *28*, 25–30. [CrossRef]
50. Benzie, I.F.F.; Strain, J.J. The Ferric Reducing Ability of Plasma (FRAP) as a Measure of "Antioxidant Power": The FRAP Assay. *Anal. Biochem.* **1996**, *239*, 70–76. [CrossRef]
51. Mesa-Vanegas, A.M.A.; Zapata-Uribe, S.n.; Arana, L.M.; Zapata, I.C.; Monsalve, Z.; Rojano, B.J. Actividad antioxidante de extractos de diferente polaridad de Ageratum conyzoides L. *Chem. Lat. Am. Caribb. Bull. Med. Aromat. Plants* **2015**, *14*, 1–10.

52. Balouiri, M.; Sadiki, M.; Ibnsouda, S.K. Methods for in vitro evaluating antimicrobial activity: A review. *J. Pharm. Anal.* **2016**, *6*, 71–79. [CrossRef] [PubMed]
53. Tsugawa, H.; Cajka, T.; Kind, T.; Ma, Y.; Higgins, B.; Ikeda, K.; Kanazawa, M.; VanderGheynst, J.; Fiehn, O.; Arita, M. MS-DIAL: Data-independent MS/MS deconvolution for comprehensive metabolome analysis. *Nat. Methods* **2015**, *12*, 523–526. [CrossRef] [PubMed]
54. Chong, J.; Xia, J. MetaboAnalystR: An R package for flexible and reproducible analysis of metabolomics data. *Bioinformatics* **2018**, *34*, 4313–4314. [CrossRef]
55. Sofina, E.V. Agricultural land-use optimization by farms based on quality management: Lines of research. *Int. J. Qual. Res.* **2019**, *13*, 915. [CrossRef]

molecules

MDPI

Article

Upgrading Common Wheat Pasta by Fiber-Rich Fraction of Potato Peel Byproduct at Different Particle Sizes: Effects on Physicochemical, Thermal, and Sensory Properties

Mohammad Namir [1], Ali Iskander [1,2], Amal Alyamani [3], Eman T. Abou Sayed-Ahmed [1], Ahmed M. Saad [4,*], Kamal Elsahy [1], Khaled A. El-Tarabily [5,*] and Carlos Adam Conte-Junior [6]

1 Food Science Department, Faculty of Agriculture, Zagazig University, Zagazig 44511, Egypt; nomairmohamed@gmail.com (M.N.); iskanderali@yahoo.com (A.I.); emansayed@yahoo.com (E.T.A.S.-A.); kamalsahy@gmail.com (K.E.)
2 General Organization for Export and Import Control, Ministry of Trade and Industry, Garden City, Cairo 11519, Egypt
3 Department of Biotechnology, Faculty of Sciences, Taif University, P.O. Box 11099, Taif 21944, Saudi Arabia; a.yamani@tu.edu.sa
4 Biochemistry Department, Faculty of Agriculture, Zagazig University, Zagazig 44511, Egypt
5 Department of Biology, College of Science, United Arab Emirates University, Al Ain 15551, United Arab Emirates
6 Center for Food Analysis (NAL), Technological Development Support Laboratory (LADETEC), Federal University of Rio de Janeiro (UFRJ), Cidade Universitária, Rio de Janeiro 21941-598, Brazil; conte@iq.ufrj.br
* Correspondence: ahmedm4187@gmail.com (A.M.S.); ktarabily@uaeu.ac.ae (K.A.E.-T.)

Abstract: Fiber-enriched food has numerous health benefits. This study develops functional fiber-enriched pasta (FEP) by partially substituting wheat flour for alcohol-insoluble residue prepared from potato peel byproducts (AIR-PPB) at various particle sizes (PS). The independent variables' effects, AIR-PPB at 2–15% substitution levels, and PS 40–250 μm were investigated in terms of chemical, cooking, thermal, and sensory properties. AIR-PPB is rich in total dietary fibers (TDF) (83%), exhibiting high water-holding capacity (WHC) and vibrant colors. Different concentrations of AIR-PPB increase TDF content in FEPs by 7–21 times compared to the control pasta (CP). Although the optimal cooking time (OCT) decreases by 15–18% compared to CP, where a lower OCT should reduce cooking time and save energy, cooking loss (Cl) increases slightly but remains within an acceptable range of 8%. Additionally, AIR-PPB altered the texture properties of FEP, with a moderate decrease in mass increase index (MII), firmness, and stickiness. AIR-PPB impairs the gluten network's structure in pasta due to AIR-PPB's WHC, which competes with starch for water binding, increasing the starch gelatinization temperature. FEPs show an increased lightness and yellowness and improved sensory properties. Highly acceptable FEPs were obtained for the following substitution levels: FEP11 (AIR-PPB at 2% and PS of 145 μm), FEP9 (AIR-PPB 4% level with PS of 70 μm), FEP6 (AIR-PPB of 4% level with 219 μm PS), and FEP1 (AIR-PPB = 8.5% with 40 μm PS), as compared to other FEPs.

Keywords: byproducts; enriched pasta; potato peel; physical properties; sensory traits

Citation: Namir, M.; Iskander, A.; Alyamani, A.; Sayed-Ahmed, E.T.A.; Saad, A.M.; Elsahy, K.; El-Tarabily, K.A.; Conte-Junior, C.A. Upgrading Common Wheat Pasta by Fiber-Rich Fraction of Potato Peel Byproduct at Different Particle Sizes: Effects on Physicochemical, Thermal, and Sensory Properties. *Molecules* **2022**, 27, 2868. https://doi.org/10.3390/molecules27092868

Academic Editor: Nazimah Hamid

Received: 14 March 2022
Accepted: 26 April 2022
Published: 30 April 2022

1. Introduction

In recent decades, there has been a surge interest in studying the relationship between food and health. Numerous studies have examined the effects of supplementing the diet with bioactive compounds [1,2], particularly dietary fibers [3]. According to AACC [4], dietary fibers are a mixture of indigestible carbohydrates, i.e., polymers of oligo- and polysaccharides from the edible remnants of plants that resist digestion in the small intestine but are completely or partially fermented in the large intestine. Dietary fibers have been shown to have various health benefits, including improving intestinal health [5] and lowering the risk of chronic diseases such as heart disease and type 2 diabetes, high blood cholesterol, insulin resistance, obesity, and cancer [6,7]. In this regard, the recommended daily intake of dietary fibers for women is 25 g and for men is 38 g [8]. On a food technology scale, dietary fibers have a variety of functional properties that affect food quality, including water and oil holding capacity, solubility, viscosity, gel-forming ability, and swelling capacity [9]. However, dietary fibers substitution levels in food formulations are still limited due to the undesirable quality characteristics such as color, texture, and taste associated with physicochemical changes caused by fiber addition [9].

Bioactive components are abundant in industrial and agricultural wastes [10–12]. Furthermore, dietary fibers are derived in large part from food processing wastes. Potato (*Solanum tuberosum*) has a global production of approximately 368 million tons per year [13,14], generating a massive amount of processing wastes, equal to about 15–40% of the fresh potato and posing a potential environmental risk [15]. Most potato processing wastes consist of peels, which contain significant amounts of antioxidants and dietary fibers [16–18]. Additionally, several findings have shown the relationship between the quality of bakery products and the addition of potato peel fibers. The incorporation of fiber fraction into bread (0.4 g fibers/100 g flour) reduced the hardness of the bread over seven-days storage period compared to the control formulation [19]. The addition of potato peel flour with a high level of dietary fibers and protein to cakes significantly increased dietary fibers content. It decreased the cake's hardness by 30.24%, but it had no effect on sensory traits compared to the control [20].

Pasta is typically made with an unleavened dough of wheat flour and water or eggs, formed into sheets or other shapes and cooked by boiling or baking. Pasta is broadly classified into two types: dried and fresh. Most dried pasta is commercially produced through an extrusion process [21,22]. Globally, the demand for pasta continues to grow due to its cost-effectiveness, ease of preparation, long shelf life, low glycemic index, low sodium and fat content, and high complex carbohydrates content [23,24]. According to the International Pasta Organization (IPO), global pasta production was estimated to be 14.3 million tons in 2019. As a result, pasta can serve as a vehicle for bioactive ingredients such as dietary fibers, antioxidants, omega-3 fatty acids, and protein, contributing to human health maintenance [25,26]. Numerous studies have examined the use of whole wheat flour rather than semolina to increase the nutritional value of pasta by increasing the proportion of active compounds such as dietary fibers, antioxidants, vitamins, and mineral salts found in the bran that are removed during the durum wheat milling process [26–31]. Incorporating bran as a dietary fiber source into wheat flour reduces dough development and strengthens the dough by disrupting the gluten network [32].

Additionally, the bran-enriched pasta exhibited poor cooking characteristics and a firm texture. The more fiber incorporation, the lower the sensory properties [23,33]. Thus, several trials have been conducted to enrich wheat flour with various ingredients, including oregano and carrot leaves [34] and other plant parts [35,36], white bean, split yellow pea, lentil [24], and chickpea flour, as well as protein isolates [37–39], quinoa flour [30], carob flour [28], and bambara groundnut flour [40]. Therefore, the challenge for food producers is to determine the most effective way to fortify pasta with dietary fibers while maintaining its quality properties.

This study aims to create functional fiber-enriched pasta (FEP) by partial substitution of wheat flour with potato peel byproducts (PPB) with different particle sizes (PS). The prepared pasta was evaluated via its proximate composition, cooking characteristics, thermal and color properties, and sensory analysis. Response surface methodology (RSM) was used to optimize the experimental processing variables for desired FEP quality attributes.

2. Materials and Methods

2.1. Materials

Potato (*Solanum tuberosum* L.) cv Spunta peel byproduct (PPB) was kindly provided by Farm Frites Factory (10th of Ramadan industrial city, Eastern Province, Egypt). Wheat flour (72% extraction rate) was purchased from a local market in Zagazig City, Zagazig, Egypt. Wheat flour contained 12.00 g/100 g, ash 0.68 g/100 g, moisture 13.41 g/100 g, total dietary fibers (TDF) 6.48 g/100 g. Following the peeling procedure, PPB samples were collected, washed with distilled water to remove impurities, and dried for 24 h in a hot air oven at 45 ± 2 °C. The dried PPB had a moisture content of 7.45 ± 0.66 g/100 g. We milled and sieved the dried samples using standard sieves. PPB fractions with PS of 40, 70, 145, 219, and 250 mm were obtained and vacuum-packed into separate plastic bags, where they were stored at 42 °C until further analysis.

2.2. Preparation of Alcohol Insoluble Residue from Potato Peel Byproduct (AIR-PPB)

The AIR-PPB was prepared as described by Latorre et al. [41] with some modifications. For 30 min, 10 g of PPB samples with specified PS was stirred in 50 mL boiling ethanol (70%, *v*/*v*). The suspension was centrifuged at 3000 rpm/10 min, and the supernatant was decanted. Subsequently, the residues were collected and extracted twice for another 30 min with boiling ethanol (95%, *v*/*v*). The AIR-PPB was filtered through Whatman No.1 filter paper, washed with acetone, and the solvents were exchanged several times before being air-dried at 30 °C.

2.3. Experimental Design and Statistical Analysis

In order to establish an appropriate statistical model for the current study, several trials were conducted to determine the maximum and minimum amounts of AIR-PPB in pasta formulations containing various PS. Statgraphics Plus for Windows (version 4.1, Centurion, XV, USA) was used to create polynomial models. A central composite rotatable design (CCRD) was used to design the test with two independent variables (AIR-PPB 2–15 g/100 g and PS 40–250 mm) coded at five levels (−1.41, −1, 0, +1, and +1.41). To demonstrate the model's applicability, the CCRD included 12 experiments with three replicates for the central point (Table 1). The experiments were conducted at random to mitigate the effects of unanticipated variability introduced by external sources. A second-order polynomial equation was used to investigate the relationship between the responses and the independent variables.

$$Y = b_0 + b_{1\times 1} + b_2X_2 + b_{11}(X_1)^2 + b_{12}X_1\,X_2 + b_{22}(X_2) \quad (1)$$

where Y response value and X_1 and X_2 denote independent variables. b_0 is a constant, b_1 and b_2 are linear coefficients, b_{11} and b_{22} are quadratic coefficients, and b_{12} is a quadratic interaction coefficient.

Table 1. Experimental design values for fiber-enriched pasta (FEP) with different combinations of alcohol insoluble residue from potato peel byproduct (AIR-PPB) (X_1) and particle size PS (X_2).

Experiments	Independent Variables	
	Coded/Real Values	
	X_1, AIR-PPB (g/100 g Wheat Flour)	X_2, PS (μm)
FEP-1	0 (8.50)	−1.41 (40)
FEP-2 (Central point)	0 (8.50)	0 (145)
FEP-3 (Central point)	0 (8.50)	0 (145)
FEP-4	+1 (13)	+1 (219)
FEP-5	0 (8.50)	+1.41 (250)
FEP-6	−1 (4)	+1 (219)
FEP-7	+1.41 (15)	0 (145)
FEP-8 (Central point)	0 (8.50)	0 (145)
FEP-9	−1 (4)	−1 (70)
FEP-10	+1 (13)	−1 (70)
FEP-11	−1.41 (2)	0 (145)
Control pasta (CP)	0	-

The coefficient of determination (R^2) and adjusted R^2 were used to assess the fit of the model, whereas analysis of variance was used to determine the model's significance ($p < 0.05$) (ANOVA). The correlation coefficient (r) was used to determine whether there was a positive (+1) or negative (−1) correlation between responses. Excel 2010 was used to calculate the means and standard deviations for the obtained results. To determine $p < 0.05$ statistically significant differences between parameters, a one-way ANOVA with Duncan's test ($p < 0.05$) was used. The experimental processing variables were then optimized for the desired fiber-enriched pasta (FEP) quality attributes, which included a high dietary fiber content, a high sensory properties score, a shorter optimum cooking time (OCT), a lower cooking loss (Cl), and the maximum desirable color changes.

2.4. Preparation of FEP and Control Pasta (CP)

The experimental FEPs are primarily composed of whole wheat flour and AIR-PPB. Firstly, a Moulinex Hm4121 stand mixer (Moulinex, France) was used to combine whole wheat flour and AIR-PPB. The mixture was then hydrated by adding warm water (30 °C), which calculated to achieve a moisture content of up to 30%, and mixed at a slower speed (120 rpm) for 12 min. Finally, the mixture was fed into a pasta extruder and extruded using strand forming technology (diameter 2 mm). For 2 h, the pasta strands were dried at 80 °C. The dried FEP samples were sealed in a zip-lock plastic bag and stored at 22 ± 2 °C until further analysis. The CP was prepared (1 kg wheat, 300 mL water), except for the amount of AIR-PPB.

2.5. Proximate composition

Moisture, protein, fat, and ash contents were estimated as described in AOAC [42]. TDF, soluble (SDF), and insoluble (IDF) dietary fibers were determined using McCleary enzymatic and gravimetric methods, McCleary et al. [43]. Differential analysis was used to determine the solubility of carbohydrates, and each analysis was conducted in triplicates.

2.6. Functional Properties

The water holding capacity (WHC) was determined as described by Namir et al. [44]. In weighted test tubes, 1 g of AIR-PPB was homogenized in 10 mL of distilled water, stirred for 30 min, and then centrifuged at 6000× *g* for 30 min. The supernatant was discarded, and the residues were weighed. WHC was calculated as mL of retained water/g of sample.

2.7. Physical Characteristics

2.7.1. Cooking Properties

OCT, min was determined in accordance with the method described by Sobota et al. [45]. The mass increase index (MII g/g) was determined by dividing the weight of cooked pasta by the weight of uncooked pasta [46]. In contrast, the Cl % was determined by measuring the solids content of the cooking water using the AACC 66-50 method [4].

2.7.2. Texture Profile Analysis

TA.XT*plus*C texture analyzer (Stable Micro Systems, Godalming, UK) was used to determine the texture. Samples were cooked for the OCT and then immersed in cold water to prevent the texture from being affected by continuous cooking heat. The pasta was drained and then placed in the center of the texture analyzer platform, where it was cut at a speed of 2 mm/s and deformation of 90% with a pasta blade (thickness = 1 mm). Firmness (*N*) was defined as the maximum force required to cut a single cooked pasta, whereas stickiness (*N*) was defined as the maximum force required to separate the probe from the sample surface (*N*). For each trial, the mean of ten replicates was calculated for each trial.

2.7.3. Differential Scanning Calorimeter (DSC)

The differential scanning calorimeter was used to determine the profile of starch gelatinization (Malvern Panalytical Ltd., Malvern, UK). Then, approximately 3 mg of dried flour was placed in an aluminum pan, followed by the addition of 1 mL of distilled water via micropipette in the DSC pan. The pan was sealed and left at room temperature condition for 12 h. The thermograms were taken at a rate of 5 °C/min between 25 and 120 °C. The thermogram was used to determine the starch gelatinization temperature, the onset temperature (T_o), the peak or melting temperature (T_p), the conclusion temperature (T_c), and the enthalpy (H) values from data recording software.

2.8. Color Measurement and Sensory Assessment

A Color Flex EZ spectrophotometer (HunterLab, Murnau, Germany) was used to determine the color of FEP. The following parameters were taken into account: Equation (2) calculated the L^*, a^*, b^*, and color difference (ΔE) between control pasta (CP) and FEP. The values given are the averages of three independent samples [47].

$$\Delta \text{E} = \sqrt{(\Delta L)^2 + (\Delta \text{a})^2 + (\Delta \text{b})^2} \tag{2}$$

The sensory characteristics of FEP were evaluated using a nine-point hedonic scale (1 = strongly dislike extremely, 5 = neither like nor dislike, and 9 = like extremely). After cooking the pasta to its OCT, it was drained and rinsed. Pasta samples were randomly coded and served on white foam plates to 25 panelists, 15 males and 10 females (aged 21–30), in a fluorescent-lit laboratory with an air-conditioning temperature of 23 ± 1 °C. Water was used to rinse the panelists' mouths between samples to avoid interfering with the results.

3. Results and Discussion

3.1. Chemical Composition and AIR-PPB Characteristics

The chemical composition, functional and color properties of AIR-PPB with various PS (40–250 μm) are listed in Table 2. AIR-PPB comprise 63.87% of the dry weight of PPB. Furthermore, TDF was the most abundant component of AIR-PPB, ranging from 82.73 to 83.02 g/100 g depending on PS, the TDF content increasing with PS increments. The TDF content of AIR-PPB was significantly higher than that of citrus peels (57 g/100 g), mango peel (51.2 g/100 g), pomegranate peel (17.53 g/100 g), and grape pomace skins (67.95 g/100 g) [48–51] but was comparable to that of passion fruit byproduct, apple, and date pomace (81.50, 82.00, and 83.70 g/100 g, respectively) [52,53]. Insoluble dietary fibers (IDF) comprise 61.39–63.10% of TDF in AIR-PPB depending on PS (40–250 μm); in contrast, soluble dietary fibers (SDF) comprise 19.92–21.34% of TDF in AIR-PPB.

The SDF content was increased in small PS due to fiber structure breakdown. Furthermore, orange byproducts, apple pomace, date pomace, and pineapple peel all demonstrated similar properties [52,54,55]. Moreover, there have been no studies conducted on the alcohol extraction of PPB. In general, fruit and vegetable byproducts are high in dietary fiber and soluble sugars but low in fat and protein [56], and the dietary fibers were combined with low protein, low fat, and soluble carbohydrates in this extraction method. Therefore, these components were identified in AIR-PPB, and their content increased as PS increased, excluding soluble carbohydrates, which exhibited the opposite trend.

The WHC of AIR-PPB was 6.73–7.75 mL/g, compared to 5.70, 4.90, and 4.12 mL/g for the date, pear, and tomato pomaces, respectively. Additionally, the WHC of the cantaloupe byproduct was 6.17 mL/g [47,57,58]. An almost identical value (7.5 mL/g) was observed in apple pomace [58]. The hydroxyl groups in polysaccharide chains may form hydrogen bonds with water, thereby increasing the water-holding capacity of fiber-rich materials [59]. The lower WHC values in AIR-PPB with smaller PS can be attributed to the degradation of polysaccharides chains, which increases the soluble fibers fraction; these chains are collapsed by centrifugation during WHC determination.

Table 2. Proximate composition, water holding capacity (WHC), and color attributes of alcohol insoluble residue from potato peel byproduct (AIR-PPB) at different particle sizes (PS).

Composition•(g/100 g)	Particles' Size (μm)				
	250	219	145	70	40
			Chemical		
Moisture (w.b) *	5.78 ± 0.03 a	5.67 ± 0.06 b	5.61 ± 0.00 bc	5.62 ± 0.04 bc	5.55 ± 0.01 c
Protein	5.46 ± 0.13 a	5.40 ± 0.18 a	5.38 ± 0.09 ab	5.26 ± 0.04 c	5.29 ± 0.21 bc
Fat	0.54 ± 0.01 a	0.50 ± 0.00 ab	0.51 ± 0.00 ab	0.45 ± 0.01 ab	0.42 ± 0.00 b
Ash	4.71 ± 0.23 a	4.69 ± 0.09 ab	4.61 ± 0.13 bc	4.64 ± 0.17 ab	4.53 ± 0.26 c
TDF••	83.02 ± 0.45 a	82.91 ± 0.31 b	82.87 ± 0.17 c	82.81 ± 0.25 cd	82.73 ± 0.21 d
IDF	63.10 ± 0.51 a	62.54 ± 0.35 b	62.15 ± 0.43 c	61.86 ± 0.41 d	61.39 ± 0.21 e
SDF	19.92 ± 0.11 e	20.37 ± 0.10 d	20.72 ± 0.09 c	20.95 ± 0.13 b	21.34 ± 0.24 a
SC•••	6.27 ± 0.31 d	6.50 ± 0.26 d	6.63 ± 0.41 c	6.84 ± 0.09 b	7.03 ± 0.81 a
			Physical		
WHC (mL/g)	7.75 ± 0.19 a	7.61 ± 0.88 b	7.21 ± 0.34 c	7.04 ± 0.56 d	6.73 ± 0.61 e
L^*	64.22 ± 0.83 e	66.14 ± 0.00 d	69.79 ± 0.63 c	74.34 ± 0.97 b	75.12 ± 0.53 a
a^*	15.52 ± 0.68 a	11.65 ± 0.56 b	9.65 ± 0.00 c	8.32 ± 0.66 d	7.24 ± 0.74 e
b^*	48.26 ± 0.28 e	51.76 ± 0.00 d	55.45 ± 0.21 c	66.31 ± 0.20 b	68.81 ± 0.37 a

*•: wet basis, •• TDF: total dietary fibers, IDF: insoluble dietary fibers, SDF: soluble dietary fibers, ••• SC: soluble carbohydrates. Values are expressed as the mean ± SD of three replicates. Means within a row with different superscript letters are significantly different at the level of $p < 0.05$. L^*: whiteness, a^*: redness, b^*: yellowness.

Table 2 demonstrates a significant increase in the values of lightness and yellowness in AIR-PPB with lower PS but a significant decrease in redness values. Thus, the color of the AIR-PPB changed from light brown to dark orange when PS was reduced from 250 to 40 mm, which will have a beneficial effect on the color of the FEP, as will be demonstrated later.

Industrial byproducts can differ considerably in their chemical composition. Additionally, it is important to consider that certain byproducts require special treatments and processing prior to further utilization due to their chemical composition. The chemical composition and structure of these byproducts significantly impact their functional properties and their applications. Additionally, byproducts are naturally colored differently. Therefore, high levels of these byproducts may cause color changes in the extrudates, which may or may not be desired by consumers [60].

According to ANOVA results, the moisture, protein, fat, ash, and soluble carbohydrates contents of FEP significantly affected by the substitution levels of AIR-PPB and its PS ($p < 0.05$). Additionally, the PS had a significant ($p < 0.05$) lowering effect on the FEP's protein and fat contents. The moisture, protein, fat, ash, and soluble carbohydrate contents of all experimental FEP were significantly decreased ($p < 0.05$) compared to the CP. FEP had a slightly lower moisture content of about 4–7% compared to CP. Furthermore, FEP4 had the lowest moisture content (9.41%). Additionally, the protein content of FEP7 and FEP10 was reduced by approximately 7% compared to the CP. Although there is a slight decrease in fat content (7–17%) as compared to the CP (Table 3), FEP7, FEP10, and FEP4 had the lowest fat content (0.97, 0.98, and 0.99 g/100 g FEP, respectively) compared to CP (1.14 g/100 g CP) (Table 3). FEP4 had the lowest ash content of all FEPs (0.77 g/100 g) as compared to CP, which had an ash content of 1.28 g/100 g (Table 3). In terms of soluble carbohydrates, CP has the highest content at 85.96%, which was significantly decreased in all experimental FEP, reaching a minimum of 12% relative to control pasta in FEP4.

Table 3. Chemical composition of fiber-enriched pasta (FEP) (g/100 g).

Chemical Composition	CP	FEP-1	FEP-2,3,8	FEP-4	FEP-5	FEP-6	FEP-7	FEP-9	FEP-10	FEP-11
Moisture	10.08 ± 0.12 ^{a}	9.67 ± 0.00 ^{c}	9.59 ± 0.11 ^{f}	9.41 ± 0.32 ^{j}	9.54 ± 0.27 ^{g}	9.62 ± 0.78 ^{e}	9.45 ± 0.14 ^{i}	9.65 ± 0.00 ^{d}	9.49 ± 0.01 ^{h}	9.72 ± 0.02 ^{b}
Protein	11.04 ± 0.27 ^{a}	10.48 ± 0.16 ^{g}	10.56 ± 0.13 ^{f}	10.32 ± 0.09 ^{h}	10.59 ± 0.12 ^{e}	10.81 ± 0.40 ^{c}	10.23 ± 0.10 ^{i}	10.79 ± 0.17 ^{d}	10.25 ± 0.22 ^{i}	10.89 ± 0.30 ^{b}
Fat	1.14 ± 0.10 ^{a}	1.01 ± 0.01 ^{d}	1.02 ± 0.04 ^{d}	0.99 ± 0.07 ^{e}	1.04 ± 0.03 ^{c}	1.05 ± 0.02 ^{c}	0.97 ± 0.00 ^{e}	1.04 ± 0.02 ^{c}	0.98 ± 0.00 ^{e}	1.07 ± 0.06 ^{b}
TDF	0.58 ± 0.09 ^{j}	7.80 ± 0.15 ^{f}	8.08 ± 0.19 ^{e}	12.07 ± 0.08 ^{a}	9.45 ± 0.12 ^{d}	5.64 ± 0.26 ^{g}	11.88 ± 0.32 ^{b}	5.55 ± 0.38 ^{h}	10.41 ± 0.29 ^{c}	4.12 ± 0.21 ^{i}
IDF	0.38 ± 0.02 ^{j}	5.54 ± 0.07 ^{f}	5.90 ± 0.13 ^{e}	9.17 ± 0.21 ^{a}	6.99 ± 0.16 ^{d}	3.95 ± 0.10 ^{g}	9.03 ± 0.14 ^{b}	3.83 ± 0.25 ^{h}	7.81 ± 0.27 ^{c}	2.76 ± 0.17 ^{i}
SDF	0.20 ± 0.00 ^{j}	2.26 ± 0.01 ^{e}	2.18 ± 0.00 ^{f}	2.90 ± 0.06 ^{a}	2.46 ± 0.01 ^{d}	1.69 ± 0.00 ^{h}	2.85 ± 0.00 ^{b}	1.72 ± 0.11 ^{g}	2.60 ± 0.00 ^{c}	1.36 ± 0.05 ^{i}
Ash	1.28 ± 0.13 ^{a}	1.09 ± 0.12 ^{e}	1.00 ± 0.17 ^{f}	0.77 ± 0.00 ^{j}	0.89 ± 0.00 ^{g}	1.13 ± 0.02 ^{d}	0.84 ± 0.00 ^{h}	1.19 ± 0.20 ^{c}	0.82 ± 0.00 ^{i}	1.22 ± 0.01 ^{b}
SC	85.96 ± 1.54 ^{a}	79.62 ± 0.81 ^{e}	79.33 ± 1.91 ^{f}	75.86 ± 1.45 ^{j}	78.03 ± 1.51 ^{g}	81.37 ± 1.20 ^{d}	76.07 ± 0.92 ^{i}	81.43 ± 0.81 ^{c}	77.53 ± 1.53 ^{h}	82.69 ± 0.40 ^{b}

Values are means ± SD. Means within a row with different superscript letters are significantly different ($p < 0.05$). FEP-2,3,8 is the center point, and their value is the average of the three central points. TDF: total dietary fibers; IDF: insoluble dietary fibers; SDF: soluble dietary fibers; SC: soluble carbohydrates, CP: control pasta.

The TDF content of FEP varies significantly ($p < 0.05$) between the experimental pastas. All FEPs contain more TDF than CP, which increased 7 to 21 times over CP (Table 3 and Figure 1A).

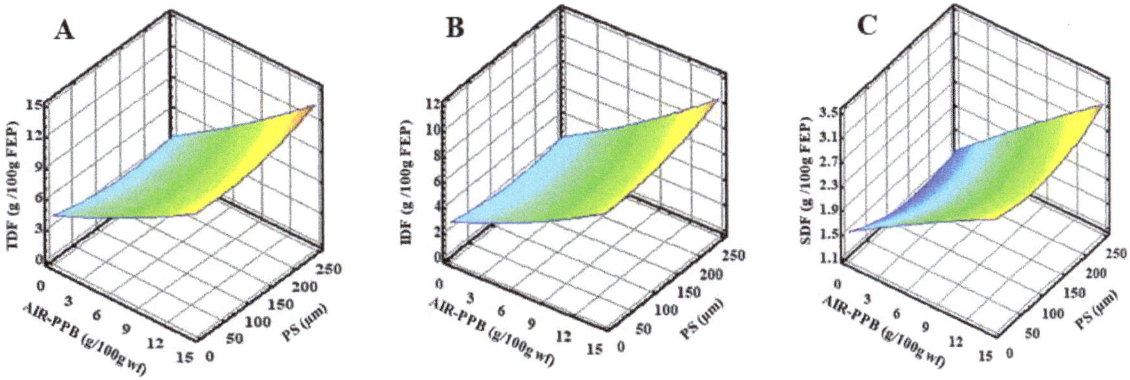

Figure 1. Response surface plots of the interaction of alcohol insoluble residue from potato peel byproduct (AIR-PPB) and particle size (PS) on (**A**) total dietary fibers (TDF) (g/100 g FEP), (**B**) insoluble dietary fibers (IDF) (g/100 g FEP), and (**C**) soluble dietary fibers (SDF) (g/100 g FEP) of FEP samples as compared to control pasta (CP) (0 AIR-PPB, 0 PS). FEP, fiber-enriched pasta.

Depending on the source of the dietary fiber, fruit and vegetable byproducts may have varying amounts of insoluble and soluble fiber, which can affect their level of inclusion in direct-expanded products [56].

In comparison to CP, FEP4 had the highest TDF content of all FEPs at 12.07%, followed by FEP7, FEP10, and FEP5. According to the European Commission Regulation [61], experimental FEPs containing more than 6 g of dietary fibers per 100 g are labeled as high in dietary fibers. IDF was the predominant fraction of dietary fibers in all experimental FEP, accounting for 67–76% of TDF when compared to CP (0.38 g/100 g CP) (Table 3). The highest IDF was observed in FEP4 (9.17 g/100 g FEP), which accounted for 75.97% of TDF (Table 3, Figure 1B). The IDF is composed of cellulose, insoluble hemicelluloses, and lignin, and it plays a critical role in human and animal health, particularly in digestive processes, by promoting intestinal peristalsis, increasing fecal volume, and adsorbing heavy metals, grease, and toxic substances, which it then quickly eliminates. Additionally, SDF refers to the fraction of fibers that dissolved in water during analysis [62,63]. SDF accounts for between 24 and 33% of TDF in all experimental FEP. FEP4 had the highest SDF content (2.90 g/100 g FEP) compared to CP (0.20 g/100 g CP) (Table 3 and Figure 1C).

Ritthiruangdej et al. [64] incorporated 30% unripe banana flour into the noodles, which increased the TDF and resistant starch content of noodles.

As illustrated in Table 3 and Figure 1, the second-order polynomial models for FEP characterization were significant ($p < 0.05$) and had a coefficient of determination (R^2) greater than 0.90, indicating that the regression model is fit and can be used to predict the responses. Due to the unique chemical composition of AIR-PPB, the partial replacement of wheat flour with AIR-PPB at certain PS resulted in significant changes ($p < 0.05$) in all measured chemical constituents of the experimental FEP as compared to CP.

3.2. Physical Properties

3.2.1. Cooking Properties

When FEPs are exposed to cooking during preparation, it is expected that heat will affect the physicochemical properties of FEP by affecting the structure of the gluten–starch network, which influences the mechanical properties of FEP. Therefore, it is critical to investigate the cooking properties of FEP in order to ascertain the effect of experimental factors on the FEP's quality. The OCT, CL, and MII were used to evaluate the cooking properties (Table 4, Figure 2A).

Table 4. Physical properties (cooking, texture, thermal) of fiber-enriched pasta (FEP).

Physical Properties	CP	FEP-1	FEP-2,3,8	FEP-4	FEP-5	FEP-6	FEP-7	FEP-9	FEP-10	FEP-11
Cooking										
OCT (min)	9.63 ± 1.21 ^{a}	8.05 ± 1.45 ^{e}	8.04 ± 0.64 ^{e}	7.90 ± 0.72 ^{i}	7.99 ± 0.56 ^{f}	8.07 ± 0.78 ^{d}	7.93 ± 0.68 ^{h}	8.09 ± 0.42 ^{c}	7.96 ± 0.81 ^{g}	8.14 ± 0.46 ^{b}
Cl (%)	4.27 ± 0.16 ^{i}	5.76 ± 0.10 ^{e}	5.83 ± 0.15 ^{d}	5.97 ± 0.06 ^{a}	5.88 ± 0.24 ^{c}	5.73 ± 0.77 ^{f}	5.92 ± 0.80 ^{b}	5.65 ± 0.81 ^{g}	5.91 ± 0.31 ^{b}	5.59 ± 0.88 ^{h}
MII (kgcp/kg up)	3.51 ± 0.86 ^{a}	2.79 ± 0.73 ^{e}	2.74 ± 0.89 ^{f}	2.55 ± 0.31 ^{i}	2.72 ± 0.10 ^{g}	2.82 ± 0.21 ^{d}	2.60 ± 0.00 ^{h}	2.98 ± 0.17 ^{c}	2.71 ± 0.30 ^{g}	3.08 ± 0.46 ^{b}
Texture										
Firmness (N)	2.69 ± 0.95 ^{a}	1.98 ± 0.36 ^{d}	1.91 ± 0.44 ^{f}	1.68 ± 0.24 ^{h}	1.92 ± 0.34 ^{ef}	2.01 ± 0.10 ^{c}	1.85 ± 0.21 ^{g}	2.16 ± 0.21 ^{b}	1.93 ± 0.67 ^{e}	2.18 ± 0.58 ^{b}
Stickiness (N)	4.73 ± 0.31 ^{a}	2.97 ± 0.28 ^{e}	2.95 ± 0.26 ^{f}	2.42 ± 0.30 ^{j}	2.88 ± 0.00 ^{g}	3.24 ± 0.93 ^{d}	2.45 ± 0.87 ^{i}	3.58 ± 0.80 ^{c}	2.83 ± 0.16 ^{h}	3.70 ± 0.01 ^{b}
Thermal properties										
To (°C)	59.44 ± 1.38 ^{j}	61.16 ± 1.09 ^{f}	61.99 ± 1.27 ^{e}	63.14 ± 1.52 ^{a}	62.14 ± 1.47 ^{d}	60.88 ± 1.22 ^{g}	62.54 ± 1.43 ^{b}	60.72 ± 1.29 ^{h}	62.33 ± 1.34 ^{c}	60.58 ± 1.71 ^{i}
Tp (°C)	63.97 ± 1.33 ^{j}	65.79 ± 1.61 ^{f}	66.78 ± 1.39 ^{e}	68.11 ± 1.23 ^{a}	66.96 ± 1.13 ^{d}	65.71 ± 1.82 ^{g}	67.47 ± 1.51 ^{b}	65.42 ± 1.57 ^{h}	67.18 ± 1.54 ^{c}	65.02 ± 1.69 ^{i}
Tc (°C)	66.54 ± 1.42 ^{j}	69.52 ± 1.37 ^{f}	70.58 ± 1.61 ^{e}	71.90 ± 1.82 ^{a}	70.72 ± 1.83 ^{d}	69.44 ± 1.76 ^{g}	71.26 ± 1.22 ^{b}	69.18 ± 1.31 ^{h}	70.96 ± 1.49 ^{c}	68.83 ± 1.71 ^{i}
H (J/g)	6.17 ± 0.98 ^{a}	5.19 ± 0.91 ^{d}	4.89 ± 0.43 ^{e}	3.54 ± 0.61 ^{i}	4.87 ± 0.28 ^{f}	5.21 ± 0.19 ^{d}	4.13 ± 0.75 ^{h}	5.28 ± 0.21 ^{c}	4.64 ± 0.19 ^{g}	5.49 ± 0.85 ^{b}

* Values are means ± SD. Means within a row with different superscript letters are significantly different ($p < 0.05$). FEP-2,3,8 is the center point, and its value is the average of the three central points. CP: control pasta, OCT: optimum cooking time (min), Cl: cooking loss (%), MII: mass increase index (kg cooked pasta/kg uncooked pasta), To: onset temperature (°C), Tp: melting temperature (°C), Tc: conclusion temperature (°C) and H: enthalpy value.

Table 4 demonstrates significant ($p < 0.05$) differences in OCT values between all experimental FEPs and CP, with FEPs showing a 15–18% decrease in OCT values when compared to CP. Additionally, the higher the level of AIR-PPB substitution at high PS, the shorter the OCT (Figure 2A). The CP had the longest OCT (9.63 min), which decreased significantly by 18% and yielded the lowest FEP4 value of 7.90 min. IDF generally distribute evenly throughout the starch matrix at low AIR-PPB concentrations, strengthen the starch matrix, and may result in increased expansion. However, at higher fiber concentrations, uniform distribution is not achieved, and fiber particles may disrupt cell walls, resulting in a reduction in expansion. Additionally, IDF may compete for water with starch, preventing the starch from fully gelatinizing, increasing the melt viscosity, increasing resistance to cell formation, and decreasing cooking time, which is beneficial because a lower OCT is desired to shorten cooking time and save energy. In contrast, SDF have no adverse effect on expansion; they may result in a slight increase in pasta volume expansion [56,65]. Previous research has corroborated these findings. Pasta made from bambara groundnut enriched fractionated whole grain wheat flour had a higher PS but a lower OCT than CP made from unfractionated flour, suggesting that fibers play a role in lowering the OCT [40]. Similarly, pasta made from olive pomace fortified durum wheat semolina had an OCT of 13.30 min for the control, which decreased to 12 min for pasta fortified with 10% olive pomace [66].

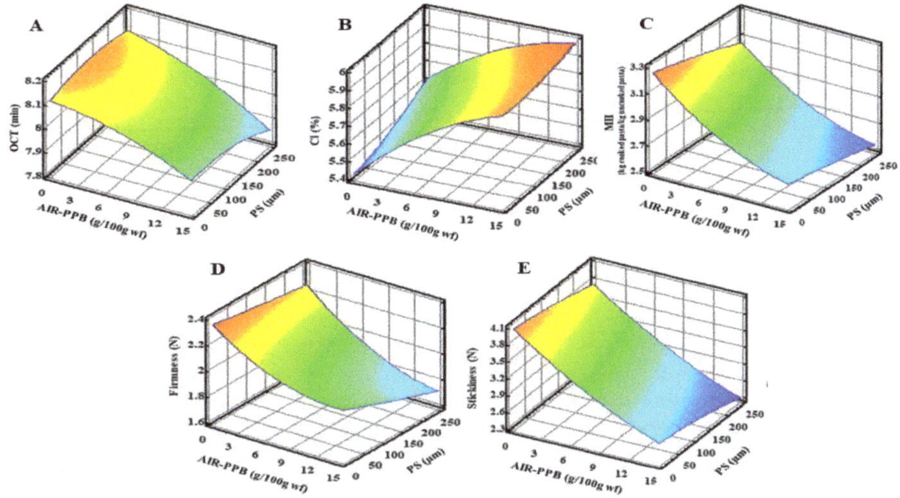

Figure 2. Response surface plots of the interaction of alcohol insoluble residue from potato peel byproduct (AIR-PPB) and Particle size (PS) on (**A**) optimum cooking time (OCT) (min), (**B**) cooking loss (Cl) (%), (**C**) mass increase index (MII) (kg cooked pasta/kg uncooked pasta), (**D**) Firmness (N), and (**E**) Stickiness (N) of fiber-enriched pasta (FEP) samples.

Furthermore, spaghetti made from durum semolina and various proportions of durum bran had a lower OCT compared to the CP [33]. Furthermore, Lončarić et al. [67] discovered that Fettuccine enriched with 10% apple peel powder had an OCT of 7.3 min, similar to the control. However, an increase above the 10% level resulted in a decrease in OCT.

One of the most critical quality indicators in pasta is Cl%, which is defined as solids remaining in the cooking water after drying that leach from the pasta during cooking. The greater the Cl concentration, the lower the quality of the FEP obtained. As shown in Table 4, the Cl values of FEPs were significantly ($p < 0.05$) higher than those of CP but within the acceptable cooking loss limit of ≤8% [68], which should not be exceeded by good-quality pasta. Increasing the AIR-PPB level with a high PS increased the Cl by 31–40% over the CP (Figure 2B). The highest Cl (5.97%) was observed in FEP4, whereas the lowest Cl (5.59%) was observed in FEP11 (4.27%). This result is explained by the interaction of AIR-PPB with the gluten–starch network, which weakens and disrupts its structure, impairing its ability to retain gelatinized starch leached from FEP into the cooking water, thereby increasing cooking loss. The addition of mango peel powder [69], carrot pomace [70], and orange dietary fiber [71] to pasta formulations revealed that at low concentrations (2.0–2.5%), these ingredients have no discernible effect on the Cl. These findings suggest that such concentrations may not significantly weaken the starch and gluten matrix. However, increasing the inclusions to 4% or 5% resulted in a significant increase in Cl in comparison to the control sample and in the treatments with a lower incorporation level.

Additionally, the results indicated a significant ($p < 0.05$) reduction in MII parameter as the concentrations of AIR-PPB and PS increased (Table 4, 6, and Figure 2C). MII was highest in CP (3.51 kg cooked pasta/kg uncooked pasta) but decreased by 12–27% in all FEPs. FEP11 had the highest MII (3.08 kg CP/kg uncooked pasta with a relative decrease of 12.45%, when compared to CP, whereas FEP4 had the lowest MII (2.55 kg cp/kg up) (Table 4, Figure 2C). The observed decrease in the MII for FEP is a result of competition for water absorption between fibers' free groups and starch, limiting the amount of available water for starch during cooking, decreasing water absorption, and consequently decreasing MII [33,72].

3.2.2. Texture

The presence of the fiber-rich material may disrupt and weaken the gluten matrix within the pasta microstructure, resulting in a loss of pasta firmness. In comparison to CP (2.69 N), the experimental FEP firmness values varied significantly ($p < 0.05$) from 1.68 N to 2.18 N (Table 4, Figure 2D). This revealed that increasing the level of AIR-PPB replacement with increasing PS decreased FEP firmness. Firmness was reduced to 1.68 N in FEP4 by 37.55% compared to CP (Table 4). FEP 11 and FEP 9 have the highest firmness values (2.18 N and 2.16 N, respectively) among all experimental FEPs (Table 4, Figure 2D).

In contrast, low post-cooking stickiness values indicate high-quality pasta. The 0.05 stickiness values of cooked FEP decreased significantly ($p < 0.05$) in parallel with the increase in the substitution level of AIR-PPB with high PS. The stickiness values decreased significantly ($p < 0.05$) in all FEPs compared to the CP, ranging from 22 to 49%. FEP11 had the highest stickiness (3.70 N), whereas FEP4 had the lowest (2.42 N) (Table 4, Figure 2E). The decrease in the stickiness values of FEPs can be explained by the decline in the total starch content, which is attributed to stickiness properties and is replaced by fibers in the experimental pasta. The addition of pomaces such as carrot pomace, apple peel powder, and flaxseed cake to pasta decreases its firmness/hardness [67,73]. This decreased firmness/hardness results from byproduct components that inhibit the formation of a strong gluten network [70].

The firmness of ziti-cut pasta containing 2% carrot pomace was significantly reduced from 5.94 N to 2.88 N [70]. However, increasing the number of byproducts in steps of 2% to 10% did not result in a further significant change.

3.2.3. Thermal Properties of FEP

When starch granules are heated (>60 °C) in water, they undergo a transition from an ordered to a disordered state, absorbing water, swelling, and losing crystallinity and amylopectin double-helical order in the process of gelatinization. In the present study, the thermal properties of FEP were investigated using differential scanning calorimetry (DSC), a technique that determines the gelatinization transition temperatures and enthalpy of gelatinization, both of which are affected by the molecular structure or crystallinity.

Table 4 and Figure 3 illustrate the effect of AIR-PPB and PS on the thermal transition parameters of FEP. All experimental FEP demonstrated an endothermic transition between 60 and 71 °C, with well-defined transition temperatures (To, Tp, and Tc). These results indicated that by increasing the replacement levels of AIR-PPB and PS in FEP, the gelatinization transition temperatures (To, Tp, and Tc) increased significantly ($p < 00.05$) from 60.58 to 63.14 °C, 65.02 to 68.11 °C, and 68.83 to 71.90 °C, respectively (Table 4 and Figure 3). Because fibers and starch compete for water absorption, starch swelling and gelation events are limited, resulting in a higher starch gelatinization temperature. The gelatinization temperatures of FEP4 and FEP7 are the highest, whereas FEP9 and FEP11 are the lowest. However, they are still higher than those of CP (Table 4), implying that FEP formulations require a higher temperature to melt their crystalline structure. Similar findings have been reported for pasta enriched with pea fibers and Kañawa flour [74].

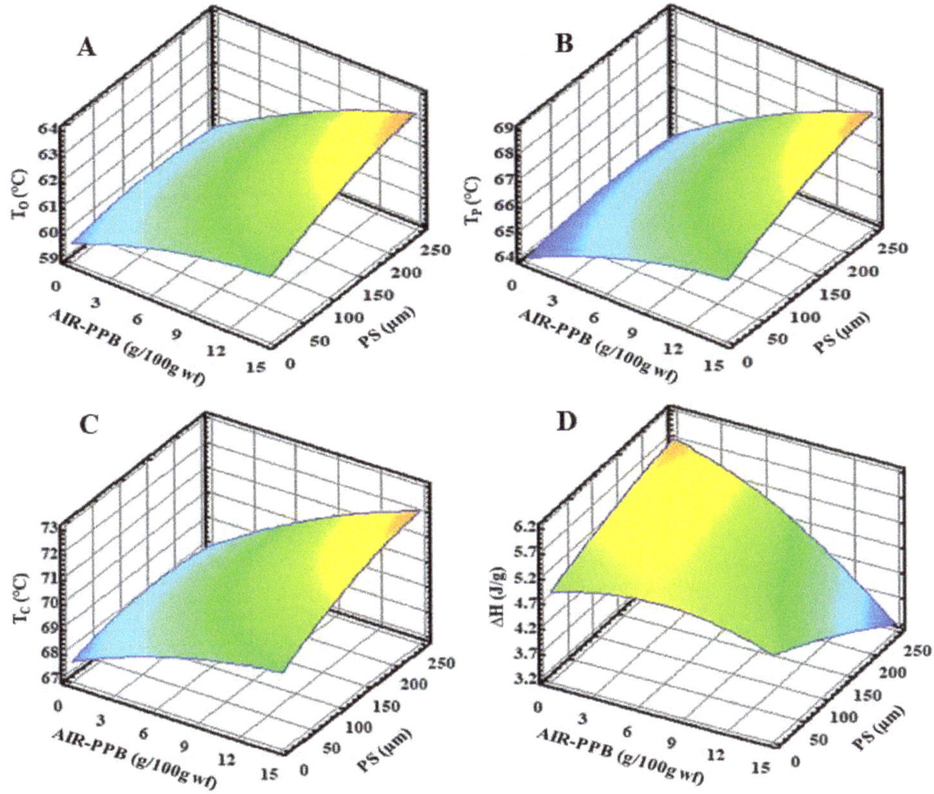

Figure 3. Response surface plots of the interaction of alcohol insoluble residue from potato peel byproduct (AIR-PPB) and particle size (PS) on (**A**) onset temperature (To) (°C), (**B**) melting temperature (Tp) (°C), (**C**) conclusion temperature (Tc) (°C), and (**D**) ΔH (J/g) of fiber-enriched pasta (FEP) samples.

The obtained results indicated that the enthalpy value (*H*) decreased significantly ($p < 0.05$) as the amount of AIR-PPB with higher PS was increased, ranging from 3.54 to 5.49 J/g, compared to the CP (6.17 J/g) (Table 4 and Figure 3D). As reported by BeMiller [75], the loss of molecular order in the starch granule is indicated by a high enthalpy of gelatinization, which serves as a general indicator of crystallinity quantity and quality. The high WHC of AIR-PPB, particularly when combined with a high PS, inhibited swelling, lowering the starch gelatinization and enthalpy values. Similarly, pasta enriched with pea fibers, bran, and Kañawa flour demonstrated similar results [33,68,74]. Lu et al. [76] discovered that adding mushroom powder increased the *H* values of semolina-only pasta. These findings indicate that the addition of mushroom powders reduced the degree of gelatinization or dextrinization of starch granules during cold extrusion and cooking. It is speculated that mushroom fiber (particularly IDF) and fat acted as a protective agent during processing, inhibiting enzyme access to starch granules within the pasta matrix and thus limiting the release of reducing sugars during starch digestion.

3.3. Color Parameters and Sensory Properties

The ability of colors to attract attention has long been used by marketers to gain customers' attention. The L^*, a^*, and b^* values of the experimental FEPs demonstrated significant variation in the color attributes of FEP compared to CP. However, regardless of the substitution level of AIR-PPB and PS, all of the experimental FEP exhibited a tendency toward a brown color from light to dark. When AIR-PPB replacement was increased, experimental FEP became darker, more reddish, and less yellowish. Although the PS had the opposite effect on the color values, increasing lightness, redness, and yellowness values as the PS was decreased. This behavior is explicable by examining the color of AIR-PPB at various PS. The findings corroborated those of Makhlouf et al. [77] who incorporated oat bran (OB), whole barley flour, and resistant starch (RS) into pasta formulations at various concentrations. They concluded that among the three fiber sources, only OB concentrations caused the apparent changes to the lightness (L^*) and red-to-green color coordinate (a^*). All three fibers' source materials altered the b^* value (blue-to-yellow coordinate), resulting in all samples having a more yellowish hue compared to the control product.

The differences in color (ΔE) results confirmed the effect of AIR-PPB replacement levels and PS on the color values of FEP. The ΔE values of FEP were significantly different ($p < 0.05$) and showed an increasing tendency as the AIR-PPB replacement level and its PS increased (Table 5). All experimental FEP demonstrated significant and observable differences in color ($p < 0.05$) according to the color difference findings. FEP4 had the highest value of color change (19.66); however, FEP 11 had the lowest value of color change (8.40) (Table 5).

Table 5. Color parameters and sensorial traits of fiber-enriched pasta (FEP).

Color Parameters	CP	FEP-1	FEP-2,3,8	FEP-4	FEP-5	FEP-6	FEP-7	FEP-9	FEP-10	FEP-11
L^*	68.82 ± 0.00 ^{a}	59.02 ± 0.02 ^{e}	58.66 ± 0.00 ^{f}	49.21 ± 0.02 ^{j}	57.14 ± 0.00 ^{g}	59.52 ± 0.01 ^{d}	49.56 ± 0.01 ^{i}	60.22 ± 0.01 ^{c}	54.32 ± 0.06 ^{h}	60.43 ± 0.00 ^{b}
a^*	0.84 ± 0.04 ^{i}	1.79 ± 0.03 ^{a}	1.61 ± 0.07 ^{c}	1.32 ± 0.03 ^{f}	1.51 ± 0.06 ^{d}	1.19 ± 0.08 ^{h}	1.38 ± 0.02 ^{e}	1.27 ± 0.01 ^{g}	1.64 ± 0.09 ^{b}	1.19 ± 0.01 ^{h}
b^*	13.24 ± 0.01 ^{a}	12.46 ± 0.08 ^{e}	12.26 ± 0.03 ^{f}	11.80 ± 0.00 ^{j}	12.24 ± 0.00 ^{g}	12.60 ± 0.00 ^{d}	11.93 ± 0.05 ^{i}	12.71 ± 0.00 ^{c}	11.97 ± 0.01 ^{h}	12.78 ± 0.04 ^{b}
ΔE	- - - -	9.87 ± 0.03 ^{f}	10.23 ± 0.01 ^{e}	19.66 ± 0.03 ^{a}	11.74 ± 0.01 ^{d}	9.32 ± 0.03 ^{g}	19.31 ± 0.02 ^{b}	8.62 ± 0.01 ^{h}	14.57 ± 0.04 ^{c}	8.40 ± 0.03 ^{i}
Sensory properties •										
Color	8.76 ± 1.96 ^{a}	8.11 ± 1.70 ^{e}	6.71 ± 1.25 ^{f}	5.86 ± 1.36 ^{j}	6.54 ± 1.22 ^{g}	8.32 ± 1.20 ^{d}	6.19 ± 1.10 ^{i}	8.56 ± 1.82 ^{c}	6.46 ± 1.70 ^{h}	8.61 ± 1.09 ^{b}
Taste	8.91 ± 1.84 ^{a}	7.94 ± 1.68 ^{e}	6.54 ± 1.19 ^{f}	5.93 ± 1.83 ^{j}	6.40 ± 1.37 ^{g}	8.09 ± 1.27 ^{d}	6.13 ± 0.98 ^{i}	8.37 ± 1.18 ^{c}	6.29 ± 1.67 ^{h}	8.41 ± 1.10 ^{b}
Flavor	8.09 ± 1.57 ^{a}	7.44 ± 1.64 ^{e}	6.03 ± 1.27 ^{f}	5.58 ± 1.56 ^{j}	5.91 ± 1.61 ^{g}	7.54 ± 1.02 ^{d}	5.69 ± 1.12 ^{i}	7.85 ± 1.77 ^{c}	5.78 ± 1.61 ^{h}	7.88 ± 1.45 ^{b}
Texture	8.93 ± 1.40 ^{a}	7.83 ± 1.82 ^{e}	6.43 ± 1.34 ^{f}	6.02 ± 1.79 ^{j}	6.33 ± 1.52 ^{g}	7.92 ± 1.17 ^{d}	6.13 ± 1.17 ^{i}	8.17 ± 1.01 ^{c}	6.22 ± 1.89 ^{h}	8.21 ± 1.27 ^{b}
Overall acceptability	8.87 ± 1.21 ^{a}	8.07 ± 1.93 ^{e}	6.67 ± 1.31 ^{f}	6.05 ± 1.83 ^{j}	6.53 ± 1.41 ^{g}	8.22 ± 1.08 ^{d}	6.26 ± 1.09 ^{i}	8.49 ± 1.89 ^{c}	6.43 ± 1.32 ^{h}	8.53 ± 1.13 ^{b}

* Values are means ± SD. Means within a row with different superscript letters are significantly different ($p < 0.05$). • Mean values of hedonic test from 25 panelists (15 males and 10 females at 21–30 years old), FEP-2,3,8 is the center point, and its value is the average of the three central points. CP: control pasta, L^*: whiteness, a^*: redness, b^*: yellowness', ΔE : color difference.

Sensory evaluation is a critical indicator for determining the optimal conditions for FEP production because it is related to the product's acceptability and marketability on a commercial scale. It is possible to determine the acceptable ranges of AIR-PPB addition and its PS, as well as the expected physicochemical properties, based on the sensory evaluation.

Table 5 compares the sensory properties of experimental FEP to those of CP. Enriching pasta with AIR-PPB at the experimental PS can result in a significant ($p < 0.05$) reduction in color, taste, flavor, texture, and overall acceptability, which could be attributed to the interaction of the macromolecules of wheat dough with AIR-PPB at the experimental PS, resulting in variation in FEP qualities. Although the CP attributes had the highest preference scores (8.76 for color, 8.91 for taste, 8.09 for flavor, 8.93 for texture, and 8.87 for overall acceptability), the values obtained from FEP1, 6, 9, and 11 were greater than 7 for all sensory attributes compared to the CP. Meanwhile, sensory scores for FEP-2, 3, 8 (center point), 4, 5, 7, and 10 were the lowest (Table 5). The vibrant yellow color of pasta is a critical factor in determining consumer acceptance. Carotenoid pigments, in particular, are responsible for the yellow color of pasta. Semolina's ash content also contributes to its coloring. Increased ash content results in the pale color of pasta. Sobotaet al. [78] showed that consumers prefer darker pasta because they think it is healthier because it contains fiber. Makhlouf et al. [77] demonstrated that FEP could be produced by adding up to 15% of dietary fiber into regular semolina-based pasta formulation, leading to acceptable products with matching characteristics of texture and color compared to commercial products. Among the three fiber sources, OB offered better characteristics of texture and taste, whereas RS featured the most desired golden color. Although all three addition levels resulted in acceptable products, the lower addition level (5 per cent) led to the highest preference from the sensory panel.

Table 6 shows that the color, taste, and flavor scores of FEP were negatively correlated with the TDF content ($r = -0.91$, -0.89, and -0.88) and the Cl content (-0.95, -0.94, and -0.94), respectively, confirming the hypothesis of dietary fibers having a negative effect on sensory properties, as well as the negative impact of micro-molecules of FEP leaching into the cooking water as a result of the gluten network's weakness. Notably, the greater the color change, the less acceptable the sensorial color ($r = -0.76$). In addition, there was a negative correlation between texture score and TDF content (r = -0.87). Conversely, we observed a positive correlation between firmness and stickiness (r = 0.86). The interaction of AIR-PPB with starch results in a decrease in the amount of gelatinized starch required to maintain the structural characteristics of dough and disrupts the starch–protein matrix [79]. Similarly, ANOVA and RSM results indicated that overall acceptability scores were negatively correlated with TDF (r = -0.89), whereas ΔE ($r = -0.73$) was negatively correlated with increasing FEP firmness and stickiness ($r = 0.88$).

Table 6. Data analysis of the effect of alcohol insoluble residue from potato peel byproduct (AIR-PPB) and particle size (PS) on chemical and physical properties of fiber-enriched pasta (FEP).

	TDF (g/100 g)		OCT (min)		Cl (%)		MII (Kg CP/kg Up)		Firmness (N)		Stickiness (N)		ΔE		To (°C)		Tp (°C)		Tc (°C)		*H* (J/g)	
Source	*F.*	*p.*	*F.*	*p.*	*F.*	*p.*	*F.*	*p.*	*F.*	*p.*	*F.*	*p.*	*F.*	*p.*	*F.*	*p.*	*F.*	*p.*	*F.*	*p.*	*F.*	*p.*
X_1	1760.54	0.0000	275.23	0.0000	336.51	0.0000	88.88	0.0002	40.92	0.0014	141.97	0.0001	328.18	0.0000	106.53	0.0001	118.31	0.0001	108.85	0.0001	54.34	0.0007
X_2	45.51	0.0011	20.04	0.0065	39.15	0.0015	11.07	0.0209	10.20	0.0242	10.77	0.0219	18.90	0.0074	13.75	0.0139	16.86	0.0093	15.85	0.0105	7.24	0.0433
$X_1 \times 1$	2.96	0.1460	4.78	0.0804	10.83	0.0217	3.11	0.1383	2.76	0.1573	1.15	0.3334	80.56	0.0003	1.78	0.2401	1.74	0.2438	1.68	0.2511	2.94	0.1473
$X_1 \times 2$	9.04	0.0299	1.82	0.2355	1.72	0.2461	0.09	0.7756	0.32	0.5945	0.00	0.9823	8.52	0.0330	2.45	0.1781	1.95	0.2217	1.98	0.2180	5.85	0.0602
$X_2 \times 2$	16.32	0.0099	5.11	0.0734	1.22	0.3189	0.06	0.8107	0.01	0.9302	0.13	0.7344	5.95	0.0587	1.36	0.2965	0.99	0.3644	1.61	0.2599	0.12	0.7424
R^2	99.74		98.5367		98.8226		95.6893		92.2817		97.0473		98.9435		96.3575		96.7292		96.452		94.1099	
R^2 (d.f.)	99.4955		97.0734		97.6453		91.3785		84.5634		94.0945		97.887		92.715		93.4583		92.904		88.2198	
MAE	0.0946195		0.0071620		0.0095813		0.0227303		0.0275063		0.0521069		0.283463		0.113512		0.121282		0.122567		0.0977168	

MAE: Mean absolute error, F: *F*-Ratio, *p*: *p*-Value, TDF: total dietary fibers, OCT: optimum cooking time (min), Cl: cooking loss (%), MII: mass increase index (kg cooked pasta/kg uncooked pasta), ΔE: color difference, To: onset temperature, Tp: melting temperature, Tc: conclusion temperature, *H*: enthalpy value.

From the previous results and RSM results, the highly acceptable FEP samples were in descending order: FEP11 (AIR-PPB = 2 g/100 g wheat flour, PS = 145 μm), FEP9 (AIR-PPB = 4 g/100 g wheat flour, PS = 70 μm), FEP6 (AIR-PPB = 4 g/100 g wheat flour, PS = 219 μm), and FEP1 (AIR-PPB = 8.50 g/100 g wheat flour, PS = 40 μm); TDF concentrations were 4.12, 5.55, 5.64, and 7.80 g/100 g FEP in these fortified samples. Daily consumption of the recommended 25 g of dietary fibers requires 606.80, 450.45, 443.26, and 320.51 g of FEP11, FEP9, FEP6, and FEP1, respectively, in comparison to the CP (4310.34 g). Notably, the OCT values at these points were 8.14, 8.09, 8.07, and 8.05 min, respectively, compared to CP. Additionally, Cl values were 5.59, 5.65, 5.73, and 5.76% higher than CP values. Firmness values were 2.18, 2.16, 2.01, and 1.98 N, respectively, whereas stickiness values were 3.70, 3.58, 3.24, and 2.97 N, corresponding to CP. The values for color changes were 8.40, 8.62, 9.32, and 9.87. The temperatures used for gelation were 65.02, 65.42, 65.71, and 65.79 °C, respectively.

4. Conclusions

Fiber supplementation improves digestion and treats a variety of recent malnutrition-related diseases. In this study, we developed a cost-effective FEP pasta by incorporating AIR-PPB into whole wheat flour. The fiber-rich fraction (AIR-PPB) was successfully isolated from PPB and used as a promising source of fibers in FEB at a variety of substitution levels and PS. The TDF, WHC, and significant color characteristics, particularly (L^*, b^*), distinguish the AIR-PPB with different PS. These unique properties of AIR-PPB correlate positively with the color and sensory quality of pasta. Additionally, there is a positive effect on the gluten network and thermal properties of starch gelatinization, resulting in improved cooking properties and texture compared to CP. The optimal processing parameters obtained via the sensory evaluation revealed that FEP 11, 9, 6, and 1 have the highest acceptability. Additionally, these formulations can help individuals meet the recommended daily intake of dietary fibers. Therefore, this study recommends using AIR-PPB at specific PS as a suitable dietary fiber source for commercial FEP.

Author Contributions: Data curation, A.I.; Formal analysis, M.N.; Funding acquisition, A.A., K.A.E.-T.; E.T.A.S.-A. and C.A.C.-J.; Investigation, A.I., E.T.A.S.-A. and C.A.C.-J.; Methodology, A.I. and A.A.; Supervision, K.E.; Writing—original draft, M.N.; Writing—review and editing, A.M.S., K.A.E.-T. and K.E. All authors have read and agreed to the published version of the manuscript.

Funding: The authors are thankful for the financial support provided by the Fundação de Amparo à Pesquisa do Estado do Rio de Janeiro (FAPERJ) Brazil—grant number [E-26/200.891/2021] and the Conselho Nacional de Desenvolvimento Científico e Tecnológico (CNPq)—grant number [313119/2020-1].

Institutional Review Board Statement: Not applicable.

Informed Consent Statement: Not applicable.

Data Availability Statement: Not applicable.

Conflicts of Interest: The authors declare no conflict of interest.

Sample Availability: Samples of the compounds are not available from the authors.

References

1. El-Saadony, M.T.; Elsadek, M.F.; Mohamed, A.S.; Taha, A.E.; Ahmed, B.M.; Saad, A.M. Effects of chemical and natural additives on cucumber juice's quality, shelf life, and safety. *Foods* **2020**, *9*, 639. [CrossRef] [PubMed]
2. Saad, A.M.; Mohamed, A.S.; El-Saadony, M.T.; Sitohy, M.Z. Palatable functional cucumber juices supplemented with polyphenols-rich herbal extracts. *LWT Food Sci. Technol.* **2021**, *148*, 111668. [CrossRef]
3. Elleuch, M.; Bedigian, D.; Roiseux, O.; Besbes, S.; Blecker, C.; Attia, H.J.F.c. Dietary fibre and fibre-rich by-products of food processing: Characterization, technological functionality and commercial applications: A review. *Food Chem.* **2011**, *124*, 411–421. [CrossRef]
4. AACC. *Approved Methods of the American Association of Cereal Chemists*; American Association of cereal Chemists, Inc.: St. Paul, MN, USA, 2007.
5. Abd El-Hack, M.E.; El-Saadony, M.T.; Shehata, A.M.; Arif, M.; Paswan, V.K.; Batiha, G.E.-S.; Khafaga, A.F.; Elbestawy, A.R. Approaches to prevent and control *Campylobacter* spp. colonization in broiler chickens: A review. *Environ. Sci. Pollut. Res.* **2021**, *28*, 4989–5004. [CrossRef]
6. Brownlee, I.A. The physiological roles of dietary fibre. *Food Hydrocoll.* **2011**, *25*, 238–250. [CrossRef]
7. Swelum, A.A.; Shafi, M.E.; Albaqami, N.M.; El-Saadony, M.T.; Elsify, A.; Abdo, M.; El-Sayed Taha, A.; Abdel-Moneim, A.-M.E.; Al-Gabri, N.A.; Mohamed, E.; et al. COVID-19 in human, animal, and environment: A review. *Front. Vet. Sci.* **2020**, *7*, 578. [CrossRef]
8. Institute of Medicine. Available online: http://www.iom.edu/Reports/2002/Dietary-Reference-Intakes-for-Energy-Carbohydrate-Fiber-Fat-Fatty-Acids-Cholesterol-Protein-and-Amino-Acids.aspx (accessed on 18 January 2011).
9. Pathania, S.; Kaur, N.J.B.C.; Fibre, D. Utilization of fruits and vegetable by-products for isolation of dietary fibres and its potential application as functional ingredients. *Bioact. Carbohydr. Diet. Fibre.* **2021**, *27*, 100295. [CrossRef]
10. Saad, A.M.; El-Saadony, M.T.; El-Tahan, A.M.; Sayed, S.; Moustafa, M.A.; Taha, A.E.; Taha, T.F.; Ramadan, M.M. Polyphenolic extracts from pomegranate and watermelon wastes as substrate to fabricate sustainable silver nanoparticles with larvicidal effect against *Spodoptera littoralis*. *Saudi J. Biol. Sci.* **2021**, *28*, 5674–5683. [CrossRef]
11. Abdel-Moneim, A.-M.E.; El-Saadony, M.T.; Shehata, A.M.; Saad, A.M.; Aldhumri, S.A.; Ouda, S.M.; Mesalam, N.M. Antioxidant and antimicrobial activities of *Spirulina platensis* extracts and biogenic selenium nanoparticles against selected pathogenic bacteria and fungi. *Saudi J. Biol. Sci.* **2022**, *29*, 1197–1209. [CrossRef]
12. El-Saadony, M.T.; Saad, A.M.; Taha, T.F.; Najjar, A.A.; Zabermawi, N.M.; Nader, M.M.; AbuQamar, S.F.; El-Tarabily, K.A.; Salama, A. Selenium nanoparticles from *Lactobacillus paracasei* HM1 capable of antagonizing animal pathogenic fungi as a new source from human breast milk. *Saudi J. Biol. Sci.* **2021**, *28*, 6782–6794. [CrossRef]
13. Açıkalın, K.J.B.C. Evaluation of orange and potato peels as an energy source: A comprehensive study on their pyrolysis characteristics and kinetics. *Biomass Convers. Biorefin.* **2021**, *12*, 501–514. [CrossRef]
14. Sampaio, S.L.; Petropoulos, S.A.; Alexopoulos, A.; Heleno, S.A.; Santos-Buelga, C.; Barros, L.; Ferreira, I.C. Potato peels as sources of functional compounds for the food industry: A review. *Trends Food Sci. Technol.* **2020**, *103*, 118–129. [CrossRef]

15. Quisperima, A.; Pérez, S.; Flórez, E.; Acelas, N.J.B.T. Valorization of potato peels and eggshells wastes: Ca-biocomposite to remove and recover phosphorus from domestic wastewater. *Bioresour. Technol.* **2022**, *343*, 126106. [CrossRef] [PubMed]
16. Albishi, T.; John, J.A.; Al-Khalifa, A.S.; Shahidi, F. Phenolic content and antioxidant activities of selected potato varieties and their processing by-products. *J. Func. Foods* **2013**, *5*, 590–600. [CrossRef]
17. Friedman, M.; Huang, V.; Quiambao, Q.; Noritake, S.; Liu, J.; Kwon, O.; Chintalapati, S.; Young, J.; Levin, C.E.; Tam, C. Potato peels and their bioactive glycoalkaloids and phenolic compounds inhibit the growth of pathogenic trichomonads. *J. Agric. Food Chem.* **2018**, *66*, 7942–7947. [CrossRef]
18. Khawla, B.J.; Sameh, M.; Imen, G.; Donyes, F.; Dhouha, G.; Raoudha, E.G.; Oumèma, N.-E. Potato peel as feedstock for bioethanol production: A comparison of acidic and enzymatic hydrolysis. *Ind. Crops Prod.* **2014**, *52*, 144–149. [CrossRef]
19. Curti, E.; Carini, E.; Diantom, A.; Vittadini, E.J.F.c. The use of potato fibre to improve bread physico-chemical properties during storage. *Food Chem.* **2016**, *195*, 64–70. [CrossRef]
20. Jeddou, K.B.; Bouaziz, F.; Zouari-Ellouzi, S.; Chaari, F.; Ellouz-Chaabouni, S.; Ellouz-Ghorbel, R.; Nouri-Ellouz, O.J.F.C. Improvement of texture and sensory properties of cakes by addition of potato peel powder with high level of dietary fiber and protein. *Food Chem.* **2017**, *217*, 668–677. [CrossRef]
21. Padalino, L.; Conte, A.; Del Nobile, M.A.J.F. Overview on the general approaches to improve gluten-free pasta and bread. *Foods.* **2016**, *5*, 87. [CrossRef]
22. Laleg, K.; Cassan, D.; Barron, C.; Prabhasankar, P.; Micard, V.J.P.O. Structural, culinary, nutritional and anti-nutritional properties of high protein, gluten free, 100% legume pasta. *PLoS ONE* **2016**, *11*, e0160721. [CrossRef]
23. Vignola, M.B.; Bustos, M.C.; Pérez, G.T.J.F.c. *In vitro* dialyzability of essential minerals from white and whole grain pasta. *Food Chem.* **2018**, *265*, 128–134. [CrossRef] [PubMed]
24. Wójtowicz, A.; Mościcki, L.J.L.-F.S. Influence of legume type and addition level on quality characteristics, texture and microstructure of enriched precooked pasta. *LWT Food Sci. Technol.* **2014**, *59*, 1175–1185. [CrossRef]
25. Oliviero, T.; Fogliano, V.J.T.i.F.S. Food design strategies to increase vegetable intake: The case of vegetable enriched pasta. *Trends Food Sci. Technol.* **2016**, *51*, 58–64. [CrossRef]
26. Saad, A.M.; El-Saadony, M.T.; Mohamed, A.S.; Ahmed, A.I.; Sitohy, M.Z. Impact of cucumber pomace fortification on the nutritional, sensorial and technological quality of soft wheat flour-based noodles. *Inter J. Food Sci. Technol.* **2021**, *56*, 3255–3268. [CrossRef]
27. Ciccoritti, R.; Taddei, F.; Nicoletti, I.; Gazza, L.; Corradini, D.; D'Egidio, M.G.; Martini, D.J.F.C. Use of bran fractions and debranned kernels for the development of pasta with high nutritional and healthy potential. *Food Chem.* **2017**, *225*, 77–86. [CrossRef]
28. Sęczyk, Ł.; Świeca, M.; Gawlik-Dziki, U.J.F.C. Effect of carob (*Ceratonia siliqua* L.) flour on the antioxidant potential, nutritional quality, and sensory characteristics of fortified durum wheat pasta. *Food Chem.* **2016**, *194*, 637–642. [CrossRef]
29. Pigni, N.B.; Aranibar, C.; Mas, A.L.; Aguirre, A.; Borneo, R.; Wunderlin, D.; Baroni, M.V.J.L. Chemical profile and bioaccessibility of polyphenols from wheat pasta supplemented with partially-deoiled chia flour. *LWT Food Sci. Technol.* **2020**, *124*, 109134. [CrossRef]
30. Tiga, B.H.; Kumcuoglu, S.; Vatansever, M.; Tavman, S. Thermal and pasting properties of Quinoa—wheat flour blends and their effects on production of extruded instant noodles. *J. Cereal Sci.* **2021**, *97*, 103120. [CrossRef]
31. Sahin, A.W.; Hardiman, K.; Atzler, J.J.; Vogelsang-O'Dwyer, M.; Valdeperez, D.; Münch, S.; Cattaneo, G.; O'Riordan, P.; Arendt, E.K. Rejuvenated Brewer's Spent Grain: The impact of two BSG-derived ingredients on techno-functional and nutritional characteristics of fibre-enriched pasta. *Innov. Food Sci. Emerg. Technol.* **2021**, *68*, 102633. [CrossRef]
32. Kaur, G.; Sharma, S.; Nagi, H.; Dar, B.N. Functional properties of pasta enriched with variable cereal brans. *J. Food Sci. Technol.* **2012**, *49*, 467–474. [CrossRef]
33. Aravind, N.; Sissons, M.; Egan, N.; Fellows, C.J.F.C. Effect of insoluble dietary fibre addition on technological, sensory, and structural properties of durum wheat spaghetti. *Food Chem.* **2012**, *130*, 299–309. [CrossRef]
34. Boroski, M.; de Aguiar, A.C.; Boeing, J.S.; Rotta, E.M.; Wibby, C.L.; Bonafé, E.G.; de Souza, N.E.; Visentainer, J.V. Enhancement of pasta antioxidant activity with oregano and carrot leaf. *Food Chem.* **2011**, *125*, 696–700. [CrossRef]
35. Abd El-Hack, M.E.; El-Saadony, M.T.; Swelum, A.A.; Arif, M.; Abo Ghanima, M.M.; Shukry, M.; Noreldin, A.; Taha, A.E.; El-Tarabily, K.A. Curcumin, the active substance of turmeric: Its effects on health and ways to improve its bioavailability. *J. Sci. Food Agric.* **2021**, *101*, 5747–5762. [CrossRef]
36. El-Saadony, M.T.; Zabermawi, N.M.; Zabermawi, N.M.; Burollus, M.A.; Shafi, M.E.; Alagawany, M.; Yehia, N.; Askar, A.M.; Alsafy, S.A.; Noreldin, A.E.; et al. Nutritional aspects and health benefits of bioactive plant compounds against infectious diseases: A review. *Food Rev. Int.* **2021**, 1–23. [CrossRef]
37. El-Sohaimy, S.A.; Brennan, M.; Darwish, A.M.; Brennan, C. Physicochemical, texture and sensorial evaluation of pasta enriched with chickpea flour and protein isolate. *Ann. Agric. Sci.* **2020**, *65*, 28–34. [CrossRef]
38. Saad, A.M.; Elmassry, R.A.; Wahdan, K.M.; Ramadan, F.M. Chickpea (*Cicer arietinum)* steep liquor as a leavening agent: Effect on dough rheology and sensory properties of bread. *Acta Period. Technol.* **2015**, *46*, 91–102. [CrossRef]
39. Saad, A.M.; Sitohy, M.Z.; Ahmed, A.I.; Rabie, N.A.; Amin, S.A.; Aboelenin, S.M.; Soliman, M.M.; El-Saadony, M.T. Biochemical and functional characterization of kidney bean protein alcalase-hydrolysates and their preservative action on stored chicken meat. *Molecules* **2021**, *26*, 4690. [CrossRef] [PubMed]

40. Oyeyinka, S.A.; Adepegba, A.A.; Oyetunde, T.T.; Oyeyinka, A.T.; Olaniran, A.F.; Iranloye, Y.M.; Olagunju, O.F.; Manley, M.; Kayitesi, E.; Njobeh, P.B.J.L. Chemical, antioxidant and sensory properties of pasta from fractionated whole wheat and Bambara groundnut flour. *LWT Food Sci. Technol.* **2021**, *138*, 110618. [CrossRef]
41. Latorre, M.E.; Narvaiz, P.; Rojas, A.M.; Gerschenson, L.N. Effects of gamma irradiation on bio-chemical and physico-chemical parameters of fresh-cut red beet (*Beta vulgaris* L. var. conditiva) root. *J. Food Eng.* **2010**, *98*, 178–191. [CrossRef]
42. AOAC. *Official Method of Analysis: Association of Analytical Chemists*, 19th ed.; AOAC: Washington, DC, USA, 2012.
43. McCleary, B.V.; DeVries, J.W.; Rader, J.I.; Cohen, G.; Prosky, L.; Mugford, D.C.; Champ, M.; Okuma, K. Determination of insoluble, soluble, and total dietary fiber (CODEX definition) by enzymatic-gravimetric method and liquid chromatography: Collaborative study. *J. AOAC Inter.* **2012**, *95*, 824–844. [CrossRef]
44. Namir, M.; Siliha, H.; Ramadan, M.F. Fiber pectin from tomato pomace: Characteristics, functional properties and application in low-fat beef burger. *J. Food Meas. Charact.* **2015**, *9*, 305–312. [CrossRef]
45. Sobota, A.; Zarzycki, P.; Rzedzicki, Z.; Sykut-Domańska, E.; Wirkijowska, A. Effect of cooking time on the texture and cooking quality of spaghetti. *Acta Geophys.* **2013**, *20*, 693–703.
46. Bonomi, F.; D'Egidio, M.G.; Iametti, S.; Marengo, M.; Marti, A.; Pagani, M.A.; Ragg, E.M. Structure–quality relationship in commercial pasta: A molecular glimpse. *Food Chem.* **2012**, *135*, 348–355. [CrossRef] [PubMed]
47. Namir, M.; Suleiman, A.R.; Hassanien, M.F.R. Characterization and functionality of alcohol insoluble solids from tomato pomace as fat substitute in low fat cake. *J. Food Meas. Charact.* **2015**, *9*, 557–563. [CrossRef]
48. Ajila, C.; Leelavathi, K.; Rao, U.P. Improvement of dietary fiber content and antioxidant properties in soft dough biscuits with the incorporation of mango peel powder. *J. Cereal Sci.* **2008**, *48*, 319–326. [CrossRef]
49. Ismail, T.; Akhtar, S.; Riaz, M.; Ismail, A. Effect of pomegranate peel supplementation on nutritional, organoleptic and stability properties of cookies. *Int. J. Food Sci. Nutri.* **2014**, *65*, 661–666. [CrossRef]
50. Chau, C.-F.; Huang, Y.-L. Comparison of the chemical composition and physicochemical properties of different fibers prepared from the peel of *Citrus sinensis* L. Cv. Liucheng. *J. Agric. Food Chem.* **2003**, *51*, 2615–2618. [CrossRef]
51. Bender, A.B.; Speroni, C.S.; Salvador, P.R.; Loureiro, B.B.; Lovatto, N.M.; Goulart, F.R.; Lovatto, M.T.; Miranda, M.Z.; Silva, L.P.; Penna, N.G. Grape pomace skins and the effects of its inclusion in the technological properties of muffins. *J. Food Sci. Technol.* **2017**, *15*, 143–157. [CrossRef]
52. Antonic, B.; Jancikova, S.; Dordevic, D.; Tremlova, B. Apple pomace as food fortification ingredient: A systematic review and meta-analysis. *J. Food Sci.* **2020**, *85*, 2977–2985. [CrossRef]
53. Martínez, R.; Torres, P.; Meneses, M.A.; Figueroa, J.G.; Pérez-Álvarez, J.A.; Viuda-Martos, M. Chemical, technological and in vitro antioxidant properties of mango, guava, pineapple and passion fruit dietary fibre concentrate. *Food Chem.* **2012**, *135*, 1520–1526. [CrossRef]
54. Ocen, D.; Xu, X. Effect of citrus orange (*Citrus sinensis*) by-product dietary fiber preparations on the quality characteristics of frozen dough bread. *Am. J. Food Technol.* **2013**, *8*, 43–53. [CrossRef]
55. Wu, M.Y.; Shiau, S.Y. Effect of the amount and particle size of pineapple peel fiber on dough rheology and steamed bread quality. *J. Food Process. Preserv.* **2015**, *39*, 549–558. [CrossRef]
56. Wang, S.; Gu, B.-J.; Ganjyal, G.M. Impacts of the inclusion of various fruit pomace types on the expansion of corn starch extrudates. *LWT Food Sci. Technol.* **2019**, *110*, 223–230. [CrossRef]
57. Namir, M.; Rabie, M.A.; Rabie, N.A. Characterization. Physicochemical, pasting, and sensory characteristics of antioxidant dietary fiber gluten-free donut made from cantaloupe by-products. *J. Food Meas. Charact.* **2021**, *15*, 5445–5459. [CrossRef]
58. Bchir, B.; Rabetafika, H.N.; Paquot, M.; Blecker, C.J. Effect of pear, apple and date fibres from cooked fruit by-products on dough performance and bread quality. *Food Bioproc. Tech.* **2014**, *7*, 1114–1127. [CrossRef]
59. Foschia, M.; Peressini, D.; Sensidoni, A.; Brennan, C.S. The effects of dietary fibre addition on the quality of common cereal products. *J. Cereal Sci.* **2013**, *58*, 216–227. [CrossRef]
60. Dey, D.; Richter, J.K.; Ek, P.; Gu, B.-J.; Ganjyal, G.M. Utilization of food processing by-products in extrusion processing: A review. *Front. Sustain. Food Syst.* **2021**, *4*, 1–18. [CrossRef]
61. European Commission Regulation, E. No 1924/2006 of the Parliament and of the Council of 20 December 2006 on nutrition and health claims made on foods. *Off. J. Eur. Union* **2007**, *12*, 3–18.
62. Makki, K.; Deehan, E.C.; Walter, J.; Bäckhed, F.J. The impact of dietary fiber on gut microbiota in host health and disease. *Cell Host Microbe.* **2018**, *23*, 705–715. [CrossRef]
63. Abd El-Hack, M.E.; El-Saadony, M.T.; Elbestawy, A.R.; Nahed, A.; Saad, A.M.; Salem, H.M.; El-Tahan, A.M.; Khafaga, A.F.; Taha, A.E.; AbuQamar, S.F.; et al. Necrotic enteritis in broiler chickens: Disease characteristics and prevention using organic antibiotic alternatives–a comprehensive review. *Poult. Sci.* **2021**, *101*, 101590. [CrossRef]
64. Ritthiruangdej, P.; Parnbankled, S.; Donchedee, S.; Wongsagonsup, R. Physical, chemical, textural and sensory properties of dried wheat noodles supplemented with unripe banana flour. *Agric. Nat. Resour.* **2011**, *45*, 500–509.
65. Masli, M.D.P.; Gu, B.J.; Rasco, B.A.; Ganjyal, G.M. Fiber-rich food processing byproducts enhance the expansion of cornstarch extrudates. *J. Food Sci.* **2018**, *83*, 2500–2510. [CrossRef] [PubMed]
66. Simonato, B.; Trevisan, S.; Tolve, R.; Favati, F.; Pasini, G.J.L. Pasta fortification with olive pomace: Effects on the technological characteristics and nutritional properties. *LWT Food Sci. Technol.* **2019**, *114*, 108368. [CrossRef]

67. Lončarić, A.; Kosović, I.; Jukić, M.; Ugarčić, Ž.; Piližota, V. Effect of apple by-product as a supplement on antioxidant activity and quality parameters of pasta. *Croat. J. Food Sci. Technol.* **2014**, *6*, 97–103. [CrossRef]
68. Tudorica, C.; Kuri, V.; Brennan, C. Nutritional and physicochemical characteristics of dietary fiber enriched pasta. *J. Agric. Food Chem.* **2002**, *50*, 347–356. [CrossRef]
69. Ajila, C.; Aalami, M.; Leelavathi, K.; Rao, U.P. Mango peel powder: A potential source of antioxidant and dietary fiber in macaroni preparations. *Innov. Food Sci. Emerg. Technol.* **2010**, *11*, 219–224. [CrossRef]
70. Gull, A.; Prasad, K.; Kumar, P. Effect of millet flours and carrot pomace on cooking qualities, color and texture of developed pasta. *LWT Food Sci. Technol.* **2015**, *63*, 470–474. [CrossRef]
71. Crizel, T.d.M.; Rios, A.d.O.; Thys, R.C.S.; Flôres, S.H. Effects of orange by-product fiber incorporation on the functional and technological properties of pasta. *Food Sci. Technol.* **2015**, *35*, 546–551. [CrossRef]
72. Brennan, M.A.; Merts, I.; Monro, J.; Woolnough, J.; Brennan, C.S. Impact of guar and wheat bran on the physical and nutritional quality of extruded breakfast cereals. *Starch-Stärke* **2008**, *60*, 248–256. [CrossRef]
73. Zarzycki, P.; Sykut-Domańska, E.; Sobota, A.; Teterycz, D.; Krawęcka, A.; Blicharz-Kania, A.; Andrejko, D.; Zdybel, B. Flaxseed enriched pasta-chemical composition and cooking quality. *Foods* **2020**, *9*, 404. [CrossRef]
74. Bustos, M.C.; Ramos, M.I.; Pérez, G.T.; Leon, A.E. Utilization of Kañawa (*Chenopodium pallidicaule* Aellen) flour in pasta making. *J. Chem.* **2019**, *2019*, 4385045. [CrossRef]
75. BeMiller, J.N. Pasting, paste, and gel properties of starch–hydrocolloid combinations. *Carbohy. Poly.* **2011**, *86*, 386–423. [CrossRef]
76. Lu, X.; Brennan, M.A.; Serventi, L.; Liu, J.; Guan, W.; Brennan, C.S. Addition of mushroom powder to pasta enhances the antioxidant content and modulates the predictive glycaemic response of pasta. *Food Chem.* **2018**, *264*, 199–209. [CrossRef] [PubMed]
77. Makhlouf, S.; Jones, S.; Ye, S.-H.; Sancho-Madriz, M.; Burns-Whitmore, B.; Li, Y.O. Effect of selected dietary fibre sources and addition levels on physical and cooking quality attributes of fibre-enhanced pasta. *Food Qual. Saf.* **2019**, *3*, 117–127. [CrossRef]
78. Sobota, A.; Rzedzicki, Z.; Zarzycki, P.; Kuzawińska, E. Application of common wheat bran for the industrial production of high-fibre pasta. *Int. J. Food Sci. Technol.* **2015**, *50*, 111–119. [CrossRef]
79. Biernacka, B.; Dziki, D.; Gawlik-Dziki, U.; Różyło, R.; Siastała, M. Physical, sensorial, and antioxidant properties of common wheat pasta enriched with carob fiber. *LWT Food Sci. Technol.* **2017**, *77*, 186–192. [CrossRef]

molecules

Article

Effects of Sphingomyelin-Containing Milk Phospholipids on Skin Hydration in UVB-Exposed Hairless Mice

Yejin Ahn [1], Min Guk Kim [1], Kyungae Jo [1], Ki-Bae Hong [2,*] and Hyung Joo Suh [1,3,*]

[1] Department of Integrated Biomedical and Life Science, Graduate School, Korea University, Seoul 02841, Korea; ahnyj708@gmail.com (Y.A.); minguk94@gmail.com (M.G.K.); kyungae11@korea.ac.kr (K.J.)
[2] Department of Food Science and Nutrition, Jeju National University, Jeju 63243, Korea
[3] BK21FOUR R&E Center for Learning Health Systems, Korea University, Seoul 02841, Korea
* Correspondence: kbhong@jejunu.ac.kr (K.-B.H.); suh1960@korea.ac.kr (H.J.S.); Tel.: +82-23-290-5639 (H.J.S.)

Abstract: Reactive oxygen species (ROS) generated by ultraviolet (UV) exposure cause skin barrier dysfunction, which leads to dry skin. In this study, the skin moisturizing effect of sphingomyelin-containing milk phospholipids in UV-induced hairless mice was evaluated. Hairless mice were irradiated with UVB for eight weeks, and milk phospholipids (50, 100, and 150 mg/kg) were administered daily. Milk phospholipids suppressed UV-induced increase in erythema and skin thickness, decreased transepidermal water loss, and increased skin moisture. Milk phospholipids increased the expression of filaggrin, involucrin, and aquaporin3 (AQP3), which are skin moisture-related factors. Additionally, hyaluronic acid (HA) content in the skin tissue was maintained by regulating the expression of HA synthesis- and degradation-related enzymes. Milk phospholipids alleviated UV-induced decrease in the expression of the antioxidant enzymes superoxidase dismutase1 and 2, catalase, and glutathione peroxidase1. Moreover, ROS levels were reduced by regulating heme oxygenase-1 (HO-1), an ROS regulator, through milk phospholipid-mediated activation of nuclear factor erythroid-2-related factor 2 (Nrf2). Collectively, sphingomyelin-containing milk phospholipids contributed to moisturizing the skin by maintaining HA content and reducing ROS levels in UVB-irradiated hairless mice, thereby, minimizing damage to the skin barrier caused by photoaging.

Keywords: milk phospholipids; sphingomyelin; skin hydration; Nrf2; hyaluronic acid

Citation: Ahn, Y.; Kim, M.G.; Jo, K.; Hong, K.-B.; Suh, H.J. Effects of Sphingomyelin-Containing Milk Phospholipids on Skin Hydration in UVB-Exposed Hairless Mice. *Molecules* **2022**, *27*, 2545. https://doi.org/10.3390/molecules27082545

Academic Editor: Smaoui Slim

Received: 29 March 2022
Accepted: 12 April 2022
Published: 14 April 2022

1. Introduction

The skin is the primary protective barrier of the human body, and it plays a role in preserving moisture in the body and protecting the skin from the external environment. The epidermal stratum corneum, the outermost layer of the skin, is involved in protecting the moisture content of the skin in a dry environment [1]. When the epidermis is repeatedly exposed to a large amount of ultraviolet (UV) light, it induces reactive oxygen species (ROS) generation and oxidative stress. UV rays can penetrate the epidermal and dermal layers and facilitate ROS generation in cells and tissues through various processes [2]. UV-induced ROS accelerate aging by inducing photooxidative damage to the skin. ROS cause an imbalance in enzymatic and non-enzymatic antioxidant defense systems of the skin and prevent normal cell functions owing to lipid peroxidation-induced cell membrane damage [3]. Increased oxidative stress in skin cells activates the expression of matrix metalloproteases, thereby reducing collagen production and elastic fiber synthesis, thereby promoting skin aging [4]. As it cannot recover from the continuous oxidation state, the skin surface becomes rough and loses its luster, leading to skin aging, which involves loss of elasticity and wrinkle formation [5].

The stratum corneum, which affects skin moisture retention, forms a lipid layer composed of ceramide, cholesterol, and free fatty acids between keratinocytes and has low permeability compared to general phospholipid biofilms, thereby inhibiting the permeation of external substances [6]. The stratum corneum produces natural moisturizing factors, such as amino acids, lactic acid, urea, citrate, and hyaluronic acid (HA), to maintain body water balance [7]. The stratum corneum of a healthy person contains approximately 10 to 30% moisture, and insufficient moisture causes abnormalities in the skin barrier and increases transdermal moisture loss, resulting in dry skin. Loss of moisture caused by abnormal skin barrier function reduces skin elasticity, thickens the epidermis, promotes wrinkle formation, causes diseases such as itching and xerosis, and worsens diseases such as psoriasis and atopic dermatitis [8]. Moisture supply and maintenance of moisture in the skin are important in terms of pathological and cosmetic aspects [7]. Active ingredients required to moisturize the skin include ceramide, hydroxy acid, glycerin, and butylene glycol, and these compounds are either applied to the skin or ingested orally [9,10].

Research on the development of commercial cosmeceuticals has been conducted through the repositioning of natural and synthetic products, and research to explore and utilize food materials with wrinkle-improving effects continues steadily [11]. This study used sphingomyelin-containing phospholipids, which are polar lipids extracted from milk whey with ethanol. Most of it contains phospholipids, mainly phosphatidylcholine and phosphatidylethanolamine, and sphingolipids, mainly sphingomyelin [12]; therefore, it is a food ingredient rich in precursors of ceramides necessary for skin moisturizing. Milk fat is a dietary source of sphingomyelin, and dietary sphingomyelin raises ceramide levels in the body. Ceramide plays a role in maintaining the moisture in the epidermis and skin barrier function [13,14].

Numerous studies have been conducted on natural diets and herbs for skin moisturizing and skin barrier function improvement, but studies on milk phospholipids are limited. Therefore, in this study, the skin moisturizing effect of milk phospholipids was evaluated by measuring the expression levels of skin hydration factors and enzymes related to the synthesis and decomposition of hyaluronic acids in hairless mice induced by photoaging. By measuring the expression level of enzymes involved in ROS removal by milk phospholipids, the potential for skin photoaging inhibition was evaluated, and the mechanism of action of ROS removal was investigated.

2. Results

2.1. Effects of Milk Phospholipids on Body Weight Changes and Plasma Biochemical Parameter

During the experimental period, all the experimental groups showed a tendency to gradually increase in body weight (Supplementary Materials Table S1). In addition, the milk phospholipid administration groups (low-dose milk phospholipids [ML]: 50 mg/kg; medium-dose milk phospholipids [MM]: 100 mg/kg; high-dose milk phospholipids [MH]: 150 mg/kg) did not show a significant difference in body weight compared to the NOR group. Plasma biochemical values are shown in Table S2. Plasma levels of glucose, aspartate transaminase and alanine transferase were not significantly different between groups. The milk phospholipid administration groups (MM and MH) showed significantly lower triglyceride levels compared to the NOR group ($p < 0.05$ and $p < 0.01$, respectively), but the total cholesterol levels were similar to the NOR group.

2.2. Effects of Milk Phospholipids on Skin Parameters

Erythema formation and skin thickness are expressed as delta values, which represent the difference in values before and after the UV treatment (Figure 1A,B). As representative phenomena of skin photoaging, the erythema index and skin thickness were significantly higher in the UVB-C group than in the normal group ($p < 0.001$). However, oral administration of milk phospholipids decreased the erythema index in a concentration-dependent manner, showing improvement in UV-induced photoaging (Figure 1A). Administration of medium (MM) and high (MH) doses of milk phospholipids significantly decreased the erythema index compared to the UVB-C group ($p < 0.05$ and $p < 0.01$, respectively). Oral milk phospholipid administration, particularly MM and MH doses, significantly decreased UV-mediated increase in skin thickness compared to the UVB-C group (Figure 1B; $p < 0.01$

and $p < 0.05$, respectively). Collectively, MM and MH groups showed positive effects on erythema and skin thickness.

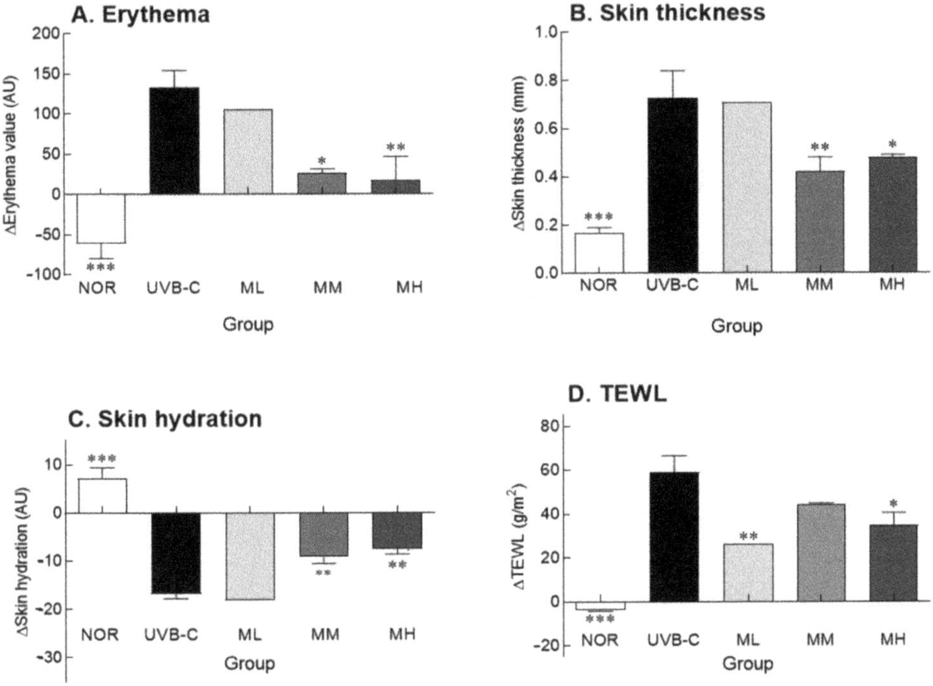

Figure 1. Effects of milk phospholipids on erythema formation (**A**), skin thickness (**B**), skin hydration (**C**) and transepidermal water loss (TEWL) (**D**) in ultraviolet (UV) B-irradiated hairless mice. NOR: oral administration of saline without UVB irradiation; UVB-C: oral administration of saline under UVB irradiation; ML: oral administration of low-dose (50 mg/kg b.w.) milk phospholipids under UVB irradiation; MM: oral administration of medium-dose (100 mg/kg b.w.) milk phospholipids under UVB irradiation; MH: oral administration of high-dose (150 mg/kg b.w.) milk phospholipids under UVB irradiation. Data are expressed as means ± standard error (n = 6). * $p < 0.05$, ** $p < 0.01$, and *** $p < 0.001$ vs. UVB-C group (Tukey's test).

Skin hydration and transepidermal water loss (TEWL) were determined to evaluate skin barrier function, which plays an important role in skin hydration. Skin moisture content and transdermal moisture loss are expressed as delta values, which represent the difference in values before and after the experiment (Figure 1C,D). There were significant differences in the delta values of skin hydration and TEWL between normal and UVB-C groups ($p < 0.001$). Furthermore, UV-induced reduction in skin hydration was reversed by milk phospholipid administration in a concentration-dependent manner (Figure 1C). Similarly, milk phospholipid administration also improved TEWL (Figure 1D). In particular, ML and MH groups showed a significant improvement in TEWL compared to the UVB-C group, but there was no dose-dependent change ($p < 0.01$ and $p < 0.05$, respectively). Compared with the UVB-C group, the MM group showed a tendency to decrease TEWL, but there was no significant difference. Taken together, milk phospholipids exhibited improving effects on the skin barrier function.

2.3. Effects of Milk Phospholipids on HA Synthesis and Degradation

HA is a compound responsible for skin moisture and is involved in inhibiting moisture loss from the epidermis and maintaining skin elasticity. HA is synthesized by hyaluronic acid synthase (HAS) and degraded by hyaluronidase (HYAL). UV irradiation significantly decreased the expression of HAS (HAS1, 2, and 3) (Figure 2A–C; $p < 0.01$, $p < 0.05$, $p < 0.05$, respectively), but significantly increased the expression of HYAL (HYAL1 and 3) (Figure 2D,E; $p < 0.01$, $p < 0.001$, respectively) compared to the normal group. However, milk phospholipids increased HAS expression, which were reduced by UV, and decreased HYAL expression, which were increased by UV, in a concentration-dependent manner (Figure 2D,E). HA content in the skin was also reduced by UV irradiation, but it was significantly increased by oral milk phospholipid administration (Figure 2F; $p < 0.001$). Hereby, milk phospholipids are thought to improve skin hydration by regulating the expression of HAS and HYAL.

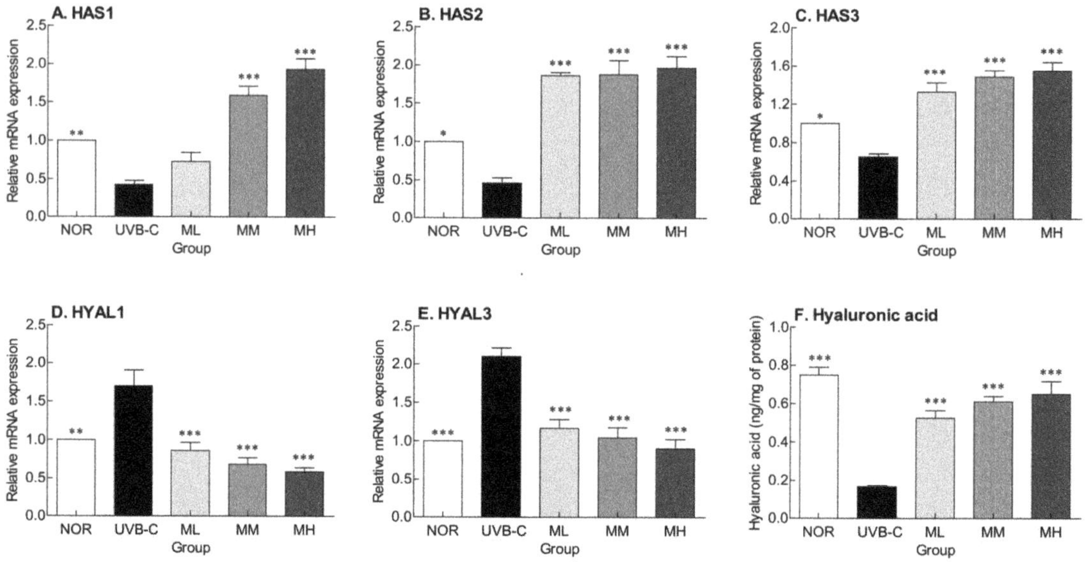

Figure 2. Effects of milk phospholipids on gene expression of HAS (**A–C**) and HYAL (**D**,**E**) and hyaluronic acid (HA) content (**F**) in UVB-irradiated hairless mice. NOR: oral administration of saline without UVB irradiation; UVB-C: oral administration of saline under UVB irradiation; ML: oral administration of low-dose (50 mg/kg b.w.) milk phospholipids under UVB irradiation; MM: oral administration of medium-dose (100 mg/kg b.w.) milk phospholipids under UVB irradiation; MH: oral administration of high-dose (150 mg/kg b.w.) milk phospholipids under UVB irradiation. Data are expressed as means ± standard error ($n = 6$). * $p < 0.05$, ** $p < 0.01$, and *** $p < 0.001$ vs. UVB-C group (Tukey's test). HAS: hyaluronan synthase; HYAL: hyaluronidase.

2.4. Effects of Milk Phospholipids on the Expression of Skin Moisture-Related Factors

UV rays damage the skin, causing abnormal skin barrier function and eventually dryness [15]. The effect of milk phospholipids on the recovery of skin barrier function damaged by UV was examined. The gene expression of involucrin and filaggrin, which are differentiation-promoting factors involved in keratinocyte membrane formation, and AQP3, a gene that encodes a protein that synthesizes the water passage in the basal outer layer of the cell membrane, were determined (Figure 3). Their expressions were significantly lower in the UVB-C group than in the normal group (Figure 3; $p < 0.01$, $p < 0.001$, and $p < 0.01$, respectively). Milk phospholipids significantly increased their expression in a concentration-dependent manner (Figure 3; $p < 0.001$). Collectively, milk phospholipids appear to be involved in the restoration of skin barrier function by suppressing UV-mediated decrease in the expression of these factors.

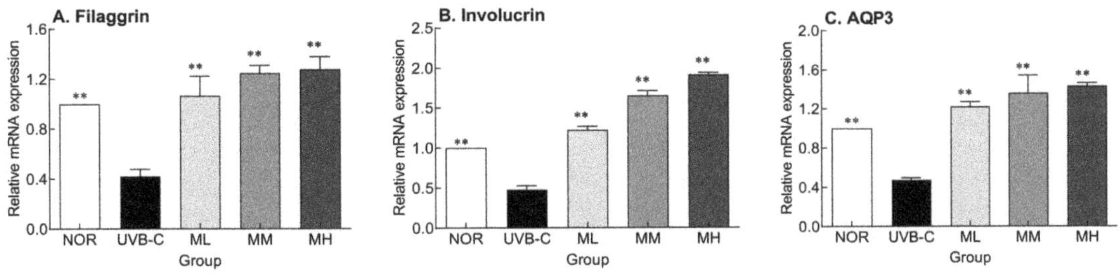

Figure 3. Effects of milk phospholipids on the expression of filaggrin (**A**), involucrin (**B**), and AQP3 (**C**) in UVB-irradiated hairless mice. NOR: oral administration of saline without UVB irradiation; UVB-C: oral administration of saline under UVB irradiation; ML: oral administration of low-dose (50 mg/kg b.w.) milk phospholipids under UVB irradiation; MM: oral administration of medium-dose (100 mg/kg b.w.) milk phospholipids under UVB irradiation; MH: oral administration of high-dose (150 mg/kg b.w.) milk phospholipids under UVB irradiation. Data are expressed as means ± standard error ($n = 6$). ** $p < 0.01$ vs. UVB-C group (Tukey's test). AQP3: aquaporin3.

2.5. Effects of Milk Phospholipids on ROS Production and Expression of Genes Encoding Antioxidant Enzymes

Figure 4 shows the inhibitory effect of milk phospholipids on ROS production and expression of genes encoding antioxidant enzymes. ROS levels were significantly higher in the UVB-C group than in the normal group (Figure 4; $p < 0.001$). Milk phospholipids significantly lowered UV-induced ROS production in a concentration-dependent manner (Figure 4A; $p < 0.01$ and $p < 0.001$, respectively). The expression of superoxide dismutase 1 (SOD1), SOD2, catalase (CAT), and glutathione peroxidase 1 (GPx1), which are involved in ROS removal, were lower in the UVB-C group than in the normal group (Figure 4B–E). Milk phospholipids suppressed UV-mediated decrease in gene expression in a concentration-dependent manner. In particular, the expression of the antioxidant enzymes was significantly increased by MM and MH doses. Altogether, oral milk phospholipid administration inhibited photoaging by suppressing ROS generation and regulating the expression of genes encoding antioxidant enzymes.

2.6. Effects of Milk Phospholipids on Nrf2-Keap1-Related Protein Expression

Oral administration of milk phospholipids suppressed UVB-induced ROS generation and the decrease in gene expression of antioxidant enzymes. Therefore, to examine the underlying mechanisms of milk phospholipids, protein expression of nuclear factor erythoride-2-related factor 2 (Nrf2) and Kelch-like ECH-associated protein 1 (Keap-1), which are affected by oxidative stress, and heme oxygenase-1 (HO-1), an antioxidant enzyme, were examined by Western blotting (Figure 5). Protein expression of Nrf2 and HO-1 ($p < 0.001$ and $p < 0.05$, respectively) were significantly lower, but that of keap1, a negative regulator of Nrf2, was significantly higher ($p < 0.01$) in the UVB-C group than in the normal group (Figure 5A,B). Milk phospholipids increased Nrf2 and HO-1 expression and decreased Keap1 expression in a concentration-dependent manner. Taken together, milk phospholipids demonstrated ROS scavenging effects by increasing the expression of the transcription factor Nrf2, contributing to an increase in ROS scavenging-related enzymes.

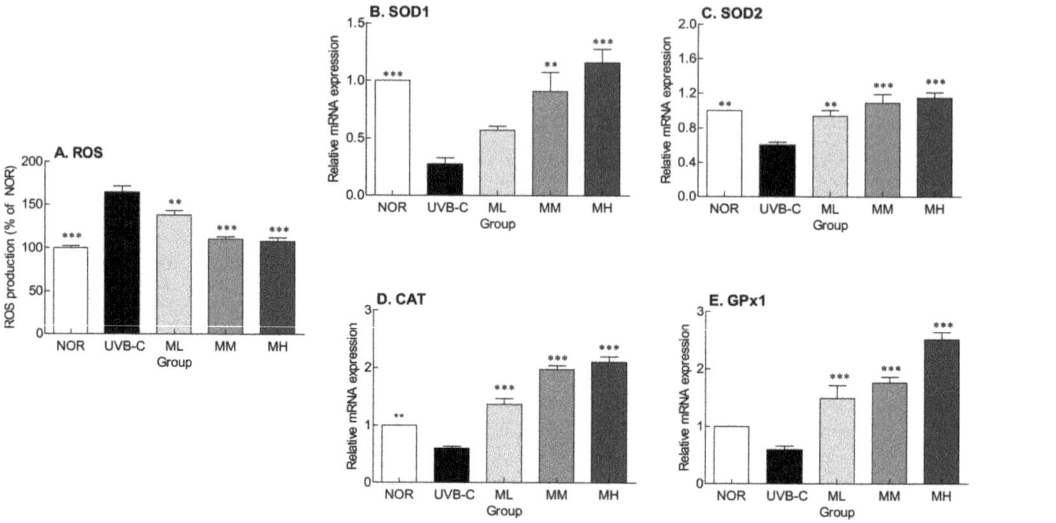

Figure 4. Effects of milk phospholipids on ROS production (**A**) and the expression of genes encoding the antioxidant enzymes SOD1 (**B**), SOD2 (**C**), CAT (**D**), and GPx1 (**E**) in UVB-irradiated hairless mice. NOR: oral administration of saline without UVB irradiation; UVB-C: oral administration of saline under UVB irradiation; ML: oral administration of low-dose (50 mg/kg b.w.) milk phospholipids under UVB irradiation; MM: oral administration of medium-dose (100 mg/kg b.w.) milk phospholipids under UVB irradiation; MH: oral administration of high-dose (150 mg/kg b.w.) milk phospholipids under UVB irradiation. Data are expressed as means ± standard error (n = 6). ** $p < 0.01$ and *** $p < 0.001$ vs. UVB-C group (Tukey's test). ROS: reactive oxygen species; CAT: catalase; SOD: superoxide dismutase; Gpx-1: glutathione peroxidase-1.

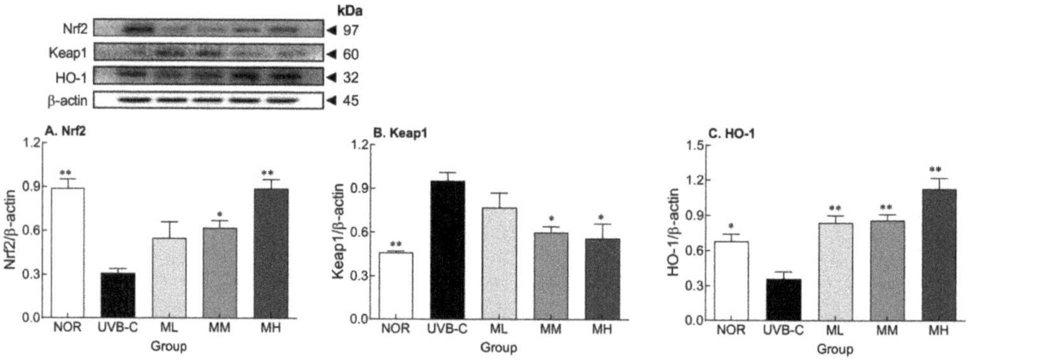

Figure 5. Effects of milk phospholipids on protein expression of Nrf2 (**A**), Keap1 (**B**), and HO-1 (**C**) in UVB-irradiated hairless mice. Western blot and protein quantifications are shown. NOR: oral administration of saline without UVB irradiation; UVB-C: oral administration of saline under UVB irradiation; ML: oral administration of low-dose (50 mg/kg b.w.) milk phospholipids under UVB irradiation; MM: oral administration of medium-dose (100 mg/kg b.w.) milk phospholipids under UVB irradiation; MH: oral administration of high-dose (150 mg/kg b.w.) milk phospholipids under UVB irradiation. Data are expressed as means ± standard error (n = 6). * $p < 0.05$ and ** $p < 0.01$ vs. UVB-C group (Tukey's test). Nrf2: Nuclear factor erythroid-2-related factor 2; Keap1: Kelch-like ECH-associated protein 1; HO-1: heme oxygenase-1.

3. Discussion

The skin has a barrier function to protect the body from environmental factors, such as chemicals, pathogens, air pollutants, and UV rays, and a moisturizing function to prevent water loss from the body. The skin is an organ that retains moisture and performs an essential barrier function to protect the body from the intrusion of external factors [16]. In particular, the stratum corneum of the epidermis acts as a skin barrier to protect the skin from the outside while retaining moisture [17]. Skin moisture is the most important factor for maintaining skin health and controlling aging. When UV-induced photoaging progresses, the optimal moisture content of the skin cannot be achieved, and enzymes that produce lipids and natural moisturizing factors are not activated, causing the stratum corneum of the skin to become dry and thick [18].

The keratinocyte membrane is formed when proteins, such as loricrin and involucrin, generated during the differentiation of keratinocytes are cross-linked by transglutaminases [19]. Genes whose expression increases as differentiation progresses include transglutaminase 1 and 3, involucrin, loricrin, cornifin, and filaggrin. Among them, transglutaminase 1 strengthens the skin barrier by crosslinking the structural proteins involucrin, loricrin, and cornifin and imparts resistance and insolubility by catalyzing a stable isotope peptide bond during keratinocyte formation [20]. UV light inhibits the production of natural moisturizing factors by reducing the expression of proteins involved in keratinocyte membrane formation, such as filaggrin, involucrin, and caspase-14 [21]. In keratinocytes, among transmembrane proteins, AQP3 is expressed; aquaporins (AQPs) specifically transport water and glycerol into cells. Glycerol is a structural component for various lipids, and has a positive effect on elasticity and wound healing by increasing the water content of the epidermal layer [22]. Interestingly, AQP3 expression decreases with age, and contributes to skin dryness [23]. As a result, UV-induced damage to the epidermis layer causes skin dryness by reducing skin moisture content [24]. We found that UV rays reduced the expression of genes encoding filaggrin, involucrin, and AQP3, which are skin moisture-related factors; however, oral milk phospholipid administration reversed this effect (Figure 3). Uncontrolled expression of inflammation-related factors causes dysfunction of the epidermal barrier, which is seen in diseases such as atopic dermatitis and psoriasis. Our previous study reported that administration of sphingomyelin-containing milk phospholipids reduces production of proinflammatory cytokines in the photoaged skin [25]. We also confirmed that milk phospholipids had similar effects to virgin coconut oil, *Centella asiatica*, and *Tamnolia vermicularis*, which are known to affect epidermal markers (filaggrin, involucrin and AQP3) responsible for keratinocyte differentiation and skin barrier function [26–28].

Although UV irradiation decreased skin moisture and increased TEWL, oral milk phospholipid administration improved the skin barrier damage (Figure 1). UV irradiation promotes the detachment of keratinocytes from the skin surface, weakening the skin hydration and skin barrier function. According to the H&E staining results of epidermis, oral administration of milk phospholipids inhibited the increase in epidermal thickness caused by UV irradiation [29]. In addition, administration of milk phospholipids lowered plasma triglyceride levels compared to the NOR group. Ref. [30] reported that milk-derived phospholipids improved plasma lipid levels, including triglycerides, in obese mice induced by a high-fat diet. UV irradiation induces the increase of TEWL and the alteration of stratum corneum lipid profile by disrupting epidermal barrier functions in skin [31]. According to our previous study [25], sphingomyelin, which is involved in the skin barrier function, decreased when irradiated with UV light, but showed a tendency to increase in the skin tissue by administration of milk phospholipids. Dietary sphingomyelin might improve skin barrier function by altering skin inflammation and covalently bound ω-hydroxy ceramides, and it is also known that dietary sphingomyelin can promote the formation of the epidermal cornified envelope by changes in inflammation-related gene expression [32].

ROS and reactive nitrogenous species (RNS), which are formed during inflammatory processes, are known to be critical for signaling, aging, and apoptosis in extrinsic or intrinsic skin. Several studies showed that hydroxyl radical or endogenous ROS affected epidermal HA catabolism by producing peroxynitrite [33,34]. Overproduction of ROS/RNS has been known to promote the degradation of HA, which has the capacity to bind and retain water molecules, resulting in the expansion of skin aging. Thus, the control of ROS/RNS in skin metabolism is essential in skin health. Furthermore, HA is a major component of the extracellular matrix and is involved in water retention, maintenance of intercellular spacing, and storage and diffusion of cell growth factors and nutrients [35]. HA content decreases with skin aging and represent a direct cause of decreases in skin elasticity and moisture content [36]. HA is synthesized by HAS and degraded by HYAL [37]. Among HAS, HAS2 and HAS3 are known to play a decisive role in HA synthesis. UV rays affect the expression of these proteins, damaging the epidermal layer and causing a decrease in skin moisture content. This eventually leads to dry skin and accelerated aging [38]. However, oral milk phospholipid administration seemed to contribute to skin hydration by maintaining HA content, which was reduced by UV irradiation, as milk phospholipids upregulated HAS gene expression and downregulated HYAL gene expression (Figure 2).

In addition, it has been reported that accumulated UV irradiation breaks the antioxidant defense system and promotes the generation of lipid oxidation products including malondialdehyde by increasing ROS production [39]. ROS generated by UVB exposure accelerate skin aging by participating in wrinkle formation and melanin generation via decomposition of binding tissue components, such as collagen and HA, and abnormal crosslinking of these components [40,41]. In this study, SOD, CAT, and GPx1 expression were decreased by UV irradiation, but milk phospholipid administration reversed this effect (Figure 4). Moreover, although sphingomyelin has been known to represent one of the main factors behind the antioxidant activity of milk and dairy products [42], the current study has demonstrated that sphingomyelin-containing milk phospholipids can upregulate the expression of antioxidant enzymes in skin tissues.

The skin barrier improvement effect of milk phospholipids appeared to be related to the activation of Nrf2-keap1, which is related to ROS removal. Milk phospholipids activated Nrf2 and increased the expression of HO-1, an antioxidant enzyme, thereby reducing ROS produced by UV irradiation (Figures 4A and 5). Nrf2, which responds sensitively to intracellular oxidative stress, is a transcription factor for some antioxidant enzymes and is known to play an important role in protecting against UV-induced skin cell death and acute skin burns [43]. In a steady state, Nrf2 levels in the cytoplasm are kept low by Keap1, which forms a complex with Nrf2 and degrades it. During oxidative stress, Nrf2 is separated from Keap1 and translocated to the nucleus [44], where it forms a dimer with the small Maf protein, binds to the antioxidant response element (ARE), and activates HO-1, an ARE-dependent antioxidant gene [45]. HO-1 is a member of the intracellular phase II enzyme family, plays an important role in ROS generation and maintenance of homeostasis against oxidative stress, and is one of the cell protection mechanisms.

We demonstrated that milk phospholipid administration improved skin hydration in a UVB-induced photoaging model. In the future, we will investigate changes in the lipid composition of the skin by administration of milk phospholipids. However, since this study evaluated the effect of milk phospholipids in a UVB-induced photoaging model, additional studies are needed to clarify the effect of milk phospholipid administration on the skin in a normal skin model. Milk phospholipids were involved in the restoration of skin barrier function damaged by UV rays and improved the skin moisture and transdermal moisture loss. This result was suspected to be because of a decrease in UV-induced ROS production following the activation of the Nrf2-Keap1 system. In addition, milk phospholipids may have improved UVB-induced skin barrier damage by supplying the skin constituent lipids containing ceramide.

4. Materials and Methods

4.1. Materials and Animals

Sphingomyelin-containing milk phospholipids were provided by Solus Advanced Materials Co., Ltd. (Yongin, Korea). Sphingomyelin-containing milk phospholipids consist of: Phospholipids 25.0 ± 5.0% (phosphatidylcholine 7.5 ± 1.5%, phosphatidylethanolamine 6.5 ± 1.5%, phosphatidylserine 1.3 ± 0.7%, sphingomyelin 6.5 ± 1.0%), lactosylceramide 1.5 ± 0.5%, glucosylceramide 0.9 ± 0.6% and GD3 ganglioside 0.3 ± 0.1%. Eight-week-old SKH-1 hairless male mice (Central Lab Animal Inc., Seoul, Korea) were acclimatized for one week before being used in the experiment. They were reared in an environment maintained at a temperature of 23 ± 2 °C, humidity of 55 ± 10%, and light/dark cycles of 12 h. They were provided with solid feed and ad libitum access to drinking water. To induce photoaging, mice were irradiated with UVB; UVB irradiation dose was 1 minimal erythemal dose (MED; 75 mJ/cm^2) in weeks 1 and 2, 2 MED in week 3, 3 MED in week 4, and 4 MED in week 5 onward, 3 times a week for a total of 8 weeks. For UV irradiation, a UV irradiator (BLX-254, Vilber Lourmat, Marne La Vallee, France) with UVB lamp (UB800, Waldman Licht Technik GmbH) was used. In addition, UV spectrum was measured using a UV light meter (UV-340, Lutron, Taipei, Taiwan) before UVB irradiation. Experimental animals were randomly divided into five groups, each group containing six mice: normal (unirradiated) group (NOR), UVB-irradiated group (UVB-C), 50 mg/kg b.w. milk phospholipid-administered group (ML), 100 mg/kg b.w. milk phospholipid-administered group (MM), and 150 mg/kg b.w. milk phospholipid-administered group (MH). Mice in the experimental groups (ML, MM, and MH) were orally administered with milk phospholipids once a day. The sample was administered intragastrically and was conducted simultaneously with UVB irradiation for a total of 8 weeks. The body weight of mice was measured once a week. After the end of the experiment period, whole blood was collected from the abdominal aorta and centrifuged (3000 rpm, 4 °C, 15 min) to separate plasma for serum biochemical analysis. The animal experiments were approved by the Korea University Institutional Animal Care and Use Committee (KUIACUC-2020-0054).

4.2. Measurement of Skin Parameters

To evaluate skin barrier function, skin hydration and TEWL were measured on the dorsal side of mice. Skin hydration content was measured using a Corneometer CM825 (Courage and Khazaka electronic GmbH, Cologne, Germany) and TEWL was measured using a Tewameter TM300 (Courage and Khazaka electronic GmbH) equipped with a Multi Probe Adapter MPA5 (Courage and Khazaka electronic GmbH). The erythema index of the mouse dorsal skin was determined using a Mexameter MX18 (Courage and Khazaka electronic GmbH). A caliper (Ozaki MFG Co., Ltd., Tokyo, Japan) was used to measure skin thickness, which was the thickness of the middle part after grabbing the skin of the lower part of the mouse tail and the neck by hand. Skin parameters (erythema, skin thickness, skin hydration, and TEWL) were expressed as delta values, which are differences from the initial values of the experiment.

4.3. Quantitative Real-Time PCR (qRT-PCR) Analysis

The qRT-PCR analysis was performed using cDNA prepared from mRNA fractions of tissue lysates as previously described [46]. Target gene expression was normalized to that of glyceraldehyde-3-phosphate dehydrogenase (GAPDH; NM_008084.3). The target gene information is as follows: HAS-1 (NM_008215.2), HAS-2 (NM_008216.3), HAS-3 (NM_008217.4), HYAL-1 (NM_008317.6), HYAL-3 (NM_178020.3), filaggrin (NM_001013804.2), involucrin (NM_008412.3), AQP3 (NM_016689.2), superoxide dismutase (SOD) 1 (NM_011434.1), SOD2 (NM_013671.3), glutathione perxodiase1 (GPx-1) (NM_008160.6), and catalase (CAT) (NM_009804.2).

4.4. Measurement of Protein Expression by Western Blot Analysis

Proteins were isolated from skin tissues using a lysis buffer, and the concentration of the isolated proteins was quantified using the Bradford assay [47]. Proteins were separated by 6–15% sodium dodecyl sulfate-polyacrylamide gel electrophoresis and then transferred to polyvinylidene fluoride membranes. After blocking the membranes with 5% skim milk solution for 1 h, anti-Nrf2 (1:1000, #12721, Cell Signaling Technology, Beverly, MA, USA), anti-Keap1 (1:1000, #8047, Cell Signaling Technology), anti-HO-1 (1:1000, SC-120745, Santa Cruz Biotechnology, Dallas, TX, USA), and anti-β-actin (1:1000, #8457, Cell Signaling Technology) were added, and incubated overnight at 4°C. The membranes were washed with 1X Tris-buffered saline (TBST) buffer, reacted with secondary antibodies (anti-rabbit IgG, 1:2000, #7074, Cell Signaling Technology) for 2 h, washed with 1X TBST buffer, and treated with enhanced chemiluminescence reagent to determine the expression of proteins. The results were normalized to the endogenous protein, β-actin.

4.5. Measurement of ROS

For ROS measurement, the skin tissues were homogenized in 40 mM Tris-HCl buffer (pH 7.4) and centrifuged [48]. Next, 10 μM 2′,7′-dichlorodihydrofluorescein diacetate (Sigma-Aldrich, St Louis, MO, USA) was added to the supernatant and reacted at 37 °C. Fluorescence (Excitation wavelength: 485 nm, emission wavelength: 535 nm) was measured after 30 min (SpectraMax Gemini EM fluorometer, Molecular Devices, Sunnyvale, CA, USA).

4.6. Statistical Analysis

Data are expressed as the mean ± standard mean error (SEM). The statistical significance was at the $p < 0.05$ level. Comparisons between treatment groups were performed using one-way ANOVA followed by the Tukey's multiple range test using the Statistical Package for the Social Science software (SPSS Version 20, SPSS Inc., Chicago, IL, USA).

Supplementary Materials: The following supporting information can be downloaded at: https://www.mdpi.com/article/10.3390/molecules27082545/s1, Table S1. Effects of milk phospholipids on body weight changes in ultraviolet (UV) B-irradiated hairless mice; Table S2. Effects of milk phospholipids on serum biochemical parameters in ultraviolet (UV) B-irradiated hairless mice.

Author Contributions: Conceptualization, K.-B.H., K.J., and H.J.S.; methodology, K.-B.H. and H.J.S.; software, M.G.K.; validation, Y.A., K.-B.H. and K.J.; formal analysis, Y.A. and M.G.K.; investigation, K.J.; data curation, Y.A.; writing—original draft preparation, Y.A. and H.J.S.; writing—review and editing, Y.A., M.G.K., K.-B.H., K.J., and H.J.S.; visualization, Y.A. and M.G.K.; supervision, K.-B.H. and H.J.S.; project administration, H.J.S.; funding acquisition, H.J.S. All authors have read and agreed to the published version of the manuscript.

Funding: This research was funded by Solus Biotech Co., Ltd. (Yongin, Korea) and Holistic Bio Co., Ltd. (Seongnam, Korea) (Q2026771).

Institutional Review Board Statement: The animal study protocol was approved by the Korea University Institutional Animal Care and Use Committee (protocol code KUIACUC-2020-0054 and date of approval: 2020.11.09.).

Informed Consent Statement: Not applicable.

Data Availability Statement: The data that support the findings of this study are available from the corresponding author upon reasonable request.

Acknowledgments: This research was supported by Solus Biotech Co., Ltd (Yongin, Korea) and Holistic Bio Co., Ltd., Korea (Seongnam, Korea). The funders had no role in the design of the study; in the collection, analyses, or interpretation of data; in the writing of the manuscript, or in the decision to publish the results.

Conflicts of Interest: The authors declare no conflict of interest.

Sample Availability: Samples of the milk phospholipids are available from the corresponding authors, Ki-Bae Hong and Hyung Joo Suh.

References

1. Parke, M.A.; Perez-Sanchez, A.; Zamil, D.H.; Katta, R. Diet and skin barrier: The role of dietary interventions on skin barrier function. *Dermatol. Pract. Concept* **2021**, *11*, e2021132. [CrossRef] [PubMed]
2. Fahad, D.; Mohammed, M.T. Oxidative stress: Implications on skin diseases. *Plant Arch.* **2020**, *20*, 4150–4157.
3. Chen, J.; Liu, Y.; Zhao, Z.; Qiu, J. Oxidative stress in the skin: Impact and related protection. *Int. J. Cosmetic Sci.* **2021**, *43*, 495–509. [CrossRef] [PubMed]
4. Liu, N.; Matsumura, H.; Kato, T.; Ichinose, S.; Takada, A.; Namiki, T.; Asakawa, K.; Morinaga, H.; Mohri, Y.; De Arcangelis, A.; et al. Stem cell competition orchestrates skin homeostasis and ageing. *Nature* **2019**, *568*, 344–350. [CrossRef] [PubMed]
5. Chavoshnejad, P.; Foroughi, A.H.; Dhandapani, N.; German, G.K.; Razavi, M.J. Effect of collagen degradation on the mechanical behavior and wrinkling of skin. *Phys. Rev. E* **2021**, *104*, 034406. [CrossRef]
6. Jang, H.H.; Lee, S.N.; Jang, H.H.; Lee, S.N. Epidermal skin barrier. *Asian J. Beauty Cosmetol.* **2016**, *14*, 339–347. [CrossRef]
7. Del Rosso, J.Q.; Levin, J. The clinical relevance of maintaining the functional integrity of the stratum corneum in both healthy and disease-affected skin. *J. Clin. Aesthet. Dermatol.* **2011**, *4*, 22.
8. Farage, M.A.; Miller, K.W.; Elsner, P.; Maibach, H.I. Structural characteristics of the aging skin: A review. *Cutan. Ocul. Toxicol.* **2007**, *26*, 343–357. [CrossRef]
9. Lipozencić, J.; Pastar, Z.; Marinović-Kulisić, S. Moisturizers. *Acta Dermatovenerol. Croat.* **2006**, *14*, 104–108.
10. Draelos, Z.D. The science behind skin care: Moisturizers. *J. Cosmet. Dermatol.* **2018**, *17*, 138–144. [CrossRef]
11. Sotiropoulou, G.; Zingkou, E.; Pampalakis, G. Redirecting drug repositioning to discover innovative cosmeceuticals. *Exp. Dermatol.* **2021**, *30*, 628–644. [CrossRef] [PubMed]
12. Lee, K.; Kim, A.; Hong, K.B.; Suh, H.J.; Jo, K. Preparation and characterization of a polar milk lipid-enriched component from whey powder. *Food Sci. Anim. Resour.* **2020**, *40*, 209–220. [CrossRef] [PubMed]
13. Parodi, P.W. Cows' milk fat components as potential anticarcinogenic agents. *J. Nutr.* **1997**, *127*, 1055–1060. [CrossRef] [PubMed]
14. Lee, K.; Kim, S.; Kim, A.; Suh, H.J.; Hong, K.B. Sphingolipid identification and skin barrier recovery capacity of a milk sphingolipid-enriched fraction (MSEF) from buttermilk powder. *Int. J. Cosmet. Sci.* **2020**, *42*, 270–276. [CrossRef]
15. Oh, M.J.; Nam, J.J.; Lee, E.O.; Kim, J.W.; Park, C.S. A synthetic C16 omega-hydroxyphytoceramide improves skin barrier functions from diversely perturbed epidermal conditions. *Arch. Dermatol. Res.* **2016**, *308*, 563–574. [CrossRef]
16. Lee, S.H.; Jeong, S.K.; Ahn, S.K. An update of the defensive barrier function of skin. *Yonsei Med. J.* **2006**, *47*, 293–306. [CrossRef]
17. Fernando, I.P.S.; Dias, M.K.H.M.; Madusanka, D.M.D.; Han, E.J.; Kim, M.J.; Jeon, Y.J.; Ahn, G. Fucoidan refined by *Sargassum confusum* indicate protective effects suppressing photo-oxidative stress and skin barrier perturbation in UVB-induced human keratinocytes. *Int. J. Biol. Macromol.* **2020**, *164*, 149–161. [CrossRef]
18. Ganceviciene, R.; Liakou, A.I.; Theodoridis, A.; Makrantonaki, E.; Zouboulis, C.C. Skin anti-aging strategies. *Derm. -Endocrinol* **2012**, *4*, 308–319. [CrossRef]
19. Woo, S.W.; Rhim, D.B.; Kim, C.; Hwang, J.K. Effect of standardized boesenbergia pandurata extract and its active compound panduratin a on skin hydration and barrier function in human epidermal keratinocytes. *Prev. Nutr. Food Sci.* **2015**, *20*, 15–21. [CrossRef]
20. Candi, E.; Schmidt, R.; Melino, G. The cornified envelope: A model of cell death in the skin. *Nat. Rev. Mol. Cell Biol.* **2005**, *6*, 328–340. [CrossRef]
21. Cheong, Y.; Kim, C.; Kim, M.B.; Hwang, J.K. The anti-photoaging and moisturizing effects of Bouea macrophylla extract in UVB-irradiated hairless mice. *Food Sci. Biotechnol.* **2018**, *27*, 147–157. [CrossRef] [PubMed]
22. Fluhr, J.W.; Darlenski, R.; Surber, C. Glycerol and the skin: Holistic approach to its origin and functions. *Br. J. Dermatol.* **2008**, *159*, 23–34. [CrossRef] [PubMed]
23. Li, J.; Tang, H.; Hu, X.; Chen, M.; Xie, H. Aquaporin-3 gene and protein expression in sun-protected human skin decreases with skin ageing. *Australas. J. Dermatol.* **2010**, *51*, 106–112. [CrossRef] [PubMed]
24. Kim, S.; Oh, H.I.; Hwang, J.K. Oral administration of fingerroot (*Boesenbergia pandurata*) extract reduces ultraviolet b-induced skin aging in hairless mice. *Food Sci. Biotechnol.* **2012**, *21*, 1753–1760. [CrossRef]
25. Ahn, Y.; Kim, M.G.; Choi, Y.J.; Lee, S.J.; Suh, H.J.; Jo, K. Photoprotective effects of sphingomyelin-containing milk phospholipids in ultraviolet B-irradiated hairless mice by suppressing nuclear factor-kappaB expression. *J. Dairy Sci.* **2022**, *105*, 1929–1939. [CrossRef]
26. Varma, S.R.; Sivaprakasam, T.O.; Arumugam, I.; Dilip, N.; Raghuraman, M.; Pavan, K.B.; Rafiq, M.; Paramesh, R. In vitro anti-inflammatory and skin protective properties of Virgin coconut oil. *J. Tradit. Complement. Med.* **2019**, *9*, 5–14. [CrossRef]
27. Shen, X.; Guo, M.; Yu, H.; Liu, D.; Lu, Z.; Lu, Y. *Propionibacterium acnes* related anti-inflammation and skin hydration activities of madecassoside, a pentacyclic triterpene saponin from *Centella asiatica*. *Biosci. Biotechnol. Biochem.* **2019**, *83*, 561–568. [CrossRef]
28. Haiyuan, Y.U.; Shen, X.; Liu, D.; Hong, M.; Lu, Y. The protective effects of beta-sitosterol and vermicularin from *Thamnolia vermicularis* (Sw.) Ach. against skin aging in vitro. *An. Acad. Bras. Cienc.* **2019**, *91*, e20181088. [CrossRef]
29. Agren, U.M.; Tammi, R.H.; Tammi, M.I. Reactive oxygen species contribute to epidermal hyaluronan catabolism in human skin organ culture. *Free Radic. Biol. Med.* **1997**, *23*, 996–1001. [CrossRef]

30. Wat, E.; Tandy, S.; Kapera, E.; Kamili, A.; Chung, R.W.; Brown, A.; Rowney, M.; Cohn, J.S. Dietary phospholipid-rich dairy milk extract reduces hepatomegaly, hepatic steatosis and hyperlipidemia in mice fed a high-fat diet. *Atherosclerosis* **2009**, *205*, 144–150. [CrossRef]
31. Bissett, D.L.; Hannon, D.P.; Orr, T.V. An animal model of solar-aged skin: Histological, physical, and visible changes in UV-irradiated hairless mouse skin. *Photochem. Photobiol.* **1987**, *46*, 367–378. [CrossRef] [PubMed]
32. Oba, C.; Morifuji, M.; Ichikawa, S.; Ito, K.; Kawahata, K.; Yamaji, T.; Asami, Y.; Itou, H.; Sugawara, T. Dietary milk sphingomyelin prevents disruption of skin barrier function in hairless mice after UV-B irradiation. *PLoS ONE* **2015**, *10*, e0136377. [CrossRef] [PubMed]
33. Al-Assaf, S.; Navaratnam, S.; Parsons, B.J.; Phillips, G.O. Chain scission of hyaluronan by peroxynitrite. *Arch. Biochem. Biophys.* **2003**, *411*, 73–82. [CrossRef]
34. Kennett, E.C.; Davies, M.J. Degradation of matrix glycosaminoglycans by peroxynitrite/peroxynitrous acid: Evidence for a hydroxyl-radical-like mechanism. *Free Radic. Biol. Med.* **2007**, *42*, 1278–1289. [CrossRef]
35. Kim, S.-H.; Nam, G.-W.; Kang, B.-Y.; Lee, H.-K.; Moon, S.-J.; Chang, I.-S. The effect of kaempferol, guercetin on hyaluronan-synthesis stimulation in human keratinocytes (HaCaT). *J. Soc. Cosmet. Sci. Korea* **2005**, *31*, 97–102.
36. Ghersetich, I.; Lotti, T.; Campanile, G.; Grappone, C.; Dini, G. Hyaluronic acid in cutaneous intrinsic aging. *Int. J. Dermatol.* **1994**, *33*, 119–122. [CrossRef]
37. Papakonstantinou, E.; Roth, M.; Karakiulakis, G. Hyaluronic acid: A key molecule in skin aging. *Derm. -Endocrinol.* **2012**, *4*, 253–258. [CrossRef]
38. Dai, G.; Freudenberger, T.; Zipper, P.; Melchior, A.; Grether-Beck, S.; Rabausch, B.; de Groot, J.; Twarock, S.; Hanenberg, H.; Homey, B. Chronic ultraviolet B irradiation causes loss of hyaluronic acid from mouse dermis because of down-regulation of hyaluronic acid synthases. *Am. J. Pathol.* **2007**, *171*, 1451–1461. [CrossRef]
39. Nakai, K.; Tsuruta, D. What are reactive oxygen species, free radicals, and oxidative stress in skin diseases? *Int. J. Mol. Sci* **2021**, *22*, 10799. [CrossRef]
40. Davies, K.J.A. Protein damage and degradation by oxygen radicals.1. General-aspects. *J. Biol. Chem.* **1987**, *262*, 9895–9901. [CrossRef]
41. Podda, M.; Traber, M.G.; Weber, C.; Yan, L.-J.; Packer, L. UV-irradiation depletes antioxidants and causes oxidative damage in a model of human skin. *Free Radic. Biol. Med.* **1998**, *24*, 55–65. [CrossRef]
42. Won, J.S.; Singh, I. Sphingolipid signaling and redox regulation. *Free Radic. Biol. Med.* **2006**, *40*, 1875–1888. [CrossRef] [PubMed]
43. Marrot, L.; Jones, C.; Perez, P.; Meunier, J.R. The significance of Nrf2 pathway in (photo)-oxidative stress response in melanocytes and keratinocytes of the human epidermis. *Pigment. Cell Melanoma Res.* **2008**, *21*, 79–88. [CrossRef] [PubMed]
44. Kaspar, J.W.; Niture, S.K.; Jaiswal, A.K. Nrf2:INrf2 (Keap1) signaling in oxidative stress. *Free Radic. Biol. Med.* **2009**, *47*, 1304–1309. [CrossRef] [PubMed]
45. Johnson, J.A.; Johnson, D.A.; Kraft, A.D.; Calkins, M.J.; Jakel, R.J.; Vargas, M.R.; Chen, P.C. The Nrf2-ARE pathway an indicator and modulator of oxidative stress in neurodegeneration. *Ann. N. Y. Acad. Sci* **2008**, *1147*, 61–69. [CrossRef] [PubMed]
46. Park, K.; Elias, P.M.; Oda, Y.; Mackenzie, D.; Mauro, T.; Holleran, W.M.; Uchida, Y. Regulation of cathelicidin antimicrobial peptide expression by an endoplasmic reticulum (ER) stress signaling, vitamin d receptor-independent pathway. *J. Biol. Chem.* **2011**, *286*, 34121–34130. [CrossRef] [PubMed]
47. Bradford, M.M. A rapid and sensitive method for the quantitation of microgram quantities of protein utilizing the principle of protein-dye binding. *Anal. Biochem.* **1976**, *72*, 248–254. [CrossRef]
48. Gupta, R.; Dubey, D.K.; Kannan, G.M.; Flora, S.J.S. Concomitant administration of *Moringa oleifera* seed powder in the remediation of arsenic-induced oxidative stress in mouse. *Cell Biol. Int.* **2007**, *31*, 44–56. [CrossRef]

molecules

Article

Flavonoids and Phenols, the Potential Anti-Diabetic Compounds from *Bauhinia strychnifolia* Craib. Stem.

Rachanida Praparatana [1], Pattaravan Maliyam [1], Louis R. Barrows [2] and Panupong Puttarak [1,3,*]

1 Department of Pharmacognosy and Pharmaceutical Botany, Faculty of Pharmaceutical Sciences, Prince of Songkla University, Hat-Yai, Songkhla 90112, Thailand; rachanida.pra@gmail.com (R.P.); 5910720014@email.psu.ac.th (P.M.)
2 Department of Pharmacology and Toxicology, University of Utah, Salt Lake City, UT 81112, USA; lbarrows@pharm.utah.edu
3 Phytomedicine and Pharmaceutical Biotechnology Excellence Center, Faculty of Pharmaceutical Sciences, Prince of Songkla University, Hat-Yai, Songkhla 90112, Thailand
* Correspondence: panupong.p@psu.ac.th; Tel.: +66-994-741598

Abstract: Bioactive compounds from medicinal plants are good alternative treatments for T2DM. They are also sources of lead molecules that could lead to new drug discoveries. In this study, *Bauhinia strychnifolia* Craib. stem, a traditional Thai medicinal plant for detoxification, was extracted into five fractions, including crude extract, BsH, BsD, BsE, and BsW, by ethanolic maceration and sequential partition with hexane, dichloromethane, ethyl acetate, and water, respectively. Among these fractions, BsE contained the highest amounts of phenolics (620.67 mg GAE/g extract) and flavonoids (131.35 mg QE/g extract). BsE exhibited the maximum inhibitory activity against α-glucosidase (IC_{50} 1.51 ± 0.01 μg/mL) and DPP-IV (IC_{50} 2.62 ± 0.03 μg/mL), as well as dominantly promoting glucose uptake on 3T3-L1 adipocytes. Furthermore, the four compounds isolated from the BsE fraction, namely resveratrol, epicatechin, quercetin, and gallic acid, were identified. Quercetin demonstrated the highest inhibitory capacity against α-glucosidase (IC_{50} 6.26 ± 0.36 μM) and DPP-IV (IC_{50} 8.25 μM). In addition, quercetin prominently enhanced the glucose uptake stimulation effect on 3T3-L1 adipocytes. Altogether, we concluded that quercetin was probably the principal bioactive compound of the *B. strychnifolia* stem for anti-diabetic, and the flavonoid-rich fraction may be sufficiently potent to be an alternative treatment for blood sugar control.

Keywords: diabetes; *Bauhinia strychnifolia*; α-glucosidase; dipeptidyl peptidase-IV; glucose uptake; flavonoids

Citation: Praparatana, R.; Maliyam, P.; Barrows, L.R.; Puttarak, P. Flavonoids and Phenols, the Potential Anti-Diabetic Compounds from *Bauhinia strychnifolia* Craib. Stem. *Molecules* **2022**, *27*, 2393. https://doi.org/10.3390/molecules27082393

Academic Editors: Smaoui Slim and Jianbo Xiao

Received: 27 February 2022
Accepted: 25 March 2022
Published: 7 April 2022

1. Introduction

Diabetes mellitus (DM) is a chronic metabolic disorder characterized by abnormally high levels of blood glucose [1]. Over 460 million people worldwide have been affected by DM, and this number has been projected to increase continuously [2,3]. From all diabetes cases, type 2 diabetes mellitus (T2DM) is the most common type, accounting for around 90–95% [4]. It is primarily caused by insulin-resistance, lack of secretion, and/or inefficient action of insulin hormone, contributing to hyperglycemia [2]. T2DM is one of the serious health problems in many countries, including Thailand, and leads to reducing life quality and increasing mortality and healthcare costs [5]. Glycemic control is the critical step of T2DM management and is very important for the prevention of long-term hyperglycemia complications, such as cardiovascular disease, neuropathy, nephropathy, retinopathy, foot damage, and hearing impairment [6,7].

Recently, many anti-diabetic drugs have been developed and consist of multiple modes of action to control blood sugar levels, such as inhibiting metabolic enzymes (such as α-amylase and α-glucosidase), blocking the dipeptidyl peptidase-IV (DPP-IV) enzyme, and enhancing glucose uptake [8,9]. However, there is no successful remedy for curing T2DM. Undesirable side effects, such as diarrhea, abdominal distention, flatulence, liver disorder, bloating, and nausea, are key and cause a lack of patient compliance and poor remedy effectiveness [10,11]. Therefore, new glucose-lowering agents from natural plants that might express less, or no, side effects would be beneficial.

Bauhinia strychnifolia Craib., a local Thai herb, commonly known as Yanang Dang, is one of the candidate plants for T2DM treatment. *B. strychnifolia* is a reddish climbing plant, mostly found in the north of Thailand. *B. strychnifolia* contains high phenolic and flavonoid compounds, which are considered likely bioactive compounds for various biological activities, including anti-diabetic [12–14]. Formerly, the leaves, roots, and stems of *B. strychnifolia* have all been used in foods, health tea drinks, tonics, and traditional Thai medicines (TTM) for immunization, nourishment, alleviating fever, reducing alcoholic toxification, allergy treatment, eliminating toxins, anti-inflammation, and anti-cancer and -diarrheal effects [14–17]. Several studies have also reported its biological and pharmacological activities, which are consistent with TTM knowledge, such as antipyretic, alleviating allergy, detoxification, antioxidants, antimicrobial, anti-cancer, anti-HIV, and anti-diabetic [14–19]. Considering anti-diabetic activity, only two studies have been reported. It was found that the ethanolic extract of the *B. strychnifolia* stem inhibited α-glucosidase in vitro and reduced glucose levels, triglycerides, and total cholesterol in the circulation of the alloxan-diabetic rats [18,19]. In addition, the two flavonoids, namely 3,5,7,3′,5′-pentahydroxy-flavanonol-3-O-a-L-rhamnopyranoside and 3,5,7-trihydroxychromone-3-O-a-L-rhamnopyranoside, that were isolated from the ethanolic extract of the *B. strychnifolia* stem also exhibited good inhibitory effect against α-glucosidase [18]. However, other glucose-lowering effects of *B. strychnifolia* stem have not been studied, with its major bioactive compounds still unknown. Herein, we determined the different anti-diabetic effects of the *B. strychnifolia* stem and explored its potent bioactive compounds. We expected that these findings could support the use of *B. strychnifolia* as an alternative treatment for T2DM or source of lead molecules for new drug discovery in the future.

2. Results and Discussions

2.1. *B. strychnifolia Stem Extraction*

Mimicking a traditional method, the dried powder of B. strychnifolia stem was extracted by ethanolic maceration to obtain crude extract. Traditionally, ethanolic crude extract was used for the treatment of various ailments, such as fever, diarrhea, cancer, and infection [14–17]. In this study, we used crude extract that has a reddish-brown appearance as a main extract for screening anti-diabetic activities. Subsequently, the crude extract was further separated into four fractions, including BsH, BsD, BsE, and BsW, according to its polarity by using hexane, dichloromethane, ethyl acetate, and water, respectively. Four partitioned fractions were then investigated for anti-diabetic effects, and the most active fraction was selected for the isolation of bioactive compounds.

Extraction is a crucial step for recovering phytochemicals from plant materials. Its efficiency depends on the solvent polarity, chemical nature of the compounds, temperature, interfering substance, times, and skill of the extractor [20]. From Table 1, the results showed that the yield of extraction ranged from 0.40% to 60%. The BsW fraction gave the highest yield (60%) by weight, followed by crude extract (10.30%), BsD (8.15%), BsE (5.90%), and BsH (0.40%) fractions, respectively. This suggested that the increasing polarity of solvent will increase extraction yield.

Table 1. Extraction yield, TPC, and TFC of crude extract and four partitioned fractions.

Extracts	% Yield Extraction	TPC mg GAE/g Extract	TFC mg QE/g Extract
Crude	10.30	562.33 ± 2.89	79.70 ± 0.72
BsH	0.40	12.33 ± 0.76	55.10 ± 0.00
BsD	8.15	210.67 ± 1.88	65.52 ± 0.72
BsE	5.90	620.67 ± 0.99 *	131.35 ± 2.50 *
BsW	60.00	392.33 ± 1.65	71.77 ± 0.72

GAE: gallic acid equivalent; QE: quercetin equivalent. * Indicates statistical difference, when compared with other fractions.

2.2. Total Phenolic Contents (TPC) and Total Flavonoid Contents (TFC)

Flavonoids are a large group of natural substances, containing variable phenolic structures, and mostly found in fruits, vegetables, nuts, tea, and herbs [21,22]. Previous studies demonstrated that most dietary flavonoids provided various medical activities, including anti-diabetic agents [21–23]. For example, rutin can reduce carbohydrate absorption by inhibiting α-glucosidase enzyme, kaempferol can enhance glucose uptake, and luteolin can inhibit lipid synthesis [21–23]. Therefore, in this study, we mainly focused on the flavonoids that were suggested as bioactive compounds. To preliminary screen for the bioactive compounds in each fraction, the TPC and TFC of crude extract and four partitioned fractions were evaluated using colorimetric assays. TPC value was obtained from gallic acid calibration curve $y = 0.0002x + 0.0492$, $R^2 = 0.999$ and expressed as the milligram gallic acid equivalent per gram extract (mg GAE/g extract), whereas TFC value was obtained from quercetin calibration curve $y = 0.0008x + 0.0392$, $R^2 = 1$, and expressed as milligram quercetin equivalent per gram extract (mg QE/g extract). From Table 1, the results showed that the highest amounts of both TPC and TFC were found in BsE, with less in crude extract, BsW, BsD, and trace in BsH. Phenolics and flavonoids in medicinal plants and foods were the principal constituents responsible for various anti-diabetic activities [22]. The higher amounts of TPC and TFC are the indication of possible therapeutic activities of plant extracts [24,25]. Thus, BsE, which has the higher amounts of both TPC and TFC, was hypothesized to exhibit greater anti-diabetic activities than other fractions. For this reason, BsE was used as a representative fraction for the further isolation of bioactive compounds (Figure 1).

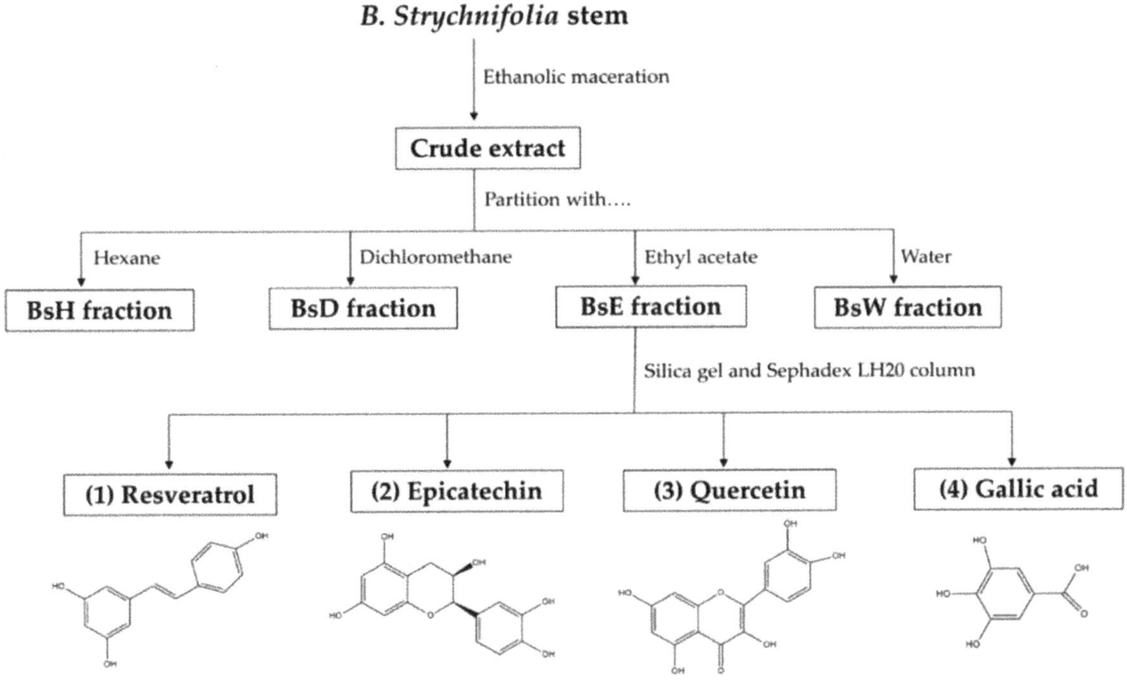

Figure 1. Schematic procedure of extraction, fractionation, and isolation of *B. strychnifolia* stem, as well as the chemical structures of compounds, which were isolated from the BsE fraction.

2.3. Liquid Chromatography Tandem Mass Spectrometry (LC-MS/MS) Analysis

Liquid chromatograph–quadrupole time-of-flight mass spectrometer (LC-QTOF/MS) was used to identify flavonoid and phenolic components in BsE fraction. Based on the retention time (RT) comparison, mass spectrometric obtained under both negative and positive electron spray ionization modes (ESI^-/ESI^+), mass error, data identification score, 25 flavonoids, and 8 phenolics were preliminarily identified. The flavonoids in BsE fraction have been reported as anti-diabetic agents, such as quercetin, kaempferol, naringenin, apigenin, and luteolin (Table 2) [26–38]. A recent systemic review showed that quercetin can reduce the serum glucose level, at the dose of 10 mg/kg of mice body weight [26]. Quercetin from berry extract was shown to induce glucose uptake via an insulin-independent 5′ adenosine monophosphate-activated protein kinase (AMPK) pathway [27]. In vitro studies revealed that kaempferol can enhance insulin secretion and glucose uptake through protein kinase C [28,29]. These results support the idea that BsE fraction contains numerous anti-diabetic flavonoids, and this (flavonoid-rich fraction) might exhibit good anti-hyperglycemia activities.

Table 2. Qualitative characterization of flavonoids and phenolics in BsE fraction.

Compounds	Formula	Retention Time (min)	Mode of Ionization (ESI^-/ESI^+)	Mass	*m/z*	Mass Error (ppm)
Flavonoids						
1. Epigallocatechin	$C_{15}H_{14}O_7$	12.54	$[M-H]^-$	306.07	305.07	0.02
2. Catechin	$C_{15}H_{14}O_6$	13.72	$[M-H]^-$	290.07	289.07	−0.05
3. Apuleirin	$C_{20}H_{20}O_9$	15.00	$[M-H]^-$	404.11	403.10	−0.40
4. Phloridzin	$C_{21}H_{24}O_{10}$	15.11	$[M-H]^-$	436.13	435.13	−0.05
5. Epicatechin	$C_{15}H_{14}O_6$	16.35	$[M-H]^-$	290.07	289.07	−0.57
6. Astilbin	$C_{21}H_{22}O_{11}$	16.71	$[M-H]^-$	450.11	449.11	−0.39
7. Rhapontin	$C_{21}H_{24}O_9$	18.86	$[M-H]^-$	420.14	419.13	−0.29
8. Quercetin3-(2-galloylglucoside)	$C_{28}H_{24}O_{16}$	19.41	$[M-H]^-$	616.11	615.10	0.18
9. Quercetin 3-galactoside	$C_{21}H_{20}O_{12}$	20.68	$[M-H]^-$	464.09	463.08	−1.01
10. Epicatechin Monogallate	$C_{22}H_{18}O_{10}$	21.12	$[M-H]^-$	442.09	441.08	−0.08
11. Taxifolin	$C_{15}H_{12}O_7$	21.70	$[M-H]^-$	304.05	303.05	−0.25
12. Kaempferol−7-o-glucoside	$C_{21}H_{20}O_{11}$	22.21	$[M-H]^-$	448.10	447.09	−1.07
13. Naringenin	$C_{15}H_{12}O_5$	23.53	$[M-H]^-$	272.07	271.06	−0.19
14. Alphitonin	$C_{15}H_{12}O_7$	24.23	$[M-H]^-$	304.06	303.05	−0.17
15. Apigenin 7-O-glucoside	$C_{21}H_{20}O_{10}$	26.12	$[M-H]^-$	432.11	431.11	−0.46
16. Phloretin	$C_{15}H_{14}O_5$	28.84	$[M-H]^-$	274.08	273.08	−0.58
17. Agehoustin C	$C_{22}H_{24}O_{10}$	31.16	$[M-H]^-$	448.13	447.13	0.31
18. Theasinensin C	$C_{30}H_{26}O_{14}$	32.04	$[M-H]^-$	610.13	609.12	3.00
19. Luteolin	$C_{15}H_{10}O_6$	32.62	$[M-H]^-$	286.04	285.04	−0.27
20. Quercetin	$C_{15}H_{10}O_7$	32.92	$[M-H]^-$	302.04	301.04	−0.12
21. Dichamanetin	$C_{29}H_{24}O_6$	33.62	$[M+HCOO]^-$	468.16	513.15	0.28
22. Viniferal	$C_{35}H_{26}O_8$	35.07	$[M+HCOO]^-$	574.17	633.18	−2.57
23. Cnidilin	$C_{17}H_{16}O_5$	35.65	$[M-H]^-$	300.10	299.09	0.09
24. Apigenin	$C_{15}H_{10}O_5$	36.89	$[M-H]^-$	270.05	269.04	0.14
25. Kaempferol	$C_{15}H_{10}O_6$	37.90	$[M-H]^-$	286.05	285.04	−0.17
Phenolics						
26. Gallic acid	$C_7H_6O_5$	4.92	$[M-H]^-$	170.22	169.01	−0.20
27. Caffeic acid	$C_9H_8O_4$	10.58	$[M-H]^-$	180.04	179.03	0.06
28. Gentisic acid	$C_7H_6O_4$	13.19	$[M-H]^-$	154.02	153.19	0.02
29. Kelampayoside A	$C_{20}H_{30}O_{13}$	13.82	$[M+HCOO]^-$	478.19	523.17	−0.08
30. Pyrocatechol	$C_{23}H_{34}O_{14}$	15.41	$[M-H]^-$	110.03	109.03	−0.04
31. Irisxanthone	$C_{20}H_{20}O_{11}$	22.63	$[M-H]^-$	436.10	435.09	−0.28
32. Resveratrol	$C_{14}H_{12}O_3$	22.70	$[M-H]^-$	228.07	227.07	−0.53
33. Isosyringinoside	$C_{23}H_{34}O_{14}$	23.43	$[M-H]^-$	534.19	533.19	0.56

Furthermore, we found that the compounds isolated from the BsE fraction (Section 2.4), which included resveratrol, epicatechin, quercetin, and gallic acid, are consistent with our hypothesis (Figure 2B, Table 2).

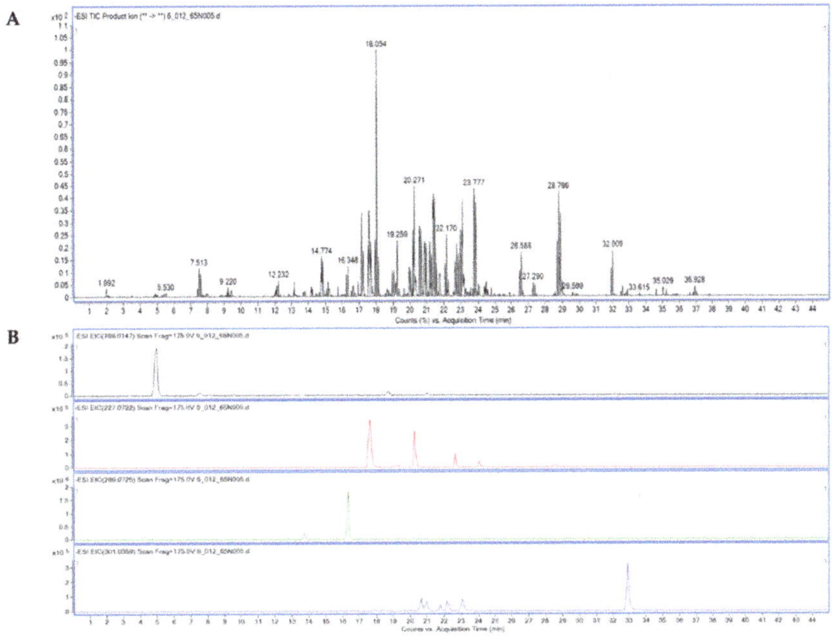

Figure 2. LC-MS/MS analysis. Total ion chromatogram (TIC) of phenolic compounds in BsE fraction (**A**) and extracted ion chromatogram (EIC) of four isolated compounds in Section 2.4, including gallic acid, resveratrol, epicatechin, and quercetin, respectively (**B**).

2.4. Isolation

Based on bioassay guide isolation, four compounds were isolated from the BsE fraction using silica gel and the Sephadex LH20 column. Their chemical structures were identified as resveratrol, epicatechin, quercetin, and gallic acid. The ^{1}H and ^{13}C nuclear magnetic resonance (NMR) spectra of each compound, shown below, also correspond with the structure of resveratrol, epicatechin, quercetin, and gallic acid in databases. Thus, it was confirmed that chemical structures in Figure 1 were resveratrol, epicatechin, quercetin, and gallic acid. To date, only two flavonoids, namely 3,5,7-Trihydroxychromone-3-O-α-L-rhamnopyranoside and 3,5,7,3′,5′-pentahydroxy-flavanonol-3-O-α-L-rhamnopyranoside were isolated from *B. strychnifolia* stem [18]. Additionally reported were β-sitosterol and stigmasterol. This is the first report that identified these four compounds from the *B. strychnifolia* stem. All isolated compounds were further investigated for anti-diabetic activities.

2.4.1. Resveratrol

^{1}H-NMR (500 MHz, CD_3OD): δ 7.35 (2H, d, J = 8.5 Hz, H2′, H6′), 6.96 (1H, d, J = 16.0 Hz, H8), 6.80 (3H, d, J = 16.0 Hz, H7), 6.44 (2H, d, J = 2.5 Hz, H2, H6), and 6.15 (1H, t, J = 2.0 Hz, H4); ^{13}C-NMR (500 MHz, CD_3OD): δ 141.3 (C1), 105.8 (C2, C6), 159.7 (C3, C5), 102.7 (C4), 129.4 (C7), 127.1 (C8), 130.5 (C1′), 128.8 (C2′, C6′), 116.5 (C3′, C5′), and 158.4 (C4′).

2.4.2. Epicatechin

^{1}H-NMR (500 MHz, CD_3OD): 4.84 (br, s, H2), 4.18 (1H, m, H3), 2.87 (1H, dd, J = 5, 17.0 Hz, H4a), 2.74 (1H,dd, J = 3, 17 Hz, H4b), 5.91 (1H, d, J = 2.5 Hz, H6), 5.94 (1H, d, J = 2 Hz, H8), 6.97 (1H, J = 2 Hz, H2′), 6.76 (1H, d, J = 8 Hz, H5′), and 6.8 (1H, dd, J = 2,8.5 Hz, H6′); ^{13}C-NMR (500 MHz, CD_3OD): δ 79.8 (C-2), 67.5 (C-3), 29.2 (C-4), 157.3 (C-5), 96.4 (C-6),

157.9 (C-7), 95.9 (C-8), 157.6 (C-9), 100.1 (C-10), 132.3 (C-1′), 115.3 (C-2′), 145.9 (C-3′), 145.8 (C-4′), 115.9 (C-5′), and 119.4 (C-6′).

2.4.3. Quercetin

^{1}H-NMR (400 MHz, CD_3OD): 12.48 (1H, s, 5-OH), 10.75 (1H, s, br, 3-OH), 9.56 (1H, s, br, 7-OH), 9.32 (2H, s, br, 2×-OH), 6.19 (1H, d, J = 2.0 Hz, H-6), 6.41 (1H, d, J = 2.0 Hz, H-8), 7.68 (1H, d, J = 2.2 Hz, H-2′), 6.89 (1H, d, J = 8.5 Hz, H-5′), 7.54 (1H, dd, J = 2.2, and 8.5 Hz, H-6′); ^{13}C-NMR (100 MHz, CD_3OD): δ 147.6 (s, C-2), 135.6 (s, C-3), 175.8 (s, C-4), 160.6 (s, C-5), 98.1 (d, C-6), 163.8 (s, C-7), 93.3 (d, C-8), 156.1 (s, C-9), 103.0 (s, C-10), 121.9 (s, C1), 115.0 (d, C-2), 145.0 (s, C-3), 146.7 (s, C-4), 115.5 (d, C-5), and 120.0 (d, C-6).

2.4.4. Gallic Acid

^{1}H-NMR (500MHz, CD_3OD): δ 7.06 (2H, s, H-galloy 1-2, 6); ^{13}C-NMR (125MHz, CD_3OD): 120.5 (C-1), 108.9 (C-2, 6), 145.0 (C-3, 5), 138.1 (C-4), and 168.9 (C-7).

2.5. α-Glucosidase Inhibitory Assay

In this study, we investigated the anti-diabetic effect of crude extract, partitioned fractions, and isolated compounds on α-glucosidase activity. As shown in Table 3, the results displayed that BsE fraction has the highest inhibitory activity, with an IC_{50} value of 1.51 ± 0.01 μg/mL, followed by crude extract (IC_{50} 2.37 ± 0.13 μg/mL), BsW (IC_{50} 2.37 ± 0.13 μg/mL), BsD (IC_{50} 10.09 ± 0.75 μg/mL), and BsH fractions (IC_{50} 10.61 ± 0.93 μg/mL), respectively. The inhibitory activity against α-glucosidase of BsE was significantly higher than acarbose, the positive control, which inhibited the enzyme with IC_{50} 329.48 ± 6.91 μg/mL. Moreover, four compounds isolated from BsE, namely resveratrol, epicatechin, quercetin, and gallic acid, were further investigated. Results demonstrated that quercetin has the highest inhibitory ability, with IC_{50} 6.26 ± 0.36 μM, followed by resveratrol, with IC_{50} 8.16 ± 10 μM, whereas epicatechin and gallic acid have no effect. IC_{50} values of both quercetin and resveratrol were also significantly lower than that of acarbose. Currently, acarbose, voglibose, and miglitol are synthetic drugs that serve as α-glucosidase inhibitors and are often prescribed for T2DM patients [39]. These drugs result in slowing down the digestion of polysaccharides from dietary into monosaccharides, followed by delaying glucose absorption and reducing postprandial hyperglycemia [40,41]. Thus, quercetin and resveratrol might be more potent than acarbose in reducing postprandial hyperglycemia. Moreover, as compared to the α-glucosidase inhibitory effect of known active flavonoids from the *B. strychnifolia* stem, namely 3,5,7-Trihydroxychromone-3-O-α-L-rhamnopyranoside (IC_{50} 540 μg/mL) and 3,5,7,3′,5′-pentahydroxy-flavanonol-3-O-α-L-rhamnopyranoside (IC_{50} 980 μg/mL) [18], it was found that quercetin and resveratrol were more potent. Overall, quercetin and resveratrol potentially inhibited α-glucosidase enzyme, showing a potential to decrease postprandial hyperglycemia.

Table 3. Inhibitory activities of the crude extract, partitioned fractions, and isolated compounds.

Samples	IC_{50}	
	α-Glucosidase (μg/mL)	**DPP-IV (μg/mL)**
Crude extract	2.37 ± 0.13 [a]	>50
BsH	10.61 ± 0.93 [a]	>50
BsD	10.09 ± 0.75 [a]	>50
BsE	1.51 ± 0.01 [a]	2.62 ± 0.03 [a]
BsW	2.42 ± 0.10 [a]	3.20 ± 0.02 [a]
Resveratrol	1.41 ± 0.02 [a] (8.16 μM)	3.22 ± 0.02 [a] (14.11 μM)
Epicatechin	>7.26 (25 μM)	>14.52 (50 μM)
Quercetin	1.89 ± 0.11[a] (6.26 μM)	2.49 ± 0.01[a] (8.25 μM)
Gallic acid	>4.25 (25 μM)	>8.50 (50 μM)
Acarbose	329.48 ± 6.91 [b] (509.60 μM)	-
Diprotin A	-	1.14 ± 0.05 [a] (3.26 μM)

Different letters (a,b) indicate statistical difference ($p < 0.05$).

2.6. Dipeptidyl Peptidase-IV (DPP-IV) Inhibitory Assay

DPP-IV is a serine protease enzyme produced from the intestine and associated with degradation of the incretin hormone (a stimulator of insulin secretion). Inhibiting DPP-IV leads to the increasing of active incretin hormone, followed by elevating insulin secretion [42–44]. Insulin regulates glucose homeostasis by inducing glucose storage in muscles, liver, and adipose tissues, stimulating triglyceride synthesis, as well as suppressing the release of free fatty acid into circulation [45]. Blocking the DPP-IV enzyme was another crucial strategy that reflected the glucose-lowering properties of crude extract, partitioned fractions, and isolated compounds. As demonstrated in Table 3, the results showed that BsE gave the highest activity, with an IC_{50} value of 2.62 ± 0.03 μg/mL, followed by BsW (IC_{50} 3.20 ± 0.02 μg/mL). Crude extract, BsD, and BsH exhibited an IC_{50} value higher than 50 μg/mL. BsE fraction showed good inhibitory activity, with low IC_{50}, which close to positive control, diprotin A (IC_{50} 1.14 ± 0.05 μg/mL). Then, four compounds, isolated from BsE, were further investigated. The results showed that quercetin possessed the maximum ability, with an IC_{50} value of 2.49 ± 0.01 μg/mL (8.25 μM), followed by resveratrol (IC_{50} 3.22 ± 0.02 μg/mL or 14.11 μM), while epicatechin and gallic acid inhibited DPP-IV activity, with an IC_{50} value higher than 50 μM. The potency of quercetin, the best compound, was similar to diprotin A. Diprotin A is a standard DPP-IV inhibitor that function as anti-diabetic drugs, such as sitagliptin, vildagliptin, and saxagliptin [42]. Although no prior studies tested the DPP-IV inhibitory activity in extracts or compounds from the *B. strychnifolia* stem, it has been reported that synthetic resveratrol and commercial quercetin could inhibit the DPP-IV enzyme; based on molecular docking, both gave high docking scores, which indicates that they could bind to the active site of DPP-IV [46,47]. These data support that quercetin and resveratrol were DPP-IV inhibitors, even if their potencies were less than anti-diabetic drugs.

2.7. Cell Viability

We used 3T3-L1 adipocytes and raw 264.7 macrophage cells as models for further examining of the cellular mechanisms of crude extract, partitioned fractions, and isolated compounds. Cytotoxicity of all samples toward 3T3-L1 adipocytes and raw 264.7 cells were assessed using MTT assay. As shown in Figure 3, the results showed that the cell viability of both cells was more than 80% at every tested concentration of all samples, indicating no significant cytotoxicity of crude extract, partitioned fractions, and isolated compounds at any tested concentration. Thus, these results, as well as the historical use of *B. strychnifolia*, suggest that the crude extract, partitioned fractions, and isolated compounds are safe and could be used for further experiments.

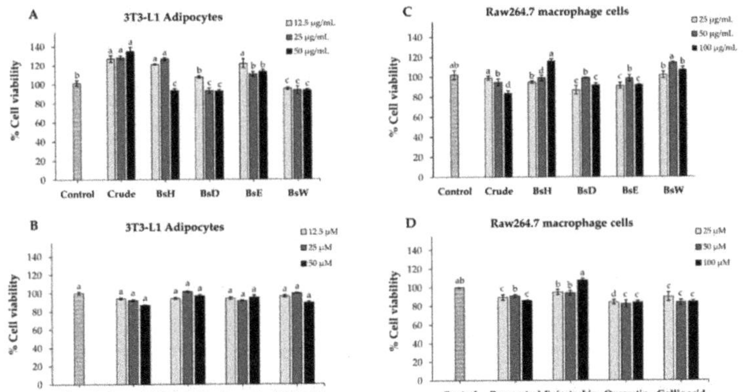

Figure 3. The percentage of cell viability of 3T3-L1 adipocytes (**A**,**B**) and raw264.7 macrophage cells (**C**,**D**) after treatment with different concentrations of crude extract, partitioned fractions, and isolated compounds. Values are expressed as the means ± SD (n = 3). Different letters (a–d) indicate statistical difference ($p < 0.05$).

2.8. Glucose Uptake in 3T3-L1 Adipocytes

To further evaluate the anti-diabetic effect of crude extract, partitioned fractions, and isolated compounds on cellular glucose uptake, we used differentiated 3T3-L1 adipocytes as a model, since adipocytes is one of the major issues involved in insulin function [45]. As demonstrated in Figure 4A, the results showed that crude extract, BsH, BsD, BsE, and BsW fractions could enhance glucose uptake in adult 3T3-L1 adipocytes in a different pattern. The BsE fraction displayed the highest percentage of glucose uptake stimulation at the lowest concentration (85.63% stimulation at 12.5 μg/mL), which is equivalent to positive controls, insulin (92.81% stimulation), and metformin (85.34% stimulation). Fuor compounds, isolated from BsE, were then investigated. The results showed that quercetin demonstrated a higher glucose uptake stimulation effect than resveratrol, gallic acid, and epicatechin, and its activity was also equivalent to positive controls, insulin (92.81% stimulation), and metformin (85.34% stimulation). Unlike most anti-diabetic drugs, metformin is derived from natural plants and used as a first-line drug for T2DM management [48]. It enhances cellular glucose uptake on muscle, liver, and adipocytes, mainly by activating the AMP-activated protein kinase (AMPK) [48]. This is the first preliminary study of extracts and their bioactive compounds from the *B. strychnifolia* stem on cellular glucose uptake. It has been reported that quercetin and 3-O-Acyl-epicatechin could promote glucose by increasing GLUT-4 translocation in skeletal muscle cells, while gallic acid also enhances glucose uptake by inducing GLUT-4 translocation in 3T3-L1 adipocytes [49–51]. At low dose, resveratrol increased glucose uptake and lipid accumulation in 3T3-L1 adipocytes via the insulin signaling pathway [52]. Altogether, the literature suggests that quercetin, resveratrol, epicatechin, and gallic acid from BsE fraction might promote glucose uptake in adipocytes via increasing GLUT-4 translocation.

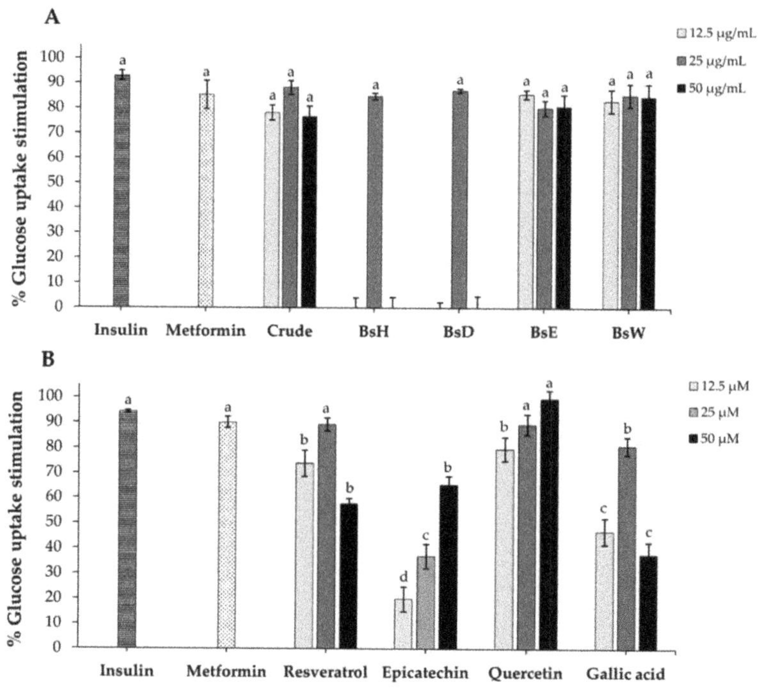

Figure 4. The percentage of glucose uptake stimulation in 3T3-L1 adipocytes after treatment with various concentrations of crude extract (**A**), partitioned fractions (**A**), and isolated compounds (**B**). Insulin (100 nM) and metformin (2 mM) were used as the positive control. Results are expressed as mean ± SD (n = 3). Different letters significantly indicate difference ($p < 0.001$).

2.9. Nitric Oxide (NO) Production

Previous studies reported that the excess nitric oxide (NO), an inflammatory cytokine, can cause insulin resistance in obesity [53]. Lipopolysaccharide (LPS)-induced raw 264.7 cells were used for investigation of the inhibitory effect of crude extract, partitioned fractions, and isolated compounds on nitric oxide production. As demonstrated in Figure 5, the results showed that BsE slightly inhibited NO-production of LPS-induced raw 264.7 cells, crude extract, and BsW slightly increased the production of NO, whereas BsD and BsH reduced NO production in a dose-dependent manner. Indomethacin, the positive control, showed potent anti-inflammatory, which was significantly higher than all fractions. Resveratrol, epicatechin, and gallic acid displayed an inhibitory effect on NO production at only high concentrations, but quercetin has no effect. These results indicated that the isolated compounds could inhibit nitric oxide production moderately. Therefore, it is possible that the low inhibitory effect of these isolated on NO production might be beneficial to improving insulin sensitivity, according to a mentioned study [54].

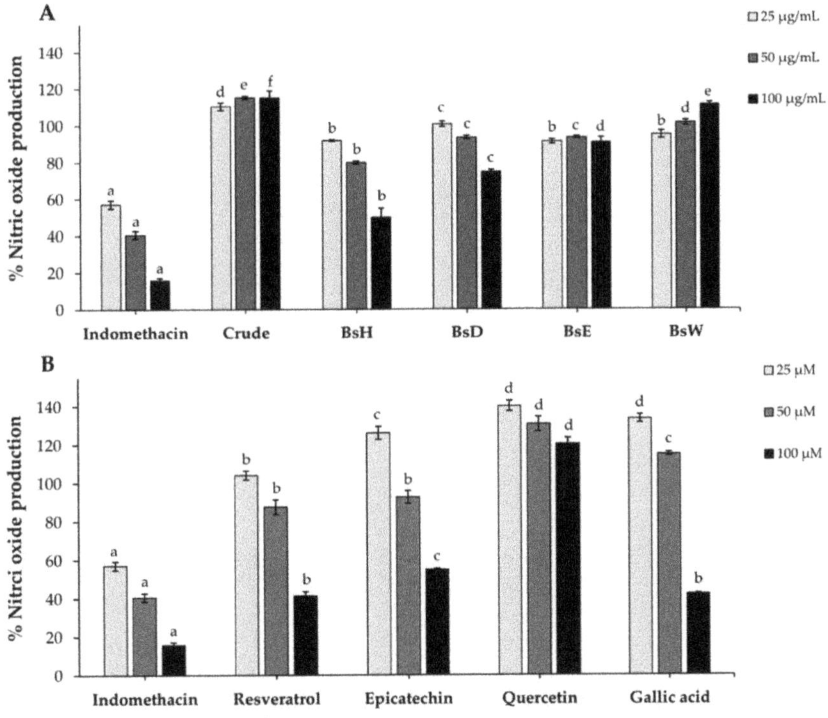

Figure 5. The percentage of nitric oxide production in LPS-induced Raw264.7 macrophage cells after 24 h of incubation with crude extract (**A**), partitioned fractions (**A**), and isolated compounds (**B**). Indomethacin was used as a positive control. Data are expressed as mean ± SD (n = 3). Different letters indicate a lack of significant difference ($p < 0.001$).

3. Materials and Methods

3.1. Chemicals and Reagents

The α-glucosidase from *Saccharomyces cerevisiae*, para-nitrophenyl-α-D-glucopyranoside (pNPG), acarbose, dipeptidyl peptidase-IV (DPP-IV) from porcine kidney, Gly-Pro-p-nitroanilide (GP-p-NA), diprotin A, lipopolysaccharide (LPS) from Escherichia coli O111: B4, dimethyl sulfoxide (DMSO), indomethacin, quercetin, gallic acid, human insulin, dexamethasone (DEX), isobutyl-methylxanthine (IBMX), metformin, and glucose (GO) assay kits were purchased from Sigma Aldrich Inc. (St. Louis, MO. USA). Folin–Ciocalteu reagent was bought from

Merck (Darmstadt, Germany). Fetal bovine serum (FBS), penicillin-streptomycin, Dulbecco's Modified Eagle Medium, high glucose (DMEM, high glucose), Roswell Park Memorial Institute (RPMI) 1640 medium, trypsin-EDTA, trypan blue dyes, and 3-(4,5-dimethylthiazol-2-yl)-2,5 diphenyltetrazolium bromide (MTT) were obtained from Thermo Fisher Scientific (Waltham, CA, USA). Ethanol, methanol, hexane, dichloromethane, ethyl acetate, and acetone (as analytical grades) were purchased from Labscan Asia Co., Bangkok, Thailand.

3.2. Plant Materials

B. strychnifolia stems were collected from the medical herbal garden at Prince of Songkla University, Thailand, in 2021. The herbarium voucher specimen was deposited in the Department of Biology, Faculty of Sciences, Prince of Songkhla University, Thailand, under the code number 0015181. Initially, plant stems were cleaned with tap water, cut into small segments, and dried in a hot air oven at 60 °C. After two days, the dried plant was ground into a fine powder and kept at RT until used.

3.3. Preparation of B. strychnifolia Extracts and Its Isolated Compounds

A total of 900 g of *B. strychnifolia* powder were macerated with 95% ethanol at RT for three days in triplicates. Subsequently, pooled ethanolic macerates were filtrated through filter papers, evaporated with a rotary evaporator at 50 °C, and freeze-dried to get crude extract. After that, 80 g of aliquoted crude extract were dissolved in 10% methanol and sequentially partitioned with hexane, dichloromethane, ethyl acetate, and water, respectively. Each partitioned solvent was then filtrated, evaporated, and freeze-dried to obtain hexane (BsH), dichloromethane (BsD), ethyl acetate (BsE), and water (BsW) fractions.

The BsE fraction that contained the highest amounts of flavonoids and exhibited good activity in enzyme assays was selected for further isolation of bioactive compounds. BsE was applied into silica gel column and eluted by a gradient system, starting with hexane in ethyl acetate, followed by mixtures of ethyl acetate and methanol, as well as mixtures of methanol and acetone. Obtained fractions were screened by TLC and re-chromatographed in silica gel with various gradient elution. Then, all compounds were re-purified by Sephadex LH20 and structurally identified by NMR analysis.

Crude extract, partitioned fractions, and isolated pure compounds were stored at 4 °C in the dark for further analysis.

3.4. Total Phenolic Contents (TPC)

In order to determine the total phenolic contents in crude extract and four partitioned fractions, including BsH, BsD, BsE, and BsW, the Folin–Ciocalteu colorimetric assay was carried out using gallic acid as a reference standard [55]. Shortly, 0.01 mL of each sample, at a concentration of 1 mg/mL, in methanol, was individually pipetted into 1.5 mL tube, followed by adding 0.20 mL of 10% Folin–Ciocalteu reagent and 1 mL of 10% sodium carbonate. After resting at RT for 20 min, each mixture was aliquoted into a 96-well plate, and we measured the absorbance at 765 nm with a spectrophotometer (DTX880 Multimode Detector, Beckman Coulter®, Wien, Austria). The results were done in triplicate and expressed as the mg of gallic acid equivalent/mg extract (mg GAE/mg extract).

3.5. Total Flavonoid Contents (TFC)

The total flavonoid contents of crude extract and four partitioned fractions were detected using quercetin as a reference standard by means of aluminum chloride colorimetric assay [56]. Briefly, 0.6 mL of each sample, at a concentration of 1 mg/mL, in methanol, was separately added into 5 mL tube, containing 1.25 mL of distilled water and 0.75 mL of 5% $NaNO_2$. Then, the solution was incubated at RT for 5 min, followed by adding 0.15 mL of 10% $AlCl_3$. Five minutes later, each reaction was thoroughly mixed with 0.5 mL of 1 M NaOH, and we measured the absorbance at 510 nm by a spectrophotometer (DTX880 Multimode Detector, Beckman Coulter®, Wien, Austria). The results were performed in triplicate and expressed as the mg of quercetin equivalent/mg extract (mg QE/mg extract).

3.6. Liquid Chromatography with Tandem Mass Spectrometry (LC-MS/MS) Analysis

Phenolic and flavonoid compounds in BsE fraction were identified by liquid chromatograph–quadrupole time-of-flight mass spectrometer (LC-QTOF MS, Agilent Technology, Santa Clara, CA, USA). The chromatographic separation was run on a Zorbax Eclipse Plus C18 column (150 × 2.1 mm, particle size 1.8 μm) at 25 °C. The injection volume was 2 μL, and the flow rate was kept at 0.2 mL/min. Elution was carried out by using two mobile phases: eluent A (0.1% formic acid in water) and eluent B (acetonitrile). The gradient program was as follows: 5% B (3 min), 23% B (22 min), 35% B (10 min), and 5% B (10 min).

MS was operated with Dual AJS source, and the instrument parameters were as follows: gas temperature, 325 °C; gas flow, 13 L/min; nebulizer, 35 psi; sheath gas temperature, 275 °C; sheath gas flow, 12 L/min; VCap, 4000 V; nozzle, 2000 V; fragmentor, 175 V; skimmer, 65 V; and octopole RF peak, 750. The scan range of the ion trap was 100–1500 m/z for MS and 50–1500 m/z for MS/MS. The mass spectra were recorded in both negative and positive ion modes. The reference mass ions purine at 112.9856 (for negative mode) and 121.0508 (for positive mode), as well as the HP921 at 1033.9881 (for negative mode) and 922.0098 (for negative mode), were used for continuously correcting any mass drift. All data were analyzed by LC/MS Data Acquisition and Qualitative Analysis Workflows software.

3.7. α-Glucosidase Inhibitory Assay

The inhibitory effects against α-glucosidase enzyme of crude extract, partitioned fractions, and all isolated pure compounds were examined according to previous assay [57]. In brief, 50 μL of each sample in 50 mM phosphate buffer, pH 6.9 (at various concentrations), was pre-mixed with 50 μL of 0.57 U/mL α-glucosidase enzyme for 10 min at 37 °C. Then, 50 μL of 5 mM pNPG substrate was added into the mixture and further incubated for 30 min, after that stopping the reaction by 50 μL of 1 M sodium carbonate solution (Na_2CO_3). A microplate reader (DTX880 Multimode Detector, Beckman Coulter®, Wien, Austria) immediately measured the absorbance at 405 nm. The following equation calculated the percentage of inhibition: $[(A_{control} - A_{sample})/A_{control}] \times 100$, where $A_{control}$ and A_{sample} were the absorbance of reaction without and with the sample, respectively. Acarbose served as a positive control. The results were reported as a concentration of sample that could inhibit enzyme activity by 50% (IC_{50}).

3.8. Dipeptidyl Peptidase-IV (DPP-IV) Inhibitory Assay

The DPP-IV inhibition activities of crude extract, partitioned fractions, and isolated pure compounds were investigated, following the procedure of Van et al. (2009) and Al-Masri et al. (2009), with some modifications [58]. Concisely, 40 μL of each sample diluted in 50 mM Tris-HCl buffer, pH 8.0 (at final concentration 50 ug/mL), was pre-incubated with 30 μL of 0.05 U/mL of DPP-IV enzyme at 37 °C. After 10 min, 100 μL of 0.2 mM GP-p-NA substrate was added to the solution and further incubated for 30 min. Then, the reaction was terminated with 30 μL of 25% acetic acid, and the absorbance at 405 nm was suddenly measured by a microplate reader (DTX880 Multimode Detector, Beckman Coulter®, Wein, Austria). The percentage of inhibition was calculated, following the formula: $[(A_{control} - A_{sample})/A_{control}] \times 100$, where $A_{control}$ and A_{sample} were the absorbances of reaction with and without sample, respectively. Diprotin A was employed as a positive control, and the results were presented as the IC_{50} value.

3.9. Cell Culture

Raw 264.7 cells (murine macrophages) were purchased from the American Type Culture Collection (Manassas, VA, USA). They were grown in RPMI 1640 medium, supplemented with 10% FBS and 1% penicillin-streptomycin at 37 °C, under a humidified atmosphere containing 5% CO_2. The medium was changed every 2–3 days and subcultured once the cells reached 80–90% confluence. The 3T3-L1 pre-adipocytes (mouse embryonic fibroblast) were generously given by the Medical Science Research and Innovation Institute,

Prince of Songkla University. They were cultured in DMEM, high glucose, supplemented with 10% FBS and 1% penicillin-streptomycin and maintained in the same condition as the RAW 264.7 cells. These cells should be subcultured once the cells were reached 70% confluence, with the medium changed every two days.

3.10. Cell Differentiation

The 3T3-L1 pre-adipocytes were induced to differentiate into adult adipocytes after 48 h of confluence in 48 wells plate. The cells were incubated with cultured media, comprising of 1 μg/mL insulin, 10 μM dexamethasone, and 0.5 mM IBMX. After 48 h of incubation, differentiate media were removed and replaced with cultured media containing 1 μg/mL insulin. Then, the maintained media were changed every two days for eight days.

3.11. Cell Viability Assay

The cytotoxicity of crude extract, partitioned fractions, and isolated compounds was assessed in 3T3-L1 adipocytes and raw264.7 macrophage cells, according to previous methods [59]. Raw264.7 cells were seeded into 96-well plates, at a density of 1×10^5 cells/well, and incubated at 37 °C for 2 h. The 3T3-L1 pre-adipocytes were seeded at a density of 5×10^3 cells/well and maintained in the same condition for 24 h. After this step, the adhered cells were treated overnight with various concentrations of samples. Then, the treated medium was changed and further incubated with the MTT solution, at a final concentration of 0.5 mg/mL for 3 h. After that, the MTT-containing medium was removed and replaced with 100 μL of DMSO to dissolve the formazan crystals. Cell viability was monitored by measuring the absorbance at 570 nm. Results were presented as the percentage of control.

3.12. Glucose Uptake Assay

The glucose uptake stimulatory effect of crude extract, partitioned fractions, and isolated compounds was determined by the previously described method [60]. Briefly, differentiated adipocytes were grown in 48-well plates, with cultured media containing 2% FBS, for 18 h. Then, the cells were washed and treated with various concentrations of samples in a complete low glucose medium. After 24 h of incubation, the treated medium of samples was individually transferred into 96-well plates and mixed with the reagent of the GO kit at 37 °C for 30 min. The reaction was terminated by 6M H_2SO_4, and the absorbance of remaining glucose at 540 nm was recorded by a microplate reader. The ability to enhance glucose uptake was measured by the amount of glucose in the media. The percentage of glucose uptake stimulation was calculated by following equation: %stimulation = $[(A_{control} - A_{sample})/A_{control}] \times 100$, where $A_{control}$ and A_{sample} were the absorbances of reaction with and without sample, respectively.

3.13. Inhibition of Nitric Oxide (NO) Production Using Griess Renitricaction Assay

The investigation of the inhibitory activities against LPS induced nitric oxide (NO) production of crude extract, partitioned fractions, and isolated compounds by Griess assay, which was accomplished as described by Kaewdana et al. [61]. Briefly, Raw264.7 cells were seeded, at the density of 1×10^5 cells/well, in 96-well plates. After 1–2 h culture, the medium was changed and co-treated with different concentrations of samples and 200 ng/mL LPS for 24 h. After that, the nitrite-containing medium was equally mixed with the Griess reagent (1% sulfanilamide in 5% phosphoric acid and 0.1% naphthyl ethylenediamine dihydrochloride (NED) in water) and measured by a microplate reader at 570 nm. Indomethacin (a nonsteroidal anti-inflammatory drug) was used as a positive control. Results were shown as the percentage of inhibition calculated, following the equation: $[(A_{control} - A_{sample})/A_{control}] \times 100$.

3.14. Statistics

All data were expressed as mean ± standard deviation and done in triplicate. Any significant difference was determined by one-way analysis of variance (one-way ANOVA), at $p < 0.05$, unless stated differently, using the GraphPad Prism 9 statistical package (GraphPad Software Inc., La Jolla, CA, USA).

4. Conclusions

This is the first study to investigate the different anti-diabetic effects of crude extract, partitioned fractions, and isolated compounds of *B. strychnifolia* stem. Using a traditional method, the ethanolic crude extract was obtained and considered for further separating into different fractions, including BsH, BsD, BsE, and BsW. The screening of TPC and TFC revealed that the BsE fraction from ethyl acetate contained the highest amounts of phenolics and flavonoids and might exert anti-diabetic effects. BsE manifested potent inhibitory activities against α-glucosidase and DPP-IV enzymes, as well as the highest rate of glucose uptake stimulation. The LC-MS/MS analysis demonstrated numerous phenolics and flavonoids in BsE fractions, and four isolated compounds from BsE, namely resveratrol, epicatechin, quercetin, and gallic acid, were also identified in both LC-MS/MS and NMR analysis. Among isolated compounds, quercetin displayed the maximum capacities in inhibition of both α-glucosidase and DPP-IV enzymes, as well as the stimulation of glucose storage in 3T3-L1 adipocytes. Our results suggested that quercetin was a promising bioactive compound, contributing to the blood glucose-lowering properties of BsE extract and the *B. strychnifolia* stem. It could be used as a bioactive marker of BsE for use as an alternative treatment T2DM. Our finding supports the safety and efficacy of the traditional use of the *B. strychnifolia* stem.

Author Contributions: Conceptualization, R.P., L.R.B. and P.P.; methodology, R.P., L.R.B. and P.P.; software, P.P.; validation, R.P., L.R.B. and P.P.; formal analysis, R.P., P.M. and P.P.; investigation, R.P., P.M. and P.P.; resources, R.P., P.M. and P.P.; data curation, R.P., L.R.B. and P.P.; writing—original draft preparation, R.P. and P.P.; writing—review and editing, R.P., L.R.B. and P.P.; visualization, R.P. and P.P.; supervision, P.P.; project administration, P.P.; funding acquisition, P.P. All authors have read and agreed to the published version of the manuscript.

Funding: The research was supported from Prince of Songkla University and Ministry of Higher Education, Science, Research, and Innovation, under the Reinventing University Project (grant number: REV64034).

Institutional Review Board Statement: Not applicable.

Informed Consent Statement: Not applicable.

Data Availability Statement: Not applicable.

Acknowledgments: We would like to thank Department of Pharmacognosy and Pharmaceutical Botany, Faculty of Pharmaceutical Sciences, Prince of Songkla University for supporting the raw material and the Medical Science Research and Innovation Institute, Prince of Songkla University for supporting 3T3 adipocyte cells.

Conflicts of Interest: The authors declare no conflict of interest.

Sample Availability: Samples of the compounds are not available from the authors.

Abbreviations

A	absorbance
ANOVA	analysis of variance
°C	degree celsius
DMEM	Dulbecco's modified Eagle's medium
DMSO	dimethyl sulfoxide
GP-p-NA	gly-pro-p-nitroanilide
IBMX	isobutyl-methylxanthine
LPS	lipopolysaccharide
MTT	3-(4,5-dimethylthiazol-2-yl)-2,5-diphenyltetra-zolium bromide
nm	nanometer

NMR	nuclear magnetic resonance
NO	nitric oxide
pNPG	para-nirophenyl-α-D-glucopyranoside
RPMI	Roswell Park Memorial Institute
UV	ultraviolet

References

1. American Diabetes Association. Diagnosis and Classification of Diabetes Mellitus. *Diabetes Care* **2007**, *30*, S42–S47. [CrossRef] [PubMed]
2. Saeedi, P.; Petersohn, I.; Salpea, P.; Malanda, B.; Karuranga, S.; Unwin, N.; Colagiuri, S.; Guariguata, L.; Motala, A.A.; Ogurtsova, K.; et al. Global and regional diabetes prevalence estimates for 2019 and projections for 2030 and 2045: Results from the International Diabetes Federation Diabetes Atlas, 9th edition. *Diabetes Res. Clin. Pract.* **2019**, *157*, 107843. [CrossRef] [PubMed]
3. Sun, H.; Saeedi, P.; Karuranga, S.; Pinkepank, M.; Ogurtsova, K.; Duncan, B.B.; Stein, C.; Stein, A.; Chan, J.C.N.; Mbanya, J.C.; et al. IDF diabetes atlas: Global, regional, and country-level diabetes prevalence estimates for 2021 and projections for 2045. *Diabetes Res. Clin. Pract.* **2022**, *183*, 109119. [CrossRef]
4. Galicia-Garcia, U.; Benito-Vicente, A.; Jebari, S.; Larrea-Sebal, A.; Siddiqi, H.; Uribe, K.B.; Ostolaza, H.; Martín, C. Pathophysiology of type 2 diabetes mellitus. *Int. J. Mol. Sci.* **2020**, *21*, 6275. [CrossRef] [PubMed]
5. Deerochanawong, C.; Ferrario, A. Diabetes management in Thailand: A literature review of the burden, costs, and outcomes. *Glob. Health* **2013**, *9*, 11. [CrossRef] [PubMed]
6. Manukumar, H.M.; Kumar, J.S.; Chandrasekhar, B.; Raghava, S.; Umesha, S. Evidences for diabetes and insulin mimetic activity of medicinal plants: Present status and future prospects. *Crit. Rev. Food Sci.* **2017**, *57*, 2712–2729. [CrossRef] [PubMed]
7. Aryangat, A.V.; Gerich, J.E. Type 2 diabetes: Postprandial hyperglycemia and increased cardiovascular risk. Vasc. *Health Risk Manag.* **2012**, *6*, 145–155. [CrossRef]
8. Rehani, P.R.; Iftikhar, H.; Nakajima, M.; Tanaka, T.; Jabbar, Z.; Rehani, R.N. Safety and Mode of Action of Diabetes Medications in comparison with 5-Aminolevulinic Acid (5-ALA). *J. Diabetes Res.* **2019**, *2019*, 4267357. [CrossRef]
9. Meneses, M.J.; Silva, B.M.; Sousa, M.; Sá, R.; Oliveira, P.F.; Alves, M.G. Antidiabetic Drugs: Mechanisms of Action and Potential Outcomes on Cellular Metabolism. *Curr. Pharm. Des.* **2015**, *21*, 3606–3620. [CrossRef]
10. Hollander, P. Safety profile of acabose, an alpha-glucosidase inhibitor. *Drugs* **1992**, *44*, 47–53. [CrossRef]
11. Chaudhury, A.; Duvoor, C.; Reddy Dendi, V.S.; Kraleti, S.; Chada, A.; Ravilla, R.; Marco, A.; Shekhawat, N.S.; Montales, M.T.; Kuriakose, K.; et al. Clinical Review of Antidiabetic Drugs: Implications for Type 2 Diabetes Mellitus Management. *Front. Endocrinol.* **2017**, *8*, 6. [CrossRef]
12. Abdullah, K.M.; Arefeen, A.; Shamsi, A.; Alhumaydhi, F.A.; Naseem, I. Insight into the In Vitro Antiglycation and In Vivo Antidiabetic Effects of Thiamine: Implications of Vitamin B1 in Controlling Diabetes. *ACS Omega.* **2021**, *6*, 12605–12614. [CrossRef]
13. Adeshara, K.A.; Bangar, N.; Diwan, A.G.; Tupe, R.S. Plasma glycation adducts and various RAGE isoforms are intricately associated with oxidative stress and inflammatory markers in type 2 diabe. *Diabetes Metab. Syndr. Clin. Res. Rev.* **2022**, 102441. [CrossRef]
14. Panchinda, C.; Ruangnoo, S.; Itharat, A. Cytotoxic activity against cancer cell lines from the ethanolic extracts and its VLC fractions of Bauhinia strychnifolia leaves. *J. Med. Assoc. Thail.* **2016**, *99* (Suppl. S4), S110–S115.
15. Kaewpiboon, C.; Lirdprapamongkol, K.; Srisomsap, C.; Winayanuwattikun, P.; Yongvanich, T.; Puwaprisirisan, P.; Svasti, J.; Assavalapsakul, W. Studies of the in vitro cytotoxic, antioxidant, lipase inhibitory and antimicrobial activities of selected Thai medicinal plants. *BMC Complemetary Altern. Med.* **2012**, *12*, 217. [CrossRef]
16. Itharat, A.; Sayompark, S.; Hansakul, P.; Dechayont, B. In vitro antioxidant activities of extracts of *Bauhinia strychnifolia* stems and leaves: Comparison with activities in green tea extracts. *Med. Aromat Plants* **2016**, *5*, 2167-0412. [CrossRef]
17. Yuenyongsawad, S.; Bunluepuech, K.; Wattanapiromsakul, C.; Tewtrakul, S. Anti-cancer activity of compounds from Bauhinia strychnifolia stem. *J. Ethnopharmacol.* **2013**, *150*, 765–769. [CrossRef]
18. Bunluepuech, K.; Tewtrakul, S.; Wattanapiromsakul, C. Alpha-glucosidase inhibitory activity of compounds from *Bauhinia strychnifolia*. *J. Chem. Pharm. Res.* **2019**, *11*, 22–26.
19. Thaina, P.; Rattanamusik, N.; Saraphee, T.; Tengyai, S.; Pattanawongsa, A.; Puttarak, P. Anti-diabetic activity of *Bauhinia strychnifolia* Craib stem extract in Alloxan-induced diabetic rat. *Thai J. Pharm. Sci.* **2022**, *46*, accepted.
20. Do, Q.D.; Angkawijaya, A.E.; Tran-Nguyen, P.L.; Huynh, L.H.; Soetaredjo, F.E.; Ismadji, S.; Ju, Y.H. Effect of extraction solvent on total phenol content, total flavonoid content, and antioxidant activity of Limnophila aromatica. *J. Food Drug Anal.* **2014**, *22*, 296–302. [CrossRef]
21. Vinayagam, R.; Xu, B. Antidiabetic properties of dietary flavonoids: A cellular mechanism review. *Nutr. Metab.* **2015**, *12*, 60. [CrossRef] [PubMed]
22. Ong, K.C.; Khoo, H.E. Effects of myricetin on glycemia and glycogen metabolism in diabetic rats. *Life Sci.* **2000**, *67*, 1695–1705. [CrossRef]
23. Al-Ishaq, R.K.; Abotaleb, M.; Kubatka, P.; Kajo, K.; Büsselberg, D. Flavonoids and Their Anti-Diabetic Effects: Cellular Mechanisms and Effects to Improve Blood Sugar Levels. *Biomolecules* **2019**, *9*, 430. [CrossRef] [PubMed]

24. Babbar, N.; Oberoi, H.S.; Sandhu, S.K.; Bhargav, V.K. Influence of different solvents in extraction of phenolic compounds from vegetable residues and their evaluation as natural sources of antioxidants. *J. Food Sci. Technol.* **2014**, *51*, 2568–2575. [CrossRef]
25. Pandey, K.B.; Rizvi, S.I. Plant polyphenols as dietary antioxidants in human health and disease. *Oxidative Med. Cell. Longev.* **2009**, *2*, 270–278. [CrossRef]
26. Bule, M.; Abdurahman, A.; Nikfar, S.; Abdollahi, M.; Amini, M. Antidiabetic effect of quercetin: A systematic review and meta-analysis of animal studies. *Food Chem. Toxicol.* **2019**, *125*, 494–502. [CrossRef]
27. Eid, H.M.; Martineau, L.C.; Saleem, A.; Muhammad, A.; Vallerand, D.; Benhaddou-Andaloussi, A.; Haddad, P.S. Stimulation of AMP-activated protein kinase and enhancement of basal glucose uptake in muscle cells by quercetin and quercetin glycosides, active principles of the antidiabetic medicinal plant Vaccinium vitis-idaea. *Mol. Nutr. Food Res.* **2010**, *54*, 991–1003. [CrossRef]
28. Zanatta, L.; Rosso, A.; Folador, P.; Figueiredo, M.S.; Pizzolatti, M.G.; Leite, L.D.; Silva, F.R. Insulinomimetic effect of kaempferol 3-neohesperidoside on the rat soleus muscle. *J. Nat. Prod.* **2008**, *71*, 532–535. [CrossRef]
29. Zhang, Y.; Liu, D. Flavonol kaempferol improves chronic hyperglycemia-impaired pancreatic beta-cell viability and insulin secretory function. *Eur. J. Pharmacol.* **2011**, *670*, 325–332. [CrossRef]
30. Campanero, M.A.; Escolar, M.; Perez, G.; Garcia-Quetglas, E.; Sadaba, B.; Azanza, J.R. Simultaneous determination of diosmin and diosmetin in human plasma by ion trap liquid chromatography-atmospheric pressure chemical ionization tandem mass spectrometry: Application to a clinical pharmacokinetic study. *J. Pharm. Biomed. Anal.* **2010**, *51*, 875–881. [CrossRef]
31. Zang, Y.; Igarashi, K.; Li, Y. Anti-diabetic effects of luteolin and luteolin-7-O-glucoside on KK-A(y) mice. *Biosci. Biotechnol. Biochem.* **2016**, *80*, 1580–1586. [CrossRef]
32. Baek, Y.; Lee, M.N.; Wu, D.; Pae, M. Luteolin Improves Insulin Resistance in Postmenopausal Obese Mice by Altering Macrophage Polarization. *Curr. Dev. Nutr.* **2019**, *13*, FS12–FS19. [CrossRef]
33. Rauter, A.P.; Martins, A.; Borges, C.; Mota-Filipe, H.; Pinto, R.; Sepodes, B.; Justino, J. Antihyperglycaemic and protective effects of flavonoids on streptozotocin-induced diabetic rats. *Phytother. Res.* **2010**, *24*, S133–S138. [CrossRef]
34. Kim, E.K.; Kwon, K.B.; Song, M.Y.; Han, M.J.; Lee, J.H.; Lee, Y.R.; Park, J.W. Flavonoids protect against cytokine-induced pancreatic beta-cell damage through suppression of nuclear factor kappaB activation. *Pancreas* **2007**, *35*, e1–e9. [CrossRef]
35. Zang, M.; Xu, S.; Maitland-Toolan, K.A.; Zuccollo, A.; Hou, X.; Jiang, B.; Cohen, R.A. Polyphenols stimulate AMP-activated protein kinase, lower lipids, and inhibit accelerated atherosclerosis in diabetic LDL receptor-deficient mice. *Diabetes* **2006**, *5*, 2180–2191. [CrossRef]
36. Priscilla, D.H.; Roy, D.; Suresh, A.; Kumar, V.; Thirumurugan, K. Naringenin inhibits alpha-glucosidase activity: A promising strategy for the regulation of postprandial hyperglycemia in high fat diet fed streptozotocin induced diabetic rats. *Chem. Biol. Interact.* **2014**, *210*, 77–85. [CrossRef]
37. Zygmunt, K.; Faubert, B.; MacNeil, J.; Tsiani, E. Naringenin, a citrus flavonoid, increases muscle cell glucose uptake via AMPK. *Biochem. Biophys. Res. Commun.* **2010**, *398*, 178–183. [CrossRef]
38. Cushnie, T.P.; Lamb, A.J. Antimicrobial activity of flavonoids. *Int. J. Antimicrob. Agents.* **2005**, *26*, 343–356. [CrossRef]
39. Martin, A.E.; Montgomery, P.A. Acarbose: An alpha-glucosidase inhibitor. *Am. J. Health Syst Pharm.* **1996**, *53*, 2277–2337. [CrossRef]
40. Ortiz-Andrade, R.R.; Garcia-Jimenez, S.; Castillo-Espana, P.; Ramirez-Avila, G.; Vila-lobosMolina, R.; Estrada-Soto, S. Alpha-glucosidase inhibitory activity of the methanolic extract from Tournefortia hartwegiana: An anti-hyperglycemic agent. *J. Ethnopharmacol.* **2007**, *109*, 48–53. [CrossRef]
41. Shang, Q.; Xiang, J.F.; Tang, Y.L. Screening α-glucosidase inhibitors from mulberry extracts via DOSY and relaxation-edited NMR. *Talanta* **2012**, *97*, 362–367. [CrossRef]
42. Dipeptidyl Peptidase IV (DPP IV) Inhibitors. Available online: https://www.ncbi.nlm.nih.gov/books/NBK542331/ (accessed on 22 January 2022).
43. Paliwal, G.; Sharma, A.; Upadhyay, N.; Das, M.; Tiwari, A. Therapeutic stimulation of glp-1 protein by implementing in silico to in vitro approach for type-2 diabetes treatment. *Middle East J. Sci. Res.* **2015**, *23*, 1005–1011.
44. Ng, V.W.S.; Glasg, F. Dipeptidyl peptidase (dpp) -iv inhibitor: A novel class of oral anti-hyperglycemic agents. *Clin. Pharmacol. Ther.* **2007**, *11*, 33–34.
45. Rahman, M.S.; Hossain, K.S.; Das, S.; Kundu, S.; Adegoke, E.O.; Rahman, M.A.; Hannan, M.A.; Uddin, M.J.; Pang, M.G. Role of Insulin in Health and Disease: An Update. *Int. J. Mol. Sci.* **2021**, *22*, 6403. [CrossRef]
46. Singh, A.-K.; Patel, P.K.; Choudhary, K.; Joshi, J.; Yadav, D.; Jin, J.-O. Quercetin and Coumarin Inhibit Dipeptidyl Peptidase-IV and Exhibits Antioxidant Properties: In Silico, In Vitro, Ex Vivo. *Biomolecules* **2020**, *10*, 207. [CrossRef]
47. Huang, P.K.; Lin, S.R.; Chang, C.H.; Tsai, M.J.; Lee, D.N.; Weng, C.F. Natural phenolic compounds potentiate hypoglycemia via inhibition of Dipeptidyl peptidase IV. *Sci. Rep.* **2019**, *9*, 15585. [CrossRef]
48. Rena, G.; Hardie, D.G.; Pearson, E.R. The mechanisms of action of metformin. *Diabetologia* **2017**, *60*, 1577–1585. [CrossRef]
49. Boesch-Saadatmandi, C.; Loboda, A.; Wagner, A.E.; Stachurska, A.; Jozkowicz, A.; Dulak, J.; Döring, F.; Wolffram, S.; Rimbach, G. Effect of quercetin and its metabolites isorhamnetin and quercetin-3-glucuronide on inflammatory gene expression: Role of miR-155. *J. Nutr. Biochem.* **2011**, *22*, 293–299. [CrossRef]
50. Ueda-Wakagi, M.; Mukai, R.; Fuse, N.; Mizushina, Y.; Ashida, H. 3-O-Acyl-epicatechins Increase Glucose Uptake Activity and GLUT4 Translocation through Activation of PI3K Signaling in Skeletal Muscle Cells. *Int. J. Mol. Sci.* **2015**, *16*, 16288–16299. [CrossRef]

51. Prasad, C.N.; Anjana, T.; Banerji, A.; Gopalakrishnapillai, A. Gallic acid induces GLUT4 translocation and glucose uptake activity in 3T3-L1 cells. *FEBS Lett.* **2010**, *584*, 531–536. [CrossRef]
52. Lee, H.; Kim, J.W. High-dose Resveratrol Inhibits Insulin Signaling Pathway in 3T3-L1 Adipocytes. *J. Lifestyle Med.* **2013**, *3*, 41–47. [PubMed]
53. Katashima, C.K.; Silva, V.R.R.; Lenhare, L.; Marin, R.M.; Carvalheira, J.B.C. iNOS promotes hypothalamic insulin resistance associated with deregulation of energy balance and obesity in rodents. *Sci. Rep.* **2017**, *7*, 9265. [CrossRef] [PubMed]
54. Aggarwal, H.; Pathak, P.; Singh, P.; Gayen, J.R.; Jagavelu, K.; Dikshit, M. Systemic Insulin Resistance and Metabolic Perturbations in Chow Fed Inducible Nitric Oxide Synthase Knockout Male Mice: Partial Reversal by Nitrite Supplementation. *Antioxidants* **2020**, *9*, 736. [CrossRef] [PubMed]
55. Hatami, T.; Emami, S.A.; Miraghaee, S.S.; Mojarrab, M. Total Phenolic Contents and Antioxidant Activities of Different Extracts and Fractions from the Aerial Parts of Artemisia biennis Willd. *Iran. J. Pharm. Res.* **2014**, *13*, 551–559.
56. Jia, Z.; Tang, M.; Wu, J. The determination of flavonoid contents in mulberry and their scavenging effects on superoxides radicals. *Food Chem.* **1998**, *64*, 555–559.
57. Bachhawat, A.; Shihabudeen, M.S.; Thirumurugan, K. Screening of fifteen Indian Ayurvedic plants for alpha-glucosidase inhibitory activity and enzyme kinetics. *Int. J. Pharm Pharm Sci.* **2011**, *3*, 267–274.
58. Al-Masri, I.M.; Mohammad, M.K.; Tahaa, M.O. Inhibition of dipeptidyl peptidase-IV (DPP-IV) is one of the mechanisms explaining the hypoglycemic effect of berberine. *J. Enzym Inhib. Med. Chem.* **2009**, *24*, 1061–1066. [CrossRef]
59. Choi, J.S.; Nurul Islam, M.; Yousof Ali, M.; Kim, E.J.; Kim, Y.M.; Jung, H.A. Effects of C-glycosylation on anti-diabetic, antiAlzheimer's disease and anti-inflammatory potential of apigenin. *Food Chem. Toxicol.* **2014**, *64*, 27–33. [CrossRef]
60. Vishwanath, D.; Srinivasan, H.; Patil, M.S.; Seetarama, S.; Agrawal, S.K.; Dixit, M.N.; Dhar, K. Novel method to differentiate 3T3 L1 cells *in vitro* to produce highly sensitive adipocytes for a GLUT4 mediated glucose uptake using fluorescent glucose analog. *J. Cell Commun. Signal.* **2013**, *7*, 129–140. [CrossRef]
61. Sudsai, T.; Wattanapiromsakul, C.; Nakpheng, T.; Tewtrakul, S. Evaluation of the wound healing property of *Boesenbergia longiflora* rhizomes. *J. Ethnopharmacol.* **2013**, *150*, 223–231. [CrossRef]

molecules

MDPI

Article

Extraction of Phenolic and Flavonoid Compounds from Mung Bean (*Vigna radiata* L.) Seed Coat by Pressurized Liquid Extraction

Benya Supasatyankul [1], Maythee Saisriyoot [2], Utai Klinkesorn [1], Kittipong Rattanaporn [3] and Sudathip Sae-Tan [1,*]

[1] Department of Food Science and Technology, Faculty of Agro-Industry, Kasetsart University, Bangkok 10900, Thailand; benya.su@ku.th (B.S.); utai.k@ku.th (U.K.)
[2] Department of Chemical Engineering, Faculty of Engineering, Kasetsart University, Bangkok 10900, Thailand; fengmts@ku.ac.th
[3] Department of Biotechnology, Faculty of Agro-Industry, Kasetsart University, Bangkok 10900, Thailand; kittipong.r@ku.th
* Correspondence: fagists@ku.ac.th; Tel.: +66-2562-5037

Abstract: Mung bean seed coat (MBC) is a by-product of the mung bean processing industry. It contains a large number of phenolic compounds with therapeutic anti-inflammatory, anti-diabetic and antioxidant properties. This research aimed to investigate the optimum conditions for phenolic and flavonoid extraction from MBC by pressurized liquid extraction (PLE). Response surface methodology (RSM) was used to study the effects of temperature (80–160 °C), pressure (1200–1800 psi) and ethanol concentration (5–95%) on total phenolic content (TPC), total flavonoid content (TFC) and 2,2′-azinobis(3-ethylbenzothiazoline-6-sulfonic acid) scavenging activity (ABTS). Scale-up extraction was also performed. The optimum conditions for extraction were 160 °C, 1300 psi and 50% ethanol. Under optimum conditions, the TPC was 55.27 ± 1.14 mg gallic acid equivalent (GAE)/g MBC, TFC was 34.04 ± 0.72 mg catechin equivalent (CE)/g MBC and ABTS scavenging activity was 195.05 ± 2.29 mg trolox equivalent (TE)/g MBC. The TFC and ABTS scavenging activity of the extracts obtained at the pilot scale (10 L) was not significantly different from the laboratory scale, while TPC was significantly increased. The freeze-dried MBC extract contained vitexin and isovitexin 130.53 ± 17.89, 21.21 ± 3.22 mg/g extract, respectively. In conclusion, PLE was able to extract phenolics, flavonoids with ABTS scavenging activity from MBC with the prospect for future scale-up for food industry.

Keywords: flavonoid extraction; mung bean; phenolic extraction; pressurized liquid extraction; response surface methodology

Citation: Supasatyankul, B.; Saisriyoot, M.; Klinkesorn, U.; Rattanaporn, K.; Sae-Tan, S. Extraction of Phenolic and Flavonoid Compounds from Mung Bean (*Vigna radiata* L.) Seed Coat by Pressurized Liquid Extraction. *Molecules* **2022**, *27*, 2085. https://doi.org/10.3390/molecules27072085

Academic Editor: Smaoui Slim

Received: 27 February 2022
Accepted: 22 March 2022
Published: 24 March 2022

1. Introduction

Mung bean (*Vigna radiata* L.) is a pulse crop widely cultivated and consumed in Asia, India and the warmer part of Europe and America, with a short growth cycle around 2–3 months [1]. Mung bean is a rich source of proteins, essential amino acids, complex carbohydrates, vitamins and minerals and it is easy to be digested [2]. Mung bean is also well known for a large number of bioactive compounds including protein, phenolic and flavonoid compounds with various health benefits [3]. The majority of by-products from the mung bean industry is mung bean seed coat and it is usually discarded. While phenolic and flavonoid compounds were reported to be abundant in mung bean seed coat for 84.2% and 83.9%, respectively [4]. Many biological activities of mung bean seed coat extract were reported. A recent study showed that mung bean seed coat water extract exerted anti-inflammatory effects by inhibiting NF-κB activation via inhibition of TAK1 phosphorylation and IκBα degradation [5]. Mung bean seed coat water extract also showed antioxidant activity [6]. While ethanolic extract of mung bean seed coat inhibited α-glucosidase activity and exerted anti-diabetic activity in *db/db* mice [7]. The proposed bioactive compounds in mung bean seed coat were vitexin and isovitexin [5]. Total phenolic content in mung bean extract is ranging from 2.03 to 79.65 mg GAE/g extract, depending on the method of extraction [8–10]. Therefore, the study on the extraction method and optimum conditions is necessary.

Conventional extraction methods for polyphenols, phenolics and flavonoids from plant materials are maceration, hydro-distillation, agitated solvent extraction and Soxhlet extraction [11]. Recently, many innovative extraction methods have been introduced to increase the efficacy of extraction including ultrasound-assisted extraction (UAE), microwave-assisted extraction (MAE) and pressurized liquid extraction (PLE). Santos et al. compared the efficacy of α-bisabolol extraction between Soxhlet extraction and PLE [12]. The results showed that the amount of α-bisabolol from both methods was not significantly different. However, PLE used only 20 min, while Soxhlet extraction used 360 min for extraction. PLE was compared with UAE in extracting phenolic compounds from the residues of *Rubus fruticosus*, *Vaccinium myrtillus* and *Eugenia brasiliensis* [13]. The results showed that PLE gave a higher yield of phenolic compounds compared to UAE. PLE was also more efficient in the extraction of terpenes, fatty acids and vitamin E from *Piper gaudichaudianum* Kunth in terms of yield and time [14]. The efficacy of PLE and MAE in extracting phenolic compounds from *Moringa oleifera* leaves was compared by Barriada-Pereira et al. [15]. The results showed that TPC and TFC from PLE were higher than those from MAE. This indicated that PLE was a potential extraction method to increase the efficacy of phenolic compounds from mung bean seed coat.

Many factors play a role in extraction. The efficiency of PLE depends on the extraction temperature, pressure, time, solvent and the ratio between extraction solvent and sample [16]. The present study was designated to investigate the effects of extraction temperature, pressure and concentration of ethanol on TPC, TFC and ABTS of mung bean seed coat. The main goal of the present study was to optimize, validate and scale up the extraction of mung bean seed coat using PLE.

2. Results

2.1. Comparison of Soxhlet and PLE Extraction

Considering the high number of bioactive compounds and therapeutic properties of MBC, the extraction of phenolic and flavonoid compounds from MBC was interesting. The total content of bioactive compounds depends on many factors including the extraction method [17]. Here, MBC was extracted with Soxhlet and PLE extraction methods. The TPC, TFC and ABTS of the MBC extract using two extraction methods were shown and compared in Figure 1. The results showed no significant difference between TPC, TFC and ABTS of the extracted solutions from Soxhlet and PLE. About the extraction time, the PLE method spent only 10 min for extraction, while the Soxhlet method spent 6 h for extraction. Both extraction methods were carried out using 95% ethanol at 80 °C, however, the pressure of 1500 psi was applied in PLE. This indicated that pressure application did not have any effect on TPC, TFC and ABTS extraction, however pressure did reduce the extraction time. The present results agreed with the previous study showed that the PLE method reduced the extraction time in TPC extraction from passion fruit rinds by 12 times [18]. The PLE method also reduced the extraction time and the solvent use in polycyclic aromatic hydrocarbon (PAH) extraction from soil (18 and 10 times lower than Soxhlet extraction method, respectively) [19]. The extraction of phenolic compounds from *Mentha piperita*, *Origanum vulgare*, *Rosmarinus officinalis* L. and *Thymusvulgaris* L. by PLE method also gave a higher yield of phenolic and volatile compounds as well as the higher antioxidant activity compared to Soxhlet extraction method [20]. This did indicate the potential of PLE on MBC extraction.

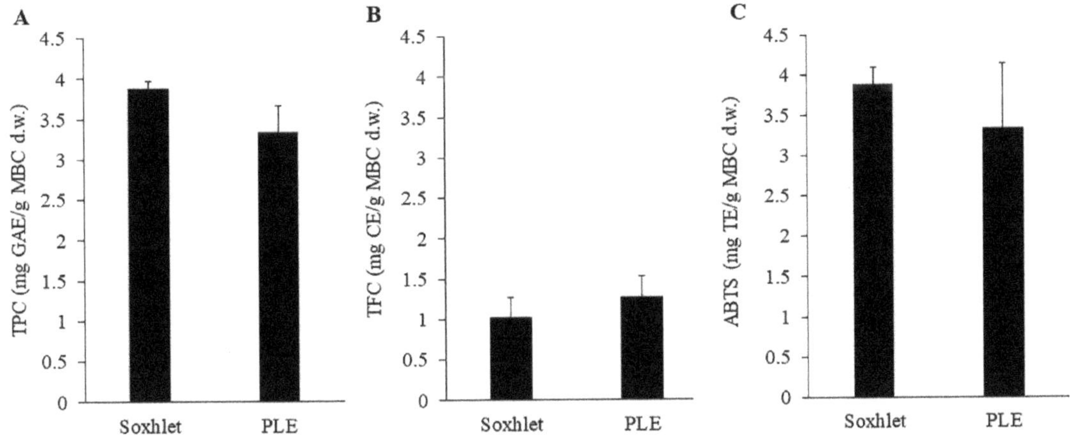

Figure 1. Total phenolic content (**A**), total flavonoid content (**B**) and ABTS (**C**) of MBC extract obtained by Soxhlet extraction (80 °C and 95% ethanol) and PLE (80 °C, 1500 psi and 95% ethanol). There was no significant difference between methods on those response variables.

2.2. Model Fitting and Analysis of Variance

Optimization of extraction conditions was performed in 45 randomized runs to study the effect of different variables on the TPC, TFC and ABTS of the MBC extract. The coded values, uncoded values and the results of dependent variables were shown in Table 1. Phenolic compounds were extracted from MBC ranged from 3.34 ± 0.32 to 52.88 ± 2.48 mg GAE/g MBC (dry weight, d.w.). Extracted flavonoids ranged from 1.27 ± 0.26 to 17.07 ± 0.53 mg CE/g MBC d.w. and ABTS ranged from 3.34 ± 0.80 to 202.71 ± 6.26 mg TE/g MBC d.w. These results showed that TPC, TFC and ABTS considerably depended on the extraction condition, which indicated that the optimization of the extraction condition was important. The optimization of the extraction process was performed by applying second-order polynomial equations. The regression coefficients for dependent variables along with the corresponding coefficient of determination value (R^2) and lack of fit test of TPC, TFC and ABTS were shown in Table 2.

Table 1. Coded and uncoded levels of independent variables in experimental design, TPC, TFC and ABTS from the MBC extract.

Treatment	Temperature (°C)		Pressure (psi)		Ethanol Concentration (%)		TPC (mg GAE/g MBC d.w.)	TFC (mg CE/g MBC d.w.)	ABTS (mg TE/g MBC d.w.)
	Code	Uncode	Code	Uncode	Code	Uncode			
1	−1	80	0	1500	+1	95	3.34 ± 0.32 [e]	1.27 ± 0.26 [g]	3.34 ± 0.80 [g]
2	0	120	−1	1200	+1	95	6.43 ± 0.31 [e]	4.93 ± 1.27 [f]	9.80 ± 1.74 [g]
3	0	120	+1	1800	+1	95	6.22 ± 0.39 [e]	5.45 ± 0.81 [e,f]	11.49 ± 1.67 [g]
4	+1	160	0	1500	+1	95	6.73 ± 0.73 [e]	6.73 ± 0.51 [e,f]	36.70 ± 2.64 [f]
5	−1	80	−1	1200	0	50	28.06 ± 2.11 [c,d]	7.22 ± 0.67 [e]	123.59 ± 5.90 [c,d]
6	−1	80	+1	1800	0	50	30.78 ± 1.94 [c]	6.80 ± 0.62 [e,f]	133.74 ± 2.01 [c]
7	+1	160	−1	1200	0	50	52.88 ± 2.48 [a]	17.07 ± 0.53 [a]	197.51 ± 7.99 [a]
8	0	120	0	1500	0	50	36.91 ± 1.63 [b]	14.13 ± 0.43 [b]	155.29 ± 9.98 [b]
9	0	120	0	1500	0	50	38.81 ± 2.40 [b]	14.21 ± 0.82 [b]	163.50 ± 9.40 [b]
10	0	120	0	1500	0	50	37.97 ± 3.74 [b]	13.90 ± 0.98 [b]	169.08 ± 8.90 [b]
11	+1	160	+1	1800	0	50	50.94 ± 0.53 [a]	13.13 ± 0.45 [b,c]	202.71 ± 6.26 [a]
12	−1	80	0	1500	−1	5	23.47 ± 0.44 [d]	7.26 ± 0.45 [e]	87.58 ± 0.62 [e]
13	0	120	−1	1200	−1	5	29.85 ± 2.72 [c]	9.76 ± 0.34 [d]	110.07 ± 2.07 [d]
14	0	120	+1	1800	−1	5	36.32 ± 1.22 [b]	11.81 ± 0.44 [c,d]	118.15 ± 1.29 [c,d]
15	+1	160	0	1500	−1	5	37.61 ± 0.80 [b]	12.96 ± 0.90 [b,c]	124.35 ± 1.52 [c,d]

Means values with the different letters in the same column are significantly different ($p \leq 0.05$).

Table 2. Regression coefficients, coefficient of determination and *p*-value of the predicted second-order polynomial models for TPC, TFC and ABTS.

Regression Term	TPC		TFC		ABTS	
	Coefficient	*p*-Value *	Coefficient	*p*-Value *	Coefficient	*p*-Value *
Constant	−51.7964	0.225	−57.4491	0.000	−10.0851	0.916
Temperature (X_1)	0.7024	0.011	0.4937	0.000	0.6657	0.270
Pressure (X_2)	0.0206	0.664	0.0443	0.000	0.0097	0.928
Ethanol (X_3)	1.2036	0.000	0.2303	0.000	4.1609	0.000
Temperature × Temperature (X_1X_1)	−0.0012	0.164	−0.0012	0.000	0.0007	0.699
Pressure × Pressure (X_2X_2)	0.0000	0.998	0.0000	0.003	0.0000	0.852
Ethanol × Ethanol (X_3X_3)	−0.0113	0.000	−0.0025	0.000	−0.0498	0.000
Temperature × Pressure (X_1X_2)	−0.0001	0.374	−0.0001	0.009	−0.0001	0.675
Temperature × Ethanol (X_1X_3)	−0.0015	0.045	0.0000	0.847	−0.0005	0.773
Pressure × Ethanol (X_2X_3)	−0.0001	0.205	0.0000	0.240	−0.0001	0.589
R^2	0.9426	-	0.9507	-	0.9809	-
R^2 adjusted	0.9279	-	0.9380	-	0.9760	-
Regression	-	0.000	-	0.000	-	0.000
Lack of fit	-	0.000	-	0.000	-	0.000

* The *p*-value more than 0.05 is not significantly different at the 5% level.

The ANOVA results indicated that three independent variables on the model were able to explain the experimental variation for TPC, TFC and ABTS as shown by the significant *p*-value for each model (Table 2). The corresponding coefficients of determination (R^2) of the model were 0.9426, 0.9507 and 0.9809 for TPC, TFC and ABTS, respectively. These values implied that more than 94.26% of the response variables variation could be explained by the regression models. Although the lack of fit was significant, the correlation between predicted values and actual values of all runs was high (R^2 = 0.9740, 0.8486 and 0.9877) for TPC, TFC and ABTS respectively (data was not shown). The predicted second-order polynomial regression equations with enter elimination method were shown in Table 3 for TPC, TFC and ABTS (Equations (1)–(3), respectively).

Table 3. Predicted second-order polynomial model equations of TPC, TFC and ABTS.

Responses	Polynomial Equations	
TPC (mg GAE/g MBC d.w.)	$Y = 51.7964 + 0.7024X_1 + 0.0206X_2 + 1.2036X_3 - 0.0012X_1^2 - 0.0113X_3^2 - 0.0001X_1X_2 - 0.0015X_1X_3 - 0.0001X_2X_3$	(1)
TFC (mg CE/g MBC d.w.)	$Y = 57.4491 + 0.4937X_1 + 0.0443X_2 + 0.2303X_3 - 0.0012X_1^2 - 0.0025X_3^2 - 0.0001X_1X_2$	(2)
ABTS (mg TE/g MBC d.w.)	$Y = 10.0851 + 0.6657X_1 + 0.0097X_2 + 4.1609X_3 + 0.0007X_1^2 - 0.0498X_3^2 - 0.0001X_1X_2 - 0.0005X_1X_3 - 0.0001X_2X_3$	(3)

2.3. *Effect of the Extraction Variables on TPC, TFC and ABTS*

The model showed a high significant value with the experimental data. ANOVA showed a positive linear effect of temperature (X_1) and ethanol concentration (X_3) was significant for TPC. The linear effect of ethanol concentration (X_3) was significant for ABTS, while all dependent variables, temperature (X_1), pressure (X_2) and ethanol concentration (X_3) were significant for TFC. The quadratic effect of temperature (X_1), pressure (X_2) and ethanol concentration (X_3) were found to be significant for TFC, while the quadratic effect of only ethanol concentration (X_3) was found to be significant for TPC and ABTS. The interaction effect of temperature (X_1) and ethanol concentration (X_3) was found to be significant for TPC, while the interaction effect of temperature (X_1) and pressure (X_2) was found to be significant for TFC. There was no interaction effect of dependent variables on ABTS.

The response surface plot of TPC using different combinations of temperature, pressure and ethanol concentration was shown in Figure 2. The TPC was higher at higher temperatures and vice versa, while pressure did not affect TPC. As shown in Table 2, TPC was significantly influenced by the interaction effect of temperature and ethanol concentration. Figure 2B showed that TPC increased with ethanol concentration increased, but this increase was up to 50% ethanol and then gradually decreased as ethanol concentration increased.

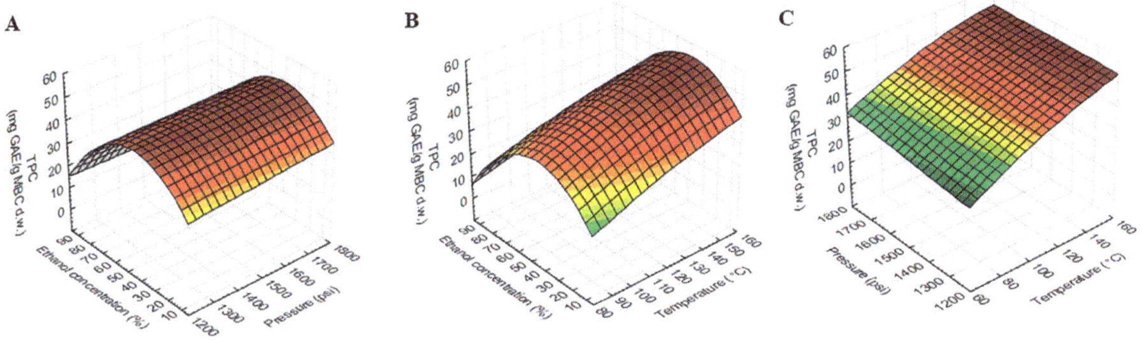

Figure 2. Response surface plots showing the interaction effects of extraction variables on TPC of MBC extract. (**A**) pressure and ethanol concentration at constant temperature (120 °C), (**B**) temperature and ethanol concentration at constant pressure 1500 psi and (**C**) temperature and pressure at constant ethanol concentration (50%).

The response surface plot of TFC using different combinations of temperature, pressure and ethanol concentration was shown in Figure 3. The TFC was higher at higher temperatures, pressure and vice versa. As shown in Table 2, TFC was significantly influenced by the interaction effect of temperature and pressure. The effect of temperature on TFC extraction seen in Figure 3C was positive but clearly lower than the effect of pressure. The effect of ethanol concentration on TFC was similar to TPC but less potent.

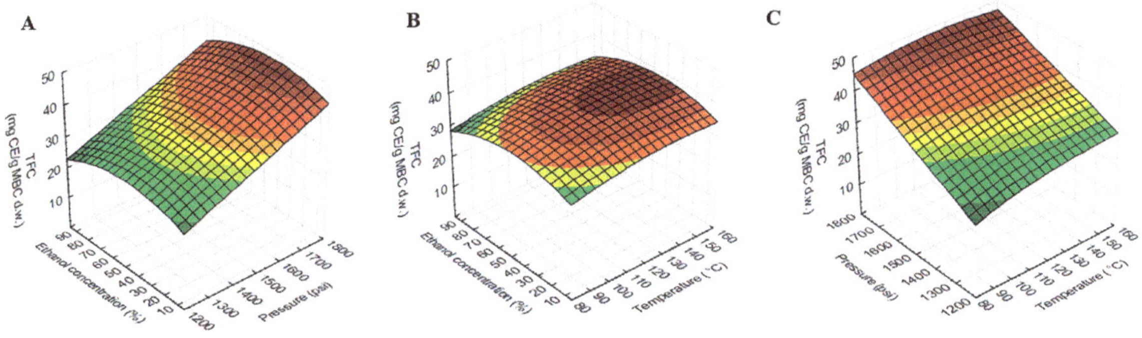

Figure 3. Response surface plots showing the interaction effects of extraction variables on TFC of MBC extract. (**A**) pressure and ethanol concentration at constant temperature (120 °C), (**B**) temperature and ethanol concentration at constant pressure 1500 psi and (**C**) temperature and pressure at constant ethanol concentration (50%).

The response surface plot of ABTS using different combinations of temperature, pressure and ethanol concentration was shown in Figure 4. Figure 4A,B showed that ABTS increased with ethanol concentration increased, but this increase was up to 50% ethanol and then gradually decreased as ethanol concentration increased (>50%).

The present results were consistent with the previous studies showing that higher extraction temperature promoted higher phenolic compounds extracted from rosemary [21], jabuticaba skins [22], cocoa bean shell [23] and rice grains [16] using PLE. The use of high temperature promotes the recovery of phenolic and flavonoid compounds because it increases the molecular motion, which in turn causes a decrease of solvent viscosity, increased diffusivity and increased solubility of target compounds. High temperature also assists in making the cell walls more permeable, enabling the target compounds to leach out into the solvent. In addition, high temperature assists in breaking down target compound-matrix interaction and promote the diffusion of the target compound to the matrix surface and mass transfer to solvent [16,24].

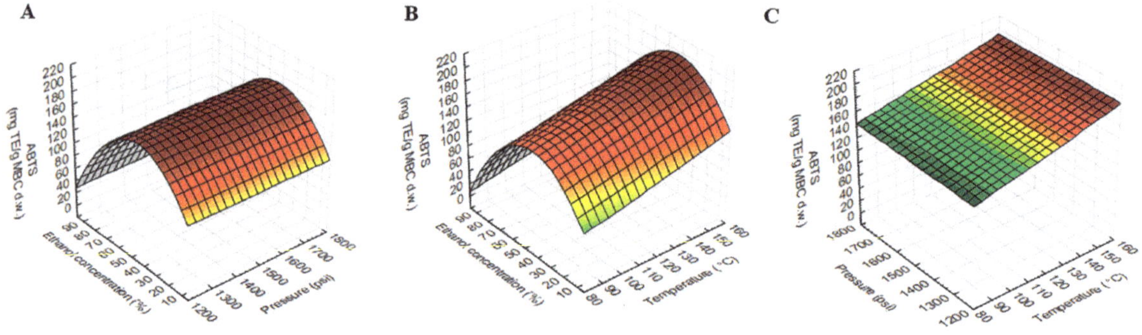

Figure 4. Response surface plots showing the interaction effects of extraction variables on ABTS of MBC extract. (**A**) pressure and ethanol concentration at constant temperature (120 °C), (**B**) temperature and ethanol concentration at constant pressure 1500 psi and (**C**) temperature and pressure at constant ethanol concentration (50%).

Although the use of elevated temperature promotes higher phenolic compounds recovery, the use of excessive extraction temperature may degrade phenolic compounds. The use of elevated temperature in potato peel extraction also increased phenolic compounds recovery, but further temperature to 190 °C resulted in a decrease in phenolic compounds yield [25]. Another study reported that the increase of temperature to a certain level (190 °C) did not affect the level of phenolic compounds [16]. The positive effect of temperature on antioxidant activity was also found in the papaya seed extraction [26]. A similar trend of TPC and ABTS in the present study was due to the highly correlated between the content of phenolic compounds and antioxidant activity as shown in other study [27].

Pressure plays an important role in PLE extraction because at high temperatures, the solvent becomes a gas state. High pressure in PLE allows the solvent to maintain a liquid state, which increases the penetration of the solvent in the sample matrix [24]. The present results indicated that pressure did not significantly affect TPC and ABTS in MBC extraction. This may be due to the lowest pressure in the present study (1200 psi) being enough to maintain the ethanol mixture in a liquid state. Therefore, the increase of pressure did not have any effect on extraction. The insignificant effect of pressure in the present study was consistent with a previous extraction study in the potato peel [25]. However, pressure showed a significant effect on TFC in the present study. Application of pressure in PLE alone without high temperature increases the viscosity of the solvent and solvents with high viscosity result in slow mass transfer, lower diffusivity and lower extraction efficacy [28]. Therefore, both high temperature and pressure need to be applied together in PLE.

Ethanol was used in the present study since it is relatively safer or less toxic. The concentration of ethanol also showed an important effect on the TPC, TFC and ABTS. The present study found that the optimum ethanol concentration for MBC extraction was 50% ethanol. The lower and higher ethanol concentration than 50% showed the lower responses, especially TPC and ABTS. The optimum ethanol concentration in the present study was consistent with previous studies. Zafari and Sharifi reported that the mixture of alcohol with 50% water gave the highest flavonoids including vitexin from *Prosopis farcta* [29]. Sixty percent of ethanol was the best concentration to extract phenolic compounds from brewers' spent grain compared to 20, 40 and 100% ethanol [30]. The vitexin and orientin yield from *Trollius chinesis* flowers extraction increased with the increase of ethanol concentration in the range of 0-60%. The highest yield was given at 60% ethanol extraction [31]. However, the yield sharply decreased when the ethanol concentration increased further. The present study and previous studies confirmed that ethanol concentration greatly impacted the phenolics and flavonoids extraction. Phenolic and flavonoid compounds are diverse with a wide range of solubilities in a single component [30]. Solvents with high polarity showed a higher ability to extract compounds with a wider polarity. Moreover, this allowed non-phenolic polar compounds such as protein and carbohydrates to dissolve during the extraction process, leading to the increased extraction yield [32]. This phenomenon was also explained by the fact that the addition of ethanol reduced the dielectric constant of extraction solvent, which facilitates the solubility and diffusion of phenolic compounds. Too high ethanol concentration, however, results in cell dehydration and denaturation of cell wall proteins, which inhibits the diffusion of phenolic compounds from plant material to extraction solvent [31]. Therefore, the mixture of ethanol and water was suitable for phenolic and flavonoid compounds extraction.

2.4. Optimization and Validation of the Extraction Condition

Multiple numerical optimizations were carried out to achieve the overall optimum PLE extraction conditions [33]. The goal was to set the optimum level of independent variables resulting in the highest TPC, TFC and ABTS. The predicted optimum extraction conditions from Minitab were: temperature of 160 °C, pressure of 1300 psi and 35% ethanol. Under recommended optimum extraction conditions, the predicted and experimental values provided residual standard error (RSE) of more than 10% (data was not shown). This indicated that the recommended optimum extraction conditions were not the actual optimum, which was indicated by the significant lack of fit of the models. Therefore, the steepest ascent method was carried out to find the actual optimum extraction condition. The procedure of the steepest ascend method is moving sequentially in the direction of the maximum response and based on the previous experiment [34]. According to Table 1, the conditions of temperature of 160 °C, pressure of 1200 psi and 50% ethanol provided the highest TPC, TFC and ABTS compared to other treatments. Therefore, 50% ethanol was selected to be a new optimum ethanol concentration. The new optimum conditions were temperature of 160 °C, pressure of 1300 psi and 50% ethanol. Under the new optimum extraction conditions, the predicted TPC, TFC and ABTS were 53.28 GAE/g MBC d.w., 32.88 mg CE/g MBS d.w. and 192.20 mg TE/g MBC d.w., respectively (Table 4). MBC was extracted under the new optimum conditions to verify the validity. The TPC, TFC and ABTS of the extract were then determined to verify the reliability of the new optimum conditions. The validation study showed the TPC, TFC and ABTS of the new MBC extracted solution were 55.267 ± 1.14 mg GAE/g MBC d.w., 34.041 ± 0.72 mg CE/g MBS d.w. and 195.046 ± 2.29 mg TE/g MBC d.w., respectively (Table 4). Under the new optimum extraction conditions, the experimental values showed that the models were in good agreement with the predicted values with RSE values of less than 4%. The present result confirmed the reliability of the new optimum extraction conditions.

Table 4. Predicted and experimental response values at the optimum extraction conditions.

Response Variables	Predicted Values	Experimental Values	RSE (%)
TPC (mg GAE/g MBC d.w.)	53.28	55.27 ± 1.14 ns	3.73
TFC (mg CE/g MBC d.w.)	32.88	34.04 ± 0.72 ns	3.54
ABTS (mg TE/g MBC d.w.)	192.20	195.05 ± 2.29 ns	1.48

2.5. Scale-Up Study

One of the main goals of this study was to scale up the extraction process of MBC. Many factors influence the upscaling extraction process such as process parameters, the kinetics of the whole process, the ratio between material and solvent [22]. The pilot-scale PLE (10 L extraction cell) was 25 times bigger than the laboratory-scale PLE (400 mL extraction cell). The optimum extraction conditions were used in the pilot-scale extraction with the same solid to solvent ratio (1:5 *w*/*v*; 2 kg MBC in 10 L ethanol). The extraction time was constant with laboratory-scale PLE for 10 min. Comparing TPC, TFC and ABTS of the laboratory-scale PLE and pilot-scale PLE (Figure 5), the results showed that TFC and ABTS from pilot-scale PLE were not significantly different from TFC and ABTS from laboratory-scale PLE. Interestingly, TPC from pilot-scale PLE was significantly higher than that from laboratory-scale PLE. The increase of TPC might be due to the change of geometric factors of laboratory-scale PLE and pilot-scale PLE. Previous study showed that geometry and dynamic factors influenced the scale-up study [35]. However, this increase should be more investigated for the explanation. These results indicated that MBC extraction with pilot-scale PLE was efficiently comparable to laboratory-scale PLE.

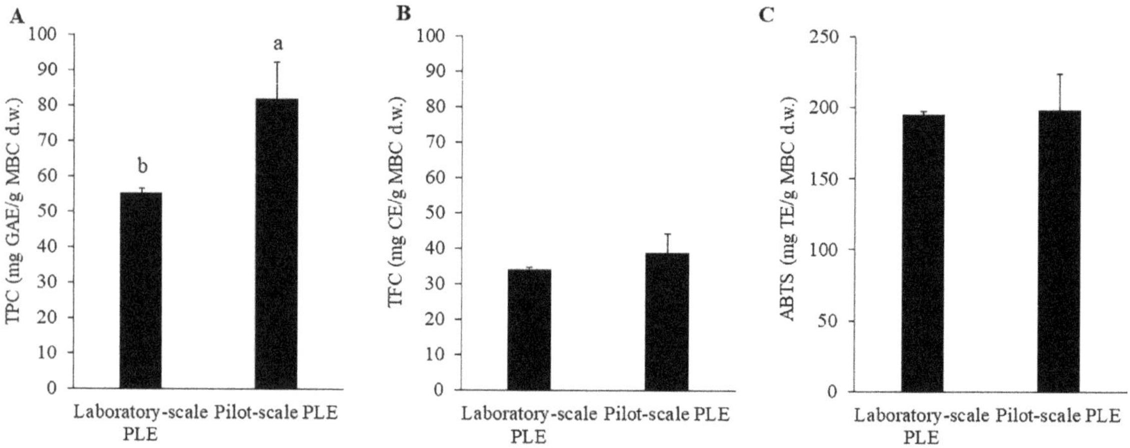

Figure 5. Total phenolic content (**A**), total flavonoid content (**B**) and ABTS (**C**) of MBC extract obtained by laboratory-scale PLE and pilot-scale PLE. There was no significant difference between laboratory-scale PLE and pilot-scale PLE. The different alphabet indicated significant difference at $p < 0.05$.

2.6. Identification of Vitexin and Isovitexin in the Extract

Vitexin and isovitexin are major flavonoids found in mung beans especially in seed coat [4]. Our previous study reported that mung bean seed coat extracted using boiling water contained vitexin and isovitexin 38.56 and 28.96 mg/g extract [5]. The present study showed that mung bean seed coat extracted using PLE contain both vitexin and isovitexin at the concentration of 130.53 ± 17.89 and 21.21 ± 3.22 mg/g freeze-dried extract (Figure 6).

In the present study, TPC and TFC of MBC extracted with the optimum conditions using scale-up PLE were 81.88 mg GAE/g MBC d.w. and 38.90 mg CE/g MBC d.w., respectively. TPC and TFC from the present study were higher than TPC and TFC of MBC extract obtained from maceration extraction method (29.58 mg GAE/g MBC and 22.08 mg CE/g MBC, respectively) [4]. UAE was also applied to increase TPC and TFC extraction. However, TPC and TFC from MBC extract obtained from UAE were only 42.22 mg GAE/g MBC and 1.13 mg CE/g MBC, respectively [9]. This indicated that PLE was a potential extraction method to extract phenolic and flavonoid compounds from MBC with high efficiency.

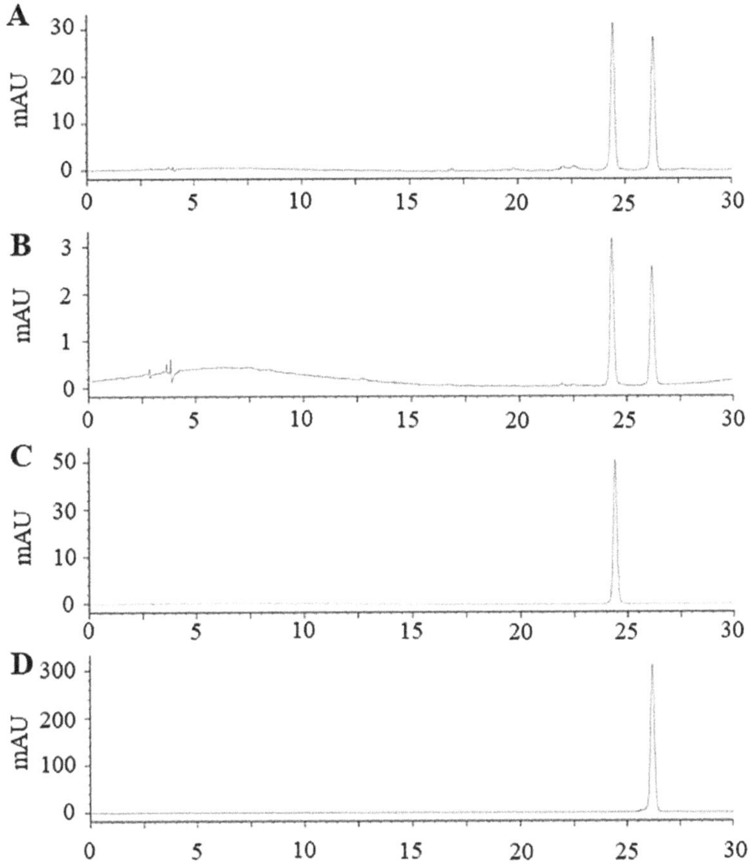

Figure 6. Representative high performance liquid chromatography-diode array detector chromatographic profile (λ = 337 nm) of (**A**) mung bean seed coat extract from laboratory scale; (**B**) mung bean seed coat freeze-dried extract from pilot scale; (**C**) standard vitexin; (**D**) standard isovitexin.

3. Materials and Methods

3.1. Materials and Chemicals

Mung bean seed coat (MBC), a co-product of the mung bean dehulled process, was received from Kittitat Co., Ltd. (Bang Khun Thian, Bangkok, Thailand) in May 2019. After foreign matters removal, MBC was kept in sealed plastic bags at 4 °C until extraction. Ethanol commercial grade (Mallinckrodt Baker Inc., Phillipsburg, NJ, USA) was used for extraction. ABTS, vitexin and isovitexin were purchased from Sigma-Aldrich (St. Louis, MO, USA). All other chemicals were of analytical grade and obtained from reputable suppliers.

3.2. Extraction Procedures

3.2.1. Soxhlet Extraction

MBC was extracted by Soxhlet extraction according to Weggler et al. [36]. Briefly, 8 g of MBC was placed in an extraction thimble. The thimble was then placed in a Soxhlet extractor and 400 mL of 95% ethanol was used as a solvent. The extraction was carried out for 6 h at 80 °C. The extract was collected in a Duran bottle and kept in a dark place at 4 °C. The extract was centrifuged at 7000 rpm, room temperature for 10 min. The supernatant was collected in a Duran bottle and kept in a dark place at 4 °C until further analysis.

3.2.2. Pressurized Liquid Extraction (PLE) at Laboratory Scale

PLE unit was assembled in the Chemical Engineering Laboratory, Department of Chemical Engineering, Faculty of Engineering, Kasetsart University, Bangkok, Thailand (Figure 7). The solvent was pumped by a preparative pump (Fluid Management System, Kalamazoo, MI, USA) into the extraction cell. The extraction cell was placed between an electrical heating system at the desired temperature until the required pressure was obtained. All connections within the system were made using stainless steel tubes.

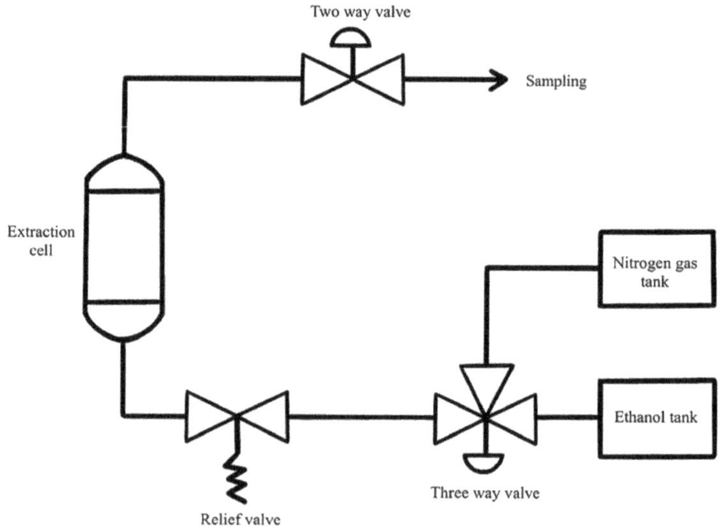

Figure 7. Pressurized liquid extraction set-up.

MBC was placed in the 400 mL extraction cell (OD 60.33 mm, length 28 mm, stainless steel 316) containing a metal filter at the top and bottom of the extraction cell. The cell containing the sample was filled with extraction solvent, heated and then pressurized. The sample was placed in the heating system for 10 min at desired temperature and pressure. After 10 min, the relief valve was carefully opened to lower pressure to atmospheric pressure. After temperature was below 70 °C, the cell was then exhaustively purged with 350 mL 50% ethanol to ensure that no residual extract solution was left in the extraction cell. The MBC extract was collected in a Duran bottle and kept in a dark place at 4 °C. The MBC extract was centrifuged at 7000 rpm, room temperature for 10 min. The supernatant was collected in a Duran bottle and kept in a dark place at 4 °C until further analysis.

3.2.3. Pressurized Liquid Extraction (PLE) at Pilot Scale

In order to scale up the extraction process, the pilot-scale PLE was used. This PLE was equipped with a 10L extraction cell (OD 143.2 mm, length 1200 mm, stainless steel 316). The optimum conditions for extraction were repeated at the pilot scale. The difference was that 2 kg of MBC was used in the extraction cell. After 10 min extraction, the MBC extract was collected in a plastic bottle and kept in a dark place at 4 °C. The MBC extract was centrifuged at 7000 rpm, room temperature for 10 min. The supernatant was collected and evaporated to remove the solvent at 60 °C, then the sample was freeze-dried and kept in an aluminum foil bag at 4 °C for further analysis.

3.3. Extract Characterization

3.3.1. Determination of Total Phenolic Content (TPC)

TPC was analyzed according to Herald et al. [37]. Briefly, 75 µL of distilled water was added to each well of the 96-well plate, followed by 25 µL of sample and 25 µL of folin ciocalteu reagent (diluted 1:1 (v/v) with distilled water). After the solutions had been mixed and left for 6 min, then 100 µL of 75 g/L Na_2CO_3 was added to each well. The solutions were mixed again, and the plates were covered and left at room temperature in a dark place for 90 min. After shaking for 60 sec, the absorbance was measured at 765 nm using a microplate reader (TECAN, Infinite 200 Pro, Manndenorf, Switzerland). Gallic acid was used as a reference standard and the results were expressed as mg GAE/g MBC (d.w.).

3.3.2. Determination of Total Flavonoid Content (TFC)

TFC was analyzed according to Herald et al. [37]. Briefly, 100 µL of distilled water was added to each well of the 96-well plate, followed by 10 µL of 50 g/L $NaNO_2$ and 25 µL of sample solution. After 5 min incubation, 15 µL of 100 g/L $AlCl_3$ was added to the mixture. Six min later, 50 µL of 1 M NaOH and 50 µL of distilled water were added. The plate was shaken for 30 sec in the plate reader before measuring absorbance at 510 nm using a microplate reader (TECAN, Infinite 200 Pro, Manndenorf, Switzerland). Catechin was used as a reference standard and the results were expressed as mg CE/g MBC (d.w.).

3.3.3. Determination of ABTS Radical Scavenging Activity (ABTS)

The 2,2′-azinobis(3-ethylbenzothiazoline-6-sulfonic acid) (ABTS) radical scavenging activity was determined according to Indracanti et al. [38]. Briefly, 10 µL of sample were mixed with 190 µL of ABTS solution (2.45 mM potassium persulfate solution and 7 mM ABTS). The mixture was incubated at room temperature for 6 min in a dark place, then the absorbance was measured at 734 nm using a microplate reader (TECAN, Infinite 200 Pro, Manndenorf, Switzerland). Trolox was used as a reference standard and the results were expressed as mg TE/g MBC (d.w.).

3.4. *Identification of Major Compounds in the Extract*

The major compounds in the extract were determined using high-performance liquid chromatography (HPLC) according to our previous method [39]. The extract was dissolved with deionized water and subjected to HPLC equipped with a diode array detector (Waters 600, Milford, MA, USA). An analytical column (C18) (4.6 × 250 mm, Inertsil ODS-3, 5 µm, GL Sciences, Tokyo, Japan) was used and kept at 30 °C while using. The spectra from 210–600 nm were recorded and UV absorbance at 337 nm was used to monitor flavonoids. The injection volume was 10 µL at a flow rate of 1 mL/min. Elution was done using two solvent gradients: solvent A (1% acetic acid in deionized water) and solvent B (1% acetic acid in methanol). The gradient program started with 10–35% B (10 min), 35–42% B (15 min), 42–75% B (10 min), 75% B (5 min), 75–10% B (5 min) and 10% B (5 min). Vitexin and isovitexin at the concentrations 20, 40, 60, 80 and 100 mg/kg were used as external standards to make standard curves and to determine the concentration of vitexin and isovitexin in the extract.

3.5. *Experimental Design and Statistical Analysis*

The effect of three independent variables: temperature (X_1; 80–160 °C), pressure (X_2; 1200–1800 psi) and ethanol concentration (X_3; 5–95%) on the response variables: TPC (Y_1), TFC (Y_2) and ABTS (Y_3) was evaluated using a three-factor-three-level Box-Behnken design (BBD). The three coded levels of temperature, pressure and ethanol concentration were incorporated into the design and were analyzed in 15 combinations (Table 1). The central point of the design was repeated three times to estimate the repeatability of the method. For each combination, three dependent variables were determined. A second-order polynomial model was used for fitting data and predicting the responses as Equation (4):

$$Y = b_0 + b_1X_1 + b_2X_2 + b_3X_3 + b_{11}X_1^2 + b_{22}X_2^2 + b_{33}X_3^2 + b_{12}X_1X_2 + b_{13}X_1X_3 + b_{23}X_2X_3 \quad (4)$$

where, Y is the response variable; $b_0, b_1, b_2, b_3, b_{11}, b_{22}, \ldots$ are regression coefficients and X_1, X_2 and X_3 are uncoded values for temperature, pressure and ethanol concentration, respectively. An analysis of variance (ANOVA) was performed at a 95% confidence level to evaluate the predicted model on the response variables and assess the effect of each factor. In addition, the regression coefficient (R^2), the p-value of the regression model, the p-value of the lack of fit were used to determine the fitness of the regression model. The graphical and numerical optimization procedures were used to determine the optimum PLE condition. Three-dimensional (3D) response surface plots were applied for the interactions. Steepest ascend approach was also applied to determine the optimum PLE conditions to provide the highest TPC, TFC and ABTS. In order to assess the adequacy of the constructed model, the actual values were compared with the predicted values and the percentage of the residual standard error (RSE) was calculated for each response [40]. All of the treatments were carried out in triplicates. The experimental design matrix, data analysis, regression coefficients and numerical optimization were analyzed using Minitab statistical software (Trial version 16.1 Minitab Inc., State College, PA, USA) and 3D graphs were provided using Statistica 10.0 (StatSoft Inc., Tulsa, OK, USA). Differences between groups of data were assessed by student t-test. Results were expressed as mean ± standard deviation.

4. Conclusions

The present study showed the possibility of obtaining phenolics, flavonoids with antioxidant activity from MBC using PLE. The effects of temperature, pressure and ethanol concentration were studied to maximize TPC, TFC and ABTS of MBC extract. The results showed that TPC, TFC and ABTS were most affected by ethanol concentration. TPC and TFC were also affected by temperature, while pressure did not affect TPC and ABTS. Temperature and ethanol concentration had an interaction effect on TPC, while temperature and pressure had an interaction effect on TFC. The optimum conditions obtained from numerical optimization and steepest ascend approach, were temperature of 160 °C, pressure of 1300 psi and 50% ethanol. The optimum conditions were performed in laboratory-scale and pilot-scale PLE. No significant decrease was observed in pilot-scale PLE. The profile of MBC extract from laboratory-scale and pilot-scale PLE also remained. This illustrated that the optimum conditions can be transferred to an industrial scale. However, industrial costs such as energy and materials are needed to be investigated. In conclusion, the PLE process appeared to be a potential technique in extracting phenolics, flavonoids with antioxidant activity from mung bean seed coat.

Author Contributions: Conceptualization, M.S., U.K., K.R., S.S.-T.; Data curation, S.S.-T.; Formal analysis, B.S. and S.S.-T.; Funding acquisition, S.S.-T.; Investigation, B.S.; Methodology, M.S., U.K., K.R. and S.S.-T.; Project administration, S.S.-T.; Resources, M.S.; Validation, M.S., U.K., K.R. and S.S.-T.; Visualization, B.S., M.S., U.K., K.R. and S.S.-T.; Writing—original draft, B.S.; Writing—review & editing, S.S.-T. All authors have read and agreed to the published version of the manuscript.

Funding: This work is financially supported by Office of National Higher Education Science Research and Innovation Policy Council through Program Management Unit for Competitiveness (Grant number C10F630076).

Data Availability Statement: The dataset generated for this research are available on request to the corresponding author.

Acknowledgments: The authors acknowledge Department of Food Science and Technology, the Faculty of Agro-Industry and Department of Chemical Engineering, Faculty of Engineering, Kasetsart University for providing laboratory facilities and research equipment. We also thank Ravipim Chaveesuk from the Department of Agro-Industrial Technology, Faculty of Agro-Industry, Kasetsart University for giving suggestions about the optimization method.

Conflicts of Interest: The authors declare no conflict of interest.

Sample Availability: Samples of the compounds are not available from the authors.

References

1. Yi-Shen, Z.; Shuai, S.; Fitzgerald, R. Mung bean proteins and peptides: Nutritional, functional and bioactive properties. *Food Nutr. Res.* **2018**, *62*. [CrossRef]
2. Anwar, F.; Latif, S.; Przybylski, R.; Sultana, B.; Ashraf, M. Chemical Composition and Antioxidant Activity of Seeds of Different Cultivars of Mungbean. *J. Food Sci.* **2007**, *72*, S503–S510. [CrossRef]
3. Hou, D.; Yousaf, L.; Xue, Y.; Hu, J.; Wu, J.; Hu, X.; Feng, N.; Shen, Q. Mung Bean (*Vigna radiata* L.): Bioactive Polyphenols, Polysaccharides, Peptides, and Health Benefits. *Nutrients* **2019**, *11*, 1238. [CrossRef]
4. Luo, J.; Cai, W.; Wu, T.; Xu, B. Phytochemical distribution in hull and cotyledon of adzuki bean (*Vigna angularis* L.) and mung bean (*Vigna radiate* L.), and their contribution to antioxidant, anti-inflammatory and anti-diabetic activities. *Food Chem.* **2016**, *201*, 350–360. [CrossRef]
5. Sae-tan, S.; Kumrungsee, T.; Yanaka, N. Mungbean seed coat water extract inhibits inflammation in LPS-induced acute liver injury mice and LPS-stimulated RAW 246.7 macrophages via the inhibition of TAK1/IκBα/NF-κB. *J. Food Sci. Technol.* **2020**, *57*, 2659–2668. [CrossRef]
6. Cao, D.; Li, H.; Yi, J.; Zhang, J.; Che, H.; Cao, J.; Yang, L.; Zhu, C.; Jiang, W. Antioxidant properties of the mung bean flavonoids on alleviating heat stress. *PLoS ONE* **2011**, *6*, e21071. [CrossRef]
7. Jang, Y.-H.; Kang, M.-J.; Choe, E.-O.; Shin, M.; Kim, J.-I. Mung bean coat ameliorates hyperglycemia and the antioxidant status in type 2 diabetic db/db mice. *Food Sci. Biotechnol.* **2014**, *23*, 247–252. [CrossRef]
8. Zhao, Y.; Du, S.-K.; Wang, H.; Cai, M. In vitro antioxidant activity of extracts from common legumes. *Food Chem.* **2014**, *152*, 462–466. [CrossRef]
9. Singh, B.; Singh, N.; Thakur, S.; Kaur, A. Ultrasound assisted extraction of polyphenols and their distribution in whole mung bean, hull and cotyledon. *J. Food Sci. Technol.* **2016**, *54*, 921–932. [CrossRef]
10. Lee, J.H.; Jeon, J.K.; Kim, S.G.; Kim, S.H.; Chun, T.; Imm, J.-Y. Comparative analyses of total phenols, flavonoids, saponins and antioxidant activity in yellow soy beans and mung beans. *Int. J. Food Sci. Technol.* **2011**, *46*, 2513–2519. [CrossRef]
11. Oreopoulou, A.; Tsimogiannis, D.; Oreopoulou, V. Chapter 15—Extraction of Polyphenols from Aromatic and Medicinal Plants: An Overview of the Methods and the Effect of Extraction Parameters. In *Polyphenols in Plants*, 2nd ed.; Watson, R.R., Ed.; Academic Press: Cambridge, MA, USA, 2019; pp. 243–259. [CrossRef]
12. Santos, K.A.; Gonçalves, J.E.; Cardozo-Filho, L.; da Silva, E.A. Pressurized liquid and ultrasound-assisted extraction of α-bisabolol from candeia (*Eremanthus erythropappus*) wood. *Ind. Crops Prod.* **2019**, *130*, 428–435. [CrossRef]
13. Machado, A.P.D.F.; Pereira, A.L.D.; Barbero, G.F.; Martínez, J. Recovery of anthocyanins from residues of *Rubus fruticosus*, *Vaccinium myrtillus* and *Eugenia brasiliensis* by ultrasound assisted extraction, pressurized liquid extraction and their combination. *Food Chem.* **2017**, *231*, 1–10. [CrossRef]
14. Péres, V.F.; Saffi, J.; Melecchi, M.I.S.; Abad, F.C.; de Assis Jacques, R.; Martinez, M.M.; Oliveira, E.C.; Caramão, E.B. Comparison of soxhlet, ultrasound-assisted and pressurized liquid extraction of terpenes, fatty acids and Vitamin E from Piper gaudichaudianum Kunth. *J. Chromatogr. A* **2006**, *1105*, 115–118. [CrossRef]
15. Barriada-Pereira, M.; González-Castro, M.J.; Muniategui-Lorenzo, S.; López-Mahía, P.; Prada-Rodríguez, D.; Fernández-Fernández, E. Comparison of pressurized liquid extraction and microwave assisted extraction for the determination of organochlorine pesticides in vegetables. *Talanta* **2007**, *71*, 1345–1351. [CrossRef]
16. Setyaningsih, W.; Saputro, I.; Palma, M.; Barroso, C. Pressurized liquid extraction of phenolic compounds from rice (*Oryza sativa*) grains. *Food Chem.* **2016**, *192*, 452–459. [CrossRef]
17. Mandal, S.C.; Mandal, V.; Das, A.K. Chapter 6—Classification of Extraction Methods. In *Essentials of Botanical Extraction*; Mandal, S.C., Mandal, V., Das, A.K., Eds.; Academic Press: Cambridge, MA, USA, 2015; pp. 83–136. [CrossRef]
18. Viganó, J.; Brumer, I.Z.; de Campos Braga, P.A.; da Silva, J.K.; Maróstica Júnior, M.R.; Reyes Reyes, F.G.; Martínez, J. Pressurized liquids extraction as an alternative process to readily obtain bioactive compounds from passion fruit rinds. *Food Bioprod. Process.* **2016**, *100*, 382–390. [CrossRef]
19. Hawthorne, S.B.; Grabanski, C.B.; Martin, E.; Miller, D.J. Comparisons of Soxhlet extraction, pressurized liquid extraction, supercritical fluid extraction and subcritical water extraction for environmental solids: Recovery, selectivity and effects on sample matrix. *J. Chromatogr. A* **2000**, *892*, 421–433. [CrossRef]
20. Rodríguez-Solana, R.; Salgado, J.M.; Domínguez, J.M.; Cortés-Diéguez, S. Comparison of Soxhlet, accelerated solvent and supercritical fluid extraction techniques for volatile (GC-MS and GC/FID) and phenolic compounds (HPLC-ESI/MS/MS) from *Lamiaceae* species. *Phytochem. Anal* **2015**, *26*, 61–71. [CrossRef]
21. Herrero, M.; Plaza, M.; Cifuentes, A.; Ibáñez, E. Green processes for the extraction of bioactives from Rosemary: Chemical and functional characterization via ultra-performance liquid chromatography-tandem mass spectrometry and in-vitro assays. *J. Chromatogr. A* **2010**, *1217*, 2512–2520. [CrossRef]
22. Santos, D.T.; Veggi, P.C.; Meireles, M.A.A. Optimization and economic evaluation of pressurized liquid extraction of phenolic compounds from jabuticaba skins. *J. Food Eng.* **2012**, *108*, 444–452. [CrossRef]
23. Okiyama, D.C.G.; Soares, I.D.; Cuevas, M.S.; Crevelin, E.J.; Moraes, L.A.B.; Melo, M.P.; Oliveira, A.L.; Rodrigues, C.E.C. Pressurized liquid extraction of flavanols and alkaloids from cocoa bean shell using ethanol as solvent. *Food Res. Int.* **2018**, *114*, 20–29. [CrossRef] [PubMed]

24. De la Guardia, M.; Armenta, S. Chapter 5—Greening Sample Treatments. In *Comprehensive Analytical Chemistry*; Guardia, M.D.L., Armenta, S., Eds.; Elsevier: Amsterdam, The Netherlands, 2011; Volume 57, pp. 87–120. [CrossRef]
25. Luthria, D.L. Optimization of extraction of phenolic acids from a vegetable waste product using a pressurized liquid extractor. *J. Funct. Foods* **2012**, *4*, 842–850. [CrossRef]
26. Samaram, S.; Mirhosseini, H.; Tan, C.P.; Ghazali, H.M.; Bordbar, S.; Serjouie, A. Optimisation of ultrasound-assisted extraction of oil from papaya seed by response surface methodology: Oil recovery, radical scavenging antioxidant activity and oxidation stability. *Food Chem.* **2015**, *172*, 7–17. [CrossRef] [PubMed]
27. Wangkiri, N.; Sarnsri, T.; Thongkanjana, T.; Sae-tan, S. Antioxidant potentials and inhibitory activities against α-amylase and α-glucosidase, and glucose uptake activity in insulin-resistance HepG2 cells of some medicinal plants. *Agric. Nat. Resour.* **2021**, *55*, 98–104.
28. Poole, C.F. Chapter 2—Solvent Selection for Liquid-Phase Extraction. In *Liquid-Phase Extraction*; Poole, C.F., Ed.; Elsevier: Amsterdam, The Netherlands, 2020; pp. 45–89. [CrossRef]
29. Zafari, S.; Sharifi, M. Optimization of Solvent Systems for the Extraction of Vitexin as the Major Bioactive Flavonoid in Prosopis farcta. *Am. J. Plant Sci.* **2020**, *11*, 595–603. [CrossRef]
30. Zuorro, A.; Iannone, A.; Lavecchia, R. Water–Organic Solvent Extraction of Phenolic Antioxidants from Brewers' Spent Grain. *Processes* **2019**, *7*, 126. [CrossRef]
31. Chen, F.; Zhang, Q.; Liu, J.; Gu, H.; Yang, L. An efficient approach for the extraction of orientin and vitexin from Trollius chinensis flowers using ultrasonic circulating technique. *Ultrason. Sonochem.* **2017**, *37*, 267–278. [CrossRef]
32. Ermi Hikmawanti, N.P.; Fatmawati, S.; Asri, A.W. The Effect of Ethanol Concentrations as The Extraction Solvent on Antioxidant Activity of Katuk (*Sauropus androgynus* (L.) Merr.) Leaves Extracts. *IOP Conf. Ser. Earth Environ. Sci.* **2021**, *755*, 012060. [CrossRef]
33. Mirhosseini, H.; Tan, C.P.; Taherian, A.R.; Boo, H.C. Modeling the physicochemical properties of orange beverage emulsion as function of main emulsion components using response surface methodology. *Carbohydr. Polym.* **2009**, *75*, 512–520. [CrossRef]
34. Montgomery, D.C. *Design and Analysis of Experiments*, 8th ed.; John Wiley & Sons, Inc.: Hoboken, NJ, USA, 2012.
35. Fernández-Ponce, M.T.; Parjikolaei, B.R.; Lari, H.N.; Casas, L.; Mantell, C.; de la Ossa, E.J.M. Pilot-plant scale extraction of phenolic compounds from mango leaves using different green techniques: Kinetic and scale up study. *Chem. Eng. J.* **2016**, *299*, 420–430. [CrossRef]
36. Weggler, B.A.; Gruber, B.; Teehan, P.; Jaramillo, R.; Dorman, F.L. Chapter 5—Inlets and sampling. In *Separation Science and Technology*; Snow, N.H., Ed.; Academic Press: Cambridge, MA, USA, 2020; Volume 12, pp. 141–203. [CrossRef]
37. Herald, T.; Gadgil, P.; Tilley, M. High-throughput micro plate assays for screening flavonoid content and DPPH-scavenging activity in sorghum bran and flour. *J. Sci. Food Agric.* **2012**, *92*, 2326–2331. [CrossRef] [PubMed]
38. Indracanti, M.; ChV, S.; Sisay, T. A 96 well-microtiter plate abts based assay for estimation of antioxidant activity in green leafy vegetables. *Biotechnol. Int.* **2019**, *12*, 22–29.
39. Saeting, O.; Chandarajoti, K.; Phongphisutthinan, A.; Hongsprabhas, P.; Sae-tan, S. Water Extract of Mungbean (*Vigna radiata* L.) Inhibits Protein Tyrosine Phosphatase-1B in Insulin-Resistant HepG2 Cells. *Molecules* **2021**, *26*, 1452. [CrossRef] [PubMed]
40. Che Sulaiman, I.S.; Basri, M.; Fard Masoumi, H.R.; Chee, W.J.; Ashari, S.E.; Ismail, M. Effects of temperature, time, and solvent ratio on the extraction of phenolic compounds and the anti-radical activity of Clinacanthus nutans Lindau leaves by response surface methodology. *Chem. Cent. J.* **2017**, *11*, 54. [CrossRef]

molecules

Review

High Hydrostatic Pressure to Increase the Biosynthesis and Extraction of Phenolic Compounds in Food: A Review

Jorge E. Navarro-Baez, Luz María Martínez, Jorge Welti-Chanes, Génesis V. Buitimea-Cantúa and Zamantha Escobedo-Avellaneda *

Escuela de Ingeniería y Ciencias, Tecnologico de Monterrey, Eugenio Garza Sada 2501, Monterrey 64700, Mexico; a00820540@itesm.mx (J.E.N.-B.); luzvidea@tec.mx (L.M.M.); jwelti@tec.mx (J.W.-C.); genesis.vidal@tec.mx (G.V.B.-C.)

* Correspondence: zamantha.avellaneda@tec.mx

Abstract: Phenolic compounds from fruits and vegetables have shown antioxidant, anticancer, anti-inflammatory, among other beneficial properties for human health. All these benefits have motivated multiple studies about preserving, extracting, and even increasing the concentration of these compounds in foods. A diverse group of vegetable products treated with High Hydrostatic Pressure (HHP) at different pressure and time have shown higher phenolic content than their untreated counterparts. The increments have been associated with an improvement in their extraction from cellular tissues and even with the activation of the biosynthetic pathway for their production. The application of HHP from 500 to 600 MPa, has been shown to cause cell wall disruption facilitating the release of phenolic compounds from cell compartments. HPP treatments ranging from 15 to 100 MPa during 10–20 min at room temperature have produced changes in phenolic biosynthesis with increments up to 155%. This review analyzes the use of HHP as a method to increase the phenolic content in vegetable systems. Phenolic content changes are associated with either an immediate stress response, with a consequent improvement in their extraction from cellular tissues, or a late stress response that activates the biosynthetic pathways of phenolics in plants.

Keywords: phenolic compounds; high hydrostatic pressure; reactive oxygen species; biosynthesis; extraction; stress response; cell wall modification

Citation: Navarro-Baez, J.E.; Martínez, L.M.; Welti-Chanes, J.; Buitimea-Cantúa, G.V.; Escobedo-Avellaneda, Z. High Hydrostatic Pressure to Increase the Biosynthesis and Extraction of Phenolic Compounds in Food: A Review. *Molecules* **2022**, *27*, 1502. https://doi.org/10.3390/molecules27051502

Academic Editor: Smaoui Slim

Received: 15 January 2022
Accepted: 21 February 2022
Published: 23 February 2022

1. Introduction

Phenolics are a group of specialized metabolites with antioxidant, antimicrobial, anti-cancer, anti-inflammatory activity, among other biological properties for human health [1,2]. Fruits and vegetables are rich sources of these compounds. Due to their association with treating and preventing some chronic diseases, a diet rich in vegetables and fruits is highly recommended [3,4].

Thermal treatments are frequently used during fruit and vegetable processing to inactivate pathogenic and spoilage microorganisms and enzymes to ensure food safety and quality [5]. However, these treatments can decrease the nutritional value by reducing thermosensitive bioactive compounds, along with modifying texture, taste, and flavor [5,6]. To overcome these adverse effects, other alternatives such as high hydrostatic pressure (HHP) have been used to preserve food. HHP consists of applying pressures normally up to 600 MPa into a chamber. The pressure is generally transmitted by water at room or moderate temperature. This treatment can inactivate microorganisms and enzymes while minimizing quality losses in the pressurized products [7]. This technology is considered an innovative technique for the nonthermal treatment of food [8].

Pressure is a physical parameter that affects the state of physical, chemical, and biological systems. Pressure can modify the chemical configuration of some food molecules changing the rate of chemical and enzymatic reactions [9]. HHP is governed by the Le

Chatelier and Pascal principles [10]. The former states that a system under pressure will adopt molecular configurations and will adjust the rate of chemical reactions to achieve the lowest volume [11]. In other words, pressure favors phenomena and reactions that are accompanied by negative volume changes [12]. While the Pascal or isostatic principle states that the pressure applied is instantly and uniformly transmitted in all directions of the fluid and its surroundings [13].

HHP to inactivate microorganisms was first proposed in the late 1980s. Since then, the use of this technology as a preservation method has increased in the food industry. The ability of HHP to inactivate microorganisms and enzymes related to food oxidation are some examples of the benefits involved [7]. HHP-treated products present better nutrient retention, flavor, and color. Moreover, HHP reduces or eliminates the use of additives and does not produce residues during processing because only water is involved in the entire process [14–16]. Nevertheless, some enzymes and bacterial spores are highly resistant to pressure requiring the combination of pressure and temperature for their inactivation. In addition, some residual enzyme activity and dissolved oxygen could cause enzymatic and oxidative degradation of specific food components, and most HHP-treated products need to be stored at low temperature [17]

In addition to food preservation, HHP has been proposed as a method to enhance bioactive compound content in foods and to improve their extraction. Some pressurized foods have shown higher contents of functional compounds such as phenolics compared with untreated products. The increments have been associated with the release of these compounds from cellular compartments, resulting in increased extractability due to the mechanical stress occurring during the pressurization, which compacts the cellular morphology, cell wall, and organelles [18]. Another mechanism for the increment is related to the immediate response of plants to generate signaling molecules that activate pathways that regulate gene expressions or with the late response associated with higher enzyme activity produced by the signaling molecules generated in the immediate response, which activate the biosynthesis of specialized metabolites [19].

This review analyzes the mechanisms for phenolic increment in vegetables and fruits immediately after HHP treatment and during storage, differentiating between increment due to cellular disruption (improvement of extraction yield) or activation of the metabolic pathways for compound biosynthesis.

2. Biosynthesis of Phenolics in Plants

The biosynthesis of phenolic in plants is achieved by the malonate and the shikimate pathways [20]. The malonate pathway is mainly focused on the synthesis of secondary fatty acids, and some phenolic compounds such as aromatic polyketides, where flavonoids are included. The precursors of this pathway are acetyl-CoA and malonyl-CoA [21]. The shikimate pathway (Figure 1) is responsible for the biosynthesis of most phenolics in plants [22]. This consists of seven steps carried out by different enzymes, and starts with two molecules, erythrose-4-phosphate (E4P) and phosphoenolpyruvate (PEP) derived from glycolysis and pentose phosphate pathways, respectively [23]. The first step consists of the condensation of erythrose-4-phosphate (E4P) and phosphoenolpyruvate (PEP) into 3-deoxy-D-arabinoheptulosonate 7-phosphate (DAHP) by the DAHP synthase [24]. In the second step, the DAHP is cyclized, forming 3-dehydroquinate (DHQ) due to the action of DHQ synthase [25]. In the third and fourth steps, DHQ is dehydrated to 3-dehydroshikimate (DHS) and then is reduced to shikimate by the bifunctional enzyme 3-dehydroquinate dehydratase/shikimate dehydrogenase [26,27]. In the fifth step, the shikimate is converted to shikimate 3-phosphate (S3P) by the shikimate kinase [28]. In the sixth step, the shikimate 3-phosphate is condensate with a second molecule of PEP into 5-enolpyruvylshikimate 3-phosphate (EPSP) [29]. Finally, the EPSP is dephosphorylated by the chorismate synthase to produce chorismite [30].

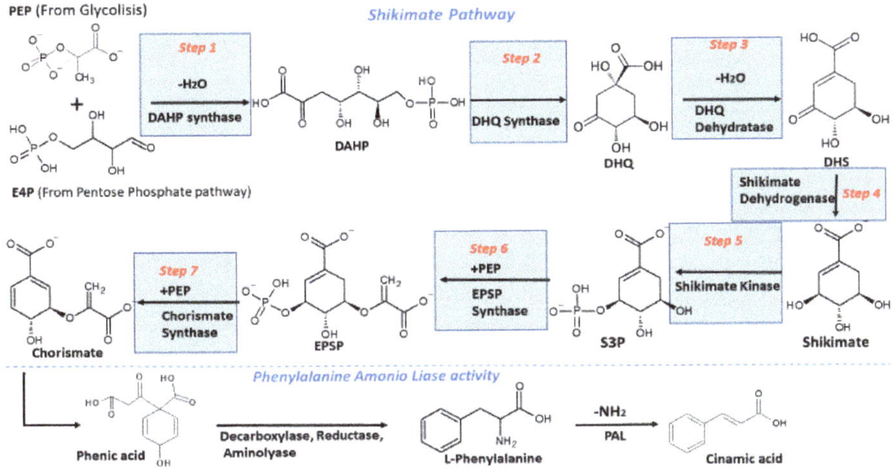

Figure 1. Shikimate pathway for the production of phenolics in plants. PEP: phosphoenolpyruvate, E4P: erythrose-4-phosphate, DAPH: 3-deoxy-D-arabinoheptulosonate 7-phosphate, DHQ: 3-dehydroquinate, S3P: shikimate 3-phosphate, EPSP: 5-enolpyruvylshikimate 3-phosphate, PAL: Phenylalanine Ammonium Lyase. Modified from [22,31].

After the production of chorismate, the phenic acid is formed by the enol-pyruvate transferase action, the phenic acid goes under a decarboxylation, and the substitution of the oxygen for an amino group results in the formation of L-phenylalanine [31]. The most important step occurs by the phenylalanine ammonium lyase (PAL). This enzyme deaminates the L-phenylalanine into cinnamic acid. L-phenylalanine is a block for the formation of secondary metabolites, so its relationship with the biosynthesis of phenolic compounds is important [31–33] (Figure 1). The production of phenolic compounds in plants is strictly related to the activity of the enzyme PAL. A study in strawberries showed that at the peak of maximum PAL activity, there was a higher anthocyanin content [34]. The formation of metabolites such as phenolic compounds is also related to the defense mechanism against biotic or abiotic stress [32].

Biotic and Abiotic Factors Influencing Biosynthesis of Phenolics

The production of specialized metabolites in plants is affected by biotic and abiotic factors, such as environmental conditions, microorganisms, insect attacks, among others [35]. Biotic stress is caused by the action of bacteria, fungi, viruses, and nematodes that attack the plants by secreting enzymes to break down tissues. Insects and vertebrates are also biotic stressors that use plants as a food source [36]. Abiotic stress is caused by external factors such as drought, soil salinization, extreme temperatures, strong winds, climate, and change of season of the year [36]. Even air pollution and the use of pesticides can act as abiotic factors [35].

Phenolic compounds are implicated in the biotic and abiotic stresses by reinforcing cell walls and scavenging of ROS (Reactive Oxygen Species) [37]. Biotic and abiotic stressors promote the production of free radicals and oxygen species in plants, inducing the synthesis of secondary products like phenolics as a mechanism to protect plants [38]. Some researchers have taken advantage of this to seek new ways to deliberately increase bioactive compounds in plants. Some techniques to increase phenolics in plants are based on controlled elicitation. One example is the use of nanoparticles (NPs) such as Cu, CdO, CeO2, CuO, Ag, and ZnO as abiotic elicitors for induction of phenolic and other bioactive compounds in plant cells [39]. NPs induce the production of ROS, leading to the transcription of secondary metabolites [40]. The effect on NPs as elicitors vary, but generally shows an increment of secondary metabolites such as phenols and flavonoids [41]. However, an important drawback of the use of NPs is their toxicity [42]. It has been suggested that the elevated production of ROS by NPs results in lipid peroxidation, which damages cell membrane, proteins, and DNA resulting in cell death [43].

Mechanical force is another important abiotic stress factor that has shown a positive effect on the production of phenolics in plants. HHP at pressure levels below 100 MPa acts as a mechanical stressor resulting in an increment of phenolic levels [44,45]. Pressure, temperature, and time of exposure are variables that have been studied for the biosynthesis of metabolites in plants under HHP [9]. A very useful advantage of this technology is the retention of cell viability at certain treatment conditions, which allows the cells to keep enzymatic activities to induce significant production of phenolics [44]. The HHP potential for the biosynthesis of phenolic compounds is presented in the following sections, differentiating it from the increment due to improvement in extraction yields.

3. HHP as a Stress Factor for the Biosynthesis of Phenolics and to Increase Their Extraction Yield

3.1. Effect of HHP on Phenolics Biosynthesis

Although the effects of HHP on the biosynthesis of phenolics have been evaluated in fruits and vegetables such as mangoes, carrots, strawberries, and suspension cultures of grapes and potato, there is not yet enough research on the mechanism(s) implied in the increment for the biosynthesis of phenolics by HHP. Several studies using pressures from 10 to 100 MPa at treatment times from 10 to 20 min at room temperature have been tested immediately after processing or during the storage at different temperatures and relative humidities; and the results have revealed a change in phenolics biosynthesis with increments up to 155%. Some studies conclude biosynthesis of phenolics due to the low-pressure levels used and the increment in their content, nevertheless, some of them do not show studies related to increment in PAL activity, ROS production, or gene expression, which are important to conclude this (Table 1).

Table 1. High hydrostatic pressure effects on the biosynthesis of phenolic in different vegetable systems.

Sample	Treatment Conditions				Storage Conditions	Analyzed Compound	Main Findings		Reference
	P (MPa)	t (min)	CUT (s)	T (°C)			Approximate Change (%)	PAL Activity (%)	
Mango *Mangifera indica* (Whole fruit)	15–60	10–20	3, 10 & 28	25	2–14 days at 25 °C and 85–90% RH	Total phenols	↓7.2 up to ↑68.4	NR	[46]
						Flavonoids	↓38.6 up to ↑36.8	NR	
Mango *Mangifera indica* (Whole fruit)	20–80	10	NR	20	1–16 days at 13 °C with 85% RH	Total phenols	↓5.2 up to ↑30	NR	[47]
						Flavonoids	↓27.6 up to ↑69.7	NR	
Vitis vinifera (Suspension culture)	40	10	NR	25	1–7 days at 25 °C	Anthocyanin	↓53.9 up to ↑53.3	NR	[48]
Carrots *Daucus carota* (Whole vegetable)	60 & 100	CUT	15.33 & 20.67	22	0–3 days at 15 °C CO_2 < [0.5 *v*/*v*]	Total phenols	↓11.8 up to ↑154.9	↓61.4 up to ↑380	[49]
Potato *Solanum tuberosum* (suspension culture)	100–200	10	NR	25	1–24 h	Polyphenols	↑54.0 up to ↑456.0	↑199	[44]
Strawberry *Seolhyang, Fragaria × ananassa* Duch (Whole fruit)	30–90	5	NR	25	NR	Total phenols	↑6.4 up to ↑23.1	NR	[50]
						Anthocyanin	↓16.9 up to ↑10.0	NR	

P: Pressure; t: time; T: Temperature; CUT: Come up time (time to achieve desired pressure); NR: Not reported; RH: Relative humidity. ↑ indicates an increment of content compared with the untreated sample; ↓ indicates decreasing of content compared with the untreated sample.

Phenolic biosynthesis due to HHP does not seem to increase proportionally with the pressure. The increments primarily depend on the type of fruit treated and the ripening stage, as well as, the storage conditions such as temperature, relative humidity, and storage time [46,47]. In addition, it has been observed that the increment does not occur immediately after processing, but during storage producing a late stress response. For example, mango treated at 60 MPa/20 min showed an increment in phenolics of about 11, 29, and 47% after two, five, and eight days of storage, respectively, compared to nontreated samples (Figure 2a) [46]. After day eight, the concentration of phenolics starts to decay [46,48]. This decrement could be associated with the activity of oxidative enzymes, which are normally inactivated at higher pressure levels (>200 MPa) than the ones used to induce stress in the vegetable systems. According to Ortega et al. [48], the initial improvement in phenolic content could be attributed to an increment in their biosynthesis due to immediate oxidative stress, while reductions at longer times could be related to the damage in cellular structures and the ripening process [46]. In another study with mango, Hu et al. [47] showed 19.6% increment in phenolics after one day of storage (Figure 2b) and 69.7% increment in flavonoids after four days. Overall, Ortega et al. [46] found greater increments in phenolics compared to Hu et al. [47], which could be attributed to differences in treatment conditions, ripening stage of the fruit, and the storage conditions. It has been proved that 25 °C is the best temperature for mango ripening, which may result in improved phenolic compounds and other metabolites such as organic acids and sugars [51]. Both studies agreed that pressures around 60 and 80 MPa have an initial increase in phenolics, but the concentration gradually decreases with storage time. Although both studies suggest biosynthesis of phenolic compounds due to HHP; the authors did not show any test to probe the biosynthesis, for example, the increment in PAL activity. Further studies need to be performed to prove the biosynthesis of phenolic compounds [32,50].

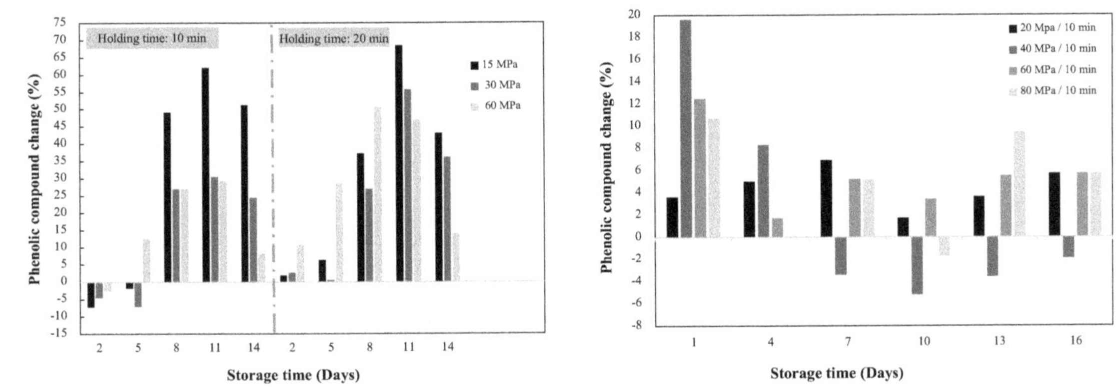

(a) (b)

Figure 2. Effect of high hydrostatic pressure in the increment of phenolic compounds of mango *(Mangifera indica)* stored at two conditions. (**a**) Fruits stored at 25 ± 1 °C with 85–90% relative humidity [46] and (**b**) fruits stored at 13 °C with ~85% humidity [46].

Ortega et al. [46] suggested that pressure as an abiotic stressor can lead to cell wall fracture or deformation causing cell wall loosening by crosslinking or depolymerizing its components [52]. Plant cells can sense the mechanical perturbation at their cell surfaces and they respond [53]. This promotes the production of ROS, like H_2O_2, which later acts, controls, and initiates enzymatic responses to repair the damaged cell wall via stress-responsive gene, oxidative burst linked with cell wall reinforcement, biosynthesis of phenolics, among others [9,48,54]. Injuries caused by mechanical stress, such as pressure, modify how plants synthesize secondary metabolites, as can be represented in Figure 3 [35]. The release of H_2O_2 is carried out in minutes, acting as the elicitor in the biosynthesis. The production of H_2O_2 at the cellular level acts as Ca+2 signaling, activates kinases, hormonal signaling, and regulates gene expression [54,55]. H_2O_2 as a signaling molecule activating metabolic pathways [56], leads to increased PAL activity, which as previously mentioned, synthesizes simple phenols derived from the cinnamic acid [19,31,32].

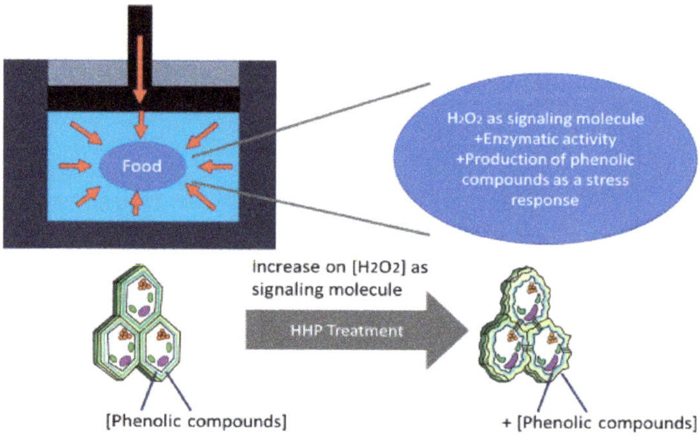

Figure 3. HHP effect in cell wall deformation from food. Modified from Gómez-Maqueo, et al. [18].

Similar to mango, for HHP-treated carrots after one day of storage, the total phenolic content increased 69.1% at 60 MPa and 154.9% at 100 MPa [49]. After day one, phenolic concentration starts to decrease showing similar values to the control (Figure 4). Viacava et al. [49], related the change in phenolic concentration with PAL activity observing an immediate effect in PAL. On day 0, PAL activity was reduced by 61.4% (100 MPa), while no significant increment was presented at day 1 at any pressure level evaluated; however, it increased at day 2 by 380.2 and 139.7% at 60 and 100 MPa, respectively. The authors attributed the low activity of PAL at day 1 and the higher concentrations of phenolic content on the same day, to the higher availability of precursors during the HHP treatment and the activation of enzymes not quantified in the study. In general, results showed that the PAL activity was higher at 60 than at 100 MPa, having greater metabolic activity because the samples were under higher oxidative stress-producing higher ROS, which resulted in higher PAL activity. It has shown that pressure has different effects on enzyme activities depending on factors such as type of product, type of enzyme, and treatment conditions. It has been stated that pressure could have favorable effects on the release of membrane-bound enzymes, or in the activation of proenzymes that require a biochemical change or a change on their configuration to expose the active site and to become active, and a direct relationship is not always observed between the increment in enzyme activity and pressure level [57]. Figure 5 shows the hypothetical model from Viacava et al. [49] explaining the immediate and late physiological response of carrots to HHP application, this hypothetical model could be applied to other vegetables and fruits. The immediate response involves cell wall deformation (mass exchange), increment in respiration rate, and production of signaling molecules such as H_2O_2, while the late response involves the biosynthesis of secondary metabolites during storage [18,19,47].

The synthesis of phenolics in cell cultures has also been evaluated. Cai et al. [48] studied the synthesis of anthocyanins during seven days of storage of a cell suspension of Vitis vinifera treated at 40 MPa for 10 min, observing the greatest increment at day 6 (53.3%). For potato (*Solanum tuberosum*) suspension culture treated at 100–200 MPa for 10 min, phenolic content increased immediately after processing by 54, 81, 267, 456, and 453% at pressures of 100, 125, 150, 175, and 200 MPa, respectively; nevertheless, these increments appear not to be associated with biosynthesis, but rather with the loss of compartmentation and subsequent release of the content of the vacuoles into the cytoplasm [44].

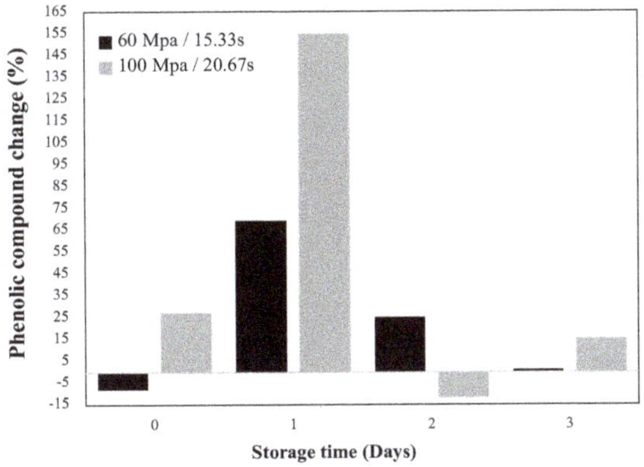

Figure 4. Effect of HHP in the biosynthesis of phenolic in carrots (*Daucus carota*). Adapted from Viacava et al. [49].

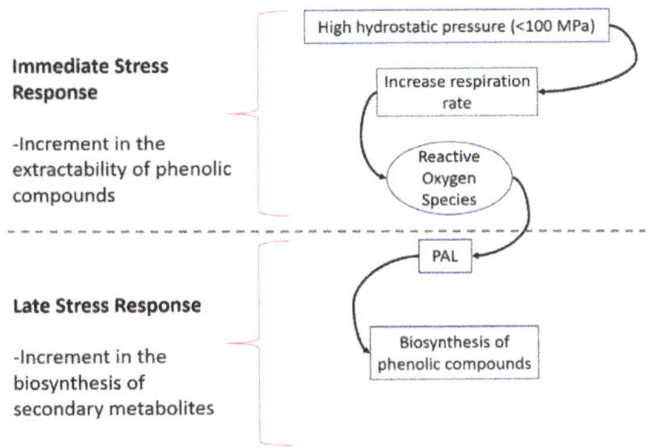

Figure 5. Immediate and late physiological response to stress produced by HHP technology to enhance extractability and biosynthesis in phenolic compounds in plants. Modified from [49].

Bioactive compounds, such as phenolics, are contained in specific organelles in the cell, which can vary depending on each type of product and variety [18]. The increment in phenolic compounds observed after HHP is not always attributed to biosynthesis. The release of phenolics by extraction from specific organelles could be responsible for the increments. The better extractability of phenolic at pressure levels above 100 MPa has been related to cell membrane disruption and release of bound phenolics, resulting in higher extractability and an improvement in bioaccessibility [58]. Making it different from biosynthesis, which is suggested to be a dual stress-response mechanism related to ATP, ROS, and the activation-deactivation of enzymes [18,19,47,50]. In the following section, some studies showing an increment of phenolics due to improvement of extraction are discussed.

3.2. *Effect of HHP on Phenolics Extraction Yield*

Several methods have been applied for phenolic extraction, such as conventional solvent extraction (CSE), and novel technologies like ultrasonic-assisted extraction (UAE), microwave-assisted extraction (MAE), and supercritical fluid extraction (SFE-CO_2) [59–61]. The application of novel technologies has resulted in favorable results on extraction yields compared with CSE [60,61].

HHP can also favor the release of bioactive compounds from cellular compartments, enhancing their extractability [18]. When fruits and vegetables are subjected to pressure levels normally higher than 100 MPa, the mechanism for phenolic biosynthesis is not promoted because the cell is inactivated before reacting to the stress caused by pressure [18]. In this case, the increment observed in phenolic can be attributed to improvement in extraction rather than to biosynthesis. HHP increases mass transfer in an immediate response due to the damage caused to the cell membrane, which increases permeability and facilitates secondary metabolites diffusion via solvent extraction [59,62]. In addition, disruption of weak interaction between phenolics and cell wall favors their release. In plants, phenolic compounds exist in both free and bound forms [60]. In dry fruits, bound phenolic content (mgGAE/100 g) ranged from 96 to 408; and free phenolic from 46 to 345. While in fresh fruits, the bound phenolic content ranged from 29 to 306; and free phenolics from 120 to 316 [63]. The main difference between free and bound phenolics is that free is solvent extractable, while the bound phenolic cannot be extracted into water or aqueous/organic solvents mixtures [60]. Based on this information, it is suggested that during the application of HHP treatments, the increment in free phenolic content would be attributed to the cell decompartmentalization, which produces phenolic release from plant tissue improving yield extraction, while the increment of bound phenolic compounds in addition to decompartmentalization is probably due to the increment of the enzyme's activity involved in the hydrolysis of proanthocyanidins, phenolic acids, and hydrolyzable tannins, which are esterified-bound and glycosylated-bound [61].

The effects on plant tissues, organelles, cell walls, and membranes, depend on the pressure level [62]. For example, in a study with prickly pears [18], the application of HHP at 100 MPa helped to release phenolics attached to cell walls by cell wall modifications (Figure 6). At this pressure level, the cell was still viable and capable to synthesize metabolites in response to the abiotic stress, but at higher pressure levels (350–600 MPa), the cell wall collapsed, enhancing the extractability of phenolic at pressures higher than 100 MPa. Higher pressure levels, and times, favored the extraction of phenolics due to the higher loss of cell wall integrity [18]. Table 2 shows studies for a variety of foods (vegetables, fruits, by products of plants, etc.) in which the authors have observed increment in phenolics (anthocyanins, flavonoids, polyphenols, and individual phenolics), suggesting improvement in their extractability at treatment conditions mainly from 300–600 MPa for 5–20 min at temperatures generally around 20–40 °C.

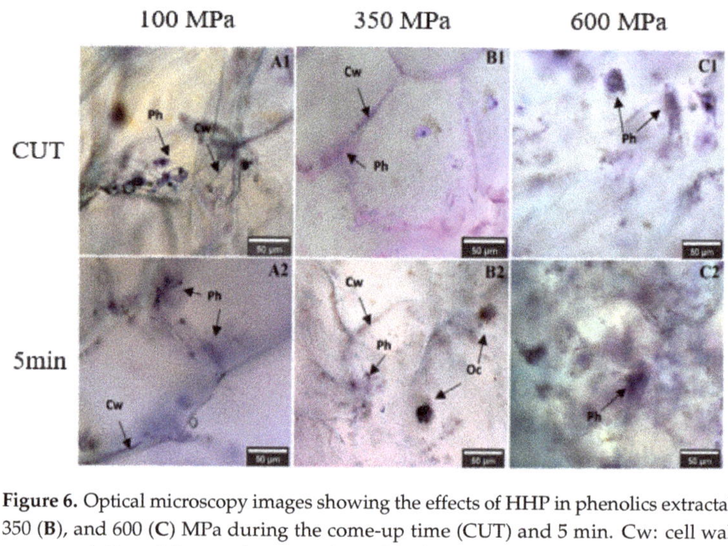

Figure 6. Optical microscopy images showing the effects of HHP in phenolics extractability at 100 (**A**), 350 (**B**), and 600 (**C**) MPa during the come-up time (CUT) and 5 min. Cw: cell wall, Ph: phenolic compound, Oc: calcium oxalate crystal. Modified from [18].

Table 2. Effect of high hydrostatic pressure on phenolic extraction yield.

Sample	Analyzed Compound	Treatment Conditions				Storage Conditions	Approximate Change (%)	Reference
		P (MPa)	t (min)	CUT (min)	T (°C)			
Apricot nectar *Prunus armeniaca* L.	TPC (Individual phenols include: Catechin, Chlorogenic acid, Neochlorogenic acid, Epicatechin, Ferulic acid, Caffeic acid, p-Coumaric acid)	300–500	5–20	2.5–4.2	34–40	2 days at 4 °C	↑2.0 up to ↑12.5	[64]
Sour cherry pomace *Prunus cerasus* L.	TPC	400 & 500	1–10	NR	20	−4 °C until analysis	↑39.5 up to ↑109.9	[65]
Grape by products (Skin, stems, and seeds) *Vitis Vinifera*	TPC	600	60	NR	70	NR	↑48.0	[66]
	Anthocyanins	600	60	NR	70	NR	↑41.4	
Jerusalem Artichoke *Helianthus tuberosus* L.	TPC (Pre-fermentation)	100	24 h	NR	50	NR	↑36.6	[67]
	TPC (Post-fermentation	100	24 h	NR	50	NR	↑61.36	
Cape gooseberry pulp *Physalis peruviana* L.	TPC	300–500	1–5	NR	25	0 and 60 days at 4 °C	↓32.3 up to ↑35.9	[68]
Grape *Vitis Vinifera*	TPPC	200–550	10	28.6 s–78.6 s	20	4 °C until fermentation(13 days)	↑55.0 up to ↑75.0	[69]
Wild Berry *Lonicera caerulea*	TPC	200–600	5–20	4–12 s	25	4 °C until analysis (48 h)	↓10.0 up to ↑14.4	[70]
	Anthocyanins	200–600	5–20	4–12 s	25	4 °C until analysis (48 h)	↓6.3 up to ↑7.9	
Açai Pulp *Euterpe oleracea Martius*	TPC	600	5	NR	25 and 65	Stored for 24 h with oxygen and light barrier	↓10.3 up to ↑11.4	[63]
Cricket *Acheta domesticus*	TPC	500	15	NR	30 and 40	NR	↑9.3 up to ↓67.3	[71]
Mealworm *Tenebrio molitor*	TPC	500	15	NR	30 and 40	NR	↓23.7 up to ↑8.6	[71]
Silvetia compressa	TPPC	400	15	2.03	35	Stored in brown glass flask at 10 °C	↓41.0	[72]
		600	5	3.07	35		↓30.0	
Ecklonia arborea	TPPC	400	15	2.03	35	Stored in brown glass flask at 10 °C	↑46.0	[72]
		600	5	3.07	35		↑20.0	

Table 2. *Cont.*

Sample	Analyzed Compound	Treatment Conditions				Storage Conditions	Approximate Change (%)	Reference
		P (MPa)	t (min)	CUT (min)	T (°C)			
Green tea *Camellia sinensis* L.	TPC	490	15	25	25	NR	↑32.6	[73]
Longan fruit pericarp *Dimocarpus longan* L.	TPC	500	2.5	NR	30	4 °C until analysis	↑43.8	[59]
Korean barberry *Berberis koreana*	TPC	500	5 & 15	NR	25	−20 °C until analysis	↑29.9 up to ↑33.1	[74]
Grape pomace	TPC	50–200	5–30	NR	25	NR	↓27.9 up to ↑18.6	[75]

P: Pressure; t: time; T: Temperature; CUT: Come up time (time to achieve desired pressure); NR: Not reported; TPC: total phenolic content. TPPC: total polyphenol content; ↑ indicates the increment in content compared with the untreated sample; ↓ indicates a decrease in content compared with the untreated sample.

In most cases presented in Table 2, HHP showed a positive effect on the extraction of phenolics. Results indicate that the higher the pressure and treatment time, the higher the extractability of phenolics. Some remarkable results are from Okur et al. [65] for sour cherry (Figure 7a), in which the increment in time of treatment from 1 to 10 min enhances the extraction yield from 39.5 up to 95% at 400 MPa and from 65 up to 109.9% at 500 MPa, respectively. Also, the increment in pressure from 400 to 500 MPa resulted in improvement of the extracted phenolics from 39.5 up to 61% [65]. The same tendency was shown for grape (Figure 7b), where the extraction of phenolic increased from 55 up to 75% by increasing the pressure treatment from 200 to 550 MPa [69]; gooseberry pulp treated at 400 and 500 MPa (during 10 min) showed an increment in the extracted phenolics of 8.3 and 22.9%, respectively [68]. In other studies, the treatment at 500 MPa of apricot nectar, showed an increment in total phenolics from 2 up to 9.6% when the holding time increases from 5 to 20 min (Figure 7c) [64]. Liu et al. [76] in wild berry demonstrated that at 200 MPa, the anthocyanin content increased from 6.3 to 8% by increasing the holding time from 5 to 10 min.

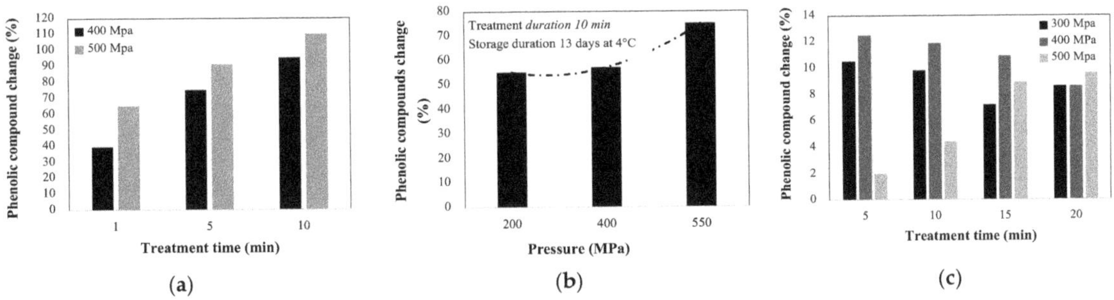

Figure 7. HHP effect in the extraction of phenolic compounds from (**a**) Sour cherry pomace (*Prunus cerasus* L.) [65], (**b**) Grape (*Vitis vinifera*) [69], and (**c**) apricot nectar (*Prunus armeniaca* L.) [64].

The levels of improvement during HHP processing depend on a variety of factors such as treatment conditions, type of compound, food physical characteristics, and composition. While, after HHP processing, the method used for phenolic extraction (i.e., type of solvent, ratio solvent: sample, contact, time, etc.) could influence the yield obtained among different studies from different authors. In addition, the storage conditions and handling after and during processing could influence results among different studies. As mentioned before, during processing, HHP could influence enzymes increasing their activity due to factors such as the release of membrane-bound enzymes or configuration changes, this last mechanism is also related to decreasing activity [57]. According to this, the extraction yields for the different samples could also be related to the residual activity of oxidative enzymes and the contact between phenolic and oxidative enzymes released from plant tissues after pressurization, which promotes oxidative reactions. The enzymatic activity could explain why at certain treatment conditions, no increment in phenolics was observed, but rather a decrease. Some examples are *Silvetia compressa*, where 400 MPa (15 min) and 600 MPa (5 min) decreased phenolics by 30 and 41%, respectively [72]. In cases where the change in the extractability yield of phenolics was not observed, it could be related to the retention of phenolics by the cell wall components. Some examples are gooseberry treated at 300 MPa for 5 min stored for 60 days [68], wild berry treated at 400 MPa for 20 min [77], and even cricket treated at 500 MPa for 15 min at a temperature of 65 °C [71].

It has been suggested that the increment in bioactive compounds by HHP could result in better bioavailability. Bioavailability determines the number of bioactive compounds that are digested, absorbed, and metabolized and therefore it determines their action in the human body [78–80]. For apple (*Granny Smith*) treated at 500 MPa, it was showing a higher absorption of minerals produced by high solubility in the intestine [79] [81], while for orange juice, higher bioavailability of vitamin C was observed. There are not enough studies showing the relationship between increment in phenolics and better bioavailability. The treatment of olives (Azeitera, Carrasqueña, Conserva de Elvas, and Morisca) at 600 MPa for 6 min at 10 °C showed an increase of bioavailability in phenols in the large intestine [79]. A review from Serment-Moreno et al. [9] concluded that HHP treatment (200–600 MPa) improved the bioavailability of phytochemical contents. This effect on the increase in bioavailability reported after HHP treatment can be related to differences in cell wall structures and improvement in the capacity of binding the phenolic compounds in the food matrix [1,4].

4. Final Remarks

This work describes two possible mechanisms for the increment of phenolics in foods after HHP treatment and during storage. The overview presented suggested that the increments could be related to phenolic biosynthesis and improvement of their extractability in food. The enhancement of the extraction of phenolic through HHP has been related to an immediate response to stress, where the HHP disrupts cellular compartments enhancing mass transfer and extractability or due to the disruption of non-covalent interactions between phenolics and cell wall. Despite a few cases, HHP has shown to be more effective at higher pressure levels, (>200 MPa); however, the level of yield achieved highly depends on the type of food and intensity of the other variables (time, temperature) as well as the extraction/analysis methods used. The use of HHP at lower values of pressure (<100 MPa) has been shown to activate biosynthesis of phenolic in a late response where the higher production of ROS such as H_2O_2 activates metabolic routes that increase the phenolic content. Despite very few cases, the use of HHP showed to be a very innovative and promising technology to improve the phenolic compounds in plants. The results presented in this review are highly relevant for the future use of HHP technology for both biosynthesis and extraction of phenolics or even other secondary metabolites with functional activities. This technology could help to generate food with better nutritional and functional value, including enhanced antioxidant activity, in addition to a better extraction could even influence the bioavailability of these compounds in the human body, increasing their beneficial effects on health.

Funding: Tecnológico de Monterrey and Consejo Nacional de Ciencia y Tecnología: A1-S-45034.

Acknowledgments: The authors acknowledge the support from CONACyT (Research Project A1-S-45034).

Conflicts of Interest: The authors declare no conflict of interest.

Sample Availability: Not applicable.

References

1. Ozcan, T.; Akpinar-Bayizit, A.; Yilmaz-Ersan, L.; Delikanli, B. Phenolics in Human Health. *Int. J. Chem. Eng. Appl.* **2014**, *5*, 393–396. [CrossRef]
2. Luna-Guevara, M.L.; Luna-Guevara, J.J.; Hernández-Carranza, P.; Ruíz-Espinosa, H.; Ochoa-Velasco, C.E. Phenolic Compounds: A Good Choice Against Chronic Degenerative Diseases. *Stud. Nat. Prod. Chem.* **2018**, *59*, 79–108. [CrossRef]
3. Osorio-Tobón, J.F. Recent advances and comparisons of conventional and alternative extraction techniques of phenolic compounds. *J. Food Sci. Technol.* **2020**, *57*, 4299–4315. [CrossRef]
4. Balasundram, N.; Sundram, K.; Samman, S. Food Chemistry Phenolic compounds in plants and agri-industrial by-products: Antioxidant activity, occurrence, and potential uses. *Food Chem.* **2005**, *99*, 191–203. [CrossRef]
5. Barba, F.J.; Putnik, P.; Kovačević, D.B.; Poojary, M.M.; Roohinejad, S.; Lorenzo, J.M.; Koubaa, M. Impact of conventional and non-conventional processing on prickly pear (*Opuntia* spp.) and their derived products: From preservation of beverages to valorization of by-products. *Trends Food Sci. Technol.* **2017**, *67*, 260–270. [CrossRef]
6. Putnik, P.; Kovačević, D.B.; Penić, M.; Fegeš, M.; Dragović-Uzelac, V. Microwave-Assisted Extraction (MAE) of Dalmatian Sage Leaves for the Optimal Yield of Polyphenols: HPLC-DAD Identification and Quantification. *Food Anal. Methods* **2016**, *9*, 2385–2394. [CrossRef]
7. Rastogi, N.K.; Raghavarao, K.S.M.S.; Balasubramaniam, V.M.; Niranjan, K.; Knorr, D. Opportunities and challenges in high pressure processing of foods. *Crit. Rev. Food Sci. Nutr.* **2007**, *47*, 69–112. [CrossRef]

8. Miao, M.; Wang, Q.; Zhang, T.; Jiang, B. Effect of high hydrostatic pressure (HHP) treatment on texture changes of water bamboo shoots cultivated in China. *Postharvest Biol. Technol.* **2011**, *59*, 327–329. [CrossRef]
9. Serment-Moreno, V.; Jacobo-Velázquez, D.A.; Torres, J.A.; Welti-Chanes, J. Microstructural and Physiological Changes in Plant Cell Induced by Pressure: Their Role on the Availability and Pressure-Temperature Stability of Phytochemicals. *Food Eng. Rev.* **2017**, *9*, 314–334. [CrossRef]
10. Perez, M.C.P. *Aplicación de Tecnologías No-Térmicas de Conservación, Pulsos Eléctricos de Alta Intensidad (PEAI), y Altas Presiones Hidrostáticas (APH), Para el Control de Cronobacter Sakazakii en Fórmula Láctea Infantil: Desarrollo de Modelos Predictivos y Valoración*; Universidad Politécnica de Valencia: Valencia, Spain, 2012.
11. Serment-Moreno, V.; Barbosa-Cánovas, G.; Torres, J.A.; Welti-Chanes, J. High-pressure Processing: Kinetic Models for Microbial and Enzyme Inactivation. *Food Eng. Rev.* **2014**, *6*, 56–88. [CrossRef]
12. Mozhaev, V.V.; Heremans, K.; Frank, J.; Masson, P.; Balny, C. Exploiting the effects of high hydrostatic pressure in biotechnological applications. *Trends Biotechnol.* **1994**, *12*, 493–501. [CrossRef]
13. Martín, J.; Asuero, A.G. High hydrostatic pressure for recovery of anthocyanins: Effects, performance, and applications. *Sep. Purif. Rev.* **2021**, *50*, 159–176. [CrossRef]
14. Cheftel, J.C. Review: High-pressure, microbial inactivation and food preservation. *Food Sci. Technol. Int.* **1995**, *1*, 75–90. [CrossRef]
15. Kimura, K.; Ida, M.; Yosida, Y.; Ohki, K.; Fukumoto, T.; Sakui, N. Comparison of Keeping Quality between Pressure-processed Jam and Heat-processed Jam: Changes in Flavor Components, Hue, and Nutrients during Storage. *Biosci. Biotechnol. Biochem.* **1994**, *58*, 1386–1391. [CrossRef]
16. Téllez-Luis, S.J.; Ramírez, J.A.; Pérez-Lamela, C.; Vázquez, M.; Simal-Gándara, J. Aplicación de la alta presión hidrostática en la conservación de los alimentos. Ciencia y Tecnología de los Alimentos. *Cienc. Tecnol. Aliment.* **2001**, *3*, 66–80. [CrossRef]
17. Srinivas, M.S.; Madhu, B.; Girijal, S. High Pressure Processing of Foods: A Review. 2018. Available online: https://www.researchgate.net/publication/328652367 (accessed on 7 February 2022).
18. Gómez-Maqueo, J.; Welti-Chanes, M.; Cano, P. Release mechanisms of bioactive compounds in fruits submitted to high hydrostatic pressure: A dynamic microstructural analysi based on prickly pear cells. *Food Res. Int.* **2019**, *130*, 108909. [CrossRef]
19. Jacobo-Velázquez, D.A.; Cuéllar-Villarreal, M.D.; Welti-Chanes, J.; Cisneros-Zevallos, L.; Ramos-Parra, P.A.; Hernández-Brenes, C. Nonthermal processing technologies as elicitors to induce the biosynthesis and accumulation of nutraceuticals in plant foods. *Trends Food Sci. Technol.* **2017**, *60*, 80–87. [CrossRef]
20. Hättenschwiler, S.; Vitousek, P.M. The role of polyphenols in terrestrial ecosystem nutrient cycling. *Tree* **2000**, *15*, 238–243. [CrossRef]
21. Bahadur, B.; Rajam, M.V.; Sahijram, L.; Krishnamurthy, K.V. *Plant Biology and Biotechnology: Plant Diversity, Organization, Function and Improvement*; Springer: Berlin/Heidelberg, Germany, 2015; Volume 1, pp. 1–827. [CrossRef]
22. Marchiosi, R.; dos Santos, W.D.; Constantin, R.P.; de Lima, R.B.; Soares, A.R.; Finger-Teixeira, A.; Mota, T.R.; de Oliveira, D.M.; Foletto-Felipe, M.D.; Abrahão, J.; et al. Biosynthesis and metabolic actions of simple phenolic acids in plants. *Phytochem. Rev.* **2020**, *19*, 865–906. [CrossRef]
23. Singh, S.A.; Christendat, D. Structure of Arabidopsis dehydroquinate dehydratase-shikimate dehydrogenase and implications for metabolic channeling in the shikimate pathway. *Biochemistry* **2006**, *45*, 10406. [CrossRef]
24. Tzin, V.; Galili, G. New Insights into the shikimate and aromatic amino acids biosynthesis pathways in plants. *Mol. Plant* **2010**, *3*, 956–972. [CrossRef]
25. Mittelstädt, G.; Negron, L.; Schofield, L.R.; Marsh, K.; Parker, E.J. Biochemical and structural characterisation of dehydroquinate synthase from the New Zealand kiwifruit Actinidia chinensis. *Arch. Biochem. Biophys.* **2013**, *537*, 185–191. [CrossRef]
26. Singh, S.A.; Christendat, D. The DHQ-dehydroshikimate-SDH-shikimate-NADP(H) complex: Insights into metabolite transfer in the shikimate pathway. *Cryst. Growth Des.* **2007**, *7*, 2153–2160. [CrossRef]
27. Herrmann, K.M.; Weaver, L.M. The shikimate pathway. *Annu. Rev. Plant Biol.* **1999**, *50*, 473–503. [CrossRef]
28. Dudareva, H.M.N. The shikimate pathway and aromatic amino acid biosynthesis in plants. *Annu. Rev. Plant Biol.* **2012**, *63*, 73–105. [CrossRef]
29. Schönbrunn, E.; Eschenburg, S.; Shuttleworth, W.A.; Schloss, J.V.; Amrhein, N.; Evans, J.N.; Kabsch, W. Interaction of the herbicide glyphosate with its target enzyme 5-enolpyruvylshikimate 3-phosphate synthase in atomic detail. *Proc. Natl. Acad. Sci. USA* **2001**, *98*, 1376–1380. [CrossRef]
30. Kitzing, K.; Auweter, S.; Amrhein, N.; Macheroux, P. Mechanism of chorismate synthase: Role of the two invariant histidine residues in the active site. *J. Biol. Chem.* **2004**, *279*, 9451–9461. [CrossRef]
31. Gordo, D.A.M. Los Compuestos Fenólicos, Un Acercamiento A Su Biosíntesis, Síntesis Y Actividad Biológica. *Rev. Investig. Agrar. Ambient.* **2018**, *9*, 81–104. [CrossRef]
32. Ardila, H.; Baquero, B.; Martinez, S. Inducción de la Actividad de la Enzima Fenilalanina Amonio Liasa en Clavel (*Dianthus caryophyllus* L.) Por Elicitores del Hongo *Fusarium oxysporum* f. sp. Dianthi raza 2. *Rev.Colomb.Quim.* **2007**, *36*, 151–167.
33. Knaggs, R. The biosynthesis of shikimate metabolites. *Nat. Prod. Rep.* **2003**, *18*, 119–136. [CrossRef]
34. Cheng, G.W.; Breen, P.J. Activity of Phenylalanine Ammonia-Lyase (PAL) and Concentrations of Anthocyanins and Phenolics in Developing Strawberry Fruit. *J. Am. Soc. Hortic. Sci.* **2019**, *116*, 865–869. [CrossRef]
35. Figueiredo, C.; Barroso, J.; Pedro, L.; Scheffer, J. Factors affecting secondary metabolite production in plants: Volatile components and essential oils. *Flavor Fragr. J.* **2008**, *23*, 213–226. [CrossRef]

36. Vidal, A.M. Respuestas Fisiológicas de Los Cítricos Sometidos a Condiciones de Estrés Biótico y Abiótico. Aspectos Comunes y Específicos. TDX (Tesis Doctorals en Xarxa). 2010, p. 213. Available online: http://repositori.uji.es/xmlui/handle/10234/29723 (accessed on 15 November 2021).
37. Cho, M.H.; Lee, S.W. Phenolic phytoalexins in rice: Biological functions and Biosynthesis. *Int. J. Mol. Sci.* **2015**, *16*, 29120–29133. [CrossRef]
38. Lattanzio, V. *Natural Products | Phenolic Compounds: Introduction*; Springer: Berlin/Heidelberg, Germany, 2013. [CrossRef]
39. Hatami, M.; Badi, H.N.; Ghorbanpour, M. Nano-elicitation of secondary pharmaceutical metabolites in plant cells: A review. *J. Med. Plants* **2019**, *18*, 6–36. [CrossRef]
40. Marslin, G.; Sheeba, C.J.; Franklin, G. Nanoparticles alter secondary metabolism in plants via ROS burst. *Front. Plant Sci.* **2017**, *8*, 832. [CrossRef]
41. Rivero-Montejo, S.d.; Vargas-Hernandez, M.; Torres-Pacheco, I. Nanoparticles as novel elicitors to improve bioactive compounds in plants. *Agriculture* **2021**, *11*, 134. [CrossRef]
42. Shalaby, T.A.; Bayoumi, Y.; Abdalla, N.; Taha, H.; Alshaal, T.; Shehata, S.; Amer, M.; Domokos-Szabolcsy, É.; El-Ramady, H. Nanoparticles, Soils, Plants and Sustainable Agriculture. *Nanosci. Food Agric.* **2016**, *20*, 283–312. [CrossRef]
43. van Breusegem, F.; Dat, J.F.; Van, F. Reactive Oxygen Species in Plant Cell Death. *Source Plant Physiol. React. Oxyg. Species* **2015**, *141*, 384–390. [CrossRef]
44. Dörnenburg, H.; Knorr, D. Evaluation of Elicitor- and High-Pressure-Induced Enzymatic Browning Utilizing Potato (*Solanum tuberosum*) Suspension Cultures as a Model System for Plant Tissues. *J. Agric. Food Chem.* **1997**, *45*, 4173–4177. [CrossRef]
45. Ramos-Parra, P.A.; García-Salinas, C.; Rodríguez-López, C.E.; García, N.; García-Rivas, G.; Hernández-Brenes, C.; de la Garza, R.I.D. High hydrostatic pressure treatments trigger de novo carotenoid biosynthesis in papaya fruit (*Carica papaya* cv. Maradol). *Food Chem.* **2018**, *277*, 362–372. [CrossRef]
46. Hu, K.; Peng, D.; Wang, L.; Liu, H.; Xie, B.; Sun, Z. Effect of mild high hydrostatic pressure treatments on physiological and physicochemical characteristics and carotenoid biosynthesis in postharvest mango. *Postharvest Biol. Technol.* **2020**, *172*, 10. [CrossRef]
47. Viacava, F.; Ortega-Hernández, E.; Welti-Chanes, J.; Cisneros-Zevallos, L.; Jacobo-Velázquez, D.A. Using High Hydrostatic Pressure Processing Come-Up Time as an Innovative Tool to Induce the Biosynthesis of Free and Bound Phenolics in Whole Carrots. *Food Bioprocess Technol.* **2020**, *13*, 1717–1727. [CrossRef]
48. Ortega, V.G.; Ramírez, J.A.; Velázquez, G.; Tovar, B.; Mata, M.; Montalvo, E. Effect of high hydrostatic pressure on antioxidant content of 'Ataulfo' mango during postharvest maturation. *Food Sci. Technol.* **2013**, *33*, 561–568. [CrossRef]
49. Yasunaga, E.; Fukuda, S.; Takata, D.; Spreer, W.; Sardsud, V.; Nakano, K. Quality changes in fresh mango fruits (*Mangifera indica* L. 'Nam Dok Mai') under actual distribution temperature profile from Thailand to Japan. *Environ. Control Biol.* **2018**, *56*, 45–49. [CrossRef]
50. Kim, T.E.; Gil, B.; Kim, C.T.; Cho, Y.J. Enrichment of Phenolics in Harvested Strawberries by High-Pressure Treatment. *Food Bioprocess Technol.* **2017**, *10*, 222–227. [CrossRef]
51. Cai, Z.; Riedel, H.; Saw NM, M.T.; Mewis, I.; Reineke, K.; Knorr, D.; Smetanska, I. Effects of elicitors and high hydrostatic pressure on secondary metabolism of Vitis vinifera suspension culture. *Process Biochem.* **2011**, *46*, 1411–1416. [CrossRef]
52. Liang, D. A salutary role of reactive oxygen species in intercellular tunnel-mediated communication. *Front. Cell Dev. Biol.* **2018**, *6*, 2. [CrossRef]
53. Yahraus, T.; Chandra, S.; Legendre, L.; Low, P.S. Evidence for a Mechanically Induced Oxidative Burst. 1995. Available online: www.plantphysiol.org (accessed on 7 February 2022).
54. Slesak, I.; Libik, M.; Karpinska, B.; Karpinski, S.; Miszalski, Z. The role of hydrogen peroxide in regulation of plant metabolism and cellular signalling in response to environmental stresses. *Acta Biochim. Pol.* **2007**, *54*, 39–50. [CrossRef]
55. Hancock, T.; Desikan, R.; Neill, S.J. Role of reactive oxygen species in cell signalling pathways. *Biochem. Soc. Trans.* **2001**, *29*, 345–350. [CrossRef]
56. Jacobo-velázquez, D.A.; Santana-Galvez, J.; Cisneros-Zevallos, L. Designing Next-Generation Functional Food and Beverages: Combining Nonthermal Processing Technologies and Postharvest Abiotic Stresses. *Food Eng. Rev.* **2021**, *13*, 592–600. [CrossRef]
57. Escobedo-Avellaneda, Z.; Pérez-Simón, I.; Lavilla-Martín, M.; Baranda-González, A.; Welti-Chanes, J. Enzymatic and phytochemical stabilization of orange-strawberry-banana beverages by high hydrostatic pressure and mild heat. *Food Sci. Technol. Int.* **2017**, *23*, 185–193. [CrossRef] [PubMed]
58. Gonzalez, M.E.; Anthon, G.E.; Barrett, D.M. Onion cells after high pressure and thermal processing: Comparison of membrane integrity changes using different analytical methods and impact on tissue textura. *J. Food Sci.* **2010**, *75*, 426–432. [CrossRef] [PubMed]
59. Prasad, N.; Yang, B.; Zhao, M.; Wei, X.; Jiang, Y.; Chen, F. High pressure extraction of corilagin from longan (*Dimocarpus longan* Lour.) fruit pericarp. *Sep. Purif. Technol.* **2009**, *70*, 41–45. [CrossRef]
60. Pérez-Jiménez, J.; Torres, J.L. Analysis of nonextractable phenolic compounds in foods: The current state of the art. *J. Agric. Food Chem.* **2011**, *59*, 12713–12724. [CrossRef] [PubMed]
61. Lou, X.; Xu, H.; Hanna, M.; Yuan, L. Identification and quantification of free, esterified, glycosylated and insoluble-bound phenolic compounds in hawthorn berry fruit (*Crataegus pinnatifida*) and antioxidant activity evaluation. *LWT* **2020**, *130*, 109643. [CrossRef]

62. Huang, H.W.; Hsu, C.P.; Yang, B.B.; Wang, C.Y. Advances in the extraction of natural ingredients by high pressure extraction technology. *Trends Food Sci. Technol.* **2013**, *33*, 54–62. [CrossRef]
63. Koubala, B.B.; Bayang, J.P.; Wangso, H.; Kolla, M.C.; Laya, A. Variation of Phenolics (Bound and Free), Minerals, and Antioxidant Activity of Twenty-Eight Wild Edible Fruits of Twenty-Three Species from Far North Region of Cameroon. *BioMed Res. Int.* **2021**, *2021*, 4154381. [CrossRef]
64. Okur, İ.; Baltacıoğlu, C.; Ağçam, E.; Baltacıoğlu, H.; Alpas, H. Evaluation of the Effect of Different Extraction Techniques on Sour Cherry Pomace Phenolic Content and Antioxidant Activity and Determination of Phenolic Compounds by FTIR and HPLC. *Waste Biomass Valoriz.* **2019**, *10*, 3545–3555. [CrossRef]
65. Morata, A.; Loira, I.; Vejarano, R.; Bañuelos, M.A.; Sanz, P.D.; Otero, L.; Suárez-Lepe, J.A. Grape Processing by High Hydrostatic Pressure: Effect on Microbial Populations, Phenol Extraction and Wine Quality. *Food Bioprocess Technol.* **2014**, *10*, 3545–3555. [CrossRef]
66. Torres-Ossandón, M.J.; Vega-Gálvez, A.; López, J.; Stucken, K.; Romero, J.; Di Scala, K. Effects of high hydrostatic pressure processing and supercritical fluid extraction on bioactive compounds and antioxidant capacity of Cape gooseberry pulp (*Physalis peruviana* L.). *J. Supercrit. Fluids* **2018**, *318*, 215–220. [CrossRef]
67. Huang, W.; Bi, X.; Zhang, X.; Liao, X.; Hu, X.; Wu, J. Comparative study of enzymes, phenolics, carotenoids and color of apricot nectars treated by high hydrostatic pressure and high temperature short time. *Innov. Food Sci. Emerg. Technol.* **2013**, *18*, 74–82. [CrossRef]
68. Liu, S.; Xu, Q.; Li, X.; Wang, Y.; Zhu, J.; Ning, C.; Chang, X.; Meng, X. Effects of high hydrostatic pressure on physicochemical properties, enzymes activity, and antioxidant capacities of anthocyanins extracts of wild Lonicera caerulea Berry. *Innov. Food Sci. Emerg. Technol.* **2016**, *36*, 48–58. [CrossRef]
69. Tapia-Salazar, M.; Arévalo-Rivera, I.G.; Maldonado-Muñiz, M.; Garcia-Amezquita, L.E.; Nieto-López, M.G.; Ricque-Marie, D.; Cruz-Suárez, L.E.; Welti-Chanes, J. The Dietary Fiber Profile, Total Polyphenol Content, Functionality of Silvetia compressa and Ecklonia arborea, and Modifications Induced by High Hydrostatic Pressure Treatments. *Food Bioprocess Technol.* **2019**, *12*, 512–523. [CrossRef]
70. Ugur, A.E.; Bolat, B.; Oztop, M.H.; Alpas, H. Effects of High Hydrostatic Pressure (HHP) Processing and Temperature on Physicochemical Characterization of Insect Oils Extracted from Acheta domesticus (House Cricket) and Tenebrio molitor (Yellow Mealworm). *Waste Biomass Valoriz.* **2020**, *12*, 4277–4286. [CrossRef]
71. Clariana, M.; Valverde, J.; Wijngaard, H.; Mullen, A.M.; Marcos, B. High pressure processing of swede (*Brassica napus*): Impact on quality properties. *Innov. Food Sci. Emerg. Technol.* **2011**, *12*, 85–92. [CrossRef]
72. Corrales, M.; Toepfl, S.; Butz, P.; Knorr, D.; Tauscher, B. Extraction of anthocyanins from grape by-products assisted by ultrasonics, high hydrostatic pressure or pulsed electric fields: A comparison. *Innov. Food Sci. Emerg. Technol.* **2008**, *9*, 85–91. [CrossRef]
73. Kim, D.; Fan, J.P.; Chung, H.C.; Han, G.D. Changes in Extractability and Antioxidant Activity of Jerusalem Artichoke (*Helianthus tuberosus* L.) Tubers by Various High Hydrostatic Pressure Treatments. *Food Sci. Biotechnol.* **2010**, *19*, 1365–1371. [CrossRef]
74. de Jesus AL, T.; Cristianini, M.; Dos Santos, N.M.; Júnior, M.R.M. Effects of high hydrostatic pressure on the microbial inactivation and extraction of bioactive compounds from açaí (*Euterpe oleracea* Martius) pulp. *Food Res. Int.* **2019**, *130*, 108856. [CrossRef]
75. Xi, J.; Wang, B. Optimization of Ultrahigh-Pressure Extraction of Polyphenolic Antioxidants from Green Tea by Response Surface Methodology. *Food Bioprocess Technol.* **2013**, *6*, 2538–2546. [CrossRef]
76. Qadir, S.A.; Kwon, M.C.; Han, J.G.; Ha, J.H.; Chung, H.S.; Ahn, J.; Lee, H.Y. Effect of different extraction protocols on anticancer and antioxidant activities of Berberis koreana bark extracts. *J. Biosci. Bioeng.* **2009**, *107*, 331–338. [CrossRef]
77. Ghasemy-piranloo, F.; Kavousi, F.; Dadashian, S. Combination of enzyme-assisted extraction and high hydrostatic pressure for phenolic compounds recovery from grape pomace. *J. Food Eng.* **2020**, *288*, 110128. [CrossRef]
78. Srinivasan, V.S. Bioavailability of nutrients: A practical approach to in vitro demonstration of the availability of nutrients in multivitamin-mineral combination products. *J. Nutr.* **2001**, *131* (Suppl. 4), 1349–1350. [CrossRef] [PubMed]
79. Sanchez-Moreno, C.; de Ancos, B.; Plaza, L.; Elez-Martinez, P.; Cano, M.P. Nutritional approaches and health-related properties of plant foods processed by high pressure and pulsed electric fields. *Crit. Rev. Food Sci. Nutr.* **2009**, *49*, 552–576. [CrossRef] [PubMed]
80. Talaverano, M.I.; Pérez-Nevado, F.; Boselli, E.; Cordeiro, A.M.; Martillanes, S.; Foligni, R.; Martín-Vertedor, D. Evaluation of phenolics and acrylamide and their bioavailability in high hydrostatic pressure treated and fried table olives. *J. Food Process. Preserv.* **2020**, *44*, 1–9. [CrossRef]
81. Briones-Labarca, V.; Venegas-Cubillos, G.; Ortiz-Portilla, S.; Chacana-Ojeda, M.; Maureira, H. Effects of high hydrostatic pressure (HHP) on bioaccessibility, as well as antioxidant activity, mineral and starch contents in Granny Smith apple. *Food Chem.* **2011**, *128*, 520–529. [CrossRef]

molecules

MDPI

Article

Identification of Carotenoids in Hairless Canary Seed and the Effect of Baking on Their Composition in Bread and Muffin Products

El-Sayed M. Abdel-Aal *, Lili Mats and Iwona Rabalski

Guelph Research and Development Centre, Agriculture and Agri-Food Canada, 93 Stone Road West, Guelph, ON N1G 5C9, Canada; lili.mats@agr.gc.ca (L.M.); iwona.rabalski@agr.gc.ca (I.R.)
* Correspondence: elsayed.abdelaal@agr.gc.ca; Tel.: +1-(226)-217-8079; Fax: +1-(226)-217-8181

Abstract: Carotenoids are essential components in the human diet due to their positive functions in ocular and cognitive health. This study investigated composition of carotenoids in hairless canary seed (HCS) as a novel food and the effect of baking on carotenoids in bread and muffin made from HCS, wheat and corn. Three bread formulations made from wheat and HCS blends were evaluated and compared with control wheat bread. In addition, three low-fat muffin recipes prepared from HCS alone or in blends with corn were assessed. The fate of carotenoid compounds in breads and muffins was monitored after dry mixing, dough/batter formation and oven baking. Carotenoids in products were quantified using UPLC and their identification was confirmed based on LC-MS/MS. Hairless canary seed and corn were fairly rich in carotenoids with a total content of 7.6 and 12.9 μg/g, respectively, compared with wheat (1.3 μg/g). Nineteen carotenoid compounds were identified, with all-*trans* lutein being the principal carotenoid in HCS followed by lutein 3-*O*-linoleate, lutein 3-*O*-oleate and lutein di-linoleate. There were significant reductions in carotenoids in muffin and bread products. It appears that batter or dough preparation causes more reductions in carotenoids than oven baking, probably due to enzymatic oxidation and degradation. Muffin-making resulted in lower lutein reductions compared with the bread-making process. The results suggest that muffins made from hairless canary seed alone or in blends with corn could boost the daily intake of lutein and/or zeaxanthin.

Keywords: hairless canary seed groats; carotenoid identification; LC-MS/MS; baked products

Citation: Abdel-Aal, E.-S.M.; Mats, L.; Rabalski, I. Identification of Carotenoids in Hairless Canary Seed and the Effect of Baking on Their Composition in Bread and Muffin Products. *Molecules* **2022**, *27*, 1307. https://doi.org/10.3390/molecules27041307

Academic Editor: Smaoui Slim

Received: 27 January 2022
Accepted: 11 February 2022
Published: 15 February 2022

1. Introduction

Carotenoids are essential components in the human diet due to their positive functions in ocular and cognitive health [1–5]. Lutein and zeaxanthin are xanthophyll carotenoids commonly found at high levels in dark green vegetables, egg yolks, einkorn wheat, and corn [1]. These pigments constitute the yellow spot in the human retina and are referred to as macular pigments [6]. They also accumulate in the brain across the lifespan [5] and have been found to improve cognitive functions [7,8]. In addition, they have been linked with a reduced risk of age-related macular degeneration and cataracts [1]. Epidemiological studies have shown that the status of macular pigment is strongly correlated with lutein intake [9]. Thus, the availability of diets rich in lutein and/or zeaxanthin is crucial for human health due to their physiological and protective attributes, particularly in the eye and brain.

Special grains such as hairless canary seed and high lutein corn could provide promising ingredients for the development of high lutein functional foods. Hairless (glabrous) canary seed (*Phalaris canariensis* L.) was developed through a successful breeding program to eliminate the potential health threats associated with the presence of tiny siliceous hairs on the surface of the hull of the kernel in the hairy (pubescent) varieties [10]. The hairless canary seed (HCS) was approved by Health Canada as a novel food [11] and received GRAS status (GRAS Notice No. GRN 000529) from the US Food and Drug Administration [12].

HCS is a true cereal grain belonging to the grass family (*Poaceae or Gramineae*), similar to wheat, barley, oat, and other cereals. The grain has a unique nutritional profile being relatively rich in protein (22.7%) and oil (7.7%) as a cereal grain, but its starch content is comparable to other cereal grains at 57.2% [13]. In addition, the protein in HCS is free from gluten, which makes it suitable for the gluten-free food market [14]. Several studies have reported that HCS groats hold great potential as a wholegrain food ingredient and/or sustainable source of starch, protein, and oil [13,15,16]. HCS is also considered a health food ingredient being a good source of bioactive peptides, carotenoids, and phenolic compounds [13,17,18]. Currently, little or no data are available on carotenoid composition in HCS and the effect of processing on carotenoids in the end products. This study was aimed at exploring the detailed composition of carotenoids in HCS. In addition, the effect of baking on carotenoids in bread and muffin prototypes made from blends of hairless canary seed, wheat, and corn was investigated to assess their potential in enhancing the consumption of carotenoids.

2. Materials and Methods

2.1. Chemicals

HPLC or MS-grade methanol, acetonitrile, methyl tert-butyl ether (MtBE), and hexane, sodium hydroxide, and hydrochloric acid were purchased from Fisher Scientific (Mississauga, ON, Canada). Suprapur formic acid (FA) was purchased from VWR (Mississauga, ON, Canada). High purity carotenoid standards, including lutein 00012453 and Zeaxanthin 00026504 were purchased from ChromaDex (Irvine, CA, USA) and β-cryptoxanthin C6368 and β-carotene C9750 from SigmaAldrich (Oakville, ON, Canada). Lutein mono- and di-esters were prepared in the lab as described in our previous publications [19,20]. The nano pure water was obtained from Milli-Q integral water purification system (Millipore Ltd., Etobicoke, ON, Canada).

2.2. Grains and Flours

The yellow hairless canary seed was kindly supplied by the Canary seed Development Commission of Saskatchewan and hard red spring wheat by the University of Saskatchewan (Saskatoon, SK, Canada). Yellow corn flour was purchased from the local market in Guelph, ON, Canada. The yellow hairless canary seeds were milled using the UDY cyclone mill (310-014, UDY Corporation, Fort Collins, CO, USA) into wholegrain flour, while the wheat grains were milled on the Brabender Junior mill (Quadrumat Junior, Brabender Instruments Inc., Duisburg, Germany) into wholemeal flour.

2.3. Preparation of Bread and Muffin Products

Two baked products were investigated, bread as a fermented food commonly consumed worldwide and muffin as a non-fermented baked food. Three low-fat muffins were prepared from HCS flour alone or in blend with cornflour at ratios of 1:1 and 1:2 (*w*/*w*). The muffin formula contained 120 g HCS flour or flour mix, 36 g sugar, 1.5 g salt, 3 g baking powder, 2.5 g egg white powder, 5 g canola oil, and 85, 80, 80 g water for 100% HCS flour, 1:1 and 1:2 HCS/corn, respectively. First, the dried ingredients were mixed for 1 min, then water and other ingredients were added and mixed for 2 min. A 25 g batter was dispensed into a paper cup, then baked for 16 min at 204 °C. The sourdough method was used for bread making according to the approved AACCI method 10-11.01 [21]. The Lallemand Inc. Florapan L62 culture starter was kindly provided by Lallemand Company (Montreal, QC, Canada). Three bread formulations were made from wheat and HCS flour blends at a ratio of 85/15, 75/25, and 50/50 (*w*/*w*) in addition to 100% wheat flour as a control treatment. The ingredients were 75 g flour, 50 g sour, 2 g salt, 5 g sugar, 3 g corn oil, 0.5 g dry yeast, and 37.5 g water. The dry ingredients and oil were pre-mixed for 2 min, then sour and water were added, and all ingredients were mixed for 3 min. The dough was placed in a proffer at 35 °C for 60 min, then punched, shaped, and panned. The second proofing was carried out for approximately 50 min until the maximum volume was reached. Bread loaves were baked at 204 °C for 21 min. All baking trials were made at least in triplicate, and subsamples were taken at three technological steps, e.g., dry mix, batter or dough, and freshly baked muffins or breads, and changes in carotenoids were monitored during the baking process. The changes in carotenoids during the baking process were calculated as % difference from their corresponding dry mixes using the following equation:

$$\% \text{ difference} = [\text{carotenoid in dry mix} - \text{carotenoid in product}]/[\text{carotenoid in dry mix}] \times 100$$

2.4. Analysis of Nutrients and Prodauct Quality

Moisture, ash, and total dietary fiber in ingredients and finished products were determined according to the approved methods of the AACCI Methods 44-15.02, 08-01.01, and 32-21.01, respectively [21]. Protein content was based on the combustion method using a Nitrogen Analyzer (Flash 2000, Thermo Fisher Scientific, Waltham, MA, USA). External and internal qualities of breads were assessed based on the scoring guidelines of the AACCI method 10-12.01 and the volume of bread loaves was measured by the rapeseed displacement method 10-05.01 [21]. The quality of the muffin products was evaluated based on the measurement of the height of the muffin, internal and external quality scores. Three trained technicians were employed to assess the quality of products. The quality score reported is the average internal quality score (crumb structure, texture, and color, flavour and mouthfeel) and external quality score (shape, smoothness, degree of breakage and shred and crust colour).

2.5. Quantification and Identification of Carotenoids

Samples of HCS flours and products (0.5 g) were extracted with 5 mL of water-saturated butanol for 30 sec using a Polytron PT 10–30 homogenizer (Kinematica, Switzerland). The butanol extracts were centrifuged at 3030 g for 10 min and the supernatants were filtered through a 0.2-µm GHP syringe filter (Pall Company, Mississauga, ON, Canada) prior to UPLC analysis. Waters Acquity UPLC H Class with eLambda PDA detection and Carotenoid C30 YMC column (10 cm × 4.6 mm × 3 µm) was used. Ten µL of carotenoid extracts were injected onto the column. The mobile phase was composed of methanol (A) and MtBE (B) using a gradient program: at 0 time 95% B, at 7 min 92% B, at 20 min 75% B, at 29 min 60% B, at 30–33 min 0% B and at 34–36 min 95% B. The column temperature was set at 35 °C. Four authentic carotenoids (all-*trans*-lutein, all-trans-zeaxanthin, all-trans-β-cryptoxanthin and all-trans-β-carotene and lab-prepared lutein mono- and di-ester compounds were used for the identification and quantification of carotenoids [20].

Carotenoid compounds were further identified and confirmed based on LC-MS/MS analysis using Thermo Scientific™ Q-Exactive™ Orbitrap mass spectrometer equipped with a Vanquish™ Flex Binary UPLC System (Waltham, MA, USA). For the identification of carotenoid compounds, extracts were subjected to further clean up to remove proteins and oil in order to concentrate carotenoids. A YMC Carotenoid S-3 µm column (150 × 4.6 mm, Chromatographic Specialties Inc., Brockville ON, Canada) was used. Both APCI and ESI methods were used in this study. The mobile phase for APCI method consisted of solvent A (MeOH) and solvent B (MtBE). The mobile phase for ESI mode consisted of solvent A (99.95% MeOH + 0.05% FA) and solvent B (99.95% MtBE + 0.05% FA). The following solvent gradient was used for both APCI and ESI studies: 0–10 min, 5% to 8% B; 10–25 min, 8% to 35% B; 25–28 min, isocratic 35% B; 28–29 min, 35% to 100% B; 29–31 min, isocratic 100% B; 31–32 min, 100% to 5% B; 32–36 min, isocratic 5% B. The column temperature was set at 35 °C. The flow rate was 0.4 mL/ min. The injection volume was 5 or 10 µL. The UV absorption wavelengths were 475 nm (UV1), 439 nm (UV2), and 450 nm (UV3). UPLC chromatograms depicting the separation of the four carotenoid standards and lutein mono- and di-esters are shown in Figure 1A,B. Positive ionization mode was used for both APCI and ESI study. The optimized APCI conditions were as follows: spray voltage, 5 kV; capillary temperature, 350 °C and auxiliary heater temperature, 400 °C. The optimized ESI conditions were as follows: spray voltage, 3.5 kV; capillary temperature,

263 °C and auxiliary heater temperature, 425 °C. Mass spectrometry data was collected using Full-MS/DDMS2 (TopN = 15) method. Data was visualized and analysed using Thermo FreeStyle™ 1.7 software.

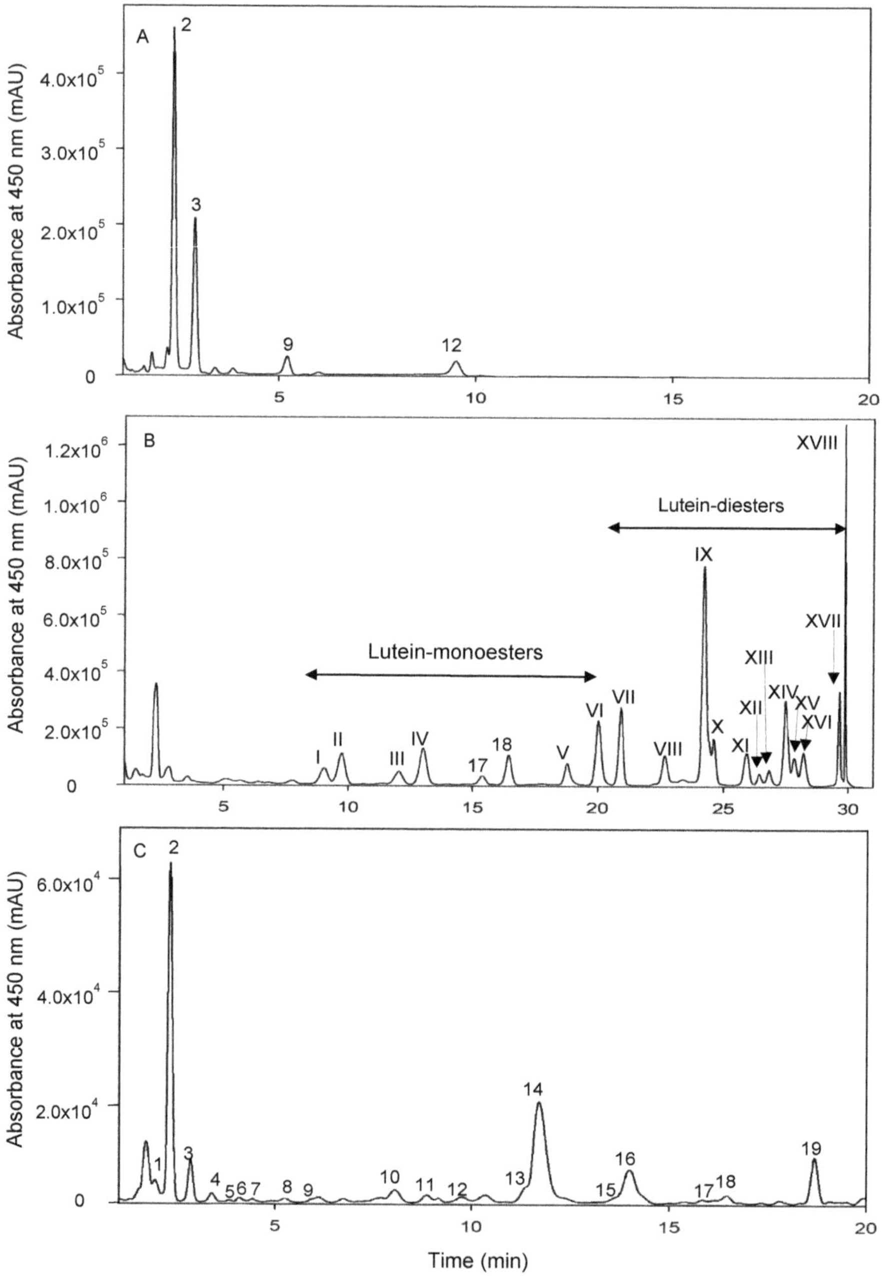

Figure 1. UPLC chromatograms of authentic carotenoid standards (**A**), lab-prepared mono- and di-esters of lutein standards (**B**) and canary seed extract (**C**). Numbers 1–19 correspond to compounds

in Table 1, and compounds I-XVIII are I (3′-*O*-laurate), II (3-*O*-laurate), III (3′-*O*-myristate), IV (3-*O*-myristate), V (3′-*O*-stearate), VI (3-*O*-stearate), VII (dilaurate), VIII (co-eluting 3-O-myristate-3′-O-laurate and 3-O-laurate-3′-O-myristate), IX (co-eluting dimyristate and 3-O-myristate-3′-O-laurate), X (3-O-laurate-3′-O-myristate), XI (co-eluting 3-O-myristate-3′-O-palmitate and 3-O-palmitate-3′-O-myristate), XII (3-O-laurate-3′-O-stearate), XIII (3-O-stearate-3′-O-laurate), XIV (dipalmitate), XV (3-O-myristate-3′-O-stearate), XVI (3-O-stearate-3′-O-myristate), XVII (co-eluting 3-O-palmitate-3′-O-stearate and 3-O-stearate-3′-O-palmitate), XVIII (distearate).

Table 1. Detailed composition of carotenoids identified in hairless canary seed and their chromatographic and spectrometric properties.

Peak Number #	Retention Time (min)	Identity of Compound	Formula	Accurate Mass (g/mol)	MS in-Source Fragmentation, *m/z* (Relative Abundance, %) [c]	UV-Vis Absorption Bands, λ_I, λ_{II} (λ_{max}) and λ_{III} (nm) [d]
1	1.9	15-*cis*-Lutein	$C_{40}H_{56}O_2$	568.43	551.43 (100), 568.43 (40)	416, 436, 464
2	2.3	all-*trans*-Lutein [a]	$C_{40}H_{56}O_2$	568.43	551.42 (100), 568.43 (38), 533.42 (13)	418, 444, 472
3	2.8	all-*trans*-Zeaxanthin [a]	$C_{40}H_{56}O_2$	568.43	569.44 (100), 551.42 (44)	424, 446, 474
4	3.4	9-*cis*-Lutein	$C_{40}H_{56}O_2$	568.43	551.43 (100), 568.43 (15)	415, 440, 464
5	3.8	9-*cis*-Zeaxanthin	$C_{40}H_{56}O_2$	568.43	569.43 (100), 551.42 (86)	420, 444, 472
6	4.1	15-*cis*-β-Cryptoxanthin	$C_{40}H_{56}O$	552.43	553.44 (100), 535.43 (14)	420, 444, 472
7	4.4	13-*cis*-β-Cryptoxanthin	$C_{40}H_{56}O$	552.43	553.44 (100), 535.43 (37)	424, 448, 476
8	5.0	13′-*cis*-β-Cryptoxanthin.	$C_{40}H_{56}O$	552.43	553.44 (100), 535.43 (37)	416, 442, 466
9	5.8	all-*trans*-β-Cryptoxanthin [a]	$C_{40}H_{56}O$	552.43	553.44 (100), 535.43 (15)	423, 450, 474
10	8.1	15-*cis*-Lutein-3-*O*-linoleate	$C_{58}H_{86}O_3$	830.66	813.65 (100), 533.41 (67), 830.66 (28), 551.41 (16)	414, 436, 464
11	8.9	13-*cis*-Lutein-3-*O*-linoleate	$C_{58}H_{86}O_3$	830.66	813.66 (100), 533.41 (64), 830.66 (28), 551.43 (14)	412, 436, 464
12	10.7	all-*trans*-β-Carotene [a]	$C_{40}H_{56}$	536.44	537.44 (100)	423, 450, 474
13	11.4	Lutein 3′-*O*-linoleate [b]	$C_{58}H_{86}O_3$	830.66	830.66 (100), 551.43 (32), 813.65 (2)	417, 444, 472
14	11.7	Lutein-3-*O*-linoleate	$C_{58}H_{86}O_3$	830.66	813.66 (100), 533.42 (63), 830.66 (24), 551.43 (12)	417, 444, 472
15	13.7	Lutein-3′-*O*-oleate [b]	$C_{58}H_{88}O_3$	832.67	832.67 (100), 551.42 (43)	419, 444, 472
16	14.0	Lutein-3-*O*-oleate [b]	$C_{58}H_{88}O_3$	832.67	832.67 (100), 815.67 (53), 533.42 (7)	419, 444, 472
17	15.9	Lutein-3′-*O*-palmitate [a]	$C_{56}H_{86}O_3$	806.66	551.43 (100), 806.66 (36), 533.41 (17), 789.65 (4)	420, 444, 472
18	16.5	Lutein 3-*O*-palmitate [a]	$C_{56}H_{86}O_3$	806.66	789.65 (100), 533.42 (78), 806.66 (56), 551.43 (19)	420, 444, 472
19	18.7	Lutein dilinoleate	$C_{76}H_{116}O_4$	1092.89	1092.88 (100), 533.41 (77)	419, 444, 472

[a] Carotenoids that are identified using authentic carotenoid standards. [b] Carotenoids that are identified using ESI positive mode. [c] Relative abundance is calculated with respect to the major in-source fragment of identified carotenoid using either APCI or ESI positive mode. [d] *cis*-Carotenoids exhibit additional peaks (λ_{cis}) around 330–340 nm. Peak numbers match those in Figure 1.

2.6. Statistical Analysis

Baking trials and analyses were carried out at least in triplicate, and the data are expressed as means ±SD (standard deviation) on a dry matter basis. One-way ANOVA was used to determine the effect of the baking process on individual carotenoid compounds, and significant differences between means were assessed using Tukey's method and considered significant at $p < 0.05$. Statistical analyses were performed using Sigma-Plot version 14.5 (Systat Software Inc., San Jose, CA, USA).

3. Results and Discussion

3.1. Composition and Identification of Carotenoids in Hairless Canary Seed

Hairless canary seed holds a promise as a novel food for the functional food industries a source of non-gluten protein, bioactive peptides, polyphenols and carotenoids [13]. Since carotenoids play significant roles in the health of the eye and brain, the current study provides detailed information on the composition of carotenoids in HCS and their behavior during the baking process. Carotenoids are made up of eight isoprene units having two classes, carotenes (purely unsaturated hydrocarbons) and xanthophyll or oxygenated carotenoids. The main carotenoids found in HCS comprised of lutein and its mono- and di-esters which belong to the oxygenated carotenoids (Table 1). We identified 19 carotenoid compounds based on their mass, UV-vis and structural properties in comparison with authentic carotenoids (Figure 1C, Table 1). Other oxygenated carotenoids in HCS were zeaxanthin, β-cryptoxanthin and their *cis* isomers. Several *cis* configurations were identified at positions C9, C13, C13′ and C15. The presence of *cis* carotenoids in wheat has been previously reported [22,23].

Both APCI+ and ESI+ modes were employed to ascertain the identity of carotenoids. APCI mode is often a preferred method of carotenoid analysis as it provides more predictable in-source fragmentation; however, ESI mode proved to be more sensitive in the analysis of certain less abundant stereoisomers, e.g., lutein mono-esters. Both methods provided similar fragmentation patterns, especially for unbound carotenoids, and are used interchangeably throughout the discussion. The mass characteristics of each carotenoid and the relative abundance percentage are given in Table 1. The MS spectra of all-*trans*-lutein, the dominant carotenoid found in HCS, are shown in APCI+ (Figure 2A) and ESI+ (Figure 2B) ionization modes. In both modes, lutein appears as a radical molecular ion at m/z 568, along with in-source fragment ions due to the loss of one or two molecules of water to yield protonated ions at m/z 551 and 533, respectively. The in-source fragmentation pattern was quite similar in both ionization modes with slight differences in their relative abundance percentages. The dominant ion for lutein was the fragment ion $[M\text{-}H_2O+H]^+$ at m/z 551 as indicated by the percent signal intensity (100%).

Despite zeaxanthin having similar molecular weight and structure as lutein, it exhibited a different fragmentation pattern. The protonated molecular ion $[M+H]^+$ at m/z 569 became the most abundant (100 %), and the water loss fragment ion $[M\text{-}H_2O+H]^+$ at m/z 551 is about half as intense. There is a subtle difference between zeaxanthin and lutein in their chemical structure (e.g., zeaxanthin has two β-ionone rings, whereas lutein has a β-ionone ring and ε-ionone ring) which could elucidate differences in the fragmentation pattern between the two molecules under the same conditions. The ε-ionone ring has a C4′-C5′ double bond adjacent to the 3′ hydroxyl group, which could encourage water loss due to allylic stabilization of the resulting cation [22]. The all-*trans*-β-carotene showed a major protonated molecular $[M+H]^+$ ion at m/z 537. Interestingly, zeaxanthin, β-cryptoxanthin and β-carotene showed the preferential formation of protonated molecular ions in APCI, whereas lutein and its esters favoured the formation of a radical molecular ion. In ESI mode, all carotenoids preferably formed non-protonated molecular ion. Ionization of carotenoid monoesters in both APCI and ESI modes produced predominantly radical non-protonated molecular ions and fragments. With our specific experimental conditions, ESI mode provided a more gentle fragmentation resulting in molecular ion often being the most abundant, whereas APCI mode resulted in more abundant in-source fragments (Supplementary Figure S2). The exact mechanism of molecular ion formation is determined by mobile phase composition and specific mass ionization conditions. Nevertheless, the fragmentation patterns either in APCI+ or ESI+ can assist in the identification of carotenoid molecules [20,24].

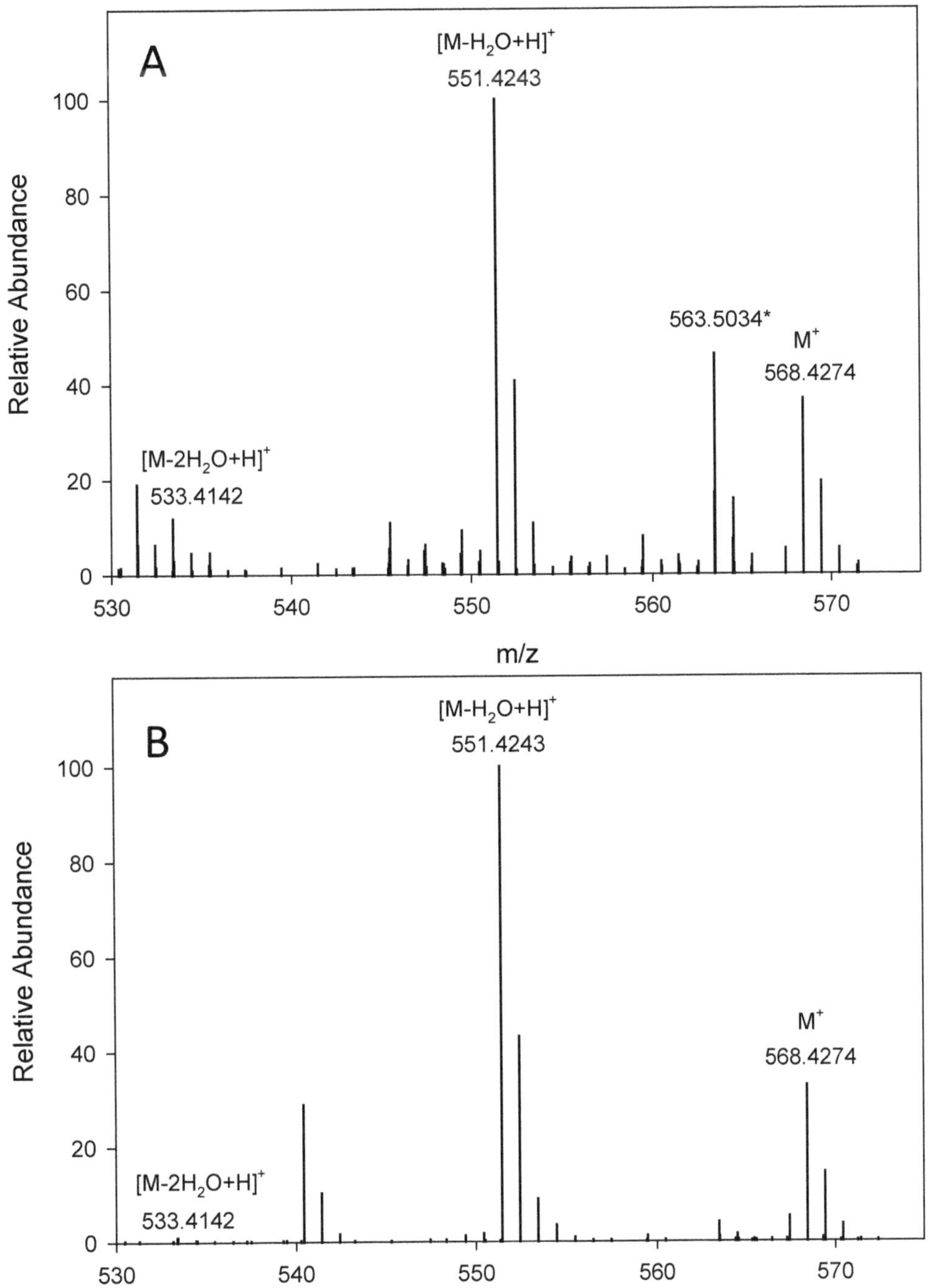

Figure 2. MS spectra of all-*trans*-lutein in two positive ionization modes, (**A**) APCI and (**B**) ESI. * Major non-carotenoid peaks.

Figure 3 shows the mass spectra of lutein mono- and di-esters in the $APCI^+$ and ESI^+ ionization modes. There were obvious differences in the mass spectra between lutein 3′-*O*-linoleate and lutein 3-*O*-linoleate (Figure 3A,B) arising from the structural differences determined by which moiety is attached to the ε-ionone ring, i.e., the acylated fatty acid at the 3′-position in the former molecule while at the 3-position in the latter one. One expects a more or less similar probability of neutral loss of a fatty acid or water molecule from either end of the lutein molecule. However, the preferred fatty acid or water loss occurs from the 3′-position due to allylic stabilization of the resulting cation generated by cleavage in the ε-ionone ring [20]. Then, subsequent losses of water or fatty acid respectively occur at the 3-position in the β-ionone ring. For that reason, the fragmentation patterns of 3′-*O*- and 3-*O*- acylated lutein molecules are different. The differences in MS fragment patterns enable facile discrimination between the 3′-*O*- and 3-*O*-monoester regioisomers of lutein. Based on the fragmentation patterns of the four isomers for lutein 3′-*O*-linoleate and lutein 3-*O*-linoleate, three of them were lutein 3-*O*-linoleate and the fourth one was lutein 3′-*O*-linoleate. The presence of *cis* peaks (λ_{cis}) around 330–340 nm along with their elution order was also employed to confirm the identity of two *cis* isomers of lutein 3-*O*-linoleate (Table 1).

In a previous study, we confirmed the identity of 3′-*O*- and 3-*O*-monoesters of lutein using ^1H NMR technique [20]. The ^1H NMR spectrum of pure lutein showed chemical shifts of 3.85 and 4.12 ppm for the C-3 and C-3′ protons on the β- and ε-ionone rings, respectively. In their study, the C-3 and C3′ proton chemical shift values for the 3-*O*-monopalmitate esters are 5.01 and 4.12 ppm and 3.85 and 5.26 ppm, respectively. This indicates that we can clearly confirm the identity of lutein 3′-*O*-and lutein 3-*O*-mono esters based on NMR data. A study has reported the presence of 4 mono-esters (lutein 3′-*O*-linoleate, lutein 3-*O*-linoleate, lutein 3′-*O*-palmitate, lutein 3-*O*-palmitate) and 4 di-esters (lutein di-linoleate, lutein 3′-*O*-linoleate- 3-*O*-palmitate, lutein 3′-*O*-palmitate-3-*O*-linoleate and Lutein di-palmitate) in the endosperm of tritordeum, a novel cereal, using LS-MS APCI positive mode [25]. As seen in Table 1 all-*trans* and *cis* carotenoids exhibited three absorption peaks λ_I, λ_{II} (λ_{max}) and λ_{III}, at approximately 412–424, 436–450 and 464–476 nm, plus λ_{cis} at 330–340 nm only in the case of *cis* isomers. In the absence of standards for *cis* isomers, these compounds were first characterized based on their λ_{cis} in addition to their relative retention times given in previous studies [20,22,26]. In general, the combined data from LC, UV-vis, MS and NMR enable us to identify and confirm the identity of free lutein and lutein mono- and di-ester isomers and other carotenoids.

Other carotenoid compounds found in HCS were lutein 3′-*O*-oleate, lutein 3-*O*-oleate, lutein 3′-*O*-palmitate, lutein 3-*O*-palmitate and lutein di-linoleate (Table 1). Once again the fragmentation pattern and UV-vis data were used in the identification of those compounds. The most abundant in-source fragment for lutein 3′-*O*-oleate was ion at m/z 551, corresponding to the loss of oleic acid on the ε-ionone ring, compared with $[M\text{-}H_2O+H]^+$ ion at m/z 816 for lutein 3-*O*-oleate. There were also similar differences in the mass spectra between lutein 3′-*O*-palmitate and lutein 3-*O*-palmitate (Supplementary Figure S2). For lutein 3′-*O*-palmitate, the most abundant fragment was $[M\text{-}FA+H]^+$ ion at m/z 551, while lutein 3-*O*-palmitate had the $[M\text{-}H_2O+H]^+$ ion at m/z 790 as the major fragment. The preferential loss of fatty acid from lutein 3′-*O*-palmitate or the loss of water from lutein 3-*O*-palmitate is from the acylated 3′-hydroxyl in the ε-ionone ring as previously reported [20]. Lutein di-linoleate appeared as the molecular ion at m/z 1093 and the fragment $[M\text{-}2FA+H]^+$ at m/z 533 under $APCI^+$ ionization mode, with the molecular fragment being the prevailing ion (Figure 3C). The loss of a fatty acid from each ε-ionone and β-ionone ring produced the fragment m/z 533, which was found at a lower intensity; we did not observe an expected $[M\text{-}FA+H]^+$ fragment, potentially due to interfering non-carotenoid peaks (Figure 3) [25]. These peaks could arise from tiny concentrations of oil impurities in the extracts. Other studies have shown the presence of triacylglycerols in seed extracts, which ionize extremely well under the same positive APCI and ESI conditions as carotenoids and can interfere with their detection [27,28]. Additionally, HCS is rich in oil that contains high amounts of oleic, linoleic and palmitic fatty acids [29]. It will be of interest if more investigations using different analytical techniques are performed to confirm the identity of fatty acids and *cis* isomers.

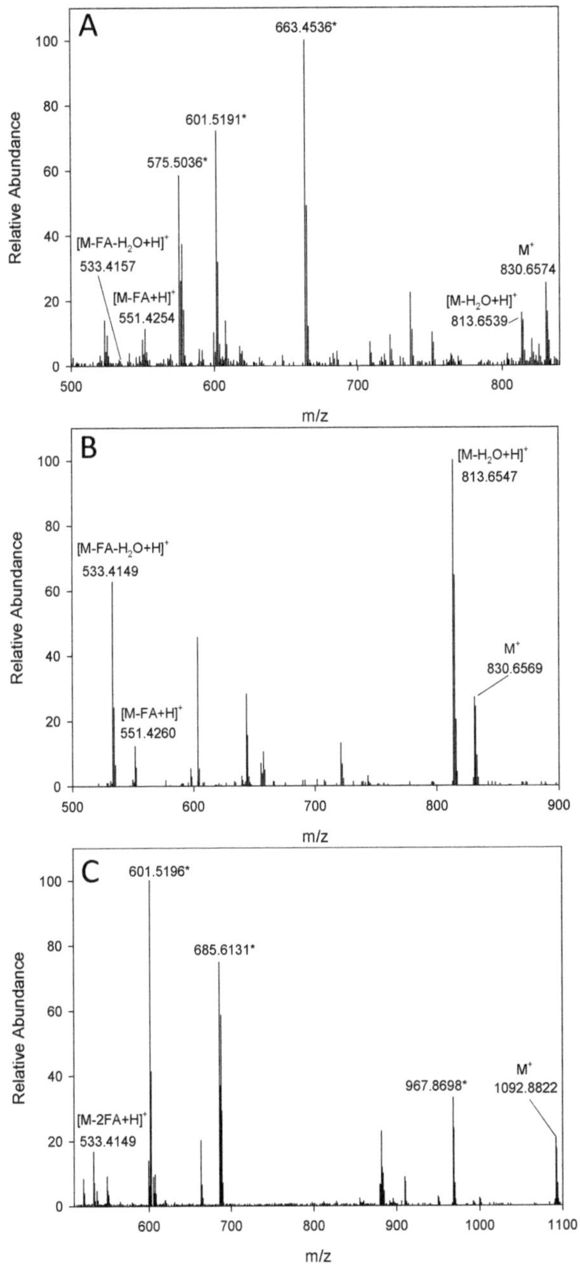

Figure 3. MS spectra of trans-lutein mono- and di-esters identified in canary seed samples in positive ionization modes, (**A**) trans-lutein-3′-O-linoleate (ESI), (**B**) trans-lutein-3-O-linoleate (APCI), (**C**) trans-lutein-dilinoleate (APCI). * Major non-carotenoid peaks.

3.2. Quality of Muffins and Breads

In the current study, prototype muffin and bread products made from HCS alone or in blends with corn and wheat were developed to determine their acceptability and the effect of the baking process on carotenoid composition. The appearance of muffins and breads is shown in Supplementary Figure S1. In order to ensure acceptability of the developed baked products and their potential market, it was necessary to evaluate the quality of the products. Three low-fat muffins (HCS, HCS/corn 1:1, and 1:2) were assessed based on their external and internal quality and nutrient content (Supplementary Table S1). Corn was incorporated into the muffin formulas due to its high concentration of carotenoids, especially lutein and zeaxanthin [22,30]. The three muffins were rated acceptable based on their height, and internal and external quality. The HCS muffin was considered a good source of protein (17.9%) and total dietary fiber (12.8%). This will provide a protein-rich functional food for the gluten-free food market.

Three bread products baked from composite flours of wheat and HCS were compared with a bread control (100% wheat) based on loaf volume, sensory properties and nutrient content (Supplementary Table S2). Replacement of wheat flour with HSC flour up to 25% by weight was acceptable and comparable with the control bread. These results are in agreement with previous research [16]. Breads made from wheat and HCS composite flour (50/50, *w*/*w*) had substantially lower loaf volume and sensory properties compared with the control and other breads. The HCS containing breads had higher protein and total dietary fiber contents compared with the control bread due to the addition of HCS wholegrain flour. These bread products could be a good source of protein, dietary fiber, and bioactive compounds.

3.3. Changes in Carotenoid Composition in Muffins

The role of lutein in the health of the eye and brain has been shown previously. Nonetheless, studies have shown a dietary gap between the daily intake (1.0–1.8 mg/day) [31] and the suggested effective dose of lutein estimated based on reducing the risk of cataracts and AMD (6 mg/day) [1]. This indicates a need for making high lutein functional foods in addition to commercially available lutein supplements, especially since no daily intake for lutein is recommended. Muffins made from HCS alone or in blends with corn could help in filling this dietary gap. Table 2 shows the main carotenoids in HCS and its blends with corn and the impact of the baking process on carotenoids. All-*trans* lutein, along with its *cis* isomers and mono- and di-esters was found to constitute the major portion of carotenoids in HCS. Altogether they made up to about 83% of the total carotenoids in HCS. All-*trans* lutein was the principal carotenoid (3.1 μg/g) followed by lutein 3-*O*-linoleate (1.8 μg/g), lutein 3-*O*-oleate (0.6 μg/g) and lutein di-linoleate (0.6 μg/g). Free or unbound lutein and lutein esters are absorbed from foods and dietary supplements, but the ester form requires prior de-esterification by intestinal enzymes [32]. A study on HCS genotypes has found β-carotene as the main carotenoid followed by lutein and zeaxanthin [33].

We detected trivial amount of β-carotene in HCS, which is in disagreement with the latter study. It is possible that β-carotene could co-elute with other carotenoids in the grain, which could lead to misidentification. In the current study, the identity of carotenoid compounds was identified and confirmed based on authentic standards, including β-carotene and lutein esters and the use of several structural properties. The corn flour used in the current study contained higher amounts of lutein (4.5 μg/g) and zeaxanthin (5.9 μg/g) compared with HCS. Since lutein and its cousin zeaxanthin constitute the macular pigments, foods that are rich in lutein, zeaxanthin, and their derivatives would boost their daily intake. Blending HCS with corn at ratios of 1:1 and 1:2 increased total unbound carotenoids by about 105% and 133%, respectively, while it decreased total bound carotenoids by 53% and 59%, respectively. These changes are due to the significant increase in the unbound or free carotenoids (e.g., lutein from 3.1 to 4.1 and 4.5 μg/g & zeaxanthin from 0.4 to 3.6 and 4.2 μg/g at ratios 1:1 and 1:2, respectively) due to the addition of cornflour. The addition of corn also resulted in an increase in total carotenoids from

7.5 μg/g in HCS to 10.3–11.3 μg/g in the blends. Currently, there are corn varieties with very high levels of lutein and/or zeaxanthin which could be used to boost lutein in food products even more.

Table 2. Content of major carotenoids in muffin formulations and products (μg/g) and the impact of baking process on carotenoids (% decrease) [x].

Carotenoids	Hairless Canary Seed (100%)			Hairless Canary Seed/Corn Blend (1:1, *w/w*)			Hairless Canary Seed/Corn Blend (1:2, *w/w*)		
	Dry Flour	Batter	Muffin	Dry Flour	Batter	Muffin	Dry Flour	Batter	Muffin
15-*cis*-Lutein	0.06 ± 0.02 [a]	0.04 ± 0.01 [a]	0.05 ± 0.01 [a]	0.03 ± 0.01 [a]	0.03 ± 0.01 [a]	0.07 ± 0.01 [a]	0.1 ± 0.03 [a]	0.1 ± 0.01 [a]	0.1 ± 0.01 [a]
all-*trans*-Lutein	3.1 ± 0.17 [a]	2.3 ± 0.15 [b]	1.5 ± 0.06 [c]	4.1 ± 0.04 [a]	2.5 ± 0.10 [b]	1.7 ± 0.04 [c]	4.5 ± 0.08 [a]	2.8 ± 0.03 [b]	2.0 ± 0.05 [c]
all-*trans*-Zeaxanthin	0.4 ± 0.02 [a]	0.3 ± 0.01 [b]	0.2 ± 0.01 [c]	3.6 ± 0.08 [a]	2.0 ± 0.05 [b]	1.5 ± 0.03 [c]	4.2 ± 0.02 [a]	2.6 ± 0.08 [b]	2.2 ± 0.05 [c]
9-*cis*-Lutein	0.1 ± 0.01 [a]	0.2 ± 0.01 [a]	0.2 ± 0.02 [a]	0.1 ± 0.01 [a]	0.1 ± 0.01 [a]	0.1 ± 0.01 [a]	0.2 ± 0.01 [a]	0.2 ± 0.01 [a]	0.1 ± 0.01 [b]
9-*cis*-Zeaxanthin	0.1 ± 0.01 [a]	0.1 ± 0.01 [a]	0.1 ± 0.01 [a]	0.2 ± 0.01 [a]	0.1 ± 0.01 [a]	0.1 ± 0.01 [a]	0.3 ± 0.01 [a]	0.2 ± 0.01 [b]	0.2 ± 0.01 [b]
15-*cis*-β-Cryptoxanthin	0.3 ± 0.02 [a]	0.2 ± 0.01 [b]	0.2 ± 0.01 [b]	0.4 ± 0.01 [a]	0.2 ± 0.01 [b]	0.2 ± 0.01 [b]	0.3 ± 0.01 [a]	0.2 ± 0.01 [b]	0.2 ± 0.01 [b]
all-*trans*-β-Cryptoxanthin	0.2 ± 0.02 [a]	0.1 ± 0.01 [b]	0.1 ± 0.01 [b]	0.3 ± 0.02 [a]	0.2 ± 0.02 [b]	0.2 ± 0.02 [b]	0.4 ± 0.01 [a]	0.2 ± 0.01 [b]	0.2 ± 0.01 [b]
Lutein-3-*O*-linoleate	1.8 ± 0.06 [a]	1.3 ± 0.07 [b]	1.0 ± 0.07 [c]	0.8 ± 0.01 [a]	0.6 ± 0.01 [b]	0.4 ± 0.01 [c]	0.8 ± 0.06 [a]	0.6 ± 0.03 [b]	0.5 ± 0.04 [c]
Lutein-3-*O*-oleate	0.6 ± 0.03 [a]	0.4 ± 0.02 [b]	0.3 ± 0.01 [b]	0.5 ± 0.05 [a]	0.4 ± 0.03 [b]	0.4 ± 0.03 [b]	0.3 ± 0.03 [a]	0.2 ± 0.01 [b]	0.2 ± 0.02 [b]
Lutein dilinoleate	0.6 ± 0.02 [a]	0.3 ± 0.02 [b]	0.2 ± 0.01 [b]	0.2 ± 0.02 [a]	0.1 ± 0.01 [a]	0.1 ± 0.01 [a]	0.2 ± 0.02	0.1 ± 0.01 [a]	0.1 ± 0.01 [a]
Total unbounds (free)	4.3	3.2	2.4	8.8	5.1	3.9	10.0	6.3	5.0
Total bounds (mono- and di-esters)	3.2	2.0	1.5	1.5	1.1	0.9	1.3	0.9	0.8
Total carotenoids	7.5	5.2	3.9	10.3	6.2	4.8	11.3	7.2	5.8
				% Decrease					
Total unbounds	-	25.6	44.2 (18.6) [y]	-	42.0	55.7 (13.7)	-	37.0	50.0 (13.0)
Total bounds	-	37.5	53.1 (15.6)	-	26.7	40.0 (13.3)	-	30.8	38.5 (7.7)
Total carotenoids	-	30.7	48.0 (17.3)	-	39.8	53.4 (13.6)	-	36.3	48.7 (12.4)

[x] For each product and compound, mean values in a row followed by a different superscript letter are significantly different at $p < 0.05$. [y] Figures between brackets are reduction percent due to oven baking.

The muffin-making process resulted in significant reductions in all carotenoid compounds but at different extents subject to carotenoid type (Table 2). For example, all *trans*-lutein had the highest reduction among carotenoids. The batter preparation resulted in higher reductions in unbound, bound, and total carotenoids than that of oven baking of muffins, e.g., 26 vs. 19%, 38 vs. 16%, and 31 vs. 17% for HCS muffin product, respectively. Similar trends were also observed for muffins made from HCS and corn blends. The total carotenoid loss had a range of 48–53% for the three muffin products. Interestingly, the reduction percent of total unbound carotenoids was much lower than that of bound carotenoids in HCS, while an opposite trend was observed in muffins made from HCS and corn blends. During the preparation of the batter, carotenoids could undergo changes due to enzyme degradation and oxidation, while during oven baking thermal degradation and isomerization could take place. Previous research has reported reductions in lutein and zeaxanthin in muffins made from einkorn wheat by about 64 and 57%, respectively, due to dough preparation and oven baking [34].

3.4. Changes in Carotenoid Composition in Breads

Bread is one of the world's most important staple food and is often consumed every day. Thus the availability of nutritious breads could make a difference in human health through healthy eating choices. The current study used HCS as a source of bioactive peptides, minerals, vitamins, carotenoids, and polyphenols to replace part of wheat flour in the bread-making process. The addition of HCS increased total carotenoids from 1.5 μg/g in wheat flour to 2.7, 3.8 and 4.9 μg/g at replacement levels of 15, 25 and 50%, respectively (Table 3). There were significant changes in carotenoids during dough formation and oven baking. Total carotenoids decreased by about 67, 48, 50 and 47% in control, 15, 25 and 50% breads, respectively. This finding indicates that degradation of carotenoids is dependent on carotenoid type; the higher amount of bound carotenoids, the more stable they remain during the baking process. The control bread, which had no bound carotenoids, exhibited the highest reduction percentage. Esterification of carotenoid with fatty acids in bound carotenoids affect their stability and behavior during processing and storage [35]. Similar to muffin, dough formation resulted in higher reductions in unbound, bound and total carotenoids compared with oven baking of breads in all bread products. It has been reported that dough preparation causes a more pronounced loss of carotenoids and tocols (51.5 and 33.0%) compared to baking (22.5 and 9.1%) in flatbread [36]. The study has also reported that carotenoids are less stable than tocols, and the presence of tocols could improve stability of carotenoids during the baking process. Carotenoids undergo several biochemical and chemical reactions during dough preparation including enzyme degradation and oxidation which contribute to their loss. During oven baking of breads, additional but smaller reductions in total carotenoids occurred, e.g., 20 vs. 47%, 11 vs. 37%, 8 vs. 42% and 8 vs. 39% for breads made form wheat, wheat/HCS (85/15), wheat/HCS (75/25) and wheat/HCS (50/50), respectively (Table 3). Carotenoids are sensitive to light and temperature through oxidation and isomerization, resulting in a partial loss of carotenoids. The replacement of wheat flour with HCS up to 25% gave acceptable breads which would increase protein, carotenoids and other bioactive compounds in the end product.

Table 3. Content of major carotenoids in bread formulations and products (μg/g) and the impact of baking process on carotenoids (% decrease) [x].

Carotenoids	Wheat (100%)			Wheat/Hairless Canary Seed (85/15, *w/w*)			Wheat/Hairless Canary Seed (75/25, *w/w*)			Wheat/Hairless Canary Seed (50/50, *w/w*)		
	Dry Flour	Dough	Bread	Dry Flour	Dough	Bread	Dry Flour	Dough	Bread	Dry Flour	Dough	Bread
15-*cis*-Lutein	0.1 ± 0.01 ^{a}	0.05 ± 0.01 ^{a}	0.04 ± 0.01 ^{a}	0.05 ± 0.01 ^{a}	0.03 ± 0.01 ^{a}	0.03 ± 0.01 ^{a}	0.05 ± 0.01 ^{a}	0.03 ± 0.01 ^{a}	0.03 ± 0.01 ^{a}	0.1 ± 0.02 ^{a}	0.06 ± 0.02 ^{a}	0.06 ± 0.02 ^{a}
all-*trans*-Lutein	1.0 ± 0.02 ^{a}	0.5 ± 0.03 ^{b}	0.3 ± 0.01 ^{c}	1.2 ± 0.04 ^{a}	0.6 ± 0.03 ^{b}	0.5 ± 0.03 ^{c}	1.7 ± 0.05 ^{a}	0.8 ± 0.02 ^{b}	0.7 ± 0.01 ^{c}	2.2 ± 0.07 ^{a}	0.9 ± 0.07 ^{b}	0.8 ± 0.03 ^{c}
all-*trans*-Zeaxanthin	0.4 ± 0.02 ^{a}	0.2 ± 0.03 ^{b}	0.2 ± 0.02 ^{b}	0.4 ± 0.02 ^{a}	0.3 ± 0.03 ^{b}	0.3 ± 0.02 ^{b}	0.4 ± 0.02 ^{a}	0.2 ± 0.01 ^{b}	0.2 ± 0.02 ^{b}	0.4 ± 0.04 ^{a}	0.3 ± 0.01 ^{b}	0.3 ± 0.01 ^{b}
Lutein-3-*O*-linoleate	nd	nd	nd	0.5 ± 0.02 ^{a}	0.3 ± 0.03 ^{b}	0.3 ± 0.03 ^{b}	0.6 ± 0.02 ^{a}	0.4 ± 0.03 ^{b}	0.4 ± 0.03 ^{b}	1.1 ± 0.06 ^{a}	0.7 ± 0.01 ^{b}	0.6 ± 0.03 ^{c}
Lutein-3-*O*-oleate	nd	nd	nd	0.4 ± 0.02 ^{a}	0.3 ± 0.03 ^{b}	0.2 ± 0.03 ^{c}	0.5 ± 0.02 ^{a}	0.4 ± 0.03 ^{b}	0.3 ± 0.03 ^{c}	0.7 ± 0.06 ^{a}	0.5 ± 0.03 ^{b}	0.4 ± 0.03 ^{c}
Lutein dilinoleate	nd	nd	nd	0.3 ± 0.02 ^{a}	0.2 ± 0.01 ^{b}	0.1 ± 0.01 ^{c}	0.5 ± 0.02 ^{a}	0.4 ± 0.02 ^{b}	0.3 ± 0.02 ^{c}	0.6 ± 0.05 ^{a}	0.5 ± 0.02 ^{b}	0.4 ± 0.02 ^{c}
Total unbounds (free)	1.5	0.8	0.5	1.5	0.9	0.8	2.2	1.0	0.9	2.5	1.3	1.2
Total bounds (mono- and di-esters)	0.0	0.0	0.0	1.2	0.8	0.6	1.6	1.2	1.0	2.4	1.7	1.4
Total carotenoids	1.5	0.8	0.5	2.7	1.7	1.4	3.8	2.2	1.9	4.9	3.0	2.6
% Decrease												
Total unbounds	-	46.7	66.7 (20.0) [y]	-	40.0	46.7 (6.7)	-	54.5	59.1 (4.6)	-	48.0	52.0 (4.0)
Total bounds	-	-	-	-	33.3	50.0 (16.7)	-	25.0	37.5 (12.5)	-	29.2	41.7 (12.5)
Total carotenoids	-	46.7	66.7 (20.0)	-	37.0	48.1 (11.1)	-	42.1	50.0 (7.9)	-	38.8	46.9 (8.1)

[x] For each product and carotenoid mean values in a row followed by a different superscript letter are significantly different at $p < 0.05$. y Figures between brackets are reduction percent due to oven baking.

4. Conclusions

Hairless canary seed is a true cereal grain with unique nutritional profile that makes it a suitable functional food ingredient. For the first time detailed composition of carotenoids in HCS has been identified based on UPLC and LC-MS/MS analyses. Hairless canary seed and corn were fairly rich in carotenoids compared with other cereal grains having a total content of 7.6 and 12.9 μg/g, respectively. HCS is rich in lutein and its mono- and di-esters especially lutein 3-*O*-linoleate, lutein 3-*O*-oleate and lutein di-linoleate, while corn is rich in free lutein and zeaxanthin. Thus, they might complement each other in their carotenoid composition in terms of lutein stability and bioavailability. Baking process resulted in significant reductions in carotenoids in muffin and bread products. It appears that batter or dough preparation causes more reductions in carotenoids than oven baking probably due to enzymatic oxidation and degradation. In addition, muffin-making resulted in lower lutein reductions compared to the bread-making process. Thus, muffins made from hairless canary seed alone or in blends with corn could boost the daily intake of lutein and/or zeaxanthin. Further research is underway to assess the bioavailability of carotenoids in HCS muffin and bread products.

Supplementary Materials: The following supporting information can be downloaded.

Author Contributions: Conceptualization, E.-S.M.A.-A.; methodology, I.R. and L.M.; validation, E.-S.M.A.-A., L.M. and I.R.; formal analysis, E.-S.M.A.-A., L.M. and I.R.; resources, E.-S.M.A.-A.; data curation, E.-S.M.A.-A.; writing—Original draft preparation, E.-S.M.A.-A.; writing—Review and editing, E.-S.M.A.-A.; L.M. and I.R.; supervision, E.-S.M.A.-A. All authors have read and agreed to the published version of the manuscript.

Funding: This research was funded by Agriculture and Agri-Food Canada grant number J-002260 And The APC was funded by A-base project J-002260.

Institutional Review Board Statement: Not applicable.

Informed Consent Statement: Not applicable.

Data Availability Statement: Not applicable.

Acknowledgments: This work is supported by the A-base project J-002260 of the Agriculture and Agri-Food Canada. The authors thank Claire Traversa for her technical support in product preparation and analysis, Honghui Zhu for her technical support in the MS analysis, Kevin Hursh at the Canary Seed Development Commission of Saskatchewan for providing the hairless canary seed sample and Pierre Hucl from the University of Saskatchewan for providing the wheat sample.

Conflicts of Interest: The authors declare no conflict of interest.

References

1. Abdel-Aal, E.-S.M.; Akhtar, H.; Zaheer, K.; Ali, R. Dietary Sources of Lutein and Zeaxanthin Carotenoids and Their Role in Eye Health. *Nutrients* **2013**, *5*, 1169–1185. [CrossRef] [PubMed]
2. Eisenhauer, B.; Natoli, S.; Liew, G.; Flood, V. Lutein and Zeaxanthin—Food Sources, Bioavailability and Dietary Variety in Age-Related Macular Degeneration Protection. *Nutrients* **2017**, *9*, 120. [CrossRef] [PubMed]
3. Johnson, E.J. Role of Lutein and Zeaxanthin in Visual and Cognitive Function Throughout the Lifespan. *Nutr. Rev.* **2014**, *72*, 605–612. [CrossRef] [PubMed]
4. Li, J.; Abdel-Aal, E.-S.M. Dietary Lutein and Cognitive Function in Adults: A Meta-Analysis of Randomized Controlled Trials. *Molecules* **2021**, *26*, 5794. [CrossRef] [PubMed]
5. Stringham, J.M.; Johnson, E.J.; Hammond, B.R. Lutein across the Lifespan: From Childhood Cognitive Performance to the Aging Eye and Brain. *Curr. Dev. Nutr.* **2019**, *3*, nzz066. [CrossRef] [PubMed]
6. Vishwanathan, R.; Schalch, W.; Johnson, E.J. Macular Pigment Carotenoids in the Retina and Occipital Cortex are Related in Humans. *Nutr. Neurosci.* **2015**, *19*, 95–101. [CrossRef]
7. Jia, Y.-P.; Sun, L.; Yu, H.-S.; Liang, L.-P.; Li, W.; Ding, H.; Song, X.-B.; Zhang, L.-J. The Pharmacological Effects of Lutein and Zeaxanthin on Visual Disorders and Cognition Diseases. *Molecules* **2017**, *22*, 610. [CrossRef]
8. Johnson, E.J. A Possible Role for Lutein and Zeaxanthin in Cognitive Function in the Elderly. *Am. J. Clin. Nutr.* **2012**, *96*, 1161S–1165S. [CrossRef]

9. Kelly, D.; Coen, R.F.; Akuffo, K.O.; Beatty, S.; Dennison, J.; Moran, R.; Stack, J.; Howard, A.N.; Mulcahy, R.; Nolan, J.M. Cognitive Function and its Relationship with Macular Pigment Optical Density and Serum Concentrations of its Constituent Carotenoids. *J. Alzheimers Dis.* **2015**, *48*, 261–277. [CrossRef]
10. Hucl, P.; Matus-Cadiz, M.; Vandenberg, A.; Sosulski, F.W.; Abdel-Aal, E.-S.M.; Hughes, G.R.; Slinkard, A.E. CDC Maria annual canarygrass. *Can. J. Plant Sci.* **2001**, *81*, 115–116. [CrossRef]
11. Health Canada. Food and Nutrition, Novel Foods. 2016. Available online: http://www.hc-sc.gc.ca/fn-an/gmf-agm/appro/canary-seed-lang-graine-alpiste-eng.php (accessed on 24 January 2022).
12. US-FDA. US Food and Drug Administration. Agency Response Letter GRAS Notice No. GRN 000529. 2015. Available online: https://www.accessdata.fda.gov/scripts/fdcc/index.cfm?set=GrASNotices&id=529 (accessed on 24 January 2022).
13. Abdel-Aal, E.-S.M. Nutritional and Functional Attributes of Hairless Canary Seed Groats and Components and their Potential as Functional Ingredients. *Trends Food Sci. Technol.* **2021**, *111*, 680–687. [CrossRef]
14. Boye, J.I.; Achouri, A.; Raymond, N.; Cleroux, C.; Weber, D.; Koerner, T.B.; Hucl, P.; Patterson, C.A. Analysis of Glabrous Canary Seeds by ELISA, Mass Spectrometry and Western Blotting for the Absence of Cross-Reactivity with major Plant Food Allergens. *J. Agric. Food Chem.* **2013**, *61*, 6102–6112. [CrossRef] [PubMed]
15. Abdel-Aal, E.-S.M.; Hucl, P.; Patterson, C.A.; Gray, D. Fractionation of Hairless Canary Seed (*Phalaris canariensis* L.) into Starch, Protein and Oil. *J. Agric. Food Chem.* **2010**, *58*, 7046–7050. [CrossRef] [PubMed]
16. Patterson, C.A.; Malcolmson, L.; Lukie, C.; Young, G.; Hucl, P.; Abdel-Aal, E.-S.M. Glabrous Canary Seed: A Novel Food Ingredient. *Cereal Foods World* **2018**, *63*, 194–200.
17. Mason, E.; L'Hocine, L.; Achouri, A.; Pitre, M.; Karboune, S. Health Promoting Bioactive Properties of Novel Hairless Canary Seed Flour in vitro Gastrointestinal Digestion. *Foods* **2020**, *9*, 932. [CrossRef] [PubMed]
18. Mason, E.; L'Hocine, L.; Achouri, A.; Karboune, S. Hairless Canaryseed: A Novel Cereal with Health Promoting Potential. *Nutrients* **2018**, *10*, 1327. [CrossRef]
19. Abdel-Aal, E.-S.M.; Rabalski, I. Composition of Lutein Ester Regioisomers in Marigold Flower, Dietary Supplement and Herbal Tea. *J. Agric. Food Chem.* **2015**, *63*, 9740–9746. [CrossRef]
20. Young, J.C.; Abdel-Aal, E.-S.M.; Rabalski, I.; Blackwell, B. Identification of Synthetic Regioisomeric Lutein Esters and their Quantification in a Commercial Lutein Supplement. *J. Agric. Food Chem.* **2007**, *55*, 4965–4972. [CrossRef]
21. AACC International. *Approved Methods of the American Association of Cereal Chemists International*, online ed.; The Association: St. Paul, MN, USA, 2016.
22. Abdel-Aal, E.-S.M.; Young, J.C.; Rabalski, I.; Frégeau-Reid, J.; Hucl, P. Identification and Quantification of Seed Carotenoids in Selected Wheat Species. *J. Agric. Food Chem.* **2007**, *55*, 787–794. [CrossRef]
23. Ziegler, J.U.; Wahl, S.; Würschum, T.; Longin, C.F.H.; Carle, R.; Schweiggert, R.M. Lutein and Lutein Esters in Whole Grain Flours made From 75 Genotypes of 5 *Triticum* Species Grown at Multiple Sites. *J. Agric. Food Chem.* **2015**, *63*, 5061–5071. [CrossRef]
24. Bijttebier, S.K.A.; D'Hondt, E.; Hermans, N.; Apers, S.; Voorspoels, S. Unravelling Ionization and Fragmentation Pathways of Carotenoids Using Orbitrap Technology: A First Step Towards Identification of Unknowns. *J. Mass Spectrom.* **2013**, *48*, 740–754. [CrossRef] [PubMed]
25. Mellado-Ortega, E.; Hornero-Méndez, D. Isolation and Identification of Lutein Esters, Including their Regioisomers, in Tritordeum (*Tritordeum Ascherson* et Graebner) Grains: Evidence for a Preferential Xanthophyll Acyltransferase Activity. *Food Chem.* **2012**, *135*, 1344–1352. [CrossRef]
26. Sander, L.C.; Sharpless, K.E.; Pursch, M. C30 Stationary Phases for the Analysis of Food by Liquid Chromatography. *J. Chromatogr. A* **2000**, *880*, 189–202. [CrossRef]
27. Gao, F.; Yang, S.; Birch, J. Physicochemical Characteristics, Fatty Acid Positional Distribution and Triglyceride Composition in Oil Extracted from Carrot Seeds Using Supercritical CO2. *J. Food Compos. Anal.* **2016**, *45*, 26–33. [CrossRef]
28. Mercadante, A.Z.; Rodrigues, D.B.; Petry, F.C.; Mariutti, L.R.B. Carotenoid esters in Foods—A Review and Practical Directions on Analysis and Occurrence. *Food Res. Int.* **2017**, *99*, 830–850. [CrossRef] [PubMed]
29. Abdel-Aal, E.-S.M.; Hernandez, M.; Rabalski, I.; Hucl, P. Composition of Hairless Canary Seed Oil and Starch-Associated Lipid and its Relationship to Pasting and Thermal Properties of Starch. *LWT-Food Sci. Technol.* **2020**, *125*, 109257. [CrossRef]
30. Zhao, X.; Liang, K.; Zhu, H. Carotenoids in Cereals and Related Foodstuffs: A Review of Extraction and Analysis Methods. *Food Rev. Int.* 2022, *Ahead-of-print*. [CrossRef]
31. Johnson, E.J. Intake of Lutein and Zeaxanthin Differ with Age, Sex, and Ethnicity. *J. Am. Diet. Assoc.* **2010**, *110*, 1357–1362. [CrossRef] [PubMed]
32. Alves-Rodrigues, A.; Shao, A. The Science Behind Lutein. *Toxicol. Lett.* **2004**, *150*, 57–83. [CrossRef]
33. Li, W.; Beta, T. An evaluation of carotenoid levels and composition of glabrous canary seed. *Food Chem.* **2012**, *133*, 782–786. [CrossRef]
34. Abdel-Aal, E.-S.M.; Young, J.C.; Akhtar, H.; Rabalski, I. Stability of Lutein in Wholegrain Bakery Products Naturally High in Lutein or Fortified with Free Lutein. *J. Agric. Food Chem.* **2010**, *58*, 10109–10117. [CrossRef] [PubMed]
35. Subagio, A.; Wakaki, H.; Morita, N. Stability of Lutein and its Myristate Esters. *Biosci. Biotechnol. Biochem.* **1996**, *63*, 1784–1786. [CrossRef] [PubMed]
36. Buresova, B.; Paznocht, L.; Kotikova, Z.; Giampaglia, B.; Martinek, P.; Lachman, J. Changes in Carotenoids and Tocols of Colored-Grain Wheat during Unleavened Bread Preparation. *J. Food Compos. Anal.* **2021**, *103*, 104108. [CrossRef]

molecules

MDPI

Article

Puree and Juice of Thai Mango and Pineapple Analyzed by High-Performance Thin-Layer Chromatography Hyphenated with Effect-Directed Assays

Gertrud E. Morlock [1,*,†], Newitchaya Wutthinithisanand [1,2,†] and Doris Rauhut [2]

1 Chair of Food Science, Institute of Nutritional Science, and Interdisciplinary Research Center, Justus Liebig University Giessen, Heinrich-Buff-Ring 26-32, 35392 Giessen, Germany; neenee250@hotmail.com

2 Department of Microbiology and Biochemistry, Hochschule Geisenheim University, Von-Lade-Str. 1, 65366 Geisenheim, Germany; Doris.Rauhut@hs-gm.de

* Correspondence: Gertrud.Morlock@uni-giessen.de

† These authors contributed equally to this work.

Abstract: The requirements for analytical tools are changing due to the global production chain, the increasing cases of adulteration, and the growing trend towards consumption of plant-based food products worldwide. The assessment of bioactivity of natural foods is currently not a quality criterion, and a paradigm shift is postulated. A non-targeted effect-directed profiling by high-performance thin-layer chromatography hyphenated with five different effect-directed assays was developed exemplarily for the puree and juice products of mango *Mangifera indica* L. (Anacardiaceae) and pineapple *Ananas comosus* (L.) Merr. (Bromeliaceae). Several bioactive compounds were detected in each sample. The additional bioactivity information obtained through effect-directed profiles improves, expands and modernizes product control. Non-target effect-directed profiling adds a new perspective to previous target analysis results that can be used not only to ensure health claims based on bioactive compounds, but also to detect unknown bioactive compounds coming from contamination or residues or changes caused by food processing.

Keywords: effect-directed analysis; healthy food; food processing; adulteration; authentication

Citation: Morlock, G.E.; Wutthinithisanand, N.; Rauhut, D. Puree and Juice of Thai Mango and Pineapple Analyzed by High-Performance Thin-Layer Chromatography Hyphenated with Effect-Directed Assays. *Molecules* **2021**, *26*, 7683. https://doi.org/10.3390/molecules26247683

Academic Editor: Smaoui Slim

Received: 2 December 2021
Accepted: 13 December 2021
Published: 19 December 2021

1. Introduction

Bioactive compounds present in natural plant-based foods and their products attract increasingly both consumers and researchers. Epidemiological evidence indicates a contribution to the preventive effects in reducing the risk of certain common diseases, e.g., cardiovascular disorder, cancer, osteoporosis, infections, cataract and diabetes [1,2]. The spectrum and content of bioactive compounds in fruits vary, among other things, depending on the cultivar and maturity stage of the product [3,4]. The most prominent bioactive compounds found in tropical fruits are ascorbic acid (vitamin C), tocopherols (vitamin E), carotenoids (provitamin A candidates), flavonoids and thiol compounds. In particular, the widely distributed polyphenols are the predominant antioxidative secondary metabolite group [5].

The pineapple *Ananas comosus* (L.) Merr. (Bromeliaceae) and the mango *Mangifera indica* L. (Anacardiaceae) are important tropical fruits popular all over the world. Global pineapple and mango production in 2019 were about 28 million tons and 56 million tons, respectively, of which Thailand accounted for about 6% and 3% [6]. Pineapple is commonly consumed as fresh fruit, or processed to conventional products including canned fruit, jam, concentrated juice, and dried chips, or used as an ingredient in exotic food [7] or as material for pineapple wine production [8]. Thailand is not only a world leader in the export volume of canned pineapple, but many local cultivars are also unique in their taste and aroma [9]. Commercial cultivars of pineapple in Thailand are Pattavia, Phuket,

Nanglae and Phulae. Pineapple is considered a very nutritious and functional fruit, as it is rich in vitamin A, vitamin B, and minerals such as calcium, phosphorus and iron [10]. It contains also several proteinases, such as bromelain, comosain and ananain, which support the digestive tract [11]. Moreover, it has several compounds with antioxidant properties, namely ascorbic acid, β-carotene and phenolic compounds such as flavonoids [12–15].

The mango fruit is widely accepted by consumers due to its succulence, sweet taste and exotic flavor. Mango flesh is consumed in both ripe and unripe stages. It is also processed to foods and drinks, such as pickled products, beverages, vinegar, chutneys, and desserts [16]. In Thailand, the major mango varieties, consumed fresh when the fruit is ripe, are Nam Dok Mai and Mahachanok. A small percentage of the production is processed to canned, dried and frozen mango [9]. Mangos contain bioactive compounds, such as ascorbic acid, carotenoids (e.g., β-carotene, violaxanthin, cryptoxanthin, neoxanthin, luteoxanthin and zeaxanthin), phenolic components, and mono-, di- and triterpenoids (e.g., ocimene, myrcene or limonene, terpinolene and careen) [2].

The domestic consumption and the export trade of pineapple and mango provide economic value to Thailand. Nevertheless, reports on bioactive compounds and antioxidant activities of these fruits from Thailand are rare. Since bioactive compounds generally occur as complex mixtures in a plant, it is a major challenge to identify the individual substances responsible for specific effects [17–19]. For such effect-directed profiling, high-performance thin-layer chromatography hyphenated with multi-imaging (UV/Vis/FLD) and effect-directed assays (EDA) was proven to be a good choice [20–24]. In this study, a non-target effect-directed profiling by HPTLC−UV/Vis/FLD−EDA was developed for puree and juice products of different Thai pineapple and mango cultivars to gain information on bioactive constituents. The obtained effect-directed profiles pointed to individual bioactive compounds and allowed a comparative evaluation with regard to their activity potential.

2. Results and Discussion

2.1. Physico-Chemical Characterization of the Produced Fruit Purees and Juices

Two cultivars each of mango and pineapple were chosen due to their importance as major commercial varieties and their widespread production in Thailand. The two pineapple cultivars Pattavia (PTV) and Nanglae (NL) were selected as sub-varieties of the smooth cayenne pineapple Pattavia variety. The two mango cultivars Nam Dok Mai (NDM) and Mahachanok (MHC) represent poly-embryonic and mono-embryonic ecotypes, respectively. At the commercial maturity stage, the fruits were harvested from the orchard at the Chiang Rai province in Northern Thailand. The regularity of fruit ripeness was controlled by the peel color observation in our study, i.e., 30–40% yellow for pineapple and fully yellow for mango. A puree and a juice sample were produced from each of the four cultivars. Thus, 8 fruit product samples were obtained. Their ripeness (content of total sugars and total acidity) and compositional values (content of total amino acids and total phenolics) were determined (Table 1). Total soluble solids, pH, and total acidity of pineapple and mango puree in both cultivars were in accordance with the literature [12,13,25–29]. The total sugar of pineapple was higher than that reported by Lu et al. [13]; however, values of natural plant-based food can vary depending on harvest time, climate influence each year, agricultural and soil management, among many other aspects. The sugar content normally increases during fruit ripening.

Table 1. Properties of the produced purees and juices of the pineapple (PTV and NL) and mango cultivars (NDM and MHC). Each value shows the mean ± standard deviation from three measurement replicates. Different letters (a–d) derived from the ANOVA and DMRT test indicate significant differences among samples.

	Pineapple Cultivar		Mango Cultivar	
	PTV	NL	NDM	MHC
Parameters	Produced Puree			
Total Soluble Solids (°Brix)	14.4 ± 0.2 [d]	15.3 ± 0.1 [c]	18.0 ± 0.2 [a]	16.4 ± 0.1 [b]
Total Sugar (g/L)	133 ± 0 [d]	142 ± 0 [c]	159 ± 0 [a]	151 ± 0 [b]
pH	4.6 ± 0.0 [a]	4.5 ± 0.0 [a]	3.7 ± 0.0 [c]	4.0 ± 0.0 [b]
Total Acidity (pH 8.1 as citric acid, g/L)	2.4 ± 0.0 [c]	2.4 ± 0.0 [c]	7.8 ± 0.2 [a]	4.1 ± 0.1 [b]
	Produced Juice			
Total Amino Acids (mg/L)	3335 ± 8 [a]	2790 ± 11 [b]	1641 ± 4 [c]	1366 ± 14 [d]
Total Phenolics (mg/L)	426 ± 3 [d]	442 ± 4 [c]	845 ± 6 [a]	616 ± 4 [b]

2.2. Development of the Effect-Directed Profiling

An effect-directed HPTLC method was developed on HPTLC plates silica gel 60 F_{254} (for the *B. subtilis* bioassay without F_{254}). For a first impression, the more complex puree sample was chosen first. The methanolic extraction was performed at a solid-liquid ratio of 40%. A highly concentrated extract solution was preferred for the analysis of bioactive secondary metabolites, usually present in trace amounts. This allows the use of small sample volumes (rapid application) and easy adjustment to the different detection capabilities of biological and enzymatic assays (dilution step is faster than concentration step). Application as area (4 mg/area) was preferred due to the expected high load of sample matrix onto the adsorbent. Six different mobile phases (which resulted in successful separations in previous projects) were investigated (Table S1 from Supplementary Materials). Solvent mixture 1 consisting of toluene–ethyl acetate–methanol–formic acid 2.4:1.8:0.7:0.06, *v*/*v*/*v*/*v*, was suited best, as it distributed the sample components well along the developing distance, especially detectable after derivatization with the anisaldehyde sulfuric acid reagent at FLD 366 nm. A large portion of each methanolic extract remained in the initial start zone area, to which the sugar content contributed most, being as high as 16% in the 8 fruit product samples (Table 1). The individual saccharides of the samples, applied at 200 µg/area, were determined using a more polar mobile phase for separation (Table S1 from Supplementary Materials), mobile phase 6: acetonitrile–water 4:1, *v*/*v*, plus 8 mg diphenylboric acid-2-aminoethylester) and the diphenylamine aniline *o*-phosphoric acid reagent for detection at white light illumination. The 8 saccharide profiles showed glucose, fructose and sucrose being present (Figure 1). Among these, the mango cultivars (Table 2, puree IDs 3,1 and 4,1; juice IDs 7,1 and 8,1) showed a lower glucose and instead a higher sucrose content than the pineapple cultivars (puree IDs 1,1 and 2,1; juice IDs 5,1 and 6,1).

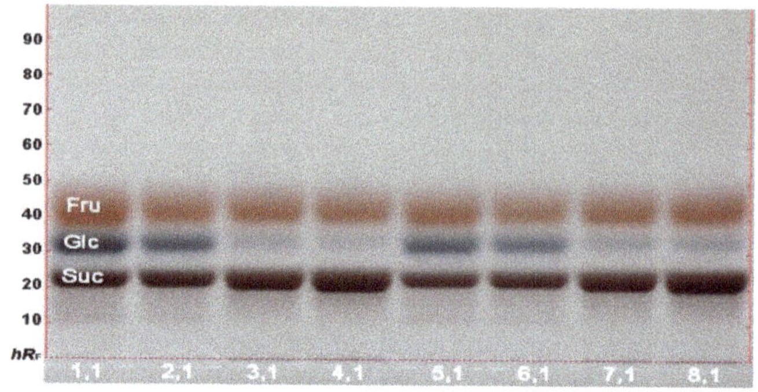

Figure 1. Saccharide analysis: HPTLC-Vis chromatogram of fructose (Fru), glucose (Glc) and sucrose (Suc) present in fruit product extracts (IDs 1–8, Table 2, 0.4 g/mL in methanol, 0.5 µL/area) on HPTLC plates silica gel 60 F_{254} with acetonitrile–water 4:1, *v*/*v*, plus 8 mg diphenylboric acid-2-aminoethylester (Table S1 from Supplementary Materials, mobile phase 6) derivatized with diphenylamine aniline *o*-phosphoric acid reagent and documented at white light illumination.

Table 2. Produced fruit products (IDs 1–8) extracted either with methanol (x,1) or ethyl acetate (x,2).

Puree Extract		Juice Extract	
ID	**Fruit Cultivar**	**ID**	**Fruit Cultivar**
1,1	Pineapple PTV	5,1	Pineapple PTV
1,2		5,2	
2,1	Pineapple NL	6,1	Pineapple NL
2,2		6,2	
3,1	Mango NDM	7,1	Mango NDM
3,2		7,2	
4,1	Mango MHC	8,1	Mango MHC
4,2		8,2	

2.3. Selection of the Extraction Solvent for Effect-Directed Profiling

Sample preparation was performed as minimally as possible to minimize any influence (alteration) in the fruit products produced. Different extraction solvents (i.e., methanol, acetone, methanol/acetone 1:1, ethanol, ethyl acetate, and acetonitrile) were comparatively studied for the intended non-targeted effect-directed profiling (Figure 2, all 4 mg/area). For polar extraction solvents, the co-extracted saccharides caused a high matrix loading on the start zone area of the adsorbent. This limited not only the amount of sample that could be applied, but also the development with more polar mobile phases (Table S1 from Supplementary Materials, mobile phases 4 and 6). In comparison, the ethyl acetate extracts discriminated the co-extraction of saccharides, as evident in the absent dark start zone area. This was found to be advantageous, in case higher sample amounts need to be used depending on the assay response. In addition, the ethyl acetate extracts contained a similar pattern of medium polar components that was almost comparable to that of the other solvents. Therefore, to cover different selectivities, not only methanol but also ethyl acetate were chosen as extraction solvent (Table 2). Apart from UV/Vis/FLD detections and the use of derivatization reagents, the compounds in the extracts were detected via the Gram-negative *Aliivibrio fischeri* bioassay (Figure 2C). This assay detects more universally any bioactive chemicals interfering with the energetic metabolism of the bacteria and is commonly used in environmental research [30]. The bioautogram revealed a similar pattern of bioactive compounds within the different samples. As the bioactivity pattern of the applied 10 µL sample (4 mg/area) was faint, it was evident that a higher sample amount was required to detect the bioactive components present at low concentrations. Therefore, a 5-fold amount of each fruit product extract (solid-liquid extract ratio of 20% in ethyl acetate) was applied (100 µL/area) in the following.

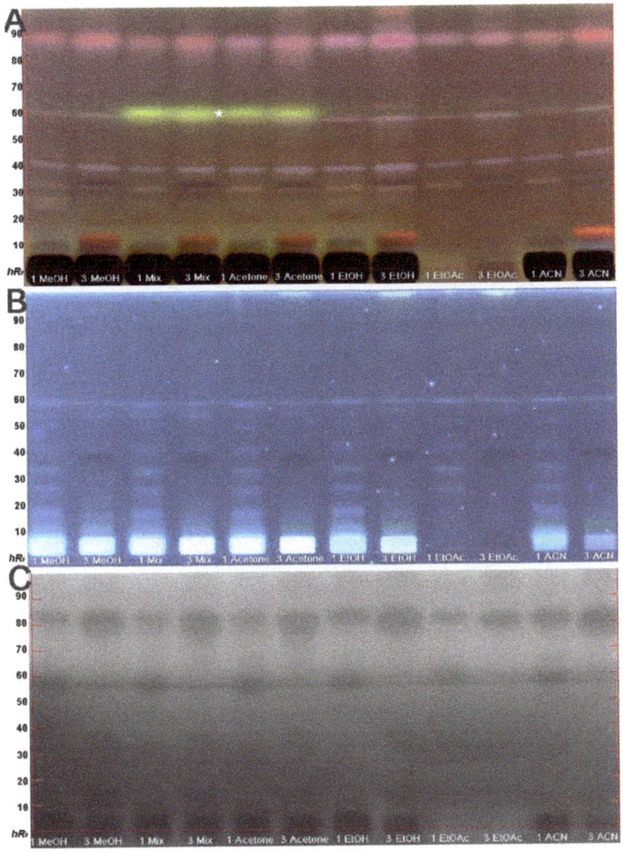

Figure 2. Investigation of different solvent selectivities for extraction: HPTLC chromatogram at FLD 366 nm after derivatization with anisaldehyde sulfuric acid reagent (**A**), at FLD 366 nm before the bioassay (**B**) and bioautogram after *Aliivibrio fischeri* bioassay, depicting bioluminescence after 30 min as grey scale image (**C**) of methanol (MeOH), mixture of methanol/acetone 1:1 (Mix), acetone, ethanol (EtOH), ethyl acetate (EtOAc) and acetonitrile (ACN) extracts of puree sample IDs 1 and 3 (Table 2, 0.4 g/mL) applied as area (6.0 × 3.0 mm, 10 µL/area), developed on HPTLC plates silica gel 60 F_{254} with toluene–ethyl acetate–methanol–formic acid 2.4:1.8:0.7:0.06, *v*/*v*/*v*/*v*, in twin-trough chamber 20 × 10 cm (* yellow acetone-soluble conta mination from plastic flask material).

2.4. Physico-Chemical Profiling of the Fruit Product Extracts

The physico-chemical profiling of the ethyl acetate extracts of the 8 different fruit products provided information on the absorbance or fluorescence of individual compounds. Close to the solvent front of the separated mango puree extracts (IDs 3,2 and 4,2), a yellow zone, most likely carotenoids, was visible (Figure 3A). UV-active compounds were comparatively more pronounced in the mango (IDs 3,2; 4,2; 7,2 and 8,2) than pineapple product extracts (Figure 3B). Among the mango products, the puree extracts (IDs 3,2 and 4,2) contained more UV-active compounds than the juice extracts, mainly retained in the start area of the adsorbent.

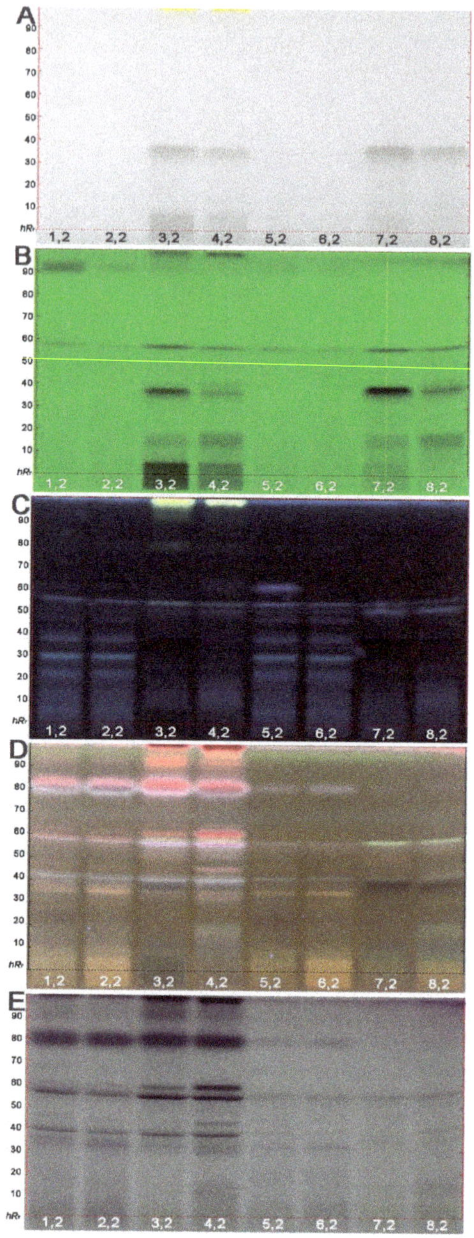

Figure 3. Physico-chemical profiling: HPTLC chromatograms at white light illumination (Vis, **A**), UV 254 nm (**B**), FLD 366 nm (**C**) and after derivatization with anisaldehyde sulfuric acid reagent (**D** FLD 366 nm and **E** Vis) of the puree and juice extracts (IDs 1–8, Table 2, 0.2 g/mL in ethyl acetate, 100 µL/area of 8.0 × 6.0 mm) developed on HPTLC plates silica gel 60 F_{254} with toluene–ethyl acetate–methanol–formic acid 2.4:1.8:0.7:0.06, *v*/*v*/*v*/*v*, in twin-trough chamber 10 cm × 10 cm.

In the FLD 366 nm chromatogram (Figure 3C), a dominantly yellow fluorescent zone was observed in the solvent front of the mango puree extracts (IDs 3,2 and 4,2). This zone was also visible and UV-active as mentioned. Tentatively, more blue fluorescent zones were observed in the pineapple (IDs 1,2; 2,2; 5,2 and 6,2) than mango products (IDs 3,2; 4,2; 7,2 and 8,2). After derivatization with the anisaldehyde sulfuric acid reagent and detection at FLD 366 nm (Figure 3D) or with white light illumination (Figure 3E), further components of the fruit products were detectable. For example, a strongly pink fluorescent (Figure 3D) or lilac (Figure 3E) compound at hR_F 80 was detected only by chemical derivatization, proving that the commonly used UV/Vis/FLD detection is not sufficient for quality control. In particular, this compound zone also proved to have multiple bioactivities, as discussed subsequently. As expected by the natural density of the product and verified by all the different detection modes, the puree extract samples (IDs 1,2–4,2) were more complex and rich in natural compounds.

2.5. *Effect-Directed Profiling of the Fruit Product Extracts*

The effect-directed profiling of the ethyl acetate extracts of the 8 different fruit products provided information on bioactive components using five different effect-directed assays (i.e., one chemical, two biological and two enzymatic assays). The DPPH• radical scavenging assay (Figure 4A) showed for both mango products (puree and juice) the strongest radical scavenging (antioxidative) compound zone at the start area and at hR_F 37 due to the ascorbic acid addition. This addition was required during juice production to prevent the enzymatic browning reaction, and thus, darkening of mango products.

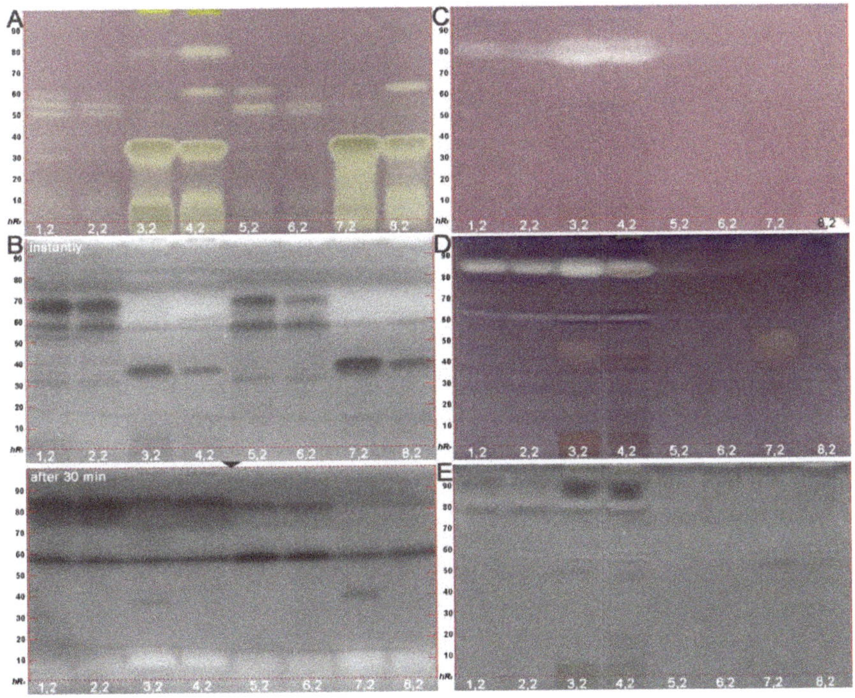

Figure 4. Effect-directed profiling: HPTLC autograms after DPPH• assay (**A**), *Aliivibrio fischeri* bioassay, instantly and after 30 min (**B**), *Bacillus subtilis* bioassay (**C**), AChE (**D,** here zones shifted upwards by 2 mm) and tyrosinase inhibition assays I detected at white light illumination (except for B: bioluminescence depicted as grey scale image) of puree and juice extracts (IDs 1–8, Table 2, 0.2 g/mL in ethyl acetate, 100 µL/area of 8.0 × 6.0 mm) developed on HPTLC plates silica gel 60 F_{254} with toluene–ethyl acetate–methanol–formic acid 2.4:1.8:0.7:0.06, *v*/*v*/*v*/*v*, in twin-trough chamber 10 cm × 10 cm (chromatograms at UV 254 nm and FLD 366 nm before the respective assay detection in Table S2 from Supplementary Materials).

Nevertheless, clear differences in the antioxidative response at hR_F 60 and hR_F 80 were evident for the two mango cultivars (IDs 3,2 versus 4,2; 7,2 versus 8,2). The comparison of the two mango products also revealed a difference. The enzymatic or heat treatment during juice production degraded the antioxidative compound zone at hR_F 80 (e.g., IDs 4,2 versus 8,2).

At the same position (hR_F 37) as the previously mentioned main antioxidative compound zone (ascorbic acid), the Gram-negative *Aliivibrio fischeri* bioautogram (Figure 4B) revealed a dominant bioactive dark zone in both mango products and cultivars which, however, substantially decreased over the 30 min imaging inspection. This indicated an acute transient effect, i.e., the energetic metabolism of the bacteria was immediately but temporarily disrupted by this compound and recovered almost completely after 30 min. In the pineapple cultivars, two further dominant antimicrobial zones at hR_F 58 and 68 were observed. The similar acute transient effect was observed for the antimicrobial zone at hR_F 68, whereas the antimicrobial zone at hR_F 58 increased in activity over the monitored 30 min period. The latter indicated a lasting effect with a delayed increase in the activity response. Over the monitoring period, a new antimicrobial compound appeared at hR_F 80 mainly in both fruit purees, as evident in the 30-min bioautogram. This indicated a delayed increase in the activity response.

The Gram-positive *Bacillus subtilis* bioassay (Figure 4C) revealed the same antimicrobial compound zone at hR_F 80 in both fruit puree extracts. This colorless zone at hR_F 80 acting against *B. subtilis* was also the main inhibitor of the AChE (Figure 4D, IDs 1,2–4,2). Another weaker AChE inhibiting zone was detected at hR_F 60. Both colorless zones were almost absent in the juice extracts, most likely degraded by the enzymatic or heat treatment during the juice production. Especially, the mango cultivar NDM (IDs 3,2 and 7,2) revealed a further AChE inhibiting compound zone at hR_F 45.

The tyrosinase inhibition assay (Figure 4E) showed three colorless inhibition zones in the puree extracts (IDs 1,2–4,2) with the main inhibition zone at hR_F 80. This multipotent compound zone showed already a pronounced activity against *B. subtilis* and AChE. Vice versa, the dark zone above (hR_F 85) in the mango puree extracts (ID 3,2 and 4,2) triggered the fruit browning. Another weaker browning-promoting dark zone at hR_F 78 was evident in all puree extracts (IDs 1,2–4,2).

Differences in the activity profiles depending on the fruit processing were evident. Both purees contained a multipotent zone at hR_F 80 (active against both bacteria, AChE and tyrosinase, Figure 4B–E), which was lost during the production of juice made out of the puree. This was a more general observation across all the different assays, as the more apolar bioactive compounds (evident in the upper autogram part) were partially or totally lost during the production of juice. Although it was expected that puree is richer in bioactive compounds than juice, the severe loss was surprising. All in all, the bioprofiling of food products turned out to be a rich source of information with regard to changes in the activity potential caused by food processing.

3. Materials and Methods

3.1. Chemicals and Reagents

Bi-distilled water was produced using a Heraeus Destamat Bi-18E (Thermo Fisher Scientific, Schwerte, Germany). Ethyl acetate (≥99.7%) and ammonia (25%) were obtained from Th. Geyer, Renningen, Germany. All amino acids, lithium citrate, all mineral salts for producing phosphate-buffered saline (PBS), acetone (≥99.5%), acetonitrile (≥99.9%), diphenylamine, aniline, 2,2-diphenyl-1-picrylhydrazyl radical (DPPH•, 97%), acetylcholinesterase (AChE) from *Electrophorus electricus* (≥245 U/mg, 10 kU/vial), bovine serum albumin (BSA, fraction V, ≥98%) and tyrosinase from mushroom (≥1000 U/mg, 25 kU/vial) were obtained from Fluka Sigma-Aldrich, Steinheim, Germany. 1-Naphthyl acetate (≥98%) and 2-naphthyl-α-D-glucopyranoside were obtained by AppliChem, Darmstadt, Germany. Toluene, ethanol, 2-propanol (all ≥ 99.9%), formic acid (≥98%), sulfuric acid (96%), diphenylboric acid-2-aminoethylester, 4-methoxybenzaldehyde (anisaldehyde,

≥97.5%), tris(hydroxymethyl)amino¬methane (TRIS, ≥99.9%) and thiazol blue tetrazolium bromide (3-(4,5-dimethylthiazol-2-yl)-2,5-diphenyl-tetrazolium bromide, MTT) were purchased from Carl Roth, Karlsruhe, Germany. Methanol (>99.8%) was purchased from VWR, Darmstadt, Germany. (2S)-2-Amino-3-(3,4-dihydroxyphenyl) propionic acid (levodopa) was obtained from Santa Cruz Biotechnology, Dallas, TX, USA. 1-Butanol was from Alfa Aesar, Karlsruhe, Germany. HPTLC plates silica gel 60 F_{254} (without F_{254} for *B. subtilis* bioassay), 20 cm × 10 cm, ascorbic acid, potassium disulfite, sodium hydroxide, Gram-positive soil bacteria *Bacillus subtilis* subsp. spizizenii (DSM-618), and Folin-Ciocalteu's phenol reagent were from Merck, Darmstadt, Germany. Gram-negative marine *A. fischeri* bacteria (NRRL-B11177, strain 7151) were bought from the Leibniz Institute DSMZ (German Collection of Microorganisms and Cell Cultures), Berlin, Germany. Fruit samples were from the orchard at Chiang Rai province, Northern Thailand.

3.2. Production of Puree Samples

Ripe pineapples were washed with water, peeled, cut into small pieces and crushed to form pineapple puree. Then 50 mg/L of sulfur dioxide was added. Ripe mangoes were also washed, peeled, cut and crushed, but after peeling the seeds were removed. The mango purees were treated with 40 mg/L of ascorbic acid and 50 mg/L of sulfur dioxide. Each fruit puree (1 kg) was placed in a polyethylene pouch and stored at −18 °C until further use.

3.3. Production of Juice Samples

The frozen puree of each cultivar was thawed at ambient temperature and treated with pectinase (400 µL/L; Trenolin®Super DF, Erbslöh, Geisenheim, Germany). The puree was incubated at 45–50 °C for 1 h. Then, it was heated at 70 °C for 5 min to inactivate the enzymes. The juice was separated from the solid material part by centrifugation at 5000× *g* at 4 °C for 10 min. Juice (100 mL) was filled in a polyethylene bottle and stored at −18 °C until use.

3.4. Physico-Chemical Characterization of the Produced Purees and Juices

The standard parameters were established as follows: total soluble solids (TSS) was measured by a digital automatic refractometer (Abbemat, Anton Paar®, St. Albans, Austria). The total saccharide content was enzymatically determined using a Konelab 20XTi analyzer (Thermo Fisher Scientific, Schwerte, Germany) and its proper kits (EnzytecTM fluid, Thermo Fisher Scientific). Total acidity and pH were analyzed by a Schott titrator (Titroline alpha plus, SI-Analytics, Texas City, TX, USA). The content of total phenols was evaluated by the Folin assay (Singleton and Rossi 1965) using an automatic analyzer (Konelab 20XTi, Thermo Fisher Scientific). Amino acids were determined by an Amino Acid Analyzer S433 (Sykam GmbH, Eresing, Germany). All measurements were made in triplicate.

3.5. HPTLC−UV/Vis/FLD−EDA Analysis of the Produced Purees and Juices

3.5.1. Extraction

For the selection of the extraction solvent, 400 mg each of pineapple (PTW) or mango (NDM) puree were placed in a 1.5 mL centrifuge tube, mixed and extracted with 1 mL solvent (methanol, acetone, ethanol, ethyl acetate, and acetonitrile) in the ultrasonic bath for 10 min. Each suspension was centrifuged (5 min, 11,600× *g*) and the supernatant transferred into an autosampler vial (1.8 mL). For effect-directed profiling, the puree (2 g) and juice samples (2 mL) were extracted with 10 mL ethyl acetate, mixed (vortexed) and centrifuged (5 min, 3000× *g*). The supernatant was removed with a Pasteur pipette, was transferred to a 10 mL glass vial and stored at −18 °C until use.

3.5.2. Sample Application and Development

Instrumentation was used from CAMAG. Plates were cut to smaller pieces using the smartCUT Plate Cutter. Solutions were sprayed as a band (8 mm) for sugar analysis or as area (6.0 × 3.0 mm for 10 µL or 8.0 × 6.0 mm for 100 µL) for other analyses on HPTLC plates with the Automatic TLC Sampler (ATS) 4. Up to 8 tracks were applied onto an HPTLC plate, 10 cm × 10 cm, with a distance of 10 mm from the lower edge, 15 mm distance from the left edge and 10 mm distance between the bands. Sample volumes ranged 10 µL for extraction solvent selection and 100 µL for effect-directed profiling. Development was performed with a mixture of with toluene–ethyl acetate–methanol–formic acid 2.4:1.8:0.7:0.06, *v*/*v*/*v*/*v* [31] up to a migration distance of 60 mm from the lower plate edge in a twin-through chamber. The chromatogram was dried in a stream of warm air (hair dryer) for 2 min.

3.5.3. Derivatization and Documentation

For derivatization, the HPTLC plate was immersed into the anisaldehyde sulfuric acid reagent (5 mL concentrated sulfuric acid was carefully added to a mixture of 500 µL anisaldehyde, 10 mL acetic acid, and 100 mL methanol) or diphenylamine aniline *o*-phosphoric acid reagent (mixture of 70 mL aniline solution, 70 mL diphenylamine solution, 2% each in acetone, and 10 mL *o*-phosphoric acid, 85%) using the TLC Chromatogram Immersion Device. The immersion speed was 3 cm/s and the immersion time 2 s. The plates were heated at 110 °C on the TLC Plate Heater for 5 min. The HPTLC chromatograms were documented at UV 254, FLD 366 nm and white light illumination (reflection and transmission mode) using the TLC Visualizer. All data obtained was processed with the software winCATS, version 1.4.7.2018.

3.5.4. Effect-Directed Detection

The HPTLC plates were immersed into the respective assays using the TLC Chromatogram Immersion Device (immersion speed 3 cm/s), heated using the TLC Plate Heater, and if not stated otherwise, documented using the TLC Visualizer or DigiStore 2 Documentation System. Before application of the biological assays, the chromatogram was neutralized with ammonia vapor for 5 min and freed from any excess vapor or acidic mobile phase traces (Automatic Developing Chamber 2; relative humidity control glass flask filled with dry molecular sieve) for 25 min.

Radical scavenging assay: The chromatogram was dipped in a 0.02% methanolic DPPH• solution [32] for 5 s (immersion time). The chromatogram was dried in the dark at room temperature for 90 s, heated at 60 °C for 30 s and documented at white light illumination (reflectance mode), repeated after a day.

A. fischeri bioassay: The nutrient medium for the bacterial suspension was prepared as described [33]. The dried, neutralized chromatogram was immersed into the suspension for 3 s. Bioluminescence images (bioautograms) were recorded and processed with the BioLuminizer software, version 1.0.2.6107. Ten images were recorded over 30 min at time intervals of 3 min, each over an exposure time of 50 s.

B. subtilis bioassay: The dried, neutralized chromatogram was immersed in the bacterial suspension for 5 s and incubated at 37 °C for 2 h [31,34]. For visualization, the plates were immersed into a 0.2% PBS-buffered MTT solution for 1 s. During the incubation for 30 min at 37 °C, the MTT was reduced to a purple formazan dye, stopped by drying the plate at 50 °C for 5 min. Bioautograms were documented at white light illumination (reflection mode).

AChE inhibition assay: The dried, neutralized plate was immersed in the enzyme solution (AChE 666 units and 100 mg BSA in 100 mL 0.05 M TRIS buffer, pH 7.8) for 5 s. The plate was incubated for 25 min at 37 °C according to Akkad and Schwack [35]; Hage and Morlock [36]. For visualization, the chromatogram was immersed in the substrate solution (25 mg α-naphthyl acetate and 50 mg Fast Blue salt B in 90 mL ethanol-water, 1:2) and documented at white light illumination (reflectance mode).

Tyrosinase inhibition assay: The dried, neutralized plate was immersed in the enzyme solution (400 U/mL mushroom tyrosinase in 0.02 M phosphate buffer, pH 6.8) for 5 s [37]. The plate was dried for about 2 min and immersed in the substrate solution (L-DOPA, 18 mM in phosphate buffer, pH 6.8) for 3 s, followed by incubation for 15 min at room temperature. The dried plate was recorded at white light illumination (reflectance mode).

4. Conclusions

The bioactivity patterns revealed multiple bioactive compounds of mango and pineapple purees and juices, self-produced from self-harvested authentic fruits. The bioactivity assessment can be used to valorize and add value to the plant-based products. It can also be used with regard to distinct health claims, e.g., based on antioxidative compounds. Some of the bioactive compounds were not detected by UV/Vis/FLD, but first with the planar biochemical or biological assays. Effect-directed profiling thus makes product control more powerful. Since it is a non-targeted method, also unknown effective compounds, which are not in the focus of current analysis, can be detected. This is of high importance due to the global food processing chain. Compared to the status quo, product changes caused by contamination, adulteration, processing, treatments, etc. can be more comprehensively and easily detected through multi-imaging exploiting an array of different detection techniques.

Supplementary Materials: The following are available online, Table S1: Investigated mobile phases 1–6, Table S2: Chromatograms at UV 254 nm and FLD 366 nm before the respective assay detection.

Author Contributions: Conceptualization, methodology, formal analysis, data curation, writing—review and editing, all; investigation, N.W. (harvest and production); writing—original draft preparation, G.E.M.; supervision, resources, project administration, G.E.M. and D.R. All authors have read and agreed to the published version of the manuscript.

Funding: This research received no external funding.

Institutional Review Board Statement: Not applicable.

Informed Consent Statement: Not applicable.

Data Availability Statement: Data are available from the authors on request.

Acknowledgments: Thanks are owed to Anja Rheinberger, Anja Giehl, Claus-Dieter Patz (Department of Beverage Research) and Birgit Krause (Department of Soil Science and Plant Nutrition), all of Hochschule Geisenheim University, for their assistance with the analysis of certain parameters (Table 1) as well as to Stefanie Krüger, Food Science, JLU Giessen, for performing HPTLC experiments.

Conflicts of Interest: The authors declare no conflict of interest.

Sample Availability: Samples of the compounds are not available from the authors.

References

1. Block, G.; Patterson, B.; Subar, A. Fruit, vegetables, and cancer prevention: A review of the epidemiological evidence. *Nutr. Cancer* **1992**, *18*, 1–29. [CrossRef] [PubMed]
2. Ribeiro, S.M.R.; Schieber, A. Chapter 34-Bioactive Compounds in Mango (*Mangifera indica* L.). In *Bioactive Foods in Promoting Health: Fruits and Vegetables*, 1st ed.; Watson, R.R., Preedy, V.R., Eds.; Academic Press: Boston, MA, USA, 2010; pp. 507–523. ISBN 978-0-12-374628-3.
3. Giuffrè, A.M. Bergamot (*Citrus bergamia*, Risso): The Effects of Cultivar and Harvest Date on Functional Properties of Juice and Cloudy Juice. *Antioxidants* **2019**, *8*, 221. [CrossRef] [PubMed]
4. Siriwoharn, T.; Wrolstad, R.E.; Finn, C.E.; Pereira, C.B. Influence of cultivar, maturity, and sampling on blackberry (*Rubus*, L. Hybrids) anthocyanins, polyphenolics, and antioxidant properties. *J. Agric. Food Chem.* **2004**, *52*, 8021–8030. [CrossRef]
5. González-Aguilar, G.; Robles-Sánchez, R.M.; Martínez-Téllez, M.A.; Olivas, G.I.; Alvarez-Parrilla, E.; Rosa, L.A. Bioactive compounds in fruits: Health benefits and effect of storage conditions. *Stewart Postharvest Rev.* **2008**, *4*, 1–10. [CrossRef]
6. Food and Agricultural Organization of the United Nations. Available online: http://www.fao.org/faostat/en/#data/QC (accessed on 16 December 2021).
7. Barretto, L.C.d.O.; Moreira, J.d.J.d.S.; Santos, J.A.B.d.; Narendra, N.; Santos, R.A.R.d. Characterization and extraction of volatile compounds from pineapple (*Ananas comosus* L. Merril) processing residues. *Food Sci. Technol (Camp.)* **2013**, *33*, 638–645. [CrossRef]

8. Pino, J.A.; Queris, O. Analysis of volatile compounds of pineapple wine using solid-phase microextraction techniques. *Food Chem.* **2010**, *122*, 1241–1246. [CrossRef]
9. Chomchalow, N.; Somsri, S.; Na Songkhla, P. Marketing and Export of Major Tropical Fruits from Thailand. *AU J.T.* **2008**, *11*, 133–143.
10. Gardner, P.T.; White, T.A.C.; McPhail, D.B.; Duthie, G.G. The relative contributions of vitamin C, carotenoids and phenolics to the antioxidant potential of fruit juices. *Food Chem.* **2000**, *68*, 471–474. [CrossRef]
11. Mhatre, M.; Tilak-Jain, J.; De, S.; Devasagayam, T.P.A. Evaluation of the antioxidant activity of non-transformed and transformed pineapple: A comparative study. *Food Chem. Toxicol.* **2009**, *47*, 2696–2702. [CrossRef]
12. Kongsuwan, A.; Suthiluk, P.; Theppakorn, T.; Srilaong, V.; Setha, S. Bioactive compounds and antioxidant capacity of phulae and nanglae pineapple. *As. J. Food Ag-Ind.* **2009**, *2*, S44–S50.
13. Lu, X.-H.; Sun, D.-Q.; Wu, Q.-S.; Liu, S.-H.; Sun, G.-M. Physico-chemical properties, antioxidant activity and mineral contents of pineapple genotypes grown in china. *Molecules* **2014**, *19*, 8518–8532. [CrossRef]
14. Freitas, A.; Moldão-Martins, M.; Costa, H.S.; Albuquerque, T.G.; Valente, A.; Sanches-Silva, A. Effect of UV-C radiation on bioactive compounds of pineapple (*Ananas comosus* L. Merr.) by-products. *J. Sci. Food Agric.* **2015**, *95*, 44–52. [CrossRef] [PubMed]
15. Ferreira, E.A.; Siqueira, H.E.; Boas, E.V.V.; Hermes, V.S.; Rios, A.D.O. Bioactive compounds and antioxidant activity of pineapple fruit of different cultivars. *Rev. Bras. Frutic.* **2016**, *38*, 527. [CrossRef]
16. Ramteke, R.S.; Vilajayalaskshmi, M.R.; Eipeson, W.E. Processing and value addition to mangoes. *Indian Food Ind.* **1999**, *18*, 155–163.
17. Morlock, G.E.; Schwack, W. Hyphenations in planar chromatography. *J. Chromatogr. A* **2010**, *1217*, 6600–6609. [CrossRef] [PubMed]
18. Morlock, G.E. Background mass signals in TLC/HPTLC-ESI-MS and practical advices for use of the TLC-MS interface. *J. Liq. Chromatogr. Relat.* **2014**, *37*, 2892–2914. [CrossRef]
19. Morlock, G.E. Bioassays–Effects-detection in chromatography. In *Reference Module in Encyclopedia of Analytical Science*, 3rd ed.; Worsfold, P.J., Poole, A., Townshend, A., Miro, M., Eds.; Elsevier Science: Amsterdam, The Netherlands, 2019; pp. 261–270. ISBN 978-0-0810-1983-2.
20. Krüger, S.; Urmann, O.; Morlock, G.E. Development of a planar chromatographic method for quantitation of anthocyanes in pomace, feed, juice and wine. *J. Chromatogr. A* **2013**, *1289*, 105–118. [CrossRef] [PubMed]
21. Cretu, G.C.; Morlock, G.E. Analysis of anthocyanins in powdered berry extracts by planar chromatography linked with bioassay and mass spectrometry. *Food Chem.* **2014**, *146*, 104–112. [CrossRef]
22. Morlock, G.E.; Klingelhöfer, I. Liquid chromatography-bioassay-mass spectrometry for profiling of physiologically active food. *Anal. Chem.* **2014**, *86*, 8289–8295. [CrossRef] [PubMed]
23. Teh, S.-S.; Morlock, G. Analysis of bioactive components of oilseed cakes by high-performance thin-layer chromatography-(bio)assay combined with mass spectrometry. *Chromatography* **2015**, *2*, 125–140. [CrossRef]
24. Ristivojević, P.M.; Morlock, G.E. Effect-directed classification of biological, biochemical and chemical profiles of 50 German beers. *Food Chem.* **2018**, *260*, 344–353. [CrossRef]
25. Joomwong, A.; Sornsrivichai, J. Morphological characteristic, chemical composition and sensory quality of pineapple fruit in different seasons. *Chiang Mai Univ. J. Nat. Sci.* **2005**, *4*, 149–164. Available online: https://cmuj.cmu.ac.th/uploads/journal_list_index/594162253.pdf (accessed on 16 December 2021).
26. Nilprapruck, P.; Pradisthakarn, N.; Authanithee, F.; Keebjan, P. Effect of exogenous methyl jasmonate on chilling injury and quality of pineapple (*Ananas comosus* L.) cv. Pattavia. *SUSTJ* **2008**, *2*, 33–42. [CrossRef]
27. Boonyaritthongchai, P.; Puthmee, T.; Wongs-Aree, C. Quality changes of fresh-cut 'Mahachanok' mango at different storage temperatures. *Acta Hortic.* **2015**, 431–434. [CrossRef]
28. Lapcharoensuk, R.; Phannote, N.; Kasetyangyunsapa, D. Physicochemical properties of pineapple at difference maturity. In Proceedings of the Conference paper of the 10th TSAE International Conference, Bangkok, Thailand, 7–9 September 2017.
29. Chuensombat, N.; Rungraeng, N.; Setha, S.; Suthiluk, P. A preliminary study of high pressure processing effect on quality changes in 'Nanglae' pineapple juice during cold storage. *JFAT* **2019**, *5*, 13–18.
30. Logemann, A.; Schafberg, M.; Brockmeyer, B. Using the HPTLC-bioluminescence bacteria assay for the determination of acute toxicities in marine sediments and its eligibility as a monitoring assessment tool. *Chemosphere* **2019**, *233*, 936–945. [CrossRef] [PubMed]
31. Jamshidi-Aidji, M.; Morlock, G.E. From bioprofiling and characterization to bioquantification of natural antibiotics by direct bioautography linked to high-resolution mass spectrometry: Exemplarily shown for *Salvia miltiorrhiza* root. *Anal. Chem.* **2016**, *88*, 10979–10986. [CrossRef]
32. Pozharitskaya, O.N.; Ivanova, S.A.; Shikov, A.N.; Makarov, V.G. Separation and free radical-scavenging activity of major curcuminoids of *Curcuma longa* using HPTLC-DPPH method. *Phytochem. Anal.* **2008**, *19*, 236–243. [CrossRef] [PubMed]
33. *Water Quality-Determination of the Inhibitory Effect of Water Samples on the Light Emission of Vibrio Fischeri (Luminescent Bacteria Test)-Part 1: Method Using Freshly Prepared Bacteria (ISO 11348-1:2007)*; Beuth Verlag: Berlin, Germany, 2009.
34. Jamshidi-Aidji, M.; Morlock, G.E. Bioprofiling of unknown antibiotics in herbal extracts: Development of a streamlined direct bioautography using *Bacillus subtilis* linked to mass spectrometry. *J. Chromatogr. A* **2015**, *1420*, 110–118. [CrossRef] [PubMed]

35. Akkad, R.; Schwack, W. Multi-enzyme inhibition assay for the detection of insecticidal organophosphates and carbamates by high-performance thin-layer chromatography applied to determine enzyme inhibition factors and residues in juice and water samples. *J. Chromatogr. B* **2010**, *878*, 1337–1345. [CrossRef]
36. Hage, S.; Morlock, G.E. Bioprofiling of Salicaceae bud extracts through high-performance thin-layer chromatography hyphenated to biochemical, microbiological and chemical detections. *J. Chromatogr. A* **2017**, *1490*, 201–211. [CrossRef] [PubMed]
37. Krüger, S.; Bergin, A.; Morlock, G.E. Effect-directed analysis of ginger (*Zingiber officinale*) and its food products, and quantification of bioactive compounds via high-performance thin-layer chromatography and mass spectrometry. *Food Chem.* **2018**, *243*, 258–268. [CrossRef] [PubMed]

molecules

MDPI

Article

Neuroprotective Effects of B-Type Cinnamon Procyanidin Oligomers on MPP$^+$-Induced Apoptosis in a Cell Culture Model of Parkinson's Disease

Qi Xu [1,2,†], Ziyu Chen [2,†], Borong Zhu [3], Yiming Li [3], Manju B. Reddy [4], Huilin Liu [1], Guodong Dang [1], Qi Jia [3,*] and Xiaojun Wu [2,*]

1 School of Public Health, Shanghai University of Traditional Chinese Medicine, 1200 Cailun Road, Shanghai 201203, China; isuxuqi@shutcm.edu.cn (Q.X.); lynnnn0303@163.com (H.L.); kyctcm@vip.sina.com (G.D.)
2 Shanghai Key Laboratory of Compound Chinese Medicine, The Ministry of Education (MOE) Key Laboratory for Standardization of Chinese Medicine, Institute of Chinese Materia Medica, Shanghai University of Traditional Chinese Medicine, 1200 Cailun Road, Shanghai 201203, China; czy714@163.com
3 School of Pharmacy, Shanghai University of Traditional Chinese Medicine, 1200 Cailun Road, Shanghai 201203, China; zhuborong1993@163.com (B.Z.); ymlius@163.com (Y.L.)
4 Department of Food Science and Human Nutrition, Iowa State University, Ames, IA 50010, USA; mbreddy@iastate.edu
* Correspondence: jq@shutcm.edu.cn (Q.J.); xiaojunwu@shutcm.edu.cn (X.W.); Tel.: +86-21-51322207 (Q.J.); 86-21-51322578 (X.W.); Fax: +86-21-51322193 (Q.J.); +86-21-51322505 (X.W.)
† These authors contributed equally to this work.

Abstract: Cinnamon procyanidin oligomers (CPOs) are water-soluble components extracted from cinnamon. This study aims to explore the neuroprotection of B-type CPO (CPO-B) against 1-methyl-4-phenylpyridinium (MPP$^+$)-mediated cytotoxicity and the molecular mechanisms underlying its protection. The results demonstrated that CPO-B showed protection by increasing cell viability, attenuating an intracellular level of reactive oxygen species, downregulating cleaved caspase-3 expression, and upregulating the Bcl-2/Bax ratio. Moreover, CPO-B completely blocked the dephosphorylation of extracellular, signal-regulated kinase 1 and 2 (Erk1/2) caused by MPP$^+$. Treatment with an Erk1/2 inhibitor, SCH772984, significantly abolished the neuroprotection of CPO-B against MPP$^+$. Taken together, we demonstrate that CPO-B from cinnamon bark provided protection against MPP$^+$ in cultured SH-SY5Y cells, and the potential mechanisms may be attributed to its ability to modulate the dysregulation between pro-apoptotic and anti-apoptotic proteins through the Erk1/2 signaling pathway. Our findings suggest that the addition of cinnamon to food or supplements might benefit patients with PD.

Keywords: Parkinson's disease; CPO-B; MPP$^+$; Erk1/2; neuroprotection

Citation: Xu, Q.; Chen, Z.; Zhu, B.; Li, Y.; Reddy, M.B.; Liu, H.; Dang, G.; Jia, Q.; Wu, X. Neuroprotective Effects of B-Type Cinnamon Procyanidin Oligomers on MPP$^+$-Induced Apoptosis in a Cell Culture Model of Parkinson's Disease. *Molecules* **2021**, *26*, 6422. https://doi.org/10.3390/molecules26216422

Academic Editor: Smaoui Slim

Received: 25 September 2021
Accepted: 20 October 2021
Published: 24 October 2021

Publisher's Note: MDPI stays neutral with regard to jurisdictional claims in published maps and institutional affiliations.

1. Introduction

Parkinson's disease (PD) is a devastating and irreversible degenerative disorder that afflicts about 1.5% of the population aged 65 years and over in the world [1]. The cardinal features of PD include tremors, rigidity, postural instability, and slowness of voluntary movement, which are the results of the premature death of dopaminergic neurons and the deficiency of dopamine thereafter in the nigrostriatal system. Mitogen-activated protein kinases (MAPKs) are a type of serine/threonine protein kinases, consisting of extracellular, signal-regulated kinase (Erk) 1 and 2 (Erk1/2), P38 isoforms, and c-Jun N-terminal kinase (JNK) [2]. They are essential components of the signaling network and function as an integral part in cellular processes, such as growth, differentiation, apoptosis, survival, and proliferation [3]. Although the exact mechanisms responsible for neuronal death in PD remain unclear, research suggests that abnormal regulation of MAPKs may contribute to the pathogenesis of PD by triggering the dysregulation of pro-apoptotic and anti-apoptotic pathways, inducing neuroinflammation, oxidative stress, and mitochondrial dysfunction [4,5]. Thus, MAPK signaling pathways are considered as a possible therapeutic solution to PD treatment.

1-Methyl-4-phenylpyridinium (MPP^+) is the ultimate metabolite of 1-methyl-4-phenyl-1,2,3,6 tetrahydropyridine (MPTP), causing the degeneration of the nigrostriatal dopaminergic neurons. MPP^+ has been used extensively in laboratories to establish in vitro models of PD. The accumulation of MPP^+ in the mitochondria interferes with the mitochondrial respiratory chain complex I, reduces mitochondrial respiration, and stimulates the generation of reactive oxygen species (ROS) [6]. It is also reported that MPTP/MPP^+-induced neurodegeneration is associated with the dysregulation of MAPK signaling cascades or other signaling pathways such as protein kinase B (AKT) both in cell culture and animal models of PD [7–9]. For example, it is demonstrated that MPTP/MPP^+-mediated suppression of Akt and Erk1/2 signaling pathways can downregulate the level of B-cell lymphoma 2 (Bcl-2)/Bcl-2-associated X protein (Bax) and promote caspases activation, leading to neuronal apoptosis [2,7,10]. We used an SH-SY5Y cell line in our experiments. When treated with neurotoxicants, the neurodegeneration recapitulated in SH-SY5Y cells serves as a model for mechanisms of dopaminergic neuronal death observed in PD [11]. Moreover, we selected undifferentiated SH-SY5Y cells since research suggests that they are suitable for the neuroprotection studies, with high vulnerability to neurotoxins such as MPP^+ [12].

Functional food has gained attention in the past few decades since people are not only concerned about the taste but also the nutritional values and health benefits. Cinnamon contains many functional food ingredients such as cinnamaldehyde, cinnamic acid, polyphenols, and flavonoids. Cinnamon is an important spice used by people worldwide and has both culinary as well as medicinal uses. The beneficial effects of cinnamon, as suggested by research, include its anti-inflammatory, antioxidant, and antidiabetic activities [13]. The extracts from cinnamon such as cinnamaldehyde and flavonoids were found to possess strong antioxidant and anti-inflammatory properties to combat chronic diseases. Cinnamon polyphenols were demonstrated to have insulin-like activities and regulate glucose metabolism [13]. Cinnamon extracts have also shown neuroprotective effects through upregulating the neurotrophic factors in PD and reducing the formation of toxic β amyloid polypeptides in Alzheimer's disease [14,15]. Active water-soluble components of cinnamon were isolated and identified as cinnamon procyanidin oligomers (CPOs). There are two main types of CPOs in plants, namely, A- and B-types, based on the interflavan linkage between the flavan-3-ol units [16]. Our previous studies also confirmed the existence of both A- and B-types of CPOs from cinnamon bark [16,17].

The beneficial effects of procyanidins have gained attention, and it has been demonstrated that procyanidins extracted from blueberries and grape seeds may alleviate neurodegeneration via the improvement of antioxidant enzymes, the enhancement of mitochondrial function, and the regulation of MAPK signaling pathways [18,19]. However, the neuroprotective effects of procyanidins extracted from cinnamon are seldomly reported. CPO-B is the major procyanidin in *Cinnamomum cassia* (L.), which is the most popular cinnamon bark in China. However, its beneficial effects in PD are not well studied. This study aims to assess whether CPO-B attenuates MPP^+-induced cytotoxicity in the dopaminergic cell SH-SY5Y, a cell model of PD, and to explore whether it exerts neuroprotection through activating the Erk1/2 signaling pathway and modulating the dysregulation between pro-apoptotic and anti-apoptotic proteins. These findings may have implications for the application of CPO-B in the therapy of PD.

2. Results

2.1. Cytotoxic Effects of MPP^+ and CPO-B

The optimal dose of MPP^+ as well as that of CPO-B used in later experiments was determined in a dose-response study of MPP^+ and CPO-B on SH-SY5Y cells using the MTS assay (Table 1). The cells were treated with increasing concentrations of MPP^+ or CPO-B for 12 h. As shown in Table 1, cell viability was reduced ($p < 0.001$) to 93%, 87%, 72%, 57%, 45%, and 34% with 0.1 mM, 0.5 mM, 1 mM, 2.5 mM, 5 mM, and 10 mM of MPP^+, respectively. CPO-B did not show any significant cytotoxicity after the 12 h incubation with a concentration less than 50 μM. Based on these results, 1 mM MPP^+ and 12.5 μM CPO-B were chosen to evaluate the neuroprotection of CPO-B in the following experiments.

Table 1. The cytotoxic effects of MPP^+ ($n = 3$) and CPO-B ($n = 6$) on SHSY-5Y cells by MTS assay.

Dose	Cell Viability (%)
MPP^+ (mM)	
0	100 ± 0.6
0.1	93 ± 1.4 ***
0.5	87 ± 1.3 ***
1	72 ± 2.5 ***
2.5	57 ± 1.7 ***
5	45 ± 1.9 ***
10	34 ± 0.2 ***
CPO-B (μM)	
0	100 ± 2.0
5	94 ± 4.8
10	91 ± 6.1
25	83 ± 2.9
50	70 ± 2.5 ***

The values (mean ± SEM) were standardized to the percentage of control; symbol * indicates the difference between control and other treatments; *** $p < 0.001$.

2.2. CPO-B Protects SH-SY5Y Cells from MPP^+-Induced Intracellular ROS Production and Apoptosis

In order to evaluate whether CPO-B suppressed MPP^+-mediated intracellular ROS generation and apoptosis, cells were incubated with 12.5 μM CPO-B for 2 h, and 1 mM MPP^+ for another 12 h. We found MPP^+ treatment stimulated intracellular ROS production by twofold ($p < 0.01$) and decreased cell viability by 55% ($p < 0.001$) (Figure 1a,b). However, 12.5 μM CPO-B showed significant protection by reducing intracellular ROS by 34% ($p < 0.05$) and increasing cell viability by 49% ($p < 0.01$). In agreement with the above results, microscopic morphological examination also indicated the protective effects of CPO-B (Figure 1c). Cells without treatment or treated with CPO-B alone were round with regular homogeneous staining of Hoechst 33258. However, MPP^+-induced apoptotic cells exhibited condensed and fragmented nuclei with a strong and bright florescence of Hoechst stain. CPO-B pretreatment markedly blocked the morphological damage caused by MPP^+ and resulted in a non-apoptotic cell phenotype in the visual fields (Figure 1c,d).

2.3. CPO-B Blocks MPP^+-Induced Alteration of Apoptosis-Associated Proteins in SH-SY5Y Cells

We further evaluated whether CPO-B provided protection against MPP^+-induced apoptosis by examining the alteration of apoptosis-associated proteins, including cleaved caspase-3, Bcl-2, and Bax (Figure 2a). MPP^+ downregulated the expression of Bcl-2 with no significant effects on the expression of Bax, resulting in a 60% decrease in the Bcl-2/Bax ratio when compared to control cells ($p < 0.01$) (Figure 2b). However, CPO-B significantly blocked the effects of MPP^+ and increased the Bcl-2/Bax ratio to 78% of the control cells ($p < 0.001$), suggesting the anti-apoptotic role of CPO-B. Moreover, MPP^+ also upregulated the expression of cleaved caspase-3 by 17% of the control ($p < 0.05$), and CPO-B pretreatment completely counteracted the effects ($p < 0.05$) and the expression of cleaved caspase-3 was downregulated to the control level (Figure 2c).

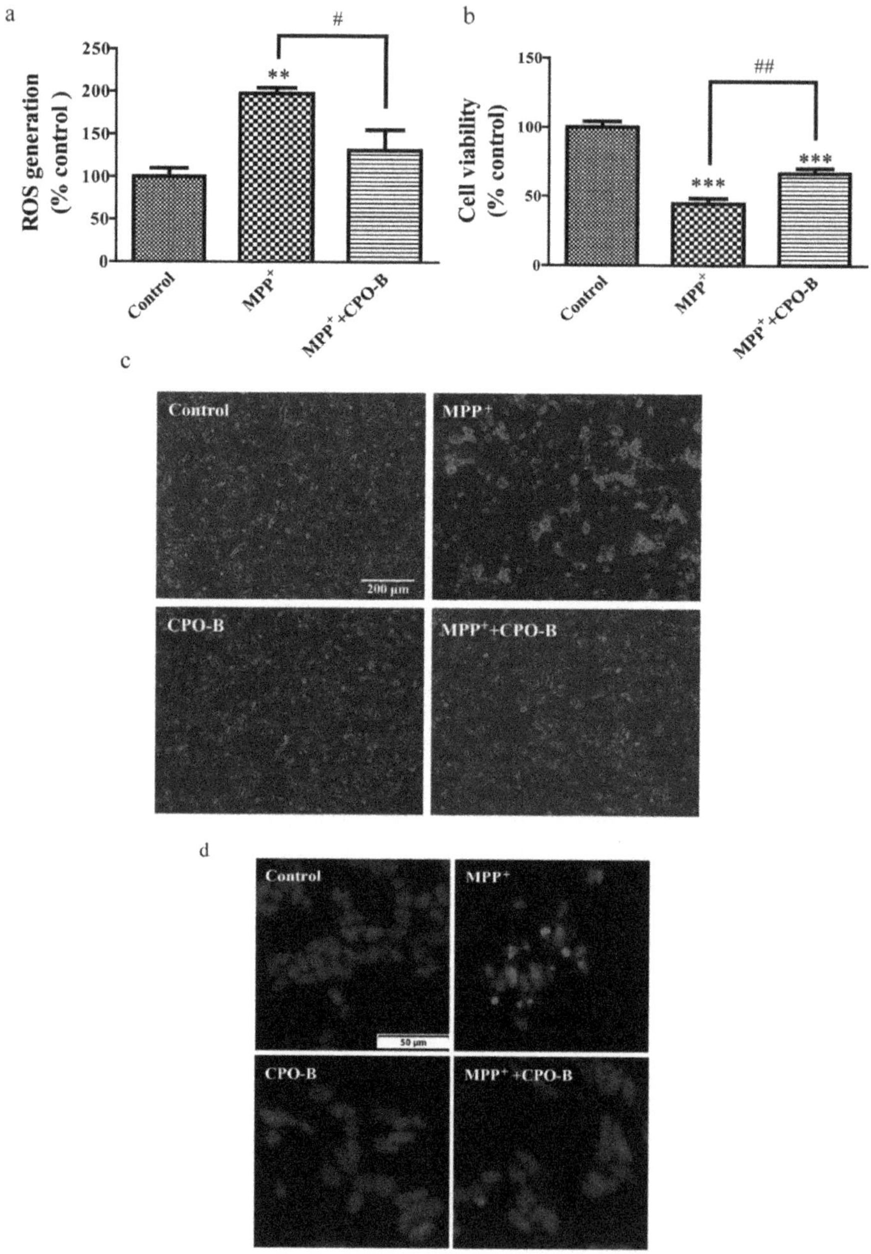

Figure 1. The effect of CPO-B on MPP^+-induced cytotoxicity measured by intracellular ROS generation (**a**, $n = 4$), cell viability (**b**, $n = 8$), microscopic examination (**c**) and Hoechst nuclear staining (**d**); the values (mean ± SEM) were standardized to the percentage of control; symbol * indicates the difference between control and other treatments; symbol # indicates the difference between MPP^+ and MPP^++CPO-B; ** $p < 0.01$, *** $p < 0.001$, # $p < 0.05$, ## $p < 0.01$; CPO-B, B-type cinnamon procyanidin oligomer; MPP^+, 1-methyl-4-phenylpyridinium.

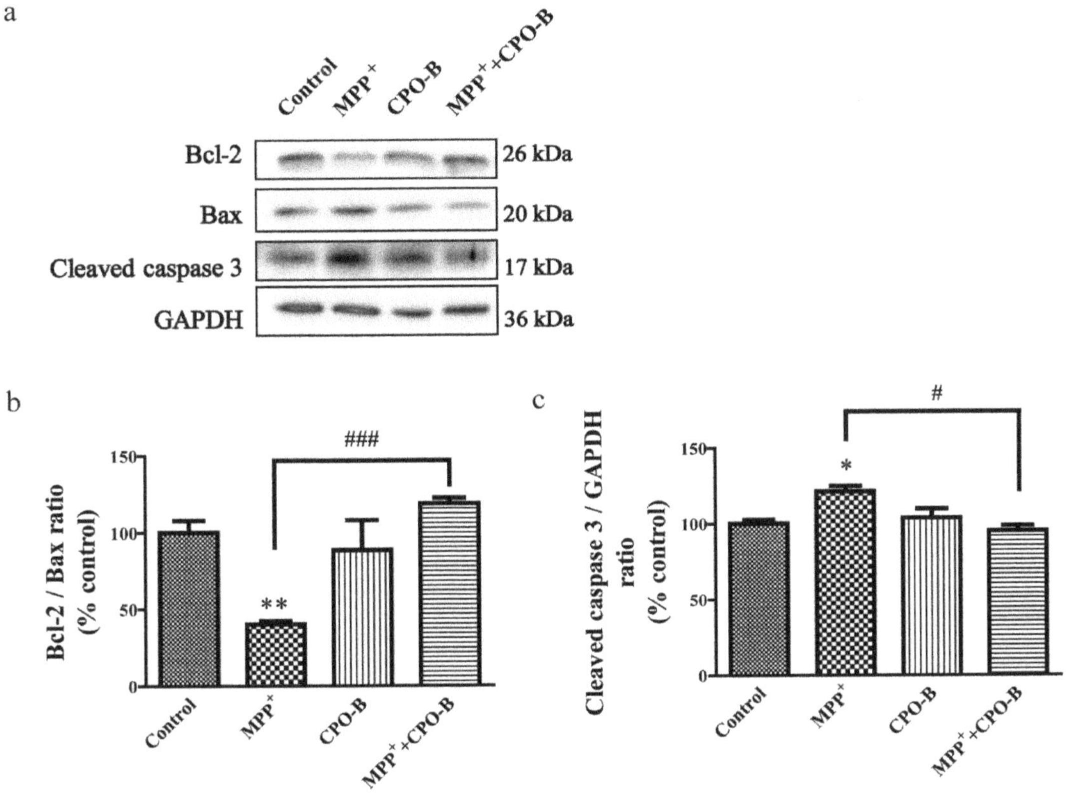

Figure 2. The effect of CPO-B on MPP$^+$-induced alteration of Bcl-2/Bax expression (**b**, n = 3) and cleaved caspase-3 expression (**c**, n = 3). The top panel shows the representative Western blots (**a**); the values (mean ± SEM) were standardized to the percentage of control; symbol * indicates the difference between control and other treatments; symbol # indicates the difference between MPP$^+$ and MPP$^+$+CPO-B; * $p < 0.05$, ** $p < 0.01$, # $p < 0.05$, ### $p < 0.001$; CPO-B, B-type cinnamon procyanidin oligomer; MPP$^+$, 1-methyl-4-phenylpyridinium; Bcl-2, B-cell lymphoma 2; Bax, Bcl-2-associated X protein; GAPDH, glyceraldehyde 3-phosphate dehydrogenase.

2.4. CPO-B Exerts Neuroprotection through the Upregulation of Phosphorylation of Erk1/2 but Not AKT Expression

For the purpose of exploring the possible molecular mechanisms behind the protective effects of CPO-B, we further examined the protein expression of Erk1/2 and Akt (Figure 3a). MPP$^+$ treatment significantly suppressed the expression of phosphorylated Erk1/2 (P-Erk1/2) but not total Erk1/2 (T-Erk1/2), resulting in a 53% decrease ($p < 0.01$) in the ratio of P-Erk1/2/T-Erk1/2 in comparison with the control (Figure 3b). CPO-B completely blocked the effects ($p < 0.05$) and upregulated the ratio of P-Erk1/2/T-Erk1/2 to the control level. However, neither MPP$^+$ nor CPO-B treatment had a significant effect on Akt phosphorylation (Figure 3c).

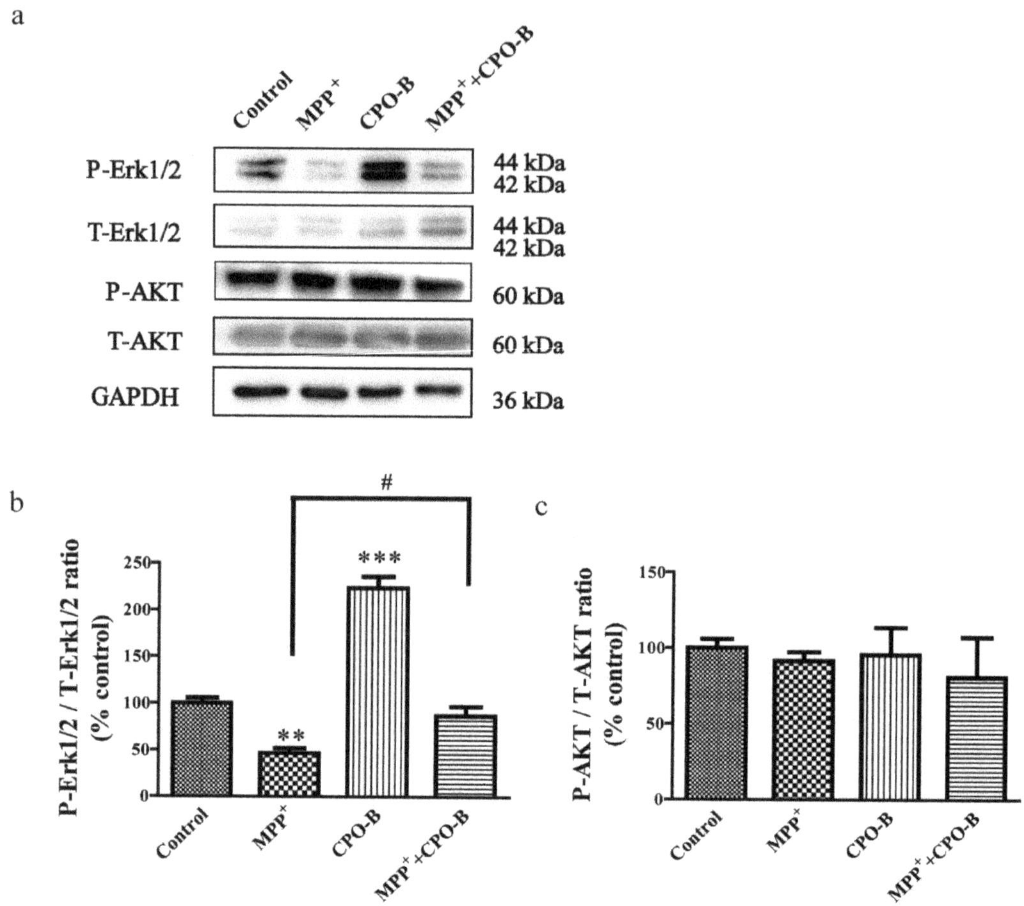

Figure 3. The effect of CPO-B on the MPP^+-induced alteration of Erk1/2 (**b**, $n = 3$) and Akt phosphorylation (**c**, $n = 3$). The top panel in the figure shows representative Western blots (**a**). The values (mean ± SEM) were normalized to control; symbol * indicates the difference between control and other treatments; symbol # indicates the difference between MPP^+ and MPP^++CPO-B; ** $p < 0.01$, *** $p < 0.001$, # $p < 0.05$; CPO-B, B-type cinnamon procyanidin oligomer; MPP^+, 1-methyl-4-phenylpyridinium; AKT, protein kinase B; P-AKT, phosphorylated AKT; T-AKT, total AKT; Erk1/2, extracellular signal regulated kinase 1 and 2; P-Erk1/2, phosphorylated Erk1/2; T-Erk1/2, total Erk1/2; GAPDH, glyceraldehyde 3-phosphate dehydrogenase.

To further examine the role of the Erk1/2 pathway in CPO-B-mediated protection, the cells were treated with 2 μM Erk1/2 inhibitor SCH772984 for 2 h, followed by the treatment of CPO-B and MPP^+. We observed that the SCH772984 treatment alone had no significant effects on cell viability (Figure 4a). However, the incubation of SCH772984 blocked the protective effects of CPO-B by inhibiting Erk1/2 phosphorylation and completely diminishing the effects of CPO-B on MPP^+-induced downregulation of Bcl-2 (Figure 4b–d). These results demonstrated that the Erk1/2 pathway is indeed involved in CPO-B-mediated neuroprotection.

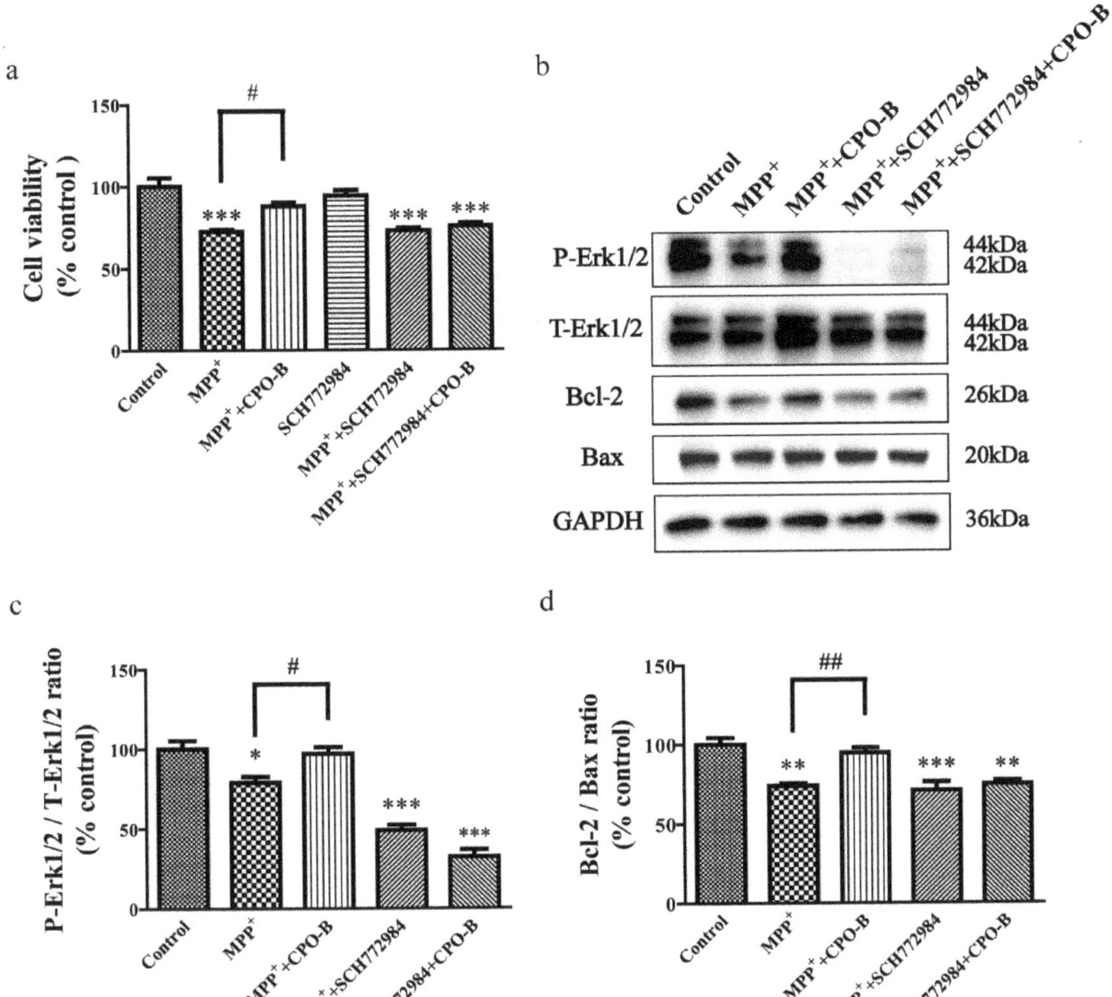

Figure 4. The effect of Erk1/2 inhibition on CPO-B-mediated neuroprotection measured by cell viability (**a**, $n = 3$), Erk1/2 phosphorylation (**c**, $n = 3$) and Bcl-2/Bax expression ratio (**d**, $n = 3$). The top panel in the figure shows representative Western blots (**b**). The values (mean ± SEM) were normalized to control; symbol * indicates the difference between control and other treatments; symbol # indicates the difference between MPP$^+$ and MPP$^+$+CPO-B; * $p < 0.05$, ** $p < 0.01$, *** $p < 0.001$, # $p < 0.05$, ## $p < 0.01$. CPO-B, B-type cinnamon procyanidin oligomer; MPP$^+$, 1-methyl-4-phenylpyridinium; Bcl-2, B-cell lymphoma 2; Bax, Bcl-2-associated X protein; Erk1/2, extracellular signal-regulated kinase 1 and 2; P-Erk1/2, phosphorylated Erk1/2; T-Erk1/2, total Erk1/2; GAPDH, glyceraldehyde 3-phosphate dehydrogenase.

3. Discussion

PD is a neurodegenerative disease that affects around 1% of the population over the age of 60 and 4% of those over the age of 80 [4]. However, there is no known cure for PD, and current treatments only alleviate the symptoms. The major goal of PD research is to identify therapeutic agents that could either delay or stop the disease's progression.

Food-derived phenolic compounds have gained particular attention as a promising therapeutic approach against neurodegenerative disorders in recent years. A number of polyphenol compounds, such as resveratrol from grape seeds or curcumin, as well as epigallocatechin gallate (EGCG) from tea, have been reported to exert neuroprotection through the scavenging of free radicals, suppressing mitochondrial dysfunction, and alleviating neuroinflammation [20,21]. Procyanidins are a class of polyphenols wildly present in many vegetables, fruits, nuts, and seeds, and are considered to be among the most potent antioxidants in nature. Research has demonstrated their beneficial effects, including reducing the risk of cerebrovascular disease, diabetes, cardiovascular disease, and cancer mortality [22]. One study also found that pretreatment with procyanidins could significantly alleviate rotenone-induced oxidative stress and reduce apoptosis in SH-SY5Y cells [19]. However, this study did not reveal where the procyanidins were extracted from, nor did it mention the degree of polymerization of the procyanidins. Since procyanidins are polymeric flavanols with a variety of linkages and subunits, the degree of polymerization and connectivity determines their bioavailability and bioactivity [23]. For example, research shows that B-type procyanidin polymers more effectively inhibited inflammation compared to monomers and oligomers in human colon cells [24]. Another study shows that the oligomeric form of procyanidins has a stronger protection ability in regard to reducing inflammation and oxidative stress than that of polymers in diabetic rats [25]. In our current study, CPO-B from cinnamon bark was shown to remarkably protect against MPP^+-induced neurotoxicity in cultured SH-SY5Y cells, suggesting the possible application of CPO-B in the therapy of PD.

The neuroprotective effects of CPO-B in our study raise issues about its absorption and the capability to cross the blood–brain barrier. Research has demonstrated that procyanidin oligomers such as dimer and trimer procyanidins are absorbable in vivo and reach maximum concentrations one hour after digestion [26]. The major components of CPO-B include procyanidin B-2, procyanidin C-1, and cassiatannin A-2. One study found that these oligomeric procyanidins were absorbed by the small intestine and detected in the different tissues without structure modification after oral ingestion [27]. A human trial demonstrated that procyanidin B-2 could be absorbed by the human intestinal tract and detected in plasma [28]. The capabilities of procyanidin metabolites to cross the blood–brain barrier and to target the brain have also been observed in several in vivo studies [29,30]. Consistent with these findings, our recent study found that A-type CPO could cross the blood–brain barrier and provide neuroprotection in an animal model of PD [31].

Our current study found 1 mM MPP^+ significantly downregulated Bcl-2, while Bax expression increased, and caspase-3 activity was activated, which was blocked by 12.5 μM CPO-B, suggesting an anti-apoptotic role of CPO-B. Although it is difficult to extrapolate the in vitro model to human physiological conditions, our study found no toxicity of CPO-B under the concentration of 50 μM. Moreover, based on the previous study showing that the absorption rate of the polyphenolic compound was 5–10% [32], we expected an adult with 5 L blood to take 500–1000 mg CPO-B daily to reach a concentration of 12.5 μM in their blood circulation. This is similar to the findings of our previous in vivo study estimating a daily dose of 720 mg for A-type CPO in the prevention of PD [31]. In addition, the data from NHANES 1999–2002 indicated that the total daily procyanidin intake in adults is around 95 mg, far below the dose used in our current study. However, our study still suggested the potential role of cinnamon procyanidin supplementation in the prevention of PD.

We further investigated the mechanisms underlying the neuroprotection of CPO-B against MPP^+-induced apoptosis by examining the role of Erk1/2 and Akt kinases. Erk1/2 is one of the major signaling cassettes of the MAPK signaling pathway and plays an essential role during cell proliferation, death, differentiation, and survival [7]. Although Erk1/2 usually functions as a mediator against apoptosis, it is also found that Erk1/2 signaling can be pro-apoptotic depending on the cell scenario and the type of cell insults. For example, one study reported that Erk1/2 is pro-apoptotic in rotenone-induced cell death, and procyanidins exerted its neuroprotection by inhibiting Erk1/2 phosphorylation. However, our study found MPP^+ reduced the phosphorylation of Erk1/2, suggesting that inhibition of Erk1/2 is involved in MPP^+-induced cell death. Moreover, CPO-B significantly blocked MPP^+-induced Erk1/2 dephosphorylation, and the Erk1/2 inhibitor SCH772984 diminished the protective effects of CPO-B. These results suggest that CPO-B exerted neuroprotection by increasing Bcl2 expression and inhibiting caspase 3 activation through Erk1/2 activation. Since Erk1/2 regulates the functions of various substrates in cells and activates multiple transcription factors associated with neuroinflammation and oxidative stress, future studies are needed to determine whether CPO-B exerts antioxidant and anti-inflammatory effects through Erk1/2 signaling pathways in experimental models of PD.

In our current study, the AKT signaling pathway was also examined, since it modulates various cellular functions, such as neuronal cell migration, proliferation, and plasticity, and provides an important signaling for neuroprotection [5]. Although studies have shown that dysregulation of AKT facilitated mitochondrial failure and contributed to MPP^+-induced cell death [33,34], we did not observe significant changes in AKT activation following treatment with MPP^+ or CPO-B. The different results may be due to the varied microenvironmental system in the experiments.

In conclusion, we extracted CPO-B from cinnamon and demonstrated that it exerted significant neuroprotection against MPP^+-induced cytotoxicity. The underlying mechanisms included its ability of activating Erk1/2 phosphorylation, enhancing the ratio of Bcl-2/Bax expression and subsequently inhibiting caspase-3 activity. While the current study focused on the effects of CPO-B in mitigating the dysregulation of cellular survival signals induced by MPP^+, future studies are underway to investigate the effects of CPO-B on dopamine system dysfunction in animal models of PD. The present findings not only provide pharmacological and mechanistic evidence to support potential therapeutic application of procyanidins in PD but also indicate that the addition of cinnamon to foods or using CPO-B as a dietary supplement might benefit patients with PD.

4. Materials and Methods

4.1. Chemicals

The SH-SY5Y cell line was supplied from Laboratory of Stem Cell Biology of Chinese Academy of Sciences (Shanghai, China). Hoechst 33258, streptomycin, penicillin, Dulbecco's modified Eagle medium (DMEM), fetal bovine serum, and 0.25% trypsin/EDTA were acquired from Thermo Fisher (Carlsbad, CA, USA). The primary antibodies against glyceraldehyde 3-phosphate dehydrogenase (GAPDH), cleaved Caspase 3, Bax, Bcl-2, Akt, Erk1/2, phospho-Akt (Thr309/Thr308/Thr305), and phospho-Erk1/2 (Thr202/Tyr204, Thr185/Tyr187) were bought from Cell Signaling Technology (Danvers, MA, USA). MPP^+ iodide was supplied from Sigma-Aldrich (St. Louis, Missouri, MO, USA). Erk1/2 inhibitor SCH772984 was purchased from Meilun (Dalian, China). One solution cell proliferation assay (MTS) was acquired from Promega (Madison, WI, USA). Before each assay, all solutions were freshly prepared.

4.2. The Extraction of Procyanidin

As described previously, CPO-B was isolated from *cassia bark* and subjected to macroresin column chromatography eluted with water and a series of different concentrations of ethanol for purification [35]. The total percentage of CPO-B was nearly 60% and identified as cassiatannin A-2 (14.8%), procyanidin C-1 (18.7%), and procyanidin B-2 (26.1%), which was confirmed by high-performance liquid chromatography.

4.3. Design of Experiments

SH-SY5Y cells at a density of 1×10^6/mL were grown in 96-well plates with DMEM medium with the supplementation of 10% fetal bovine serum and 1% penicillin/streptomycin in a humid chamber at 37 °C. The cells were treated with six concentrations of MPP^+

(0.1 mM, 0.5 mM, 1 mM, 2.5 mM, 5 mM, 10 mM) and four concentrations of CPO-B (5 μM, 10 μM, 25 μM, 50 μM) to select the desired concentration of MPP^+ and CPO-B for the subsequent experiments. To examine whether CPO-B provided protection against MPP^+-induced toxicity, cells were treated with 1 mM MPP^+ for 12 h with or without the pretreatment of CPO-B for 2 h. To further explore the role of Erk1/2 activation in CPO-B mediated neuroprotection, cells were pretreated with Erk1/2 inhibitor SCH772984 for 2 h followed by MPP^+ with or without CPO-B treatments.

4.4. Cell Viability

An MTS assay was conducted to evaluate cell viability following the manufacturer's recommendations. Following the treatment, the cells were exposed with 20 μL MTS reagent at 37 °C for 4 h in 96-well plates. The tetrazolium reagents were reduced by the cellular hydrogenases to yield the soluble formazan, which was quantified by Thermo Fisher Scientific Varioskan Flash at 490 nm (Waltham, MA, USA).

4.5. Measurement of Intracellular ROS

The intracellular ROS level was assessed by the cell permeable ROS sensor 5-(and-6)-chloromethyl-2′,7′-dichlorodihydrofluorescein diacetate (CM-H_2DCFDA). After treatments, cells were loaded with 100 μL PBS containing 10 μM CM-H_2DCFDA for 30 min in the dark. The fluorescence was quantified with Thermo Fisher Scientific Varioskan Flash (Waltham, MA, USA) at 490 nm excitation/530 nm emission.

4.6. Nuclear Staining with Hoechst 33258

Hoechst 33258 dye has been used extensively for the qualitative identification of nuclear apoptotic morphology [36]. Cells were plated on coverslips for 24 h before the treatments, fixed with 4% paraformaldehyde, and counterstained with 10 μg/mL Hoechst dye 33258.

4.7. Western Blotting Analysis

After the experimental treatments, cell pellets were lysed with a CelLytic™ MT cell lysis buffer. The cell lysates were analyzed by SDS-polyacrylamide gel electrophoresis (10% or 12%) and blotted onto polyvinylidene fluoride membranes. The proper primary antibodies at 1/1000 dilution and the secondary antibodies at 1/5000 dilution were used following the manufacturer's recommendations. GAPDH at 1/5000 dilution was used as an internal reference for an equal loading of protein.

4.8. Statistical Analysis

The values were standardized to the percentage of control and expressed as mean ± SEM in each experiment. The data were analyzed with a one-way ANOVA test with Tukey's or Dunnett's analysis using the GraphPad Prism 5 package (Graph Software, San Diego, CA, USA). The differences among treatments were considered statistically significant at $p < 0.05$.

5. Conclusions

In conclusion, we extracted CPO-B from cinnamon and demonstrated that it exerted significant neuroprotection against MPP^+-induced cytotoxicity. The underlying mechanisms included its ability to activate Erk1/2 phosphorylation, enhancing the ratio of Bcl-2/Bax expression and subsequently inhibiting caspase-3 activity. While the current study focused on the effects of CPO-B in mitigating the dysregulation of cellular survival signals induced by MPP^+, future studies are underway to investigate the effects of CPO-B on dopamine system dysfunction in animal models of PD. The present findings not only provide pharmacological and mechanistic evidence to support potential therapeutic application of procyanidins in PD but also indicate that the addition of cinnamon to foods or using CPO-B as a dietary supplement might benefit patients with PD.

Author Contributions: Conceptualization, Q.X. and X.W.; methodology, Z.C. and B.Z.; validation, Q.J. and Y.L.; formal analysis, H.L. and G.D.; investigation, Q.X. and X.W.; writing—original draft preparation, Q.X. and Z.C.; writing—review and editing, M.B.R. and X.W.; supervision, Q.J. and X.W.; funding acquisition, Q.X. and Y.L. All authors have read and agreed to the published version of the manuscript.

Funding: This research was funded by Funding of Shanghai Health Commission (2020JP006), Funding of Shanghai University of Traditional Chinese Medicine (GJ202101), and National Natural Science Foundation of China (81773791).

Institutional Review Board Statement: Not applicable.

Informed Consent Statement: Informed consent was obtained from all subjects involved in the study.

Data Availability Statement: All data used to support the findings of this study are included within the article and they are also available from the corresponding author upon request.

Conflicts of Interest: The authors declare that there are no competing interests.

Sample Availability: Not applicable.

References

1. Blesa, J.; Przedborski, S. Parkinson's disease: Animal models and dopaminergic cell vulnerability. *Front. Neuroanat.* **2014**, *8*, 155. [CrossRef] [PubMed]
2. Farzaei, M.H.; Tewari, D.; Momtaz, S.; Arguelles, S.; Nabavi, S.M. Targeting ERK signaling pathway by polyphenols as novel therapeutic strategy for neurodegeneration. *Food Chem. Toxicol.* **2018**, *120*, 183–195. [CrossRef] [PubMed]
3. Mebratu, Y.; Tesfaigzi, Y. How ERK1/2 activation controls cell proliferation and cell death: Is subcellular localization the answer? *Cell Cycle* **2009**, *8*, 1168–1175. [CrossRef] [PubMed]
4. Dexter, D.T.; Jenner, P. Parkinson disease: From pathology to molecular disease mechanisms. *Free Radic. Biol. Med.* **2013**, *62*, 132–144. [CrossRef]
5. Jha, S.K.; Jha, N.K.; Kar, R.; Ambasta, R.K.; Kumar, P. p38 MAPK and PI3K/AKT Signalling Cascades inParkinson's Disease. *Int. J. Mol. Cell. Med.* **2015**, *4*, 67–86.
6. Subramaniam, S.R.; Chesselet, M.F. Mitochondrial dysfunction and oxidative stress in Parkinson's disease. *Prog. Neurobiol.* **2013**, *106-107*, 17–32. [CrossRef]
7. Teng, L.; Kou, C.; Lu, C.; Xu, J.; Xie, J.; Lu, J.; Liu, Y.; Wang, Z.; Wang, D. Involvement of the ERK pathway in the protective effects of glycyrrhizic acid against the MPP+-induced apoptosis of dopaminergic neuronal cells. *Int. J. Mol. Med.* **2014**, *34*, 742–748. [CrossRef]
8. Zeng, W.; Zhang, W.; Lu, F.; Gao, L.; Gao, G. Resveratrol attenuates MPP(+)-induced mitochondrial dysfunction and cell apoptosis via AKT/GSK-3beta pathway in SN4741 cells. *Neurosci. Lett.* **2017**, *637*, 50–56. [CrossRef]
9. Wang, G.; Ma, W.; Du, J. beta-Caryophyllene (BCP) ameliorates MPP+ induced cytotoxicity. *Biomed. Pharmacother.* **2018**, *103*, 1086–1091. [CrossRef]
10. Hu, M.; Li, F.; Wang, W. Vitexin protects dopaminergic neurons in MPTP-induced Parkinson's disease through PI3K/Akt signaling pathway. *Drug Des. Dev. Ther.* **2018**, *12*, 565–573. [CrossRef] [PubMed]
11. Kovalevich, J.; Langford, D. Considerations for the use of SH-SY5Y neuroblastoma cells in neurobiology. *Methods Mol. Biol.* **2013**, *1078*, 9–21. [CrossRef] [PubMed]
12. Cheung, Y.T.; Lau, W.K.; Yu, M.S.; Lai, C.S.; Yeung, S.C.; So, K.F.; Chang, R.C. Effects of all-trans-retinoic acid on human SH-SY5Y neuroblastoma as in vitro model in neurotoxicity research. *Neurotoxicology* **2009**, *30*, 127–135. [CrossRef]
13. Rao, P.V.; Gan, S.H. Cinnamon: A multifaceted medicinal plant. *Evid. Based Complement. Altern. Med.* **2014**, *2014*, 642942. [CrossRef]
14. Frydman-Marom, A.; Levin, A.; Farfara, D.; Benromano, T.; Scherzer-Attali, R.; Peled, S.; Vassar, R.; Segal, D.; Gazit, E.; Frenkel, D.; et al. Orally administrated cinnamon extract reduces beta-amyloid oligomerization and corrects cognitive impairment in Alzheimer's disease animal models. *PLoS ONE* **2011**, *6*, e16564. [CrossRef]
15. Jana, A.; Modi, K.K.; Roy, A.; Anderson, J.A.; van Breemen, R.B.; Pahan, K. Up-regulation of neurotrophic factors by cinnamon and its metabolite sodium benzoate: Therapeutic implications for neurodegenerative disorders. *J. Neuroimmune Pharmacol.* **2013**, *8*, 739–755. [CrossRef]
16. Chen, L.; Yang, Y.; Yuan, P.; Yang, Y.; Chen, K.; Jia, Q.; Li, Y. Immunosuppressive Effects of A-Type Procyanidin Oligomers from Cinnamomum tamala. *Evid. Based Complement. Altern. Med.* **2014**, *2014*, 365258. [CrossRef]
17. Sun, P.; Wang, T.; Chen, L.; Yu, B.W.; Jia, Q.; Chen, K.X.; Fan, H.M.; Li, Y.M.; Wang, H.Y. Trimer procyanidin oligomers contribute to the protective effects of cinnamon extracts on pancreatic beta-cells in vitro. *Acta Pharmacol. Sin.* **2016**, *37*, 1083–1090. [CrossRef]
18. Chen, H.; Xu, J.; Lv, Y.; He, P.; Liu, C.; Jiao, J.; Li, S.; Mao, X.; Xue, X. Proanthocyanidins exert a neuroprotective effect via ROS/JNK signaling in MPTPinduced Parkinson's disease models in vitro and in vivo. *Mol. Med. Rep.* **2018**, *18*, 4913–4921. [CrossRef] [PubMed]

19. Ma, J.; Gao, S.S.; Yang, H.J.; Wang, M.; Cheng, B.F.; Feng, Z.W.; Wang, L. Neuroprotective Effects of Proanthocyanidins, Natural Flavonoids Derived From Plants, on Rotenone-Induced Oxidative Stress and Apoptotic Cell Death in Human Neuroblastoma SH-SY5Y Cells. *Front. Neurosci.* **2018**, *12*, 369. [CrossRef]
20. Ding, Y.; Xin, C.; Zhang, C.W.; Lim, K.L.; Zhang, H.; Fu, Z.; Li, L.; Huang, W. Natural Molecules From Chinese Herbs Protecting Against Parkinson's Disease via Anti-oxidative Stress. *Front. Aging Neurosci.* **2018**, *10*, 246. [CrossRef]
21. Strathearn, K.E.; Yousef, G.G.; Grace, M.H.; Roy, S.L.; Tambe, M.A.; Ferruzzi, M.G.; Wu, Q.L.; Simon, J.E.; Lila, M.A.; Rochet, J.C. Neuroprotective effects of anthocyanin- and proanthocyanidin-rich extracts in cellular models of Parkinsons disease. *Brain Res.* **2014**, *1555*, 60–77. [CrossRef] [PubMed]
22. Blade, C.; Aragones, G.; Arola-Arnal, A.; Muguerza, B.; Bravo, F.I.; Salvado, M.J.; Arola, L.; Suarez, M. Proanthocyanidins in health and disease. *BioFactors* **2016**, *42*, 5–12. [CrossRef] [PubMed]
23. Neilson, A.P.; O'Keefe, S.F.; Bolling, B.W. High-Molecular-Weight Proanthocyanidins in Foods: Overcoming Analytical Challenges in Pursuit of Novel Dietary Bioactive Components. *Annu. Rev. Food Sci. Technol.* **2016**, *7*, 43–64. [CrossRef] [PubMed]
24. Bitzer, Z.T.; Glisan, S.L.; Dorenkott, M.R.; Goodrich, K.M.; Ye, L.; O'Keefe, S.F.; Lambert, J.D.; Neilson, A.P. Cocoa procyanidins with different degrees of polymerization possess distinct activities in models of colonic inflammation. *J. Nutr. Biochem.* **2015**, *26*, 827–831. [CrossRef]
25. Lee, Y.A.; Kim, Y.J.; Cho, E.J.; Yokozawa, T. Ameliorative effects of proanthocyanidin on oxidative stress and inflammation in streptozotocin-induced diabetic rats. *J. Agric. Food Chem.* **2007**, *55*, 9395–9400. [CrossRef]
26. Serra, A.; Macia, A.; Romero, M.P.; Valls, J.; Blade, C.; Arola, L.; Motilva, M.J. Bioavailability of procyanidin dimers and trimers and matrix food effects in in vitro and in vivo models. *Br. J. Nutr.* **2010**, *103*, 944–952. [CrossRef] [PubMed]
27. Wang, L.; Yamashita, Y.; Komeda, S.; Saito, A.; Ashida, H. Absorption, metabolism, distribution and faecal excretion of B-type procyanidin oligomers in mice after a single oral administration of black soybean seed coat extract. *Food Funct.* **2018**, *9*, 5362–5370. [CrossRef]
28. Zeng, Y.X.; Wang, S.; Wei, L.; Cui, Y.Y.; Chen, Y.H. Proanthocyanidins: Components, Pharmacokinetics and Biomedical Properties. *Am. J. Chin. Med.* **2020**, *48*, 813–869. [CrossRef]
29. Serra, A.; Macia, A.; Rubio, L.; Angles, N.; Ortega, N.; Morello, J.R.; Romero, M.P.; Motilva, M.J. Distribution of procyanidins and their metabolites in rat plasma and tissues in relation to ingestion of procyanidin-enriched or procyanidin-rich cocoa creams. *Eur. J. Nutr.* **2013**, *52*, 1029–1038. [CrossRef]
30. Serra, A.; Macia, A.; Romero, M.P.; Angles, N.; Morello, J.R.; Motilva, M.J. Distribution of procyanidins and their metabolites in rat plasma and tissues after an acute intake of hazelnut extract. *Food Funct.* **2011**, *2*, 562–568. [CrossRef]
31. Xu, Q.; Chen, Z.; Zhu, B.; Wang, G.; Jia, Q.; Li, Y.; Wu, X. A-Type Cinnamon Procyanidin Oligomers Protect Against 1-Methyl-4-Phenyl-1,2,3,6-Tetrahydropyridine-Induced Neurotoxicity in Mice Through Inhibiting the P38 Mitogen-Activated Protein Kinase/P53/BCL-2 Associated X Protein Signaling Pathway. *J. Nutr.* **2020**, *150*, 1731–1737. [CrossRef]
32. Catalkaya, G.; Venema, K.; Lucini, L.; Rocchetti, G.; Delmas, D.; Daglia, M.; Filippis, A.D.; Xiao, H.; Quiles, J.L.; Xiao, J.; et al. Interaction of dietary polyphenols and gut microbiota: Microbial metabolism of polyphenols, influence on the gut microbiota, and implications on host health. *Food Front.* **2020**, *1*, 109–133. [CrossRef]
33. Hashimoto, R.; Yu, J.; Koizumi, H.; Ouchi, Y.; Okabe, T. Ginsenoside Rb1 Prevents MPP(+)-Induced Apoptosis in PC12 Cells by Stimulating Estrogen Receptors with Consequent Activation of ERK1/2, Akt and Inhibition of SAPK/JNK, p38 MAPK. *Evid. Based Complement. Altern. Med.* **2012**, *2012*, 693717. [CrossRef] [PubMed]
34. Pariyar, R.; Lamichhane, R.; Jung, H.J.; Kim, S.Y.; Seo, J. Sulfuretin Attenuates MPP(+)-Induced Neurotoxicity through Akt/GSK3beta and ERK Signaling Pathways. *Int. J. Mol. Sci.* **2017**, *18*, 2753. [CrossRef]
35. Chen, L.; Sun, P.; Wang, T.; Chen, K.; Jia, Q.; Wang, H.; Li, Y. Diverse mechanisms of antidiabetic effects of the different procyanidin oligomer types of two different cinnamon species on db/db mice. *J. Agric. Food Chem.* **2012**, *60*, 9144–9150. [CrossRef] [PubMed]
36. Janhom, P.; Dharmasaroja, P. Neuroprotective Effects of Alpha-Mangostin on MPP(+)-Induced Apoptotic Cell Death in Neuroblastoma SH-SY5Y Cells. *J. Toxicol.* **2015**, *2015*, 919058. [CrossRef] [PubMed]

molecules MDPI

Article

Effects of Airflow Ultrafine-Grinding on the Physicochemical Characteristics of Tartary Buckwheat Powder

Qinglian Xu [1], Faying Zheng [1], Xiaotong Cao [1], Ping Yang [1], Yage Xing [1,*], Ping Zhang [2], Hong Liu [1], Guangchao Zhou [2], Xiaocui Liu [1] and Xiufang Bi [1]

[1] Key Laboratory of Grain and Oil Processing and Food Safety of Sichuan Province, College of Food and Bioengineering, Xihua University, Chengdu 610039, China; xuqinglian01@163.com (Q.X.); zfyasdzxc2021@163.com (F.Z.); cxc442278103@163.com (X.C.); yangp960119@163.com (P.Y.); gyliuhong@126.com (H.L.); xiaocuiliu777@126.com (X.L.); bxf1221@163.com (X.B.)

[2] Huantai Biotechnology Co., Ltd., Chengdu 610225, China; htzp2019@163.com (P.Z.); zhou_423yan@163.com (G.Z.)

* Correspondence: xingyg@mail.xhu.edu.cn or xingyage1@163.com

Abstract: Five different ultrafine milled flours (UMFs) were prepared from Tartary buckwheat via airflow ultrafine-grinding at different grinding pressures. The airflow ultrafine-grinding resulted in marked differences in particle size (from 100 to 10 μm). The UMFs were all brighter in appearance (higher L*) than Tartary buckwheat common flour (TBCF). Illustrated by the example of 70 °C, the UMFs were also found to have a greater water holding capacity (from 4.42 g/g to 5.24 g/g), water solubility (from 12.57% to 14.10%), and water solubility index (from 5.11% to 6.10%). Moreover, as the particle sizes reduced, the moisture content decreased (from 10.05 g/100 g DW to 7.66 g/100 g DW), as did the total starch content (from 68.88 g/100 g DW to 58.24 g/100 g DW) and the protein content (from 13.16% to 12.04%). However, the grinding process was also found to have negative effects on the mineral content of the Tartary buckwheat. Additionally, several substantial variations were found in their hydration properties along with grinding pressure changes in the differently ground UMFs. Consequently, fine Tartary buckwheat powders of a bright yellow color, with superior food processing properties, were prepared in this study by airflow ultrafine-grinding.

Keywords: Tartary buckwheat powder; airflow ultrafine-grinding; grinding pressure; particle size; chemical compositions; morphology

Citation: Xu, Q.; Zheng, F.; Cao, X.; Yang, P.; Xing, Y.; Zhang, P.; Liu, H.; Zhou, G.; Liu, X.; Bi, X. Effects of Airflow Ultrafine-Grinding on the Physicochemical Characteristics of Tartary Buckwheat Powder. *Molecules* **2021**, *26*, 5841. https://doi.org/10.3390/molecules26195841

Academic Editor: Smaoui Slim

Received: 3 September 2021
Accepted: 22 September 2021
Published: 26 September 2021

1. Introduction

Buckwheat is a gluten-free pseudocereal belonging to the Polygonaceae family of plants. It contains protein of a high nutritional value, dietary fiber, vitamins, and minerals [1–3]. Tartary buckwheat (or bitter buckwheat) is the most commonly cultivated species, which has a higher concentration of bioactive compounds, such as flavonoids and phenolics [4–6]. Tartary buckwheat is, therefore, an important coarse cereal and is expected to become the target of many future planting industries [7]. In China, Tartary buckwheat is cultivated at high altitude, mountainous regions, and harsh climatic conditions [8]. In the Liangshan Yi Autonomous Prefecture, located in the Yunnan-Guizhou Plateau, the cultivation of Tartary buckwheat is an important source of income for local farmers in the high mountains and hilly areas, where other major crops may fail. By 2019, Tartary buckwheat plantations covered more than 2660 square kilometers in the Liangshan Prefecture.

In comparison to buckwheat, Tartary buckwheat has attracted more food science research interest due to its unique chemical composition and high efficiency as a functional food, including its anti-oxidative, anti-cancer, anti-hypertension, and anti-diabetic properties [9]. Tartary buckwheat is the crop that provides seven nutrients, with quantities of sugar, protein, fat, minerals, fiber, vitamins, and water, and its outstanding health and nutritional properties, thus endow it with a high edible value [10]. Furthermore, Tartary buckwheat extract exhibits high levels of antioxidant activity. As an exogenous antioxidant, it can promote or interact with endogenous antioxidants to form a cooperative network of cellular antioxidants [11]. One study found that the extract could increase the contribution of exercise-induced oxidative stress, thereby improving the physiological condition of humans [12]. The development of Tartary buckwheat as a beneficial nutritional supplement has thus far been suppressed by its bitter flavor and poor palatability, however, with the application of new food processing methods, products such as Tartary buckwheat-enriched biscuits, noodles, and beverages have been introduced and are increasing in popularity among consumers [13–17].

Superfine grinding technology, which is developing rapidly in food processing, can produce powder with superior properties to conventional particles and it is, thus, being used increasingly with a variety of food materials to improve the quality of powder [18]. Micronization is the process of reducing the average diameter of a solid material's particles [19,20]. Superfine grinding methods including airflow grinding, liquid flow grinding, low-temperature grinding, ball milling, ultrasonic disintegrator grinding, etc. [21]. To date, this ultrafine grinding technology has been applied in biotechnology and to achieve various foodstuffs, such as ginger powder, mushroom powder, green tea powder, hull-less barley bran, Agrocybe chaxingu Huang powder, and whole-wheat powder [22–27]. Wu et al. found that ultrafine grinding improved the solubility, oil-holding capacity, and brightness of Panax notoginseng powder. The contents of total saponins, minerals, phenols, and flavonoids were highest, and the antioxidant activity was the best, in the smallest particles of the Panax notoginseng powder [28]. In another study, the ultrafine powder of pomegranate peel showed strong specific surface energy, significant fluidity, and superior water-holding capacity, water solubility, polyphenols and flavonoids release, and DPPH radical scavenging activity, and it could be used as a new value-added product to provide health benefits in food processing because of its excellent dispersibility and dissolution [29]. These studies suggest that ultrafine grinding is a useful tool for producing ultrafine powders with good surface properties such as dispersion and solubility [30]. Moreover, it has been revealed that the physical and functional properties of buckwheat powder are affected differently by different milling methods, which thus affect the properties of products [31,32]. However, very limited information is available on the effects of airflow ultrafine-grinding on the physical characteristics of Tartary buckwheat powder.

The aim of this work is, thus, to investigate the application of airflow ultrafine-grinding technology on Tartary buckwheat. In this paper, five different ultrafine milled flours (UMFs) were prepared from Tartary buckwheat via airflow ultrafine-grinding at different grinding pressures. The particle size analysis and morphological observation were firstly conducted. Furthermore, the physicochemical characteristics of obtained different particles were analyzed, including chemical composition, hydration properties, and mineral content, in comparison to those of commonly ground Tartary buckwheat particles.

2. Results and Discussion

2.1. Particle Size Analysis and Morphological Observation

Analysis of the distribution of the UMF particle sizes obtained via the airflow ultrafine-grinding methods (Figure 1) showed them to be between 10 μm and 80 μm, demonstrating them to be within the size range of ultrafine powder [33].Furthermore, the Dv values of the Tartary buckwheat samples are summarized in Table 1. Generally, D10, D50, and D90 are used to characterize particle size distributions, and are the equivalent volume diameters at 10%, 50% and 90% of the cumulative volume, respectively [34]. It was found that the mean particle size of the TBCF was 100 μm, which was almost three times, five times, and two times that of the UMF for D10, D50, and D90, respectively. The D50 values of the UMFs and the TBCF were 10.97 μm, 11.19 μm, 12.24 μm, 13.18 μm, 13.07 μm, and 61.84 μm. Additionally, with increases in grinding pressure, the particle sizes of the ultrafine powder decreased; however, there were no significant ($p > 0.05$) differences between the samples at a grinding pressure at 0.1 MPa (No. 1 and No. 2), and when the grinding pressure changed by 0.2 MPa, the particle sizes of the UMF showed a significant ($p < 0.05$) decline. These results indicated that the increase in grinding pressure produced a finer UMF comprised of small particles, while the uniformity of the powder was also improved.

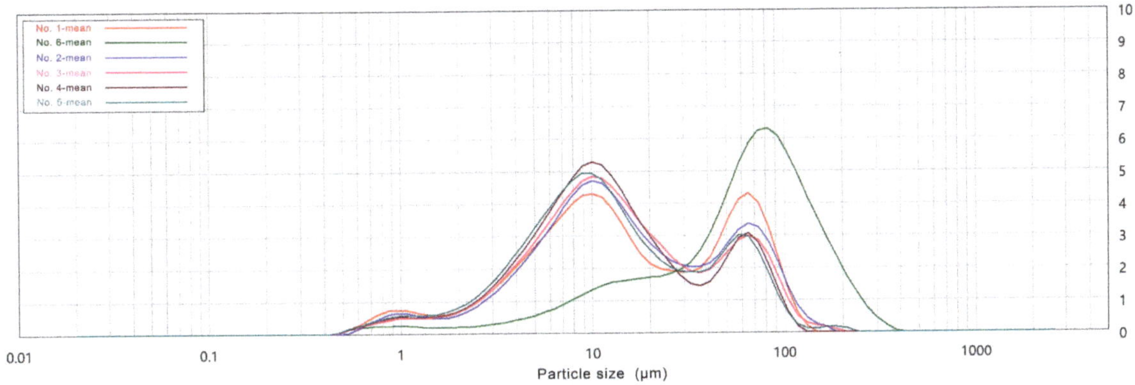

Figure 1. Particle size distribution of different Tartary buckwheat flour via airflow ultrafine-grinding at different grinding pressures. (No. 1: Tartary buckwheat ultrafinemilled flour No. 1; No. 2: Tartary buckwheat ultrafinemilled flour No. 2; No. 3: Tartary buckwheat ultrafinemilled flour No. 3; No. 4: Tartary buckwheat ultrafinemilled flour No. 4; No. 5: Tartary buckwheat ultrafinemilled flour No. 5; No. 6: Tartary buckwheat common flour No. 6).

To further understand the morphological properties of Tartary buckwheat powders, the samples in this study were analyzed using SEM and differences were found in the shapes, sizes, and granule surfaces of the powders with and without the ultrafine milling treatment. As shown in Figure 2, the treated Tartary buckwheat flour particles were spherical, oval, regular, and polygonal in shape, and were also highly uniform. In addition, the surface roughness of No. 1 (grinding pressure at 0.8 MPa) was found to be greater than that of the other five superfine powders (Figure 2A), which were seriously damaged because of mechanical force, with rough surfaces and a large number of residual fibrous fragments and protein fragments. However, the images in Figure 2F(b) reveal that the powder particles of the TBCF were uniform, arranged in an orderly manner with more pores. Moreover, the particle sizes of the TBCF sample were significantly larger than those of the UMFs, as verified by morphology observation (Figure 2F). This suggested that airflow ultrafine-grinding influenced the morphological characteristics of the Tartary buckwheat flour, increasing its surface area and leading to the increase in surface properties. Moreover, the changes in the particle size of the Tartary buckwheat flour had significant effects on its physico-chemical properties.

Ultrafine grinding technology can improve the structural characteristics of Tartary buckwheat powders by reducing their particle size, and the morphology of Tartary buckwheat flour is also changed by different grinding pressures [35]. In our study, the results indicated that superfine pulverization using a mini-type airflow pulverization instrument could reduce the sizes of the Tartary buckwheat powder particles. The ultrafine pulverization process produces more energy with higher reduction efficiency, making it easier for Tartary buckwheat flour to be shaped into a powder of smaller particle sizes, while the rough surface of the powder is created by the strong grinding force, which destroys the integrity of the starch granules during the ultrafine grinding process, producing starch granules with smaller fragments [36]. The UMF, with its narrower particle size distribution and more even particle size obtained via airflow ultrafine-grinding treatments, is more easily processed to significantly improve the quality and taste of products. Similar results were observed in red grape pomace powders (Zhao et al., 2015) and whole-wheat flour (Niu et al., 2014), while contrary results were discovered in red grape pomace powders (Zhao et al., 2015), in which superfine grinding was found to smooth the surface [27,37].

Such differences may be caused by varying experimental conditions, such as working frequency, feeding velocity, feeding pressure, and grinding pressure. In our experiment, the series of grinding pressures were set before experimentation began.

Ultrafine grinding was applied in this study. In general, the higher the pressure, the smaller the particle size [26]. As expected, the particle size decreased with increasing grinding pressure; however, there were no significant differences in the particle sizes of the No. 1 and No. 2 samples, indicating that grinding pressure may not have had much effect on the smaller powders. This result is consistent with the findings of Zhang et al., in which the powder of Agrocybe chaxingu Huang was prepared using ultrafine grinding [26]. The results showed that the particle size of the powder decreased from 110 μm to 20 μm when the grinding pressure was increased from 0.3 MPa to 0.4 MPa. However, when the grinding pressure continued to increase from 0.1 MPa to 0.5 MPa, the particle sizes of the Agrocybe chaxingu Huang hardly changed. While a substantial increase in grinding pressure did indeed change the particle size of the Tartary buckwheat flour in this study, it should also be noted that the smaller the particles, the less significant the impact of any further increase in grinding pressure on their size.

Table 1. Particle size distribution of Tartary buckwheat flour with different particle size [1].

Sample	D10 (μm)	D50 (μm)	D90 (μm)
No. 1	3.03 ± 0.01c	10.97 ± 0.04c	60.64 ± 0.44d
No. 2	3.21 ± 0.05bc	11.19 ± 0.15c	61.07 ± 0.59d
No. 3	3.33 ± 0.05bc	12.24 ± 0.18bc	67.49 ± 1.15c
No. 4	3.42 ± 0.04b	13.18 ± 0.07b	72.27 ± 0.38b
No. 5	2.96 ± 0.02c	13.07 ± 0.06b	71.83 ± 0.46b
No. 6	8.93 ± 0.48a	61.84 ± 2.36a	151.97 ± 2.20a

[1] Values in the same column with different letters are significantly different at $p < 0.05$. (Abbreviations: No. 1: Tartary buckwheat ultrafinemilled flour No. 1; No. 2: Tartary buckwheat ultrafinemilled flour No. 2; No. 3: Tartary buckwheat ultrafinemilled flour No. 3; No. 4: Tartary buckwheat ultrafinemilled flour No. 4; No. 5: Tartary buckwheat ultrafinemilled flour No. 5; No. 6: Tartary buckwheat common flour No. 6).

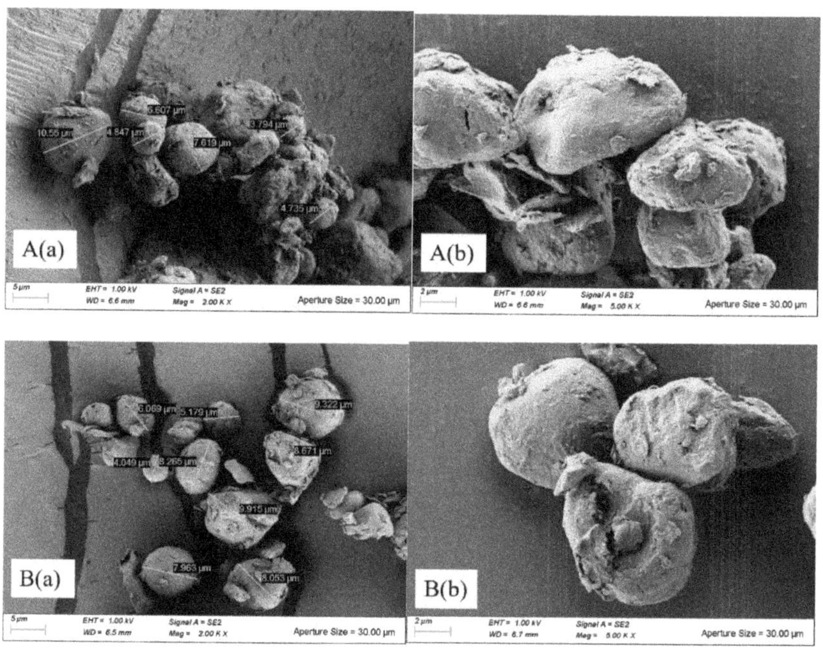

Figure 2. *Cont.*

Figure 2. Scanning electron microscopy observation of different Tartary buckwheat flour via airflow ultrafine-grinding at different grinding pressures. (Tartary buckwheat ultrafinemilled flour No. 1 (**A**); Tartary buckwheat ultrafinemilled flour No. 2 (**B**); Tartary buckwheat ultrafinemilled flour No. 3 (**C**); Tartary buckwheat ultrafinemilled flour No. 4 (**D**); Tartary buckwheat ultrafinemilled flour No. 5 (**E**); Tartary buckwheat common flour No. 6 (**F**); (**a**) is magnified 2000 times and (**b**) 5000 times.)

2.2. Color Difference Analysis

Color and particle size are important qualities in food powders. Color is an important parameter for most food products and is usually a consideration for consumers in their evaluation and selection of products. The investigation by Hu et al. indicated that particle size had a significant effect on green tea powders [24]. The influence of various particle sizes on the color of Tartary buckwheat flour in this study is presented in Table 2, in which 'L' expresses brightness, while 'a' and 'b' are chromaticity coordinates, with +a indicating the red direction, −a the green direction, +b the yellow direction, and −b the blue direction [38]. As illustrated by No. 5 in Table 2, compared to the TBCF sample, 'L' values increased significantly ($p < 0.05$) from 88.95 to 93.37, whereas 'a' and 'b' values decreased markedly ($p < 0.05$) from 2.36 to 1.70, and 22.07 to 17.29, respectively. These results indicate that ultrafine grinding improved the color of the Tartary buckwheat flour, making it appear brighter, while the appearance of the TBCF was a dim yellow in comparison. Moreover, as shown in Table 3, 'L' increased via the treatment No. 5 to No. 3 but decreased from treatment No. 3 to No. 2, then finally increased from No. 2 to No. 1. By contrast, 'a' and 'b' showed the opposite trend. In addition, with increasing grinding pressure, No. 1 expressed the highest level of brightness, indicating that the parameter of grinding pressure had an effect on the brightness of the Tartary buckwheat.

In this investigation work, the brightness of Tartary buckwheat flour was negatively correlated with particle size. The smaller the particle size, the greater the relative surface area, the better the reflective effect, and the greater the brightness value. Therefore, Tartary buckwheat flour with the highest level of brightness was obtained by the airflow ultrafine-grinding treatments. Moreover, the grinding treatments' improvement of the color of the Tartary buckwheat may be due to the fact that surface area increased with the decrease in particle size, and the internal structure of the cellulose and hemicellulose were exposed, which affected the color of the powder [39]. The same results were obtained in the experiments of Hu and Li, in which the colors of green tea powder and soybean residue powder, respectively, changed significantly with decreases in particle size during superfine grinding [24,36]. Huang et al. also found that the superfine powder of the Moringa oleifera leaf was obviously brighter than the Moringa leaf powder of millimeter grade, indicating that particle size and superfine grinding technology had a significant effect on the brightness of those powders [40]. Notably, in this study, neither the brightness nor the chromaticity coordinates of the Tartary buckwheat flour were positively correlated with grinding pressure. Color changes could be related to the heat in the UMF process. Since the smaller particles are subjected to more mechanical and high temperature damage during grinding, the pigment in the Tartary buckwheat powder was degraded, resulting in increased whiteness. Overall, UMF of smaller particle sizes was relatively brighter, and either more green or more blue in color. Importantly, since the color of a powder has a great influence on its processed products, a compound product of Tartary buckwheat may be more easily accepted by consumers because of its brightness [41].

Table 2. Color difference of different Tartary buckwheat flour obtained from airflow ultrafine-grinding [1,2].

Sample	L*	a*	b*
No. 1 (UMF)	95.52 ± 0.07a	1.02 ± 0.09d	14.90 ± 0.21d
No. 2 (UMF)	93.91 ± 0.07c	1.74 ± 0.04b	17.29 ± 0.11b
No. 3 (UMF)	95.32 ± 0.06a	1.47 ± 0.04c	15.57 ± 0.13c
No. 4 (UMF)	94.80 ± 0.03b	1.44 ± 0.03c	15.61 ± 0.20c
No. 5 (UMF)	93.37 ± 0.15d	1.70 ± 0.04b	17.29 ± 0.26b
No. 6 (TBCF)	88.95 ± 0.26e	2.36 ± 0.12a	22.07 ± 0.37a

[1] Data are expressed as the mean ± standard deviation. Values in the same column with different letters are significantly different at $p < 0.05$. [2] The 'L' is the indicator of lightnesse-darkness; the 'a' is the indicator of greenness (when it is a minus value) and redness (plus value); the 'b' is the indicator of blueness (minus value) and yellowness (plus value).

Table 3. Effects of airflow ultrafine-grinding on the basic components in Tartary buckwheat flour [1].

Nutrient Contents	No. 1 (UMF)	No. 2 (UMF)	No. 3 (UMF)	No. 4 (UMF)	No. 5 (UMF)	No. 6 (TBCF)
Moisture (g/100gDW)	7.66 ± 0.02c	8.22 ± 0.04b	7.42 ± 0.03d	8.15 ± 0.04b	8.19 ± 0.02b	10.05 ± 0.11a
Total starch (g/100gDW)	58.24 ± 0.38e	62.19 ± 0.42d	65.4 ± 0.47c	67.12 ± 0.41b	67.64 ± 0.40b	68.88 ± 0.54a
Protein (%)	12.32 ± 0.10d	12.91 ± 0.15b	12.59 ± 0.05c	12.04 ± 0.03e	13.16 ± 0.07a	13.03 ± 0.03ab
Crude fat (g/100gDW)	1.18 ± 0.12a	1.16 ± 0.14a	1.17 ± 0.14a	1.26 ± 0.11a	1.27 ± 0.22a	1.24 ± 0.02a
Crude fibre (%)	—	—	—	—	—	0.30 ± 0.10
Ca (mg/L)	13.41 ± 0.15c	14.14 ± 0.09b	13.95 ± 0.24b	13.37 ± 0.07c	12.16 ± 0.07d	14.65 ± 0.08a
Fe (mg/L)	1.85 ± 0.03a	1.48 ± 0.02c	1.16 ± 0.01d	0.63 ± 0.00e	0.39 ± 0.01f	1.80 ± 0.03b
Mn (mg/L)	0.54 ± 0.01ab	0.47 ± 0.00c	0.52 ± 0.02b	0.56 ± 0.02a	0.54 ± 0.00ab	0.54 ± 0.01ab
Zn (mg/L)	1.21 ± 0.02a	1.18 ± 0.01b	1.16 ± 0.03b	1.15 ± 0.02b	0.99 ± 0.00c	1.24 ± 0.01a
Cu (mg/L)	0.20 ± 0.00a	0.20 ± 0.00a	0.19 ± 0.00b	0.20 ± 0.01a	0.19 ± 0.00b	0.20 ± 0.00a

[1] data are expressed as the mean ± standard deviation. Values in the same row with different letters are significantly different ($p < 0.05$).

2.3. Chemical Composition of Tartary Buckwheat Flour

The effects of airflow ultrafine-grinding on the chemical compositions of the flour samples are summarized in Table 3, in which it is evident that they varied greatly. The maximum UMF moisture content (8.22 g/100 g) was significantly ($p < 0.05$) lower than that of the TBCF (10.05 g/100 g). A significant difference ($p < 0.05$) was also observed in the protein content among all samples. The crude fiber content of the UMF could not be determined because of their particle size; thus, there was no significant difference ($p > 0.05$) in the crude fat content of the UMF and TBCF. Moreover, the total starch content, which accounted for the proportion of TBCF, was approximately 84.55% in the UMF. Furthermore, changes in the mineral content of the Tartary buckwheat flour were evaluated after the airflow ultrafine-grinding treatments. As shown in Table 3, Ca content in the UMF was significantly ($p < 0.05$) lower than that in the TBCF, as was the content of Fe, with the exception of No. 1. There were no obvious increasing or decreasing trends in the Mn, Zn, or Cu contents among the UMFs treated by different grinding pressures. However, the contents of protein and total starch decreased with increasing grinding pressure, from 13.16 to 12.04% and from 67.64 g/100 g DW to 58.24 g/100 g DW, respectively.

The variations in the composition of the samples may be attributed to different operating conditions. Tartary buckwheat flour is influenced by the heat produced by the machine during milling and it is, therefore, likely that there would be changes in the contents of fat, starch, protein, and moisture during the grinding process. These results are consistent with a study by Liu et al., who found that the quality of protein and lipids could be changed by high temperatures during milling [32]. In addition, by studying the effects of superfine grinding on the quality characteristics of whole-wheat flour and its raw noodle product, Niu found that grinding treatments could induce the physical transformation of starch granules, including a reduction in the starch crystal area and an increase in starch damage, leading to changes in the properties of the starch [27]. Here, in summary, the mineral content of the Tartary buckwheat powder decreased after airflow ultrafine-grinding, and the Ca and Fe contents changed significantly, possibly because the friction of the materials bouncing off each other caused heat. Similar results were found by Liu et al. in their study of the effects of grinding methods on the chemical composition and antioxidant capacity of Tartary buckwheat powder, in which they compared the effects of four different grinding methods on the mineral content. The results showed that the mineral content of the Tartary buckwheat powder obtained by stone grinding was significantly higher than that of the ultrafine powder, almost triple that of roller milling and quadruple that of wet grinding [32].

2.4. Hydration Properties

2.4.1. Water Holding Capacity Analysis

The water holding capacity (WHC) of the five different particle-sized UMFs and TBCF sample are shown in Figure 3. It was initially discovered that the WHC increased with the decreasing size of Tartary buckwheat particles. The WHC increased at first and then decreased within the measured temperature range (from 50 °C to 80 °C) and reached its highest level at 70 °C, at which it was recorded in the following order: 5.24 g/g (No. 2) > 4.97 g/g (No. 3) > 4.75 g/g (No. 4) > 4.65 g/g (No. 5) > 4.54 g/g (No. 6) > 4.42 g/g (No. 1). With the exception of No. 1, the WHC was higher in all UMFs than that of the TBCF at all temperatures, indicating that the WHC of Tartary buckwheat flour was improved during airflow ultrafine-grinding. In addition, with increasing grinding pressure, the WHC of the ultrafine powders increased at the same temperature except for sample No. 1. This result may have been because, when the particle size decreased to a certain extent, the gaps between them also decreased, resulting in a lower ability of the powder to retain water. More importantly, with the increase in temperature, the WHC of the superfine powder increased significantly, indicating that it had improved hydrophilic ability, which can prevent the loss of water and delay the aging of starch in the hot process.

WHC refers to the quantity of water that is bound to the fibers without the application of any external force (except for gravity and atmospheric pressure) [42]. Airflow ultrafine-grinding technology produces powders with a greater WHC than TBCF because the grinding process lowers the interfacial tension and exposes more polar groups, surface area, and water-binding sites to the surrounding water [43]. The significant improvement in the WHC of the UMF may be due to changes in the surface properties of the superfine powders, such as the increase in surface area, gaps, and liquid holding space. Zhao et al. found that superfine grinding could alter or destroy the internal structure of ginger, thus resulting in the exposure of its hydrophilic groups and subsequent easier integration with water, which eventually increased its WHC [22]. Similar results were found in the studies of Astragalus membranaceus powder, pear pomace powder, and Lycium ruthenicum Murray powder [35,44–46].

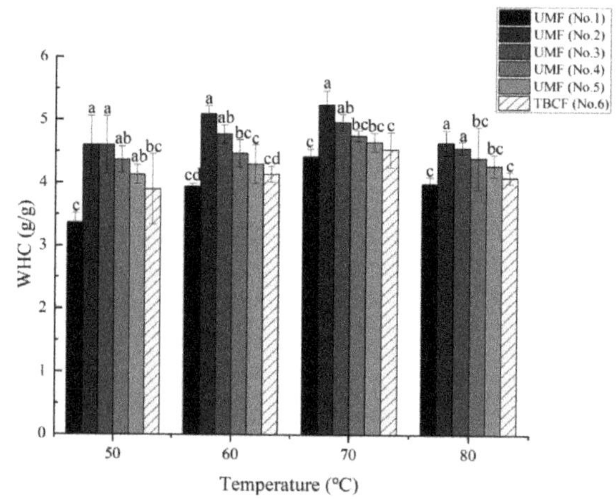

Figure 3. Effects of airflow ultrafine-grinding on the water holding capacity in Tartary buckwheat flour. The values represented by the different letters are significantly different at $p < 0.05$ compared with each other among the same temperature.

2.4.2. Swelling Power and Solubility Analysis

Solubility and the degree of expansion during steaming treatment reflect the quality of Tartary buckwheat flour. The swelling power of TBCF and ultrafine Tartary buckwheat flour at different temperatures (from 50 °C to 90 °C) are shown in Figure 4. At the same temperature, almost all UMF samples exhibited a higher SPI than that of the TBCF. When the temperature reached 70 °C, swelling power reached its highest levels, with indexes recorded in the following order: 6.10% (No. 2) > 5.77% (No. 3) > 5.45% (No. 4) > 5.33% (No. 5) > 5.19% (No. 6) > 5.11% (No. 1). Furthermore, for the samples from No. 5 to No. 2, with increasing grinding pressure, the swelling power of the samples increased at the same temperature. The results showed that the superfine milling significantly changed the swelling power of the Tartary buckwheat flour and the prevention of water dispersion incapacitation was also enhanced. Similar results were also obtained for WS, shown in Figure 5. The WS of the Tartary buckwheat flour increased at first and then decreased slightly within the measured temperature range (from 50 °C to 80 °C), reaching its highest level at 70 °C, at which WS was recorded in the following order: 14.10% (No. 2) > 13.90% (No. 3) > 13.43% (No. 1) > 12.87% (No. 4) > 12.77% (No. 5) > 12.57% (No. 6). These results, thus, also showed that the WS of the Tartary buckwheat flours were improved by airflow ultrafine-grinding technology.

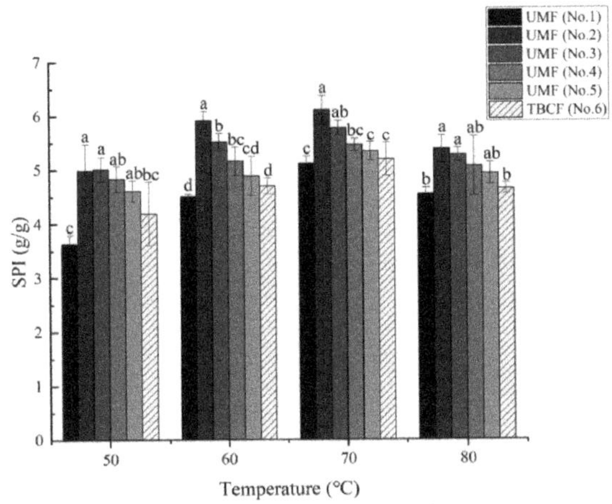

Figure 4. Effects of airflow ultrafine-grinding on the swelling power in Tartary buckwheat flour. The values represented by the different letters are significantly different at $p < 0.05$ compared with each other among the same temperature.

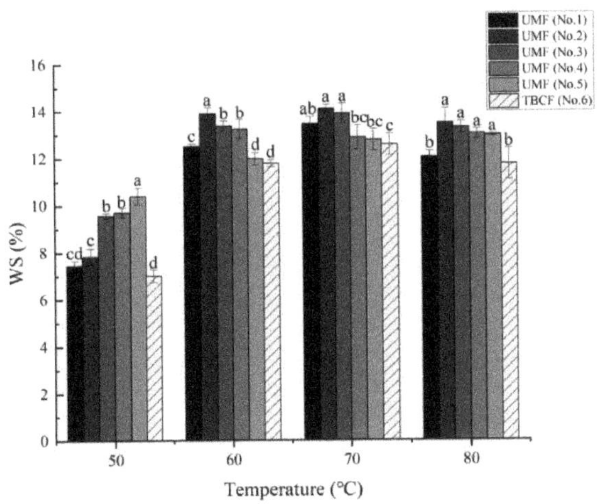

Figure 5. Effects of airflow ultrafine-grinding on the water solubility in Tartary buckwheat flour. The values represented by the different letters are significantly different at $p < 0.05$ compared with each other among the same temperature.

Water solubility and swelling power provide a measure of the degree of reciprocal action between starch chains in the amorphous and crystalline domains [47–49]. In our study, the swelling power of the ultrafine powder was found to be generally higher than that of the TBCF, possibly indicating that the starch structure of the Tartary buckwheat powder was destroyed, and that the content of amylopectin had increased in the process of airflow ultrafine-grinding [50]. This is similar to the results of Huang et al., in which the double helix structure of starch was destroyed by ultrafine grinding and the molecular structure of amylose was depolymerized [51]. In addition, water molecules can easily enter the inner regions of starch particles after ultrafine grinding, which could reduce the interaction between amylose and amylopectin molecules [52]. This may explain the higher swelling power of Tartary buckwheat after ultrafine grinding compared with that of common buckwheat. Similar observations have been made in superfine powders of buckwheat starch [51]. In terms of WS, the index increased as the size of the Tartary buckwheat particles decreased, which may be attributed to the larger surface area and high charge density [53]. Furthermore, the porosity of the powders was increased after superfine grinding, thus raising the hydration rate and bioavailability of components [48]. Similar phenomena were observed in the study of Vaccinium bracteatum Thunb leaves powder. However, these results also indicate that the paste phenomenon is more obvious in UMF processing.

It is worth noting that not all samples showed higher hydration properties after superfine grinding at the same temperature, with the exception of a No. 1 sample (prepared under the maximum pressure of 0.8 MPa). Generally speaking, with the increase in grinding pressure, the particle sizes in the ultrafine powder decreased. This is mainly because, at a certain range, the increase in grinding pressure improves the velocity of airflow at the outlet of the nozzle, causing the material to gain more kinetic energy. When particle sizes decrease to a certain value, the particles agglomerate with the increase in the powder surface energy. This may explain why the hydration properties of No. 1 sample were almost the same as those of the TBCF after ultrafine grinding. Therefore, the surface properties of powder may change dramatically after superfine grinding under high crushing pressure [54].

3. Materials and Methods

3.1. Main Material

The Tartary buckwheat powder used in this work was produced by Huantai Biotechnology Co., Ltd. (Sichuan, China). Sodium hydroxide, petroleum ether, sulfuric acid, hydrochloric acid, anhydrous ethanol, and nitric acid were purchased from Cologne Co., Ltd. (Chengdu, China), among which the nitric acid was of guaranteed grade. Iron (Fe), calcium (Ca), zinc (Zn), copper (Cu), and other standard solutions were obtained from the National Research Center for Standard Materials.

3.2. Preparation of Tartary Buckwheat Ultrafine Milled Flours

Tartary buckwheat common flour (TBCF) was passed through a 120-mesh sieve, and then dried in a constant temperature blast oven at 50 °C for 1 h. The TBCF was further pulverized using an airflow pulverization instrument (YQ50-1, Saishan Powder Machinery Manufacturing Co., Ltd., Shanghai, China), and the grinding pressure was variously regulated to 0.4 MPa, 0.5 MPa, 0.6 MPa, 0.7 MPa, and 0.8 MPa, finally resulting in five different ultrafine milled flours (UMF), correspondingly labeled No. 5, No. 4, No. 3, No. 2, and No. 1, respectively. The TBCF was used as the control sample, labeled No. 6.

3.3. Determination of the Chemical Compositions

Moisture, protein, and crude fiber in all samples were determined using the method described by Pandord, with some modifications [55]. Moisture content was determined after drying the samples at 105 °C for 4 h in an air oven. Proximate analysis of protein content was performed using the Kjeldahl method, with a conversion factor of N $\times$ 6.25. Crude fiber was determined by the filtration method, in which defatted samples were boiled in concentrated sulfuric acid for 30 min, in potassium hydroxide (instead of sodium hydroxide) for 30 min, then dried and, finally, reduced to ash in a muffle furnace. In addition, the crude fat content was assayed as described by Weber, with slight modifications [56]. Crude fat was obtained via extraction from the samples with petroleum ether (instead of hexane) for 6 h in a Soxhlet apparatus, then dried in a water bath for 1 h at 100 °C, after which the weight differences of the samples were determined.

The starch in Tartary buckwheat (which accounts for approximately 70% of its content) is particularly important in determining the textural properties of its products [57]. In this study, the starch content was analyzed using the methods described by Senanayake with some modifications [58]. Briefly, a flour sample of 0.3 g was washed, first, with petroleum ether and then with ethanol, and the residue was subsequently filtrated in a solution of 250 mL water and 30 mL hydrochloric acid. The mixture was then heated for 2 h in a conical flask fitted with a reflux condenser, after which it was cooled and neutralized with sodium hydroxide (NaOH). The volume was increased to 500 mL with distilled water and the sugar formed was determined as dextrose using the Lane and Eynon method for the estimation of reducing sugars. The dextrose multiplied by 0.9 was taken as starch.

Minerals play an important role in the physiological functioning of the body, especially in the growth and metabolic regulation process [59]. The mineral content in the Tartary buckwheat flour was assayed as described by Özcan, with slight modifications [60]. Approximately 0.4 g dried and ground sample was placed in a digestion tank, to which 7 mL pure nitric acid (HNO3) was added. This mixture was then counteracted in a microwave digestion instrument at different temperatures. Mineral concentrations were determined via inductively coupled plasma (Avio 200 ICP Optical Emission Spectrometer; PerkinElmer Instrument Co., Ltd., Singapore).

3.4. Morphologies Observation of Tartary Buckwheat Flour

The particle size distribution of the Tartary buckwheat flour samples was determined using a laser particle size analyzer (Bettersizer 2600; Dandong Baxter Instrument Co., Ltd., Liaoning, China) in the wet method mode. In addition, the morphological characterization of Tartary buckwheat flour particles was performed on images acquired using a scanning electron microscope (SEM) (ZEISS Gemini 500; Xiangyan Co., Shanghai, China) at an accelerating voltage of 1.00 kV. A small amount of Tartary buckwheat flour adhered to the surface of the double-sided conductive carbon adhesive tape and an aurilave was used to remove those particles. The sample was then subjected to a gold spray treatment under vacuum conditions prior to testing.

3.5. Color Difference Analysis

The colors of the different Tartary buckwheat flour samples were determined using the VeriVide DigiEye system (Yunding International Trade Co., Ltd., Shanghai, China), and the device was calibrated with a standard white surface calibration plate. Six varieties of Tartary buckwheat flour were placed on a plate and then selected randomly. The L*, a*, and b* values for each sample, indicating lightness, redness (+)/(−) greenness, and yellowness (+)/(−) blueness, respectively, were recorded.

3.6. Determination of Hydration Properties

The water holding capacity (WHC), water solubility (WS) and water solubility index (SPI) of the Tartary buckwheat flour were measured using the method reported by Tsai, Li, and Lii [61]. Approximately 0.1 g of Tartary buckwheat flour and 10 mL of deionized water were placed together in a beaker and mixed thoroughly by an ultrasonic instrument. The beaker was kept in a water bath (at temperatures of 50 °C, 60 °C, 70 °C, and 80 °C) for 30 min, respectively, and then centrifuged for 20 min at 3000 r/m). The weighing bottle (m_1) was weighed, and the collected supernatant was transferred to the pre-weighed weighing bottle, dried at 105 °C until the mass difference between the two was no more than 2 mg, and then weighed (m_2). It was finally weighted (m_3) after the supernatant had been discarded. The WHC, WS, and SPI were calculated by the following equation:

$$WHC = m_3/0.1 \tag{1}$$

$$WS = (m_2 - m_1)/0.1 \times 100\% \tag{2}$$

$$SPI = m_3/0.1 \times (1 - WS) \tag{3}$$

3.7. Statistical Analysis

All data were presented as the mean standard deviation of at least three replicates and analysis performed was one-way analysis of variance (ANOVA) followed by the Duncan' s multiple comparison test using SPSS version 25.0 software.

4. Conclusions

In this study, airflow ultrafine-grinding treatment was found to exert significant influence on the physicochemical characteristics and mineral content of Tartary buckwheat powder. Smaller and brighter Tartary buckwheat powders, with greater food processing properties, were obtained after superfine grinding. With increases in grinding pressure, the particle sizes of the Tartary buckwheat powder decreased and reached equilibrium finally. Moreover, the superfine grinding process decreased moisture content, crude fat content, and protein content, which is beneficial for health and weight loss. The ultrafine Tartary buckwheat powder was found to retain the essential nutritional characteristics of the original Tartary buckwheat; however, its mineral content was affected and it is, therefore, important to consider both nutrition and the cost of the product when processing ultrafine Tartary buckwheat powder. Hydration properties were significantly improved in the superfine Tartary buckwheat powder and the increase in swelling power could benefit satiation; however, the higher water solubility also suggests that the superfine powder is more likely to create a soup paste during cooking. Further studies could be conducted on the polyphenol and lipid content obtained from Tartary buckwheat powders with different particle sizes. Superfine grinding technology could be used to broaden the application of Tartary buckwheat powders in functional foods.

Author Contributions: Conceptualization, P.Y., X.C. and F.Z.; methodology, X.C. and F.Z.; software, X.C.; validation, X.C., X.B. and X.L.; formal analysis, X.C.; investigation, Y.X.; data curation, X.C.; writing-original draft preparation, F.Z., X.C., P.Y. and Q.X.; visualization, P.Y.; supervision, H.L.; project administration, Y.X.; funding acquisition, Y.X.; resources, Y.X., P.Z. and G.Z. All authors have read and agreed to the published version of the manuscript.

Funding: This research was funded by the Science and Technology Program of Sichuan Province (2018NZ0097 and 19ZDYF1593), the Key project of Education Department of Sichuan Province (18ZA0456), and the Key Laboratory Project of Grain and Oil Engineering and Food Safety in Xihua University (szjj2017-105).

Institutional Review Board Statement: Not applicable.

Informed Consent Statement: Not applicable.

Data Availability Statement: Not applicable.

Conflicts of Interest: All of the authors of this study have no relevant conflicts of interest to disclose.

Sample Availability: Not applicable.

References

1. Morishita, T.; Yamaguchi, H.; Degi, K. The contribution of polyphenols to antioxidative activity in common buckwheat and Tartary buckwheat grain. *Plant Prod. Sci.* **2007**, *10*, 99–104. [CrossRef]
2. Sanchez, A.; Schuster, T.M.; Burke, J.M.; Kron, K.A. Taxonomy of Polygonoideae (Polygonaceae): A new tribal classification. *Taxon* **2011**, *60*, 151–160. [CrossRef]
3. Wijngaard, H.H.; Arendt, E.K. Buckwheat. *Cereal Chem.* **2006**, *83*, 391–401. [CrossRef]

4. Kim, S.J.; Zaidul, I.S.M.; Suzuki, T.; Mukasa, Y.; Hashimoto, N.; Takigawa, S.; Yamauchi, H. Comparison of phenolic compositions between common and Tartary buckwheat (Fagopyrum) sprouts. *Food Chem.* **2008**, *110*, 814–820. [CrossRef] [PubMed]
5. Ahmed, A.; Khalid, N.; Ahmad, A.; Abbasi, N.A.; Latif, M.S.Z.; Randhawa, M.A. Phytochemicals and biofunctional properties of buckwheat: A review. *J. Agric. Sci.* **2014**, *152*, 349–369. [CrossRef]
6. Bhinder, S.; Singh, B.; Kaur, A.; Singh, N.; Kaur, M.; Kumari, S.; Yadav, M.P. Effect of infrared roasting on antioxidant activity, phenolic composition and Maillard reaction products of Tartary buckwheat varieties. *Food Chem.* **2019**, *285*, 240–251. [CrossRef]
7. Ruan, J.; Zhou, Y.; Yan, J.; Zhou, M.; Woo, S.H.; Weng, W.; Zhang, K. Tartary buckwheat: An under-utilized edible and medicinal herb for food and nutritional security. *Food Rev. Int.* **2020**, 1–15. [CrossRef]
8. Bonafaccia, G.; Marocchini, M.; Kreft, I. Composition and technological properties of the flour and bran from common and Tartary buckwheat. *Food Chem.* **2003**, *80*, 9–15. [CrossRef]
9. Zhu, F. Chemical composition and health effects of Tartary buckwheat. *Food Chem.* **2016**, *203*, 231–245. [CrossRef]
10. Ren, Q.; Li, Y.; Wu, C.; Wang, C.; Jin, Y.; Zhang, J. Metabolism of secondary metabolites isolated from Tartary buckwheat and its extract. *Food Chem.* **2014**, *154*, 134–144. [CrossRef]
11. Jin, H.M.; Wei, P. Anti-fatigue properties of Tartary buckwheat extracts in mice. *Int. J. Mol. Sci.* **2011**, *12*, 4770–4780. [CrossRef]
12. Powers, S.K.; Deruisseau, K.C.; Quindry, J.; Hamilton, K.L. Dietary antioxidants and exercise. *J. Sports Sci.* **2004**, *22*, 81–94. [CrossRef] [PubMed]
13. Ma, Y.J.; Guo, X.D.; Liu, H.; Xu, B.N.; Wang, M. Cooking, textural, sensorial, and antioxidant properties of common and Tartary buckwheat noodles. *Food Sci. Biotechnol.* **2013**, *22*, 153–159. [CrossRef]
14. Rufa, L. The development and utilization of Tartary buckwheat resources. In Proceedings of the 9th International Symposium on Buckwheat, Prague, Czech Republic, 18–22 August 2004; pp. 252–258.
15. Zhang, M.; Chen, H.; Li, J.; Pei, Y.; Liang, Y. Antioxidant properties of Tartary buckwheat extracts as affected by different thermal pro-cessing methods. *LWT-Food Sci. Technol.* **2010**, *43*, 181–185. [CrossRef]
16. Baljeet, S.Y.; Ritika, B.Y.; Roshan, L.Y. Studies on functional properties and incorporation of buckwheat flour for biscuit making. *Int. Food Res. J.* **2010**, *17*, 1067–1076.
17. Giménez-Bastida, J.A.; Piskula, M.K.; Zielinski, H. Recent advances in processing and development of buckwheat derived bakery and non-bakery products-a review. *Pol. J. Food Nutr. Sci.* **2015**, *65*, 9–20. [CrossRef]
18. Niu, M.; Hou, G.; Lee, B.; Chen, Z. Effects of fine grinding of millfeeds on the quality attributes of reconstituted whole-wheat flour and its raw noodle products. *LWT-Food Sci. Technol.* **2014**, *57*, 58–64. [CrossRef]
19. Duodu, K.G.; Nunes, A.; Delgadillo, I.; Parker, M.L.; Mills, E.N.C.; Belton, P.S.; Taylor, J.R.N. Effect of grain structure and cooking on sorghum and maize in vitro protein digestibility. *J. Cereal Sci.* **2002**, *35*, 161–174. [CrossRef]
20. Skrabanja, V.; Kreft, I.; Golob, T.; Modic, M.; Ikeda, S.; Ikeda, K.; Kosmelj, K. Nutrient content in buckwheat milling fractions. *Cereal Chem.* **2004**, *81*, 172–176. [CrossRef]
21. Zhao, X.Y.; Ao, Q.; Yang, L.W.; Yang, Y.F.; Sun, J.C.; Gai, G.S. Application of superfine pulverization technology in biomaterial industry. *J. Taiwan Inst. Chem. Eng.* **2009**, *40*, 337–343. [CrossRef]
22. Wu, G.C.; Zhang, M.; Wang, Y.Q.; Mothibe, K.J.; Chen, W.X. Production of silver carp bone powder using superfine grinding technology: Suitable production parameters and its properties. *J. Food Eng.* **2012**, *109*, 730–735. [CrossRef]
23. Zhao, X.; Yang, Z.; Gai, G.; Yang, Y. Effect of superfine grinding on properties of ginger powder. *J. Food Eng.* **2009**, *91*, 217–222. [CrossRef]
24. Hu, J.; Chen, Y.; Ni, D. Effect of superfine grinding on quality and antioxidant property of fine green tea powders. *LWT-Food Sci. Technol.* **2012**, *45*, 8–12. [CrossRef]
25. Zhu, F.; Du, B.; Xu, B. Superfine grinding improves functional properties and antioxidant capacities of bran dietary fibre from Qingke (hull-less barley) grown in Qinghai-Tibet Plateau, China. *J. Cereal Sci.* **2015**, *65*, 43–47. [CrossRef]
26. Zhang, M.; Zhang, C.; Shrestha, S. Study on the preparation technology of superfine ground powder of Agrocybe chaxingu Huang. *J. Food Eng.* **2005**, *67*, 333–337. [CrossRef]
27. Niu, M.; Hou, G.G.; Wang, L.; Chen, Z. Effects of superfine grinding on the quality characteristics of whole-wheat flour and its raw noodle product. *J. Cereal Sci.* **2014**, *60*, 382–388. [CrossRef]
28. Wu, Z.; Ameer, K.; Jiang, G. Effects of Superfine Grinding on the Physicochemical Properties and Antioxidant Activities of Sanchi (Panax notoginseng) Flower Powders. *J. Food Sci. Technol.* **2020**, *58*, 62–73. [CrossRef]
29. Zhong, C.; Zu, Y.; Zhao, X.; Li, Y.; Ge, Y.; Wu, W.; Guo, D. Effect of superfine grinding on physicochemical and antioxidant properties of pomegranate peel. *Int. J. Food Sci. Technol.* **2016**, *51*, 212–221. [CrossRef]
30. Tkacova, K.; Stevulova, N. Selected problems of the dispersity analysis of milled ultrafine powders. *Freib. Forsch. A (Partik.) A* **1998**, *841*, 14–25.
31. Villada, J.A.; Sánchez-Sinencio, F.; Zelaya-Ángel, O.; Gutiérrez-Cortez, E.; Rodríguez-García, M.E. Study of the morphological, structural, thermal, and pasting corn transformation during the traditional nixtamalization process: From corn to tortilla. *J. Food Eng.* **2017**, *212*, 242–251. [CrossRef]
32. Liu, F.; He, C.; Wang, L.; Wang, M. Effect of milling method on the chemical composition and antioxidant capacity of Tartary buckwheat flour. *Int. J. Food Sci. Technol.* **2018**, *53*, 2457–2464. [CrossRef]
33. Orumwense, O.A.; Forssberg, E. Superfine and ultrafine grinding—A literature survey. *Miner. Process. Extr. Metullargy Rev.* **1992**, *11*, 107–127. [CrossRef]

34. Li, G.; Guo, W.; Gao, X.; Wang, Y.; Sun, S. Effect of superfine grinding on physicochemical and antioxidant properties of soybean residue powder. *Food Sci. Nutr.* **2020**, *8*, 1208–1214. [CrossRef] [PubMed]
35. Zhang, J.; Dong, Y.; Nisar, T.; Fang, Z.; Wang, Z.C.; Guo, Y. Effect of superfine-grinding on the physicochemical and antioxidant properties of Lycium ruthenicum Murray powders. *Powder Technol.* **2020**, *372*, 68–75. [CrossRef]
36. Li, W.; Cao, F.; Fan, J.; Ouyang, S.; Luo, Q.; Zheng, J.; Zhang, G. Physically modified common buckwheat starch and their physicochemical and structural properties. *Food Hydrocoll.* **2014**, *40*, 237–244. [CrossRef]
37. Zhao, X.; Zhu, H.; Zhang, G.; Tang, W. Effect of superfine grinding on the physicochemical properties and antioxidant activity of red grape pomace powders. *Powder Technol.* **2015**, *286*, 838–844. [CrossRef]
38. Liu, K. Particle size distribution of distillers dried grains with solubles (DDGS) and relationships to compositional and color properties. *Bioresour. Technol.* **2008**, *99*, 8421–8428. [CrossRef]
39. Zhao, X.; Ao, Q.; Du, F.; Zhu, J.; Liu, J. Surface characterization of ginger powder examined by X-ray photoelectron spectroscopy and scanning electron microscopy. *Colloids Surf. B Biointerfaces* **2010**, *79*, 494–500. [CrossRef]
40. Huang, X.; Liang, K.H.; Liu, Q.; Qiu, J.; Wang, J.; Zhu, H. Superfine grinding affects physicochemical, thermal and structural properties of Moringa Oleifera leaf powders. *Ind. Crops Prod.* **2020**, *151*, 112472. [CrossRef]
41. Xu, Q.; Huang, R.; Yang, P.; Wang, L.; Xing, Y.; Liu, H.; Zhang, P. Effect of different superfine grinding technologieson the physicochemical and antioxidant properties of Tartary buckwheat bran powder. *RSC Adv.* **2021**, *11*, 30898–30910. [CrossRef]
42. Raghavendra, S.N.; Rastogi, N.K.; Raghavarao, K.S.M.S.; Tharanathan, R.N. Dietary fiber from coconut residue: Effects of different treatments and particle size on the hydration properties. *Eur. Food Res. Technol.* **2004**, *218*, 563–567. [CrossRef]
43. Ting, Y.; Jiang, Y.; Ho, C.T.; Huang, Q. Common delivery systems for enhancing in vivo bioavailability and biological efficacy of nutraceuticals. *J. Funct. Foods* **2014**, *7*, 112–128. [CrossRef]
44. Jiang, L.; Xu, Q.X.; Qiao, M.; Ma, F.F.; Thakur, K.; Wei, Z.J. Effect of superfine grinding on properties of Vaccinium bracteatum Thunb leaves powder. *Food Sci. Biotechnol.* **2017**, *26*, 1571–1578. [CrossRef] [PubMed]
45. Zhao, X.; Du, F.; Zhu, Q.; Qiu, D.; Yin, W.; Ao, Q. Effect of superfine pulverization on properties of Astragalus membranaceus powder. *Powder Technol.* **2010**, *203*, 620–625. [CrossRef]
46. Yan, L.; Li, T.; Liu, C.; Zheng, L. Effects of high hydrostatic pressure and superfine grinding treatment on physicochemical/functional properties of pear pomace and chemical composition of its soluble dietary fibre. *LWT-Food Sci. Technol.* **2019**, *107*, 171–177. [CrossRef]
47. Li, W.; Guo, H.; Wang, P.; Tian, X.; Zhang, W.; Saleh, A.S.; Zhang, G. Physicochemical characteristics of high pressure gelatinized mung bean starch during re-crystallization. *Carbohydr. Polym.* **2015**, *131*, 432–438. [CrossRef] [PubMed]
48. Jin, Z.; Hsieh, F.; Huff, H.E. Effects of soy fiber, salt, sugar and screw speed on physical properties and microstructure of corn meal extrudate. *J. Cereal Sci.* **1995**, *22*, 185–194. [CrossRef]
49. Kusumayanti, H.; Handayani, N.A.; Santosa, H. Swelling power and water solubility of cassava and sweet potatoes flour. *Procedia Environ. Sci.* **2015**, *23*, 164–167. [CrossRef]
50. Tester, R.F.; Morrison, W.R. Swelling and gelatinization of cereal starches. I. Effects of amylopectin, amylose, and lipids. *Cereal Chem.* **1990**, *67*, 551–557.
51. Huang, Y.; Sun, X.; Guo, H.; He, X.; Jiang, J.; Zhang, G.; Li, W. Changes in the thermal, pasting, morphological and structural characteristic of common buckwheat starch after ultrafine milling. *Int. J. Food Sci. Technol.* **2021**, *56*, 2696–2707. [CrossRef]
52. Xia, W.; Yang, H.U.; Ji-Hua, L.I.; Wei, X.Y.; Wang, F.; Lin, Y.Y. Effects of superfine grinding on retrogradation properties of tapioca starch. *Sci. Technol. Food Ind.* **2017**, *24*, 44–47.
53. Zhao, Y.; Wu, X.; Wang, Y.; Jing, R.; Yue, F. Comparing physicochemical properties of hawthorn superfine and fine powders. *J. Food Process. Preserv.* **2017**, *41*, e12834. [CrossRef]
54. Viuda-Martos, M.; Ruiz-Navajas, Y.; Martin-Sánchez, A.; Sánchez-Zapata, E.; Fernández-López, J.; Sendra, E.; Pérez-Álvarez, J.A. Chemical, physico-chemical and functional properties of pomegranate (*Punica granatum* L.) bagasses powder co-product. *J. Food Eng.* **2012**, *110*, 220–224. [CrossRef]
55. Pandord, J.A.; Williams, P.C.; DeMan, J.M. Analysis of oilseeds for protein, oil, fiber and moisture by near-infrared reflectance spectroscopy. *J. Am. Oil Chem. Soc.* **1988**, *65*, 1627–1634. [CrossRef]
56. Weber, C.W.; Gentry, H.S.; Kohlhepp, E.A.; McCrohan, P. The nutritional and chemical evaluation of chia seeds. *Ecol. Food Nutr.* **1991**, *26*, 119–125. [CrossRef]
57. Liu, H.; Lv, M.; Peng, Q.; Shan, F.; Wang, M. Physicochemical and textural properties of Tartary buckwheat starch after heat–moisture treatment at different moisture levels. *Starch-Stärke* **2015**, *67*, 276–284. [CrossRef]
58. Senanayake, S.A.; Ranaweera, K.K.D.S.; Gunaratne, A.; Bamunuarachchi, A. Comparative analysis of nutritional quality of five different cultivars of sweet potatoes (I pomea batatas (L) Lam) in S ri L anka. *Food Sci. Nutr.* **2013**, *1*, 284–291. [CrossRef]
59. Leghari, M.H.; Sheikh, S.A.; Kumbhar, M.B.; Baloch, A.F. Mineral content in dehydrated mango powder. *J. Basic Appl. Sci.* **2013**, *9*, 21. [CrossRef]
60. Özcan, M.; Arslan, D.; Ünver, A. Effect of drying methods on the mineral content of basil (*Ocimum basilicum* L.). *J. Food Eng.* **2005**, *69*, 375–379. [CrossRef]
61. Tsai, M.L.; Li, C.F.; Lii, C.Y. Effects of granular structures on the pasting behaviors of starches. *Cereal Chem* **1997**, *74*, 750–757. [CrossRef]

MDPI
St. Alban-Anlage 66
4052 Basel
Switzerland
Tel. +41 61 683 77 34
Fax +41 61 302 89 18
www.mdpi.com

Molecules Editorial Office
E-mail: molecules@mdpi.com
www.mdpi.com/journal/molecules

www.ingramcontent.com/pod-product-compliance
Lightning Source LLC
Chambersburg PA
CBHW061543220926
51850CB00005B/237
* 9 7 8 3 0 3 6 5 6 7 9 3 8 *